73 Springer Series in Solid-State Sciences
Edited by Peter Fulde

Springer Series in Solid-State Sciences

Editors: M. Cardona, P. Fulde, K. von Klitzing, H.-J. Queisser

Managing Editor: H. K. V. Lotsch

Volumes 1–49 are listed on the back inside cover

50 **Multiple Diffraction of X-Rays in Crystals**
By Shih-Lin Chang

51 **Phonon Scattering in Condensed Matter**
Editors: W. Eisenmenger, K. Laßmann, and S. Döttinger

52 **Superconductivity in Magnetic and Exotic Materials.** Editors: T. Matsubara and A. Kotani

53 **Two-Dimensional Systems, Heterostructures, and Superlattices**
Editors: G. Bauer, F. Kuchar, and H. Heinrich

54 **Magnetic Excitations and Fluctuations**
Editors: S. Lovesey, U. Balucani, F. Borsa, and V. Tognetti

55 **The Theory of Magnetism II**
Thermodynamics and Statistical Mechanics
By D. C. Mattis

56 **Spin Fluctuations in Itinerant Electron Magnetism.** By T. Moriya

57 **Polycrystalline Semiconductors**
Physical Properties and Applications
Editor: G. Harbeke

58 **The Recursion Method and Its Applications**
Editors: D. G. Pettifor and D. L. Weaire

59 **Dynamical Processes and Ordering on Solid Surfaces.** Editors: A. Yoshimori and M. Tsukada

60 **Excitonic Processes in Solids**
By M. Ueta, H. Kanzaki, K. Kobayashi, Y. Toyozawa, and E. Hanamura

61 **Localization, Interaction, and Transport Phenomena.** Editors: B. Kramer, G. Bergmann, and Y. Bruynseraede

62 **Theory of Heavy Fermions and Valence Fluctuations.** Editors: T. Kasuya and T. Saso

63 **Electronic Properties of Polymers and Related Compounds.** Editors: H. Kuzmany, M. Mehring, and S. Roth

64 **Symmetries in Physics.** Group Theory Applied to Physical Problems. By W. Ludwig and C. Falter

65 **Phonons: Theory and Experiments II**
Experiments and Interpretation of Experimental Results. By P. Brüesch

66 **Phonons: Theory and Experiments III**
Phenomena Related to Phonons
By P. Brüesch

67 **Two-Dimensional Systems: Physics and New Devices**
Editors: G. Bauer, F. Kuchar, and H. Heinrich

68 **Phonon Scattering in Condensed Matter V**
Editors: A. C. Anderson and J. P. Wolfe

69 **Nonlinearity in Condensed Matter**
Editors: A. R. Bishop, D. K. Campbell, P. Kumar, and S. E. Trullinger

70 **From Hamiltonians to Phase Diagrams**
The Electronic and Statistical-Mechanical Theory of sp-Bonded Metals and Alloys. By J. Hafner

71 **High Magnetic Fields in Semiconductor Physics.** Editor: G. Landwehr

72 **One-Dimensional Conductors**
By S. Kagoshima, H. Nagasawa, and T. Sambongi

73 **Quantum Solid-State Physics**
By S. V. Vonsovsky and M. I. Katsnelson

74 **Quantum Monte Carlo Methods** in Equilibrium and Nonequilibrium Systems. Editor: M. Suzuki

75 **Electronic Structure and Optical Properties of Semicoconductors**
By M. L. Cohen and J. R. Chelikowsky

76 **Electronic Properties of Conjugated Polymers**
Editors: H. Kuzmany, M. Mehring, and S. Roth

77 **Fermi Surface Effects**
Editors: J. Kondo and A. Yoshimori

78 **Group Theory and Its Applications in Physics**
By T. Inui, Y. Tanabe, and Y. Onodera

79 **Elementary Excitations in Quantum Fluids**
Editors: K. Ohbayashi and M. Watabe

80 **Monte Carlo Simulation in Statistical Physics**
An Introduction. By K. Binder and D. W. Heermann

81 **Core-Level Spectroscopy in Condensed Systems.** Editors: J. Kanamori and A. Kotani

82 **Introduction to Photoemission Spectroscopy**
By S. Hüfner

83 **Physics and Technology of Submicron Structures.** Editors: H. Heinrich, G. Bauer, and F. Kuchar

84 **Beyond the Crystalline State**
An Emerging Perspective
By G. Venkataraman, D. Sahoo, and V. Balakrishnan

85 **The Fractional Quantum Hall Effect**
Properties of an Incompressible Quantum Fluid
By T. Chakraborty and P. Pietiläinen

86 **The Quantum Statistics of Dynamic Processes**
By E. Fick, G. Sauermann

S. V. Vonsovsky M. I. Katsnelson

Quantum Solid-State Physics

With 151 Figures

Springer-Verlag Berlin Heidelberg New York
London Paris Tokyo HongKong

Academican Serghey V. Vonsovsky
Dr. Mikhail I. Katsnelson

Institute of Metal Physics, Ural Science Research Center of the USSR Academy of Sciences, 18, S. Kovalevskaya Street, SU-620219 Sverdlovsk, GSP-170/USSR

Series Editors:
Professor Dr., Dres. h. c. Manuel Cardona
Professor Dr., Dr. h. c. Peter Fulde
Professor Dr., Dr. h. c. Klaus von Klitzing
Professor Dr. Hans-Joachim Queisser

Max-Planck-Institut für Festkörperforschung, Heisenbergstraße 1
D-7000 Stuttgart 80, Fed. Rep. of Germany

Managing Editor: Dr. Helmut K. V. Lotsch

Springer-Verlag, Tiergartenstraße 17,
D-6900 Heidelberg, Fed. Rep. of Germany

Title of the original Russian edition:
Kvantovaja fizika tverdogo tela, "Nauka" Publishing House, Moscow 1983

ISBN 978-3-642-50166-1 ISBN 978-3-642-50164-7 (eBook)
DOI 10.1007/978-3-642-50164-7

Vonsovskiĭ, S. V. (Sergeĭ Vasil 'evich) [Kvantovai͡a fizika tverdogo tela. English] Quantum solid-state physics/S. V. Vonsovsky, M. I. Katsnelson. p. cm. – (Springer series in solid-state sciences : 73). Translation of: Kvantovai͡a fizika tverdogo tela. Bibliography: p. Includes indexes.
ISBN 978-3-642-50166-1
1. Solid state physics. 2. Quantum theory. I. Kat͡snel 'son, M. I. (Mikhail Iosifovich) II. Title. III. Series. QC176.V6613 1989, 530.4'1–dc 19 89-5948

Softcover reprint of the hardcover 1st edition 1989

2154/3155-543210 – Printed on acid-free paper

Preface

The quantum theory of solids occupies a peculiar and important place in the general structure of modern theoretical physics. There are currently no grounds for questioning the statement that all properties of solids can, in principle, be accounted for on the basis of firmly established principles of quantum and statistical mechanics. Nevertheless, these properties of real solids and the condensed state of matter in general, are so complicated and diverse that it is well nigh impossible, at least at present, to explain rigorously and fully from first principles the observed characteristics of crystals, even those which are close to perfect, – let alone explain the fact that they exist! Therefore, alongside the mathematical methods and physical concepts applied in other more fundamental areas of theoretical physics, solid-state theory has developed approaches of its own to account for the most important properties of the various substances. Significantly, these approaches now have a profound reciprocal effect on not only statistical physics but also on particle physics, and even on astrophysics and cosmology. Apart from this, the tremendous and ever-increasing applied significance of solid-state theory must be noted. Suffice it to mention here the theory of semiconducting devices, the theory of strength and plasticity, the theory of magnetic properties of materials, etc. In this respect, modern solid-state theory employs with great practical success a sufficiently simple and, at the same time, adequate theoretical background, based on a purely phenomenological approach and microscopic models that are comparatively simple in terms of mathematics and very lucid physically.

As stated above, a quantum theory of solids that realizes the "first-principles" program in its entirety, i. e., a theory in which all properties of solids are derived from those of individual constituent atoms, does not exist. However, it may well be assumed that, for example, the indubitable and sizable success of the pseudopotential method that now enjoys wide use in the theory of simple (normal) metals is an important step toward the construction of such a physically consistent first principles theory. Rather than choosing the deductive method of presentation, we have therefore opted, in this text, for a method based on a treatment and analysis of simple empirically established properties of solids, resorting to more complicated models only where necessary. In a way, such an exposition reproduces the evolution of this important province of modern theoretical physics (differing in this respect from the diverse monographs and textbooks devoted to the problem concerned) and, in our view, is most appropriate to initiate the reader into the subject. We have also assumed that a detailed treatment of a number of classical topics such as the one-dimensional

Schrödinger equation with a periodic potential, the metal-insulator criterion, the effect of electric and magnetic fields on electronic states, and other similar problems would be very instructive. At the same time, the book presents a number of up-to-date topics: the scattering of neutrons by the crystal lattice, plasma and Fermi liquid effects, elements of pseudopotential theory, fundamentals of the theory of disordered systems, etc.

It is assumed that the essentials of quantum and statistical mechanics are a sufficient theoretical background for reading this text. As far as possible we have tried to outline the body of mathematics in conjunction with those specific problems in which it is immediately exploited. Thus, for instance, the secondary quantization method is expounded for the first time in connection with the problem of neutron scattering on phonons, the resolvent (Green's-function) method is outlined in connection with the problem of electron localization on impurities, etc. The presentation of a number of problems, in which allowance for correlation effects in electronic systems (superconductivity, the properties of transition metals, and the like) is essential, has turned out to be extremely concise, for we did not succeed in finding a simplified enough version of the mathematical sophistication needed for a more rigorous exposition. In those (and some other) cases we had to confine ourselves to simple model problems and often to purely qualitative lines of argument.

Although the present book outlines sufficiently general properties of solids, as well as some specific problems of the physics of semiconductors, ionic crystals, etc., it is still primarily the metal that we view as the model of "a solid in general." This is partly because the metal is in essence a gigantic molecule and demonstrates most dramatically the features peculiar to the electronic properties of crystals, which are not reducible to a mere "sum" of the properties of individual constituent atoms or molecules.

An acquaintance with this text should prepare the reader for a more detailed study of particular areas of solid-state theory, which at present are outlined fully enough at an up-to-date level in the abundant body of special literature. The text is intended to appeal to experimental physicists, chemists, engineers concerned with problems of solid-state theory and wishing to become conversant with the pertinent body of mathematics, theoretical physicists in other fields, and undergraduate and graduate students.

Throughout the book the numerical values of physical quantities are given in both the units normally employed by physicists and SI units. For the equations of electrodynamics the CGSE system is used throughout.

This monograph evolved largely from notes written for a course offered in the Department of Theoretical Physics of the Ural Institute for Physics and Engineering by Prof. S. P. Shubin as far back as the thirties and from lectures delivered for many years at the A. M. Gorky Ural State University by one of us (S. V. Vonsovsky).

Very encouraging assistance during the development of the manuscript was rendered by Dr. L. A. Shubina who executed a lot of systematizing work, helped to improve the text, and skillfully typed the manuscript in its entirety. We are deeply indebted to her for this invaluable help and express keen sorrow at her not being able to see the book published.

During the development of the English edition, numerous alterations were introduced in the text. New material has been added: Sects 1.7 (and a large portion of Sect. 1.8), 1.10, 2.5, 4.2.5, 4.4.3, 4.6.7, 4.9, 5.3.2 (and a large portion of Sect. 5.3.3), 5.7. The inaccuracies noticed in the Russian edition have been corrected, a number of new figures added, and the list of references extended substantially.

It is our pleasant duty to thank Dr. H. Lotsch for his offer to have the book published and for making it a reality. We are sincerely grateful to Mr. A. P. Zavarnitsyn for the English translation and the assistance rendered during the development of this project. We would like to express our sincere appreciation to Mrs. M. L. Katsnelson for her enthusiastic help in preparing the manuscript and to Mr. V. G. Reprintsev for typing the rough draft.

Sverdlovsk, Spring 1989

Serghey V. Vonsovsky
Mikhail I. Katsnelson

Contents

1. Introduction. General Properties of the Solid State of Matter

This first chapter deals with the most general problems of solid state physics such as crystal structure and crystal symmetry, chemical bonding, classification of solids, and the amorphous state. With reference to simple examples, some concepts of group theory and fundamentals of the quantum chemistry of solids are treated briefly. An overview will be provided of the properties of the various types of crystalline and amorphous solids, including quantum crystals and metallic glasses.

1.1 General Thermodynamic Description of the Solid State

Along with the liquid and gaseous states, the solid state is one of the three thermodynamically stable states of aggregation of matter. Sometimes we talk of a fourth state of aggregation, that is, the plasma state. But the term is inaccurate, for no phase transition occurs between the plasma and the gas; gaseous plasma matter is a highly ionized gas.

To begin with we consider pure, homogeneous, and one-constituent substances that consist of atoms of the same species. Three equilibrium conditions are known from thermodynamics: the constancy of temperature T over the entire volume of the solid,

$$T = \text{constant} \ , \tag{1.1.1}$$

the constancy of pressure p,

$$p = \text{constant} \ , \tag{1.1.2}$$

and the constancy of the chemical potential $\mu(p, T)$,

$$\mu(p, T) = \text{constant} \ . \tag{1.1.3}$$

Even in the absence of external fields, a one-constituent solid in equilibrium is not necessarily homogeneous. For example, given an energy E and volume V, a solid in equilibrium may break down into two or more spatially separated, homogeneous (if there are no external fields) parts that touch, i.e., phases that are in different states of aggregation or in different allotropic states. The equilibrium conditions for these phases in a one-constituent solid are similar to (1.1.1)–(1.1.3). For two phases they have the form

$$T_1 = T_2 \, , \quad p_1 = p_2 \, , \tag{1.1.4}$$

$$\mu_1(p, T) = \mu_2(p, T) \, , \tag{1.1.5}$$

where the subscripts 1 and 2 number the phases. Condition (1.1.5) is the equation of the equilibrium curve for phases 1 and 2; $p = f_{1,2}(T)$. At each point of the curve both phases are in equilibrium. At points that lie along both its sides either one phase or the other exists.

The equilibrium conditions for three phases are

$$T_1 = T_2 = T_3 \, , \quad p_1 = p_2 = p_3 \, , \tag{1.1.6}$$

$$\mu_1(p, T) = \mu_2(p, T) = \mu_3(p, T) \, . \tag{1.1.7}$$

The system of two equations (1.1.7) with unknown p and T in the case of real and positive solutions yields an isolated triple point p_{tr}, T_{tr} (point O in Fig. 1.1). At this point all the three phases of the solid are in equilibrium. Point O is the intercept of three equilibrium curves. The OC line is a sublimation curve, OB a melting or solidification (crystallization) curve, and OA a boiling or condensation curve. They delineate in the p, T plane the regions of existence of three phases: solid (I), liquid (II), and gaseous (III). If the state of a one-constituent solid is defined by two quantities, p and T, more than three phases cannot be in equilibrium at once. This follows from the Gibbs phase rule, which relates the number of possible phases n to the number of constituents k and thermodynamic variables r that are required for a complete determination of the equilibrium state of a system,

$$n \leqslant k + r \, . \tag{1.1.8}$$

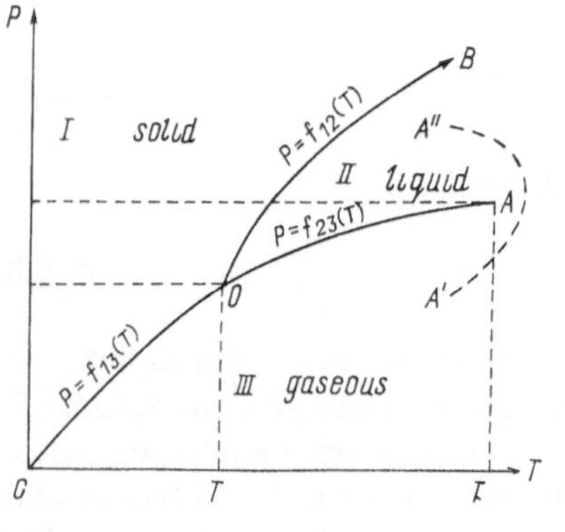

Fig. 1.1. Phase diagram for three phases of a one-constituent material whose state is determined by two parameters—pressure p and temperature T: (I) melting (solidification) curve $p = f_{12}(T)$; (II) boiling (condensation) curve $p = f_{23}(T)$; (III) sublimation curve $p = f_{13}(T)$

In the case of a one-constituent system ($k = 1$) described by two variables ($r = 2$) it follows from (1.1.8) that $n \leqslant 1 + 2 = 3$.

Note that the sublimation curve OC normally arises at the origin of coordinates ($T = 0$, $p = 0$). Experience shows that the sublimation pressure tends to zero when $T \to 0$ (for an exception to this rule see Fig. 1.2) and the other end of the OC curve lies at the triple point O. The latter is also the origin of the boiling curve OA which terminates at a critical point $A(p_{cr}, T_{cr})$. This term dates back to D. I. Mendeleyev (1860).

The existence of a critical point indicates that the transition from the liquid to the gaseous state and vice versa can also occur continuously, in bypassing the boiling curve with abrupt variations in the magnitudes of specific volume, entropy, etc. In Fig. 1.1 this is indicated by a dashed line $A'A''$. The melting curve OB begins at the triple point and terminates nowhere, extending to infinity. On account of difficulties associated with experimental verification (which necessitates superhigh pressures and temperatures), this conclusion is currently a hypothesis. However, the OB curve cannot, at any rate, terminate at a critical point, for the latter is possible for phases that differ only quantitatively, for example, in the magnitude of interparticle-interaction energy, as is the case for a gas and a liquid.

The absence of a critical point on the melting curve indicates that the transition between the liquid phase and the solid phase always occurs in jumps because of the difference in their symmetry (disordered liquid and ordered crystal, Sect. 1.2). This phenomenon excludes from the solids in thermal equilibrium the amorphous solids that should be ranked among the liquids in the metastable (supercooled) state. However, in spite of fundamental differences between the liquid and the crystal, a liquid at temperatures not very close to the boiling point becomes in many ways similar to a solid [1. 1].

The phase diagrams of one-constituent substances may be more complicated than those depicted in Fig. 1.1, if the solid in these phases has two or more allotropic modifications.

The phase diagram of helium isotopes He^4 and He^3 differs substantially from those of other substances. Figure 1.2 [1.2] presents a phase diagram of

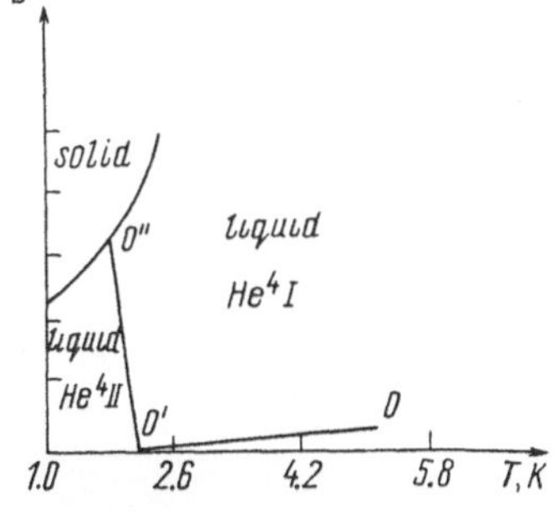

Fig. 1.2. Phase diagram for He^4 isotope with two liquid phases: normal He^4I and superfluid He^4II

He^4. There is no equilibrium between the solid phase and the gas here, and therefore there is no triple equilibrium point for the gas, liquid, and crystal simultaneously. There is a triple point O'' between the solid phase and the two liquid phases, normal (He^4 I) and superfluid (He^4 II). While the lowest-temperature phase in normal diagrams is the solid phase, even at low pressures, the counterpart in helium is the liquid phase (Sect. 1.2, 1.7).

In the case of multiconstituent systems the situation is more complicated. Two variables, T and p, do not suffice here to determine the state. For example, for a binary alloy one needs to know three quantities: T, p, and the concentration c ($0 \leqslant c \leqslant 1$), which is equal to the ratio of the number of particles or one of the constituents to the total number of particles. According to (1.1.8), the maximum number of phases in equilibrium may not exceed four, since the state here is determined by a point in three-dimensional space T, p, and c. The interface of two phases is the surface $p = f_{12}(T, c)$. Three phases will be in equilibrium at the points of the line of intersection of three surfaces $p = f_{12}(T, c)$, $p = f_{13}(T, c)$, and $p = f_{23}(T, c)$, i.e., the triple-point line. Finally, to the equilibrium of four phases there correspond isolated points—intercepts of triple-point lines. In physical metallurgy use is made of two-dimensional diagrams. The systems of coordinates T, c (with p = const.) and c, p (with T = const.) may be used for solid–gas (sublimation diagram), liquid–gas (boiling diagram), and liquid–crystal (melting diagram) transitions, and also for polymorphous transformations of crystals. These phase diagram lines, generated by interceptions of $p = f(T, c)$ surfaces by p = const. or T = const. planes, are called equilibrium curves.

To illustrate this, Fig. 1.3 presents a diagram of equilibrium curves for a gold–platinum alloy (at p = const.). The shaded area of the diagram corre-

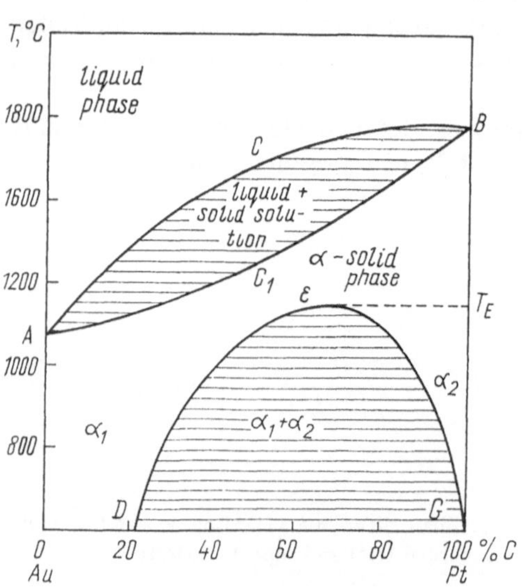

Fig. 1.3. Schematic phase diagram of a binary Au-Pt system

sponds to the regions of equilibrium of two phases (liquid phase with the α-solid phase or two solid phases α_1 and α_2). The unshaded domain refers to the regions of equilibrium of one phase (liquid phase or α, α_1, α_2 solid phases). Point *A* is the melting point of pure Au, *B* that of pure Pt. The liquidus line *ACB* separates the two-phase (liquid phase and α-solid phase) region from the pure-liquid region. The solidus line AC_1B separates the same two-phase region from the α-solid phase region. The *DEG* line indicates phase transformations that occur in the solid phase when the alloy is cooled below the critical point T_E. The solid α solution breaks down into two phases: Au–rich α_1 and Pt-rich α_2. Thus the binary alloy Au–Pt consists of two constituents that are miscible in all proportions in the liquid state at all temperatures above the liquidus line *ACB* and in the solid state below the solidus line AC_1B up to the temperature T_E (in the real case of Au–Pt the critical point $T_E > T_A$, and therefore in the solid state there is no region of one solid α phase over the entire composition range).

Figure 1.4 illustrates a silver–copper alloy. Again, this is a binary mixture with complete solubility of the constituents in the liquid state above the liquidus line *AEB* (one-phase liquid solution) which in this case has a break at point *E*. In the solid state, below the broken solidus line *ACEDB*, we deal with restricted solubility: a substitutional solid solution of Cu in Ag in the α-phase region (*ACF*) and a substitutional solid solution of Ag in Cu in the β-phase region (*BDG*). Comparison of these regions shows that copper does not dissolve silver so well as silver dissolves copper.

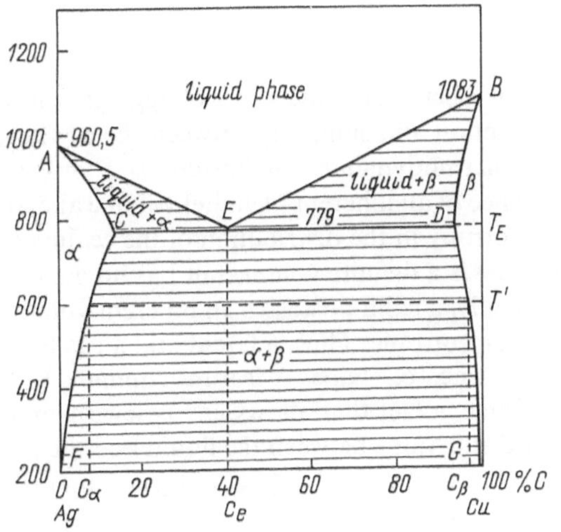

Fig. 1.4. Schematic phase diagram of a binary Ag-Cu system

The shaded regions *ACE* and *BDE* correspond to the equilibrium of the liquid and solid α, β phases respectively. The region *FCEDG* is the domain of existence of two solid phases α + β. The mixture consists of two solid solutions with concentrations c_α and c_β that correspond to a given temperature T' (see the points of intersection of the dashed straight line with the FC and DG lines).

Of particular significance is the point E. Here three phases—a liquid phase and two solid α and β phases—coexist in equilibrium. It is called the eutectic point, and the diagram itself and the related system of alloys are said to belong to the eutectic type. An alloy of eutectic composition c_E at the crystallization point T_E forms a mechanical mixture of crystals of both phases α and β. Examples of these diagrams do not exhaust all the types possible [1.3].

To provide a theoretical explanation for the various diagrams, it is necessary to know the dependence of, for example, the thermodynamic potential ϕ on the variables T, p, c, etc. This dependence can be obtained from experiment or microtheory [1.4]. However, thermodynamics, too, allows us to establish some systematics of all types of diagrams [1.4]. A complete theoretical solution of this problem would be attained if we could write the equations of the state of a substance starting from first principles. This has not yet been achieved. For the current status of this problem, see [1.5].

To sum up, we may say that, from the thermodynamic standpoint, the solid state is the phase for all condensed matter at low enough temperatures, except for helium. In order to explain this fact in terms of atomistic concepts, we have to recall the major points of what is now known about the structure of crystals.

1.2 Crystal Structure of Solids

From a thermodynamic consideration (Sect. 1.1) of the states of aggregation of matter we can perceive traits of distinction and similarity between them. Thus, subject to the condition that the thermodynamic equilibrium is complete, features peculiar to a gas are the absence of an inherent (definite) volume and an inherent shape and the presence of isotropy in the distribution of the centers of gravity of atoms. A liquid phase possesses a definite volume but has no definite shape and, like a gas, is isotropic. Finally, a solid possesses an inherent volume and an inherent shape and is always anisotropic. Thus, a liquid and a gas have features in common: isotropy and absence of shape. A feature common to a liquid and a solid is the inherent volume, a fact that allows us to view them as condensed phases. Above T_{cr}, p_{cr} all distinctions between the liquid and the gas disappear. Elsewhere, the transitions between crystal, liquid and gaseous states are always abrupt.

Mention must be made of one more type of condensed state of matter: liquid crystals (mesophases), which exist in the temperature interval between the crystal and the amorphous liquid. Such states have been found to occur only in organic matter with molecules of extended linear shape. Two types of liquid crystals are distinguished: nematic (from the Greek “nema”: thread) and smectic (from the Greek "smegma": soap, since, in its properties, this mesophase resembles soap) [1.6].

In a nematic liquid the molecules are so arranged that their long axes are mutually parallel over the entire volume of the sample (i.e., a long-range order of the orientational type is present). However, the coordination long-range order of the centers of gravity of molecules is absent (just as in a normal amorphous liquid). There is one more variety of nematic liquid crystal, viz, cholesteric crystals (the term comes from the name of pure cholesterol ether, one of the first-discovered representatives of this mesophase). These crystals have uniaxial orientational order consisting of mutually parallel planes filled with lamellar molecules that lie in these planes. The centers of gravity of the molecules are randomly distributed in the planes.

A smectic liquid has a system of equidistant planes in which the ends of the molecules are fastened with the long axes parallel to those planes. The coordination and orientation order in the directions perpendicular to the long axis is absent (like in nematic liquids). The three principal structures of liquid–crystal order are schematically depicted in Fig. 1.5a–c. For more details, see [1.4, 1.7].

Thus, a specific feature of the solid phase is its anisotropy, which manifests itself in the crystal structure of all solids in equilibrium. In experiment, this

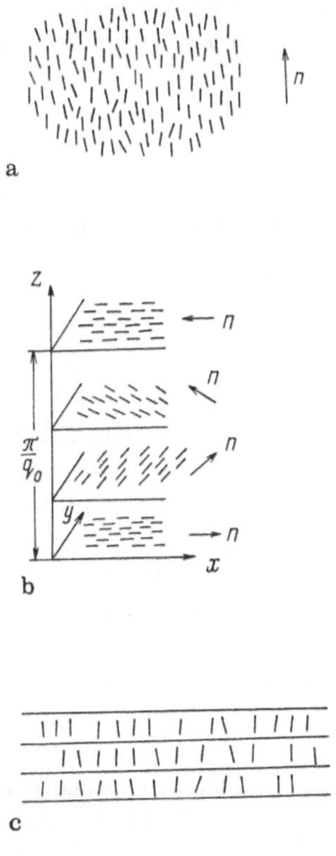

Fig. 1.5. Liquid-crystal structure types: (**a**) nematic, (**b**) cholesteric, and (**c**) smectic

reveals itself in the fact that, when the equilibrium conditions for a phase transition from a gas or liquid are thoroughly fulfilled, a solid always forms as single crystals. However, during solidification of a liquid under normal conditions there is always a large number of crystallization centers that give rise to polycrystals of solids.

According to atomistic concepts, in a gas and a liquid the atoms are in a state of chaotic thermal motion whose intensity is determined by temperature, the atoms not being associated with regularly space-distributed equilibrium positions. We introduce two times that are characteristic of thermal motion: the mean free time $\bar{\tau}$ between two successive collisions and the mean time of the interaction of atoms during collision—the impact time τ_{im}. Then the difference of the gas from the liquid is defined by the magnitude of the dimensionless ratio of these times:

$$\xi_\tau = \tau_{\text{im}}/\bar{\tau} \ . \qquad (1.2.1)$$

In a gas $\xi_\tau \ll 1$, and in a liquid $\xi_\tau \lesssim 1$. The larger value of ξ_τ in the liquid makes it a condensed phase, but the atomic-interaction energy is too small to overcome the randomness of thermal motion, and the liquid remains isotropic.

During crystallization of the liquid, long-range order arises in the distribution of the centers of equilibrium of the atoms that perform thermal oscillations. One more dimensionless parameter can be introduced which defines the dynamics of the crystal and is equal to the ratio of the mean amplitude of the atomic displacement from the equilibrium position of the atoms during thermal oscillations $\langle u \rangle_T$ to the principal geometric crystal parameter, viz., the crystal lattice constant d which is equal to the spacing between nearest atomic equilibrium centers. At temperatures not very close to the melting point ϑ_{mel}, the amplitude of thermal vibrations of atoms will be small:

$$\xi_T = \langle u \rangle_T/d \ll 1 \ . \qquad (1.2.2)$$

Proceeding from classical concepts, it might be assumed that at $T \to 0$ K the quantity $\langle u \rangle_T$ also tends to zero. However, according to the Heisenberg uncertainty principle, at 0 K an atom executes "zero" vibrations of finite amplitude $\langle u \rangle_0$. In most of the crystals the quantity $\langle u \rangle_0$ as well as $\langle u \rangle_T$ is very small when $T \ll \vartheta_{\text{mel}}$, i.e., $\xi_0 = \langle u \rangle_0/d \ll 1$. Only with helium isotopes is $\langle u \rangle_0$ nearly as high as 30% of d, and for this reason at pressures below 25 atm (or 2.5×10^6 Pa in SI units) this substance does not crystallize even at 0 K (Fig. 1.2).

In treating crystalline solids we first of all are confronted with their symmetry properties. A knowledge of these properties, in turn, allows important general conclusions to be drawn concerning the nature of many physical properties of solids, as will be seen more than once later in the text, especially in Chap. 4 (band theory of metals).

In the late nineteenth century scientists began to develop the microscopic theory of crystal symmetry [1.8, 1.9]. Now we have at our disposal an elaborate science—crystallography [1.10]—which is based on symmetry theory and group theory and makes extensive use of concepts of the atomic structure of crystals. For convenience, we here supply some indispensable information on this point. At the outset we acquaint ourselves with some results of the mathematical group theory, which are widely used in physical studies of the structure of crystalline solids, i.e., in solid-state theory. In presenting this material we do not quote complicated analytical derivations but confine ourselves to references to special literature on group theory [1.10–1.14].

Thus we begin with the main property of crystals: the symmetric spatial arrangement of their atoms. Since the atoms are in thermal motion, they do not occupy exact positions in space. To describe crystals, we introduce a radius-vector-dependent function $\varrho(\boldsymbol{r})$ which defines the particle-distribution probability density:

$$\boldsymbol{r} = x\boldsymbol{i} + y\boldsymbol{j} + z\boldsymbol{k} \ , \tag{1.2.3}$$

with x, y, z being coordinates, and $\boldsymbol{i}, \boldsymbol{j}, \boldsymbol{k}$ unit vectors along the coordinate axes. The probability of a particle being contained in a volume element $d\boldsymbol{r} = dxdydz$ is equal to $\varrho(\boldsymbol{r})d\boldsymbol{r}$. The symmetry of spatial atomic arrangement is determined by the various coordinate transformations (such as translations of origin, rotations of axes, reflections in coordinate planes, etc.) under which the function $\varrho(\boldsymbol{r})$ remains invariant. The set of all symmetry transformations of a given type of crystal constitutes its symmetry group [1.11].

Let us find the symmetry group of a simple crystal model such as a plane square $ABCD$ with its center at point O (Fig. 1.6a, b). Its symmetry group contains eight symmetry elements that map the square into itself: E, the identity transformation; C_{4z}^{+}, counterclockwise rotation through an angle of 90° about the z axis which passes through the center O and is perpendicular to the plane of the square; C_{4z}^{-}, clockwise rotation through an angle of 90° about the same z axis; C_{2z}, rotation through an angle of 180° about the z axis; σ_x, reflection in the line passing in the plane of the square through its center O normal to the x axis; σ_y, reflection in the line passing in the plane through the center O normal to the y axis; σ_1, reflection in the diagonal AC of the square; σ_2, reflection in the diagonal BD of the square.

The product (i.e., successive application) of any two elements of the group also is an element of the latter. This is illustrated by the multiplication rules summarized in Table 1.1. The symmetry operators do not necessarily commute with one another, for instance,

$$C_{4z}^{-}\sigma_x = \sigma_2 \neq \sigma_x C_{4z}^{-} = \sigma_1 \ .$$

If all the elements of the group commute, the group is said to be Abelian. Each element α of the group has an inverse α^{-1} which possesses the properties

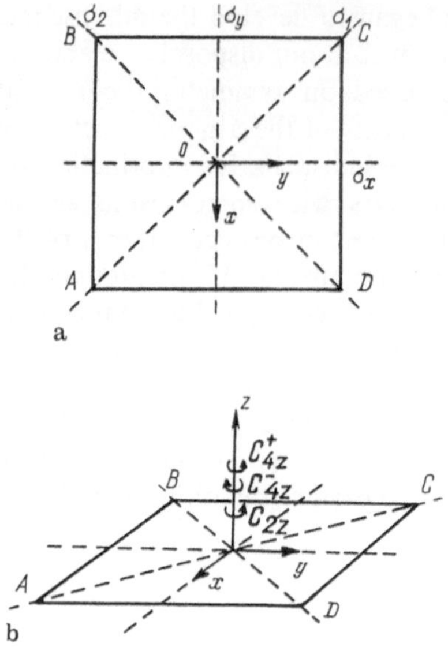

Fig. 1.6. Symmetry operations of the square: (x, y) plane—reflection lines σ_x, σ_y; σ_1, σ_2 (**a**); (x, y, z) space—rotation axes C_{4z}^+, C_{4z}^-, C_{2z} (**b**)

Table 1.1. Products of square group symmetry operations. The symmetry symbol in each box is equal to the product of the operations whose symbols stand first in the row and column that intersect in this box

E	C_{4z}^+	C_{2z}	C_{4z}^-	σ_x	σ_y	σ_1	σ_2
C_{4z}^+	C_{2z}	C_{4z}^-	E	σ_1	σ_2	σ_y	σ_x
C_{2z}	C_{4z}^-	E	C_{4z}^+	σ_y	σ_x	σ_2	σ_1
C_{4z}^-	E	C_{4z}^+	C_{2z}	σ_2	σ_1	σ_x	σ_y
σ_x	σ_2	σ_y	σ_1	E	C_{2z}	C_{4z}^-	C_{4z}^+
σ_y	σ_1	σ_x	σ_2	C_{2z}	E	C_{4z}^+	C_{4z}^-
σ_1	σ_x	σ_2	σ_y	C_{4z}^+	C_{4z}^-	E	C_{2z}
σ_2	σ_y	σ_1	σ_x	C_{4z}^-	C_{4z}^+	C_{2z}	E

$\alpha\alpha^{-1} = \alpha^{-1}\alpha = E$, i.e., performs an operation that destroys the effect of the element α. In the group of the square these are C_{4z}^+ and C_{4z}^-, etc. If an element A of the group is applied successively n times so that $A^n = E$, the integer n is called the order of element A. In the group of the square, as is seen, for example, from Table 1.1, the elements $C_{4z}^{\pm}$ are fourth order ($n = 4$), and the others are second order ($n = 2$). The elements α and β of the group belong to the same class, if in the group there is an element γ such that $\gamma\alpha\gamma^{-1} = \beta$. The elements α and β are called conjugate elements, and we therefore talk of a class of conjugate elements. The identity transformation always forms a separate class. The group of the square may be broken up into five classes C_i ($i = 1, \ldots, 5$):

$$E(C_1);\quad C_{2z}(C_2);\quad C_{4z}^+, C_{4z}^-(C_3)\ ;$$

$$\sigma_x, \sigma_y(C_4);\quad \sigma_1, \sigma_2(C_5)\ .$$

The number of elements in a class C_i, denoted by h_i, is called the order of the class. For the five classes ($r = 5$) of the group of the square we have $h_1 = h_2 = 1$ and $h_3 = h_4 = h_5 = 2$. Also, we define the coefficients of the product of the classes: $c_{ij,s}$ is the sth coefficient of the product $C_i C_j$ of the classes i and j. In other words, from Table 1.1 we have, for example,

$$C_i C_j = \sum_s c_{ij,s} C_s \tag{1.2.4}$$

or symbolically: $C_3 \cdot C_4 = 2C_5$ and $C_5 \cdot C_5 = 2C_1 + 2C_2$. As a result, we obtain the following nonzero coefficients:

$$\begin{array}{lllll}
c_{11,1} = 1 & c_{22,1} = 1 & c_{33,1} = 2 & c_{44,1} = 2 & c_{55,1} = 2 \\
c_{12,1} = 1 & c_{23\ 3} = 1 & c_{33,2} = 2 & c_{44,2} = 2 & c_{55,2} = 2 \\
c_{13,3} = 1 & c_{24,4} = 1 & c_{34,5} = 2 & c_{45,2} = 2 & \\
c_{14,4} = 1 & c_{25,5} = 1 & c_{35,4} = 2 & & \\
c_{15,5} = 1 & & & &
\end{array}$$

		C_4	
		σ_x	σ_y
C_3	C_{4z}^+	σ_1	σ_2
	C_{4z}^-	σ_2	σ_1

		C_5	
		σ_1	σ_2
C_5	σ_1	E	C_{2z}
	σ_2	C_{2z}	E

In the International System the above group of the square is referred to as 4 mm. The first index denotes a 4-fold axis that is perpendicular to the square, the second and third indices m the reflection operations with respect to straight lines of two types interrelated by the symmetry operation (σ_x, σ_y and σ_1, σ_2). A pictorial representation of reflection axes and planes (lines) of different symmetry is given in Fig. 1.7, which represents the point group of the square.

In group-theory applications, representing the operators of a group by matrices plays an important part (evidence for this is provided later in Chap. 4). We illustrate this with the example of a square. Let there be on its surface a vector $\boldsymbol{P}$ (Fig. 1.8) which starts from the center O with the coordinates of the end being x and y. When a transformation, for example, C_{4z}^+, is applied, the vector $\boldsymbol{P}$

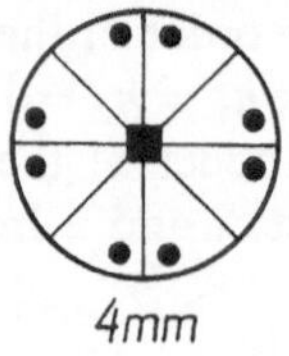

Fig. 1.7. The two-dimensional group of the square

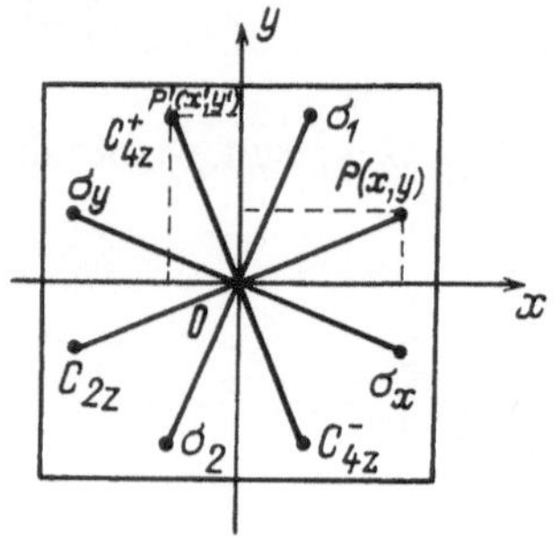

Fig. 1.8. Influence of square-group symmetry operations on point (vector) position $P(x, y)$

will be transformed to $\boldsymbol{P}'$, and the coordinates of its end, x' and y', are related to x and y by the equations

$$x' = 0 \cdot x - 1 \cdot y \ , \qquad y' = 1 \cdot x - 0 \cdot y \ ,$$

or, symbolically,

$$\boldsymbol{P}'(x'y') = \begin{vmatrix} 0 & -1 \\ 1 & 0 \end{vmatrix} \boldsymbol{P}(x, y) \ ,$$

where the matrix $\begin{vmatrix} 0 & -1 \\ 1 & 0 \end{vmatrix}$ represents the element C_{4z}^{+}. Constructing matrices for all square-group transformations is easy:

$$E \rightarrow \begin{vmatrix} 1 & 0 \\ 0 & 1 \end{vmatrix} , \quad C_{4z}^{+} \rightarrow \begin{vmatrix} 0 & -1 \\ 1 & 0 \end{vmatrix} , \quad C_{4z}^{-} \rightarrow \begin{vmatrix} 0 & 1 \\ -1 & 0 \end{vmatrix} ,$$

$$C_{2z} \rightarrow \begin{vmatrix} -1 & 0 \\ 0 & -1 \end{vmatrix} , \quad \sigma_x \rightarrow \begin{vmatrix} -1 & 0 \\ 0 & 1 \end{vmatrix} , \quad \sigma_y \rightarrow \begin{vmatrix} 1 & 0 \\ 0 & -1 \end{vmatrix} , \qquad (1.2.5)$$

$$\sigma_1 \rightarrow \begin{vmatrix} 0 & -1 \\ -1 & 0 \end{vmatrix} , \quad \sigma_2 \rightarrow \begin{vmatrix} 0 & 1 \\ 1 & 0 \end{vmatrix} .$$

These satisfy all the relations between the elements of the group of the square (Table 1.1) and are one of the possible representations of this group. An infinite number of such representations exist. An important representation is what is known as a regular representation. The matrix can be constructed by replacing any one element of Table 1.1 by unity and then replacing all other elements by zero, repeating this process for all the elements. This gives regular representation matrices corresponding to all the elements. Although these matrices are 8×8 and higher-order matrices [as compared with the 2×2 order in (1.2.5)], they are far from being the largest possible.

The infinite number of possible representations includes a particular type known as irreducible representations. Its matrices for the elements of a group have the same block-diagonal form. By means of a unitary transformation, every initial reducible matrix may be reduced to this block-diagonal form. Group theory proves that the number of possible irreducible representations is finite and is equal to the number of classes (r) of a group.

An element of the group G may be treated as a unitary operator $\hat{G}$, acting on functions φ_i ($i = 1, \ldots, f$):

$$\hat{G}\varphi_i = \sum_{k=1}^{f} G_{ki}\varphi_k \; , \tag{1.2.6}$$

which we can always take to be orthonormalized. In this case the unitary transformation matrix G_{ki} coincides with the operator matrix $\hat{G}$:

$$G_{ki} = \int \varphi_i^* \hat{G}\varphi_k \, d\tau \; .$$

Recall that the condition for the unitarity of G_{ki} matrices is

$$\sum_k G_{ki}^* G_{kj} = \sum_k G_{ik}^* G_{jk} = \delta_{ij}$$

[Ref. 1.12, Chap. 12). The functions $\varphi_1, \ldots, \varphi_f$ that define these matrices (the set of which gives the matrix representation for all the elements of the group) form the basis of the representation, their number $f = d_k$ defining the dimension of the representation d_k.

If, with the aid of the linear unitary transformation $\hat{S}$, we pass from the basis $\varphi_1, \ldots, \varphi_{d_k}$ to a new basis $\varphi'_1, \ldots, \varphi'_{d_k}$ with $\varphi'_i = \hat{S}\varphi_i$, then the latter gives a new representation of the same dimension: G'_{ik}. Such representations are called equivalent representations; there is no substantial difference between them. They are related to each other by the operator formula

$$\hat{G}' = \hat{S}^{-1}\hat{G}\hat{S} \; .$$

Consider an arbitrary group representation of dimension d_k. With the help of a linear transformation (1.2.6), the functions of the basis may be broken up into several sets of $d_{k1}, d_{k2}, \ldots$ functions ($d_{k1} + d_{k2} + \ldots = d_k$) such that when acted upon by the elements of the group the functions of each of these sets transform only through functions of the same set. Under these conditions the initial representation turns out to be reducible. If there does not exist a single linear transformation capable of decreasing the number of functions that transform through each other, then the representation carried out is irreducible. Thus, every reducible representation may be broken up into irreducible ones, which are of prime importance in all physical applications of group theory.

Of particular significance is the trace of the representation matrices, i.e., the sum of their diagonal elements, which is called the character. All matrices that represent elements of a given class possess equal characters (1.2.5). The characters of the matrices of equivalent representations coincide, too. By the characters we therefore distinguish equivalent representations from all others. The characters of the elements of the ith class of irreducible representation k are denoted by the symbol χ as follows:

$$\chi_i^k \to \chi_i^{(1)}, \chi_i^{(2)}, \dots, \chi_i^{(r)} .$$

The unit element of group E is always represented by a unit matrix whose character $\chi(E)$ is equal to the dimension of the representation d_k.

If it is necessary to find irreducible representations, there is no special need to determine the corresponding block matrices. It is sufficient to know the general orthogonality relations between the characters of the irreducible representations. These relations are derived from group theory (for detailed information see any monograph on the theory of groups, for example, [1.10, 1.13]) and have the form

$$h_i h_j \chi_i^k \chi_j^k = d_k \sum_{s=1}^{r} c_{ij,s} h_s \chi_s^k , \tag{1.2.7}$$

$$\sum_{s=1}^{r} h_s (\chi_s^k)^* \chi_s^l = g\delta_{kl} , \tag{1.2.8}$$

where g is the number of elements in a group (its order and the other symbols were introduced earlier). As an example, we derive the character table for the square-group. For this purpose we use (1.2.7), substituting there the known orders of the classes h_i and the coefficients $c_{ij,s}$ from (1.2.4). Using $c_{11,1}$ we find $\chi_1^k \chi_1^k = d_k \chi_1^k$ or $\chi_1^k = d_k$; using $c_{22,1}$ we find $\chi_2^k \chi_2^k = d_k \chi_2^k = d_k^2$ or $\chi_2^k = \pm d_k$; using $c_{33,1}$ and $c_{35,2}$ we find $4\chi_3^k \chi_3^k = d_k(2\chi_1^k + 2\chi_2^k)$ or $\chi_3^k = \pm d_k$, if $\chi_2^k = d_k$ and $\chi_3^k = 0$, if $\chi_2^k = -d_k$; using $c_{44,1}$ and $c_{44,2}$ we find in the same way $\chi_4^k = \pm d_k$, if $\chi_2^k = d_k$ and $\chi_4^k = 0$, if $\chi_2^k = -d_k$; and, finally, using $c_{34,5}$ we find $4\chi_3^k \chi_4^k = d_k 4\chi_5^k$, $\chi_5^k = \chi_3^k \chi_4^k / d_k$.

As a result, we obtain a table of characters for five classes and five irreducible representations of the group of the square (Table 1.2). Thus, the characters have been found in terms of the dimension of irreducible representations. Next, using

Table 1.2. Characters χ for five classes C_i^1. The five irreducible representations of the group of the square are specified as a function of the dimensions d_k of those representations

	C_1	C_2	C_3	C_4	C_5
χ^1	d_k	d_k	d_k	d_k	d_k
χ^2	d_k	d_k	d_k	$-d_k$	$-d_k$
χ^3	d_k	d_k	$-d_k$	d_k	$-d_k$
χ^4	d_k	d_k	$-d_k$	$-d_k$	d_k
χ^5	d_k	$-d_k$	0	0	0

the fact that the characters are real numbers, we obtain from (1.2.8) the following expression for the group of the square ($g = 8$):

$$\sum_{s=1}^{r} h_s(\chi_s^k)^2 = 8 \ .$$

Hence, we immediately find $d_k = 1$ for the first four irreducible representations and $d_k = 2$ for the fifth representation (Table 1.3).

Table 1.3. Characters for the five classes of the group of the square. The table specifies the first four irreducible representations $d_k = 1$: A_1, A_2, B_1, B_2 and the last fifth, $d_k = 2$: E

	$\{E\}$	$\{C_{2z}\}$	$\{C_{4z}^+ C_{4z}^-\}$	$\{\sigma_x \sigma_y\}$	$\{\sigma_1 \sigma_2\}$
A_1	1	1	1	1	1
A_2	1	1	1	-1	-1
B_1	1	1	-1	1	-1
B_2	1	1	-1	-1	1
E	2	-2	0	0	0

We have considered the symmetry of a two-dimensional entity. Real crystals are three-dimensional. (It can be proved that there are no ordered systems in which the density $\varrho(x)$ depends only on one coordinate [Ref. 1.4, Sect. 163], but two-dimensional systems with $\varrho(x, y)$ can, in principle, exist.) An ideal three-dimensional crystal is an infinitely extended solid which is constructed by infinite replication in space of the same structural element known as the unit cell. Finite crystals are imperfect due to the presence of boundary surfaces. Apart from this, as a rule, they also possess internal structural defects that arise during the crystallization process. Ignoring this, we concentrate here on infinite perfect three-dimensional crystals.

The example of the square (unit cell of a plane square lattice) shows that typical characteristics of a crystal are its symmetry properties. An important intrinsic feature of all ideal crystals is their translational symmetry, i.e., the property of the crystal to map wholly into itself when translated in some direction by certain translation vectors. The minimum value of such a vector gives the periodicity in a given direction. The translation periods may turn out to be equal to the distances between the nearest neighboring sites of the crystal. The translation group is given by a set of translation vectors

$$\boldsymbol{R}_m = m_1 \boldsymbol{a}_1 + m_2 \boldsymbol{a}_2 + m_3 \boldsymbol{a}_3 \,, \tag{1.2.9}$$

where $m_i = \pm 1, \pm 2, \ldots$ are positive and negative integers, including zero, and $i = 1, 2, 3$; $\boldsymbol{a}_1, \boldsymbol{a}_2, \boldsymbol{a}_3$ are noncoplanar primitive vectors that define the primitive lattice periods. The ends of the vectors (1.2.9), drawn from an arbitrary initial lattice site, give the positions of other lattice sites, which are defined by sets of three integers m_1, m_2, m_3 or, more briefly, m.

A parallelepiped constructed on vectors $\boldsymbol{a}_1$ is called a unit (primitive) cell. Its volume is equal to $\boldsymbol{a}_1 \cdot (\boldsymbol{a}_2 \times \boldsymbol{a}_3)$. Thus, a lattice in its entirety may be built of a set of identical unit cells, just as a wall can be built of bricks. The choice of the vectors $\boldsymbol{a}_i$ and the cell constructed on them is ambiguous, being dictated by convenience in the description of a given crystal lattice. (This ambiguity is not dangerous subject to the condition that a particular choice permits location of all lattice sites. It is convenient to take the vectors $\boldsymbol{a}_i$ to be the unit vectors of the system of coordinates that is connected with the crystal symmetry axes.) The choice is often made in such a way that the lattice sites are only at the vertices of a cell (primitive cell). Since each site at the vertex of the cell pertains simultaneously to eight adjacent cells, each of them has one site, which is defined by (1.2.9). But if, in addition, there are atoms on the edges and faces of the cell or inside it, the vectors (1.2.9) no longer suffice to determine their position and it is necessary to specify the vectors, defining the position of these "nonvertex" sites, which specify the basis of the lattice. If there are σ sites in the cell that do not lie at the vertices, their position may be defined by the vectors

$$\boldsymbol{R}_s = \xi_1^s \boldsymbol{a}_1 + \xi_2^s \boldsymbol{a}_2 + \xi_3^s \boldsymbol{a}_3 \,, \tag{1.2.10}$$

where $s = 1, 2, \ldots \sigma$; $0 \leqslant \xi_{(i)}^s \leqslant 1$ $(i = 1, 2, 3)$ are the coordinates of an sth lattice site in the cell, counted off from one of the vertices. The position of a site in a lattice with a basis is given by the sum of the vectors (1.2.9, 10):

$$\boldsymbol{R}_{ms} = \boldsymbol{R}_m + \boldsymbol{R}_s \,. \tag{1.2.11}$$

Hence two methods of constructing an atomic lattice follow: We begin by constructing a basis of σ atoms with the help of (1.2.10) and then translate it by means of (1.2.9); using (1.2.9), we first construct a lattice for each site σ of the basis and then insert them into one another by displacing them by (1.2.10).

Like atoms (equivalent sites) reside at the vertices of the unit cells. Each of the atoms can be moved into any other by translation by one of the lattice periods. A set of equivalent sites generates a Bravais lattice of a crystal. It stands to reason that this lattice does not incorporate all lattice sites or even all equivalent sites, since there may also exist equivalent sites that can be moved into each other only under symmetry transformations such as rotation and reflection. Generally speaking, a real crystal can be made up of more than one Bravais lattice. One may be inserted into another, with each referring to a particular atomic species and arrangement.

Crystallography enables us to establish all the possible Bravais lattice symmetry types. The above translation operations do not affect crystals externally. But there are symmetry properties, also mentioned above, that are liable to manifest themselves in the external shape of the crystal, e.g., in the regular relative position of its external faces. A definite set of symmetry transformations can be referred to each type of crystal. These are symmetry transformations that are not translation operations but include only those operations associated with the rotation of the crystal about its symmetry axes and with reflection in its planes. In group theory, a set of such crystal symmetry elements is known as a point group. An example of such a group for a plane square lattice was examined in the foregoing. We saw that all the point-group symmetry elements mentioned there leave the point O, i.e., the center of the square, fixed (Fig. 1.6). This is what explains the term "point group." Adding translations to a point group and also allowing for combinations of translations (even not by complete periods) with rotations and reflections (screw axes, glide-reflection planes), we come to the concept of the crystal symmetry space group.

A theorem exists that establishes possible crystal symmetry axes on the basis of constraints imposed on the point group by the requirement of translational invariance. Let point A (Fig. 1.9) be a Bravais lattice site through which a symmetry axis passes perpendicular to the plane of the drawing. If a similar axis passes through point B, then the latter is separated from A by a translation period d. If we rotate the lattice about the axis passing through A by an angle $\alpha = 2\pi/n$ (where n is the multiplicity of the axis—we denote the rotation operation by C_n according to *Schöenflies* [1.9], and by n according to the International System), the point B and the axis that passes through it will be carried into point B'. A similar rotation about B will transfer A with the C_n axis to point A'. Both new points A' and B' can, by symmetry, be superposed by translation along the straight line $A'B'$. If d is the minimum period in a specified

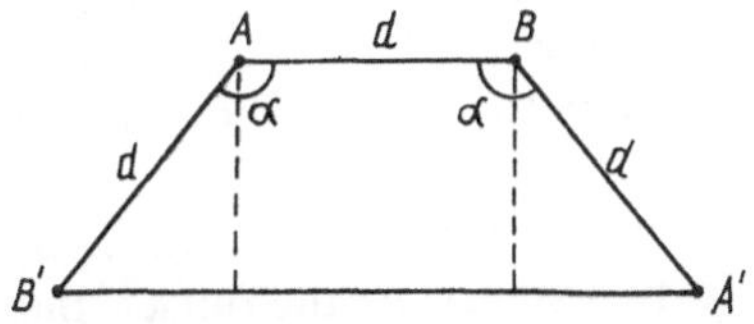

Fig. 1.9. Determination of possible crystal symmetry axes

direction (by construction, AB is parallel to $A'B'$), the distance $A'B'$ will be a multiple of the length d, i.e., $A'B' = kd$, with k being an integer. It follows from Fig. 1.9 that

$$d + 2d\sin(\alpha - \pi/2) = d - 2d\cos\alpha = kd \ ,$$

whence $\cos\alpha = (1-k)/2$. From the condition $0 \leqslant |\cos\alpha| \leqslant 1$ we have $k = -1$, 0, 1, 2, 3. Hence we find the unique values of the angles of rotation about the symmetry axes C_n: $2\pi/n = 60°, 90°, 120°, 180°$, and $360° = 0°$, i.e., $n = 6, 4, 3, 2, 1$. Thus, only five types of symmetry axes are permissible. In each site a limited number of rotation axes may intersect. A consideration similar to the above treatment [1.11] gives six possible combinations of rotation axes: 2 2 2, 2 2 3, 2 2 4, 2 2 6, 2 3 3, and 2 3 4. A possible-spatial arrangement thereof is illustrated in Fig. 1.10.

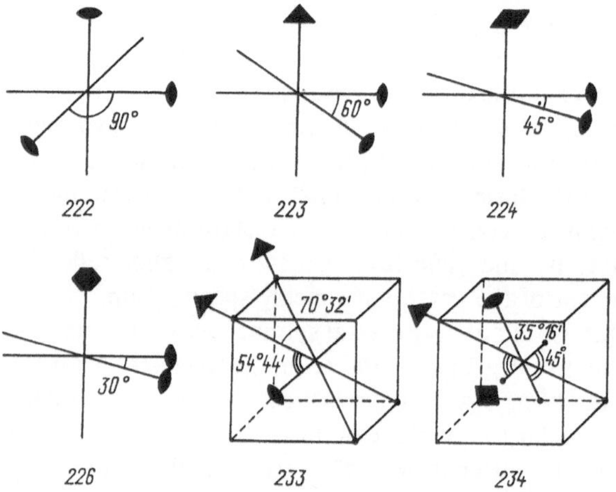

Fig. 1.10. Six possible crystal symmetry axis types

The translations and rotations are called the proper symmetry operations. There also exist improper symmetry operations: (1) inversion—a reflection at a point (center of inversion C_i or $\bar{1}$); (2) reflection in the plane (reflection plane m); (3) inversion rotation—a product of an inversion and rotation, which may be only five in number, with the symbols 1, 2 (≡ m), 3, 4, 6. If there is a symmetry axis n, it may be parallel to the plane (vertical) or perpendicular to the plane (horizontal). These axes are denoted by nm and n/m, respectively. For example, 4m denotes a 4-fold axis and a parallel (vertical) symmetry plane m that passes through it; $6/m$ stands for a 6-fold axis and a plane perpendicular (horizontal) to it.

Crystallographic methods permit the determination of the total number of allowed crystal lattices. The counting is done in two stages. At the outset, the possible point-symmetry types are determined. These are divided into seven crystal systems, with a total of 32 crystal classes or groups (Hessel, 1830). The crystal point groups are listed in Table 1.4, which specifies the numbers of Bravais lattices, their symmetry, and the major characteristics of the cells for each crystal system. The locations of the unit cell axes ***a***, ***b***, and ***c*** and the angles between them (α, β, and γ) are depicted in Fig. 1.11. The 14 Bravais lattice types themselves, are portrayed in Fig. 1.12. Note that the Bravais unit cells do not

Table 1.4. Point groups of crystal lattices

Crystal system	Number of elements	Crystal classes: Schönflies notation	Crystal classes: International system (SI)	Number of Bravais lattices	Characteristics of unit cell
Triclinic	1	C_1	1	1	$a \neq b \neq c$
	2	$C_i(S_2)$	$\bar{1}$		$\alpha \neq \beta \neq \gamma$
Monoclinic	2	C_2	2	2	$a \neq b \neq c$
	2	$C_s(C_{1h})$	m		
	4	C_{2h}	$2/m$		$\alpha = \gamma = \pi/2 \neq \beta$
Rhombic	4	$D_2(V)$	2 2 2	4	$a \neq b \neq c$
	4	C_{2v}	$m\ m\ 2$		
	8	$D_{2h}(V_h)$	$m/2\ m/2\ m/2$		$\alpha = \beta = \gamma = \pi/2$
Tetragonal	4	C_4	4	2	
	4	S_4	$\bar{4}$		$a = b \neq c$
	8	C_{4h}	$4/m$		
	8	D_4	4 2 2		
	8	C_{4v}	$4\ m\ m$		$\alpha = \beta = \gamma = \pi/2$
	8	$D_{2d}(V_d)$	$\bar{4}\ 2\ m$		
	16	D_{4h}	$4/m\ 2/m\ 2/m$		
Trigonal	3	C_3	3	1	$a = b = c$
	6	$C_{3i}(S_6)$	$\bar{3}$		
	6	D_3	3 2		$\alpha = \beta = \gamma$
	6	C_{3v}	$3\ m$		$< 120°$, $\neq \pi/2$
	12	D_{3d}	$\bar{3}\ 2/m$		
Hexagonal	6	C_6	6	1	$a = b \neq c$
	6	C_{3h}	$\bar{6}$		
	12	C_{6h}	$6/m$		$\alpha = \beta = \pi/2$
	12	D_6	6 2 2		$\gamma = 120°$
	12	C_{6v}	$6\ m\ m$		
	12	D_{3h}	$\bar{6}\ m\ 2$		
	24	D_{6h}	$6/m\ 2/m\ 2/m$		
Cubic	12	T	2 3	3	$a = b = c$
	24	T_h	$2/m\ 3$		
	24	O	4 3 2		$\alpha = \beta = \gamma = \pi/2$
	24	T_d	$\bar{4}\ 3\ m$		
	48	O_h	$4/m\ \bar{3}\ 2/m$		

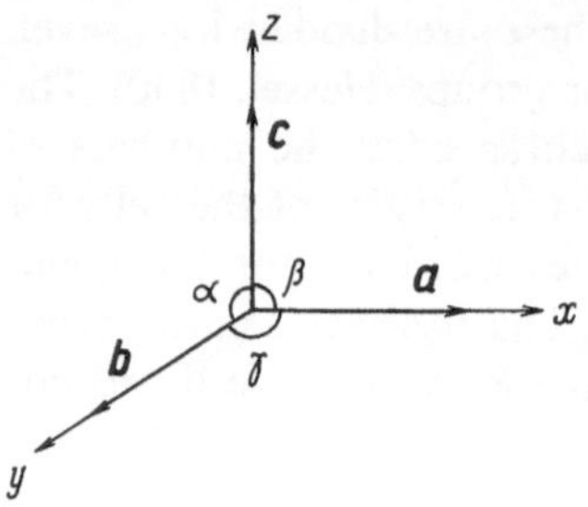

Fig. 1.11. Crystal axes ***a***, ***b***, ***c*** and angles α, β, γ between them

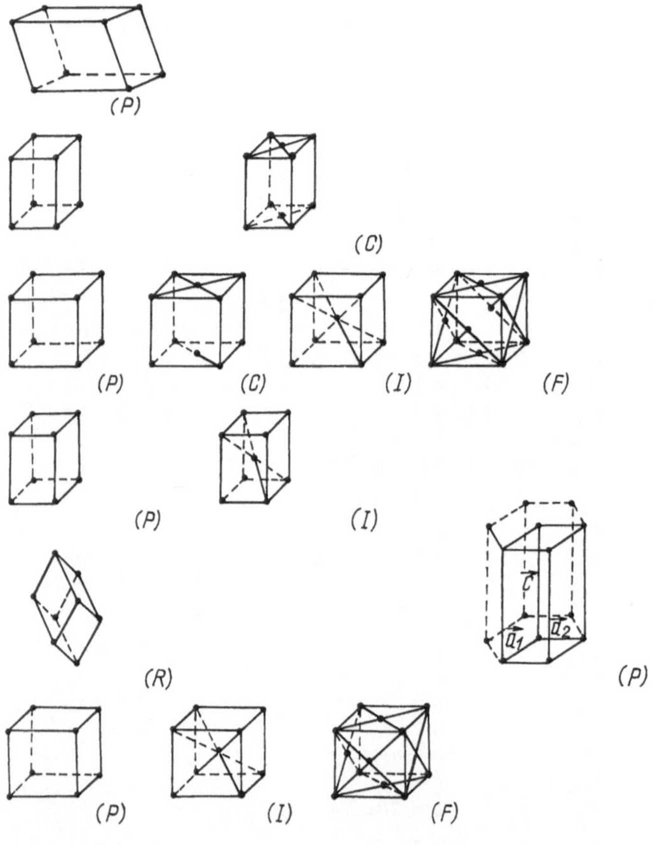

Fig. 1.12. Fourteen Bravais lattices of different crystal systems with specification of lattice symbol: *P* is a primitive cell, *C* is a base-centered cell, *I* is a body-centered cell, *F* is a face-centered cell, and *R* is a trigonal (primitive) cell. Crystal systems: triclinic (upper row), monoclinic (second row), rhombic (third row), tetragonal (fourth row), trigonal on the left and hexagonal on the right (fifth row), cubic (sixth row)

coincide with the primitive cell (P) in all the cases; a cell which is a multiple of the primitive cell often demonstrates the crystal symmetry elements more straightforwardly (Figs 1.14, 15 below).

In Table 1.4 the Schönflies notation and the International System are used for the point groups. We clarify them here (partly as a reminder): $C_1 (S_2)$ and $\bar{1}$ is an inversion operation with respect to the center of symmetry; C_n ($n = 1, 2, 3, 4, 6$ and $\bar{1}, \bar{2}, \bar{3}, \bar{4}, \bar{6}$) is an n-fold rotation axis; C_{nh} ($n = 2, 3, 4, 6$), $C_s = C_{1h}$ and $2/m, \bar{6}, 4/m, 6/m, m$ is an n-fold axis with a symmetry plane perpendicular to it; C_{nv} ($n = 2, 3, 4, 6$) and $mm2$, $3m$, $4mm$, $6mm$ stand for n-fold axes and symmetry planes that pass through them; S_n ($n = 2, 4, 6$) and $\bar{1}, \bar{4}, \bar{3}$ is an n-fold rotoflection axis; D_n ($n = 2, 3, 4, 6$) ($D_2 = V$) $-$ 2 2 2, 3 2 2, 4 2 2, 6 2 2 stands for an n-fold axis and a family of 2-fold axes perpendicular to it; D_{nh} ($n = 2, 3, 4, 6$) ($D_{2h} = V_h$) and $2/m\ 2/m\ 2/m$, $\bar{6}m2$, $4/m\ 2/m\ 2/m$, $6/m\ 2/m\ 2/m$ is the same as D_n, with the addition of symmetry planes passing through the 2-fold axis; D_{nd} ($n = 2, 3$) ($D_{2d} = V_d$) and $\bar{4}2m$, $\bar{3}2/m$ is the same as D_n, with the addition of vertical symmetry planes passing through the n-fold axis in the middle between every two adjacent horizontal 2-fold axes.

The groups T and 2 3 are formed by adding to the system of group D_n axes four inclined 3-fold axes, rotations about these axes carrying the 2-fold axes into each other. The group T_d or $\bar{4}\ 3\,m$ is formed from T by adding symmetry planes, each of which passes through one 2-fold axis and two 3-fold axes. The group T_h is formed from T by adding three mutually perpendicular symmetry planes passing through every two 2-fold axes, and the 3-fold axes become rotoflection axes; O and 4 3 2 is a system of cube symmetry axes. In the group O_h and $4/m\ \bar{3}2/m$ symmetry planes are added to O.

In determining the classes of point groups, the following procedure is normally used. If there is some point group G and, besides, a group C_i (consisting of two elements: a unit element E and an inversion element I), then the direct product of these groups, $G \times C_i$, is also a group which has twice as many elements as the G group has. One half of the elements of the new group coincide with those of the G group, and the other half consists of G elements multiplied by the inversion I. The number of classes in the group also doubles. To each class A of the G group the class AI is added. Table 1.4 shows that the number of groups of the 32 classes may be reduced to 12 types: C_n, S_{2n}, C_{nh}, C_{nv}, D_n, D_{nd}, T, T_d, $T_h = T \times C_i$, O, and $O_h = O \times C_i$. Apart from this, there also exist icosahedron point groups Y and $Y_h = Y \times C_i$, which occur in nature as groups of molecules in exceptional cases.

Each of the 32 point groups has a definite number of symmetry elements (1, 2, 3, 4, 6, 8, 12, 24, and 48). And in each class (Table 1.4) the last symmetry element for a given crystal system contains all elements of the corresponding system. These final symmetry elements are called holohedral elements. As an example, Fig. 1.13 presents elements of the group O_h or $4/m\ \bar{3}\ 2/m$ of a cube. To start with, we have here a system of cubic axes: three 4-fold axes passing through the centers of the opposite faces (C_4), four 3-fold axes passing through the

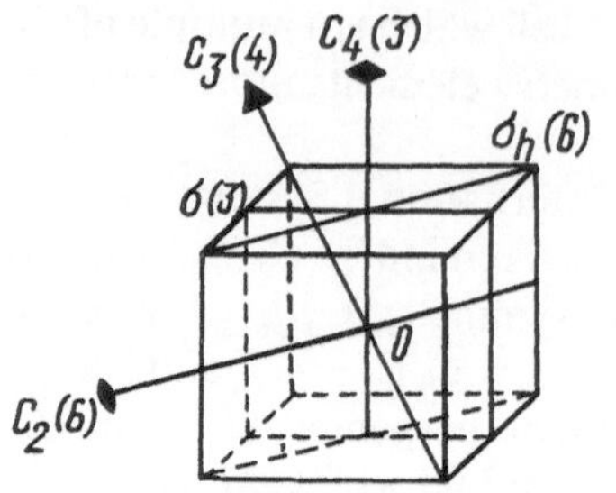

Fig. 1.13. Point-symmetry (group O_h) elements of the cube: axes C_2, C_3, and C_4; inversion center O_v; reflection planes σ_h and σ_v (the total number of symmetry elements is indicated in brackets)

opposite vertices (C_3), and six 2-fold axes passing through the middles of the opposite edges (C_2). This group is O or 4 3 2. It contains 24 elements and 5 classes: a unit element E, eight rotations C_3 and C_3^2, six rotations C_4 and C_4^3, three rotations C_4^2 and, finally, six rotations C_2. When we pass from the O group to the group of all cube-symmetry transformations, O, the center of symmetry is added. This transforms the 3-fold axes of the O group to 6-fold rotoflection axes (body diagonals of the cube) and, besides, leads to the occurrence of six more symmetry planes that pass through each pair of opposing edges and to the occurrence of three planes parallel to the cube faces (σ_h and σ_v in Fig. 1.13). There is a total of 48 elements and 10 classes in the O_h group. To the 24 elements and 5 classes of the O group we have added 24 elements in 5 classes which are made up of an inversion I, eight rotoflection transformations S_6 and S_6^5, six rotoflection transformations $C_4\sigma_h$, $C_4^3\sigma_h$ about the 4-fold axes, three reflections σ_h in the planes horizontal with respect to the 4-fold axes, and six reflections σ_v in the planes vertical with respect to these axes.

The rest of the 30 point groups may be treated in a similar fashion. In Table 1.5 we restrict our attention to the geometric and atomic characteristics of sc, bcc, and fcc lattices. Figure 1.14 furnishes Bravais unit cells and ordinary primitive cells for these crystals; Fig. 1.15 a supplies analogous data for the case

Table 1.5. Major geometric and atomic characteristics for three cubic crystal lattice types

Major characteristics	Type of cubic lattice		
	sc(P)	bcc(I)	fcc(F)
Bravais unit-cell volume	a^3	a^3	a^3
Number of lattice sites per cell (basis)	1	2	4
Number of sites per unit volume	a^{-3}	$2a^{-3}$	$4a^{-3}$
Nearest-neighbor distance	a	$\sqrt{3}a/2$	$a/\sqrt{2}$
Number of nearest neighbors (in the first coordination zone)	6	8	12
Next-nearest-neighbor distance	$\sqrt{2}a$	a	a
Number of neighbors in the second coordination zone	12	6	6

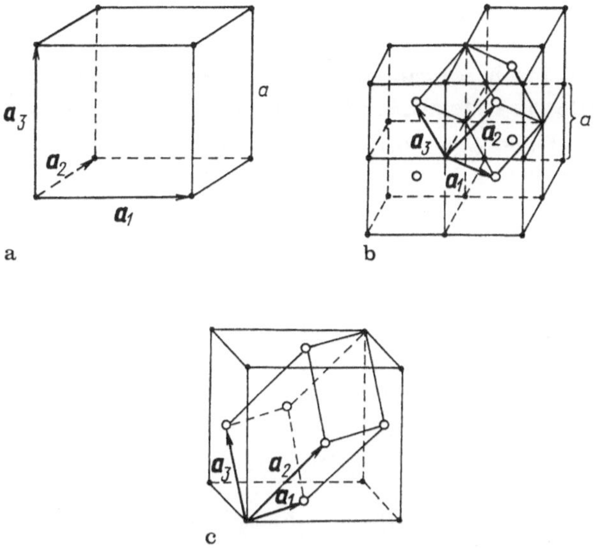

Fig. 1.14. Primitive translation vectors $\boldsymbol{a}_1$, $\boldsymbol{a}_2$, $\boldsymbol{a}_3$ and primitive (one cubic and two rhombohedral) Bravais cells for three cubic lattices: sc (**a**); bcc (circlets indicate sites at the center of the elementary cube) (**b**); fcc (circlets indicate sites at the center of the faces of the elementary cube) (**c**)

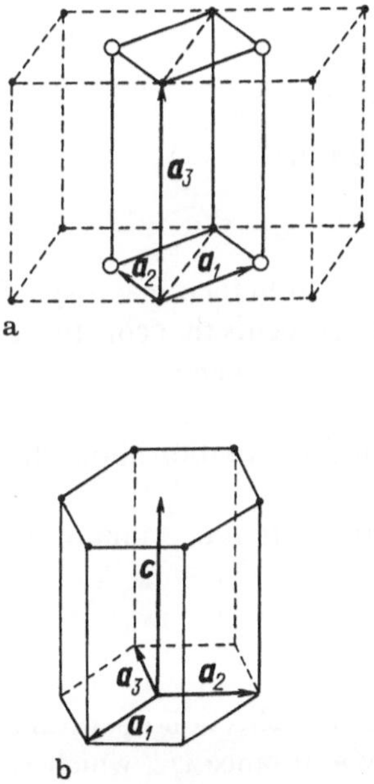

Fig. 1.15. Primitive translation vectors $\boldsymbol{a}_1$, $\boldsymbol{a}_2$, $\boldsymbol{a}_3$ and primitive *C*-type cell of the monoclinic system (**a**); a hexagonal crystal described by a coordinate system of four axes $\boldsymbol{a}_1$, $\boldsymbol{a}_2$, $\boldsymbol{a}_3$, and $\boldsymbol{c}$ (**b**)

of crystals of a monoclinic system such as C; and Fig. 1.15b presents similar data for a hexagonal lattice.

As seen from Figs. 1.14, 15a, b, we can conveniently express the primitive vectors $\boldsymbol{a}_i$ in terms of a Cartesian system x, y, z (with the unit vectors $\boldsymbol{e}_1, \boldsymbol{e}_2, \boldsymbol{e}_3$), of which the components are equal to a_{ix}, a_{iy}, a_{iz}. For each lattice there is a set of nine such components that generate a 3×3 matrix A: (1.2.9) then assumes the form

$$\boldsymbol{R} = \begin{pmatrix} a_{1x} & a_{2x} & a_{3x} \\ a_{1y} & a_{2y} & a_{3y} \\ a_{1z} & a_{2z} & a_{3z} \end{pmatrix} \begin{pmatrix} m_1 \\ m_2 \\ m_3 \end{pmatrix} = \boldsymbol{A}\boldsymbol{m} \ .$$

Accordingly, for cubic and hexagonal lattices,

$$\boldsymbol{A}_{\rm sc} = \begin{pmatrix} a & 0 & 0 \\ 0 & a & 0 \\ 0 & 0 & a \end{pmatrix} , \quad \text{i.e.,} \quad \boldsymbol{a}_i = a\boldsymbol{e}_i \ ,$$

$$\boldsymbol{A}_{\rm bcc} = \begin{pmatrix} -a/2 & a/2 & a/2 \\ a/2 & -a/2 & a/2 \\ a/2 & a/2 & -a/2 \end{pmatrix} , \quad \begin{aligned} \boldsymbol{a}_1 &= (a/2)(-\boldsymbol{e}_1 + \boldsymbol{e}_2 + \boldsymbol{e}_3) \ , \\ \boldsymbol{a}_2 &= (a/2)(\boldsymbol{e}_1 - \boldsymbol{e}_2 + \boldsymbol{e}_3) \ , \\ \boldsymbol{a}_3 &= (a/2)(\boldsymbol{e}_1 + \boldsymbol{e}_2 - \boldsymbol{e}_3) \ , \end{aligned}$$

$$\boldsymbol{A}_{\rm fcc} = \begin{pmatrix} 0 & a/2 & a/2 \\ a/2 & 0 & a/2 \\ a/2 & a/2 & 0 \end{pmatrix} , \quad \begin{aligned} \boldsymbol{a}_1 &= (a/2)(\boldsymbol{e}_1 + \boldsymbol{e}_2) \ , \\ \boldsymbol{a}_2 &= (a/2)(\boldsymbol{e}_2 + \boldsymbol{e}_3) \ , \\ \boldsymbol{a}_3 &= (a/2)(\boldsymbol{e}_3 + \boldsymbol{e}_1) \ , \end{aligned}$$

$$\boldsymbol{A}_{\rm hcp} = \begin{pmatrix} a & -a/2 & 0 \\ 0 & a\sqrt{3}/2 & 0 \\ 0 & 0 & c \end{pmatrix} , \quad \begin{aligned} \boldsymbol{a}_1 &= (a/2)\boldsymbol{e}_1 + (a\sqrt{3}/2)\boldsymbol{e}_2 \ , \\ \boldsymbol{a}_2 &= -(a/2)\boldsymbol{e}_1 + (a\sqrt{3}/2)\boldsymbol{e}_2 \ , \\ \boldsymbol{a}_3 &= c\boldsymbol{e}_3 \ . \end{aligned}$$

The true microscopic symmetry of crystals, as distinct from their macrosymmetry considered above, is defined by a set of symmetry elements that constitute its space group. In addition to the familiar point-group elements, the space groups comprises screw axes and glide planes.

An n-fold screw axis is a symmetry element that includes a rotation about the n-fold axis and a displacement parallel to that axis.

The displacement is an integral multiple of the n-th fraction of a minimum translation τ in a given direction, i.e., by

$$\alpha\tau/n \ (n = 2, 3, 4, 6; \alpha = 1, 2, \ldots, (n-1)) \ .$$

The glide plane is a symmetry element that includes a reflection operation and a displacement parallel to the reflection plane by a distance $\tau/2$, which is equal to the half-period of an ordinary minimum translation in a given direction. These two constituent symmetry operations are schematically represented in Figs. 1.16, 17. Tables 1.6, 7 summarize the most important graphic symbols of the principal axis and plane symmetry operations according to the International System.

Table 1.6. Designation of some of the most important symmetry elements

Name of symmetry element	International system (SI)	Graphic representation according to SI: vertical	Graphic representation according to SI: horizontal
Vertical symmetry plane	*m*		
Horizontal symmetry plane	/*m*		(this symbol is placed in the right-hand upper corner in SI drawings)
Rotation axes of			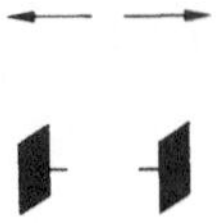
1-fold symmetry	1		
2-fold symmetry	2		
3-fold symmetry	3		
4-fold symmetry	4		
6-fold symmetry	6		
Inversion axes of			
1-fold symmetry, center of inversion	$\bar{1}$		
2-fold symmetry, symmetry plane	$\bar{2}$		
3-fold symmetry	$\bar{3}$		
4-fold symmetry	$\bar{4}$		
6-fold symmetry	$\bar{6}$		

Table 1.7. Designation of screw axes

Screw axes	Value of the translation of one component	Designation of a screw axis	Graphic representation according to the International system (SI): Vertical screw axes	Vertical screw axes + center of inversion	Horizontal screw axes
2	$\frac{1}{2}\tau$	2_1			
3	$\frac{1}{3}\tau$	3_1			
	$-\frac{2}{3}\tau$	3_2			

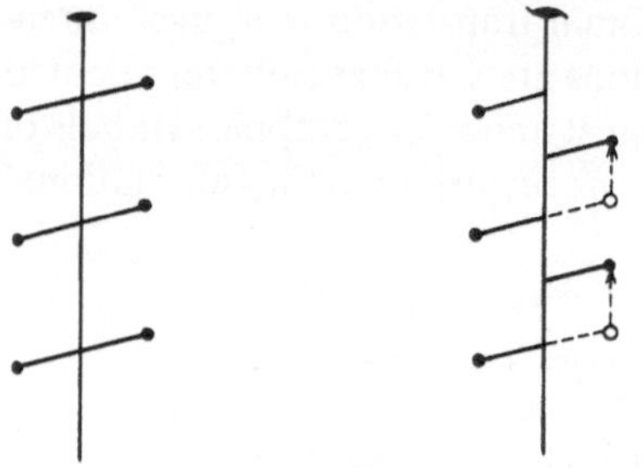

a b

Fig. 1.16. Usual two-fold axis 2 (**a**); double screw axis 2_1 (**b**)

a b

Fig. 1.17. Usual reflection plane *m* (**a**); glide reflection plane *a* (along the *a* axis) (**b**)

Allowing tor all the translation operations, one finds that to each of the 32 crystalline classes there corresponds one of the 28 space groups; the total number of groups is equal to 230 [1.8], [1.9]. For a derivation of these groups, we refer to special literature [1.13].

1.3 Reciprocal Lattice

To describe the physical characteristics of crystals, it is convenient to use what is known as the reciprocal lattice, a concept first introduced by J.W. Gibbs. All these characteristics have the same period as the crystal lattice, i.e., for each quantity $A(\boldsymbol{r})$ we have

$$A(\boldsymbol{r}+\boldsymbol{R}_m)=A(\boldsymbol{r}) \ , \tag{1.3.1}$$

where the vector $\boldsymbol{R}_m$ is given by (1.2.9). The periodic function $A(\boldsymbol{r})$ may be expanded in a Fourier series

$$A(\boldsymbol{r})=\sum_{b_g} A_{b_g} \exp(2\pi \mathrm{i}\, \boldsymbol{b}_g \cdot \boldsymbol{r}) \ , \tag{1.3.2}$$

where the sum is taken over all possible values of the vector $\boldsymbol{b}_g$ that satisfy the periodicity condition (1.3.1), when the replacement of $\boldsymbol{r}$ by $\boldsymbol{r}+\boldsymbol{R}_m$ does not alter the exponential cofactors in (1.3.1). The condition hence follows that the scalar product

$$(\boldsymbol{b}_g\cdot\boldsymbol{R}_m)=m_1\boldsymbol{b}_g\cdot\boldsymbol{a}_1+m_2\boldsymbol{b}_g\cdot\boldsymbol{a}_2+m_3\boldsymbol{b}_g\cdot\boldsymbol{a}_3$$

always be an integer, i.e., that the three products

$$\boldsymbol{b}_g\cdot\boldsymbol{a}_1=g_1;\quad \boldsymbol{b}_g\cdot\boldsymbol{a}_2=g_2;\quad \boldsymbol{b}_g\cdot\boldsymbol{a}_3=g_3$$

be integers (positive or negative, including zero) g_1, g_2, g_3. This can be obtained if we introduce the reciprocal-lattice vector

$$\boldsymbol{b}_g=g_1\boldsymbol{b}_1+g_2\boldsymbol{b}_2+g_3\boldsymbol{b}_3\ , \tag{1.3.3}$$

where the noncoplanar vectors $\boldsymbol{b}_i$ $(i=1,2,3)$ are related to the direct-lattice vectors by the equations

$$\begin{aligned}\boldsymbol{b}_1&=\bar{\omega}^{-1}(\boldsymbol{a}_2\times\boldsymbol{a}_3)\ ,\quad &\boldsymbol{b}_2&=\bar{\omega}^{-1}(\boldsymbol{a}_3\times\boldsymbol{a}_1)\ ,\\ \boldsymbol{b}_3&=\bar{\omega}^{-1}(\boldsymbol{a}_1\times\boldsymbol{a}_2)\ ,\quad &\bar{\omega}&=\boldsymbol{a}_i(\boldsymbol{a}_2\times\boldsymbol{a}_3)\ ,\end{aligned} \tag{1.3.4}$$

and $\bar{\omega}$ is the unit-cell volume. It follows from (1.3.4) that

$$\delta_{ij}=\boldsymbol{a}_i\cdot\boldsymbol{b}_j=\begin{cases}0, & \text{if } i\neq j\\ 1, & \text{if } i=j,\ i,j=1,2,3.\end{cases} \tag{1.3.5}$$

Inspection of (1.3.3–5) indicates that the summation in (1.3.2) is carried out over all possible values of the integers g_1, g_2, g_3. Since the vector products $(\boldsymbol{a}_1\times\boldsymbol{a}_2)$ etc. yield the areas of the corresponding unit-cell faces, it follows from (1.3.4) that the dimension of the vectors $\boldsymbol{b}_i$ is the reciprocal length and they are equal in magnitude to the reciprocal values of the heights of the parallelepiped in the direct-lattice unit cell. The vectors $\boldsymbol{b}_i$ (1.3.3) may be viewed as primitive periods of a reciprocal lattice. It can be readily verified that the reciprocal lattices of 1-fold primitive Bravais lattices are 1-fold primitive lattices of the same system. In particular, the fcc and bcc lattices "interchange" during transformation from the direct to the reciprocal lattice. The reciprocal lattices of base-centered structures preserve the same type. It will be left as an exercise for the reader to verify this statement and also to prove that the unit-cell volume $\bar{\omega}'$ for a reciprocal lattice is equal to the reciprocal value of the direct-lattice cell volume, i.e.,

$$\bar{\omega}'=\boldsymbol{b}_1\cdot(\boldsymbol{b}_2\times\boldsymbol{b}_3)=(\boldsymbol{a}_i\cdot(\boldsymbol{a}_2\times\boldsymbol{a}_3))^{-1}=\bar{\omega}^{-1}\ . \tag{1.3.6}$$

If we refer the vectors $\boldsymbol{b}_i$ to a Cartesian coordinate system, their components will be $\boldsymbol{b}_{1x}$, $\boldsymbol{b}_{1y}$, $\boldsymbol{b}_{1z}$, and (1.3.3) will become

$$\boldsymbol{b}_g = g\boldsymbol{B} = (g_1 g_2 g_3)\begin{pmatrix} b_{1x} & b_{1y} & b_{1z} \\ b_{2x} & b_{2y} & b_{2z} \\ b_{3x} & b_{3y} & b_{3z} \end{pmatrix} .$$

For example, the $\boldsymbol{B}_{\text{bcc}}$ matrix for a direct bcc lattice has the same form as $\boldsymbol{A}_{\text{fcc}}$

$$\boldsymbol{B}_{\text{bcc}} = 2\pi\begin{pmatrix} 0 & 1/a & 1/a \\ 1/a & 0 & 1/a \\ 1/a & 1/a & 0 \end{pmatrix} .$$

One often deals with a system of parallel crystal planes. If we choose the origin of coordinates to be at one of the Bravais lattice sites, we take a definite reciprocal lattice vector $\boldsymbol{b}_g$, and write an arbitrary direct-lattice vector $\boldsymbol{r}$ as $\boldsymbol{r} = m_1\boldsymbol{a}_1 + m_2\boldsymbol{a}_2 + m_3\boldsymbol{a}_3$. Then the equation for the system of planes will be

$$\boldsymbol{b}_g \cdot \boldsymbol{r} = m_1 g_1 + m_2 g_2 + m_3 g_3 = n \ , \tag{1.3.7}$$

with n being a given constant. For these planes to be crystalline, i.e., filled with direct Bravais lattice sites, (1.3.7) must be satisfied by a set of three numers: m_1, m_2, m_3 (1.2.9). If g_1, g_2, g_3 are assigned and the constant n is chosen as a numerical series, (1.3.7) gives a family of parallel crystal planes. Thus, each reciprocal lattice vector $\boldsymbol{b}_g$ corresponds to an appropriate family of crystal planes. The numbers g_i in (1.3.7) may always be chosen in such a way that they have no common divisor, except for unity; i.e., they are reciprocally simple. They are referred to as the Miller indices of a given family of planes and are denoted by sets of three numbers enclosed in round brackets (g_1, g_2, g_3).

The plane defined by (1.3.7) intersects the axes of coordinates chosen along the vectors $\boldsymbol{a}_i$ at points $(n/g_1)\boldsymbol{a}_1, (n/g_2)\boldsymbol{a}_2, (n/g_3)\boldsymbol{a}_3$. Then the ratio of the lengths of these segments in terms of a_i is equal to $g_1^{-1}:g_2^{-1}:g_3^{-1}$, i.e., is reciprocal to the Miller indices. For planes parallel to the coordinate planes (for which $g_1^{-1}:g_2^{-1}:g_3^{-1} = 1:\infty:\infty$; $\infty:1:\infty$ and $\infty:\infty:1$), these indices therefore have the form (100), (010) and (001). If a plane intersects an axis in the region of negative values of coordinates, a minus sign is put above the index: $(\bar{1}00)$, $(0\bar{1}0)$, $(00\bar{1})$. Planes that pass through the diagonals of opposing faces of an elementary cube have the indices (110), (101), (011), etc. A plane that cuts segments a_i off from three coordinate axes of a cube has the indices (111). If one is concerned with crystal planes equivalent in the type of symmetry, their indices are enclosed in curly brackets, for instance, $\{100\}$.

To denote crystal axes, indices which give a set of smallest numbers are also used and are related as the components of a vector parallel to a given direction in the corresponding system of coordinates. Such indices are collected in square brackets. For example, for the axis along the vector $\boldsymbol{a}_1$ we have [100], etc. The equivalent directions are written as $\langle g_1 g_2 g_3 \rangle$. In cubic crystals, the direction of a normal to a plane has the same indices, e.g., (100) and [100], etc. In crystals of other systems this does not occur in the general case. For hexagonal lattices, a four-digit system of indices is used, corresponding to the four axes $\boldsymbol{a}_1, \boldsymbol{a}_2, \boldsymbol{a}_3, \boldsymbol{c}$ (Fig. 1.15b). A hexagonal axis will thus be indexed [0001].

Two more properties of the reciprocal lattice which will be needed in the subsequent presentation must be noted. First, the reciprocal-lattice vector $\boldsymbol{b}_g$ indicating the point g_1, g_2, g_3 in the reciprocal lattice is perpendicular to the plane $(g_1 g_2 g_3)$ of the direct lattice. Indeed, the vector $(\boldsymbol{a}_1/g_1 - \boldsymbol{a}_2/g_2)$ lies in the $(g_1 g_2 g_3)$ plane. According to (1.3.3) and (1.3.5), the scalar product of this vector by the vector $\boldsymbol{b}_g$ therefore is

$$\boldsymbol{b}_g \cdot (\boldsymbol{a}_1/g_1 - \boldsymbol{a}_2/g_2) = (g_1 \boldsymbol{b}_1 + g_2 \boldsymbol{b}_2 + g_3 \boldsymbol{b}_3) \cdot (\boldsymbol{a}_1/g_1 - \boldsymbol{a}_2/g_2) = 0 \; ;$$

i.e., these vectors are mutually perpendicular. Second, the length of the vector $\boldsymbol{b}_g$ is equal to the reciprocal value of the distance between the adjacent planes of the direct lattice family $(g_1 g_2 g_3)$. Let $\boldsymbol{n}_0$ be the unit vector of the normal to these planes. The scalar product $\boldsymbol{a}_1 \boldsymbol{n}_0 / g_1$ then gives the value of the interplanar distance. But since, according to the condition $\boldsymbol{n}_0 = \boldsymbol{b}_g / |\boldsymbol{b}_g|$, then, in view of (1.3.5),

$$d(g_1 g_2 g_3) = \boldsymbol{n}_0 \boldsymbol{a}_1 / g_1 = \boldsymbol{b}_g \cdot \boldsymbol{a}_1 / g_1 |\boldsymbol{b}_g| = 1/|\boldsymbol{b}_g| \; . \tag{1.3.8}$$

This formula holds provided that the components of the vector $\boldsymbol{b}_g$ have no common divisor. In the opposite case,

$$d(g_1 g_2 g_3) = n/|\boldsymbol{b}_g| \; , \tag{1.3.9}$$

where n is an integer, the greatest common divisor of the components of the vector $\boldsymbol{b}_g$ with respect to the three reciprocal-lattice axes.

1.4 Examples of Simple Crystal Structures

Before proceeding to a consideration of actual crystals, we shall recall two types of closest spherical packing. There exists only one closest-packed arrangement of spheres of equal diameter in one plane, when a central sphere is surrounded by six spheres whose centers lie at the vertices of a regular hexagon (Fig. 1.18a). The situation is much more complicated when we deal with the stacking of such layers onto one another. For the packing to be closer, the spheres of each next layer should be placed into the hollows formed between the spheres of a preceding layer. Then the spacing c between the centers of the spheres of the first and the third layer is defined by the ratio $c/a = 1.633$, with a being the diameter of the sphere (Fig. 1.18b), rather than $c/a = 2$, when the spheres of a successive layer are placed not into the hollows but onto the spheres of a previous layer (Fig. 1.18c). Figure 1.18a shows that there are six hollows at the base of the unit cell (dark circles) and only three spheres at a time can be placed into them, thus occupying only half the number of hollows.

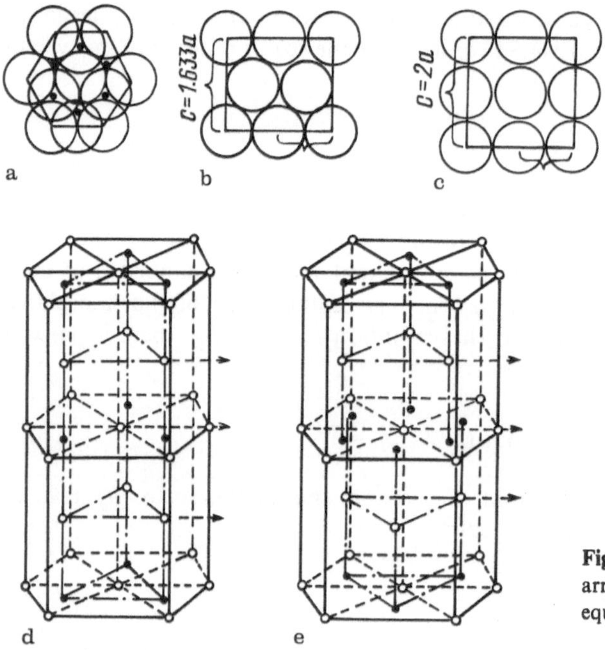

Fig. 1.18. Two closest-packing arrangement types of spheres of equal radius a

Two possibilities arise here. First, the three spheres of the upper (third) layer occupy the hollows of the second layer exactly above the spheres of the first. Thus the first and the third layer turn out to be identical to one another. This is the case of a hexagonal close packing (hcp). The unit cell of a hexagonal close-packed structure is depicted in Fig. 1.18d. Second, the three spheres of the third layer occupy the hollows of the second layer, marked in Fig. 1.18e with black circlets, and the three spheres of the first layer occupy the three other hollows of the second layer, which are indicated by rimmed black circlets in Fig. 1.18e. Consequently, it is only in the fifth layer that we in this case obtain a replication of the first layer. This results in a rhombohedral unit cell that is composed of four layers (of which three are portrayed in Fig. 1.18e). If, in this case, the axial ratio c/a is equal to $1.633 + 0.817 = 2.45$, we have a face-centred cubic (fcc) lattice. But if $c/a \neq 2.45$, the lattice is not cubic but rhombohedral. The rhombohedral axis coincides with the cubic three-fold rotation axis C_3. Thus, this packing corresponds to a closest-packed cubic arrangement.

The concept of the crystal as a system of close-packed spheres permits determination of the atomic (ionic) radii from a consideration of the unit cells [1.15, 16]. Although this is a very crude description of complex entities such as atoms (in a crystal, the electronic cloud of an atom or of an ionic core is, as a rule, not even spherically symmetric), it is very convenient in practice, and in studies of the various modifications of chemical elements, their alloys, and compounds one obtains the same values for the radii, the spread in values being very small. This is particularly true of the ions in alkali-halide compounds where the interatomic distances d satisfy, to an accuracy of several hundredths of Å, the rule that the difference of the quantities d in these compounds for NaF and KF, NaCl and KCl, NaBr and KBr, NaI and KI is a quantity constant to within the accuracy specified. Thus, we can assume that each ion may be approximated by a sphere of specified radius:

$$d(\mathrm{NaF}) - d(\mathrm{KF}) = 0.36\,\text{Å} \;,$$

$$d(\mathrm{NaCl}) - d(\mathrm{KCl}) = 0.33\,\text{Å} \;,$$

$$d(\mathrm{NaBr}) - d(\mathrm{KBr}) = 0.31\,\text{Å} \;,$$

$$d(\mathrm{NaI}) - d(\mathrm{KI}) = 0.30\,\text{Å} \;.$$

Figures 1.19–22 present unit cells of some chemical elements for the fcc, bcc, and hcp lattices, and also for a cubic diamond lattice. The latter is constructed in the following fashion. We begin by generating an fcc cell of α atoms (Fig. 1.22) and then place one β atom on each of the four principal cube diagonals at distances of 1/4 their length from the adjacent vertices α (in diamond, α atoms

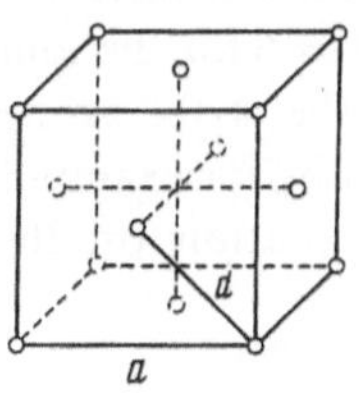

Fig. 1.19

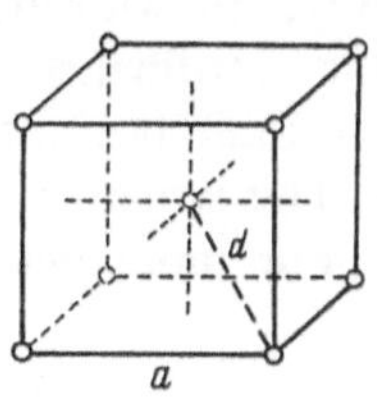

Fig. 1.20

Fig. 1.19. Atomic unit cell of fcc crystals (e.g., copper, silver, gold). Space group $O_h^5 - Fm3m$; basis $\sigma = 4$; number of neighbors in the first coordination zone $z_1 = 12$; number of neighbors in the second zone $z_2 = 6$; a is an edge of the elementary cube; d is half the cube diagonal; $d = (\sqrt{2}/2)a$

Fig. 1.20. Atomic unit cell of bcc crystals (e.g., α, β, and δ iron). Space group $O_h^9 - Jm3m$; basis $\sigma = 2$; number of neighbors in the first coordination zone $z_1 = 8$; number of neighbors in the second zone $z_2 = 6$; a is an edge of the elementary cube; d is half the cube diagonal; $d = (\sqrt{3}/2)a$

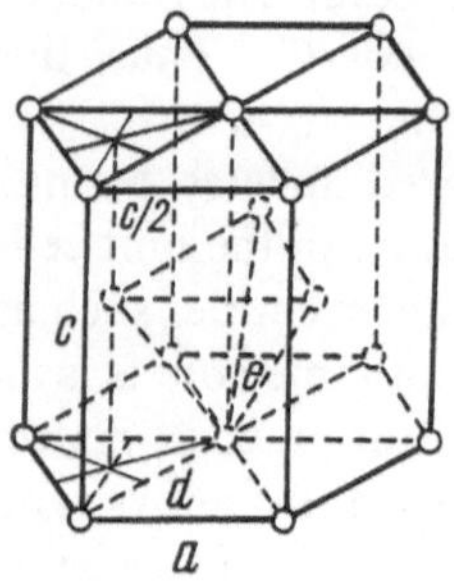

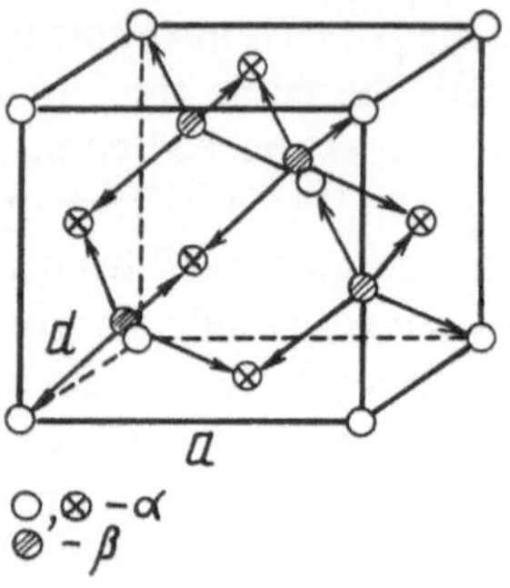

Fig. 1.21 Fig. 1.22

Fig. 1.21. Atomic unit cell of hcp crystals (e.g., zinc and cadmium). Space group $D^4_{6h}-C6/mmc$; basis $\sigma=2$; number of neighbors in the first coordination zone $z_1=6$; number of neighbors in the second zone $z_2=8$; a is side of the equilateral hexagon at the base of the elementary prism; d is half the hexagón diagonal; e is the nearest-site distance for the adjacent spheres of the hcp lattice; c is the height of the elementary prism $d=a$, $e=(a^2/3+c^2/4)^{1/2}$

Fig. 1.22. Atomic unit cell of cubic (diamond-type) carbon, silicon, germanium, and α-tin lattices. Space group O^7_h-Fd3m; basis $\sigma=8$; a is an edge of the elementary cube; d is the spacing between α-type atoms lying at the cube vertices or cube centers and β-type nearest atoms located inside the elementary cube; $d=(\sqrt{3}/4)a$

and β atoms are alike). An fcc cell of β atoms can also be formed which turns out to be displaced relative to the one constructed on α atoms. Essential information on the crystal structure of the majority of chemical elements with the most typical lattices is contained in Table 1.8.

As an illustration of simple ionic structures in compounds, we present data for alkali halides such as LiF, LiCl, NaCl, etc. (Fig. 1.23a), as well as for other compounds such as AgBr, MgO, UO_5, etc. with an fcc lattice that have the space group $O_h-Fm\,3m$, $\sigma=4$ and the coordination number $z=6$ for the ions of both signs. Figure 1.23b depicts an ionic bcc lattice with the space group O_h-Pm3m and with $\sigma=1$ and $z=8$. This structure is inherent in, for example, $\alpha=$ RbCl, CsCl, CsBr, CsSeH, etc. For more detailed information on the structure of crystals, see [1.17].

1.5 Experimental Techniques for Determining the Periodic Atomic Structure of Solids [1.18]

An experimental proof of the atomic structure of crystals was obtained in the second decade of the twentieth century by W. Bragg, G.W. Wulf, P. Debye, and M. Laue, using X-ray diffraction. Later, neutrons, electrons, etc. were used for

Table 1.8. Some of the most important data on the crystal structure of a number of chemical elements

Element	Density, [g/cm^3] (10^3 kg/m^3 at 20°C)	Lattice constant at 20°C, [nm]	Atomic volume [cm^3/mol] (10^{-3} m^3/kmol)	Nearest-neighbor distance, [nm]
	fcc-type lattice			
Aluminum	2.70	0.404	9.99	0.286
Argon	—	0.543 (20 K)	—	0.383
Gold	19.32	0.407	10.2	0.288
Calcium	1.55	0.556	25.9	0.393
Lanthanum	6.15	0.529	—	0.373
Copper	8.96	0.361	7.09	0.255
Nickel	8.90	0.352	6.59	0.249
Palladium	12.0	0.388	8.89	0.274
Platinum	21.45	0.392	9.10	0.277
Lead	11.34	0.494	18.27	0.349
Silver	10.49	0.408	10.28	0.288
Strontium	2.6	0.605	34.0	0.430
Cerium	6.9	0.514	20.00	0.364
	bcc type lattice			
Barium	3.5	0.501	39.0	0.434
Vanadium	6.0	0.303	8.5	0.263
Tungsten	19.3	0.316	8.53	0.273
Iron (α)	7.87	0.286	7.1	0.248
Lithium	0.53	0.350	13.0	0.303
Molybdenum	10.2	0.314	9.41	0.272
Sodium	0.97	0.428	24.0	0.371
Rubidium	1.53	0.562	55.9	0.487
Tantalum	16.6	0.330	10.9	0.285
Chromium	7.19	0.283	7.23	0.249
	hcp-type lattice			
Beryllium	1.82	0.227 0.359	4.96	0.222
Gadolinium	7.95	0.362 0.575	19.7	0.355
Cobalt	8.9	0.251 0.407	6.6	0.250
Magnesium	1.74	0.320 0.520	14.0	0.319
Thallium	11.85	0.345 0.551	17.24	0.340
Titanium	4.54	0.295 0.473	10.6	0.291
Zinc	7.13	0.266 0.494	9.17	0.316
	diamond-type lattice			
Germanium	5.36	0.563	13.5	0.244
Silicon	2.33	0.543	12.0	0.235
Tin (gray)	5.75	0.646	—	—
Carbon (diamond)	3.51	0.356	—	0.154

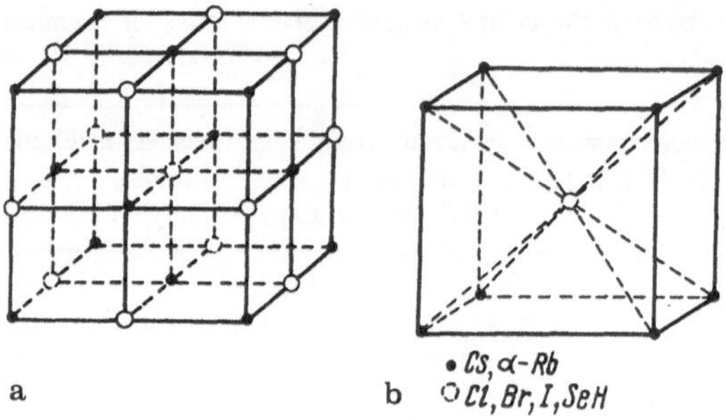

Fig. 1.23. (**a**) Atomic unit cell of cubic crystals (e.g., halogenides and hydrides of alkali metals M with F, Cl, Br, I, H, D): space group $O_h^5 - Fm3m$; basis $\sigma = 4$; number of neighbors in the first coordination zone $z_1 = 6$; a is an edge of the elementary cube; d is the distance between metallic ions and haloid ions; $d = a/2$. (**b**) Atomic unit cell of cubic crystals (e.g., halogenides and hydrides of alkali metals α-RbCl, CsCl, CsBr, CsI, CsSeH): space group $O_h^7 - Pm3m$; basis $\sigma = 4$, number of neighbors in the first coordination zone $z_1 = 8$; a is an edge of the elementary cube; d is half the cube diagonal; $d = (\sqrt{3}/2)a$

the same purpose. Nowadays, sophisticated techniques are available for studying the crystal structure.

We recall briefly the fundamentals of determining the crystal structure by X-ray diffraction. As is known (Table 1.8), the mean distance between the nearest neighbor atoms in a crystal is on the order of several angstroms (1 Å $= 10^{-8}$ cm $= 10^{-10}$ m). The diffraction of radiation therefore will manifest itself at wavelengths $\lambda \gtrsim 1$ Å. According to Planck, X-ray quanta of this wavelength possess the energy

$$\varepsilon = h\nu = hc/\lambda \ , \tag{1.5.1}$$

where ν is the frequency in s^{-1} (or Hz). At $\lambda \sim 10^{-8}\,\mathrm{cm} = 10^{-10}\,\mathrm{m}$ $\nu = c/\lambda \sim 10^{18}\,\mathrm{s}^{-1}$, the velocity of light in vacuum $c \sim 3 \times 10^8$ m/s, and the Planck constant $h = 6.625 \times 10^{-27}$ erg s $= 6.625 \times 10^{-34}$ J s. Substitution of these quantities into (1.5.1) yields $\varepsilon \sim 10^4$ eV (1 eV $= 1.6 \times 10^{-19}$ J). According to the de Broglie formula we have for the electrons or neutrons

$$\lambda = h/p = h/\sqrt{2m\varepsilon} \ , \tag{1.5.2}$$

with p being the momentum and m the particle mass. For a neutron $m_n = 1.674 \times 10^{-24}$ g, and for an electron $m_e = 9.107 \times 10^{-28}$ g. Then (1.5.2) yields the result that the energies of waves of length ~1 Å will be on the order of 10^2 eV for an electron and 10^{-1} eV for a neutron. Here we limit ourselves to a consideration of X-ray diffraction. However, many of the results obtained may be used in both neutron-diffraction and electron-diffraction studies.

Consider a small single crystal of some substance as a cube with an edge l. A monochromatic beam of rays of wavelength λ will be passed through this cube.

Each atom located at the sites of the crystal then becomes a source of secondary waves. The problem is to find the overall effect of these waves at distances R that largely exceed the size of the crystal ($R \gg l$). Some simplifying assumptions will be adopted which are not exploited in a rigorous theory:

1. We shall assume that the primary beam propagates in the crystal with the same velocity c, as it does in vacuum (i.e., the index of refraction of the crystal is taken to be approximately equal to 1). The errors thus entailed may be shown to play no particularly substantial role in locating the diffraction peaks.
2. We shall disregard the multiple scattering of a secondary wave from a given atom by the atoms located in the other lattice sites. It also may be shown here that when $R \gg l \gg 1$ Å, the error is negligible.
3. We shall completely neglect the absorption of radiant energy in the crystal (a transparent crystal). Subject to the condition that $R \gg l \gg \lambda$, this is a quite acceptable simplification.
4. We shall ignore the effects of thermal motion, using a purely static crystal model.
5. We shall not allow for the atomic scattering factor describing the interference effects within a scattering atom which arise from the fact that its size is commensurate with the wavelength.

Let us calculate the phase difference of the secondary waves from two lattice atoms residing at the sites A_1 and A_2, which are connected by the vector $\boldsymbol{r}$ (Fig. 1.24). Denote the unit vectors of the normal of the incident and scattered waves by $\boldsymbol{t}_0$ and $\boldsymbol{t}$, respectively. Find the phase difference of the secondary waves at point Q, at a large distance R ($R \gg |\boldsymbol{r}|, l$), so that the A_1Q and A_2Q lines may with high accuracy be considered parallel to one another and to the normal $\boldsymbol{t}$. The path difference of two scattered rays is equal to

$$A_2B_2 - A_1B_1 = \boldsymbol{r}\cdot(\boldsymbol{t}-\boldsymbol{t}_0) = \boldsymbol{r}\cdot\boldsymbol{q} \ . \tag{1.5.3}$$

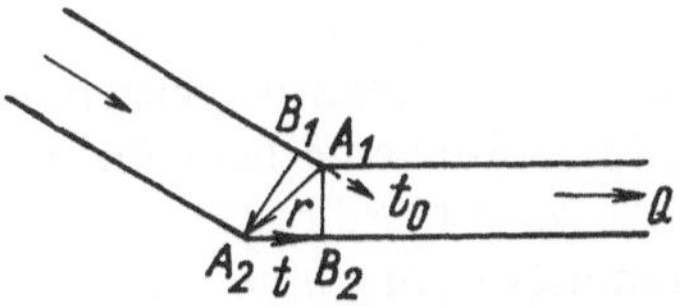

Fig. 1.24. Determination of the scattered-ray path difference by two lattice sites, A_1 and A_2

The vector $(\boldsymbol{t}-\boldsymbol{t}_0) = \boldsymbol{q}$, as seen from Fig. 1.25, is normal to a plane which, by convention, plays the role of a reflection plane for an incident ($\boldsymbol{t}_0$) and a reflected ($\boldsymbol{t}$) ray. If we denote the angle for the incident and reflected rays on this plane by ϑ, the scattering angle (the angle between $\boldsymbol{t}_0$ and $\boldsymbol{t}$) is equal to 2ϑ and, consequently, we find from Fig. 1.25

$$|\boldsymbol{q}| = |\boldsymbol{t}-\boldsymbol{t}_0| = 2\sin\vartheta \ . \tag{1.5.4}$$

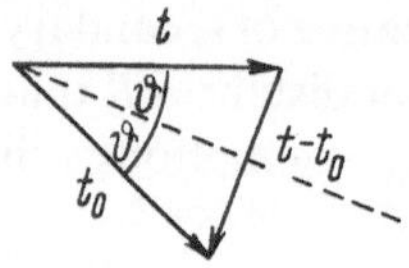

Fig. 1.25. Construction of *a* normal to the X-ray glide plane in the crystal

If we introduce a wave vector $\boldsymbol{k} = 2\pi \boldsymbol{t}/\lambda$, of which the modulus is equal to

$$|\boldsymbol{k}| = 2\pi/\lambda \ , \tag{1.5.5}$$

the path difference (1.5.3) yields for the phase difference

$$\Delta\varphi = k\boldsymbol{r}\cdot(\boldsymbol{t} - \boldsymbol{t}_0) = k(\boldsymbol{r}\cdot\boldsymbol{q}) \ . \tag{1.5.6}$$

The amplitude of a scattered wave at point Q will be maximal for directions in which the phase difference $\Delta\varphi$ is a multiple of the quantity 2π, when the amplitudes of the waves scattered from the atoms at sites A_1 and A_2 add constructively. Recall now that $\boldsymbol{r}$ is one of the translation vectors (1.2.9). Then the Laue conditions should hold for the diffraction maximum

$$\varphi_{\boldsymbol{a}_i} = \frac{2\pi}{\lambda}(\boldsymbol{a}_i\cdot\boldsymbol{q}) = 2\pi m_i \quad (i = 1, 2, 3) \ , \tag{1.5.7}$$

where m_i are integers. Label the direction cosines of the vector $\boldsymbol{q}$ with respect to the $\boldsymbol{a}_i$ axes by α_i. Then (1.5.7) becomes

$$(\boldsymbol{a}_i\cdot\boldsymbol{q}) = 2a_i \sin\vartheta\cdot\alpha_i = m_i\lambda \ . \tag{1.5.8}$$

This results in a selective diffraction picture of X-ray reflection from the crystal, because (1.5.8) is capable of solution only for some angles ϑ and wavelengths λ with a given lattice $(\boldsymbol{a}_i)$ and for the choice of the reflection plane (α_i).

From (1.5.8) we now obtain the Wulf–Bragg relations for selective reflection from a given family of parallel crystal planes with a distance d between the neighboring planes. It follows from (1.5.8) that the α_i in the direction of the diffraction maximum are proportional to the quantities m_1/a_1, m_2/a_2, m_3/a_3. The successive lattice planes $(m_1\,m_2\,m_3)$ intercept the $\boldsymbol{a}_i$ axes, according to the definition of the Miller indices, at the respective distances a_1/m_1, a_2/m_2, a_3/m_3. It follows from elementary geometry that the direction cosines to the planes $(m_1\,m_2\,m_3)$ are also proportional to a_i/m_i; i.e., these planes are parallel to the reflection plane. The distance $d\,(m_1\,m_2\,m_3)$ between the adjacent planes of the family $(m_1\,m_2\,m_3)$ is equal to

$$d(m_1\,m_2\,m_3) = a_1\alpha_1/m_1 = a_2\alpha_2/m_2 = a_3\alpha_3/m_3 \ , \tag{1.5.9}$$

and then the Laue equations take the form

$$2d(m_1 m_2 m_3) \sin \vartheta = \lambda \ . \tag{1.5.10}$$

Observe that the integers m_1, m_2, and m_3 in (1.5.10) are not merely Miller indices of the different crystal planes, for they may contain an integral common multiplier n (which cancels out due to Miller indexing). Therefore, if $m_i' = m_i/n$ in (1.5.10) are held to be Miller indices, we obtain

$$2d(m_1' m_2' m_3') \sin \vartheta = n\lambda \ . \tag{1.5.11}$$

This is the famous Wulf–Bragg formula in which the integer n gives the order of reflection. [The relation (1.5.11) was first derived by the Russian scientist G.W. Wulf in 1912, and a year later by the Englishman W. Bragg who verified it experimentally in 1913.]

The Laue equations and the Wulf–Bragg formula are due only to the fundamental property of the crystal, that is, the periodicity of its atomic structure, and are associated with neither its chemical composition nor the atomic arrangement in the reflection planes. The latter factors influence the magnitude of peak intensity and are therefore very important in determining relative peak intensities. It is also seen from (1.5.11) that the condition for the occurrence of a diffraction peak is $\lambda \leqslant 2d$.

To describe X-ray diffraction in crystals, we make use of the concept of the reciprocal lattice introduced in Sect. 1.3. Instead of $d(m_1' m_2' m_3')$ from (1.3.8) we may introduce into (1.5.11) the quantity $b(m_1' m_2' m_3')^{-1}$. Then, if g_i and m_i' contain a common multiplier n, the nth site of the reciprocal lattice (in the array of sites, counting from the origin of the reciprocal lattice) with the components g_i (or m_i) refers to an nth-order peak of an X-ray reflected from the corresponding crystal planes. Each reciprocal-lattice site therefore refers to a possible reflection peak. *Ewald* (1913) (see [1.19]) has offered a simple geometric interpretation for this fact (Fig. 1.26).

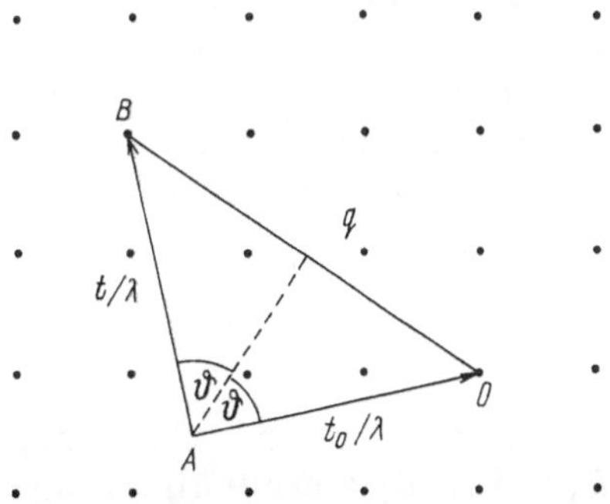

Fig. 1.26. Geometric interpretation (after Ewald) of the relation between reflection peak and reciprocal-lattice site: O is the origin of coordinates of the reciprocal lattice; AO the incident wave vector; AB the reflected wave vector; OB the normal to reflection plane

Let the $\boldsymbol{AO}$ segment be equal to the vector $\boldsymbol{t}_0/\lambda$ of length λ^{-1} oriented along the normal to the incident radiation wave front. In precisely the same way, the $\boldsymbol{AB}$ segment is equal to the vector $\boldsymbol{t}/\lambda$ of the same length, parallel to the direction along which the scattering is considered. Then the vector $\boldsymbol{OB}$ will be parallel to the scattering vector $\boldsymbol{q}$ (1.5.3) and normal to one of the lattice planes, e.g., g_i. The length of $\boldsymbol{OB}$, according to (1.5.4), is equal to $2\sin(\vartheta/\lambda)$, which, according to (1.5.11), is equal to $1/d(g_1g_2g_3)$, if the diffraction condition is fulfilled. Thus $\boldsymbol{OB}$ lies in the direction of the reciprocal-lattice vector and is equal to it in modulus. Therefore, if the point O coincides with the origin of the reciprocal lattice, the point B should occur at site g_i.

Then, following Ewald, we are in a position to execute the following construction. If O is the origin, we should draw a vector $\boldsymbol{AO}$ in the incident direction of length λ^{-1}, terminating at the origin, and construct a sphere of radius $\lambda^{-1} = |\boldsymbol{AO}|$ with its center at A. In the case under consideration, this sphere should pass through point B. Therefore, the Laue (1.5.7) or Wulf–Bragg (1.5.11) conditions are equivalent to the condition that a diffraction peak cannot arise unless at least one more of the reciprocal lattice sites B joins this sphere. The latter is referred to as the propagation sphere. This method, replacing the consideration of planes in direct space by points in reciprocal space, considerably simplifies the solution of all crystal diffraction problems.

Since this method will later be used on more than one occasion, we present one more vector modification of the Wulf–Bragg conditions (1.5.11). Let

$$\boldsymbol{b}^* = 2\pi\boldsymbol{b}_g \, , \tag{1.5.12}$$

where $\boldsymbol{b}_g$ is the reciprocal-lattice vector (henceforth the reciprocal-lattice vector will be referred to as $\boldsymbol{b}_g^*$) and $\boldsymbol{k}$ is the wave vector of an incident wave (1.5.5). From the Ewald construction (Fig. 1.26) we find $(\boldsymbol{AO} + \boldsymbol{OB})^2 = (\boldsymbol{BA})^2$, or $(\boldsymbol{k} + \boldsymbol{b}^*)^2 = (\boldsymbol{k})^2$. Hence (1.5.11) becomes

$$2(\boldsymbol{k} \cdot \boldsymbol{b}^*) + \boldsymbol{b}^{*2} = 0 \, . \tag{1.5.13}$$

We now calculate the so-called atomic scattering factor [Ref. 1.12, Sect. 126]. Strictly speaking, to do this we need to use the methods of quantum mechanics. However, here we confine ourselves to a quasi-classical method. For simplicity, we assume that the scattering process is elastic when the state of the scattering atom does not change and so the potential energy of the atom may be averaged over its wave function. According to the Born approximation, the differential scattering cross section is proportional to the quantity

$$|\textstyle\int \psi_{\boldsymbol{k}}^* V(\boldsymbol{r})\psi_{\boldsymbol{k}_0} d\boldsymbol{r}|^2 = |V_{\boldsymbol{k}\boldsymbol{k}_0}|^2 \, ,$$

where $\psi_{\boldsymbol{k}}$ and $\psi_{\boldsymbol{k}_0}$ are respectively the wave functions of an incident and an elastically scattered particle ($|\boldsymbol{k}| = |\boldsymbol{k}_0|$), i.e., $\psi_{\boldsymbol{k}} = \exp((\mathrm{i}/\hbar)\boldsymbol{p}\cdot\boldsymbol{r})$ and

$\psi_{k_0} = \exp((i/\hbar)\boldsymbol{p}_0 \cdot \boldsymbol{r})$. Here $\boldsymbol{p} = \hbar\boldsymbol{k}$ and $\boldsymbol{p}_0 = \hbar\boldsymbol{k}_0$ are the momenta of the incident and scattered particles. From the expression for V_{kk_0} we can readily obtain the Wulf–Bragg formula (1.5.11). To this end, we expand the periodic potential $V(\boldsymbol{r})$ in a Fourier series (1.3.2)

$$V(\boldsymbol{r}) = \sum_{\boldsymbol{b}_g^*} V_{\boldsymbol{b}_g^*} \exp(i\boldsymbol{b}_g^* \cdot \boldsymbol{r}) \ ,$$

see also (1.5.12). For V_{kk_0} with wave functions as plane waves we then have

$$V_{kk_0} = \sum_{\boldsymbol{b}_g^*} \int \exp[i(\boldsymbol{k}_0 + \boldsymbol{b}_g^* - \boldsymbol{k}) \cdot \boldsymbol{r}] d\boldsymbol{r}$$

$$= \begin{cases} V_{\boldsymbol{b}_g^*} \text{ at } \boldsymbol{k}_0 + \boldsymbol{b}_g^* - \boldsymbol{k} = 0 \\ 0 \text{ in all the other cases} \ . \end{cases} \tag{1.5.14}$$

Thus, a scattered ray may be observed only in directions with the wave vectors $\boldsymbol{k} = \boldsymbol{k}_0 + \boldsymbol{b}_g^*$. Apart from this, because of the elastic nature of the scattering, we have $|\boldsymbol{k}| = |\boldsymbol{k}_0|$. From Fig. 1.26 it then follows immediately (setting $\boldsymbol{t} = \boldsymbol{k}$, $\boldsymbol{t}_0 = \boldsymbol{k}_0$) that $|\boldsymbol{b}_g^*| = 2|\boldsymbol{k}| \sin\vartheta$. Next, using (1.5.4), (1.5.5), and (1.5.11) we arrive at (1.5.12).

Returning to the scattering cross section, we see that here it is equal to the square of the modulus of the integral (1.5.5, 7):

$$\int V(\boldsymbol{r}) e^{i\boldsymbol{q}\cdot\boldsymbol{r}} d\boldsymbol{r} \ . \tag{1.5.15}$$

The average potential energy of an atom is equal to $V(\boldsymbol{r}) = e\varphi(\boldsymbol{r})$, with $\varphi(\boldsymbol{r})$ being the atomic field potential and e the electronic charge. If we denote the density of the electric charge of an atom by $\varrho(\boldsymbol{r})$, the $\varphi(\boldsymbol{r})$ and $\varrho(\boldsymbol{r})$ are related to each other by the Poisson equation $\Delta\varphi(\boldsymbol{r}) = -4\pi\varrho(\boldsymbol{r})$. The same equation should be satisfied by all the Fourier components of the potential $\varphi_{\boldsymbol{q}} \exp(i\boldsymbol{q}\boldsymbol{r})$, that is,

$$\Delta[\varphi_{\boldsymbol{q}} \exp(i\boldsymbol{q}\cdot\boldsymbol{r})] = q^2 \varphi_{\boldsymbol{q}} \exp(i\boldsymbol{q}\cdot\boldsymbol{r}) = 4\pi\varrho_{\boldsymbol{q}} \exp(i\boldsymbol{q}\cdot\boldsymbol{r}) \ ;$$

this immediately gives $\varphi_{\boldsymbol{q}} = (4\pi/q^2)\varrho_{\boldsymbol{q}}$. Substituting the explicit expressions for the Fourier coefficients into this equation yields

$$\int \varphi(\boldsymbol{r}) \exp(-i\boldsymbol{q}\cdot\boldsymbol{r}) d\boldsymbol{r} = 4\pi q^{-2} \int \varrho(\boldsymbol{r}) \exp(-i\boldsymbol{q}\cdot\boldsymbol{r}) d\boldsymbol{r} \ . \tag{1.5.16}$$

The atomic charge density $\varrho(\boldsymbol{r})$ involves a nuclear point charge $+Ze\delta(\boldsymbol{r})$ [with Z being the serial element number, $\delta(\boldsymbol{r})$ the Dirac delta function] and an electronic charge Z of spatial density $\varrho(\boldsymbol{r})$. For the integral on the right-hand side of (1.5.16) we then obtain

$$\int \varrho(\boldsymbol{r})\exp(-\mathrm{i}\boldsymbol{q}\cdot\boldsymbol{r}) = -e\int n(\boldsymbol{r})\exp(\mathrm{i}\boldsymbol{q}\cdot\boldsymbol{r})d\boldsymbol{r} + Ze \ .$$

Thus the integral (1.5.15) has the form

$$\int V(\boldsymbol{r})\exp(\mathrm{i}\boldsymbol{q}\cdot\boldsymbol{r})d\boldsymbol{r} = 4\pi e^2 q^{-2}[Z - F(\boldsymbol{q})] \ , \tag{1.5.17}$$

where the quantity

$$F(\boldsymbol{q}) = \int n(\boldsymbol{r})\exp(\mathrm{i}\boldsymbol{q}\cdot\boldsymbol{r})d\boldsymbol{r} \tag{1.5.18}$$

is called the atomic form factor. It gives the magnitude of the ratio of the amplitude of a wave scattered by the "smeared" electronic density of the atom, $n(\boldsymbol{r})$, to the amplitude of a wave scattered on a point electron; $F(\boldsymbol{q})$ is a function of the scattering angle ϑ and of the wave vector of the particle being scattered. If $n(\boldsymbol{r})$ is a spherically symmetric function, i.e., $n(\boldsymbol{r}) = n(r)$, we have

$$\int n(\boldsymbol{r})\exp(\mathrm{i}\boldsymbol{q}\cdot\boldsymbol{r})d\boldsymbol{r} = \int_0^\infty\int_0^\pi\int_0^{2\pi} n(r)\exp(\mathrm{i}qr\cos\vartheta)$$

$$\times\, r^2\sin\vartheta\, d\vartheta\, d\varphi\, dr = 4\pi\int_0^\infty r^2 n(r)\sin qr(qr)^{-1}dr$$

Regarding $4\pi r^2 n(r)dr = w(r)dr$ as the probability that the scattering electron resides in the layer between the spheres of radii r and $r + dr$, we get in place of (1.5.18)

$$F(q) = \int_0^\infty w(r)\sin qr(qr)^{-1}dr \ . \tag{1.5.19}$$

Tables are available for $F(q)$ which are calculated according to the Thomas–Fermi method or the Hartree–Fock method [Ref. 1.12, Sect. 13] using the values of $n(r)$ or $w(r)$. Figure 1.27 presents plots of the neutron (N) and X-ray (X) form factors for iron as functions of $q = 4\pi\sin\vartheta/\lambda$ [1.20]. With neutrons the quantity $n(r)$ should signify the density of noncompensated magnetic moments, i.e., the charge-density difference for electrons with positive (+) and negative (−) spin projections: $\Delta n(r) = n_+(r) - n_-(r)$. Only the spin-dependent part of the neutron form factor is implied here. Since the spin density is related chiefly to the outer portion of the electronic shell of the atom (3d shell in the atom of iron), the spin-density radial distribution maximum $\Delta n(r)$ lies farther from the center of the nucleus than does the charge-density maximum. Equation (1.5.19) shows that the magnitude of the magnetic form factor falls off faster with decreasing q

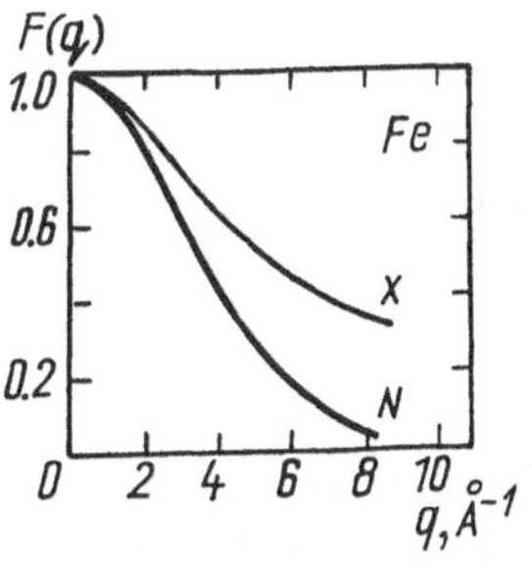

Fig. 1.27. Neutron (N) and x-ray (X) formfactors $F(q)$ of iron as a function of the wave number q determined by the change of momentum in an atom-scattered particle

than that of the X-ray form factor (Fig. 1.27). An immediate observation from (1.5.19) is that the value of the form factor for $q = 0$ is equal to the number of electrons in the atom, Z (for the X-ray case), or to the magnitude of the noncompensated spin moment (in the neutron case).

As has already been noted, the Laue formula as well as the Wulf–Bragg formula (with the assumptions that have been made) only allow us to determine the position of the peaks. But is it also important to know the magnitude of the relative intensity of the various peaks, their shape, and their temperature dependence. That information would enable us to determine the type of crystal unit cell, the number of and the arrangement of atoms in it, etc. To this end, it is necessary to determine the amplitude of a wave scattered in some direction by all unit-cell atoms. Just as with the atomic scattering factor, we call the ratio of the amplitude of a reflected wave to the amplitude of a wave scattered on a point electron (provided that the particle being scattered is of the same wavelength) the structure amplitude $F(g_1 g_2 g_3)$ for a g_i-type reflection. The expression for $F(g_1 g_2 g_3)$ has the form

$$F(g_1 g_2 g_3) = \sum_s F_s \exp(\mathrm{i}\Delta\varphi_s) = \sum_s F_s \exp(\mathrm{i}(2\pi/\lambda)\boldsymbol{R}_s \cdot \boldsymbol{q}) \; , \tag{1.5.20}$$

where the summation is taken over all unit-cell atoms, $\Delta\varphi_s$ is the phase of a wave scattered by an s-type atom ($s = 1, 2, \ldots, \sigma$), $\boldsymbol{R}_s$ is the vector defined from (1.2.10), and F_s is the atomic factor of an atom s (1.5.18, 19). In keeping with (1.5.6, 10), we have

$$\boldsymbol{R}_s \boldsymbol{q} = \lambda(g_1 \xi_1^{(s)} + g_2 \xi_2^{(s)} + g_3 \xi_3^{(s)}) \; , \tag{1.5.21}$$

and therefore it follows from (1.5.20) that

$$F(g_1 g_2 g_3) = \sum_s F_s \exp[\mathrm{i}2\pi(g_1 \xi_1^{(s)} + g_2 \xi_2^{(s)} + g_3 \xi_3^{(s)})] \tag{1.5.22}$$

and also that

$$|F(g_1g_2g_3)|^2 = \left[\sum_s F_s \cos 2\pi(g_1\xi_1^{(s)} + g_2\xi_2^{(s)} + g_3\xi_3^{(s)})\right]^2 + \left[\sum_s F_s \sin 2\pi(g_1\xi_1^{(s)} + g_2\xi_2^{(s)} + g_3\xi_3^{(s)})\right]^2 . \quad (1.5.23)$$

If all of the unit-cell atoms are alike, the factors F_s are the same for all s, and they may be taken outside the summation sign. Instead of (1.5.22) we then obtain

$$F(g_1g_2g_3) = FS , \quad (1.5.24)$$

where the quantity S is called the structure factor and has the form

$$S = \sum_s \exp[i2\pi(g_1\xi_1^{(s)} + g_2\xi_2^{(s)} + g_3\xi_3^{(s)})] . \quad (1.5.25)$$

For example, in an fcc lattice made up of like atoms there are four atoms in a cell which lie at points 000, 0 1/2 1/2, 1/2 0 1/2, and 1/2 1/2 0. In this case

$$S = 1 + \exp[i\pi(g_2 + g_3)] + \exp[i\pi(g_1 + g_3)] + \exp[i\pi(g_1 + g_2)] . \quad (1.5.26)$$

If the sum of two indices g_i is even, then the corresponding summand in (1.5.26) is equal to $(+1)$. If this sum is odd, the summand is (-1). For instance, for $\{100\}$-type planes we have $g_2 + g_3 = 0$, which is even, and $g_1 + g_2 = 1$, $g_1 + g_3 = 1$, which is odd, and therefore, $S = 0$; i.e., there is no reflection. Conversely, for $\{111\}$-type planes we have $g_2 + g_3 = g_1 + g_2 = g_1 + g_2 = 2$, which is an even sum, and $S = 4$; i.e., reflection takes place.

A simple physical explanation can be provided for this. In the fcc lattice the $\{100\}$ planes are reflection planes, and in the case of reflection from two adjacent parallel cube faces the phase difference is equal to 2π. But the atoms that reside in the middle of the faces generate an intermediate plane which produces the phase difference π and thus suppresses the contribution to the scattering amplitude by the adjacent planes. This is schematically illustrated in Fig. 1.28. It stands to reason that such suppression occurs when all four atoms in the cell are alike. In the case of a bcc lattice $S = 1 + \exp[i\pi(g_1 + g_2 + g_3)]$. Similarly, the

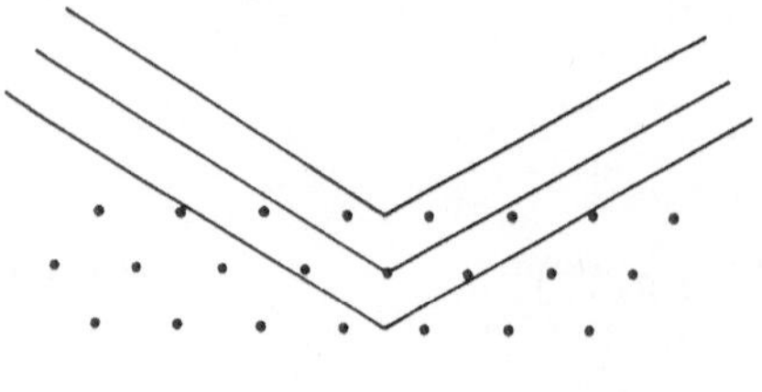

Fig. 1.28. Explanation of the absence of reflection from (100)-type planes in fcc lattices composed of like atoms

{100}-planes do not reflect here. However, for a bcc lattice with unlike atoms, for example CsCl, we have $F = F_{Cs} + F_{Cl}\{\exp[i\pi(g_1 + g_2 + g_3)]\}$, and the reflections from {100} will not be suppressed.

In practice, three methods are used in X-ray diffraction studies:

1. The *Bragg method* or rotating crystal method. The crystal investigated is rotated about a fixed axis, usually perpendicular to the specified direction of the incident monochromatic (λ = const) X-ray beam. With the crystal being rotated through an angle ϑ, when (1.5.11) is satisfied, the beam is diffracted and the reflected ray is recorded on a photographic plate.

2. The *Laue method.* In this case the crystal remains fixed with respect to the X-ray beam. However, the X-ray beam is not monochromatic but "white," with the wavelength continuously varying over a sufficiently wide range. Each system of planes, spaced a distance $d(g_1 g_2 g_3)$ apart, then in a sense picks out from the "white" light the component of wavelength λ which produces a selective reflection at an angle ϑ satisfying (1.5.11).

3. The *powder method* or Debye–Sherrer method. The sample of fine-grained polycrystalline structure or of compacted fine powder exposed to a monochromatic beam of X-ray light is fixed. In this method the monochromatic beam of wavelength λ = const., selects out crystallites in the polycrystal or powder, with planes whose spacing and orientation conform to the condition (1.5.11). At appropriate angles ϑ reflections are observed which are recorded as rings on a photographic plate (Debye powder patterns).

1.6 Qualitative Concepts of the Electronic and Nuclear Crystal Structure

To gain insight into the atomic structure of close-packed crystals, we recall the experimental fact that the lattice parameter d is close to the mean atomic size $2r_{at}$ (Table 1.8). The atoms in a solid "touch" one another just as spheres do. At the same time we should remember that the electronic shell of an atom is a highly complex entity. Roughly, the electronic shell may be thought of as composed of subshells, which are specified by two quantum numbers: the main quantum number $n = 1, 2, 3, \ldots$ and the orbital quantum number $l = 1, 2, \ldots, (n-1)$; $2l+1$ electrons in a shell with a given n and l are numbered by magnetic quantum numbers $-l \leqslant m \leqslant l$. In addition to this, there may be two electrons with these three quantum numbers—n, l, and m—in keeping with the two possible spin orientations (the spin quantum number $\sigma = \pm 1/2$). Thus the number of electrons in a shell with quantum numbers n and l is equal to 2 $(2l+1)$. For states with different orbital numbers the following notation has been adopted: $s(l=0)$, $p(l=1)$, $d(l=2)$, $f(l=3)$, etc. The filling of a shell, i.e., its electronic configuration, is designated as nl^m, e.g., $1s^2$, $3d^5$, etc.

Table 1.9. Successive filling of electron shells

Configuration with given n and l								Total number of electrons in shell	Symbol of shell
n	$l=s$	p	d	f	g	h	k		
1	$1s^2$							2	K
2	$2s^2$	$2p^6$						8	L
3	$3s^2$	$3p^6$	$3d^{10}$					18	M
4	$4s^2$	$4p^6$	$4d^{10}$	$4f^{14}$				32	N
5	$5s^2$	$5p^6$	$5d^{10}$	$5f^{14}$	$5g^{18}$			50	O
6	$6s^2$	$6p^6$	$6d^{10}$	$6f^{14}$	$6g^{18}$	$6h^{22}$			P
7	$7s^2$	$7p^6$	$7d^{10}$	$7f^{14}$	$7g^{18}$	$7h^{22}$	$7k^{26}$	98	Q

A successive filling of shells with electrons is shown in Table 1.9. However, in reality, the filling follows a somewhat different scheme. Sometimes it appears more advantageous to start completing a shell with a larger value of the main quantum number n but with a smaller value of the orbital l. For example, after ^{18}Ar of $1s^2 2s^2 2p^6 3s^2 3p^6$ configuration it would be reasonable to start completing the $3d$ shell with its ten sites, but the configurations of the next two periodic table elements actually are ^{19}K ($1s^2 2p^6 3s^2 3p^6 \ldots 4s$) and ^{20}Ca ($1s^2 2s^2 2p^6 3s^2 3p^6 \ldots 4s^2$). It is only from scandium ^{21}Sc that the unfilled $3d$ shell starts to fill. This tardy completion encompasses seven more elements from ^{22}Ti to ^{28}Ni and terminates at copper ^{29}Cu whose atomic shell is filled up "correctly" (according to the scheme presented in Table 1.9: $1s^2 2s^2 2p^6 3s^2 3p^6 3d^{10} 4s$). Then the completion of the $4s$ and $4p$ shells goes on without disturbance up to krypton ^{36}Kr, but again there is a blank for rubidium ^{37}Rb and strontium ^{38}Sr: the $5s$ shell fills, whereas the $4d$ and $4f$ shells remain empty. The $4d$ shell starts to fill from yttrium ^{39}Y to palladium ^{46}Pd, and the filling of the $4f$ shell commences with cerium ^{58}Ce and terminates with ytterbium ^{70}Yb. Table 1.10 presents equilibrium atomic configurations for all the 92 elements of the periodic table.

Elements said to be normal are those in which all shells, except perhaps for the outermost shell, are successively filled from the beginning to the end or in which there are completely vacant shells. Transition elements are those in which

Table 1.10. Equilibrium electron shell configurations for the elements of the periodic table

	^{1}H	^{2}He						
$1s$	1	2						
	^{3}Li	^{4}Be	^{5}B	^{6}C	^{7}N	^{8}O	^{9}F	^{10}Ne
$2s$	1	2	2	2	2	2	2	2
$2p$			1	2	3	4	5	6

Table 1.10 (continued)

	^{11}Na	^{12}Mg	^{13}Al	^{14}Si	^{15}P	^{16}S	^{17}Cl	^{18}Ag
3*s*	1	2	2	2	2	2	2	2
3*p*			1	2	3	4	5	6

	^{19}K	^{20}Ca	$^{21}_{(1)}$Sc	^{22}Ti	^{23}V	^{24}Cr	^{25}Mn	^{26}Fe	^{27}Co	^{28}Ni	^{29}Cu
3*d*			1	2	3	5	5	6	7	8	10
4*s*	1	2	2	2	2	1	2	2	2	2	1

	^{30}Zn	^{31}Ga	^{32}Ge	^{33}As	^{34}Se	^{35}Br	^{36}Kr
4*s*	2	2	2	2	2	2	2
4*p*		1	2	3	4	5	6

	^{37}Rb	^{38}Sr	$^{39}_{(2)}$Y	^{40}Zr	^{41}Nb	^{42}Mo	^{43}Tc	^{44}Ru	^{45}Rh	^{46}Pd
4*d*			1	2	4	5	6	7	8	10
5*s*	1	2	2	2	1	1	1	1	1	

	^{47}Ag	^{48}Cd	^{49}In	^{50}Sn	^{51}Sb	^{52}Te	^{53}I	^{54}Xe
5*s*	1	2	2	2	2	2	2	2
5*p*			1	2	3	4	5	6

	^{55}Cs	^{56}Ba	$^{57}_{(3)}$La	^{58}Ce	^{59}Pr	^{60}Nd	^{61}Pm	^{62}Sm
4*f*				1	3	4	5	6
5*d*			1	1				
6*s*	2	2	2	2	2	2	2	2
	^{63}Eu	^{64}Gd	^{65}Tb	^{66}Dy	^{67}Ho	^{68}Ez	^{69}Tm	^{70}Yb
4*f*	7	7	8	10	11	12	13	14
5*d*		1	1					
6*s*	2	2	2	2	2	2	2	2

	$^{71}_{(4)}$Lu	^{72}Hf	^{73}Ta	^{74}W	^{75}Re	^{76}Os	^{77}Ir	^{78}Pt	^{79}Au
5*d*	1	2	3	4	5	7	7	9	10
6*s*	2	2	2	2	2	1	2	1	1

	^{80}Hg	^{81}Tl	^{82}Pb	^{83}Bi	^{84}Po	^{85}At	^{86}Rn
6*s*	2	2	2	2	2	2	2
6*p*		1	2	3	4	5	6

	^{87}Fr	^{88}Ra	$^{89}_{(5)}$Ac	^{90}Th	^{91}Pa	^{92}U
5*f*					2	3
6*d*			1	2	1	2
7*s*	1	2	2	2	2	2

the previously unfilled shells are filled up. Of the 92 elements (up to the transuranium elements—Table 1.10), there are 40 transition elements. They are divided into three groups: 24 *d* elements that constitute three subgroups of eight elements with an incomplete 3*d* shell (iron group), 4*d* shell (palladium group), and 5*d* shell (platinum group); 12 *f* elements with incomplete 4*f* shells (rare-earth lanthanide group); and four mixed *d*–*f* elements with incomplete 6*d* and 5*f* shells (actinide group).

All transition-metal crystals are metals. Figure 1.29 presents radial electron densities $P^2(r) = 4\pi n(r)r^2$ in different shells of the Gd^+ ion, calculated according to the Hartree–Fock method [1.21]. It also shows the magnitude of half the lattice parameter. As seen from the figure, only the outer 6*s* shells of the adjacent atoms will superpose and undergo substantial deformation as compared with the shell of an isolated atom. An individual wave function $\psi_\alpha(\boldsymbol{r})$ may be approximately introduced for each electron in the atomic shell (the exact many-electron wave function depends on the coordinates and spins of all atomic-shell electrons). The parameter characterizing the effect of the interatomic interaction in the crystal on the motion of an electron, described by the wave function $\psi_\alpha(\boldsymbol{r})$, may therefore be taken to be the product of the wave functions of two electrons of adjacent atoms

$$S_{\alpha\alpha'}^{(n)}(\boldsymbol{r}) = \psi_\alpha^*(\boldsymbol{r})\psi_{\alpha'}(\boldsymbol{r} + n\boldsymbol{a}) \,, \tag{1.6.1}$$

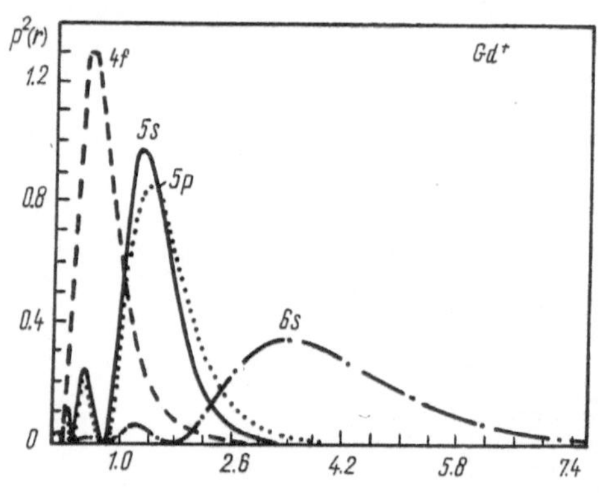

Fig. 1.29. Radial charge densities for 4*f*, 5*s*, and 6*s* electrons in a Gd^+ ion as a function of the distance from the nuclear center

with $n = 1$ for the nearest neighbors and with $\alpha = \alpha'$ for the same states. If the function $S_{\alpha\alpha'}^{(n)}(\boldsymbol{r})$ describing the overlap of the wave functions of the neighboring atoms is small in the entire space, the interaction between the corresponding electrons will be weak and, in the crystal, they will move in nearly the same way as in isolated atoms. If this function is different from zero at least in some regions of space, the interaction is liable to be large enough and the electron motion may alter substantially in character. The wave functions of outer- (valence-) shell electrons overlap more strongly than those of inner-shell electrons. For each shell, a mean effective radius r_α may be introduced which corresponds to the maximum of its radial density. Then the dimensionless parameter

$$\xi = 2r_\alpha/d \tag{1.6.2}$$

for electrons that practically do not change the character of their motion during condensation from a gas into a crystal will be $\xi \ll 1$. For electrons that undergo substantial changes the dimensionless parameter will be $\xi \gtrsim 1$. This criterion provides ample reason to permit atoms in a crystal to be "divided" into two parts: the atomic nucleus with inner electrons, for which $\xi \ll 1$ is the ion core, and electrons with $\xi \gtrsim 1$, which "separate" from their former "hosts", i.e., the isolated atoms, and gain access to the crystal as a whole.

Following Frenkel, we shall call the latter itinerant electrons. Thus the crystal may be broken up into two more or less self-contained subsystems (note that the problem as to the degree of self-containedness appears to be not so simple, for these subsystems may be strongly interrelated). One of them is the crystal lattice of ion cores, which holds a lion's share of the mass of the crystal „($M_{\text{ion}} \geqslant 1836\ \text{m}_{\text{el}}$), the ion cores performing small oscillations near the lattice sites. The mean amplitude of these displacements $\langle u \rangle_T$ at temperatures not very close to the melting point is small compared to the lattice parameter d [except for helium—see (1.2.2)]. The other subsystem consists of the itinerant electrons, for which the character of collective behavior may be quite diverse—from the formation of quasi-molecular orbits to that of a conduction-electron "gas" or "liquid" in metals, which flows around the lattice of ion cores.

The difference between the ion core mass and the electron mass allows us to introduce one more small dimensionless parameter, which is equal to the ratio of these masses

$$\eta = m_{\text{el}}/M_{\text{ion}} \ll 1 \ . \tag{1.6.3}$$

The large mass of the cores sometimes permits the system of these cores to be described quasi-classically. The smallness of the electron mass, as follows from the Heisenberg uncertainty relation, always necessitates allowance for the quantum nature of the electronic subsystem. This accounts for the failures ("catastrophes") which the classical Lorentz electron theory has experienced (Sect. 3.3). At very low temperatures, because of the presence of zero vibrations, the quantum effects in the ionic subsystem must be considered as well (for the case of quantum crystals, see Sect. 1.8.7).

The smallness of the parameter η enables us to take advantage of the adiabatic approximation [1.22], which is widely used in the quantum theory of molecules and solids (Sect. 1.9).

1.7 Fundamental Concepts of the Chemical Bonding in Solids

As has already been stated, the collective behavior of valence electrons in crystals may be very varied and thus is responsible for the enormous variety of crystal structures of pure elements, let alone binary, ternary, and more complicated compounds. Along with simple NaCl and CsCl, Li and Na, and Cu and Fe lattices, there exist complicated structures of α Se and α Mn, of molecular crystals of organic compounds and complicated biological substances—DNA (deoxyribonucleic acid), RNA (ribonucleic acid), etc. In the following we confine ourselves to simple structures, although we do not yet have at our disposal any quantitative theory that is rigorous enough to explain completely even the atomic arrangement in them.

The crystal structure is primarily determined by the structure of the outermost electronic subshell of atoms, i.e., by the atomic number Z of the relevant chemical element. Thus, for example, crystals constructed of atoms whose outermost electronic subshell is completely filled or has only one electron or, finally, is short of only one electron to be complete are quite different in their physicochemical properties. Examples of crystals of the first type are the crystals of rare (inert) gases, to the second type belong the crystals of alkali metals, and to the third the crystals of halides.

It may be expected that the electronic shells in the crystals of rare gases will be the least modified. In the crystals of alkali elements the atoms will readily supply the only outer-shell electron to the "pool" of the crystal. As for halide crystals constructed from halogen atoms, the latter will readily accept an electron from the atoms of the other constituents of the crystal (e.g., alkali-metal atoms) in order to fill the only empty place in the valence shell. Although very crude, this picture offers a quantitatively correct description of the electronic nature of chemical bonds. We make use of this picture to classify the main types of bonding forces between atoms (ions) or molecules in solids.

Naturally, the problem arises as to what forces are responsible for crystalline bonds. Atomic quantum theory shows that the structure of the atomic shell is determined by electric and magnetic electron interactions. Let us estimate the order of magnitude of these interactions. To this end we consider the energies of the electric, ε_{el}, and magnetic, ε_{mag}, interactions of two elementary charges e and two elementary magnetic moments (Bohr magnetons μ_B) that are separated by an interatomic spacing r_{at}. From the Coulomb and Biot–Savart laws we find for the energies of these interactions

$$\varepsilon_{el} \sim e^2/r_{at} \quad \text{and} \quad \varepsilon_{mag} \sim \mu_B^2/r_{at}^3 \ .$$

Since $r_{at} \sim 10^{-8}$ cm, $e \sim 10^{-10}$ erg$^{1/2}$ cm$^{1/2} \approx 10^{-19}$ C and $\mu_B \sim 10^{-20}$ erg$^{3/2}$ cm$^{3/2}$ J T^{-1}, we obtain

$$\varepsilon_{el} \sim 10^{-12}\,\text{erg} = 10^{-19}\,\text{J} \quad \text{and} \quad \varepsilon_{mag} \sim 10^{-16}\,\text{erg} = 10^{-23}\,\text{J} \ .$$

The electric-interaction energy thus is four orders of magnitude higher than the magnetic-interaction energy. From the melting temperatures of typical solids, $\vartheta_{\mathrm{mel}} \sim 10^3$–$10^4$ K, we find the estimate of the mean atomic binding energy ε_{b} in the crystal

$$\varepsilon_{\mathrm{b}} \sim k_{\mathrm{B}}\vartheta_{\mathrm{mel}} = 10^{-13}\text{–}10^{-12}\ \mathrm{erg} = 10^{-20}\text{–}10^{-19}\ \mathrm{J} \ ,$$

with $k_{\mathrm{B}} = 1.38 \times 10^{-16}$ erg/K $= 1.38 \times 10^{-23}$ J/K being the Boltzmann constant. A comparison of the last two formulas shows that the observed binding energies of actual crystals arise chiefly from electric forces. Weaker magnetic forces play a less important part in crystalline bonds. But in the case of heavy elements with uncompensated magnetic moments in their electronic shells ($\sim 10\ \mu_{\mathrm{B}}$) the magnetic spin–orbit interactions are liable to make an appreciable contribution.

1.7.1 Interaction Between Atoms (Ions) with Filled Electron Shells

We begin our treatment of the various chemical bond types in a solid with the relatively simple case of atoms (ions) having completely filled electron shells. These may be neutral atoms of rare gases, singly negatively charged halide ions, singly positively charged alkali-metal ions, etc.

We shall consider atoms (ions) that are separated by a spacing R which is larger compared to their effective radii r_0 (Fig. 1.30). The Hamiltonian of the electrostatic interaction between them has the form

$$\hat{H}_{\mathrm{int}} = \sum_{ij} e_i e'_j |\boldsymbol{r}_i - \boldsymbol{r}'_j - \boldsymbol{R}|^{-1} \ , \tag{1.7.1}$$

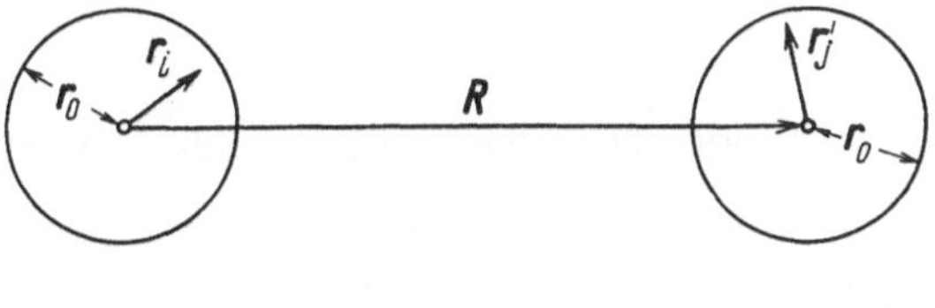

Fig. 1.30. Derivation of the van der Waals interaction

where e_i denotes the charges of the particles (of the electrons and the nucleus) that constitute the first ion, and $\boldsymbol{r}_i$ the radii of the vector with respect to the nucleus.The prime stands for similar quantities that belong to the second ion (the indices i and j run over the numbers of all the electrons and of the nucleus of the corresponding ion), and $\boldsymbol{R}$ is a vector that connects two nuclei. We expand (1.7.1) in a power series of a small dimensionless parameters r_0/R. This gives

$$|\boldsymbol{r}_i-\boldsymbol{r}'_j-\boldsymbol{R}|^{-1}=\frac{1}{R}+(r_{i\alpha}-r'_{j\alpha})\frac{\partial}{\partial R_\alpha}\frac{1}{R}$$
$$+\tfrac{1}{2}(r_{i\alpha}-r'_{j\alpha})(r_{i\beta}-r'_{j\beta})\frac{\partial^2}{\partial R_\alpha \partial R_\beta}\frac{1}{R}+\dots, \tag{1.7.2}$$

where the recurrent Cartesian indices α, $\beta = x, y, z$ imply summation. Substitute (1.7.2) into (1.7.1). In doing so, we introduce the notation

$$\sum_i e_i = q \ , \quad \sum_j e'_j = q' \tag{1.7.3}$$

for the charges of two ions (for neutral atoms $q=q'=0$), and

$$\sum_i e_i r_{i\alpha} = d_\alpha \ , \quad \sum_j e'_j r'_{j\alpha} = d'_\alpha \tag{1.7.4}$$

for the components of their dipole moments. Then

$$\sum_{ij} e_i e'_j (r_{i\alpha}-r'_{j\alpha}) = d_\alpha q' - d'_\alpha q \ ,$$

$$\sum_{ij} e_i e'_j (r_{i\alpha}-r'_{j\alpha})(r_{i\beta}-r'_{j\beta}) = d_\alpha d'_\beta - d_\beta d'_\alpha \tag{1.7.5}$$

$$+q'\sum_i e_i r_{i\alpha} r_{i\beta} + q\sum_j e'_j r'_{j\alpha} r'_{j\beta} \ . \tag{1.7.6}$$

Next, we calculate the derivatives of R^{-1} in (1.7.2):

$$\frac{\partial}{\partial R_\alpha}\frac{1}{R} = -\frac{n_\alpha}{R^2} \ , \qquad \frac{\partial^2}{\partial R_\alpha \partial R_\beta}\frac{1}{R} = \frac{1}{R^3}(3n_\alpha n_\beta - \delta_{\alpha\beta}) \ , \tag{1.7.7}$$

where $n_\alpha = R_\alpha/R$.

With allowance for (1.7.5–7), substitution of (1.7.2) into (1.7.1) yields the final result

$$\hat{H}_{\text{int}} = qq'/R + R^{-2}[q(\boldsymbol{d}'\cdot\boldsymbol{n}) - q'(\boldsymbol{d}\cdot\boldsymbol{n})]$$
$$+\tfrac{1}{2}R^{-3}\left\{q'\sum_i e_i[3(\boldsymbol{r}_i\cdot\boldsymbol{n})^2 - r_i^2] + q\sum_j e'_j[3\boldsymbol{r}'_j\cdot\boldsymbol{n})^2 - r_j'^2]\right\} \tag{1.7.8}$$
$$-R^{-3}(3n_\alpha n_\beta - \delta_{\alpha\beta})d_\alpha d'_\beta + \dots \ .$$

To calculate the correction to the energy of the system considered, which comes from the perturbation (1.7.8), we need to determine the set of zero-approximation wave functions. The latter can be chosen as the product of the wave functions describing the electronic systems of the first ($|m\rangle$) and second ($|n'\rangle$) atom (ion)

$$|mn'\rangle = |m\rangle|n'\rangle \ , \tag{1.7.9}$$

where m and n' are the respective many-electron states of the shells of the first and second atom (ion). The atomic (ionic) nuclei, in virtue of their large mass, are treated classically as electric field sources (adiabatic approximation—Sect. 1.9). Specifically, the wave function (1.7.9) of the ground state of the system involved is $|00'\rangle = |0\rangle|0'\rangle$, i.e., the product of the ground-state wave functions of each of the atoms (ions). Strictly speaking, the choice of (1.7.9) is not altogether correct, since, even in the absence of a dynamic interaction between two-atom (ion) shell electrons, it is necessary to take into account, in accordance with the Pauli principle, their statistical interaction, i.e., the antisymmetry of the wave functions with respect to the permutation of electrons (more exactly, the permutation of the spatial and spin coordinates of every single electron pair) belonging to different atoms (ions) (the wave functions of each individual ion may be assumed to be correct in this sense). However, it may be shown (more details will be given later) that such antisymmetrization leads only to the occurrence, in the interaction energy, of contributions that are exponentially small for $R \to \infty$.

According to conventional quantum-mechanical perturbation theory, the first-order correction for perturbation (1.7.8) to the ground-state energy is equal to the diagonal matrix element of the operator (1.7.8) which in this case has the form

$$E_1(R) = \langle 00'|\hat{H}_{\text{int}}|00'\rangle = qq'/R \ . \tag{1.7.10}$$

In fact, since the electron-density distribution in the closed electron shell of atoms (ions) is spherically symmetric, not only the dipole moments are absent—

$$\langle 0|\boldsymbol{d}|0\rangle = \langle 0'|\boldsymbol{d}'|0'\rangle = 0 \tag{1.7.11}$$

but also the quantity

$$\langle 0|n_\alpha n_\beta Q_{\alpha\beta}|0\rangle \equiv \langle 0|\sum_i e_i[3(\boldsymbol{r}_i\cdot\boldsymbol{n})^2 - r_i^2]|0\rangle \tag{1.7.12}$$

becomes identically equal to zero. A quadrupole moment tensor is introduced here:

$$Q_{\alpha\beta} = \sum_i e_i(3r_{i\alpha}r_{i\beta} - r_i^2\delta_{\alpha\beta}) \ . \tag{1.7.13}$$

We wish to prove the validity of (1.7.12). In virtue of the aforementioned spherical symmetry of the shells of isolated atoms (ions) with closed shells, the quantities (1.7.12) should not depend on the direction of the vector $\boldsymbol{n}$. Averaging (1.7.12) over spherical angles then yields a zero value. Indeed, for any vector $\boldsymbol{a}$ we have

$$\begin{aligned}\int d\omega[3(\boldsymbol{a}\cdot\boldsymbol{n})^2-a^2] &= \int_0^{2\pi} d\varphi \int_0^{\pi} \sin\vartheta d\vartheta(3a^2\cos^2\vartheta-a^2)\\ &= \pi a^2 \int_{-1}^{1} dx(3x^2-1)=0\ ,\end{aligned} \tag{1.7.14}$$

where the polar axis is aligned with the vector $\boldsymbol{a}$ and the substitution $\cos\vartheta = x$ is used. Since (1.7.12) does not depend on the vector $\boldsymbol{n}$, it is equal to its mean value, i.e., zero, as follows from (1.7.14). This is the case provided

$$\langle 0|Q_{\alpha\beta}|0\rangle = \langle 0'|Q'_{\alpha\beta}|0'\rangle = 0\ . \tag{1.7.15}$$

Allowing for (1.7.11) and (1.7.15), we obtain the result (1.7.10) for the diagonal matrix element of the operator (1.7.8). Moreover, this result remains valid also when we take into account the arbitrary-order terms in the small parameter r_0/R in (1.7.8). To see this, we calculate the electric potential $\varphi(R)$ produced by a nonpointlike ion with spherically symmetric distribution of the electron density $\varrho(r)$ in its shell. Let the ion be circumscribed by a sphere of radius R with its center coinciding with the center of the nucleus of the ion. Then, the electric field intensity $\boldsymbol{E}(R) = -(\partial\varphi(R)/\partial R)\boldsymbol{n}$ is oriented along the vector $\boldsymbol{R}$ and its flux through the above sphere is equal to $-4\pi R^2\partial\varphi(R)/\partial R$. According to the Gauss–Ostrogradskii theorem, we now have

$$-4\pi R^2\partial\varphi(R)/\partial R = 4\pi q(R)\ , \tag{1.7.16}$$

where

$$q(R) = \int_0^R 4\pi r^2\varrho(r)\,dr \tag{1.7.17}$$

is the charge contained within the sphere of radius R. Substituting (1.7.17) into (1.7.16) and performing integration by parts, we find

$$\begin{aligned}\varphi(R) &= \int_R^\infty drr^{-2}q(r) = -\int_R^\infty d(1/r)q(r)\\ &= q(R)/R - \int_R^\infty drr^{-1}dq(r)/dr = q(R)/R - 4\pi\int_R^\infty drr\varrho(r)\ .\end{aligned} \tag{1.7.18}$$

From (1.7.18) it follows that if $\varrho(r)$ decays exponentially with $r \to \infty$, as is the case for free atoms and ions, the correction to the asymptotic value of the potential $q(\infty)/R$ is exponentially small, too. Thus the quantity $E_1(R)$ is nothing but the electrostatic-interaction energy of nonpointlike ions that are in the ground state. Expression (1.7.10) therefore holds to within exponentially small terms. When taking account of the latter, however, in the zero approximation we must allow for the antisymmetry of the zero-approximation wave functions with respect to the permutation of the electrons of two ions. A quantum-mechanical calculation [1.23] thus leads to

$$\Delta E(R) = A(R)\exp(-\gamma R) \ , \tag{1.7.19}$$

where γ is a constant for a given pair of atoms (ions) and $A(R)$ is a function R which is smooth compared with the exponent and which also is normally replaced by a constant. An interaction of the type (1.7.19) is referred to as the Born–Mayer repulsion. It arises from both the purely classical effect of the electrostatic interaction of nonpointlike atoms (ions) and the Pauli principle, as well as from the overlap of the wave functions of different atoms (ions).

Now we proceed to a calculation of the second-order correction of perturbation theory respect to $\hat{H}_{\text{int}}$ from (1.7.8). For this correction, according to the general theory, we have

$$E_2(R) = \sum_{n=0;n' \neq 0'} \frac{|\langle 00'|\hat{H}_{\text{int}}|nn'\rangle|^2}{E_0 + E_{0'} - E_n - E_{n'}} \ , \tag{1.7.20}$$

where E_n and $E_{n'}$ are the energies of the corresponding states of noninteracting atoms (ions). In virtue of the orthogonality of the excited states to the ground state,

$$\langle 0|n\rangle = \langle 0'|n'\rangle = 0 \ , \tag{1.7.21}$$

the contribution to (1.7.20) is made only by the last term in the operator (1.7.8). This is so because only this term contains products of the operators that act upon the states of the first and the second ion. Denoting the ion excitation energies by $\varepsilon_n \equiv E_n - E_0$, $\varepsilon_{n'} \equiv E_{n'} - E_{0'}$ and substituting (1.7.8) into (1.7.20), we obtain

$$\begin{aligned} E_2(R) = &-(3n_\alpha n_\beta - \delta_{\alpha\beta})(3n_\gamma n_\eta - \delta_{\gamma\eta})R^{-6} \\ &\times \sum_{nn'} \langle 0|d_\alpha|n\rangle\langle n|d_\gamma|0\rangle\langle 0'|d'_\beta|n'\rangle\langle n'|d'_\eta|0'\rangle/(\varepsilon_n + \varepsilon_{n'}) \ . \end{aligned} \tag{1.7.22}$$

As a consequence of spherical symmetry, upon summation over states with the same energy but with different directions of the total angular momentum of an atom (ion), all off-diagonal tensor elements $\langle 0|d_\alpha|n\rangle\langle n|d_\gamma|0\rangle$ will go to zero and the diagonal elements will be equal to each other. In (1.7.22) we therefore may set

$$\langle 0|d_\alpha|n\rangle\langle n|d_\gamma|0\rangle \to \delta_{\alpha\gamma}|\langle 0|d_z|n\rangle|^2 \ ,$$
$$\langle 0'|d'_\beta|n'\rangle\langle n'|d'_\eta|0'\rangle \to \delta_{\beta\eta}|\langle 0'|d_z|n'\rangle|^2 \ . \tag{1.7.23}$$

Substituting (1.7.23) into (1.7.22) and allowing for the fact that

$$(3n_\alpha n_\beta-\delta_{\alpha\beta})(3n_\gamma n_\eta-\delta_{\gamma\eta})\delta_{\alpha\gamma}\delta_{\beta\eta}=(3n_\alpha n_\beta-\delta_{\alpha\beta})(3n_\alpha n_\beta-\delta_{\alpha\beta})$$
$$=9(n^2)^2-6n^2+3=6 \ ,$$

we obtain the final expression

$$E_2(R)=-6R^{-6}\sum_{nn'}|\langle 0|d_z|n\rangle|^2|\langle 0'|d'_z|n'\rangle|^2/(\varepsilon_n+\varepsilon_{n'}) \ . \tag{1.7.24}$$

The interaction (1.7.24) is called the van der Waals attraction. Its origin may be clarified as follows (for simplicity, we shall talk of neutral atoms). Assume that the first atom has passed virtually into an excited state with a nonzero dipole moment. Then, the latter produces an electric field of intensity F that decreases with distance as R^{-3}, which, in turn, induces on the second atom, owing to finite polarizability, a dipole moment whose magnitude is proportional to the field F, i.e., is again proportional to $\sim R^{-3}$. The induced dipole moment is oriented parallel to the field vector $\boldsymbol{F}$, leading to an energy gain proportional to $\sim R^{-3}R^{-3}=R^{-6}$. The van der Waals interaction thus is wholly determined by atomic (ionic) polarizability. It stands to reason that, in the case of neutral atoms, (1.7.24) gives only a leading term in the interaction energy expansion in r_0/R. The subsequent terms diminish as $\sim R^{-8}$, $\sim R^{-10}$, $\sim R^{-12}$, etc.

Thus, aside from the normal electrostatic energy which is absent in the case of neutral atoms, the forces of interaction in the case of atomic systems with filled electron shells have both a contribution that corresponds to attraction (1.7.24) and a contribution that corresponds to repulsion (1.7.19), the attraction prevailing at large distances. This attraction at large distances is due to fluctuations of the dipole moment (and also of the quadrupole and higher moments). The repulsion (1.7.19) is determined by the overlap of the wave functions (and, consequently, of the electron charge-density distribution tails) on different atoms (ions) and increases drastically with decreasing distance R. Expression (1.7.19) works well only in the limit of large R. In order to describe approximately the interaction of two neutral atoms with filled electron shells (rare gases) at any R, the Lennard–Jones potential

$$V(R)=4\varepsilon[(\sigma/R)^{12}-(\sigma/R)^6] \tag{1.7.25}$$

with fitting parameters ε and σ is normally used. The behavior of this potential is qualitatively shown in Fig. 1.31. Although (1.7.25) cannot be substantiated by quantum mechanics, it gives a fairly good fit to the various experimental data.

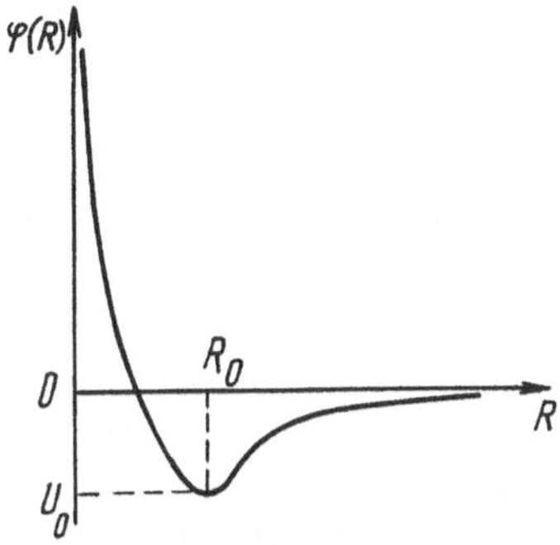

Fig. 1.31. The Lennard–Jones potential (schematic)

The interactions that we have considered here are universal; they contribute to the binding energy of both NaCl-type ionic crystals and solid rare gases. They play a certain role in metallic crystals too [1.24]. And still, they are relatively weak and, as a rule, cannot compare in order of magnitude with the bonding forces that are due to the electrons of the outer partially filled atomic or ionic shells. These interactions, as distinct from the van der Waals and Born–Mayer forces, arise from a substantial rearrangement of the electronic structure of the atomic (ionic) shell when they are united into a solid. They are subdivided into ionic bonding, when the outer-shell electrons of some atoms migrate into the shells of other atoms; covalent bonding, when the outer electrons are shared among the neighboring atoms; and metallic bonding, when the outermost electrons of the atoms are shared by all the atoms of the crystal. Like any classification, such a subdivision of solids is possible only approximately, and there an intermediate (hybridized) bond often occurs.

In subsequent sections we consider the covalent bond with the example of molecules and later on we will discuss how the "dominating" type of chemical bond governs the properties of the crystal.

1.7.2 Molecular Orbitals

To ascertain the nature of a chemical bond when a molecule or a crystal consists of atoms with partially filled outer electron shells, we will consider a simple model problem first. Let there be n noninteracting atoms, on each of which the valence electron may be in one orbital state with a wave function $\varphi_i(\boldsymbol{r})$ and energy ε_i $(i=1, 2, \ldots, n)$; the atoms may be different. Then the interaction $\hat{V}$ of valence electrons with "foreign" ion cores (by an ion core we understand a system composed of an atomic nucleus and all shell electrons, except for the valence electrons of the outer incomplete shell) comes into play; the interaction of the valence electrons themselves with each other will be disregarded for the time being. The Hamiltonian of the system being considered will thus become

$$\hat{\mathscr{H}} = \sum_{i=1}^{n} \hat{\mathscr{H}}_i + \hat{V} , \tag{1.7.26}$$

with $\hat{\mathscr{H}}_i$ being the Hamiltonian of a valence electron on the ith atom (ion). Therefore, the Schrödinger equations that satisfy the function $\varphi_i(\boldsymbol{r})$ and energy ε_i are of the form

$$\hat{\mathscr{H}}_i \varphi_i(\boldsymbol{r}) = \varepsilon_i \varphi_i(\boldsymbol{r}) . \tag{1.7.27}$$

On the assumption that the perturbation $\hat{V}$ is small compared with the energy spacing from ε_i to the energy of the next state of a given atom, but, possibly, not small compared with the energy difference $|\varepsilon_i - \varepsilon_j|$ for two different atoms $(i \neq j)$, we should make use of the perturbation theory for a degenerate level and seek the eigenfunctions $\psi(\boldsymbol{r})$ of the Hamiltonian (1.7.26) as an expansion in $\varphi_i(\boldsymbol{r})$ with coefficients c_i:

$$\psi(\boldsymbol{r}) = \sum_i c_i \varphi_i(\boldsymbol{r}) . \tag{1.7.28}$$

The functions $\psi(\boldsymbol{r})$ will be called the molecular orbitals of the problem considered.

We also include in our consideration the case of like atoms, when all ε_i are equal and the usual perturbation theory, which disregards this degeneracy, is not meaningful whatever $\hat{V}$ may be. Substituting (1.7.28) into the Schrödinger equation with the Hamiltonian (1.7.26) and taking into account (1.7.27), we have

$$(\hat{\mathscr{H}} - E)\psi(\boldsymbol{r}) = 0 \tag{1.7.29}$$

$$\sum_i c_i(\varepsilon_i - E)\varphi_i(\boldsymbol{r}) + \sum_i c_i \hat{V} \varphi_i(\boldsymbol{r}) = 0 . \tag{1.7.30}$$

Premultiplying (1.7.30) by $\varphi_j^*(\boldsymbol{r})$ and integrating over the entire configurational space $\boldsymbol{r}$, we obtain a system of linear algebraic equations to define the coefficients

$$\sum_i [(\varepsilon_i - E) S_{ji} + V_{ji}] c_i = 0 , \tag{1.7.31}$$

where

$$V_{ji} = \int d\boldsymbol{r}\, \varphi_j^*(\boldsymbol{r}) \hat{V} \varphi_i(\boldsymbol{r}) , \quad S_{ji} = \int d\boldsymbol{r}\, \varphi_j^*(\boldsymbol{r}) \varphi_i(\boldsymbol{r}) \tag{1.7.32}$$

are respectively the matrix elements of the perturbation operator $\hat{V}$ and the overlap integrals of the functions $\varphi_i(\boldsymbol{r})$ (at $i \neq j$) or their normalizations (at $i=j$). The condition for the system of equations (1.7.27) to be solvable is

$$\det \| \varepsilon_i S_{ji} + V_{ji} - E S_{ji} \| = 0 \ . \tag{1.7.33}$$

The secular equation (1.7.33) defines n, generally distinct eigenvalues E_i. Assuming that the functions $\varphi_i(\boldsymbol{r})$ of different atoms do not overlap to any great extent, and using the condition for approximate orthonormality of these functions

$$S_{ji} \approx \delta_{ji} \equiv \begin{cases} 1, & i=j \\ 0, & i \neq j \ , \end{cases} \tag{1.7.34}$$

equation (1.7.33) has the standard form of an eigenvalue problem. When the interatomic spacing is large, S_{ji} differs very little from δ_{ji} and can be written as

$$S_{ji} = \delta_{ji} + \lambda \sigma_{ji} \ , \tag{1.7.35}$$

with σ_{ji} being some matrix, and λ a formal small parameter proportional to the overlap of the wave functions of valence electrons of different atoms. The reciprocal matrix S^{-1} may then be defined as an expansion in the small quantity λ:

$$S_{ij}^{-1} = (1+\lambda\sigma)_{ij}^{-1} = \delta_{ij} - \lambda\sigma_{ij} + \lambda^2(\sigma^2)_{ij} - \ldots \ . \tag{1.7.36}$$

Multiplying (1.7.31) by S_{kj}^{-1} from (1.7.36) and summing over k, we can readily verify that E_i are the eigenvalues of a matrix with the elements

$$\tilde{H}_{ik} = \varepsilon_i \delta_{ik} + \sum_k S_{kj}^{-1} V_{ji} \approx \varepsilon_i \delta_{ik} + V_{ki} - \lambda \sum_j \sigma_{kj} V_{ji} + \ldots \ . \tag{1.7.37}$$

As is known, the sum of all the eigenvalues of any matrix is equal to its trace, i.e., the sum of all the diagonal elements. Therefore, to within small terms of the order of λV, we have

$$\sum_{i=1}^{n} E_i = \sum_{i=1}^{n} (\varepsilon_i + V_{ii}) \ . \tag{1.7.38}$$

The mean energy value of our "molecular" problem thus is equal to the mean value of the energy of an atom (with the natural correction for the potential energy of the interaction with the other atoms, V_{ii}). This signifies that if all atomic levels were filled (and thus all "molecular" levels with energies E_i also turned out to be filled), the energy of the system during the transition of the electrons to the molecular orbitals would not alter [the small variation on the order of $\sim \lambda^2$ arising from the overlap of wave functions on different sites can be shown to correspond to the Born–Mayer repulsion (1.7.19) between the filled shells]. But since we deal with partially filled atomic states and since the filling of molecular orbitals begins with lowest-energy states, we are bound, by virtue of (1.7.38), to obtain an energy gain in the case of partial filling as compared with the mean energy of an electron in the atom. This gain is very easy to understand in another way: when an electron passes from the atomic state on to the molecular orbital, the probability of its being located on "foreign" sites ($c_i \neq 0$ for any i) increases. In consequence, the indeterminacy of the coordinate of the electron increases and, therefore, by the Heisenberg uncertainty principle, the indeterminacy of its momentum, i.e. its kinetic energy, decreases.

To gain a better understanding of the results obtained, we examine in more detail the case of a diatomic molecule with one electron per atom. In doing so, we adopt for simplicity (1.7.34) and also set $V_{11} = V_{22} = 0$: (1.7.38) then immediately yields

$$E_{1,2} = (\varepsilon_1 + \varepsilon_2)/2 \mp [(\varepsilon_1 - \varepsilon_2)^2/4 + |V_{12}|^2]^{1/2} . \tag{1.7.39}$$

Each of the energy states $E_{1,2}$ of the molecule may be occupied by two electrons with opposite spin projections. For the relation of the probability of an electron being located at core 1 and core 2 in the state with the lowest energy E_1 corresponding to the minus sign in (1.7.39), (1.7.31) yields

$$\left|\frac{c_1}{c_2}\right|^2 = \frac{\left(\sqrt{(\varepsilon_1 - \varepsilon_2)^2 + 4[V_{12}]^2} + \varepsilon_2 - \varepsilon_1\right)^2}{4|V_{12}|^2} . \tag{1.7.40}$$

If $|V_{12}|^2 \ll |\varepsilon_1 - \varepsilon_2|^2$ and (for certainty) $\varepsilon_2 > \varepsilon_1$, it follows from (1.7.40) that

$$\left|\frac{c_1}{c_2}\right|^2 \approx \frac{(\varepsilon_2 - \varepsilon_1)^2}{|V_{12}|^2} \gg 1 ; \tag{1.7.41}$$

that is, both electrons in this case will with high probability be localized in the vicinity of atom 1, i.e., the lower-energy atom. This is the simplest model of the ionic bond, of which the transition of an electron from one atom to another is typical. But if $\varepsilon_1 = \varepsilon_2$, then, in view of (1.7.40),

$$\left|\frac{c_1}{c_2}\right|^2 = 1 ; \tag{1.7.42}$$

i.e., the electrons are located with equal probability on the two atoms. This case models a purely covalent bond, when both electrons are shared between two equal atoms. For $|V_{12}| \sim |\varepsilon_1 - \varepsilon_2|$, an intermediate case will take place.

Molecular orbitals with an energy smaller than the mean energy of valence electrons in atoms are called bonding orbitals, and those with an energy exceeding the mean energy of valence electrons are referred to as antibonding orbitals. These terms derive from the fact that the passage of a valence electron to the bonding molecular orbital decreases the total energy of the molecule, whereas the filling of antibonding orbitals weakens the chemical bond. The bonding and antibonding orbitals also differ drastically in electron-density distribution.

Let us consider the case of a purely covalent bond with $\varepsilon_1 = \varepsilon_2$, when $|c_1|^2 = |c_2|^2$. For simplicity, we take the functions φ_i to be real (which can always be done in view of the Hamiltonian being real in the absence of a magnetic field); then the matrix element V_{12} is real, too. The electron on atom 1 is attracted to the ion core of atom 2, so $V_{12} < 0$. But then it follows immediately from (1.7.31, 39) that for a bonding orbital (with lower energy)

$$\frac{c_1}{c_2} = -\frac{V_{12}}{|V_{12}|} = 1 \ ,$$

and for an antibonding orbital (with higher energy) $c_1 = -c_2$. Thus, the wave function in the first case is symmetric with respect to the rearrangement of atoms 1 and 2, and in the second case it is antisymmetric (and therefore goes to zero in the middle between the atoms). For an antibonding orbital, the probability of finding an electron between the atoms therefore is small, whereas the probability of finding it near the atoms is large. Conversely, for a bonding orbital, the electron-density maximum is reached in the space enclosed between the atoms.

As a result, we see that the concepts of the covalent and ionic bonds may already be introduced to consider diatomic molecules. As for the metallic bond, it is characterized by a complete sharing of valence electrons throughout the crystal. As will be shown in Chap. 4, the wave function of an electron in a metal may be regarded as a special molecular orbit that encompasses the entire crystal. Surprisingly, by neglecting the interaction between electrons, the state of any electron in any solid may, in a way, be described in this fashion. True, in this case formulating a criterion that would distinguish a metal from a nonmetal and, thus, a metallic bond from a covalent bond is not so simple. This problem will be treated in detail in Chap. 4. We should, however, state here that it cannot be regarded as ultimately resolved because of the necessity to take into account the electron–electron interaction, which is normally disregarded in the treatment of this problem (Chap. 5).

Let us emphasize once again that no sharply delineated borderlines exist between the different types of bond. For example, a number of metals reveal covalent-bond effects and some metallic alloys even show the presence of an ionic type of bond (see below).

1.7.3 The Heitler–London Method

A consideration of the molecular orbitals shows that even one electron is capable of stabilizing a molecule, i.e., of creating a chemical bond, if this electron occupies the bonding orbital (for example, a simple molecular ion of hydrogen, H_2^+, which consists of two protons and only one valence electron, is quite stable). Each molecular orbital may, according to the Pauli principle, be occupied by two electrons with opposite spins and therefore chemical bonding is usually effected by a pair of such electrons. However, the question arises of what impact the interelectron interaction might have on the energy gain in the case of electrons filling molecular orbitals (Sect. 1.7.2). After all, if each of the two valence electrons in a diatomic molecule, say, in a hydrogen molecule H_2, is situated with equal probability on any one of the atoms, there is one fourth the probability that both electrons will be on the same atom, a situation which is obviously energetically unfavorable from the point of view of their repulsive Coulomb interaction energy. Therefore, a simple molecular orbital description is applicable, at least qualitatively, only with interatomic distances that are not too large, when the energy gain due to the delocalization of an electron, on the order of $|V_{12}|$, exceeds the repulsive Coulomb interaction energy due to the localization of electrons on one atom.

In the opposite case, a more adequate approach to the problem of chemical bonding is to use the earlier method of W. Heitler and F. London (1927). The Heitler–London method disregards from the very outset the polar configurations, when the number of electrons near the nucleus differs from the number of electrons in a free atom (in a hydrogen molecule these are two and one); this leads to minimization of the electrons' Coulomb repulsion, but the electrons are deprived of the opportunity of lowering their kinetic energy owing to delocalization. It turns out, however, that the occurrence of interatomic bonding forces can be understood in terms of this approach as well.

In a nonrelativistic approximation, the Hamiltonian of a many-electron system does not depend on the spin variables of the electrons $s_i = \pm 1/2$. This immediately leads to two conclusions: (1) The wave function Ψ of the problem of interest may be represented as the product of two functions, one that depends only on the Cartesian coordinates $\Phi(\boldsymbol{r}_1, \boldsymbol{r}_2)$ and one that depends only on the spin variables $\chi(s_1, s_2)$:

$$\Psi(\boldsymbol{r}_1, s_1; \boldsymbol{r}_2, s_2) = \Phi(\boldsymbol{r}_1, \boldsymbol{r}_2)\chi(s_1, s_2) \ . \tag{1.7.43}$$

(2) The states of the system may be classed according to the values of their total spin (integral of motion) S; in the present case we deal with only two values $S = 0; 1$ (singlet state and triplet state respectively). In a relativistic treatment, a spin–orbit interaction arises which mixes the spin and orbit variables. Then, the representation of the function Ψ in the form of (1.7.43) is no longer correct and,

in addition, the total spin ceases to be an integral of the motion; only the total mechanical moment is an integral of the motion. However, the corresponding relativistic effects are very small since they are proportional to the ratio $\sim (v/c)^2$, where v is the mean velocity of the electrons in atoms. It is worth noting that, in a number of cases, the spin–orbit interaction can significantly affect the properties of molecules and solids, especially their magnetic behavior.

The spin function $\chi(s_1, s_2)$ is constructed from one-electron-spin functions $\alpha(s)$ and $\beta(s)$, where

$$\alpha(s) = \begin{pmatrix} 1 \\ 0 \end{pmatrix}, \quad \beta(s) = \begin{pmatrix} 0 \\ 1 \end{pmatrix}, \tag{1.7.44}$$

i.e., $\alpha(+1/2) = 1$, $\alpha(-1/2) = 0$, etc. The functions α and β refer to an electron with the spin projection oriented "upward" and "downward" respectively. The functions $\chi(s_1, s_2)$, which correspond to the total spin $S = 1$ and to its z projection equal to ± 1, can be written immediately

$$\begin{aligned} \chi_{1,1} &= \alpha(s_1)\alpha(s_2) \\ \chi_{1,-1} &= \beta(s_1)\beta(s_2) \end{aligned} \tag{1.7.45}$$

(in the first case both z-spin projections are directed upward, in the second both are oriented downward). The state with the spin $S = 1$ and z projection $S_z = 0$ may be obtained from $\chi_{1,1}$ by the action of the operator S^-, which lowers the spin projection by unity [Ref. 1.12, Sects. 27, 55]:

$$S^- \chi_{S,S_z} = \sqrt{(S + S_z)(S - S_z + 1)}\, \chi_{S,S_z - 1} . \tag{1.7.46}$$

But in this case $S^- = (\sigma_1^- + \sigma_2^-)/2$, where $\sigma^- = \sigma^x - i\sigma^y$ are the corresponding Pauli matrices acting on the variables s_1 and s_2:

$$\sigma^- \alpha(s) = 2\beta(s) , \quad \sigma^- \beta(s) = 0 . \tag{1.7.47}$$

From (1.7.46, 47) we then find

$$\chi_{1,0}(s_1, s_2) = \frac{1}{\sqrt{2}}[\alpha(s_1)\beta(s_2) + \alpha(s_2)\beta(s_1)] . \tag{1.7.48}$$

Finally, the function $\chi_0(s_1, s_2)$ describing the singlets $S = 0$ and $S_z = 0$ should also, similar to $\chi_{1,0}$, be constructed from the products of $\alpha(s_1)\beta(s_2)$ and $\alpha(s_2)\beta(s_1)$ $(S_z = 0)$ but should be orthogonal to $\chi_{1,0}(s_1, s_2)$:

$$\sum_{s_1, s_2} \chi_0^*(s_1, s_2)\chi_{1,0}(s_1, s_2) = 0 . \tag{1.7.49}$$

As is easy to verify, these conditions are satisfied by the function

$$\chi_0(s_1, s_2) = \frac{1}{\sqrt{2}}[\alpha(s_1)\beta(s_2) - \alpha(s_2)\beta(s_1)] \ . \tag{1.7.50}$$

From (1.7.45, 48, 50) it follows that the singlet state is described by an antisymmetric spin function, and the triplet state by a symmetric function with respect to the permutation of the spin coordinates: $s_1 \rightleftarrows s_2$. In accordance with the Pauli principle, the wave function (1.7.43) should be antisymmetric with respect to the simultaneous permutation of spin and orbital variables. Therefore, for the orbital parts of the total function we have

$$\begin{aligned} \phi_0(\boldsymbol{r}_1, \boldsymbol{r}_2) &= \phi_0(\boldsymbol{r}_2, \boldsymbol{r}_1) \ , \\ \phi_1(\boldsymbol{r}_1, \boldsymbol{r}_2) &= -\phi_1(\boldsymbol{r}_2, \boldsymbol{r}_1) \ , \end{aligned} \tag{1.7.51}$$

where the subscript of the functions denotes the magnitude of the total spin.

Up to this point, the treatment has been absolutely general in character. In order to obtain concrete results, we need to exploit perturbation theory and, to this end, to represent the total Hamiltonian in the form of (1.7.26). In doing so, the question naturally arises as to where we should enter the interelectron interaction: into the zero part of the Hamiltonian $\hat{\mathscr{H}}_0$ or into the perturbation $\hat{V}$? If we completely disregard this interaction, we return to the problem of molecular orbitals, which is outlined in Sect. 1.7.2. The wave functions of a one-electron problem in this case will have the form

$$\psi_{s,a}(\boldsymbol{r}) = c_{s,a}[a(\boldsymbol{r}) \pm b(\boldsymbol{r})] \ , \tag{1.7.52}$$

where we have introduced the notation $\varphi_1(\boldsymbol{r}) = a(\boldsymbol{r})$, $\varphi_2(\boldsymbol{r}) = b(\boldsymbol{r})$, and the subscripts s and a stand for the bonding and antibonding orbitals, respectively. The constants $c_{s,a}$ are defined by the normalization conditions. Hence we have

$$c_{s,a} = \frac{1}{\sqrt{2(1 \pm |S_{12}|^2)}} \ , \tag{1.7.53}$$

where S_{12} is the matrix element (1.7.32) if a and b are normalized. By virtue of (1.7.51, 52), the lowest-energy singlet-state function may be taken to be

$$\phi_0(\boldsymbol{r}_1, \boldsymbol{r}_2) = \psi_s(\boldsymbol{r}_1)\psi_s(\boldsymbol{r}_2) \ , \tag{1.7.54}$$

and the triplet function chosen as

$$\phi_1(\boldsymbol{r}_1, \boldsymbol{r}_2) = \frac{1}{\sqrt{2}}[\psi_s(\boldsymbol{r}_1)\psi_a(\boldsymbol{r}_2) - \psi_a(\boldsymbol{r}_1)\psi_s(\boldsymbol{r}_2)] \ . \tag{1.7.55}$$

The function (1.7.54) describes the state with two electrons with opposite z-spin projections on the bonding orbital, and (1.7.55) two electrons with parallel z-spin projections, one of which occupies the bonding orbital and the other the antibonding orbital. In the molecular-orbitals method, perturbation theory then allows the electron–electron interaction and other effects to be taken into account.

As stated previously, this approach is unsatisfactory in the case of large interatomic distances, when the interelectron interaction in one atom considerably exceeds the quantity $|V_{12}|$. In the Heitler–London scheme, the zero-approximation functions are chosen so that the energy of the atoms (including the interelectron interaction) is minimized. As for the interaction of electrons on different atoms, it is included, together with the interaction with "foreign" nuclei, in the perturbation operator $\hat{V}$. For a hydrogen molecule, this means that the zero-approximation function may involve only the product of the functions $a(\boldsymbol{r}_1)\,b(\boldsymbol{r}_2)$ and $a(\boldsymbol{r}_2)b(\boldsymbol{r}_1)$, but not the product of $a(\boldsymbol{r}_1)\,a(\boldsymbol{r}_2)$ and $b(\boldsymbol{r}_1)\,b(\boldsymbol{r}_2)$, since a configuration with two electrons on one atom is disadvantageous in terms of interelectron repulsion. Then we arrive at

$$\phi_0(\boldsymbol{r}_1, \boldsymbol{r}_2) = \frac{a(\boldsymbol{r}_1)b(\boldsymbol{r}_2) + a(\boldsymbol{r}_2)b(\boldsymbol{r}_1)}{\sqrt{2(1 + |S_{12}|^2)}} , \tag{1.7.56}$$

$$\phi_1(\boldsymbol{r}_1, \boldsymbol{r}_2) = \frac{a(\boldsymbol{r}_1)b(\boldsymbol{r}_2) - a(\boldsymbol{r}_2)b(\boldsymbol{r}_1)}{\sqrt{2(1 + |S_{12}|^2)}} , \tag{1.7.57}$$

with the coefficients being chosen from the normalization condition. To a zero approximation with respect to V, the energies of these states are equal. The energy correction in the first order of perturbation theory is of the form

$$\Delta E_0 = \int d\boldsymbol{r}_1\, d\boldsymbol{r}_2\, \phi_0^*(\boldsymbol{r}_1, \boldsymbol{r}_2)\, \hat{V} \phi_0(\boldsymbol{r}_1, \boldsymbol{r}_2) = (1 + |S_{12}|^2)^{-1}(K_{12} + J_{12}) , \tag{1.7.58}$$

$$\Delta E_1 = (1 - |S_{12}|^2)^{-1}(K_{12} - J_{12}) , \tag{1.7.59}$$

where

$$K_{12} \equiv \int d\boldsymbol{r}_1\, d\boldsymbol{r}_2\, a^*(\boldsymbol{r}_1)b^*(\boldsymbol{r}_2)\, \hat{V} a(\boldsymbol{r}_1)b(\boldsymbol{r}_2) , \tag{1.7.60}$$

$$J_{12} \equiv \int d\boldsymbol{r}_1 d\boldsymbol{r}_2 a^*(\boldsymbol{r}_1)b^*(\boldsymbol{r}_2)\, \hat{V} a(\boldsymbol{r}_2)b(\boldsymbol{r}_1) . \tag{1.7.61}$$

Calculating the corrections ΔE_0 and ΔE_1 as a function of interatomic spacing R for a hydrogen molecule leads to the result schematically portrayed in Fig. 1.32: in the triplet state the atoms repel each other, and in the singlet state there is a minimum which corresponds to a stable bound state of two atoms.

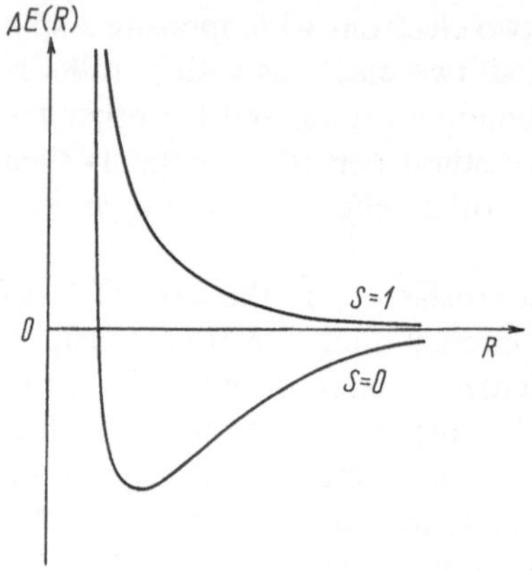

Fig. 1.32. Interatomic-interaction energy in H_2 molecule as a function of the magnitude of the total spin according to the Heitler–London method

Note that, even in the triplet state, when the distances are very large, the repulsion is replaced by an attraction that arises from van der Waals forces, which, however, are not taken into account here.

Thus, both the molecular-orbitals method and the Heitler–London method leads to the conclusion that the chemical bond in a hydrogen molecule is due to a pair of electrons with opposite spin projections. Quantitatively, neither of these methods provides a description that is altogether satisfactory; more sophisticated computational techniques are outlined by Slater [1.23].

The Pauli principle leads to the dependence of the energy of a many-electron system on the total spin, although in a nonrelativistic approximation the Hamiltonian itself is independent of spin. From (1.7.58, 59) we find the energy difference of the triplet and singlet states

$$\Delta E_1 - \Delta E_0 = 2(1-|S_{12}|^4)^{-1}\,(|S_{12}|^2 K_{12} - J_{12}) \equiv -2I_{12}\ , \tag{1.7.62}$$

where the quantity J_{12} is called the exchange integral (sometimes the quantity J_{12} is referred to as the exchange integral, but this is not quite accurate). Inspection of Fig. 1.32 shows that $I_{12} < 0$. Formula (1.7.62) may be treated as resulting from some spin-dependent electrostatic interaction, which is said to be an exchange interaction. Since the total spin $\boldsymbol{S} = \boldsymbol{s}_1 + \boldsymbol{s}_2$, we have the identity

$$\boldsymbol{S}^2 = (\boldsymbol{s}_1 + \boldsymbol{s}_2)^2$$

or

$$S(S+1) = 2\cdot\tfrac{1}{2}(\tfrac{1}{2}+1) + 2\boldsymbol{s}_1\boldsymbol{s}_2 = \tfrac{3}{2} + 2\boldsymbol{s}_1\boldsymbol{s}_2\ .$$

Thus the eigenvalues of the operator

$$\hat{P}_{12} = \tfrac{3}{4} + \boldsymbol{s}_1\boldsymbol{s}_2 \tag{1.7.63}$$

are equal to unity in the triplet state and equal to zero in the singlet state, and (1.7.62) may therefore be represented as

$$E = \text{const} - 2I_{12}\mathbf{s}_1\mathbf{s}_2 \ , \qquad (1.7.64)$$

where an energy contribution independent of the total spin is introduced into the constant.

The exchange interaction plays a most important part in the theory of the magnetic properties of solids (Chap. 5), for, by its nature, it is electrostatic and is normally much stronger than "truly magnetic" relativistic interactions. The concept "exchange interaction" owes its origin to *Heisenberg* [1.25] who in 1926 introduced it into the theory of the helium atom. The exchange interaction in the form of (1.7.64) was introduced by *Dirac* [1.26].

Note that the dependence of the total energy on the total spin S is certainly present in the one-electron approximation as well, where it is related to the Pauli principle: if $S \neq 0$, then at least part of the energy levels are singly electron-occupied, which is energetically unfavorable. Therefore, if the number of electrons in the system is even, then, in the absence of an interaction (and external magnetic fields), $S = 0$ in the ground state. In most of the molecules this is so also when the interelectron interaction is taken into account. However, there are exceptions, too—for example, $S = 1$ in an oxygen molecule O_2 in the ground state. This is a remote (molecular) forerunner of the ferromagnetism of metals such as Fe, Co, and Ni, when the entire crystal possesses a nonzero total spin in the ground state. This phenomenon is treated in more detail in Chap. 5.

1.7.4 Covalent Bond

The preceding treatment shows that the number of valence bonds in which an atom may participate is, as a rule, equal to the number of singly occupied states; i.e., it is equal to the number of unpaired electrons, $2S$, where S is the total spin of the atom. It stands to reason that this simple rule does not work well in all cases. For example, for a Eu^{++} ion, a half-filled $4f$ shell has a spin $S = 7/2$ (Table 1.10). Nevertheless, these seven unpaired electrons are not usually shared; i.e., they do not participate in the formation of chemical bonds. The point is that the interaction between $4f$-shell electrons on one site is much stronger than the energy gain from this collective behavior, because of the extremely weak overlap of the $4f$ functions with neighboring lattice sites (Sect. 1.6). In terms of interelectron–interatomic repulsion, the $4f^7$ configuration is, however, very stable. Broadly speaking, the problem as to the valence of d and f elements and as to the degree to which their electrons are collectivized is a very complicated one (Chap. 5).

For s and p elements, the assertion that the valence is equal to the number of unpaired electrons is almost always correct, but let it be emphasized that this statement does not necessarily apply to the number of unpaired electrons in the ground state of the atom. Since the formation of covalent bonds involves an energy gain, it is often energetically favorable to produce as many valence bonds as possible by sacrificing some of the electrons, which are transferred to excited states of the atom. As an illustration, we consider the behavior of valence bonds in the carbon atom [1.27].

The configuration of the valence electron shell of the carbon atom with the principal quantum number $n = 2$ in the ground state is $2s^2\,2p^2$, both p electrons having the same spin projection. This is a consequence of Hund's rules, by which configurations with maximum values of S are energetically favored. For a more elaborate treatment of this rule refer to Chap. 5. The energy of the excited configuration $2s^1\,2p^3$, which now has not two (as in the $2s^2\,2p^2$ configuration) but four unpaired electrons, $s^\uparrow\,(p^\uparrow)^3$, is also relatively low. Normally, in carbon, the energy loss resulting from the promotion of one s electron into a p state, is abundantly compensated for by the possibility of producing two additional bonds, and therefore carbon is most frequently quadrivalent.

Following Pauling, we consider the most important property of covalent bonds: their spatial orientation. In order to maximize the energy gained during the formation of a covalent bond, the overlap of the wave functions with those of neighboring atoms should also be maximum. This is possible if the neighboring atoms are situated in those directions from the central atoms in which the atomic wave functions take on maximum values. The bond here is stronger the larger these values are. Since carbon is quadrivalent, the bonding energy must be larger than the energy difference between the levels of the $2s$ and $2p$ states. There are four basis functions corresponding to the spherical harmonics [Ref. 1.12, Sect. 28]:

$$Y_{0,0}(\vartheta, \varphi) = \frac{1}{\sqrt{4\pi}}\,, \qquad Y_{1,0}(\vartheta, \varphi) = \mathrm{i}\sqrt{\frac{3}{4\pi}}\cos\vartheta\,,$$
$$Y_{1,\pm 1}(\vartheta, \varphi) = \mp\mathrm{i}\sqrt{\frac{3}{8\pi}}\sin\vartheta\exp(\pm\mathrm{i}\varphi)\,, \tag{1.7.65}$$

where ϑ and φ are polar angles. Rather than take the functions $Y_{1,m}(\vartheta, \varphi)$ to be the basis functions, it is more convenient to choose their orthonormalized linear combinations of the form

$$\frac{\mathrm{i}}{\sqrt{2}}[Y_{1,1}(\vartheta, \varphi) - Y_{1,-1}(\vartheta, \varphi)] = \sqrt{\frac{3}{4\pi}}\sin\vartheta\cos\varphi\,,$$
$$\frac{\mathrm{i}}{\sqrt{2}}[Y_{1,1}(\vartheta, \varphi) + Y_{1,-1}(\vartheta, \varphi)] = \sqrt{\frac{3}{4\pi}}\sin\vartheta\sin\varphi$$
$$-\mathrm{i}\,Y_{1,0}(\vartheta, \varphi) = \sqrt{\frac{3}{4\pi}}\cos\vartheta\,, \tag{1.7.66}$$

which transform under rotations as the Cartesian coordinates x, y, and z, respectively. Recall that the radial components of the s and p functions in our approximation are equal in magnitude and may be omitted together with the constant factor $1/\sqrt{4\pi}$ which is not important here. Then the angular dependence of the four basis functions which we will introduce in lieu of $Y_{l,m}(\vartheta, \varphi)$ can be represented as

$$|s\rangle = 1 \ ,$$

$$|x\rangle = \sqrt{3}\sin\vartheta\cos\varphi \ , \quad |y\rangle = \sqrt{3}\sin\vartheta\sin\varphi \ , \quad |z\rangle = \sqrt{3}\cos\vartheta \ . \tag{1.7.67}$$

We now seek linear combinations of $|\psi\rangle$ functions (1.7.67) that will assure maximum overlap with the functions of the adjacent atoms. This requires that the value of $\alpha = \max\limits_{\vartheta,\varphi} \psi$ be a maximum. With the normalization that we have chosen, $\alpha = 1$ for the s states, and $\alpha = \sqrt{3}$ for the p functions of $|x\rangle$, $|y\rangle$, and $|z\rangle$. We then represent the function $|\psi\rangle$ as

$$|\psi\rangle = a|s\rangle + b_1|x\rangle + b_2|y\rangle + b_3|z\rangle \ , \tag{1.7.68}$$

where a and b_i are real-valued coefficients that satisfy the normalization condition

$$a^2 + b_1^2 + b_2^2 + b_3^2 = 1 \ . \tag{1.7.69}$$

The function $|\psi\rangle$, then, is normalized in the same way as (1.7.67). This follows from their mutual orthogonality

$$\int d\omega |\psi(\vartheta, \varphi)|^2 \equiv \langle\psi|\psi\rangle = a^2\langle s|s\rangle + b_1^2\langle x|x\rangle + b_2^2\langle y|y\rangle + b_3^2\langle z|z\rangle = 4\pi \ ,$$

with $d\omega$ being an element of solid angle. For the time being, the orientation of the axes in our case is arbitrary. Let us assume that in one of the functions ψ, for which α is a maximum, this maximum value is reached in the direction along the diagonal of the cube (1, 1, 1), with the carbon atom at its center and with the coordinate axes parallel to its edges (Fig. 1.33). Then $b_1 = b_2 = b_3 = b$. The (1, 1, 1) direction is given by angles ϑ and φ such that

$$\sin\varphi = \cos\varphi = 1/\sqrt{2} \ , \quad \cos\vartheta = 1/\sqrt{3} \ , \quad \sin\vartheta = \sqrt{2/3} \ ,$$

so that

$$|x\rangle = |y\rangle = |z\rangle = 1 \ .$$

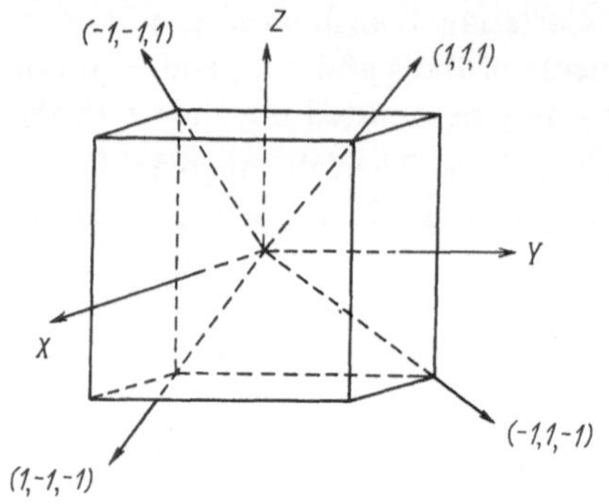

Fig. 1.33.

In addition,

$$\alpha = a + 3b = a + \sqrt{3(1 - a^2)} \tag{1.7.70}$$

where we have used the conditions (1.7.67). The maximum of α as a function of a is reached for $a = 1/2$ and is equal to 2. The quantity b in this case is equal to 1/2. Thus the first orbital with maximum values in the axes of coordinates that we have chosen is of the form

$$|1\rangle = \tfrac{1}{2}(|s\rangle + |x\rangle + |y\rangle + |z\rangle) \ . \tag{1.7.71}$$

It can be readily shown that the functions

$$\begin{aligned} |2\rangle &= \tfrac{1}{2}(|s\rangle + |x\rangle - |y\rangle - |z\rangle) \ , \\ |3\rangle &= \tfrac{1}{2}(|s\rangle - |x\rangle + |y\rangle - |z\rangle) \ , \\ |4\rangle &= \tfrac{1}{2}(|s\rangle - |x\rangle - |y\rangle + |z\rangle) \end{aligned} \tag{1.7.72}$$

correspond to the same value $\alpha = 2$. The functions $|i\rangle$ $(i = 1, 2, 3, 4)$ are mutually orthogonal. They take on their maximum values along the (1, 1, 1), $(1, \bar{1}, \bar{1})$, $(\bar{1}, 1, \bar{1})$, and $(\bar{1}, \bar{1}, 1)$ axes, i.e., along the axes of the tetrahedron, and, therefore, the maximum gain in chemical-bonding energy corresponds to the tetrahedral environment of the carbon atom. In spite of being qualitative, the treatment that we have performed above nevertheless explains the character of the crystal structure of the periodic table group-IV elements (diamond-type lattice, Fig. 1.22) as well as the shape of the methane molecule, which is very close to being tetrahedral. The above treatment may be approximated to reality by allowing in some manner for the interelectron interaction, specifically the electrostatic repulsion of charges on the bonds, which may somewhat alter the equilibrium values of the angles between them.

Thus, when considering covalent bonds, it is reasonable to proceed not from the atomic functions directly but from their linear combinations such as (1.7.71, 72), which are called hybrid orbitals. In constructing these orbitals, group-theory methods [1.23, 28] are very helpful in complicated cases.

The interatomic–interaction forces arising from the covalent bond may not be viewed as either paired or central, since the character of the interaction of any two atoms in a molecule in this case is largely determined by the position of the other atoms. The covalent bond possesses the property of saturation; the number of nearest neighbors Z of an atom in covalent-bond-stabilized crystals or molecules is fixed and is equal to the valence of that atom. Finally, the spatial electron-density distribution in the case of a covalent bond is strongly anisotropic: the covalent bond has maxima along certain directions.

All this distinguishes the covalent bond from the metallic bond, where the valence electrons collectively belong to the whole crystal. In typical metals such as sodium, electrons completely lose contact with their original ion cores and, as they do so, are distributed in space more or less uniformly. Nor do the metals possess the property of saturation. Thus, sodium is univalent but occurs in two modifications (bcc and hcp) with rather close packing (with the number of nearest neighbors at the lattice site being $Z = 8$ and $Z = 12$, respectively). As the valence increases, i.e., as we move to the right of the periodic table, the character of bonding changes from an almost purely metallic type of bond in the first group to an almost purely covalent bond in the fourth. The tendency for covalence weakens as we move downward from the top of the periodic table. Thus, carbon as diamond, germanium, and silicon are almost ideal examples of substances with a covalent type of bond, whereas in lead the bond is metallic rather than covalent. The difference in the properties of covalent-bonded metals and crystals will be discussed in more detail in Sect. 1.8.

1.7.5 Electrostatic Bonding Energy of Ionic Crystals

Basically, the ionic bond appears to be very simple and clear: it stems chiefly from the electrostatic point-ion interaction energy (1.7.10). Of course, quantum mechanics is an indispensable tool for both explaining why a given element exists in a given compound as an ion with a given charge and for taking into account the fact that the ions are not point charges (Sect. 1.7.1). But, by its origin, the fundamental contribution to the interaction energy is purely classical (unless the question is posed as to the origin of the ions). Therefore, a quantitative quasi-classical calculation of the electrostatic energy of ionic crystals can be carried out completely and accurately, in contrast to other types of interaction, where one has to solve fairly complicated problems of the quantum theory of many interacting particles.

Let us examine a crystal made up of positively and negatively charged point ions. The set of vectors (1.2.9), $\{\boldsymbol{R}_m\}$ forms the Bravais lattice of the crystal, and the position of the ions in each unit cell is given by the vectors (1.2.10), $\boldsymbol{R}_s$ ($s = 1, 2, \ldots, \sigma$), the ion charges q_s satisfying the neutrality condition

$$\sum_{s=1}^{\sigma} q_s = 0 \,. \tag{1.7.73}$$

We need to calculate the electrostatic energy of this system of charges:

$$E_c = \frac{1}{2} \sum_{m,s,s'}{}' \frac{q_s q_{s'}}{|\boldsymbol{R}_m + \boldsymbol{R}_s - \boldsymbol{R}_{s'}|} \,, \tag{1.7.74}$$

where the prime on the summation denotes that the term with $\boldsymbol{R}_m + \boldsymbol{R}_s - \boldsymbol{R}_{s'} = 0$ is canceled out. The difficulties in the calculation of (1.7.74) arise from the fact that the function R^{-1} falls very slowly when $R \to \infty$. Calculations of this kind were first carried out by E. Madelung (1918). Here we outline, with some minor differences from the standard form, an elegant method proposed by *Ewald* [1.29]. This method is based on the use of the reciprocal lattice (Sect. 1.3). By analogy with conventional Fourier series theory, we show that any function $f(\boldsymbol{r})$ possessing the translational symmetry of a given Bravais lattice can be expanded as a Fourier series in reciprocal-lattice vectors $\boldsymbol{b}_g^*$:

$$f(\boldsymbol{r}) = \sum_g \tilde{f}_g \exp(\mathrm{i}\, \boldsymbol{b}_g^* \cdot \boldsymbol{r}) \,, \tag{1.7.75}$$

where $\boldsymbol{b}_g^* = 2\pi \boldsymbol{b}_g$. Here the vector $\boldsymbol{b}_g$ is given by (1.3.3), and the expansion coefficients $\tilde{f}_g$ are equal to

$$\tilde{f}_g = \frac{1}{\bar{\omega}} \int_{\Omega} \varphi(\boldsymbol{r}) \exp(-\mathrm{i}\boldsymbol{b}_g^* \cdot \boldsymbol{r})\, d\boldsymbol{r} \,, \tag{1.7.76}$$

where the integral is taken over the unit cell, and $\bar{\omega}$ is the unit-cell volume. In particular, using the properties of the Dirac delta function, according to which the sum $\sum_m \delta(\boldsymbol{r} - \boldsymbol{R}_m)$ does not vary when $\boldsymbol{r}$ is shifted by any Bravais lattice vector, we have, by virtue of (1.7.75, 76),

$$\sum_m \delta(\boldsymbol{r} - \boldsymbol{R}_m) = \bar{\omega}^{-1} \sum_g \exp(\mathrm{i}\, \boldsymbol{b}_g^* \cdot \boldsymbol{r}) \,. \tag{1.7.77}$$

For any function φ and any vector $\boldsymbol{\varrho}$ we therefore find

$$\sum_{\boldsymbol{R}} \varphi(\boldsymbol{R} + \boldsymbol{\varrho}) \equiv \sum_{\boldsymbol{R}} \int d\boldsymbol{r} \delta(\boldsymbol{r} - \boldsymbol{R} - \boldsymbol{\varrho}) \varphi(\boldsymbol{\varrho}) = \bar{\omega}^{-1} \sum_g \varphi_g \exp(-\mathrm{i}\, \boldsymbol{b}_g^* \cdot \boldsymbol{\varrho}) \,, \tag{1.7.78}$$

where, in contrast to (1.7.76), the integration is performed over the entire space. Unfortunately, the Fourier transform of the function R^{-1} also falls off very slowly for $|\boldsymbol{b}_g^*| \to \infty$. The idea of Ewald's method is to break the function $|\boldsymbol{R}+\boldsymbol{\varrho}_i-\boldsymbol{\varrho}_j|^{-1}$ down into two summands, of which one decreases rapidly when $R \to \infty$ and the other possesses a Fourier transform that decreases rapidly when $|\boldsymbol{b}_g^*| \to \infty$. We use the identity

$$\int_0^\infty dx e^{-x^2} = \sqrt{\pi}/2 \ ,$$

from which it follows that the function may be partitioned:

$$\frac{1}{R} \equiv \varphi^{(1)}(R) + \varphi^{(2)}(R) \ ,$$

where

$$\varphi^{(1)}(R) = \frac{2}{\sqrt{\pi}\,R}\int_{\lambda R}^{\infty} dx e^{-x^2} \ , \qquad \varphi^{(2)}(R) = \frac{2}{\sqrt{\pi}\,R}\int_0^{\lambda R} dx\, e^{-x^2} \ . \tag{1.7.79}$$

We represent (1.7.1) as

$$E_c = \frac{1}{2}\sum_{ijR}{}' q_i q_j \varphi^{(1)}(|\boldsymbol{R}+\boldsymbol{\varrho}_i-\boldsymbol{\varrho}_j|) + \frac{1}{2}\sum_{ijR}{}' q_i q_j \varphi^{(2)}(|\boldsymbol{R}+\boldsymbol{\varrho}_i-\boldsymbol{\varrho}_j|) \ . \tag{1.7.80}$$

The first sum in (1.7.80) converges very well, provided λ is not too small. We proceed to transform the second sum. To start with, we calculate the Fourier transform of the function $\varphi^{(2)}(R)$:

$$\begin{aligned}\varphi_{\boldsymbol{q}}^{(2)} &= \int d\boldsymbol{r} \exp(-\mathrm{i}\boldsymbol{q}\cdot\boldsymbol{r})\,\varphi^{(2)}(r) = 4\pi\int_0^\infty r^2\,dr\,\frac{\sin qr}{qr}\,\frac{2}{\sqrt{\pi}\,r}\int_0^{\lambda r} dx\, e^{-x^2}\\ &= \frac{8\sqrt{\pi}}{q^2}\int_0^\infty(-d\cos qr)\int_0^{\lambda r} dx\, e^{-x^2} = \frac{8\lambda\sqrt{\pi}}{q^2}\int_0^\infty dr\cos qr\, e^{-\lambda^2 r^2}\\ &= \frac{4\pi}{q^2}\exp(-q^2/4\lambda^2) \ .\end{aligned} \tag{1.7.81}$$

We can now make use of (1.7.78), since the quantity $\phi_q^{(2)}$ decreases rapidly as $q \to \infty$, provided λ is not too large. Unfortunately, the $\phi_q^{(2)}$ diverges with $q \to 0$. However, as a result of electroneutrality condition (1.7.73), the infinite constant equal to $\int d\boldsymbol{r}\phi^{(2)}(r) \equiv \phi_0^{(2)}$ may be subtracted from $\phi^{(2)}(|\boldsymbol{R}+\boldsymbol{\varrho}_i-\boldsymbol{\varrho}_j|)$. This is equivalent to the exclusion of the term with $\boldsymbol{g} = 0$ in the summation in (1.7.75). Finally, for the second sum in (1.7.80) to be brought into the form of (1.7.78), we need to add and to subtract the missing term with $\boldsymbol{R}+\boldsymbol{\varrho}_i-\boldsymbol{\varrho}_j = 0$, allowing for the identity

$$\phi^{(2)}(0) = \frac{2\lambda}{\sqrt{\pi}} , \tag{1.7.82}$$

which is proved by evaluating the indeterminate forms in (1.7.79) for $R \to 0$. The result is

$$E_c = \frac{1}{2}\sum_{ijR}' q_i q_j \phi^{(1)}(|\boldsymbol{R} + \boldsymbol{\varrho}_i - \boldsymbol{\varrho}_j|) + \frac{1}{2\varpi}\sum_{ij} q_i q_j \sum_{\boldsymbol{b}_g^*} \exp[\mathrm{i}\boldsymbol{b}_g^* \cdot (\boldsymbol{\varrho}_j - \boldsymbol{\varrho}_i)] \times \frac{4\pi}{(\boldsymbol{b}_g^*)^2} \exp(-(\boldsymbol{b}_g^*)^2/4\lambda^2) - \frac{\lambda}{\sqrt{\pi}}\left(\sum_i q_i\right)^2 . \tag{1.7.83}$$

If the magnitude of λ is taken to be on the order of the reciprocal-lattice period, then both sums in (1.7.83) will converge very rapidly. The quantity E_c is normally represented as

$$E_c = -\frac{\alpha_M (Ze)^2}{d} , \tag{1.7.84}$$

where $\pm Ze$ is the charge of the lattice ions (with e being the charge of an electron), d is the spacing between nearest neighbors, and α_M is the Madelung constant. For an NaCl-type lattice (Fig. 1.23a) $\alpha_M = 1.7476$, and for a CsCl-type lattice (Fig. 1.23b) $\alpha_M = 1.7627$. Thus, with the same interatomic distance, a CsCl-type lattice is somewhat more favorable in terms of electrostatic energy than an NaCl-type lattice. However, it should be emphasized that the corresponding energy difference is very small.

Thus, by referring to simple model problems, we have seen in this section how the total lattice energy can be calculated with allowance for electrostatic, van der Waals, covalent, and other bonding forces. In actual solids, all these forces play a more or less significant part. Nevertheless, they may be approximately classified according to the "predominant" type of bond into five groups, which will be treated in Sects. 1.8.1–5. In addition, because of the uniqueness of many of their physical properties, low-dimensional (one- and two-dimensional) and quantum crystals are classified into two individual groups (VI and VII), which will be considered in Sects. 1.8.6, 7.

1.8 Types of Crystalline Solids

The differences in the electrical, optical, magnetic, and mechanical properties of crystals are determined chiefly by the difference in the type of chemical bond. Certainly it is not simple to trace how, for example, the sharing of valence electrons in metals is related to their characteristic luster or to the difference in the magnetic properties of EuO and metallic europium. The interrelation between the observed properties and the electronic structure of solids constitutes one of the domains of the quantum theory of solids. Prior to describing these interrelations in detail, it would, however, be helpful to give an empirical classification of solids that is based on a classification of the main types of chemical bonding in them.

By convention, seven major types of solids may be singled out: (I) ionic crystals, (II) covalent-bonded crystals and semiconductors, (III) metallic crystals, (IV) molecular crystals, (V) hydrogen-bonded crystals, (VI) quasi-two-dimensional and quasi-one-dimensional crystals, and (VII) quantum crystals. Although the substances of the last two types are sometimes also referred to one of the preceding groups, their properties are so peculiar that it is reasonable to single them out into special classes of solids.

1.8.1 Ionic Crystals

As has already been noted in Sect. 1.7, from the standpoint of classical electrostatic concepts, the ionic type of bond is the easiest to understand. Simple examples of ionic compounds are binary compounds (ionic crystals of pure elements do not exist). Typical representatives of ionic compounds are crystals that consist of positive alkali-metal ions and negative halide ions. The ions are so arranged at the lattice sites that an ion of a specified sign has only ions of the opposite sign as its nearest neighbors. Examples of such sc (simple cubic) and bcc lattices are given in Fig. 1.23a, b. The electrical attraction of nearest ions of different signs proves to be stronger than the repulsion of next nearest ions of the same charge.

The ionic bond turns out to be, in the main, a purely electrostatic heteropolar bond. In many cases, just as in alkali-halide crystals, the electron shells of ions with charges of both signs resemble closely the atomic shells of rare gases with spherical charge distibution. For example, the electronic configuration of a Na^+ ion is $1s^2 2s^2 2p^6$; i.e., it coincides with the atomic configuration of neon.

The family of ionic crystals is very vast. In addition to halides of alkali and other metals, this class comprises hydrates, oxides, sulfides, and selenides of the various metals, and also binary, ternary, and more complicated compounds. At low temperatures ionic crystals are very bad conductors of electricity (insulators), and at high temperatures they become ionic conductors. They also absorb strongly in the infrared. The bonding energy of ionic crystals at room temperature is normally very high ranging from 81 kcal/mol ($\approx$ 325 J/kmol) for KH to 667 kcal/mol ($\approx$ 2790 J/kmol) for Al_2O_3.

The lattices of ionic crystals, unlike metals, do not generally belong to the type with close-packed atomic arrangements. For this reason, the overlap parameter (1.6.1) for the outer shells of adjacent ions in these crystals is very small. Practically, the construction of shells in ionic crystals boils down to the transition of the valence electron of the metal atom to the non-metal atom. However, because of the finiteness of the small parameter (1.6.1), no purely ionic bond exists in real crystals, since they always have an "admixture" of quantum valence bonds (Sect. 1.7.4).

The structure and properties of ionic crystals are primarily determined by the competition between the electrostatic energy, for which the method of computation is discussed in Sect. 1.7.5, and the Born–Mayer repulsive interaction forces between ions (1.7.19). The van der Waals attraction also plays a certain role. Thus, with neglect of the weak covalence effects, the interaction between ions may be described with the help of central pairwise interaction forces. At large distances these forces decrease very slowly (in accordance with the Coulomb law), and at small distances they increase sharply, so that the ions may roughly be regarded as solid spheres of determinate radius (Sect. 1.4 and Fig. 1.31).

1.8.2 Covalent Crystals and Semiconductors

By contrast with the ionic heteropolar bond, the covalent homopolar bond of covalent and semiconducting crystals requires a quantum description from the very outset. Typical representatives of covalent-bonded crystals are diamond, one of the modifications of solid carbon, and silicon carbide SiC, a carborundum (Fig. 1.22). The same diamond-type lattice and the same type of bond are characteristic of the semiconducting elements Si, Ge, and α Sn. All of them are situated in the fourth periodic table column and are normal elements with the same valence electron configuration ns^2np^2 ($n = 2, 3, 4, 5$); i.e., they belong to the group of valence-four elements. The four valence electrons of the atom are "hybridized", giving pairwise directional and spin-saturated bonds to the four nearest neighbors in the tetrahedral environment of the diamond lattice sites. Figure 1.34 schematically presents the plane projection of the spatial distribution of ns^2np^2 lattice electrons, where the shadowed areas correspond to the maximum density of the electrons that participate in the bonds [Ref. 1.30,

Fig. 1.34. Electronic structure of Ge with a purely covalent chemical bond (diamond-type lattice in a plane projection)

Chap. 4]. Each "bridge" of the directional spin-saturated covalent bond contains a pair of former s and p electrons of next nearest atoms with opposite spin projections. This bond possesses saturated valence, which is a direct consequence of the Pauli principle. As already discussed (Sect. 1.7.4), the covalent bond produces a diamond-type structure which is less dense than the more closely packed fcc and hcp structures of metals.

Valence crystals (not semiconductors) are good insulators over the entire temperature range of their existence. The semiconducting elements and compounds of this group of solids reveal low electronic conductivity (at $T > 0$ K), which increases as the temperature is raised. When small amounts of impurities are added to a compound, when it is exposed to radiation, and/or when there is a departure from a rigorous stoichiometric composition, the compound may exhibit purely p-type (hole) conductivity as well as a mixed (complex) electron-hole conductivity. Semiconductors are quite hard but possess low cohesive energy, with bonding forces approximately the same as those of ionic compounds. Thus, the bonding energy of Ge is 85 kcal/mol ($\approx$355 J/kmol), that of diamond 170 kcal/mol ($\approx$ 700 J/kmol), and that of quartz as high as 406 kcal/mol ($\approx$ 1650 J/kmol).

Ionic heteropolar crystals and covalent homopolar crystals are, in a way, opposites in their electronic structure. In the case of binary compounds, crystals with an ionic heteropolar bond are, as a rule, made up of the ions of the first and seventh periodic table columns $A^{I+}B^{VII-}$. Typical covalent-bonded crystals, by contrast, are pure elements of the fourth column, that is, A^{IV}. Between these cases we have intermediate ones, those of hybrid (mixed) and covalent bonds. Found here, for example, are compounds of the elements of the third and fifth columns $A^{III}B^{V}$, specifically Ga and As with their valence shell configurations $4s^2 4p^1$ and $4s^2 4p^3$. Similar to Ge, which is situated between Ga and As in the same row, this compound possesses a diamond lattice with saturated two-electron bonds (Fig. 1.35). For four neighboring atoms we again have eight p valence electrons of the shells $4s^2 4p$ and $4s^2 4p^3$. Of these, the larger share (5/8) consists of As atoms (B^V), and the smaller share (3/8) of Ga atoms (A^{III}). Therefore, the shaded covalent-bond bridges in Fig. 1.35 are not symmetric, as compared to those in Fig. 1.34, but are thick near As ions with a core charge $+5$e and thin near Ga ions where the charge of the core is equal to $+3e$.

Fig. 1.35. Electronic structure of $A^{III}B^{V}$ (GaAs) compounds with hybrid covalent and ionic bonds (diamond-type lattice in a plane projection)

Gallium arsenide is a typical semiconductor of the class $A^{III}B^{V}$ and features a hybrid bond in which the covalent bond prevails.

Consider a binary compound that consists of elements contained in the second and sixth columns $A^{II}B^{VI}$—for example, ZnSe with the respective valence shell configurations $4s^2$ and $4s^2 4p^4$, and a diamond lattice. Again, two-electron bridges (Fig. 1.36) arise between the nearest neighbors. The asymmetry of these bridges is more pronounced than that in the case of GaAs, because the ionic core of zinc has a charge $+2e$, and the core of selenium $+6e$.

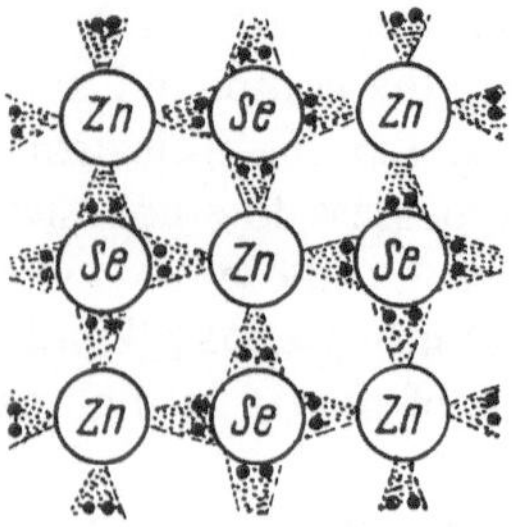

Fig. 1.36. Electronic structure of $A^{II}B^{VI}$ (ZnSe) compounds with hybrid covalent and ionic bonds (diamond-type lattice in a plane projection)

Finally, in $A^{I+}B^{VII-}$ compounds such as $Cu^{+}Br^{-}$ we are dealing with a typical ionic bond. There is no diamond lattice in this case. Nor are there any traces of a homeopolar bond in these compounds. As can be seen from the above examples, the drastic differences between covalent and ionic bonds that have been adopted for their classification are, in many cases (e.g., in $A^{III}B^{V}$ and $A^{II}B^{VI}$ compounds), of a relative nature only.

1.8.3 Metals, Their Alloys, and Compounds

In the solid state, about 70 of the 92 stable elements are metals, accounting for nearly 75% of all the periodic table elements. In the case of normal elements, all the elements of the first and second columns are metals, i.e., the alkali metals and the alkaline-earth metals, respectively. In the third column all the elements, except for boron, are metals. The fourth column contains only one metal, namely, lead. Tin and carbon are metals only when in certain modifications, and Si and Ge are semiconductors. The normal elements of columns V through VIII are all nonmetals, and the transition elements are all metals, including the noble metals Cu, Ag and Au and the elements Zn, Cd, and Hg.

The most remarkable of the physical properties of metals is their high electrical and thermal conductivity. As the temperature is lowered, the specific electrical resistivity of metals falls off, tending to a minimum value when

$T \rightarrow 0$ K. This leads to the assumption that the valence electrons of a metal form a highly mobile system of conduction electrons, or an "electronic liquid", which is easy to accelerate in an electric field. The nature of the metallic bond, as well as of the homopolar bond, could not be understood in terms of classical physics (that was another of its "catastrophes"), and only quantum physics has introduced theoretical enlightenment. Energetically, the metallic bond is weaker than the ionic and covalent bonds. This is particularly true of normal metals. In these, the bonding energy at room temperature ranges from 15 kcal/mol ($\approx$ 62.5 J/kmol) (Hg) to 92 kcal/mol ($\approx$ 386 J/kmol) (Au). In transition metals the bonding energy is higher. For example, for W it is equal to 210 kcal/mol ($\approx$ 875 J/kmol), which is due to the contribution that the electrons of the closed *d* and *f* shells make to the bonding energies. At least in part, this bond is of a covalent nature.

The difference between the metallic bond and the covalent bond arises primarily from the magnitude of the electron–ion interaction. For typical representatives of metallic-bonded crystals, i.e., alkali metals, this interaction is relatively weak. Therefore, as a first approximation, it may be assumed that the metal lattice is stabilized by the energy of the free-electron gas, and the role of the ionic cores is, chiefly, to assure electroneutrality (see the "plasma model of the metal", Sect. 5.1). The energy of typical metals consists of a large structure-independent contribution and a relatively small structure-dependent contribution. The energy difference for the various structures with the same mean electron–ion density therefore is not large and that is why it is very typical of metals to exhibit polymorphy. Thus, lithium and sodium exist at atmospheric pressure in low-temperature hcp and high-temperature bcc phases. Typically, metals possess very closely packed bcc, fcc, and hcp lattices. As the charge of the ionic core (i.e., the valence) builds up, the electron–ion interaction strengthens, and the electron-density distribution becomes nonuniform. A tendency appears for piling up charges on the bonds, and we proceed to the case of covalent crystals. For polyvalent metals, the bond is normally of an intermediate type; structures of very low symmetry occur quite often in which the number of nearest neighbors is small (for example, three nearest neighbors in the lattices of gallium and bismuth). Very complicated types of structure occur in transition metals such as manganese, tungsten, etc. For a more detailed treatment of the chemical bonding in metals, see [1.31].

Chemical bonds in alloys and intermetallic compounds are even more varied. Here we deal with a multitude of behavior types—from the practically complete insolubility of one type of metal in other types to the formation of a continuous series of solid solutions. The problem of alloy formation is an extremely complicated one and so far we have to confine ourselves to mainly empirical and semiempirical rules [1.32]. We just note here that in the case of alloys or compounds of metals whose valence differs substantially one may talk of a bond that is intermediate between the metallic and ionic bonds. Thus, a plurality of intermediate forms exist between the three types of "tight" chemical bonds (ionic, covalent, and metallic), of which electron transfer between the structural units of the crystal is typical.

Now we proceed to a consideration of the type of solids with weak chemical bonds.

1.8.4 Molecular Crystals

In molecular crystals the bonding forces within the molecules located at the lattice sites are appreciably stronger than the crystalline binding forces between the molecules. Since the electron shells of the molecules are, as a rule, closed, the intermolecular forces in the crystal are due to polarization (van der Waals forces). Note that neutral atoms or molecules polarize not only in the presence of a charged particle. Polarization also occurs for two neighboring neutral and isotropic particles. The potential energy of these van der Waals forces is inversely proportional to the sixth power of the interparticle distance (r^{-6}) for sufficiently far-off particles, when $r \gg 2r_{at}$. Since this condition is not satisfied in crystals, the calculations for them are highly complicated. Van der Waals forces are not only essential in molecular crystals—they introduce a small correction into the energy of ionic crystals, and also into the energy of metals. Strictly speaking, they should not be regarded as purely pairwise interaction forces, because the polarization of a pair will be affected by other neighbors as well. Roughly, however, they may be viewed as pairwise-interaction, central, and short-range forces.

Typical representatives of molecular crystals are the normal elements of the eighth column of the periodic table—rare gases; molecular hydrogen H_2; oxygen O_2; nitrogen N_2; the halides Cl_2, I_2, etc.; molecular NH_3, CO_2, CH_4, and a huge number of organic compounds, including the most complicated biological systems. As a rule, the structure of these crystals is determined by the requirement of a close-packed arrangement of molecules. Since the latter normally have a complicated form, the structure of the crystal turns out to be complicated too [1.33]. The bonding forces in molecular crystals are hundreds of times smaller than those in ionic crystals. For example, in helium they amount to 0.053 kcal/mol ($\approx$ 0.218 J/kmol); for argon they are as small as 1.77 kcal/mol ($\approx$ 7.37 J/kmol); and for CH_4—2.4 kcal/mol ($\approx$ 10 J/kmol).

1.8.5 Hydrogen-Bonded Crystals

Hydrogen-bonded crystals are also molecular-type crystals. However, the bond in them differs considerably from both the ionic type of bonding and van der Waals bonding, and is characterized by a shared proton (ion of the hydrogen atom). To some extent this type of bond may be thought of as being of an ionic nature—hydrogen transfers its electron to a neighboring atom and thus makes it a negative ion. The small size of the proton reduces the number of its nearest neighbors to a minimum, i.e., two neighbors. The bonding energy in these crystals is as low as 5 kcal/mol (≈ 20.5 J/kmol). Hydrogen bonds and van der Waals bonding are vitally important to biology in the formation of the structure of proteins and nucleic acids [1.34].

The most common hydrogen bonds are OH . . . O and NH . . . O. Very frequently the bond turns out to have two minima, i.e., the proton involved in the bonding may be in two equivalent or almost equivalent energy positions (i.e., a two-well potential for the proton [1.35]). Proton ordering on bonds of this type is one of the mechanisms of structural phase transitions in solids [1.36].

1.8.6 Quasi-One-Dimensional and Quasi-Two-Dimensional Crystals

A very interesting class of substances is represented by long polymeric molecules. Strong covalent bonds exist between the constituent atoms, whereas the interaction of different molecules arises from relatively weak van der Waals or hydrogen-bonding forces. If there are conjugated bonds in the molecule, electron transfer along the chain is also possible. Thus, according to the conductivity type, the $(SN)_x$ system (Fig. 1.37) is metallic, as are a number of organic polymers, although semiconductive behavior in such systems is actually much more common—the result of a phenomenon treated in Sect. 4.4.2. Of the pure elements, selenium and tellurium form "chain" structures with relatively weak bonds between the chains. In connection with some biological problems, studying electron motion in quasi-one-dimensional systems is particularly interesting because such chains are a constituent part of many biologically active molecules (chlorophyll, vitamin A, etc.).

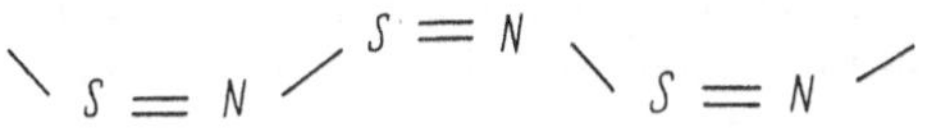

Fig. 1.37. Inorganic polymer $(SN)_x$

Not only because of their different electronic properties do we distinguish this type of solid from molecular crystals. In a classification according to chemical bond types, crystalline polymers also should come under a special heading, since the chain structure depends on the strength of the covalent bond, and the packing of the chains is determined by van der Waals bonding or hydrogen bonds.

Now we briefly examine quasi-two-dimensional crystals. A number of substances possess a layered (or quasi-two-dimensional) structure. Examples include graphite, boron nitride, and compounds with a formula of the MX_2 type (where M is a transition metal of groups IV–VII, X = S, Se, Te). As an

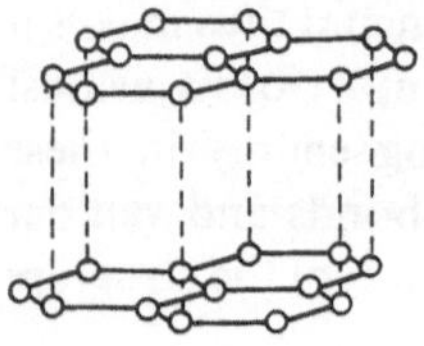

Fig. 1.38. The graphite lattice

illustration, Fig. 1.38 presents the structure of graphite. In such substances the chemical bond of the atoms is much stronger in the layers than between the layers. The interlayer spacing may be made larger (and the bond weaker) by introducing foreign molecules into the space between the layers (intercalation).

Low-dimensional systems are highly specific in their lattice and electronic properties. Normally, the interelectron interaction in these systems plays a more important part, with electronic phase transitions often occurring (e.g., metal–semiconductor transitions).

1.8.7 Quantum Crystals

In Sect. 1.2 we mentioned the quantum "zero-point" oscillations of atoms (ions) in solids with a mean amplitude $\langle u \rangle_0$, which is different from zero even when $T = 0$ K. We also noted that in most of the crystals the quantity $\langle u \rangle_0$ is very small compared to the crystal lattice parameter d, i.e., $\langle u \rangle_0/d \ll 1$. However, there is only one exceptional case of crystals of two helium isotopes He^4 and He^3, when this "rule" does not apply and $\langle u \rangle_0$ amounts to $\sim 30\%$ of the quantity d. That is why, in the low-temperature limit, the state of aggregation in these substances at low pressures (below 25 atm or 2.5×10^6 Pa, for example, for the isotope He^4) is the liquid phase. Yet even in the case of high pressures, when what becomes a state of thermodynamic equilibrium is the solid state, the latter acquires a number of unique features, because of the comparatively large quantity $\langle u_0 \rangle$ (in comparison with normal crystals). For this reason we say that solid helium-4 and helium-3 are quantum crystals and single them out into a special class of solids. We will try to give some, at least very approximate, quantitative description of the most important physical parameters of these substances. We may always specify the atomic vibrations in a solid by some set of equivalent oscillators (Chap. 2). Then, for a quantum oscillator of natural frequency ω, its ground-state energy will be equal to $\hbar\omega/2$. Proceeding from the quasi-classical approximation, the mean amplitude $\langle u \rangle_0$ of this oscillation can be evaluated from the relation

$$\tfrac{1}{2}\kappa \langle u \rangle_0^2 \sim \tfrac{1}{2}\hbar\omega \ , \qquad (1.8.1)$$

where the left-hand side represents the expression for the elastic potential energy (κ is the elasticity factor). Thus, if, in addition, we make use of the expression for the frequency ω in terms of κ and atomic mass M—$\omega = (\kappa/M)^{1/2}$—relation (1.8.1) for the mean amplitude $\langle u \rangle_0$ yields

$$\langle u \rangle_0 \sim \frac{\hbar^{1/2}}{(\kappa M)^{1/4}} . \tag{1.8.2}$$

Subject to the condition that $\langle u \rangle_0 \ll d$, the relation gives

$$d \gg \frac{\hbar^{1/2}}{(\kappa M)^{1/4}} .$$

This inequality holds for nearly all crystals. It fails only when M and κ decrease, i.e., for light atoms with relatively small interaction energies (small κ). Specifically, this may be expected for the crystals of the helium isotopes. Helium belongs to the group of rare gases (from He with $Z = 2$ to radon, Rn, with $Z = 86$) whose atoms have an ideal spin-saturated closed electron shell. The interaction between such atoms has been considered in Sect. 1.7.1. This interaction may be approximated by the Lennard–Jones potential (1.7.25). Then the parameter σ (dimensions of length) responsible for the spatial extent of the interaction turns out to be on the order of the lattice parameter ($\sigma \sim d$), and ε, which is an energy parameter, determines the magnitude of the interaction. De Boer [1.37] has introduced a dimensionless parameter specifying the ratios of the square of the amplitude of zero-temperature oscillations, $\langle u \rangle_0$, to the lattice parameter d:

$$\Lambda \sim \left(\frac{\langle u \rangle_0}{d} \right)^2 ,$$

which, in virtue of the estimate $\kappa \cong (d^2E/dR^2)_{R=\sigma} \sim \varepsilon/\sigma^2$ and $d \sim \sigma$, may be taken to be

$$\Lambda = \frac{\hbar}{\sigma\sqrt{M\varepsilon}} . \tag{1.8.3}$$

The condition $\Lambda \ll 1$ corresponds to small zero-point oscillations. Again, we see from (1.8.3) that the smaller the atomic mass M and the smaller the interatomic-interaction energy ε, the less applicable is the condition of small Λ.

As is seen from Table 1.11, the condition for the smallness of the parameter Λ works well for the heavy rare gases Xe and Kr, somewhat worse for Ar and Ne, and quite badly for both helium isotopes He^4 and He^3. If we assess $\langle u \rangle_0$ more accurately and pass from the de Boer parameter Λ on to the usual relation $\xi_0 = \langle u \rangle_0/d$, the result for the isotope He^3 will be $\langle u \rangle_0 \sim 1$ Å and $d \sim 3.77$ Å and, hence, as has already been stated above, $\xi_0 \sim 0.3$, i.e., zero-temperature oscillations produce atomic displacements that amount to nearly 30% of the

Table 1.11. Parameters of rare gases and their crystals

Element	M, [a.u.]	Z	a_0 atomic radius, [Å]	ε, [K] from (1.7.25)	σ, [Å]	d, [Å]	Λ
He^4; He^3	4; 3	2	0.93	10.2	2.56	3.77	0.4; 0.5
Ne	20	10	1.17	36.3	2.82	3.16	0.09
Ar	40	18	1.54	119.3	3.45	3.76	0.03
Kr	84	36	1.69	159	3.60	4.33	0.02
Xe	131	54	1.9	228	3.97	3.99	0.01

lattice parameter. As for Kr, the quantity $\langle u \rangle_0 \sim 0.32$ Å, $d \sim 4.33$ Å, and $\xi_0 \sim 0.08$. Thus these estimates clearly indicate that in the case of the helium isotopes the effects of quantum zero-point oscillations should be particularly pronounced. The atomic mass of the hydrogen isotopes is smaller even than that of helium. However, the atoms of the hydrogen isotopes possess no closed electron shell, and this results in the magnitude of the energy parameter ε being much larger. The quantum effects for the hydrogen isotopes may turn out to be considerable when the latter are impurities, for example, in crystals of rare gases. It is the combination of the smallness M and smallness ε that makes both helium isotopes quantum crystals.

As noted, one of the major features peculiar to helium is that at pressures below 25 atm for He^4 and below 30 atm for He^3 the two substances cannot exist in the solid phase and always remain liquid whatever the temperature. Helium crystals may form only at pressures exceeding the specified values.

The necessity to take into account the quantum zero-point oscillations in helium largely complicates the entire problem of describing the condensed state in comparison with normal substances for which the condition $\xi_0 \ll 1$ or $\Lambda \ll 1$ is satisfied. With these low-temperature liquid phases at normal atmospheric pressure we are dealing with quite unique properties. He^3, for example, is a quantum Fermi liquid, since, although the electron shell of the atom has a zero (integral) spin, the nucleus of He^3, which consists of three nucleons (two protons and one neutron), possesses a half-integral spin. The isotope He^4, on the other hand, is an example of a quantum Bose liquid, because both the electron shell and the atomic nucleus (composed of two protons and two neutrons) have a zero spin. The most striking property of these quantum liquids is superfluidity [1.38]. Similarly, "quantum behavior" might be expected to manifest itself in the physical properties of quantum crystals as well. As will be seen later, a number of the properties of solid helium are, in fact, of a specific quantum nature.

If we treat the motion of the atoms of the crystal near the equilibrium position approximately as quantum oscillators, their wave function has the form of a normal Gaussian: $A \exp(-\alpha^2 u^2/2)$, with a maximum at the coordinate origin (for $u = 0$ we have a center of equilibrium, that is, a crystal lattice site). An important characteristic of the Gaussian is its half-width $w = \alpha^{-1}$. The mean kinetic energy of the oscillator is directly proportional to the square of the uncertainty of the momentum $\delta p \sim \hbar/w$, i.e., $\sim w^{-2}$. The mean potential energy of the oscillator is directly proportional to the square of the half-width of the Gaussian $\sim w^2$. The actual half-width of the Gaussian therefore should correspond to a minimum of the total energy.

True potential-energy curves for the atoms of the crystal (with allowance for the repulsion at short distances) have a shape like that shown in Fig. 1.31. The same is true of rare gases, but the depth of the potential well of heavier atoms, for example, argon atoms, is nearly a factor of 10 (i.e., an order of magnitude) higher than the depth of the potential well of helium. Therefore, the well of the argon atom is deep enough for its mean kinetic energy to be incapable of "throwing" it out of the potential well (i.e., incapable of melting the crystal). This also holds for the case of sufficiently strong localization (i.e., when the quantity w^2 is sufficiently small). Conversely, the attempt of the helium atom to be localized in a comparatively shallow well raises its mean kinetic energy to the extent that the atom can no longer be tied to the lattice site and the crystal melts. For the atoms to be "driven" into the potential wells, the "intervention" of an external pressure is required.

Even in the case of solid helium, because of the large magnitude of the quantity $\langle u \rangle_0$, the atoms may not be regarded as particles that vibrate independently near the lattice sites. In the quantum crystal of helium it is necessary to take into account the correlation of the motions of atoms at close distances. Experimental evidence for the existence of this correlation is that helium crystals are very "soft". Their "softness" manifests itself in that helium crystals prove to be "self-releasing" entities, i.e., all crystal lattice structure defects in them are always cured by self-treatment, even at the lowest temperatures.

Up to this point we have explored the properties of quantum crystals in their ground state at $T = 0\,\mathrm{K}$. It is also of interest to consider their weakly excited states at finite temperatures. In this context, we may immediately ask whether we are entitled to exploit the concept of phonons (Chap. 2) for quantum crystals, i.e., to use the treatment which has been successful in describing thermal motion in normal crystals, at least at temperatures not very close to the melting point. As touchstones for examining this problem, we point out two typical phenomena of two different classes of macroscopic properties of crystals. One of them relates to the thermodynamic equilibrium properties that provide information on the characteristics of the energy spectrum (dispersion relation) of elementary excitations such as phonons. It is most convenient to represent these properties by the heat capacity of the crystal lattice. The other phenomenon, which pertains to kinetic properties, provides information concerning the character of the interaction between elementary excitations (phonons). As the representative of the above properties we choose, for example, the thermal conductivity in the crystal.

Detailed heat-capacity measurements for solid helium (almost in the entire temperature region of the occurrence of the solid phase and down to pressures of 300 atm) have shown that, for example, the temperature dependence of the heat capacity is qualitatively the same as that for the solid phases of heavy rare gases and, in general, all other nonquantum crystals, i.e., the Debye T^3–law holds good, etc. (Sect. 2.3). The temperature dependence of the thermal conductivity in a helium crystal can also be explained in terms of conventional concepts of phonon collisions. These two examples demonstrate that, at least in some cases, relatively large zero-temperature oscillations of the atoms of solid helium do not give rise to any qualitative peculiarities in its microscopic properties (although a quantitative allowance for these oscillations is absolutely indispensable).

However, quantum crystals also exhibit physical properties in which large zero-point oscillations do play a leading part—for example in their magnetic properties. Of the two helium isotopes, the isotope He^4 is entirely nonmagnetic since, as has been noted, its electron shell and atomic nucleus possess a zero spin. For the same reason, the isotope He^3 has no electron magnetism, but the nucleus has a spin 1/2 and therefore is magnetic. And here the large zero-temperature oscillation amplitudes lead us to expect an anomalously large exchange interaction (Sect. 5.6). Indeed, straightforward measurements of the nuclear magnetic susceptibility of the helium-3 crystal have revealed obvious departures from the Curie law (typical of normal paramagnets) at a temperature $\vartheta_M \sim 0.001$ K. This allows an estimation of the magnitude of the nuclear exchange energy per lattice site $A_{ex} \sim k_B\vartheta_M \sim 10^{-16} \times 10^{-3} \sim 10^{-19}$ erg. At the same time, estimates for normal crystals yield $\vartheta_M \sim 10^{-6}$ K or $A_{ex} \sim 10^{-22}$ erg. On the whole, solid and liquid helium-3 possesses unique nuclear magnetic properties, which are entirely due to its quantum nature.

One more specific quantum effect, which is a direct consequence of the relatively large magnitudes of the quantity $\langle u \rangle_0$, arises from the fact that the large magnitude of zero-point oscillations enables significant quantum atom tunneling. In quantum crystals, even in the region of temperatures close to 0 K (for example, in the presence of vacant sites), the tunneling of atoms into these sites results in the propagation of vacancy waves or vacancy quasiparticles over the crystal. This happens because of the translational invariance of the crystal (Sect. 4.2). In quantum crystals that contain small impurities of atoms of another isotope or another element, the cooperative tunneling of the impurity atom and the atom of the nearest neighbor of the matrix also leads to the wave propagation of specific impurity quasiparticles or "mass fluctuation waves" [1.39]. Because of the quantum effects, the temperature dependence of the diffusion coefficient of defects in quantum crystals differs from the usual temperature dependence.

1.9 Formulation of the General Quantum-Mechanical Problem of the Crystal

In general terms, the theory of the crystal is a problem of many interacting bodies, that is, atomic particles. If we are interested in processes involving energy variations that are much smaller than the rest energy of an electron, $m_{el}c^2$, we may exploit nonrelativistic quantum mechanics. This is particularly important in many-particle problems, where a relativistic generalization presents fundamental difficulties (the velocity of propagation of the interactions is finite). However, one of the conclusions of the relativistic theory should still be taken into account. Namely, allowance must be made for the presence of the intrinsic mechanical moment, that is, the electronic spin. If we fail to do this, we are not in a position to take into account correctly the symmetry of the wave functions of a many-particle system, a situation that may lead to fundamental errors even in the nonrelativistic approximation. On these grounds, the presence of the electronic spin and its influence on the statistical properties of the system will henceforth always be taken into account. Relativistic spin or dynamic magnetic interactions will always be considered to be weak and viewed as a small perturbation.

Thus, solid-state theory will be treated as a particular case of the problem of many bodies that interact with each other. It is their interaction that counts here. If it did not exist at all or were very weak (as in a rarefied gas), then this would essentially be not a many-body problem but a sum of one-body problems. In a quantum problem of many bodies, even identical ones, and in the absence of a dynamic interaction between particles, we must always take into account the statistical correlation which follows from the symmetry properties of the wave function with respect to the permutation of the coordinates of individual identical particles (symmetric and antisymmetric wave functions of the system of bosons and fermions respectively). However, at the outset the calculation can be carried out for one particle and then the wave functions of the system can be symmetrized as the product of one-particle wave functions. Thus a central point of a many-particle problem is to allow for the interparticle interaction.

Like every physical problem, where an attempt is made on the basis of atomic concepts to find a theoretical explanation of the various macroscopic properties of solids, the problem may be broken up into two stages, quantum mechanical and quantum statistical. The first stage is aimed at finding the energy spectrum of the system and the eigenstate wave functions that correspond to this spectrum. In the second stage, when considering the equilibrium properties, we determine the partition function and then some thermodynamic potential as a function of the corresponding thermodynamic variables. In the case of kinetic problems we have to solve Boltzmann-type or more complicated kinetic equations.

Consider the first stage, that is, the solution of a quantum-mechanical problem. One of the methods of solving quantum-mechanical problems is by using the Schrödinger equation, which for eigenstate problems has the form

$$\hat{\mathscr{H}}\psi = E\psi \ , \tag{1.9.1}$$

where $\hat{\mathscr{H}}$ is the Hamiltonian operator (Hamiltonian) of the system, ψ the wave function of the dynamic variables of all particles, and E the eigenstate energy of the system. To solve (1.9.1), we need to determine the Hamiltonian of the system. Assuming a crystal composed of N atomic nuclei or ion cores and n electrons of all the electrons or n itinerant electrons only, and using a nonrelativistic approximation that takes into account only the pairwise interparticle interactions, the Hamiltonian has the following coordinate representation:

$$\begin{aligned}\hat{\mathscr{H}} &= -\sum_{i=1}^{N} \hbar^2 \Delta_i/2M_i - \sum_{j=1}^{n} \hbar^2 \Delta_j/2m_{\mathrm{el}} \\ &+ \sum_{i,j=1}^{N,n} G(\boldsymbol{R}_i - \boldsymbol{r}_j) + \sum_{i<i'=1}^{N} V(\boldsymbol{R}_i - \boldsymbol{R}_{i'}) + \sum_{j,j'} W(\boldsymbol{r}_j - \boldsymbol{r}_{j'}) \ .\end{aligned} \tag{1.9.2}$$

(Broadly speaking, $n \neq N$, the relation between n and N being given by the neutrality condition).

For bare nuclei that have charge $+Ze$ when the neutrality $n = ZN$, we should take into account the three-particle and many-particle bonds in (1.9.2). However, since these bonds are weaker than pairwise bonds they are disregarded. This is natural for a gas, where the probability of interactions on the order of n (i.e., the collision of n particles) decreases drastically with n. In (1.9.2), $i = 1, 2, \ldots, N$ and $j = 1, 2, \ldots, n$ stands for the numbers of nuclei or ion cores of mass M_i with radius vectors $\boldsymbol{R}_i$ and of itinerant electrons of mass m_{el} with radius vectors $\boldsymbol{r}_j$, respectively; Δ_i and Δ_j are the Laplacian operators of the ith nucleus or ion core and of the jth electron; $G(\boldsymbol{R}_i - \boldsymbol{r}_j)$ is the operator of the potential energy of the interaction between the ith nucleus or core and the jth electron; $V(\boldsymbol{R}_i - \boldsymbol{R}_{i'})$ is the same between the ith and the i'th nuclei or ion cores; $W(\boldsymbol{r}_j - \boldsymbol{r}_{j'})$ is the same between the jth and the j'th electrons (in the general case these energies may also depend on the momenta of both the particles). In the simplest case, G, V, and W have the form of Coulomb attraction and repulsion energies

$$-Ze^2/|\boldsymbol{R}_i - \boldsymbol{r}_j|, \ Ze^2/|\boldsymbol{R}_i - \boldsymbol{R}_{i'}|, \ e^2/|\boldsymbol{r}_j - \boldsymbol{r}_{j'}| \ .$$

The wave function ψ in (1.9.1) in the coordinate representation is a composite function of the coordinates and spins of all $N + n$ particles:

$$\psi(\boldsymbol{R}_1 s_1, \boldsymbol{R}_2 s_2, \ldots, \boldsymbol{R}_N s_N; \boldsymbol{r}_1 \sigma_1, \boldsymbol{r}_2 \sigma_2, \ldots, \boldsymbol{r}_n \sigma_n) = \psi(\boldsymbol{R}, s; \boldsymbol{r}, \sigma) \ , \tag{1.9.3}$$

with s_i $(i = 1, 2, \ldots, N)$ and σ_j $(j = 1, 2, \ldots, n)$ being the spin coordinates of nuclei or ion cores and electrons. The $\boldsymbol{R}$, $\boldsymbol{r}$ and s, σ imply a set of all N radius vectors and spin coordinates of nuclei or ion cores and a set of all n radius vectors and spin coordinates of electrons.

Dividing the crystal into subsystems of ion cores and itinerant electrons enables us to do without the coordinates of all ZN electrons, because the coordinates of the electrons included into the ion cores are unnecessary in the description of the state of a crystal. Next, a normal procedure is to employ the adiabatic approximation already mentioned above, according to which the ion cores in the zero approximation may be thought of as being at rest. Then the wave function of the system will depend on $3n$ coordinates $\boldsymbol{r}_j$ and n of electronic spins σ_j, contain $3N$ parameters $\boldsymbol{R}_i$, and be represented as $\psi_{\boldsymbol{R}}^{(0)}(\boldsymbol{r})$. The energy of the system also will depend on the parameters $\boldsymbol{R}_i$ and will have the form of $E_0(\boldsymbol{R})$. If the motion of nuclei in a crystal is described by a wave function $\varphi(\boldsymbol{R})$, the wave function of the entire system has the form

$$\psi(\boldsymbol{r}, \boldsymbol{R}) = \varphi(\boldsymbol{R})\psi_{\boldsymbol{R}}^{(0)}(\boldsymbol{r}) \ . \tag{1.9.4}$$

It follows from (1.9.4) that the state of electrons at any instant of time is described by a function $\psi_{\boldsymbol{R}}^{(0)}(\boldsymbol{r})$, in which $\boldsymbol{R}$ is taken to represent the positions of nuclei at exactly the same instant of time. This is the gist of the adiabatic approximation. The function $\psi_{\boldsymbol{R}}^{(0)}(\boldsymbol{r})$ describes the electronic system for an adiabatic change of the parameters $\boldsymbol{R}$ determining the system of cores or nuclei.

The procedure for deriving an applicability criterion for (1.9.4) is treated in detail in Sect. 5.9. This criterion has the form

$$\hbar\langle v_{\text{nucl}}\rangle/\Delta R \ll \Delta E_{\text{el}} \ , \tag{1.9.5}$$

where $\langle v_{\text{nucl}}\rangle$ is the velocity of a nucleus or core, ΔR is the nuclear displacement that causes the function $\psi_{\boldsymbol{R}}^{(0)}(\boldsymbol{r})$ to vary appreciably, and ΔE_{el} is the energy gap between the ground-state electronic energy and the first excited level (at a given $\boldsymbol{R}$). This criterion usually holds good for inner-shell electrons. For valence electrons, it holds true in ionic, covalent-bonded, and molecular crystals, where ΔE_{el} is always large ($\gtrsim 1$ eV) and considerably exceeds the left-hand side of (1.9.5). This is valid for the ground state of core electrons. For excited states, (1.9.5) normally fails. In the case of metals, this criterion does not hold for former valence electrons, in which, as will be seen later, the energy spectrum has no gap ($\Delta E_{\text{el}} = 0$). However, the adiabatic approximation turns out to be applicable for inner electrons, just as it is in nonmetallic crystals (Chap. 5).

Thus, in the most general case this approximation may always be applied to ion cores, and therefore we may assume to a zero approximation that the first summand in the Hamiltonian (1.9.2)—that is, the kinetic energy of nuclei or cores—is equal to zero, and the summand next to the last—that is, the nuclear or core interaction potential—is equal to a constant. The Hamiltonian of the electronic subsystem in this approximation is

$$\hat{\mathscr{H}}_{\mathrm{el}} = -\sum_{j=1}^{n} \hbar^2 \Delta_j / 2m_{\mathrm{el}} + \sum_{i,j=1}^{N,n} G(\boldsymbol{R}_i, \boldsymbol{r}_j) + \sum_{j<j'=1}^{n} W(\boldsymbol{r}_j, \boldsymbol{r}_{j'}) \ , \tag{1.9.6}$$

where $\boldsymbol{R}_i$ stands for the positions of fixed nuclei or cores, that is, the sources of the field acting on the electrons. However, in this approximation, too, the solution of the Schrödinger equation meets with huge mathematical difficulties. The major difficulty lies in the third term in (1.9.6), which interrelates the electronic coordinates and does not allow the Fourier variable separation method to be used, i.e., does not allow a many-body problem to be reduced to a sum of one-particle problems. Nor is it possible to employ the theory of small perturbations. The small dimensionless parameter normally cannot be separated out in the last of the sums in (1.9.6) because practically all of the interactions are strong, especially the Coulomb interaction. Even if we formally regard it as small and apply perturbation theory, the result obtained for second-order and higher-order energy corrections will be at variance with the magnitude of first-order energy corrections.

An obvious possibility is to get rid of strong interactions and not to consider them at all. However, if we do not consider them, we will probably lose some of the essential properties of the system with which we are concerned. Therefore, attempts have been made to develop methods that allow for an interaction and at the same time reduce its effect to some quasi-external field, thus permitting a one-particle approach to the solution of a many-body problem. One such method was proposed by Hartree (1930) and refined by Fock (1932) in connection with problems of many-electron atoms. In the Hartree–Fock method a many-particle wave function is sought in the form of an antisymmetrized product of one-particle functions with the pairwise interaction being averaged over these functions. As a result, a linear many-particle problem reduces to a nonlinear one-particle problem. This method has been successfully employed in problems of many-electron atoms and molecules. Its utilization in crystals is associated with great difficulties. Sometimes it is simply assumed that the averaged Hartree–Fock interaction is involved in some way in the potential energy of an electron with respect to ion cores. In what follows we will make use of this crude technique. In addition to the Hartree–Fock method, many other methods have been developed which find successful applications in quantum treatments of the states of molecular and crystalline systems. For an acquaintance with some of the basic treatments, see [1.40].

In addition to the Hartree–Fock method, another method enjoys wide use, the so-called unitary transformation method. The idea of this method is that the Hamiltonian (1.9.6) is subjected to a unitary transformation by proceeding from initial coordinates to new coordinates to assure that the term involving the interaction in (1.9.6) becomes small. Bare particles then do not coincide with initial ones, but are quasiparticles in which the interaction of initial particles is included, at least in part. We illustrate this method with the example of two solids, where it can be used quite precisely. In this case the Hamiltonian has the form

$$\hat{\mathscr{H}} = -\hbar^2 \Delta_1/2m_1 - \hbar^2 \Delta_2/2m_2 + W(\boldsymbol{r}_1 - \boldsymbol{r}_2)\ . \tag{1.9.7}$$

Proceed from the coordinates $\boldsymbol{r}_1$ and $\boldsymbol{r}_2$ to the relative coordinate $\boldsymbol{r}$ and center-of-mass coordinate $\boldsymbol{R}$:

$$\boldsymbol{r} = \boldsymbol{r}_1 - \boldsymbol{r}_2$$
$$(m_1 + m_2)\boldsymbol{R} = m_1 \boldsymbol{r}_1 + m_2 \boldsymbol{r}_2\ .$$

As a result, instead of (1.9.7) we obtain

$$\hat{\mathscr{H}}' = -\hbar^2 \Delta_R/2(m_1 + m_2) - \hbar^2 \Delta_r/2\left(\frac{m_1 m_2}{m_1 + m_2}\right) + W(\boldsymbol{r}) = \hat{\mathscr{H}}'_R + \hat{\mathscr{H}}'_r\ .$$

Thus, a complete separation of the variables has been achieved in the new coordinates. We obtain two Schrödinger equations for describing the motion of two quasiparticles: one with effective mass $(m_1 + m_2)$ and center-of-mass coordinates $\boldsymbol{R}$, and the other with effective mass equal to the reduced mass $m_1 m_2/(m_1 + m_2)$ and with the relative coordinate $\boldsymbol{r}$. In this fashion, the unitary transformation method enables us, in this case, to reduce a two-particle problem to a sum of two one-particle problems for quasiparticles. The latter sometimes differ only slightly from initial particles. An example is the case when an initial particle interacts with its environment weakly, As it moves, it repels or attracts other particles and becomes surrounded by a "cloud" of excited neighbors. A quasiparticle is nothing else but a sum of an initial particle and an interaction-created cloud. The latter may screen an initial particle from other particles, thus leading to a decrease in the interaction between quasiparticles. Later we will deal with examples of such particles.

Besides quasiparticles whose genesis from initial particles can be traced clearly, there may be quasiparticles that are altogether unlike initial particles and are said to be collective excitations. These entities arise not from the motion of an individual initial particle and its cloud but from the collective motion of the entire system as a whole. Typical examples of such quasiparticles are the phonon—a quantum of sound (elastic) oscillations of the ionic lattice (Chap. 2)—and also the plasmon—a quantum of oscillations of the electron density in the crystal (Chap. 5).

1.10 Properties of Disordered Condensed Systems

1.10.1 General Remarks

Up to this point we have been concerned with ideal crystals, to which this book is mainly devoted. However, noncrystalline condensed substances as well as disordered alloys in the crystalline state, whose properties are in some respects close to those of ideal crystals and in some respects drastically different from them, also play an important role in modern solid-state physics.

It had long been held that the ideal periodicity of the crystal lattice was the determinative property of solids (crystals), which sharply distinguished them from liquids and gases. However, as far back as 1933 *Shubin* [1.41] noted that the electronic states in liquid and solid metals were in many ways alike. Later *Frenkel* [1.1] indicated that in their atomic and thermodynamic properties liquids were closer to solids than to gases. Since the nineteen-fifties, investigations of amorphous materials (glasses), which, just as crystals, may be semiconductors, metals, and dielectrics, have developed intensively. It turns out that among the glasses are many materials that exhibit unique electrical, magnetic, and mechanical properties. Finally, research has also concentrated on the properties of disordered alloys and solid solutions which possess no ideal spatial periodicity either. This large body of information on the properties of liquids, glasses, and alloys can be understood in part in terms of concepts borrowed from the theory of crystalline solids. On the other hand, a need also arose to develop a number of new concepts and methods for a theoretical study of what is now known as disordered materials. The work of *Anderson*, *Mott*, *Lifshitz*, and others was of great importance here.

At present the physics of disordered materials has become a rapidly developing branch of the physics of the condensed state of matter as a whole. The latter now incorporates crystalline solids, disordered alloys and solid solutions, liquids and glasses—all of which have turned out to possess a number of features in common. Broadly speaking, the relationship of the condensed state to the gaseous phase has proved to be more profound than the above-noted similarities between crystals and liquids. Although the present book deals almost exclusively with crystalline solids, here and in the subsequent chapters we will briefly describe some properties of disordered materials.

Disordered materials may be divided into three classes:

1. *Structurally disordered materials*, i.e., liquids and amorphous materials (glasses). There is no crystalline long-range order in these systems, but the short-range order in the arrangement of atoms is maintained. In terms of thermodynamics, glasses may be regarded as supercooled liquids. These are metastable, i.e., nonequilibrium, states. In a "frozen" atomic configuration glasses possess some properties that differ drastically both from those of crystals and from those of liquids (Sect. 1.10.2).

2. *Disordered alloys and solid solutions.* These are substances in which the crystal lattice is preserved, whereas the spatial periodicity is perturbed by the random distribution of the atoms or ions of the different constituents of the alloy in the sites of this lattice (crystallochemical disorder). If the structure of an alloy may be represented as resulting from the substitution of part of the ions of a metal for those of another metal, the alloy is said to be substitutional. But if the ions (atoms) of one of the constituents of an alloy occupy the interstitial crystal lattice sites, the alloy is said to be interstitial. Interstitial alloys are normally composed of metals and light nonmetals with small-sized atoms such as those of hydrogen, carbon, and nitrogen.

3. *Substances disordered by neutron and other irradiation, heavy external deformation, etc.* In this case a multitude of internal defects are present in the substance. Such disordered systems will not be considered at all in this book; their properties have been described in detail in a number of monographs [1.42–43].

Irrespective of the particular nature of disordered systems, the quantum-mechanical problem per se for these systems should be posed in a fashion quite different from the way it is formulated for a crystal. The point is that the crystal structure of any element or chemical compound, in principle, should be unambiguously determined from the condition of the minimum of free energy. Possibly, this is not an absolute minimum. For example, the diamond structure is energetically less favorable than is graphite, but, at any rate, there exists a finite number of structures of a given compound and the properties of each structure are prescribed unambiguously. Therefore, the problem "to calculate the properties of a perfect diamond crystal" is equally as well defined as the problem "to calculate the properties of a carbon atom". A quite different thing is that neither the first nor the second problem can be solved exactly—yet we seek approximately those same characteristics of the system which we would have sought if the solution were exact, i.e., the eigenstates and their energies.

May the problem of "calculating the properties of a disordered alloy for a given concentration of its constituents" be thought of as equally well defined? Evidently, it may not. This is so because the constituent atoms may have different arrangements which depend on the way the alloy is synthesized. Generally, the atoms in a liquid intermix and—although the Schrödinger equation can, in principle (only in principle!), be solved in an adiabatic approximation for any nuclear configuration—it is quite unclear which configuration should be of interest to us.

The problem here is about the same as that in classical statistical mechanics, where it is not only practically impossible but also unnecessary to solve equations of motion for each atom in order to describe the macroscopic properties of a substance. The latter are characterized by special parameters such as temperature, entropy, volume, etc. As in the theory of disordered systems, there is no point in trying to solve a Schrödinger equation for an arbitrary nuclear configuration, where some characteristics should be sought which refer not to the instantaneous state of a liquid and not to one particular lump of alloy or glass but to any instant of time for a liquid of specified temperature and density and to any lump of alloy or glass of specified composition produced using a given (in outline) method of synthesis. Thus, the structure of a glass or liquid is characterized not by a meaningless enumeration of the coordinates of all nuclei but by the pairwise-interaction distribution function, i.e., by the probability of finding two nuclei at a certain distance. In themselves, the electron wave functions in a disordered system are random quantities that depend on a given realization of the potential, i.e., on a particular nuclear arrangement in a substance. Some characteristics, however, possess the property of self-averaging; i.e., they lose their stochastic nature and become reliable when we pass to infinitely large systems, which is a normal procedure in statistical physics. This permits essentially exact quantitative predictions, and also necessitates the development of sophisticated methods of searching for and ascertaining such quantities [1.49]. These problems are treated in some detail in Chap. 4.

As already noted, disordered systems are often (but not always) away from thermodynamic equilibrium. This state manifests itself most directly in that their properties, generally speaking, depend on prehistory and may, because of the slow diffusion of atoms, vary with time. The characteristic scales of these temporal variations, however, are frequently very large. Therefore, it may be thought that glasses and nonequilibrium solid solutions are describable with the help of ordinary relations of equilibrium thermodynamics, although this has not been proved rigorously. The problem of the thermodynamic properties of glasses will be considered in Sect. 2.5.

Disordered materials feature a broad spectrum of properties. For lack of space we are unable to describe them in detail and therefore restrict our attention to the properties of metallic glasses.

1.10.2 Metallic Glasses (Example of Amorphous Solids)

In the most general case the amorphous condensed state of matter should be a condensed state in which there is absolutely no long-range order in the spatial arrangement of atoms (ions) or molecules of a substance. Two cases must be distinguished here: when this state corresponds to thermodynamic equilibrium and when there is no such equilibrium; i.e., a nonequilibrium metastable state takes place. The first case occurs, for example, in all liquids in the temperature interval between thermodynamic equilibrium melting (crystallization) temperatures and boiling (condensation) temperatures, whereas the second case is typical, for example, of a supercooled or superheated liquid.

What are known as amorphous solids (AS) should also be placed among supercooled liquids. On what physical grounds is it possible to single out this type of disordered state of condensed matter? The point is that because of thermal motion a substantial interchange of neighboring atomic particles occurs in liquids in both equilibrium and nonequilibrium states. The extent of interchange depends substantially on the viscosity of the liquid. When the coefficient of viscosity reaches such high values that the above interchange of atomic particles no longer occurs, a nonequilibrium supercooled liquid may be regarded as an AS. Put another way, an AS may be treated as a liquid that has solidified instantaneously, and the thermal motion in that liquid will be of the same character as that in a solid in thermodynamic equilibrium; i.e., in the form of small oscillations about equilibrium positions, which, however, are devoid of long-range crystalline order. The absence of the latter will show itself macroscopically in an AS, primarily in the isotropy of its physical properties, by contrast with the pronounced anisotropy of crystals. Further, there is no clearly marked melting point for an AS.

In itself, the AS family is not uniform and may be broken up into subclasses: amorphous dielectrics (AD), amorphous semiconductors (ASC), and metallic glasses (MG). A typical representative of AD is the usual glass (SiO_2), which everyone knows as window glass. Schematically shown in Fig. 1.39 is the temperature dependence of specific volume V for a normal soda-lime glass (AD), which does not crystallize when rapidly cooled from the liquid state to low temperatures. Two critical temperature points are marked in the plot: T_{melt}, the thermodynamic equilibrium temperature of melting (solidification) during sufficiently slow cooling from the liquid state into a crystalline solid state, and the temperature T_g, which corresponds to the transition of a metastable AS (dielectric glass) to a crystalline state of thermodynamic equilibrium (the so-called devitrification temperature) during sufficiently slow heating.

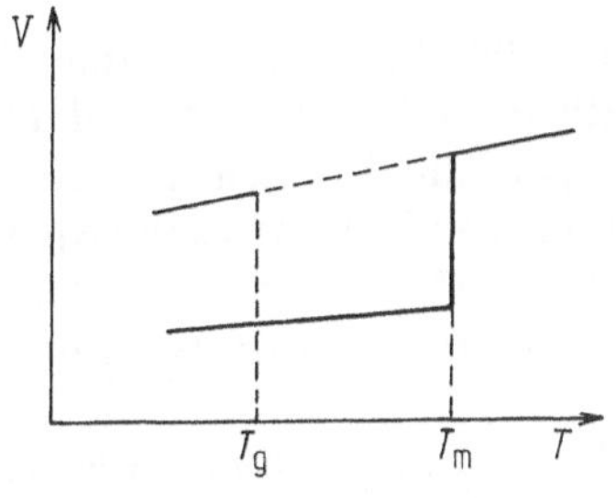

Fig. 1.39. Temperature dependence of specific volume for soda-lime glass (melting and glass-formation temperatures, T_m and T_g, are indicated)

Much consideration has recently been given to amorphous solids possessing semiconducting properties, that is, ASC. Included here are the usual pure (intrinsic) semiconductors, viz., germanium and silicon; $A^{III}B^{V}$- and $A^{II}B^{VI}$-type semiconducting compounds; oxide glasses (for example, V_2O_5, etc.); chalcogenide glasses of elements such as As, Se, Te; and many other types of ASC (Chap. 4).

Let us dwell in somewhat more detail on metallic glasses (MG). Just as AS, they are metastable solids, the state of which may be fixed by a very high rate of cooling (10^5–10^{10} or even 10^{15} K s^{-1}). The superfast heat abstraction which is required here is only possible provided that one of the dimensions of the samples produced is very small. Currently, a number of methods are used to produce MG. By slapping melt drops on a metallic anvil, a circle 15–25 mm in diameter and 40–70 mm in thick is produced. Cooling on a quickly rotating disc or rolling a melt jet between two rolls produces a ribbon 3–6 mm wide and 40–100 μm thick. Also, a wire of cylindrical shape is produced by squeezing out a melt into a cooling liquid. Many MG production processes already exist on an industrial scale.

Typical MGs are binary, ternary, and more complicated many-component systems. They consist of two types of chemical elements. Normally these are about 80 at. % of transition metals (Mn, Fe, Co, Ni, Zr, Pd, La, Pt, . . .) or noble metals (Cu, Au, Ag), and about 20 at.% of polyvalent glass-forming elements, usually metalloids (B, C, N, Si, Ge, etc.) and sometimes the metal Al. As examples, we may cite some compositions of commercial grades of MGs: 2826–$Fe_{40}Ni_{40}P_{14}B_6$, 2826A–$Fe_{32}Ni_{36}Cr_{14}P_{12}B_6$, 2826B–$Fe_{29}Ni_{49}P_{14}B_6Si_2$, 2605–$Fe_{80}B_{20}$, 2605A–$Fe_{78}Mo_2B_{20}$. These compositions usually correspond to a eutectic, although this is not necessarily so.

The MGs attract attention because, in the first place, they add a fresh page to solid-state physics and, in the second place, they are a new class of engineering materials that exhibit a peculiar complex of physical and technological properties. They are relatively simple and cheap to manufacture, and in their physical properties they are especially attractive for combining properties that cannot be combined in other materials. For example, MGs combine high strength with remarkable plasticity; electrically, they are electronic conductors that simultaneously combine high specific resistivity and a weak temperature dependence of the resistivity. They are also ideal soft magnetic materials because of the vanishingly small magnetic anisotropy.

In contrast to the usual soda-lime glasses (Fig. 1.39), the specific volume of the amorphous state in the case of MGs differs little from that of the crystal. In addition, when heated slowly the MGs at a temperature T_g often display a reversible transition first to the supercooled liquid state, and only at a somewhat higher temperature T'_g does crystallization occur. This is shown in Fig. 1.40 with the example of the temperature dependence of the specific heat capcity C_p of a typical MG.

The most direct information on the spatial distribution of atoms in MGs may be extracted from X-ray, neutron, and electron scattering experiments. Unfortunately, these do not allow the three-dimensional atomic distribution to be determined unambiguously (plane projection). Therefore, one has to use some general modeling principles which apply to the atomic structure of all MGs and ASs. Two major models compete here: (1) the microcrystalline model (MCM), according to which an AS is a conglomeration of tiny crystallites with a random distribution of the orientation of their crystallographic axes, and (2) the model of a liquid that has solidified instantaneously, or the model of a random close-packed arrangement of solid spheres (CPASS) according to *Bernal* [1.46].

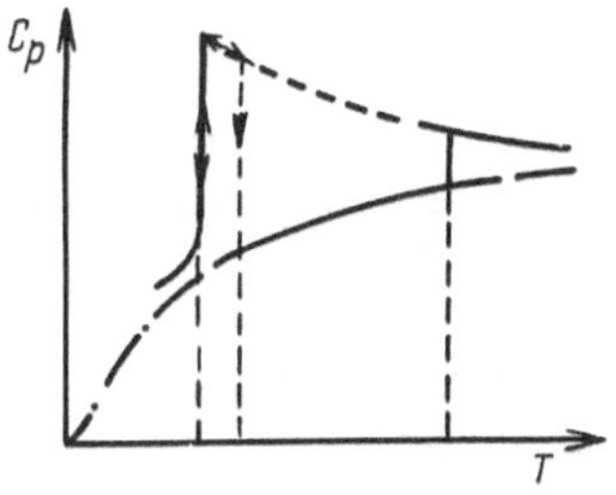

Fig. 1.40. Temperature dependence of specific heat capacity C_p for metallic glass

To be able to decide on the spatial atomic structure of a solid, we need to construct a spatial atomic distribution function, which can then be compared with experiments, for example, with X-ray scattering data. The distribution of N particles, with positions $\boldsymbol{r}_k$ $(k = 1, 2, 3, \ldots, N)$ in the vicinity of a given particle at $\boldsymbol{r}_l$, is given by the function

$$\varrho'(r, r_l) = \sum_{k=1}^{N} \delta[r-(r_k-r_l)] \ , \tag{1.10.1}$$

with $\delta(x)$ being a Dirac delta function. For a particle-density function independent of the position of some separated particle with the radius vector of its center of gravity $\boldsymbol{r}_l$, we sum (1.10.1) over all N particles with the numbers l and divide by the total number N of particles to obtain

$$\varrho''(r) = \frac{1}{N}\sum_{l=1}^{N} \varrho'(r, r_l) = \frac{1}{N}\sum_{l,k=1}^{N} \delta[r-(r_k-r_l)] \ . \tag{1.10.2}$$

For a sufficiently large system we may formally remove from (1.10.2) the limit $N \to \infty$. Thus we determine the pairwise atomic distribution function $\varrho(r)$, which signifies that the quantity $\varrho(r)dr$ gives us the mean number of particles present in the volume element dr, if we know exactly that only one particle is at the origin. If the medium is isotropic, as is the case with liquids and amorphous solids, we may carry out integration over polar angles and thereby immediately obtain the mean number of particles in a spherical layer of thickness dr between spheres of radii r and $r+dr$; $4\pi r^2 \varrho(r)dr$. The function $4\pi\varrho(r)r^2$ is normally referred to as the radial distribution function (RDF)

$$f(r) = 4\pi\varrho(r)r^2 \ . \tag{1.10.3}$$

Particle scattering experiments allow us to find not the distribution function itself but its Fourier transform, and using this, the function (1.10.3) can be constructed from experimental data.

The RDF curves for the density of particles may, in principle, be used as the criterion for choosing between the two aforementioned amorphous-state models. If the crystallite model is valid, the RDF curves should be smooth for distances larger than 10 Å, and, if the Bernal model is valid, the RDF curves for

MGs should be close to the corresponding curves for a liquid. Figure 1.41 presents RDF curves for amorphous and liquid cobalt [1.46]. Comparison of these curves shows that they are very close to each other. However, a substantial difference also may be noted. A typical (although thus far imperfectly understood) signature of the RDFs of ASs is the double splitting of the second peak on the curve. A number of very exact theoretical calculations of RDF curves have recently been carried out on the basis of Bernal-type models involving the use of a computer [1.47]. Several reviews on the structure and various physical properties of amorphous materials have appeared in recent years [1.48–50].

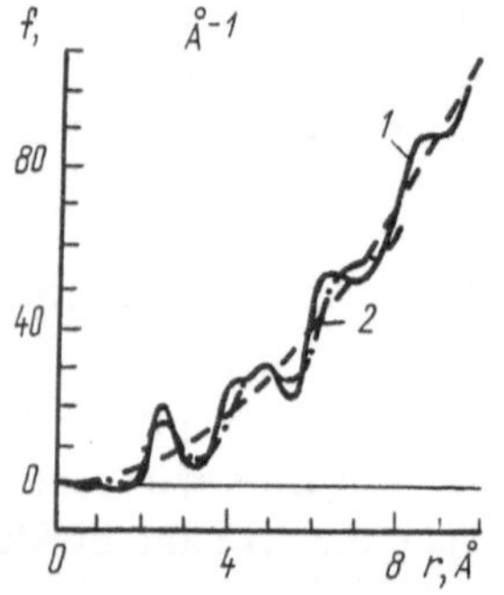

Fig. 1.41. Radial distribution function for amorphous (*1*) and liquid (*2*) cobalt

In the first chapter we have considered the most important general features of solid-state physics. The chapters that follow deal with specific prbperties of solids. In the second chapter we concentrate on the dynamics of the ionic lattice of solids, and in the subsequent three chapters we focus attention on the quantum theory of the electronic properties of solids, especially in the metallic state.

Note Added in Proof

Perhaps the most important development in crystallography since the times of Fedorov and Schönflies has been the discovery of quasicrystals. D. Shechtman, I. Bloch, D. Gratias, and J. W. Cahn (Phys. Rev. Lett. 53, L551 (1984)) have detected an icosahedral short-range order (with 5-fold symmetry axes) in Al–Mn alloy. A large number of such entities have already been discovered by now. They possess a certain long-range order, but of a more complete type than that in usual crystals. In a way, quasicrystals may be viewed as three-dimensional cross-sections of multidimensional crystals or as Penrose-tile type entities (see review by D. Gratias in La Recherche 178, 788 (1986) and the references cited therein).

2. Dynamic Properties of the Crystal Lattice

This chapter is devoted to the theory of crystal lattice vibrations. Simple one-dimensional models of both perfect and point-defect crystals are considered at length. We discuss general properties of the crystal lattice vibration spectrum, lattice thermodynamics, and optical properties of ionic crystals in the infrared region. A very detailed treatment is given to the theory of scattering on a vibrating lattice, including an exposition of the requisite mathematical machinery (second quantization, devices and techniques for handling second-quantized operators). A discussion of the theory of the Mössbauer effect is provided. Consideration is given to features peculiar to the low-temperature thermodynamics of glasses ("two-level center" model).

2.1 The Dynamics of the Ionic Lattice

Our objective now is to consider the motion of ions in the ionic lattice. As has been stated, their thermal motion in the crystal has the character of small oscillations around the lattice sites and, in keeping with quantum mechanics, the ions at 0 K possess the energy of zero-point oscillations, which is essential to quantum crystals. In other solids the amplitudes of these oscillations are so small that the crystallization process occurs well before the effect of quantum zero-point fluctuations comes into play. In the solid state of inert gases (except for helium) from neon to xenon, the zero-point oscillation energy also manifests itself noticeably, chiefly at low temperatures and normal pressure. Therefore, the crystals of these elements may be thought of as being close to quantum crystals (Sect. 1.8.7). To sum up, for an overwhelming majority of solids (except for quantum crystals) at temperatures not very close to the melting point, (1.2.2) always holds true, thus allowing us to exploit perturbation theory in solving dynamic problems.

2.1.1 A Linear Monatomic Array

To get a better understanding of the character of the ionic motion, we begin by considering a simple classical model of a crystal with one atom in the unit cell as a one-dimensional linear array (chain) composed of N equidistant atoms, with parameter d, of mass m (Fig. 2.1). This model was first considered by *Bernoulli* in

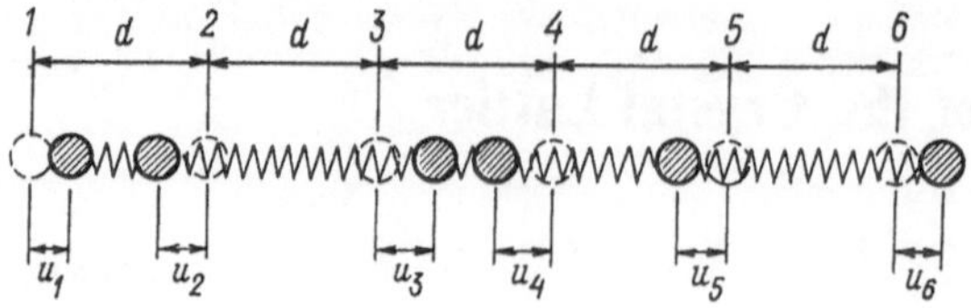

Fig. 2.1. One-dimensional model of a monatomic crystal made up of atoms of mass m

1728 and later solved by Lagrange; see also [2.1]. Such an array may model, for example, the crystal of any element whose atoms are bound by weak dispersive forces. Let us label the displacement of the lth atom along the array by u_l and the force constant by α (l is the number of the atom in terms of d: $l=0d$, $1d$, $2d$, $3d, \ldots$). Since these displacements are small ($u_l/d \ll 1$), the potential energy may be expanded in a power series of the differences of these displacements for atomic pairs. Because the oscillations take place near equilibrium positions that correspond to minimum potential energy, the expansion should begin with quadratic terms. We will restrict ourselves to allowance for these terms (harmonic approximation). The potential energy V of the array in the nearest-neighbors approximation to within the additive constant has the form

$$V = \sum_l \frac{\alpha}{2}(u_l - u_{l+d})^2 \ . \tag{2.1.1}$$

The equations of motion then will be

$$m\ddot{u}_l = -\alpha(2u_l - u_{l+d} - u_{l-d}) \ , \tag{2.1.2}$$

where $\ddot{u}_l$ is the second time derivative u_l. The system of difference equations (2.1.2) can be solved in the form

$$u_l = U_q(t)\exp(iql) \ ; \tag{2.1.3}$$

the possible values of q are determined from the boundary conditions, and $U_q(t)$ is the required temporal function. Substitution of (2.1.3) into (2.1.2) yields

$$\begin{aligned} m\ddot{U}_q &= -\alpha[2 - \exp(iqd) - \exp(-iqd)]\,U_q \\ &= -4\alpha\sin^2(qd/2)\,U_q \ . \end{aligned} \tag{2.1.4}$$

This is the differential equation of a harmonic oscillator with the frequency

$$\omega_q = 2\sqrt{\alpha/m}\,\sin(qd/2) \ . \tag{2.1.5}$$

It follows from dispersion relation (2.1.5) that the frequency is a periodic function of the wave number q, and the solution of (2.1.4) has the form

$$U_q(t) = A_q \exp(-\mathrm{i}\omega_q t) \ , \tag{2.1.6}$$

where A_q is a complex q-dependent amplitude of oscillation.

To determine the possible values of q, we must ascertain the conditions at the ends of the array. Strictly speaking, (2.1.1) is valid for an infinitely long array since the forces acting on the atoms located at the edges (and close to the edges) differ from those inside the array. This difference will become important if we progress beyond the nearest-neighbor approximation. The difficulty is eliminated by forming the array into a loop in such a way that the last Nth atom is separated from the first by a distance d ($l = 0$). For $N \gg 1$ the properties of the ring will differ little from those of a finite linear array. This technique was proposed by Born and *von* Karman [2.2]. In this case we exclude from consideration any oscillations, which are only determined by "correct" boundary conditions, but the spectrum of the other modes remains almost unchanged. For the ring we have $u_{l+Nd} = u_l$ or, with (2.1.3), $U_q \exp[\mathrm{i}q(l+Nd)] = U_q \exp(\mathrm{i}ql)$; this gives $\exp(\mathrm{i}qNd) = 1$. This condition is satisfied if $qNd = k2\pi$, where $k = 0, \pm 1, \pm 2, \ldots$, and for the possible values of q we obtain

$$q = k2\pi/Nd \ . \tag{2.1.7}$$

Since $Nd \gg 2\pi$, $q_{k+1} - q_k = 2\pi/Nd \ll 1/d$; i.e., the number q for sufficiently long arrays varies quasi-continuously. Because of the periodic dependence of ω_q on q, obeying (2.1.5), we are interested in the values of q only in the limits

$$-\pi/d < q \leqslant \pi/d \quad \text{and} \quad -N/2 < k \leqslant N/2 \ ; \tag{2.1.8}$$

i.e., the number of k values is equal to N, the total number of degrees of freedom of a linear array made up of N atoms. Thus all solutions of (2.1.2) have been found with the help of substitution (2.1.3). Condition (2.1.8) determines the so-called Brillouin zone for possible values of the wave number q. The values of q, which lie outside the interval (2.1.8), in virtue of the periodicity of the dependence (2.1.5), lead to the recurrence of already known ion-oscillation frequencies.

Consider the dispersion relation (2.1.5) for small wave numbers ($qd \ll 1$), when the sine may be replaced by its argument to give a linear dependence

$$\omega_q \approx (\sqrt{\alpha/m}\, d)q = v_s 2\pi/\lambda \ . \tag{2.1.9}$$

Instead of q, the wavelength λ is introduced here,

$$\lambda = 2\pi/q \ , \tag{2.1.10}$$

and v_s stands for the velocity of propagation of sound oscillations

$$v_s = \sqrt{\alpha/m}\,d \ . \tag{2.1.11}$$

In this approximation the quantity v_s is a constant.

As seen from (2.1.9), the linear dispersion relation holds for both long waves ($\lambda \gg d$) and normal elastic waves of a continuous medium. Therefore, when $\lambda \gg d$, a linear array of atoms behaves, for example, like a massive string. This is what allowed Debye [2.3] to develop his theory of thermal oscillations of a solid, regarding the crystal as a continuum with frequency-independent elastic constants. He used two major assumptions: (1) the propagation of elastic acoustic oscillations in a solid coincides with the thermal oscillations of its atoms and (2) the wave method may be applied to a calculation of thermal oscillations since it yields good results in acoustic problems. Debye's theory met with a difficulty in the calculation of the number of possible frequencies. In continuum theory the existence of an infinite number of oscillation overtones is allowed (for example, in a string or rod there are no restrictions on the maximum frequencies $\omega_{q\max}$ or minimum wavelengths $\lambda_{\min}$). In the theory of a discrete crystal the quantity $\lambda_{\min}$ is determined by its atomic structure, i.e., the value of the lattice parameter d ($\lambda_{\min} \sim d$). Therefore, Debye had to "cut off" the spectrum at $\lambda_{\min}$.

Let us now consider the asymptote of the dispersion relation at large q, i.e., near the boundary of the zone (2.1.8). The linear dispersion relation does not hold here; the velocity (2.1.11) is no longer constant but depends on q. At the edges of the zone $q = \pm\pi/d$ and the quantity λ reaches a minimum, and ω_q a maximum

$$\lambda_{\min} = 2\pi/|q_{\max}| = 2d \quad \text{and} \quad \omega_{\max} = 2\sqrt{\alpha/m} \ ; \tag{2.1.12}$$

the ω_q curve has a tangent parallel to the q axis (Fig. 2.2).

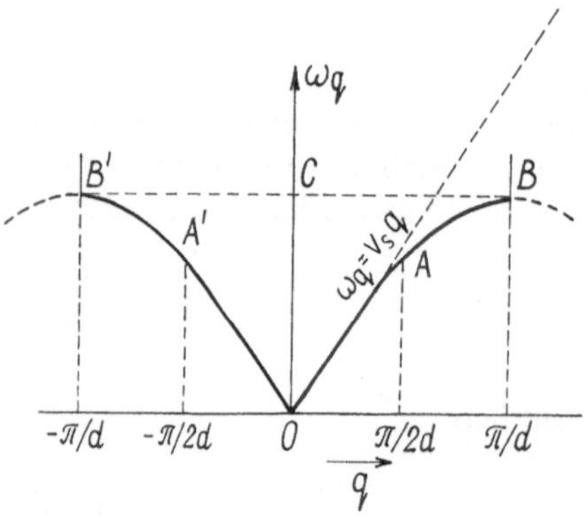

Fig. 2.2. Dispersion relation for the oscillation frequency of a monatomic array

In many problems where the spectrum of crystal oscillations is used, for example, in heat-capacity calculations, we need to know the spectral frequency distribution. In the one-dimensional case we denote the number of normal oscillations for a unitary interval of q values by $w(q)$ and call this function the density of states in q space. It follows from (2.1.7) that one oscillation falls on the interval $\Delta q = 2\pi/Nd$. Therefore,

$$w(q) = Nd/2\pi \ . \tag{2.1.13}$$

We will be interested in the number of states or the spectral density $D(\omega)\,d\omega$ for the frequency interval $d\omega$ with frequency ω. The functions $w(q)$ and $D(\omega)$ are related as follows:

$$w(q)\frac{dq}{d\omega}\,d\omega = D(\omega)\,d\omega \ . \tag{2.1.14}$$

Thus, to determine $D(\omega)$, we need to know $dq/d\omega$. It is found from (2.1.5), which is written in the form

$$q = (2/d)\arcsin(\omega/\omega_{max}) \ . \tag{2.1.15}$$

Consequently, in our case

$$dq/d\omega = (2/d)(\omega_{max}^2 - \omega^2)^{-1/2} \ , \tag{2.1.16}$$

and, in virtue of (2.1.13, 14),

$$D(\omega) = w(q)\frac{dq}{d\omega} = \frac{1}{\pi}N(\omega_{max}^2 - \omega^2)^{-1/2} \ . \tag{2.1.17}$$

As follows from (2.1.17), the function $D(\omega)$ at the band boundaries ($\omega = \omega_{max}$) shows a singularity (tends to ∞). The $D(\omega)$ is much more difficult to determine in two- and three-dimensional lattices.

2.1.2 A Linear Diatomic Array

Let us generalize the problem under consideration by adding to the unit cell another ion of mass M and by placing this ion exactly in the middle between the ions of mass m. The parameter of this lattice will be $2d$ (Fig. 2.3). We label the displacements with U_l and u_l, and the force constant, again, with α; we confine ourselves to the approximation of nearest neighbors (in this case between

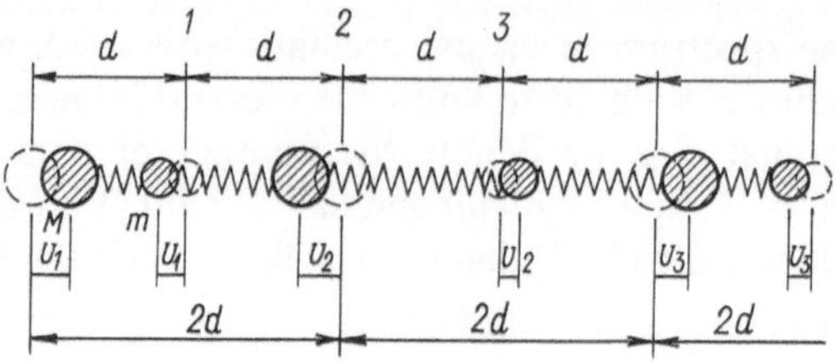

Fig. 2.3. One-dimensional model of a diatomic crystal composed of unlike ions of different mass (m and M)

different ions). The equations of motion, similar to (2.1.4), have the form

$$M\ddot{U}_q = -2\alpha U_q + 2\alpha\cos(qd)u_q ,$$
$$m\ddot{u}_q = -2\alpha u_q + 2\alpha\cos(qd)U_q , \tag{2.1.18}$$

and

$$U_q = A_q\exp(-\mathrm{i}\omega_q t) , \qquad u_q = B_q\exp(-\mathrm{i}\omega_q t) . \tag{2.1.19}$$

Substitution of (2.1.19) into (2.1.18) yields

$$\begin{aligned}(2\alpha - M\omega_q^2)A_q - 2\alpha\cos(qd)B_q &= 0 ,\\ -2\alpha\cos(qd)\,A_q + (2\alpha - m\omega_q^2)B_q &= 0 .\end{aligned} \tag{2.1.20}$$

Equations (2.1.20) have nonzero solutions at the frequencies which make the determinant of the system vanish

$$\begin{vmatrix} 2\alpha - M\omega_q^2 & -2\alpha\cos qd \\ -2\alpha\cos qd & 2\alpha - m\omega_q^2 \end{vmatrix} = 0 . \tag{2.1.21}$$

The roots of (2.1.21) are equal to

$$\omega_q^2(\pm) = \alpha\left(\frac{1}{M} + \frac{1}{m}\right) \pm \alpha\left[\left(\frac{1}{M} + \frac{1}{m}\right)^2 - \frac{4\sin^2 qd}{Mm}\right]^{1/2} . \tag{2.1.22}$$

Again, it is seen from dispersion relation (2.1.22) that the frequency is a periodic function of q. Because of the doubling of the parameter, the zone of possible q's now will be

$$-\pi/2d < q \leqslant \pi/2d , \tag{2.1.23}$$

instead of (2.1.8). The plot of (2.1.22) is presented in Fig. 2.4. Inspection of the plot shows that the diatomic array has two frequency branches. One of them,

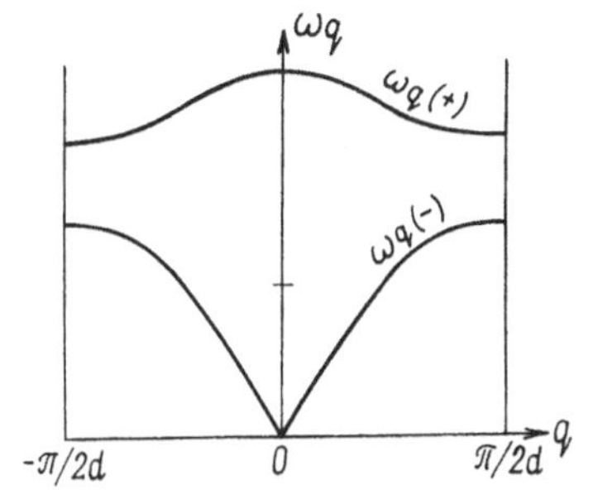

Fig. 2.4. Dispersion relation for the oscillation frequencies of the acoustic $\omega_{q(-)}$ and the optical $\omega_{q(+)}$ branches of a linear diatomic array

$\omega_{q(-)}$, when $q = 0$, has $\omega_{0(-)} = 0$ and is known as the acoustic branch. At small q's ($qd \ll 1$) a linear dispersion relation holds:

$$\omega_{q(-)} \approx qd[2\alpha/(m+M)]^{1/2} \; , \tag{2.1.24}$$

which is similar to (2.1.9) for a monatomic array (Fig. 2.2). For small q's it is analogous to the dispersion relation of the Debye continuum. The frequency at the zone edges ($q = \pm\pi/2d$), by (2.1.22), is equal to

$$\omega_{q(-)\max} = \alpha^{1/2}\left[\left(\frac{1}{M}+\frac{1}{m}\right)-\left|\frac{1}{M}-\frac{1}{m}\right|\right]^{1/2} = \begin{cases} \text{at } m > M & (2\alpha/m)^{1/2} \\ \text{at } M > m & (2\alpha/M)^{1/2} \end{cases} . \tag{2.1.25}$$

The other branch, $\omega_{q(+)}$, is termed the optical branch. At small q values, the two branches are separated in frequency (Fig. 2.4). At the zone edges, with $q = \pm\pi/2d$, they draw closer to each other, but a "gap" still remains between them:

$$\Delta\omega = \omega_{q(+)}(\pm\pi/2d) - \omega_{q(-)}(\pm\pi/2d) = \pm(2\alpha)^{1/2}(1/\sqrt{M} - 1/\sqrt{m}) \tag{2.1.26}$$

(a plus sign for $M < m$ and a minus sign for $M > m$). The value of the limiting frequency of the optical branch at $q = 0$ is equal to

$$\omega_{q(+)\max} = (2\alpha)^{1/2}\left(\frac{1}{M}+\frac{1}{m}\right)^{1/2} . \tag{2.1.27}$$

The proof that acoustic-branch oscillations in a two-atom array proceed at small q in the same way as they do in a monatomic array (two adjacent unlike ions oscillate in phase) is left as an exercise for the reader. In the case of optical-branch oscillations, adjacent positive and negative ions vibrate in antiphase; i.e., optically active oscillations of the electrical dipole moment occur in each cell. For transverse oscillations this is seen in Fig. 2.5. In real ionic crystals the

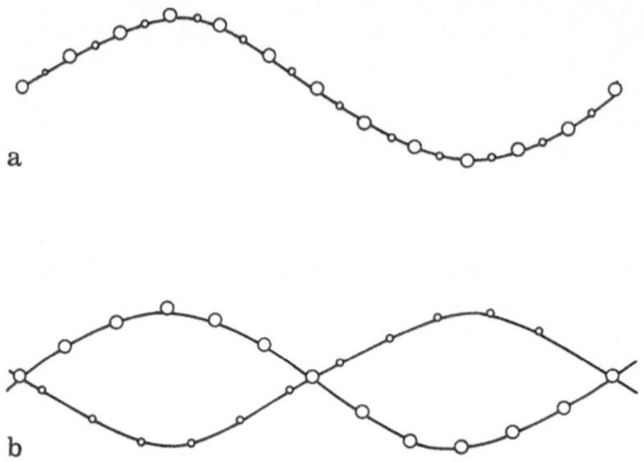

Fig. 2.5. Model of the acoustic (**a**) and the optical (**b**) branch of oscillations in a linear diatomic array (for equal wavelengths)

frequencies of the optical branch lie in the infrared region. Therefore, these crystals readily absorb infrared light (Sect. 2.6).

Let us compare the dispersion relation of a monatomic array (Fig. 2.2) and that of a diatomic array (Fig. 2.4). In the first array, the parameter may formally be doubled. The zone in q space will then decrease in size by a factor of 2 (from $-\pi/2d$ to $\pi/2d$). By displacing the segments AB and $A'B'$ of the ω_q curve to the left and to the right, respectively, along the q axis by the quantity π/d (or $-\pi/d$), we may represent the dispersion relation of a monoatomic array as two branches (Fig. 2.6). In contrast to Fig. 2.4, at points $q = \pm\pi/2d$ there is no gap between the acoustic and optical branches. Both branches, as portrayed in Fig. 2.2, pass continuously into each other at points A and A'. The curves presented in Fig. 2.4 will now be modified. Taking advantage of the circumstance that the functions $\omega_{q(\pm)}$ are periodic, we consider one more period on the q axis (from $-\pi/2d$ to $-\pi/d$ and from $\pi/2d$ to π/d), as illustrated in Fig. 2.7. Comparison with Fig. 2.2 shows that in going from the monatomic to the diatomic array new zone boundaries appear at $q = \pm\pi/2d$. The frequency spectrum at these boundaries splits up and a gap $\Delta\omega$ appears. The representation of the frequency spectrum as shown in Fig. 2.4 is called the reduced-zone representation, and the same in Fig. 2.6 the extended-zone representation.

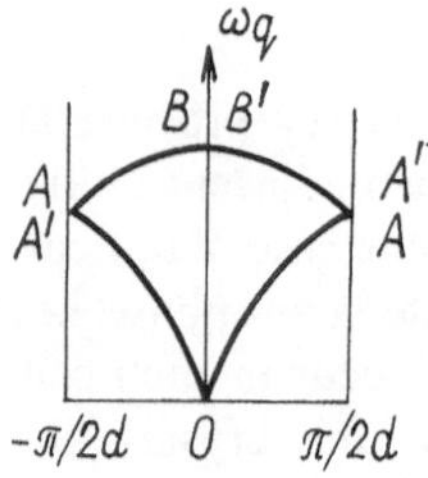

Fig. 2.6. Schematic drawing of two branches in the oscillation spectrum of a monatomic array after doubling its period ($d \to 2d$)

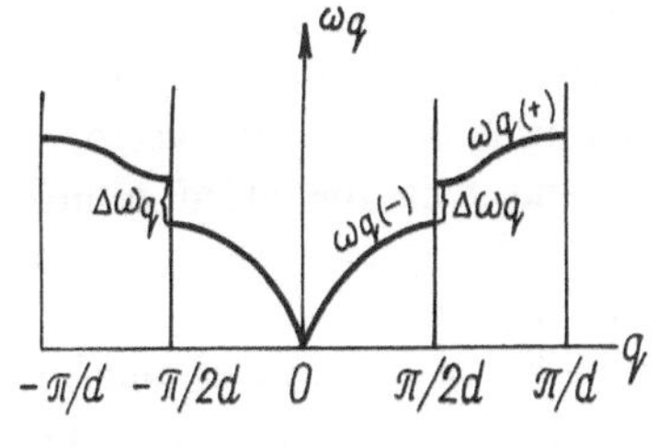

Fig. 2.7. Dispersion relation for oscillation frequencies of a linear diatomic array in the extended-zones representation

2.1.3 The Three-Dimensional Crystal Case

Passing to a two-dimensional and a three-dimensional lattice renders the problem much more complicated. Since the number of vector $\boldsymbol{u}_l$ components in the three-dimensional case is equal to three, the equation for determining the ω_q^2 of the acoustic branch will be cubic rather than linear. The meaning of its three roots can be understood in the limiting case of small q's, by proceeding to the Debye elastic continuum approximation. It is known from elasticity theory that three branches of acoustic oscillations, with different velocities and polarizations, may occur in a medium. If the medium is isotropic, one of the oscillations has longitudinal polarization; i.e., the displacement vector is parallel to the direction of propagation of the corresponding wave travelling with a velocity $v_s^{(l)}$. The two other oscillations exhibit transverse polarization with equal velocity $v_s^{(s)}$ (most frequently $v_s^{(s)} < v_s^{(l)}$) and with the displacement vector normal to the propagation vector. The constant-frequency surfaces $\omega_q^{(l)}$ and $\omega_q^{(s)}$ in $\boldsymbol{q}$ space (with the q's being small), where the linear dispersion relation holds good, have the form of concentric spheres (Fig. 2.8a, b). All this becomes more complicated when the values of $\boldsymbol{q}$ are close to the boundaries of the zones, and also when we take into account the anisotropy of actual crystals. Specifically, constant-frequency surfaces may differ from spheres substantially (Fig. 2.8b).

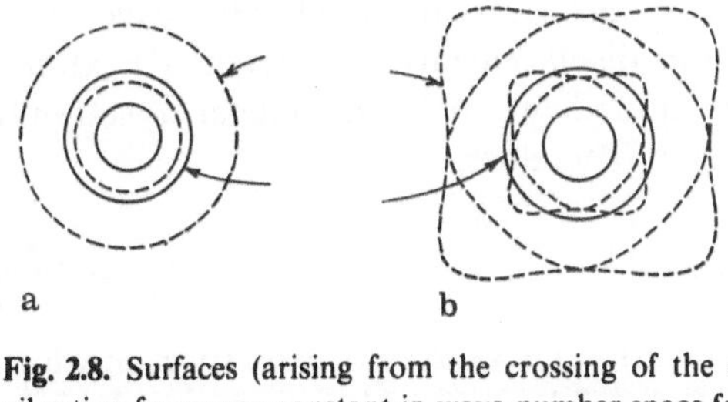

a b

Fig. 2.8. Surfaces (arising from the crossing of the extended zones by the plane) of the lattice vibration frequency constant in wave-number space for the isotropic case (**a**), for a real anisotropic crystal (**b**): solid lines indicate a longitudinal wave, dashed lines a transverse wave

Let us consider in brief the pattern of calculation for normal atomic vibrations in a three-dimensional crystal. For lattices with a basis (1.2.11) the displacement of atoms from the equilibrium position is given by the vector $\boldsymbol{u}_{ls}$, where $\boldsymbol{l}$ is the "number" (vector) of the cell, and s is the atom number ($s = 1, 2, 3, \ldots, \sigma$). Assuming the displacements to be small ($|\boldsymbol{u}_{ls}|/d \ll 1$), we expand the potential energy V in a power series of the displacement vector components $u_{ls}^{(j)}$ ($j = x, y, z$) [see (2.1.1) in the one-dimensional case] and limit ourselves to terms that are not higher than the quadratic one (harmonic approximation). Then we have

$$V = V_0 + \frac{1}{2} \sum_{\substack{(ls, l's') \\ jj'}} [\partial^2 V / \partial u_{ls}^{(j)} \partial u_{l's'}^{(j')}]_0 \, u_{ls}^{(j)} u_{l's'}^{(j')} + \ldots . \tag{2.1.28}$$

In (2.1.28) the linear term is absent for the same reason as in (2.1.1), and the "0" subscript of the second-order derivatives denotes that differentiation is carried out for the condition $u_{ls}^{(j)} = 0$. In view of (2.1.28), the equations of motion for displacements have the form

$$m_s \ddot{u}_{ls}^{(j)} = - \sum_{l's'j'} [\partial^2 V / \partial u_{ls}^{(j)} \partial u_{l's'}^{(j')}]_0 \, u_{l's'}^{(j')} , \tag{2.1.29}$$

with $\boldsymbol{l}$ running over all N unit cells of the crystal.

Thus we have obtained a system of $3N\sigma$ coupled differential equations. We start by analyzing the coefficients on the right-hand sides of these equations, which are Cartesian components of the tensor of rank two:

$$G_{ls,l's'}^{jj'} = [\partial^2 V / \partial u_{ls}^{(j)} \partial u_{ls,l's'}^{(j')}]_0 \; . \tag{2.1.30}$$

In vector form (2.1.29) then becomes

$$m \ddot{\boldsymbol{u}}_{ls} = - \sum_{l's'} G_{ls,l's'} \, \boldsymbol{u}_{l's'} \; . \tag{2.1.31}$$

The separate terms on the right-hand side of (2.1.31) are equal to the force which acts on the atom $\boldsymbol{l}$, s and arises from displacements of different $\boldsymbol{l}'s'$ atoms. Regardless of the nature of the interaction, they depend in an ideal crystal only on the relative position of the unit cells. Therefore,

$$G_{ls,l's'} = G_{ss'}(\boldsymbol{l}' - \boldsymbol{l}) = G_{ss'}(\boldsymbol{n}) = G_{ss'}(-\boldsymbol{n}) \; . \tag{2.1.32}$$

Here $\boldsymbol{n} = \boldsymbol{l}' - \boldsymbol{l}$ and the last equality in (2.1.32) is valid only for lattices in which atom is the center of symmetry. In virtue of (2.1.32), (2.1.31) will assume the form

$$m_s \ddot{u}_{ls} = -\sum_{ns'} G_{ss'}(n) u_{l+n,s'} \ . \tag{2.1.33}$$

The solution of this system of equations will be tried in the form

$$u_{ls}(t) = A_{sq} \exp[\mathrm{i}(ql - \omega_q t)] \ . \tag{2.1.34}$$

Substituting (2.1.34) into (2.1.33) and reducing the result by $\exp[\mathrm{i}(ql - \omega_q t)]$, we find

$$\begin{aligned} m_s \omega_q^2 A_{sq} &= \sum_{s'} \left[\sum_n G_{ss'}(n) \exp(\mathrm{i} q \cdot n) \right] A_{s'q} \\ &= \sum_{s'} G_{ss'}(q) A_{s'q} \ , \end{aligned} \tag{2.1.35}$$

where the notation

$$G_{ss'}(q) \equiv \sum_n G_{ss'}(n) \exp(\mathrm{i} q \cdot n) \tag{2.1.36}$$

is introduced for the Fourier transform of the tensor $G_{ss'}(n)$. Comparing (2.1.35) with input equations (2.1.29, 31), we see that they have decreased in number from $3N\sigma$ to 3σ. This is a direct consequence of the translational invariance of the crystal lattice. The number of degrees of freedom is $3N\sigma$, as before. Therefore, to determine the motion of the entire system, we must know the number of degrees of freedom for each of the possible values of the vector q. As we have seen in the one-dimensional problem (2.1.7, 8), the boundary conditions give N such values and an individual calculation should be carried out for each of them [since A_{sq} and $G_{ss'}(q)$ depend on q]. In practice, calculations are normally carried out for a special limited set of q values taken from the Brillouin zone; solutions for the other q's are found using the interpolation method.

It is suitable to bring (2.1.35) into a more compact form

$$\sum_{s'j'} [G_{ss'}^{jj'}(q) - \omega^2 m_s \delta_{ss'} \delta_{jj'}] A_{s'q}^{j'} = 0 \ . \tag{2.1.37}$$

This system is an example of a problem in determining the eigenvalues of ω^2 and the eigenvectors $A_{s'q}^{j'}$. The condition for obtaining nonzero solutions to the homogeneous system (2.1.37) is the equality to zero of the determinant involved in the coefficients of the system. From the resulting equation of degree 3σ we find the roots—the branches of the spectrum ω_q^2. For crystals without a basis ($\sigma = 1$) we find from the cubic equation three acoustic branches ($\omega_q \to 0$ when $q \to 0$), which are depicted in Fig. 2.9 for some direction q. These branches give a linear

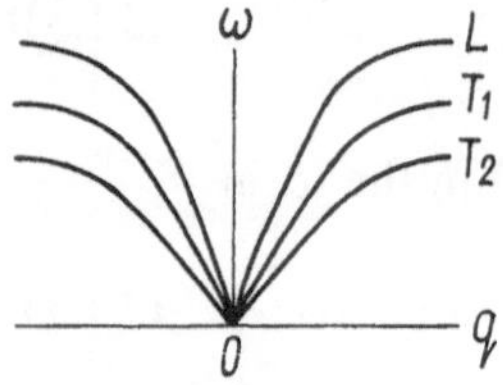

Fig. 2.9. Dependence of the frequency ω of a monatomic three-dimensional lattice (for some direction in wave-vector $\boldsymbol{q}$ space) for longitudinal L and transverse T_1, T_2 oscillations within one Brillouin zone

dependence of $\omega_{\boldsymbol{q}}$ on $\boldsymbol{q}$ in the vicinity of $\boldsymbol{q} = 0$, the slope of the curves being equal to the velocity of propagation of the corresponding oscillation.

We wish to elaborate on this important problem. The potential energy (2.1.28) should not vary during shifting of the entire crystal as a whole, i.e., when $u^j_{ls} \equiv u^j = \text{const}$. We leave as an exercise for the reader the proof that this fact entails the identity $\sum_{ss'} G_{ss'}(\boldsymbol{q} = 0) = 0$. It thus follows from (2.1.37) that with $\boldsymbol{q} \to 0$, $\omega_{\boldsymbol{q}} \to 0$ and three s'-independent solutions exist for $j = x, y, z$. The existence of acoustic branches for which $\omega_{\boldsymbol{q}} = 0$ at $\boldsymbol{q} = 0$ thus comes from the fact that the lattice is translationally invariant. This is a particular case of *Goldstone*'s fundamental theorem [2.4], which relates the symmetry properties of a system to the existence of low-frequency modes in it and has numerous applications in the theory of magnetism, superconductivity, and phase transitions.

As $|\boldsymbol{q}|$ increases and approaches the boundary of the zone, the linearity is disturbed; at the boundary itself, by contrast with the one-dimensional case (Fig. 2.2), the point of contact with the horizontal tangent may not be reached. However, the $\omega_{\boldsymbol{q}}$ curves in the general case remain periodic, the period being related to the reciprocal-lattice vectors; namely, the points $\boldsymbol{q}$ and $\boldsymbol{q} + \boldsymbol{b}^*$ (with $\boldsymbol{b}^*$ being a reciprocal-lattice vector) are equivalent (i.e., possess equal values: $\omega_{\boldsymbol{q}} = \omega_{\boldsymbol{q}+\boldsymbol{b}^*}$).

For a diatomic ionic crystal with $\sigma = 2$, the equation for $\omega^2_{\boldsymbol{q}}$ is of sixth degree, and in the general case we obtain six spectral branches ($\omega^{(i)}_{\boldsymbol{q}}$, $i = 1, 2, 3, \ldots, 6$), of which three are acoustic and three are optical. This is diagrammatically represented (for some direction of $\boldsymbol{q}$) in Fig. 2.10. The extended-zone diagram in

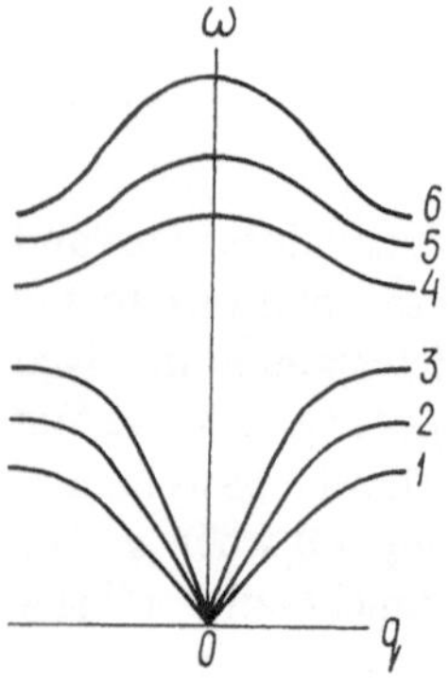

Fig. 2.10. Dispersion relation of acoustic (1–3) and optical (4–6) branches for a diatomic lattice

Fig. 2.11. Surfaces of the frequency constant $\omega_{\boldsymbol{q}}^{(i)} = \text{const}$ in wave-vector $\boldsymbol{q}$ space for one of the oscillation branches in the extended-zones scheme

Fig. 2.11 shows (for some definite cross sections in $\boldsymbol{q}$ space) the picture of periodic constant-frequency surfaces for one of the spectral branches $\omega_{\boldsymbol{q}}^{(i)}$. In the general case, the number of spectral branches will be equal to 3σ:

$$\omega_{\boldsymbol{q}}^{(s)}; \quad (s = 1, 2, \ldots, \sigma) \ . \tag{2.1.38}$$

2.1.4 Quantization of Ionic-Lattice Vibrations

Prior to considering the physical properties of ionic lattices, we wish to give a quantum description of their vibrations. As already stated in Sect. 1.9, an efficient way of solving many-particle problems is by using the unitary transformation technique, which allows the interaction term in the Hamiltonian to be brought into diagonal form so that the complicated motion of the system can be described, approximately, as the motion of an ideal quasiparticle gas. This is what we actually did when considering the classical problem of the oscillations of arrays and a three-dimensional crystal. In fact, the solution of (2.1.3 or 6) is nothing but a unitary transformation, which diagonalizes the oscillation array. In this classical problem an elementary excitation is a sine wave.

Let us see how this situation changes in the quantum case. We add to the generalized-coordinate operators $\hat{u}_l$ the conjugate-momentum operators $\hat{p}_l$. Then the kinetic-energy operator will be $\hat{T} = \sum_l \hat{p}_l^2/2m$, where the summation is taken over all N array sites, and the potential-energy operator in the nearest-neighbor approximation is given by (2.1.1). The Hamiltonian of the system. $\hat{\mathscr{H}}$, which is equal to the sum of the operators $\hat{T}$ and $\hat{U}$, will be

$$\begin{aligned}\hat{\mathscr{H}} &= \hat{T} + \hat{U} = \sum_l (\hat{p}_l^2/2m + \alpha\hat{u}_l^2) - \frac{\alpha}{2}\sum_l (\hat{u}_l\hat{u}_{l+d} + \hat{u}_l\hat{u}_{l-d}) \\ &= \sum_l \hat{p}_l^2/2m + \alpha\hat{u}_l^2 - \alpha\sum \hat{u}_l\hat{u}_{l+d} \ ,\end{aligned} \tag{2.1.39}$$

with the operators $\hat{u}_l$ and $\hat{p}_l$ satisfying the commutation relations

$$[\hat{u}_l, \hat{p}_l]_- = i\hbar\delta_{ll'} , \quad [\hat{u}_l, \hat{u}_{l'}]_- = [\hat{p}_l, \hat{p}_{l'}]_- = 0 . \tag{2.1.40}$$

The symbol $[\hat{a}, \hat{b}]_- = \hat{a}\hat{b} - \hat{b}\hat{a}$ stands for the commutator. The first sum of the right-hand side of (2.1.40) is the energy operator of N independent harmonic oscillators, and the second sum takes into account the interaction of nearest neighbors in the array. The unitary transformation diagonalizing (2.1.39) has the form

$$\hat{u}_l = N^{-1/2} \sum_q \left[\hat{U}_q \cos ql - \frac{1}{m\omega_q} \hat{P}_q \sin ql \right] ,$$

$$\hat{p}_l = N^{-1/2} \sum_q [m\omega_q \hat{U}_q \sin ql + \hat{P}_q \cos ql] , \tag{2.1.41}$$

where ω_q is defined by (2.1.5), and the operators $\hat{U}_q$ and $\hat{P}_q$ satisfy, by virtue of (2.1.40, 41), the commutation relations

$$[\hat{U}_q, \hat{P}_{q'}]_- = i\hbar\delta_{qq'} , \quad [\hat{U}_q, \hat{U}_{q'}]_- = [\hat{P}_q, \hat{P}_{q'}]_- = 0 , \tag{2.1.42}$$

the proof of which is left as an exercise for the reader. Using (2.1.41, 42), the Hamiltonian (2.1.39) may be shown to assume the diagonal form in the new variables

$$\hat{\mathscr{H}}' = \sum [\hat{P}_q^2/2m + (m\omega_q^2/2)\hat{U}_q^2] . \tag{2.1.43}$$

We also leave as an exercise for the reader the proof that this is so. The operator (2.1.43) is equal to the sum of the Hamiltonians of linear harmonic oscillators with frequencies ω_q. The energy of the system is equal to the sum of the energies of such oscillators

$$\mathscr{E} = \sum_q (n_q + \tfrac{1}{2})\hbar\omega_q , \quad n_q = 0, 1, 2, \ldots . \tag{2.1.44}$$

Thus the oscillation of the array is represented as a "gas" of independent effective oscillators, that is, quantized sonic waves with frequencies ω_q. A quantized sonic wave with a wave vector q, frequency ω_q, and energy $(n_q + 1/2)\hbar\omega_q$ may be regarded as a set of n_q quanta with an energy $\hbar\omega_q$ each and with a zero-point oscillation energy $\hbar\omega_q/2$. These quanta are referred to as phonons. Sometimes, not altogether correctly, they are said to be quantized sonic waves. In fact, a sonic wave is associated with n_q phonons plus the zero-point vibrational energy. For this reason the phonon should be called the sound quantum. The quantity $\hbar\omega_q$ is the smallest portion of energy above the ground-state level $\hbar\omega_q/2$. The phonon, therefore, is the collective excitation of the ionic lattice as a whole.

One more transformation may be carried out on the Hamiltonian (2.1.43):

$$\hat{b}_q = (2m\hbar\omega_q)^{-1/2}\hat{P}_q - \mathrm{i}(m\omega_q/2\hbar)^{1/2}\hat{U}_q \ , \tag{2.1.45}$$

$$\hat{b}_q^+ = (2m\hbar\omega_q)^{-1/2}\hat{P}_q - \mathrm{i}(m\omega_q/2\hbar)^{1/2}\hat{U}_q \ . \tag{2.1.46}$$

It can be readily verified, by means of (2.1.42), that the operators $\hat{b}_q$ and $\hat{b}_q^+$ obey the commutation rules

$$[\hat{b}_q, \hat{b}_{q'}^+]_- = \delta_{qq'} \ , \quad [\hat{b}_q^+, \hat{b}_{q'}^+]_- = [\hat{b}_q, \hat{b}_{q'}]_- = 0 \tag{2.1.47}$$

and the Hamiltonian (2.1.43) will take a more compact form

$$\hat{\mathscr{H}}' = \sum_q (\hat{b}_q^+ \hat{b}_q + 1/2)\hbar\omega_q$$

$$= \tfrac{1}{2}\sum_q \hbar\omega_q + \sum_q \hat{b}_q^+ \hat{b}_q \hbar\omega_q = \mathscr{E}_0 + \sum_q \hat{n}_q \hbar\omega_q \ . \tag{2.1.48}$$

Comparison of (2.1.48) with (2.1.44) shows that the operator $\hat{b}_q^+ \hat{b}_q$ is a quasiparticle number operator; i.e., $\hat{n}_q$ is a phonon number operator. The operators $\hat{b}_q^+$ and $\hat{b}_q$ are called the Bose operators, and phonons are Bose particles, or bosons, similar to light quanta, or photons. The wave function of the system of bosons is symmetric with respect to permutations of the coordinates of any pair of particles. In a statistical description of bosons the Bose–Einstein statistics are used (hence the name of these particles). Specifically, the mean with respect to the number of bosons $\bar{n}_q$ is given by the well-known formula (Sect. 2.7.2):

$$\bar{n}_q = [\exp(\hbar\omega_q/k_B T) - 1]^{-1} \ , \tag{2.1.49}$$

and the mean energy of the oscillator is equal to

$$\bar{\varepsilon}_q = (\bar{n}_q + 1/2)\hbar\omega_q \ . \tag{2.1.50}$$

In the three-dimensional case it is necessary to replace q by the vector $\boldsymbol{q}$, and also to take into account the polarization for the acoustic and optical branches.

2.2 The Specific-Heat Capacity of the Lattice

As far back as 1819 Dulong and Petit discovered an empirical law according to which the atomic heat capacity in the case of a constant volume is practically constant (independent of T) for an overwhelming majority of solids and is equal to

$$C_V = 6\,\mathrm{kcal\,deg^{-1}\,mol^{-1}} \approx 25\,\mathrm{J\,mol^{-1}\,K^{-1}}\ . \tag{2.2.1}$$

This is easy to explain using the general laws of classical statistical mechanics and the concepts about the oscillatory nature of the thermal motion of crystals. A mole of a monatomic crystal contain N_A atoms (N_A is Avogadro's number $\sim 6.02 \times 10^{23}\,\mathrm{mol^{-1}}$). Therefore, the number of degrees of freedom is equal to $3N_A$. Acording to the law of equipartition of energy, each degree of freedom has on the average $k_B T/2$ of kinetic energy. If we visualize the crystal as an assembly of $3N_A$ harmonic oscillators, their total energy is equal to the sum of the kinetic ε_k and the potential ε_p energy. It is also known from statistics that the mean values of these energies for a harmonic oscillator are equal to each other and each of them is equal to $3k_B T/2$, and their sum is $3k_B T$. (We leave it as an exercise for the reader to prove all this.) Therefore, the thermal energy of a mole of a monatomic crystal is equal to $\mathscr{E} = 3N_A k_B T = 3RT$, where the gas constant $R \sim 2\,\mathrm{kcal/mol} = 8.31\,\mathrm{J\,mol^{-1}\,K^{-1}}$, and for the heat capacity we have

$$C_V = \left(\frac{\partial \mathscr{E}}{\partial T}\right)_V = 3R \approx 6\,\mathrm{kcal\,deg^{-1}\,mol^{-1}} \approx 25\,\mathrm{J\,kmol^{-1}\,K^{-1}}\ . \tag{2.2.2}$$

Thus the theoretical formula (2.2.2) confirms Dulong's and Petit's experimental law (2.2.1).

Originally, the experiments of *Dulong* and *Petit* related to a comparatively narrow temperature interval (close to room temperature, $\sim 20°\mathrm{C}$). As that interval was extended, two tendencies were noticed: the quantity C_V increased with increasing T, as the melting point was approached (Sect. 2.3.2), and C_V decreased during cooling, in which case a general law was revealed—$C_V \to 0$ when $T \to 0$ (Fig. 2.12). After *Nernst* discovered in 1906 the thermal theorem (third law of thermodynamics), that property of the heat capacity proved to be

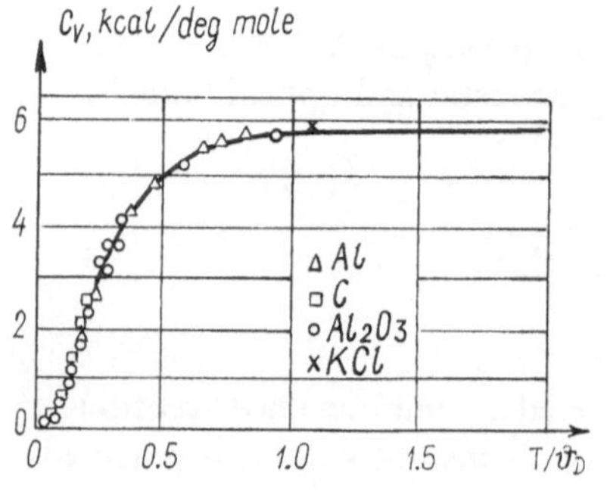

Fig. 2.12. Comparison of experimental data for the molecular heat capacity C_v (in kcal/deg mole) of some solids as a function of reduced temperature T/ϑ_D with the Debye theoretical curve (2.2.16)

an immediate corollary of the thermal theorem. In 1906 *Einstein* was the first to explain qualitatively the curve (Fig. 2.12) for $C_V(T)$, using quantum theory. He suggested that the crystal be treated as a totality of $3N_A$ linear quantum oscillators with equal frequency ω, the energy of which is always a multiple of the minimum quantity $\hbar\omega$: $n\hbar\omega$ ($n = 1, 2, \ldots$); i.e., there is a spectrum of discrete nondegenerate states ($g_n = 1$).

Using the standard formula for the partition function Z, we obtain

$$Z(T) = \sum_{n=0}^{\infty} \exp[-(n+1/2)\hbar\omega/k_B T]$$
$$= \sum_{n=0}^{\infty} [\exp(-\hbar\omega/k_B T)]^{(n+1/2)} \tag{2.2.3}$$
$$= \frac{\exp(-\hbar\omega/2k_B T)}{1-\exp(-\hbar\omega/k_B T)} \; .$$

By definition, the mean energy of a linear oscillator is equal to

$$\bar{\varepsilon} = \frac{\hbar\omega}{2} + \frac{1}{Z(T)} \sum_{n=0}^{\infty} n\hbar\omega \exp(-n\hbar\omega/k_B T)$$
$$= \frac{\hbar\omega}{2} - \frac{\partial}{\partial(1/k_B T)} \ln\left[\sum_n \exp(-n\hbar\omega/k_B T)\right] \tag{2.2.4}$$
$$= \frac{\partial}{\partial(1/k_B T)} \ln Z(T) + \frac{\hbar\omega}{2} \; .$$

Employing (2.2.3) for $Z(T)$, we find

$$\bar{\varepsilon} = \frac{\hbar\omega}{2} + \frac{1}{\partial(1/k_B T)} \ln[1-\exp(-\hbar\omega/k_B T)]$$
$$= \frac{\hbar\omega}{2} + \frac{\hbar\omega}{\exp(\hbar\omega/k_B T)-1} \; . \tag{2.2.5}$$

Equation (2.2.5) yields

$$\bar{\mathscr{E}} = \frac{3N_A\hbar\omega}{\exp(\hbar\omega/k_B T)-1} + \frac{3N_A\hbar\omega}{2} \tag{2.2.6}$$

for the molar energy and

$$C_V = \left(\frac{\partial\bar{\mathscr{E}}}{\partial T}\right)_V = 3N_A k_B(\hbar\omega/k_B T)^2 \frac{\exp(\hbar\omega/k_B T)}{[\exp(\hbar\omega/k_B T)-1]^2} \; . \tag{2.2.7}$$

This is the celebrated Einstein formula for the atomic heat capacity of a crystal. Let us consider its asymptote at high and low T.

At high temperatures $k_B T \gg \hbar\omega$, or $\hbar\omega/k_B T \ll 1$, so in the denominator of the fraction on the right-hand side of (2.2.7), we can use the expansion $\exp(\hbar\omega/k_B T) \approx 1 + \hbar\omega/k_B T + \dots$, and in the numerator replace $\exp(\hbar\omega/k_B T)$ by unity. As a result, we obtain $C_V \approx 3N_A k_B = 3R$, i.e., the classical formula (2.2.2) corresponding to experiment.

At low temperatures $\hbar\omega/k_B T \gg 1$, the unity in the denominator of the fraction in (2.2.7) may be neglected in comparison with $\exp(\hbar\omega/k_B T)$. Then the approximate result for C_V will be

$$C_V \approx 3N_A k_B(\hbar\omega/k_B T)^2 \exp(-\hbar\omega/k_B T) \ . \qquad (2.2.8)$$

In the limit $T \to 0$ the exponential factor tends to zero faster than the power factor $(\hbar\omega/k_B T)^2$ increases, and therefore, by (2.2.8), $C_V \to 0$ when $T \to 0$ K as $\exp(-\hbar\omega/k_B T)$, which agrees with the Nernst theorem and is qualitatively consistent with experiment. However, the experimental curve in the vicinity of 0 K obeys the power law ($\sim T^3$) rather than an exponential law. This difference apparently is due to Einstein's assumption of the value of ω being the same for all atomic oscillators, which is too crude an approximation.

A further refinement of the theory of the heat capacity of solids was made by *Debye* [2.3]. He took account of the fact that the crystal had an oscillation frequency spectrum (2.1.37). Confining himself to the continuum approximation, he used only one acoustic branch, assuming the optical branches to be absent and the three acoustic branches to coincide. Furthermore, he assumed the dispersion relation to be linear. To take into account the discreteness of the crystal and the correct number of degrees of freedom , Debye, in his formulas for the thermal energy of the crystal, extended the integration not over the Brillouin zone but over a sphere in q space of finite radius q_{max} (thus cutting the spectrum off at a minimum wavelength $\lambda_{min} = 2\pi/q_{max}$). Taking account of the fact that the density of allowed values of $\boldsymbol{q}$ in $\boldsymbol{q}$ space is equal to $V/(2\pi)^3$, with V being the volume of a solid [(2.1.13) for a one-dimensional array], the value of q_{max} can be determined from the equation

$$(4\pi/3)\, q_{max}^3 (V/8\pi^3) = N \quad \text{or} \quad q_{max} = (6\pi^2 N/V)^{1/3} = (6\pi^2/v_c)^{1/3} \ , \qquad (2.2.9)$$

where v_c is the volume per lattice site, which may be represented as a sphere of equal volume of radius r_s (Wigner–Seitz sphere):

$$v_c = \tfrac{4}{3}\pi r_s^3 \ , \qquad q_{max} = (9\pi/2)^{1/3}/r_s$$

and

$$\lambda_{min} = 2\pi q_{max} \approx 2,6\, r_s \ . \qquad (2.2.10)$$

Then the maximum frequency according to Debye is

$$\omega_{max} = v_s\, q_{max} \;. \tag{2.2.11}$$

In addition, since the Debye model involves not one frequency, as in the case of the Einstein model, a simple multiplication by $3N_A$ does not permit us to proceed from (2.2.5 to 6), so we have to integrate over the Debye sphere:

$$\bar{\mathscr{E}} = \int_0^{\omega_{max}} \frac{\hbar\omega\, D(\omega)d\omega}{\exp(\hbar\omega/k_B T) - 1} \;. \tag{2.2.12}$$

Calculating (2.2.12) requires a knowledge of the spectral density $D(\omega)$ The quantity $D(\omega)d\omega$ is equal to the product of $3N_A$ by the ratio of the volume of the spherical layer $4\pi q^2/dq$ to the volume of the entire Debye sphere $(4\pi/3)\, q_{max}^3$, i.e.,

$$D(\omega)\, d\omega = 3N_A\, 4\pi q^2\, dq/(4\pi/3)q_{max}^3 = 9\, N_A\, \omega^2\, d\omega/\omega_{max}^3 \;. \tag{2.2.13}$$

Here we have employed (2.2.11) and the linear dispersion relation $\omega = v_s\, q$. As seen from (2.2.13), the spectral density according to Debye has the form of a parabolic law $D(\omega) \approx \omega^2$ (Fig. 2.13). Substitution of (2.2.13) into (2.2.12) yields

$$\bar{\mathscr{E}} = \frac{9\, N_A \hbar}{\omega_{max}^3} \int_0^{\omega_{max}} \frac{\omega^3\, d\omega}{\exp(\hbar\omega/k_B T) - 1} \;,$$

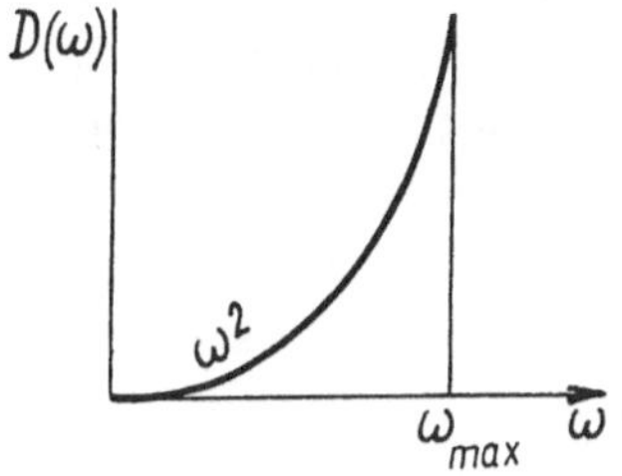

Fig. 2.13. Spectral density after Debye

and for the heat capacity we then have

$$C_V = \left(\frac{\partial \bar{\mathscr{E}}}{\partial T}\right)_V = \frac{9N_A k_B}{\omega_{max}} \int_0^{\omega_{max}} \frac{(\hbar\omega/k_B T)^2 \omega^2 \exp(\hbar\omega/k_B T) d\omega}{[\exp(\hbar\omega/k_B T) - 1]^2} \;. \tag{2.2.14}$$

Introduce a new variable $x = \hbar\omega/k_B T$ and determine the Debye temperature

$$\vartheta_D = \hbar\omega_{max}/k_B \;. \tag{2.2.15}$$

Ultimately, we find for C_V

$$C_V = 9R(T/\vartheta_D)^3 \int_0^{\vartheta_D/T} (e^x - 1)^{-2} x^4 e^x \, dx \ . \tag{2.2.16}$$

This is the celebrated Debye heat-capacity formula. At high temperatures ($T \gg \vartheta_D$) the upper limit of the integral in (2.2.16) is very small ($\vartheta_D/T \ll 1$), and therefore the integrand may be expanded in a power series of x. To a first approximation, this gives

$$\int_0^{\vartheta_D/T} (x^4/x^2) \, dx = \tfrac{1}{3}(\vartheta_D/T)^3 \ .$$

Then it is simple to see that (2.2.16) assumes the form of $C_V \approx 3R$; i.e., we come again to the Dulong and Petit law. At low temperatures the upper limit of the integral in (2.2.16) may be replaced as $\vartheta_D/T \to \infty$, and we arrive at the tabulated integral

$$\int_0^\infty x^4 e^x (e^x - 1)^{-2} \, dx = 4\pi^4/15 \ ,$$

and then find for C_V

$$C_V \approx (12\pi^4/5) R (T/\vartheta_D)^3 \ ; \tag{2.2.17}$$

i.e., the "Debye T^3 law". Figure 2.12 shows that it gives a very good fit to experimental curves in the region of low temperatures. The temperature ϑ_D is determined from experimental data and tabulated. It will suffice here to quote some of its values:

Mg	406 K
Cr	402 K
Fe	467 K
Cu	399 K
Ag	225 K

At very low temperatures, and also in some other cases, metals exhibit departures from the T^3 law. Explanations will be furnished in Chap. 3.

In spite of the excellent agreement with experiment, Debye's theory cannot be regarded as rigorous, because of the simplifications on which it is based. One of the major simplifications is the choice of a quadratic dependence for the spectral density $D(\omega)$, which may differ radically from its true form (except in the range of very small frequencies). As an illustration, Fig. 2.14 presents $D(\omega)$

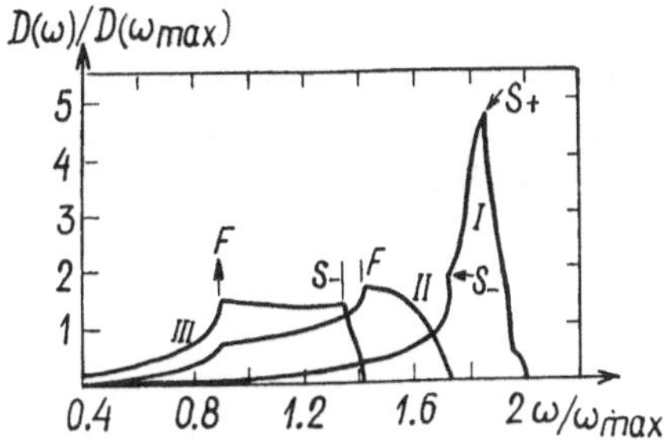

Fig. 2.14. Spectral density $D(\omega)/D(\omega_{\max})$ as a function of $\omega/\omega_{\max}$ for a copper crystal, calculated by the root-fitting method for a model taking account of the interaction between atoms up to the second coordination zone [Ref. 2.1, Fig. 9]: (*1*) branch of longitudinal oscillations, (*2, 3*) two branches of transverse oscillations; F stands for features arising from nonsingular critical points, and S for those due to the usual saddle points

functions calculated by *Owerton* [2.5] using the root-fitting method. In the figure curves are shown for three acoustic branches of copper, the letters F and S marking the singular points of the function $D(\omega)$. From the curves depicted in Fig. 2.14, it is seen that their initial segment is well approximated by the Debye parabola, showing that Debye's theory is in agreement with experiment for small frequencies.

Any attempt to compute the function $D(\omega)$ reduces to a problem of numerical computation. This compels one to proceed from the general equations (2.1.37). However, qualitative assumptions as to the form of the function $D(\omega)$ can also be made on the basis of purely topological theorems, without numerical computation. Specifically, the existence of critical points in the family of surfaces $\omega_{\boldsymbol{q}} = \text{const}$ in $\boldsymbol{q}$ space and of a singular function $D(\omega)$ can be shown to be a necessary consequence of lattice periodicity. This problem was first explored by *van* Hove [2.6] (van Hove's theorem).

We do not intend to enlarge upon this problem, but we illustrate the method for the case of one spectral branch, following *Ziman's* treatment [1.30]. According to the definition of the function $D(\omega)$ (2.1.14, 2.2.13), we may set

$$D(\omega)\, d\omega = (v_c/8\pi^3) \int d\boldsymbol{q} \ . \tag{2.2.18}$$

The integration in (2.2.18) is performed over the volume of the layer in $\boldsymbol{q}$ space, where the frequency $\omega_{\boldsymbol{q}}$ is contained in a narrow interval $\omega \leqslant \omega_{\boldsymbol{q}} \leqslant \omega + d\omega$. The integral in this equation may be transformed if the integration element is chosen not in the form of $dq_x\, dq_y\, dq_z$ but as an infinitely small cylinder whose lateral surface is perpendicular to the surface $\omega_{\boldsymbol{q}} = \omega$, with the area of the base being dS_ω and the height

$$dq_\perp = d\omega/|\text{grad}\ \omega_{\boldsymbol{q}}| = d\omega/|v_{\boldsymbol{q}}| \ . \tag{2.2.19}$$

Here grad $\omega = \{(\partial\omega_q/\partial q_x), (d\omega_q/\partial q_y), (\partial\omega_q/\partial q_z)\}$ is a frequency gradient, that is, a vector which has the dimensions of velocity (in the Debye model this is the constant v_s), the direction of the vector coinciding with the normal to the surface $\omega_q = \omega$. As a result of this transformation, (2.2.18) becomes

$$D(\omega)\,d\omega = (v_c/8\pi^3)\int\int dS_\omega\, dq_\perp \ . \tag{2.2.20}$$

Substituting (2.2.19) into (2.2.20), we find

$$D(\omega) = (v_c/8\pi^3)\int (dS_\omega/|v_q|) \ . \tag{2.2.21}$$

Formulas such as (2.2.21) will ascertain the singularities of the function $D(\omega)$, and later on they will be employed to calculate the density of electronic states in metals (Chap. 4).

Singularities evidently should arise at some critical points q_{cr} for which the quantity v_q goes to zero; i.e., the frequency ω_q at these points turns out to be a locally "planar" function. Consider this function in the vicinity of a critical point, expanding it there in a power series of the difference $y = q - q_{cr}$. There is no linear term in this expansion since, by the definition of the critical point, at such a point $v_q = \nabla_q\omega_q = 0$. The quadratic terms, after being brought into the normal form (upon transformation to the principal axes), will reduce to a sum of squares

$$\omega_q = \omega_{cr} + \alpha_1 y_1^2 + \alpha_2 y_2^2 + \alpha_3 y_3^2 + \dots , \tag{2.2.22}$$

with $\alpha_i = \frac{1}{2}\partial^2\omega_q/\partial y_i^2$ $(i = 1, 2, 3)$. If all $\alpha_i < 0$, the function ω_q at point q_{cr} has a maximum and the surfaces $\omega_q = \text{const}$, according to (2.2.22), have the form of a family of ellipsoids. The volume of an ellipsoid of this family for the frequency $\omega_q = \omega$ with the surface surrounding the point q_{cr} is equal to

$$(4\pi/3)(\omega_{cr} - \omega)^{3/2}/|\alpha_1\alpha_2\alpha_3|^{1/2} \ .$$

The differential of this volume when multiplied by $v_c/8\pi^3$ gives the spectral density

$$D(\omega) = (v_c/4\pi^3)(\omega_{cr} - \omega)^{1/2}\,|\alpha_1\alpha_2\alpha_3|^{1/2} \ , \tag{2.2.23}$$

where $\omega < \omega_{cr}$. It is clear from (2.2.23) that this singularity does not perturb the continuity of the function $D(\omega)$ itself, but its derivative $\partial D(\omega)/\partial\omega$ has a discontinuity, tending to $-\infty$ when $\omega \to \omega_{cr}$ from below. A similar situation occurs for the minimum ω_q. Figure 2.14 portrays the different singularities of $D(\omega)$ for the acoustic branch of the spectrum of copper.

If one of the α_i has a sign opposite to that of the two others, a saddle point appears for which the singularity of the derivative $\partial D(\omega)/\partial\omega$ arises again. We demonstrate this for a frequency spectrum by reference to a planar lattice model.

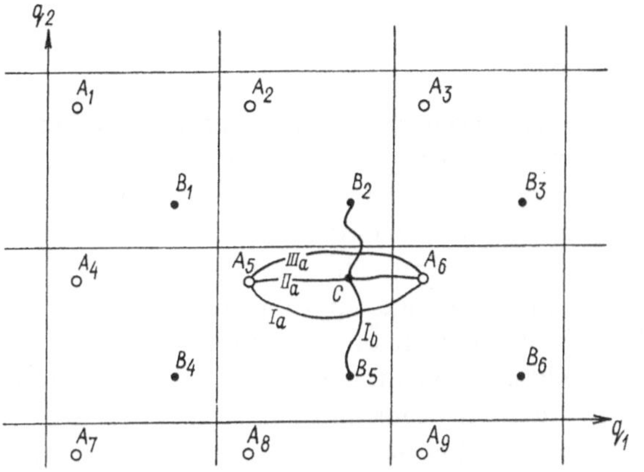

Fig. 2.15. Searching for the saddle point in wave-number space

Shown in Fig. 2.15 are several unit cells, that is, zones in $\boldsymbol{q}$ space (q_1, q_2). The function $\omega_{\boldsymbol{q}}$ is considered to be periodic and continuous. It may therefore be expected that each cell will have at least one maximum (clear circlet A_1, A_2, $A_3, \ldots$) and one minimum (dark circlet B_1, B_2, $B_3, \ldots$). If we connect the maxima of closely adjacent cells, for example A_5 and A_6, by a curve (I_a in Fig. 2.15), the latter will have at least one point at which the function ω_q takes the smallest value for this curve. Similar points will be present on any other curve that connects the maxima A_5 and A_6 (for example, on curves II_a, III_a).

The locus of all these points generates a continuous curve $B_2B_5(I_b)$ that connects two adjacent minima of the surface ω_q. On this curve there will be a point C at which the ω_q assumes the largest value on this curve. This point should be a saddle point, since on the curve I_b, when we move from B_2 to B_5, it will correspond to a relative maximum, and, if we move along curve II_a from A_5 to A_6, it will correspond to a relative minimum. The same applies to the curves which connect points A_5 and A_7, B_4 and B_5. Therefore, the function ω_q has at least two saddle points in each cell. This line of argument certainly becomes more complicated if we take into account all the branches of the spectrum, but the gross features of the treatment remain the same.

For small q the dispersion relation is linear: $\omega \sim q$ i.e., $\partial\omega_j/\partial q \equiv v_j = \text{const}$. Then it follows from (2.2.21) that when $\omega \ll \omega_{\max}$ the density of states $D(\omega) \sim \omega^2$. At temperatures $T \ll \hbar\omega_{\max}/k_B$ these frequencies contribute to the integral (2.2.12). Then, repeating the same computations as those used in the derivation of (2.2.17), we obtain $C \sim T^3$. Hence it follows that the Debye T^3 law for heat capacity at low temperatures is not a consequence of model concepts but will hold in a rigorous calculation as well, merely resulting from the acoustic nature of the phonon spectrum at small wave vectors.

2.3 Allowance for Anharmonic Terms

Up to this point lattice dynamics have been treated in a harmonic approximation. However, many phenomena exist for which the higher-order (anharmonic) terms of the expansion of potential energy in degrees of displacement, in principle, may not be neglected. Besides, at high temperatures it is also necessary to take into account the anharmonic terms to provide a better explanation even for those phenomena for which the harmonic approximation suffices, for example, for the heat capacity. Finally, with quantum crystals—in which the mean amplitudes of zero-point oscillations constitute an appreciable part of the lattice parameter—and apparently also with substances near the melting temperature, even the anharmonic approximation that takes into account the finite number of highest-order expansion terms may fail to yield the result desired. Special methods, not based on perturbation theory, have then to be developed.

First and foremost, we here briefly survey three typical problems from the dynamics of crystals, for which it is sufficient in the first approximation to take into account the first two anharmonic expansion terms (cubic and quartic): (1) the thermal expansion of solids, (2) the temperature trend of the lattice heat capacity at high temperatures, and (3) the thermal conductivity of the ionic lattice (in a purely qualitative treatment).

2.3.1 Thermal Expansion of Crystals

Thermal expansion cannot be explained in the harmonic approximation. This is evident from the elementary definition of the mean displacement u. In fact, in the harmonic approximation (2.1.1) we have for the mean displacement

$$\bar{u} = \int_{-\infty}^{\infty} u \exp(-\alpha u^2/2k_B T)\, du \left[\int_{-\infty}^{\infty} \exp(-\alpha u^2/2k_B T) du \right]^{-1} = 0 \,, \qquad (2.3.1)$$

because the integrand function in the numerator is odd. For the anharmonic terms in (2.1.28) to be taken into account rigorously, we need to add the following terms

$$\begin{aligned} &\frac{1}{3!} \sum_{\substack{m m' m'' \\ s s' s'' \\ j j' j''}} \left[\frac{\partial^3 V}{\partial u_{ms}^{(j)} \partial u_{m's'}^{(j')} \partial u_{m''s''}^{(j'')}} \right]_0 u_{ms}^{(j)} u_{m's'}^{(j')} u_{m''s''}^{(j'')} \\ &+ \frac{1}{4!} \sum_{\substack{m m' m'' m''' \\ s s' s'' s''' \\ j j' j'' j'''}} \left[\frac{\partial^4 V}{\partial u_{ms}^{(j)} \partial u_{m's'}^{(j')} \partial u_{m''s''}^{(j'')} \partial u_{m'''s'''}^{(j''')}} \right]_0 \qquad (2.3.2) \\ &\times u_{ms}^{(j)} u_{m's'}^{(j')} u_{m''s''}^{(j'')} u_{m'''s'''}^{(j''')} \,. \end{aligned}$$

For simplicity, we restrict ourselves to a more elementary one-dimensional case in order to allow for the anharmonic terms [2.7]. The potential energy as a function of displacement u at $T = 0$ K has the form

$$V(u) = V_0 + (\alpha/2)u^2 - \beta u^3 - \gamma u^4 , \tag{2.3.3}$$

where the term with u^3 describes the asymmetry of the curve in Fig. 1.32, and the term with u^4 the general "softening" of oscillations at large amplitudes. By developing the exponent up to linear terms in β and γ, we will have for the mean displacement

$$\begin{aligned} \bar{u} &\approx \int_{-\infty}^{\infty} \exp\left(-\frac{\alpha u^2}{2k_B T}\right)\left(u + \frac{\beta u^4}{k_B T} + \frac{\gamma u^5}{k_B T}\right) du \left[\int_{-\infty}^{\infty} \exp\left(-\frac{\alpha u^2}{2k_B T}\right) du\right]^{-1} \\ &\approx 3\beta k_B T/\alpha^2 . \end{aligned} \tag{2.3.4}$$

The quantity $\partial \bar{u}/\partial T = 3\beta k_B/\alpha^2$ gives the temperature coefficient of thermal expansion α_p in the first anharmonic approximation ($\beta \neq 0$), i.e., at sufficiently high temperatures. From (2.3.4) we see that $\bar{u}$ is proportional to the mean thermal energy $k_B T$ and, therefore, the temperature coefficient is proportional to the specific heat capacity. According to the Dulong and Petit law, the temperature coefficient is a constant (Grüneisen's rule). If we represent (2.3.4) as

$$\bar{u} = \frac{3\beta}{\alpha^2} \bar{\mathscr{E}} \tag{2.3.5}$$

and extend it to include the low-temperature range [by replacing it by the quantum expression (2.2.5)], then the thermal expansion coefficient will cease to be constant and will decrease sharply at $T \ll \vartheta$, tending to zero for $T \to 0$ K, in keeping with Nernst's theorem and experiment [2.8]. Remember that these derivations are approximate to within the cubic terms in the expansion of $V(u)$, as well as the simplifications used in the calculation of the integrals in (2.3.4).

Let us also outline the formal thermodynamic derivation of the Grüneisen formula (2.3.5). Write the differential of the thermodynamic potential of the crystal, ϕ, as

$$d\phi = -SdT + Vdp . \tag{2.3.6}$$

Since

$$\frac{\partial^2 \phi}{\partial T \partial p} \equiv \left(\frac{\partial V}{\partial T}\right)_p = \frac{\partial^2 \phi}{\partial p \partial T} \equiv -\left(\frac{\partial S}{\partial p}\right)_T ,$$

the thermal expansion coefficient, by definition, will be

$$\alpha_p \equiv \frac{1}{V}\left(\frac{\partial V}{\partial T}\right)_p = -\frac{1}{V}\left(\frac{\partial S}{\partial p}\right)_T . \tag{2.3.7}$$

In the Debye model, the oscillation spectrum of the crystal is given by one parameter $\vartheta_D(p)$, for which $S = S(T/\vartheta_D(p))$ from dimensional considerations. Then

$$\left(\frac{\partial S}{\partial p}\right)_T = -T\left(\frac{\partial S}{\partial T}\right)_p \frac{1}{\vartheta_D^2(p)} \frac{\partial \vartheta_D(p)}{\partial p} \equiv -\frac{C_p}{\vartheta_D^2}\frac{\partial \vartheta_D}{\partial p} , \tag{2.3.8}$$

with C_p being the heat capacity at constant pressure. It is known from mechanics that the dependence of the oscillation spectrum on the oscillation amplitude, determined by pressure, is absent for a harmonic oscillator and thus $\partial \vartheta_D/\partial p \neq 0$ only when the terms (2.3.2) in the potential energy of the crystal are taken into account. Substituting (2.3.8) into (2.3.7), we find $\alpha_p \sim C_p$, which is equivalent to (2.3.5) (the difference $C_p - C_V$ is small for solids and may be neglected).

2.3.2 Heat-Capacity Term Linear in Temperature

The deviation of the heat capacity from the Dulong and Petit law (2.2.1) at high temperatures is wholly determined by the anharmonic terms in (2.3.3). It stands to reason that they make a contribution at lower temperatures, too, but they are so small that they practically do not affect the heat capacity observed. As the temperature is raised, they begin to play an increasingly larger role. Even allowance for the finite number of anharmonic terms near the melting point probably does not suffice any longer to describe experimental facts. Using (2.3.3), the result for the distribution function is

$$Z(T) = (2\pi m k_B T)^{1/2} \int_{-\infty}^{\infty} du \exp(-(\alpha u^2/2 - \beta u^3 - \gamma u^4)/k_B T). \tag{2.3.9}$$

Expanding the integrand in a power series of small quantities β and γ and maintaining the first-order terms in γ and the first- and second-order terms in β, we have

$$\begin{aligned} Z(T) = (2\pi m k_B T)^{1/2} &\int_{-\infty}^{\infty} du \exp(-\alpha u^2/2k_B T) \\ &\times (1 + \beta u^3/k_B T + \gamma u^4/k_B T + \beta^2 u^6/2k_B^2 T^2) , \end{aligned} \tag{2.3.10}$$

where the integral with u^3, because it is odd, is equal to zero, and the three other integrals are readily evaluated so that

$$Z(T) = 2\pi \left(\frac{m}{\alpha}\right)^{1/2} k_B T \left(1 + \frac{3\gamma}{\alpha^2} k_B T + \frac{15\beta^2}{2\alpha^3} k_B T\right) . \tag{2.3.11}$$

Using (2.2.4) and multiplying it by $3N_A$, we find the mean molar energy of a monatomic crystal,

$$\bar{\mathscr{E}} = 3N_A k_B T + \frac{9N_A \gamma}{\alpha^2} (k_B T)^2 + \frac{45 N_A \beta^2}{2\alpha^3} (k_B T)^2 , \tag{2.3.12}$$

and obtain the heat capacity,

$$C_V = 3R \left[1 + (6k_B T/\alpha)\left(\frac{5\beta^2}{2\alpha^2} + \frac{\gamma}{\alpha}\right)\right] . \tag{2.3.13}$$

Equation (2.3.13) demonstrates that the anharmonic terms lead to the occurrence in the Dulong and Petit formula of an extra term which is linearly dependent on T. Each expansion coefficient V (i.e., the displacement u derivative of the potential) may be assumed to differ from the next coefficient (i.e., the next derivative) by a factor on the order of the interatomic spacing d. Thus, $d \sim \alpha/\beta \sim \beta/\gamma \sim \ldots$ etc., and therefore the two last terms in round brackets (2.3.13) are of the same order d^{-2}. In (2.3.12) the order of these terms to within the factor N_A will be $(k_B T)^2/\alpha d^2$. Therefore, these terms are small compared to the first term $3k_B T$ as long as the temperature is small compared to the temperature for which the mean-square amplitude is equal to d. Experiment qualitatively confirms (2.3.13) and a more rigorous derivation may be found in [2.7].

2.3.3 Thermal Conductivity of an Ionic Lattice

In the harmonic approximation the Hamiltonian of the crystal reduces to that of an ideal Bose gas of phonons (with the chemical potential equal to zero). As in any perfect gas, the establishment of thermodynamic equilibrium is impossible here: the phonons must be capable of exchanging energy and momentum with each other or with some external scatterers. In an ideal crystal, the sole mechanism that assures normal thermodynamic and kinetic behavior is the collision of phonons, and also the processes of merging of two phonons into one (by contrast with a normal gas, the number of phonons is not conserved), i.e., anharmonic effects.

If we impart some amount of thermal energy to the crystal at some point, these processes will assure its redistribution among phonons and the occurrence of an ordered heat flux. Interphonon interactions, however, do not alter the total momentum of the phonon system and, by themselves, do not entail equalization of the temperature gradient produced; i.e., they do not lead to the value of thermal conductivity being finite. To achieve this, the phonon flux must be hindered by some external forces; the momentum of the phonon flux can be transmitted only to the lattice of an ideal crystal as a whole. The lattice, because of its translational invariance, cannot have an arbitrary momentum but only one which is equal to some reciprocal-lattice vector multiplied by Planck's constant (just as in X-ray scattering, Sect. 1.5). The interphonon-interaction processes accompanied by such "loss" of momentum are referred to as the umklapp processes [2.7, 9].

2.4 Localization of Phonons on Point Defects

Up to now, we have considered the oscillations of an ideal lattice. Actual crystals always have various kinds of structural imperfections, viz., point, linear, and plane defects. Even if it becomes possible to grow an absolutely ideal crystal of any natural chemically pure substance, the crystal will still contain ions of different isotopes of the same element; these ions will display practically no difference in electronic properties but will differ in mass. (In addition, even in this case the crystal will be finite in dimension and its outer surface also will be a "defect", the effect of which we neglect here). Consideration of the simple case of a point defect allows us to ascertain some general properties of the oscillation spectrum of a real crystal. (A similar problem was first considered by *Lifshitz* [2.10]).

Thus, let us consider a lattice, constructed of ions of mass m, in which an isotopic ion of mass $m(1+\Delta)$ is placed at site 0. The potential energy of the ion interaction is determined by the charge and electronic properties of the ions and therefore is almost the same for all isotopes of a given element. For simplicity, we will deal with a lattice that has no basis, i.e., contains one ion per unit cell and thus possesses acoustic oscillation branches only. Finally, to render the problem as simple as possible, we restrict our attention to the one-dimensional case. Then the equations for the displacements u_l will take the form

$$m(1+\Delta\delta_{l,0})\ddot{u}_l = -\alpha(2u_l - u_{l+d} - u_{l-d}) \, , \tag{2.4.1}$$

with $\delta_{l,0}$ being Kronecker symbols. We use time-dependent solutions in the form $u_l(t) = \exp(-\mathrm{i}\omega t)u_l$. In addition, the u_l will be treated as a set of Fourier coefficients of some function $U(q)$ defined in the interval $(-\pi/d, \pi/d)$:

$$U(q) = \sum_l u_l \exp(iql) , \quad u_l = \frac{d}{2\pi} \int_{-\pi/d}^{\pi/d} dq\, U(q) \exp(-iql) . \tag{2.4.2}$$

Multiply (2.4.1) termwise by exp(iql) and calculate the sum over l. This gives the following:

$$m \sum_l (1 + \Delta\delta_{l,0}) \ddot{u}_l \exp(iql) = -m\omega^2 \left[U(q) + \frac{d\Delta}{2\pi} \int_{-\pi/d}^{\pi/d} dq\, U(q) \right] \tag{2.4.3}$$

$$-\alpha \sum_l (2u_l - u_{l+d} - u_{l-d}) \exp(iql)$$

$$= -\alpha \sum_l [2 - \exp(iqd) - \exp(-iqd)] u_l \exp(iql) = -m\omega_q^2 U(q) . \tag{2.4.4}$$

Equating (2.4.3) to (2.4.4), we find

$$(\omega_q^2 - \omega^2) U(q) = \left(\frac{\omega^2 d\Delta}{2\pi} \right) \int_{-\pi/d}^{\pi/d} dq\, U(q) . \tag{2.4.5}$$

Consider two limiting cases. First, let

$$\int_{-\pi/d}^{\pi/d} dq\, U(q) = 0 . \tag{2.4.6}$$

Then $U(q) \neq 0$ only if $\omega^2 = \omega_q^2 \equiv (4\alpha/m)\sin^2(qd/2)$. For $\omega^2 > 4\alpha/m$ such solutions do not exist at all, and for $\omega^2 < 4\alpha/m$ there are two solutions: $q = \pm q_0$, $q_0 \equiv (2/d)\arcsin[(m\omega^2/4\alpha)^{1/2}]$. Condition (2.4.6) yields

$$U(q) = A[\delta(q + q_0) - \delta(q - q_0)] , \tag{2.4.7}$$

where A is an arbitrary constant, and $\delta(q)$ a Dirac δ function.

As is clear from (2.1.5), the presence of a defect under (2.4.6) does not in any way change the spectrum of these oscillations in comparison with an ideal lattice. Let now, in contrast to (2.4.6),

$$(d/2\pi) \int_{-\pi/d}^{\pi/d} dq\, U(q) \equiv C \neq 0 . \tag{2.4.8}$$

Then (2.4.5) immediately yields

$$U(q) = \omega^2 \Delta C / (\omega_q^2 - \omega^2) , \tag{2.4.9}$$

$$(\omega^2 d\Delta / 2\pi) \int_{-\pi/d}^{\pi/d} (\omega_q^2 - \omega^2)^{-1} dq = 1 . \tag{2.4.10}$$

Introducing the notation $\omega^2 \equiv 4\alpha\Omega^2/m$, $qd = x$, rewrite (2.4.10) in the form

$$(\Omega^2\Delta/2\pi)\int_{-\pi}^{\pi}(\sin^2(x/2)-\Omega^2)^{-1}dx = 1 \; . \tag{2.4.11}$$

In solving (2.4.11) for $\Omega^2 < 1$, we encounter a difficulty which is associated with the nonintegrable divergence of the integrand for $\sin(x/2) = \pm\Omega$. To eliminate the divergence, we add to the right-hand side of (2.4.1) the frictional force (proportional to the velocity)

$$f_l = -\gamma m\dot{u}_l = \mathrm{i}\omega\gamma m u_l \; . \tag{2.4.12}$$

This leads to the replacement $\omega^2 \to \omega^2 + \mathrm{i}\omega\gamma$ or $\Omega^2 \to \Omega^2 + \mathrm{i}\Omega\eta$ ($\eta = \gamma(4\alpha/m)^{-1/2}$). Then we should let η approach zero and make use of the identity [2.11]:

$$\int(\varphi(x)-\mathrm{i}\eta)^{-1}dx|_{\eta\to 0} = ⨍ dx/\varphi(x) + \mathrm{i}\pi\int dx\delta[\varphi(x)] \; , \tag{2.4.13}$$

where $⨍$ denotes the integral in terms of the principal value. Calculation yields

$$-(1/2\pi)⨍_{-\pi}^{\pi}(\sin^2(x/2)-\Omega^2)^{-1}dx = (2\pi\Omega(1-\Omega^2))^{-1/2}$$

$$\times\ln\left|\frac{\mathrm{tg}(x/2)-\Omega(1-\Omega^2)^{-1/2}}{\mathrm{tg}(x/2)+\Omega(1-\Omega^2)^{-1/2}}\right|\Bigg|_{-\pi}^{+\pi} = 0 \; , \tag{2.4.14}$$

$$(\mathrm{i}/2)\int_{-\pi}^{\pi}dx\delta(\sin^2(x/2)-\Omega^2) = \mathrm{i}\Omega(1-\Omega^2)^{1/2} \; . \tag{2.4.15}$$

We assume here that $\Omega^2 < 1$. Substituting (2.4.15) into (2.4.11), we find

$$\mathrm{i}\Omega\Delta/(1-\Omega^2)^{1/2} = 1 \; , \quad \Omega^2 = (1-\Delta^2)^{-1} \; , \tag{2.4.16}$$

which contradicts the initial assumption $\Omega^2 < 1$. Then we try a solution of (2.4.11) for $\Omega^2 > 1$. The integral in (2.4.11) is easy to calculate, and we obtain

$$(1/2\pi)\int_{-\pi}^{+\pi}[\sin^2(x/2)-\Omega^2]^{-1}dx = -1/\Omega(\Omega^2-1)^{1/2} \; ,$$

$$\Delta\Omega + (\Omega^2-1)^{1/2} = 0 \; .$$

When $\Delta > 0$ this equation is incapable of solution, and for $\Delta < 0$ (i.e., for a lighter isotope)

$$\Omega = (1-\Delta^2)^{-1/2} \; . \tag{2.4.17}$$

Thus, in addition to the solutions with $\Omega^2 < 1$ which existed in an ideal crystal, a split-off level has arisen which possesses the frequency $\omega_0 = \omega_{max}(1 - \Delta^2)^{-1/2} > \omega_{max}(\omega_{max} = 2(\alpha/m)^{1/2})$. It follows from formula (2.4.9) that $U(q)$ at $\omega = \omega_0$ does not display singularities (as a function of q) anywhere on the real axis. As is known from the Fourier theory [2.12], the integral (2.4.2) for u_l decreases quickly to zero when $l \to \infty$. Thus, (2.4.17) yields an expression for the frequency of a localized phonon, i.e., a lattice vibration mode in which the displacements u_l are localized near a defect. The possibility of the occurrence of such vibrational modes is one of the most important features peculiar to the spectrum of a real crystal as compared to that of an ideal one.

2.5 Heat Capacity of Glasses at Low Temperatures

As already noted in Sect. 1.10, modern solid-state physics is by no means concerned with crystals only. Of great interest are the properties of amorphous solids (glasses). Naturally, the question arises: Which of the above results may be applied to glasses and, primarily, are there phonons in them? This problem is nontrivial. As has been emphasized in Sect. 1.10, it is altogether pointless to describe the states of glasses in the language of wave functions—some averaged characteristics are needed—and it is therefore not even clear how the problem of elementary excitations in glasses and other disordered systems should be posed at all. An adequate mathematical device is the resolvent (Green's function) method, which will be considered in Chap. 4.

All these difficulties, however, arise in attempting to consider excitations similar to localized phonons (Sect. 2.4) or short-wavelength phonons (which in a disordered system, generally speaking, may even not be specified by a definite quasimomentum $\boldsymbol{q}$). Everything is much simpler for long-wavelength phonons. The point is that when considering them we may approximate the real atomic structure of a glass as an isotropic elastic continuum [2.13]. In the latter, as can be shown purely phenomenologically, longitudinal and transverse sound waves exist with definite frequency and wave vector, ω and $\boldsymbol{q}$, with $\omega \sim q$. In keeping with the general principles of quantum mechanics, one can assign quasiparticles to these waves with a momentum $\hbar\boldsymbol{q}$ and energy $\hbar\omega$. As with crystals, it is quite natural to call such quasiparticles phonons. They also obey the Bose–Einstein statistics. An immediate consequence might seem to be that at low temperatures, when excitations with small ω and q are substantial, crystals and glasses should not differ in thermodynamic properties and the Debye T^3 law for the heat capacity C_V should hold for glasses. However, experiment shows that at very low temperatures practically all glasses exhibit a linear T dependence of C_V.

As will be shown in Chap. 3, the same linear temperature dependence in metals is due to conduction electrons. This linear term in the lattice heat capacity can be separated out directly in dielectric and semiconducting glasses, but in metallic glasses it can only be separated out at temperatures below the superconducting transition temperature if this exists. As follows from the above arguments, this dependence does not have to be due to phonons. This implies that there may exist in glasses some extra specific excitations that are absent in crystals. A model for such excitations, called two-level centers (or tunnel states), was proposed by *Anderson* et al. [2.14], and also *Phillips* [2.15].

As already mentioned in Sect. 1.10, in contrast to a crystal, a glass has no uniquely defined structure. *Anderson* et al. and *Phillips* [2.14, 15] suggest that in glasses there exist atoms (or groups of atoms) which can occupy two positions almost equivalent in energy. This double-well character of interatomic potentials, which arises quite legitimately, say, in hydrogen-bonded crystals (Sect. 1.7), may "fortuitously" occur for some atoms, whatever the type of bond. For a detailed discussion of this assumption see [2.16]. Let it be noted that the shape of the potential with two minima (as a function of one of the x coordinates) differs drastically from a parabolic one. Thus, we deal with regions in which the anharmonic effects are anomalously large; i.e., the structure is highly softened. Three-well and more complicated potentials are apparently less probable than two-well potentials and, therefore, do not seem to make any substantial contribution to the thermodynamic properties of glasses.

The problem of the energy spectrum of a quantum system with a potential having two minima was solved in Sect. 1.7. If E_1 and E_2 are the energies of the lowest levels in the first and the second well, the distance between the two lowest energy levels of a two-well system is equal to (Sect. 1.7)

$$\Delta E = [(E_1 - E_2)^2 + |V_{12}|^2]^{1/2} , \qquad (2.5.1)$$

where V_{12} is the matrix element of the Hamiltonian describing the tunneling of an atom from one well into the other. Since the atomic mass is very large, the tunneling probability and, hence, the quantity V_{12} are extremely small. The energies E_1 and E_2 may take different values, but we are concerned with the case $E_1 \approx E_2$ and therefore the quantity ΔE is small, too.

Let us consider the problem of the contribution which a two-level system with an excitation energy $\Delta E \sim k_B T$ makes to the thermodynamic quantities (higher excited states possess energies that are much larger than $k_B T$ and their contribution is exponentially small). According to the Gibbs distribution, the mean energy of a two-level system with energy levels ε_i $(i = 1, 2)$ is

$$\bar{\varepsilon} = \frac{\sum\limits_i \varepsilon_i \exp(-\varepsilon_i/k_B T)}{\sum\limits_i \exp(-\varepsilon_i/k_B T)} = \frac{\Delta E}{\exp(\Delta E/k_B T) + 1} , \qquad (2.5.2)$$

where the lowest level is taken to be the zero of energy $\varepsilon_1 = 0$; then $\varepsilon_2 = \Delta E$. The relevant contribution to the heat capacity is then equal to

$$C_V(\Delta E) = d\bar{\varepsilon}/dT = \frac{(\Delta E)^2}{k_B T^2} \frac{\exp(\Delta E/k_B T)}{(\exp(\Delta E/k_B T)+1)^2} . \qquad (2.5.3)$$

Thus from (2.5.3), the contribution to the heat capacity from a two-level defect (Schottky contribution) goes exponentially to zero for $k_B T \ll \Delta E$, falls off as T^{-2} for $k_B T \gg \Delta E$, and has a maximum at intermediate temperatures. But in glasses, according to the model discussed, there exist two-level centers with diverse excitation energies, including very small ones. To get a final answer, we need, in addition, to average formula (2.5.3) over all values of ΔE with the corresponding density of states $D(\Delta E)$:

$$C_V = \int_0^\infty d(\Delta E) D(\Delta E) C_V(\Delta E) . \qquad (2.5.4)$$

It is assumed that $D(0) \neq 0$; i.e., there is a fairly large number of two-level centers with excitation energies as small as desired (to be more exact, with $\Delta E \ll k_B T$). Of course, as follows from (2.5.1), the quantity ΔE may not be less than the minimum value $|V_{12}|$; however, as noted above, the latter is apparently very small.

If $k_B T$ is small compared to the representative splitting value $(\Delta E)_0$ assigning the scale of decay of the function $D(\Delta E)$ with increasing ΔE, the $D(\Delta E)$ in (2.5.4) may be replaced by $D(0)$. Then, upon replacement of the variables $\Delta E \equiv k_B T x$ we have

$$C_V \cong k_B^2 T D(0) \int_0^\infty \frac{dx x^2 e^x}{(e^x+1)^2} = \frac{\pi^2}{6} D(0) k_B^2 T . \qquad (2.5.5)$$

Integration by parts allows the integral involved in (2.5.5) to be reduced to the form

$$I \equiv \int_0^\infty \frac{dx x^2 e^x}{(e^x+1)^2} = \int_0^\infty dx x^2 \left(-\frac{d}{dx}\right) \frac{1}{e^x+1} = 2\int_0^\infty \frac{dx x}{e^x+1} .$$

Taking account of the fact that

$$\frac{1}{e^x+1} = \frac{e^{-x}}{1+e^{-x}} = \sum_{n=1}^\infty (-1)^{n+1} e^{-nx}$$

and using the replacement of the variables $nx = y$, we have

$$I = 2 \sum_{n=1}^\infty (-1)^{n+1} \int_0^\infty dx x e^{-nx} = 2\int_0^\infty dy e^{-y} y \cdot \sum_{n=1}^\infty \frac{(-1)^{n+1}}{n^2} .$$

Since

$$\int_0^\infty dy e^{-y} y = 1 \, , \quad \sum_{n=1}^{\infty} \frac{(-1)^{n+1}}{n^2} = \frac{\pi^2}{12} \, ,$$

we obtain the result that

$$\int_0^\infty \frac{dx x^2 e^x}{(e^x+1)^2} = \frac{\pi^2}{6} \, . \tag{2.5.6}$$

Thus we have obtained the required linear temperature dependence of heat capacity for $k_B T \ll (\Delta E)_0$. It stands to reason that, in addition to the contribution (2.5.5), there is also a phonon contribution to the heat capacity $\sim T^3$, which at very low temperatures is smaller than the contribution of the two-level centers.

According to Anderson et al. [2.14], a contribution to the heat capacity is made only by those two-level centers which have enough time to get into thermodynamic equilibrium during the time of the experiment t_0, i.e. by those centers for which the tunneling time $\tau_{12} \simeq \hbar/|V_{12}|$ falls short of t_0. If there is a center with $E_1 = E_2$ and a very high barrier between the wells, i.e., with a very small $|V_{12}|$, then $\Delta E \to 0$ for such a center. Also, if $\tau_{12} > t_0$, one does not have to allow for the contribution of these centers to $D(0)$, since during the experiment the atom may be regarded as localized in one of the wells; it simply does not have the time to become "smeared" over the two wells. As is known from quantum mechanics, the quantity τ_{12} depends exponentially on the spacing d between the wells and on the height U of the barrier:

$$\tau_{12} \sim \exp(\sqrt{2MU}d/\hbar) \, , \tag{2.5.7}$$

with M being the mass of a tunneling atom. Here U and d are distributed approximately with a constant density, and the number of centers for which $\tau_{12} < t_0$ depends logarithmically on t_0. Thus far the prediction that the heat capacity depends logarithmically on measurement time has not been corroborated experimentally. If this effect exists, it clearly supports the description of glass as a nonequilibrium system whose thermodynamic characteristics depend on time!

Two-level centers manifest themselves not only in the heat capacity, but also in the thermal conductivity of glasses; in anomalies in ultrasonic absorption; and they make a substantial contribution to electrical resistivity and other electronic characteristics of metallic glasses.

2.6 High-Frequency Permittivity of Ionic Crystals

As shown in Sect. 2.1.2, optical oscillations of the ionic lattice should interact strongly with an electromagnetic field. This coupling is the strongest at $\omega \sim \omega_+ \sim v_s/d$ (2.1.11, 27). The corresponding wavelength (and frequency) of the electromagnetic field is $\lambda = 2\pi c/\omega \sim (c/v_s)2\pi d \gg d$ (with c being the velocity of light), so that the photon wave vector may to a high degree of precision be set equal to zero; i.e., the electric field $\boldsymbol{E}$ may be assumed to be homogeneous.

Let us consider a linear two-ion array with ions of mass M and charge $+e$, and with ions of mass m and charge $-e$. The equations of motion will then take the form (2.1.18)

$$\begin{aligned} M\ddot{\boldsymbol{U}} &= -2\alpha(\boldsymbol{U}-\boldsymbol{u}) + e\boldsymbol{E}\exp(-\mathrm{i}\omega t) \ , \\ m\ddot{\boldsymbol{u}} &= -2\alpha(\boldsymbol{u}-\boldsymbol{U}) - e\boldsymbol{E}\exp(-\mathrm{i}\omega t) \ , \end{aligned} \tag{2.6.1}$$

where, in accord with what has been stated above, $\boldsymbol{q} = 0$; this corresponds to $\omega_{\max}$ (2.1.27). According to the well-known electrodynamic relation,

$$\boldsymbol{E} = \boldsymbol{D} - 4\pi\boldsymbol{P} \ , \tag{2.6.2}$$

where $\boldsymbol{D}$ is the electric induction,

$$\boldsymbol{P} = ne(\boldsymbol{U}-\boldsymbol{u}) \tag{2.6.3}$$

is the polarization vector of the medium, and n is the number of unit cells in unit volume.

The first term in (2.6.2) describes the field produced by external forces, and the second the long-range (Coulomb) contribution of the ion interaction which is taken into account in the mean-field approximation (2.6.2). On the other hand, the short-range contribution of the interionic interaction is described by the term that contains α in (2.6.1). Trying a solution of (2.6.1) in the form

$$\boldsymbol{U}(t) = \boldsymbol{U}\mathrm{e}^{-\mathrm{i}\omega t} \ , \quad \boldsymbol{u}(t) = \boldsymbol{u}\mathrm{e}^{-\mathrm{i}\omega t} \ .$$

and combining (2.6.1), we find

$$M\boldsymbol{U} + m\boldsymbol{u} = 0 \ . \tag{2.6.4}$$

Substituting (2.6.4) into (2.6.1) yields

$$\boldsymbol{U} = e\boldsymbol{E}/M(\omega_0^2 - \omega^2) \ , \tag{2.6.5}$$

where ω_0 is defined in (2.1.27). The polarization $\boldsymbol{P}$, according to (2.6.3, 5), is equal to

$$\boldsymbol{P} = ne\boldsymbol{U}\left(1+\frac{M}{m}\right) = ne^2\left(\frac{1}{M}+\frac{1}{m}\right)\frac{\boldsymbol{E}}{\omega_0^2-\omega^2} . \tag{2.6.6}$$

The permittivity $\varepsilon(\omega)$ is determined (2.6.2) by the formulas

$$\boldsymbol{D} = \varepsilon(\omega)\boldsymbol{E} \quad \text{or} \quad \boldsymbol{P} = [\varepsilon(\omega)-1]\boldsymbol{E}/4\pi . \tag{2.6.7}$$

Comparing (2.6.6, 7), we find

$$\varepsilon(\omega) = 1 + \omega_p^2/(\omega_0^2-\omega^2) . \tag{2.6.8}$$

The notation introduced here is

$$\omega_p^2 = 4\pi ne^2\left(\frac{1}{M}+\frac{1}{m}\right) . \tag{2.6.9}$$

The $\varepsilon(\omega)$ relation is schematically represented by a solid line in Fig. 2.16a. The expression for $\varepsilon(\omega)$ is inapplicable in the vicinity of ω_0, where $\varepsilon(\omega) \to \infty$.

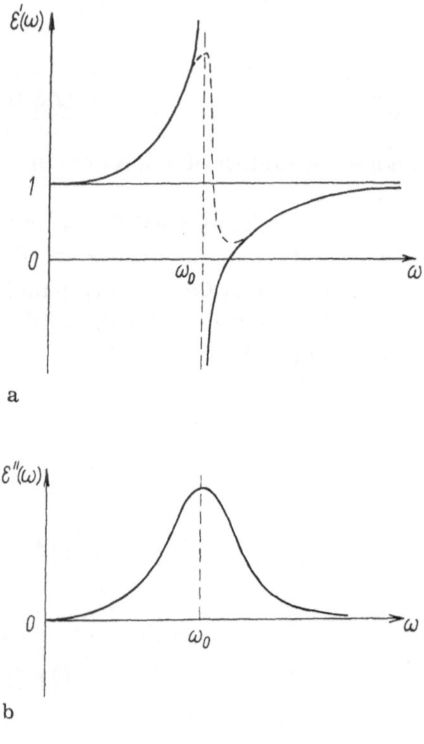

Fig. 2.16. Schematic frequency ω dependence of (**a**) the real ε' and (**b**) the imaginary ε'' part of permittivity

Allowance for the frictional force (2.4.12) leads to the replacement $\omega^2 \to \omega^2 + i\gamma\omega$ (Sect. 2.4). The result for the real and imaginary parts $\varepsilon(\omega) = \varepsilon'(\omega) + i\varepsilon''(\omega)$ (Fig. 2.16) then will be

$$\begin{aligned} \varepsilon'(\omega) &= 1 + \omega_p^2(\omega_0^2 - \omega^2)/[(\omega_0^2 - \omega^2)^2 + \gamma^2\omega^2] \, , \\ \varepsilon''(\omega) &= \gamma\omega\omega_p^2/[(\omega_0^2 - \omega^2)^2 + \gamma^2\omega^2] \, . \end{aligned} \tag{2.6.10}$$

The maximum of absorption [determined by $\varepsilon''(\omega)$] [2.17] at small γ corresponds to the optical-phonon frequency ω_0 (which justifies the attribute "optical").

As is known from the electrodynamics of continuous media [2.17], longitudinal (L) and transverse (T) electromagnetic waves may propagate in a system. The frequencies of these waves are determined by the equations (for $\gamma = 0$)

$$\varepsilon(\omega_L) = 0 \, , \quad \omega_L = (\omega_p^2 + \omega_0^2)^{1/2} \, , \tag{2.6.11}$$

$$\omega_T^2 \varepsilon(\omega_T)/c^2 = q^2 \, , \quad c^2 q^2 = \omega_T^2[1 + \omega_p^2/(\omega_0^2 - \omega_T^2)] \, . \tag{2.6.12}$$

Since the thermal motion of ions is neglected, the frequency of longitudinal waves turns out to be q independent. If we turn off the interaction of the lattice with the electromagnetic field ($e \to 0$), (2.6.12) yields solutions (Fig. 2.17a)

$$\omega_L = \omega_0 \, , \quad \omega_T = cq \tag{2.6.13}$$

describing a photon and an optical phonon that do not interact. An interaction leads to mixing (hybridization) of the modes (Fig. 2.17b). For $\omega \ll \omega_0$ we have from (2.6.12) the ordinary solution

$$\omega_T = cq/n_0 \tag{2.6.14}$$

for a photon in a refracting medium with a refractive index

$$n_0 = (1 + \omega_p^2/\omega_0^2)^{1/2} \, . \tag{2.6.15}$$

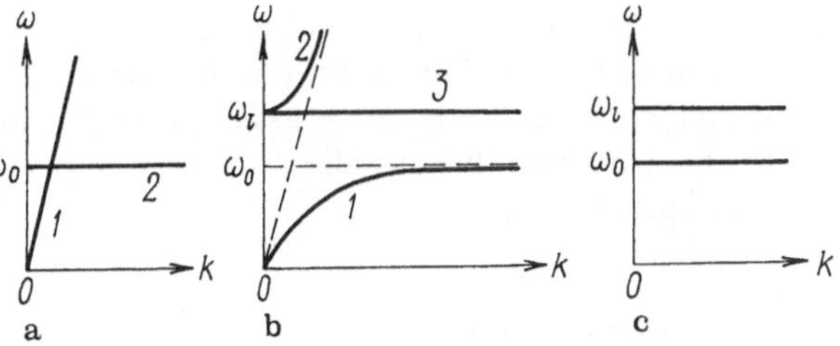

Fig. 2.17. Dispersion relations (**a**) of a photon (*1*) and an optical phonon (*2*), (**b**) of two branches of transverse waves (*1*, *2*) resulting from hybridization and of one longitudinal wave (*3*), (**c**) with $c = \infty$

When $\omega \gg \omega_0$, the lattice does not have enough time to respond to the variation of the electromagnetic field, and $n_0 \to n_\infty = 1$. It is also easy to see that the maximum frequency of the lower branch of transverse oscillations (reached when $q \to \infty$) is equal to ω_0, and the minimum frequency of the upper branch (reached for $q = 0$) coincides with ω_L (Curves 1, 2, Fig. 2.17b). A characteristic relation holds (Fig. 2.17c),

$$(\omega_L/\omega_0)^2 = (n_0/n_\infty)^2 = \varepsilon(0)/\varepsilon(\infty) \ , \tag{2.6.16}$$

in which we do not replace $\varepsilon(\infty)$ by unity since there may be a contribution from the polarizability of the ion cores themselves (in this sense $\omega = \infty$ denotes $\omega_0 \ll \omega \ll \omega_e$, where ω_e is the characteristic oscillation frequency of electrons in atoms: $\omega_e \sim 10^{15}$–$10^{16}\,\mathrm{s}^{-1}$, whereas $\omega_0 \sim 10^{12}$–$10^{13}\,\mathrm{s}^{-1}$).

2.7 Lattice Scattering and the Mössbauer Effect

2.7.1 Scattering Probability and the Correlation Function

The properties of the lattice may be explored with the help of macroscopic quantities such as heat capacity, sound velocities, thermal expansion coefficients, etc. The information thus derived is far from giving us complete picture of the major characteristics of the crystal (structure, phonon spectrum, etc.). As will be shown later, a widely used method for a straightforward investigation of these characteristics is the scattering of particles—neutrons, electrons, photons, i.e., light rays and X rays. As a concrete example we will discuss neutron scattering.

Let the crystal at the time zero be in a state $|\Psi_l\rangle$, and the free neutron have a wave vector $\boldsymbol{k}$ and, accordingly, a wave function

$$\psi_{\boldsymbol{k}}(\boldsymbol{r}) = (2\pi)^{-3/2}\exp(\mathrm{i}\boldsymbol{k}\cdot\boldsymbol{r}) \tag{2.7.1}$$

(normalized to a δ function). As a result of the interaction of the neutron with the crystal, the latter passes to a state $|\Psi_n\rangle$, and the former to a state $|\psi_{\boldsymbol{k}'}\rangle$. The probability of this process $P_{l\boldsymbol{k}\to n\boldsymbol{k}'}$ may be calculated from the time-dependent quantum-mechanical perturbation theory (Sect. 1.5), on the assumption that the potential of the interaction of the neutron with the crystal, V, is small compared to the characteristic excitation energies [1.12],

$$\begin{aligned} dP_{l\boldsymbol{k}\to n\boldsymbol{k}'} &= \frac{1}{\hbar^2}\int_{-\infty}^{\infty} dt \exp\left[\frac{\mathrm{i}}{\hbar}(E_n - E_l + \varepsilon_{\boldsymbol{k}} - \varepsilon_{\boldsymbol{k}'})t\right] \\ &\quad \times |\langle \Psi_n \psi_{\boldsymbol{k}'}|V|\Psi_l \psi_{\boldsymbol{k}}\rangle|^2 d\nu_{\boldsymbol{k}'} \ , \end{aligned} \tag{2.7.2}$$

where E and ε are the energies of the corresponding states of the crystal and neutron, and $d\nu_{k'}$, is the density-of-final-states differential of the neutron. Represent V as

$$V = \sum_j v(\boldsymbol{r} - \boldsymbol{r}_j) , \tag{2.7.3}$$

where

$$\boldsymbol{r}_j = \boldsymbol{R}_j + \boldsymbol{u}_j \tag{2.7.4}$$

are the coordinates of the jth ion, $\boldsymbol{R}_j$ is its equilibrium position, and $\boldsymbol{u}_j$ is the displacement vector. For simplicity assume that all ions in the lattice are alike and the form of the function $v(\boldsymbol{r} - \boldsymbol{r}_j)$ does not depend on j. Substituting (2.7.1, 3) into the matrix element V, we find

$$\begin{aligned}\langle \Psi_n \psi_{k'} | V | \Psi_l \psi_k \rangle &= \sum_j \int \frac{d\boldsymbol{r}}{(2\pi)^3} \langle \Psi_n | v(\boldsymbol{r} - \boldsymbol{r}_j) \exp[\mathrm{i}(\boldsymbol{k} - \boldsymbol{k}') \cdot (\boldsymbol{r} - \boldsymbol{r}_j + \boldsymbol{r}_j)] | \Psi_l \rangle \\ &= \frac{1}{(2\pi)^3} \sum_j \langle \Psi_n | \exp(-\mathrm{i}\boldsymbol{q} \cdot \boldsymbol{r}_j) | \Psi_l \rangle v_{\boldsymbol{q}} \\ &\equiv \langle \Psi_n | n_{\boldsymbol{q}} | \Psi_l \rangle v_{\boldsymbol{q}} ,\end{aligned} \tag{2.7.5}$$

where the scattering vector $\boldsymbol{q} = \boldsymbol{k}' - \boldsymbol{k}$ and the notation $v_{\boldsymbol{q}} = \int d\boldsymbol{R} v(\boldsymbol{R}) \cdot \exp(-\mathrm{i}\boldsymbol{q} \cdot \boldsymbol{R})$, $n_{\boldsymbol{q}} = (2\pi)^{-3} \sum_j \exp(-\mathrm{i}\boldsymbol{q} \cdot \boldsymbol{r}_j)$ are introduced. The operator $n_{\boldsymbol{q}}$ is the Fourier component of the ionic-density operator in the crystal

$$n(\boldsymbol{r}) = \sum_j \delta(\boldsymbol{r} - \boldsymbol{r}_j) , \quad n_{\boldsymbol{q}} = ((2\pi)^{-3}) \int d\boldsymbol{r} n(\boldsymbol{r}) \exp(-\mathrm{i}\boldsymbol{q} \cdot \boldsymbol{r}) . \tag{2.7.6}$$

Next, we make use of the identity

$$\begin{aligned}\exp\left[\frac{\mathrm{i}}{\hbar}(E_n - E_l)t\right] \langle \Psi_n | n_{\boldsymbol{q}} | \Psi_l \rangle &= \langle \Psi_n | \exp(\mathrm{i}\hat{\mathscr{H}} t/\hbar) n_{\boldsymbol{q}} \exp(-\mathrm{i}\hat{\mathscr{H}} t/\hbar) | \Psi_l \rangle \\ &\equiv \langle \Psi_n | n_{\boldsymbol{q}}(t) | \Psi_l \rangle ,\end{aligned} \tag{2.7.7}$$

where $\hat{\mathscr{H}}$ is the Hamiltonian of the crystal and $n_{\boldsymbol{q}}(t)$ is the Heisenberg representation of the operator $n_{\boldsymbol{q}}$ [1.12]. Here we have taken into account that $\hat{\mathscr{H}} | \Psi_m \rangle = E_m | \Psi_m \rangle$. Substitution of (2.7.5, 7) into (2.7.2) yields

$$
\begin{aligned}
dP_{lk\to nk'} &= \frac{1}{\hbar^2}\int_{-\infty}^{\infty} dt \exp(\mathrm{i}\omega t)|\langle \Psi_n|n_q(t)|\Psi_l\rangle|^2 |v_q|^2 dv_{k'} \\
&= \frac{1}{\hbar^2}\int_{-\infty}^{\infty} dt \int_{-\infty}^{\infty} dt' \exp(\mathrm{i}\omega(t-t'))\langle \Psi_l|n_{-q}(t)|\Psi_n\rangle \\
&\quad \times \langle \Psi_n|n_q(t')|\Psi_l\rangle |v_q|^2 dv_{k'} ,
\end{aligned} \tag{2.7.8}
$$

where we have introduced the notation $\omega = \hbar^{-1}(\varepsilon_{k'} - \varepsilon_k)$ and extended the square of the modulus of the matrix element allowing for the readily verifiable relation

$$
\langle \Psi_n|n_q|\Psi_l\rangle^* = \langle \Psi_l|n_{-q}|\Psi_n\rangle .
$$

A direct observable is the transition probability per unit time, summed over all final states of the crystal (with $\sum_n|\Psi_n\rangle\langle\Psi_n| = 1$ by the completeness condition) and averaged together with the Gibbs distribution function ϱ_l with respect to the initial states

$$
\varrho_l = \frac{1}{Z}\exp(-\beta E_l) , \quad \beta = (k_B T)^{-1} , \quad Z = \sum_n \exp(-\beta E_n) . \tag{2.7.9}
$$

According to (2.7.8), this probability is equal to

$$
\begin{aligned}
dw_{k\to k'} &= \lim_{\tau\to\infty} \frac{|v_q|^2}{\hbar^2}\sum_{nl} \varrho_l \frac{1}{\tau}\int_{-\tau/2}^{\tau/2} dt \int_{-\tau/2}^{+\tau/2} dt' \exp(\mathrm{i}\omega(t-t'))\langle \Psi_l|n_{-q}(t)|\Psi_n\rangle \\
&\quad \times \langle \Psi_n|n_q(t')|\Psi_l\rangle dv_{k'} \\
&= \lim_{\tau\to\infty} \frac{|v_q|^2}{\hbar^2}\frac{1}{\tau}\int_{-\tau/2}^{+\tau/2} dt \int_{-\tau/2}^{+\tau/2} dt' \exp(\mathrm{i}\omega(t-t'))\langle n_{-q}(t) n_q(t')\rangle dv_{k'} \\
&= \frac{|v_q|^2}{\hbar^2}\int_{-\infty}^{\infty} dt \exp(\mathrm{i}\omega t)\langle n_{-q}(t) n_q\rangle dv_{k'} ,
\end{aligned} \tag{2.7.10}
$$

where the notation

$$
\langle A\rangle = \sum_l \langle l|A|l\rangle \varrho_l \tag{2.7.11}
$$

is introduced for the quantities averaged over the canonical ensemble, and it is taken into account that, in virtue of time homogeneity,

$$
\langle n_{-q}(t) n_q(t')\rangle = \langle n_{-q}(t-t') n_q\rangle . \tag{2.7.12}
$$

The easiest way to prove (2.7.12) is to introduce the operator

$$\varrho = \frac{1}{Z}\exp(-\beta\hat{\mathscr{H}}) \tag{2.7.13}$$

with the eigenvalues (2.7.9); then (2.7.11) becomes

$$\langle A\rangle = \mathrm{Tr}(A\varrho) \ . \tag{2.7.14}$$

In this case

$$\begin{aligned}
&\mathrm{Tr}(n_{-q}(t)n_q(t')\exp(-\beta\hat{\mathscr{H}})) = \mathrm{Tr}(\exp(\mathrm{i}\mathscr{H}t/\hbar) \\
&\quad\times n_{-q}\exp(-\mathrm{i}\hat{\mathscr{H}}t/\hbar)\exp(\mathrm{i}\mathscr{H}t'/\hbar)n_q\exp(-\mathrm{i}\mathscr{H}t'/\hbar)\exp(-\beta\hat{\mathscr{H}})) \\
&\quad= \mathrm{Tr}[\exp(\mathrm{i}\mathscr{H}t/\hbar)n_{-q}\exp(-\mathrm{i}\mathscr{H}(t-t')/\hbar)n_q\exp(-\beta\hat{\mathscr{H}}) \\
&\quad\times \exp(-\mathrm{i}\hat{\mathscr{H}}t'/\hbar)] = \mathrm{Tr}[\exp(-\mathrm{i}\hat{\mathscr{H}}t'/\hbar)\exp(\mathrm{i}\mathscr{H}t/\hbar) \\
&\quad\times n_{-q}\exp(-\mathrm{i}\mathscr{H}(t-t')/\hbar)n_q\exp(-\beta\hat{\mathscr{H}})] \ ,
\end{aligned}$$

which proves (2.7.12). Here we have used the familiar identities $\exp(\alpha A)\exp(\beta A) = \exp(\beta A)\exp(\alpha A) = \exp(\alpha+\beta)A$, where α and β are c numbers, and the properties of the trace $\mathrm{Tr}\,AB = \mathrm{Tr}\,BA$. We have examined this identity closely because later on we will have to carry out similar transformations, for example, when proving Wick's theorem in Sect. 2.7.2.

Introduce the notation

$$S(\boldsymbol{q},\omega) = \int_{-\infty}^{\infty} dt\exp(\mathrm{i}\omega t)\langle n_{-q}(t)n_q\rangle \ . \tag{2.7.15}$$

Then (2.7.10) will become

$$dw_{k\to k'}/dv_{k'} = |v_q|^2 S(\boldsymbol{q},\omega)/\hbar^2 \ . \tag{2.7.16}$$

In this formula the properties of the crystal are characterized by the quantity $S(\boldsymbol{q},\omega)$, which is the quantity directly measured in scattering experiments [2.18]. The quantity $\langle n(\boldsymbol{r},t)n(\boldsymbol{r}')\rangle$ is called the density correlation function, and the function $S(\boldsymbol{q},\omega)$, which is its Fourier transform in space and time variables, is called the dynamic form factor.

The quantity

$$S(\boldsymbol{q}) = \int_{-\infty}^{\infty} \frac{d\omega}{2\pi} S(\boldsymbol{q},\omega) = \langle n_{-q}n_q\rangle \tag{2.7.17}$$

is referred to as the structure factor [in deriving (2.7.17) we have used the relation $(1/2\pi)\int_{-\infty}^{\infty} dt\,\exp(i\omega t)=\delta(t)$], the static form factor, or sometimes simply the form factor. Performing the Fourier transformation, we find

$$\langle n(\boldsymbol{r})n(\boldsymbol{r}')\rangle = \int (2\pi)^{-3} S(\boldsymbol{q})\exp(\mathrm{i}\boldsymbol{q}\cdot(\boldsymbol{r}'-\boldsymbol{r}))d\boldsymbol{q}\ . \qquad (2.7.18)$$

Thus scattering experiments provide complete information on the crystal structure. Further it will be seen that the factor $S(\boldsymbol{q},\omega)$ contains also a complete description of the phonon spectrum. But prior to this we must establish how to handle the phonon operators and how to calculate the averages (2.7.14) with the operator ϱ from (2.7.13), which may be expressed in terms of the crystal Hamiltonian.

2.7.2 Some Properties of the Phonon Operators and of the Averages Containing Them

First we need to express the displacement $u_l(t)$ in terms of the phonon operators $\hat{b}^+$ and $\hat{b}$, similar to the way this was done in (2.1.4) for linear array. In these variables the Hamiltonian takes a particularly simple form. We then determine in what fashion these operators in the Heisenberg representation depend on time, since the quantity $\hat{n}_{\boldsymbol{q}}(t)$ involves, according to (2.7.4, 6), the operators $\hat{u}_l(t)$. Thereupon we have to ascertain how the averages (2.7.14) should be calculated in this problem, and then we will be able to find the dynamic form factor of the crystal.

For a linear array, the relation of $\hat{u}_l$ to $\hat{b}^+$ and $\hat{b}$ is given by (2.1.41, 45, 46). On finding the quantities $\hat{u}_q$ and $\hat{\mathscr{P}}_q$ from (2.1.45, 46) and substituting these quantities into (2.1.41), we obtain

$$\hat{u}_l = N^{-1/2}\mathrm{i}\sum_q (\hbar/2m\omega_q)^{1/2}(\hat{b}_q-\hat{b}^+_{-q})\exp(\mathrm{i}ql)\ . \qquad (2.7.19)$$

For a three-dimensional crystal, just as for a linear array, the crystal oscillations in the harmonic approximation may be represented in terms of an assembly of independent oscillators specified by the numbers $\nu=\boldsymbol{q}, s$, where $\boldsymbol{q}$ is the wave vector, $s=1,2,\ldots,3\sigma$ is the branch number, and σ is the number of atoms (ions) per unit cell (2.1.3). Here we introduce the operators $\hat{b}^+_\nu$ and $\hat{b}_\nu$, which satisfy commutation relations such as (2.1.47)

$$[\hat{b}_\nu,\hat{b}^+_{\nu'}]_- = \delta_{\nu\nu'}\ ,\qquad [\hat{b}_\nu,\hat{b}_{\nu'}]_- = [\hat{b}^+_\nu,\hat{b}^+_{\nu'}]_- = 0\ , \qquad (2.7.20)$$

and are related to the Hamiltonian $\hat{\mathscr{H}}$ and the displacement vector $\hat{\boldsymbol{u}}_j$ in the same way as they are in (2.1.46, 48):

$$\hat{\mathscr{H}} = \sum_\nu \hbar\omega_\nu(\hat{b}^+_\nu\hat{b}_\nu+\tfrac{1}{2}) \equiv E_0+\sum_\nu \hbar\omega_\nu\hat{b}^+_\nu\hat{b}_\nu \qquad (2.7.21)$$

$$\hat{\boldsymbol{u}}_j = \mathrm{i}\sum_{\nu} (\hbar/2Nm\omega_\nu)^{1/2} \boldsymbol{e}_\nu (\hat{b}_{\boldsymbol{q}s} - \hat{b}^+_{-\boldsymbol{q},s}) \exp(\mathrm{i}\boldsymbol{q}\cdot\boldsymbol{R}_j) \ , \tag{2.7.22}$$

where $\boldsymbol{e}_\nu$ is the corresponding matrix eigenvector G (2.1.35) normalized to unity $|\boldsymbol{e}_\nu| = 1$ (called the polarization vector). Since we assume that the lattice is made up of like atoms (ions), $m_s = m$ is independent of s. It is no trouble to derive formulas (2.7.20–22), which are written by analogy with a linear array (we leave it as an exercise for the reader to prove this). More important is the manipulation of such Hamiltonians and operators since they arise in a wide variety of solid-state theory problems that might seem to have no bearing on lattice dynamics (Chap. 5).

To start with, we calculate

$$\hat{b}^+_\nu(t) = \exp(\mathrm{i}\mathscr{H}t/\hbar)\hat{b}^+_\nu \exp(-\mathrm{i}\hat{\mathscr{H}}t/\hbar) \ . \tag{2.7.23}$$

Since, according to (2.7.20), $[\hat{b}^+_\nu, \hat{b}^+_{\nu'}\hat{b}_{\nu'}]_- = 0$, $[\hat{b}^+_\nu \hat{b}_\nu, \hat{b}^+_{\nu'}\hat{b}_{\nu'}]_- = 0$ when $\nu \neq \nu'$, we obtain

$$\begin{aligned}
\hat{b}^+_\nu(t) &= \exp\left(\frac{\mathrm{i}t}{\hbar}\sum_{\nu'\neq\nu} \hbar\omega_{\nu'}\hat{b}^+_{\nu'}\hat{b}_{\nu'}\right)\exp(\mathrm{i}\omega_\nu t\hat{b}^+_\nu\hat{b}_\nu) \\
&\quad \times \hat{b}^+_\nu \exp(-\mathrm{i}\omega_\nu t\hat{b}^+_\nu\hat{b}_\nu)\exp\left(-\frac{\mathrm{i}t}{\hbar}\sum_{\nu'\neq\nu} \hbar\omega_{\nu'}\hat{b}^+_{\nu'}\hat{b}_{\nu'}\right) \\
&= \exp\left(\frac{\mathrm{i}t}{\hbar}\sum_{\nu'\neq\nu} \hbar\omega_{\nu'}\hat{b}^+_{\nu'}\hat{b}_{\nu'}\right)\exp\left(-\frac{\mathrm{i}t}{\hbar}\sum_{\nu'\neq\nu} \hbar\omega_{\nu'}\hat{b}^+_{\nu'}\hat{b}_{\nu'}\right) \\
&\quad \times \exp(\mathrm{i}\omega_\nu t\hat{b}^+_\nu\hat{b}_\nu)\hat{b}^+_\nu \exp(-\mathrm{i}\omega_\nu t\hat{b}^+_\nu\hat{b}_\nu) \\
&= \exp(\mathrm{i}\omega_\nu t\hat{b}^+_\nu\hat{b}_\nu)\hat{b}^+_\nu \exp(-\mathrm{i}\omega_\nu t\hat{b}^+_\nu\hat{b}_\nu) \ .
\end{aligned} \tag{2.7.24}$$

Thus, we need to calculate

$$\hat{B}^+_\nu(\alpha) = \exp(\alpha\hat{b}^+_\nu\hat{b}_\nu)\hat{b}^+_\nu \exp(-\alpha\hat{b}^+_\nu\hat{b}_\nu) \ . \tag{2.7.25}$$

Differentiating $\hat{B}^+_\nu(\alpha)$ with respect to α, we find

$$\begin{aligned}
d\hat{B}^+_\nu(\alpha)/d\alpha &= \exp(\alpha\hat{b}^+_\nu\hat{b}_\nu)\hat{b}^+_\nu\hat{b}_\nu\hat{b}^+_\nu \exp(-\alpha\hat{b}^+_\alpha b_\nu) \\
&\quad - \exp(\alpha\hat{b}^+_\nu\hat{b}_\nu)\hat{b}^+_\nu\hat{b}^+_\nu\hat{b}_\nu \exp(-\alpha\hat{b}^+_\nu\hat{b}_\nu) \\
&= \exp(\alpha\hat{b}^+_\nu\hat{b}_\nu)[\hat{b}^+_\nu\hat{b}_\nu, \hat{b}^+_\nu]_- \exp(-\alpha\hat{b}^+_\nu\hat{b}_\nu) \\
&= \exp(\alpha\hat{b}^+_\nu\hat{b}_\nu)\{\hat{b}^+_\nu[\hat{b}_\nu, \hat{b}^+_\nu]_- + [\hat{b}^+_\nu, \hat{b}^+_\nu]_-\hat{b}_\nu\} \\
&\quad \times \exp(-\alpha\hat{b}^+_\nu b_\nu) = \hat{B}^+_\nu(\alpha) \ ,
\end{aligned} \tag{2.7.26}$$

where we have made use of the identity

$$[AB, C]_- \equiv A[B, C]_- + [A, C]_- B \ , \tag{2.7.27}$$

which is proved by direct expansion of the right-hand and left-hand sides, and employs (2.7.20). Equation (2.7.26) may be integrated with the initial condition $\hat{B}_\nu^+(0) = \hat{b}_\nu^+$. Then, by virtue of (2.7.25),

$$\hat{B}_\nu^+(\alpha) = \hat{b}_\nu^+ \exp\alpha \ . \tag{2.7.28}$$

The possibility of extending the usual methods of solving equations for functions to operators is, generally speaking, nontrivial. In the present case it may be achieved as follows:

$$\begin{aligned}\hat{B}_\nu^+(\alpha) &= \hat{B}_\nu^+(0) + \alpha \left.\frac{d\hat{B}_\nu^+}{d\alpha}\right|_{\alpha=0} + \tfrac{1}{2}\alpha^2 \left.\frac{d^2\hat{B}_\nu^+}{d\alpha^2}\right|_{\alpha=0} + \dots \\ &= \hat{B}_\nu^+(0)(1 + \alpha + \tfrac{1}{2}\alpha^2 + \dots) = \hat{b}_\nu^+ \exp\alpha \ ,\end{aligned}$$

where the function of the operator is expanded in a Taylor series.

Comparing (2.7.23, 25) and allowing for (2.7.28), we find

$$\hat{b}_\nu^+(t) = \exp(\mathrm{i}\omega_\nu t)\hat{b}_\nu^+ \ . \tag{2.7.29}$$

Similarly, we may find

$$\hat{b}_\nu(t) = \exp(-\mathrm{i}\omega_\nu t)\hat{b}_\nu \ . \tag{2.7.30}$$

Substitution of (2.7.29, 30) into (2.7.31) yields

$$\begin{aligned}\hat{\boldsymbol{u}}_j(t) &= \mathrm{i}\sum_\nu (\hbar/2Nm\omega_\nu)^{1/2} \boldsymbol{e}_\nu \exp(\mathrm{i}\boldsymbol{q}\cdot\boldsymbol{R}_j) \\ &\quad \times [\hat{b}_{\boldsymbol{q},s}\exp(-\mathrm{i}\omega_\nu t) - \hat{b}^+_{-\boldsymbol{q},s}\exp(\mathrm{i}\omega_\nu t)] \ .\end{aligned} \tag{2.7.31}$$

Note also that for any operator A we have

$$\begin{aligned}A^n(t) &= \exp(\mathrm{i}\hat{\mathscr{H}}t/\hbar)A^n\exp(-\mathrm{i}\hat{\mathscr{H}}t/\hbar) \\ &= \exp(\mathrm{i}\hat{\mathscr{H}}t/\hbar)A\exp(-\mathrm{i}\hat{\mathscr{H}}t/\hbar)\exp(\mathrm{i}\hat{\mathscr{H}}t/\hbar)A \\ &\quad \times \exp(-\mathrm{i}\hat{\mathscr{H}}t/\hbar)\exp(\mathrm{i}\hat{\mathscr{H}}t/\hbar)\dots A\exp(-\mathrm{i}\hat{\mathscr{H}}t/\hbar) = [A(t)]^n \ ,\end{aligned}$$

and hence for any function $f(A)$ prescribed by a Taylor series

$$\exp(\mathrm{i}\hat{\mathscr{H}}t/\hbar)f(A)\exp(-\mathrm{i}\hat{\mathscr{H}}t/\hbar) = f(\exp(\mathrm{i}\hat{\mathscr{H}}t/\hbar)\,A\exp(-\mathrm{i}\hat{\mathscr{H}}t/\hbar)) \ . \tag{2.7.32}$$

Now we are aware of what $n_q(t)$ is. Let us turn to the problem of calculating the mean values of the products of the operator $\hat{b}^+$, $\hat{b}$ according to (2.7.14), where ϱ is defined according to (2.7.13), and $\mathscr{H}$ according to (2.7.21). It follows from (2.7.13, 25, 28) that

$$\varrho^{-1}\hat{b}_\nu^+\varrho = \hat{B}^+(\beta\hbar\omega_\nu) = \hat{b}_\nu^+\exp(\beta\hbar\omega_\nu) \ ,$$

i.e.,

$$\hat{b}_\nu^+\varrho = \exp(\beta\hbar\omega_\nu)\varrho\hat{b}_\nu^+ \ . \tag{2.7.33}$$

Similarly,

$$\hat{b}_\nu\varrho = \exp(-\beta\hbar\omega_\nu)\varrho\hat{b}_\nu \ . \tag{2.7.34}$$

Then

$$\begin{aligned}\mathrm{Tr}([\hat{b}_\nu,\hat{b}_\nu^+]_-\varrho) &= \mathrm{Tr}(\hat{b}_\nu\hat{b}_\nu^+\varrho) - \mathrm{Tr}(\hat{b}_\nu^+\hat{b}_\nu\varrho)\\ &= \mathrm{Tr}(\hat{b}_\nu\varrho\hat{b}_\nu^+)\exp(\beta\hbar\omega_\nu) - \mathrm{Tr}(\hat{b}_\nu^+\hat{b}_\nu\varrho)\\ &= \mathrm{Tr}(\hat{b}_\nu^+\hat{b}_\nu\varrho)\exp(\beta\hbar\omega_\nu) - \mathrm{Tr}(\hat{b}_\nu^+\hat{b}_\nu\varrho)\\ &= \mathrm{Tr}(\hat{b}_\nu^+\hat{b}\varrho)(\exp(\beta\hbar\omega_\nu)-1) \ ,\end{aligned} \tag{2.7.35}$$

where, just as in the derivation of (2.7.12), we have used the freedom to perform a cyclic permutation of the operators under the trace sign. But $[\hat{b}_\nu^+,\hat{b}_\nu]_- = -1$, $\mathrm{Tr}\varrho = 1$, and $\mathrm{Tr}(\hat{b}_\nu^+\hat{b}_\nu\varrho) = \langle\hat{b}_\nu^+\hat{b}_\nu\rangle$, and therefore we find from (2.7.35)

$$\langle\hat{b}_\nu^+\hat{b}_\nu\rangle = (\exp(\beta\hbar\omega_\nu)-1)^{-1} \ . \tag{2.7.36}$$

Thus, we have derived the formula (2.1.49) for the mean of the occupation number in the Bose–Einstein statistics. A similar technique will be used for calculating the average of an arbitrary number of phonon operators A_1, $A_2, \ldots, A_n$ (each operator A_i is either $\hat{b}_\nu^+$ or $\hat{b}_\nu$ [2.19]). Then, according to (2.7.20), $[A_i, A_j]_-$ is a c number (i.e., commutes with any one of the operators $\hat{b}_\nu^+, \hat{b}_\nu$). Therefore,

$$\begin{aligned}[A_1, A_2A_3\ldots A_n]_- &\equiv A_1A_2\ldots A_n - A_2A_3\ldots A_nA_1\\ &= [A_1,A_2]_-A_3\ldots A_n + A_2A_1A_3\ldots A_n - A_2A_3\ldots A_nA_1\\ &= [A_1,A_2]_-A_3\ldots A_n + [A_1,A_3]_-A_2A_4\ldots A_n\\ &\quad + A_2A_3A_1\ldots A_n - A_2A_3\ldots A_nA_1 = \ldots\\ &= \sum_{i=2}^{n}[A_1,A_i]_-A_2\ldots A_{i-1}A_{i+1}\ldots A_n \ .\end{aligned} \tag{2.7.37}$$

Now we average the right-hand and left-hand sides of (2.7.37):

$$\begin{aligned}\mathrm{Tr}([A_1, A_2 \ldots A_n]_- \varrho) &= \mathrm{Tr}(A_1 A_2 \ldots A_n \varrho) - \mathrm{Tr}(A_2 A_3 \ldots A_n A_1 \varrho) \\ &= \mathrm{Tr}(A_1 A_2 \ldots A_n \varrho) \\ &\quad - \exp(\pm\beta\hbar\omega_1)\mathrm{Tr}(A_1 A_2 \ldots A_n \varrho) \\ &= (1-\exp(\pm\beta\hbar\omega))\langle A_1 A_2 \ldots A_n\rangle ,\end{aligned} \tag{2.7.38}$$

where the plus or minus signs are put depending on whether A_1 is the operator $\hat{b}_1^+$ or $\hat{b}_1$. Similarly,

$$\begin{aligned}&\mathrm{Tr}([A_1, A_i]_- A_2 \ldots A_{i-1} A_{i+1} \ldots A_n \varrho) \\ &= [A_1, A_i]_- \langle A_2 \ldots A_{i-1} A_{i+1} \ldots A_n\rangle \\ &= (1-\exp(\pm\beta\hbar\omega_1))\langle A_1 A_i\rangle\langle A_2 \ldots A_{i-1} A_{i+1} \ldots A_n\rangle .\end{aligned} \tag{2.7.39}$$

Substituting (2.7.38, 39) into (2.7.37), we obtain

$$\begin{aligned}\langle A_1 A_2 \ldots A_n\rangle &= \langle A_1 A_2\rangle\langle A_3 \ldots A_n\rangle \\ &\quad + \langle A_1 A_3\rangle\langle A_2 A_4 \ldots A_n\rangle + \ldots \\ &\quad + \langle A_1 A_n\rangle\langle A_2 A_3 \ldots A_{n-1}\rangle .\end{aligned} \tag{2.7.40}$$

The averages involved in (2.7.40) may be transformed further proceeding in the same way:

$$\begin{aligned}\langle A_1 A_2 A_3 A_4\rangle &= \langle A_1 A_2\rangle\langle A_3 A_4\rangle + \langle A_1 A_3\rangle\langle A_2 A_4\rangle \\ &\quad + \langle A_1 A_4\rangle\langle A_2 A_3\rangle ,\end{aligned}$$

etc.

We have arrived at the result that an average such as that involved in (2.7.40) (with n being even) is equal to the sum of all the products of the averages of the pairs of operators involved in the product $A_1, A_2, \ldots, A_n$, the operators in each pair being taken in the same order as in the initial product. For an odd n it may be shown in an analogous fashion that $\langle A_1 A_2 \ldots A_n\rangle = 0$ (since $\langle A_i\rangle = 0$). The statements formulated above also hold true in the case when the A_i are arbitrary linear combinations of the operators $\hat{b}_\nu^+$ and $\hat{b}_\nu$. The result obtained (normally referred to as Wick's theorem) was proved by *Wick* for the averaging over the ground state [2.20], and by *Bloch* and *de Dominicis* for a canonical ensemble [2.21]. Two important relationships were used here: (1) the commutator of any two operators $\hat{b}^+$, $\hat{b}$ is a c number; (2) the commutator of any operator $\hat{b}^+$, $\hat{b}$ with the Hamiltonian is the same operator (to within the factor $\pm\hbar\omega$), which results in identities (2.7.33, 34). When the anharmonic terms in the

Hamiltonian are taken into account, this is invalid and Wick's theorem does not hold any longer.

Now we are in a position to calculate the dynamic form factor, having learned in passing some operator handling procedures which will be employed in Chap. 5.

2.7.3 Calculating the Dynamic Form Factor in the Harmonic Approximation

We assume that the crystal is described by a harmonic Hamiltonian (2.7.21). We need to calculate the quantity

$$S(\boldsymbol{q},\omega)=(2\pi)^{-6}\int_{-\infty}^{\infty}dt\exp(\mathrm{i}\omega t)$$

$$\times\sum_{jj'}\langle\exp[-\mathrm{i}\boldsymbol{q}\cdot\boldsymbol{u}_j(t)]\exp(\mathrm{i}\boldsymbol{q}\cdot\boldsymbol{u}_{j'})\rangle\exp[\mathrm{i}\boldsymbol{q}\cdot(\boldsymbol{R}_{j'}-\boldsymbol{R}_j)] \qquad (2.7.41)$$

[we have substituted (2.7.4, 6) into (2.7.15)]. The expression to be averaged in (2.7.41) can be readily represented as the exponent of some operator (if $[A,B]_-\neq 0$, then $\mathrm{e}^A\mathrm{e}^B\neq\mathrm{e}^{A+B}$). Find the commutator of the operators $-\mathrm{i}\boldsymbol{q}\hat{\boldsymbol{u}}_j(t)$ and $\mathrm{i}\boldsymbol{q}\hat{\boldsymbol{u}}_{j'}$. Inserting (2.7.31) and using (2.7.20), we obtain

$$[-\mathrm{i}\boldsymbol{q}\cdot\hat{\boldsymbol{u}}_j(t),\mathrm{i}\boldsymbol{q}\cdot\hat{\boldsymbol{u}}_{j'}]_-=\frac{1}{N}\sum_{\substack{\boldsymbol{k}\boldsymbol{k}'\\ss'}}(\hbar/(2m(\omega_{\boldsymbol{k}s}\omega_{\boldsymbol{k}'s'})^{1/2})$$

$$\times\exp[\mathrm{i}(\boldsymbol{k}\boldsymbol{R}_j+\boldsymbol{k}'\cdot\boldsymbol{R}_{j'})](\boldsymbol{q}\cdot\boldsymbol{e}_{\boldsymbol{k}s})(\boldsymbol{q}\cdot\boldsymbol{e}_{\boldsymbol{k}'s'})[\hat{b}_{\boldsymbol{k}s}\exp(-\mathrm{i}\omega_{\boldsymbol{k}s}t)$$

$$-\hat{b}^+_{-\boldsymbol{k}s}\exp(\mathrm{i}\omega_{\boldsymbol{k}s}t),\hat{b}_{\boldsymbol{k}'s'}-\hat{b}^+_{-\boldsymbol{k}'s'}]_-$$

$$=\frac{1}{N}\sum_{\substack{\boldsymbol{k}\boldsymbol{k}'\\ss'}}\hbar(2m)^{-1}(\omega_{\boldsymbol{k}s}\omega_{\boldsymbol{k}'s'})^{-1/2}$$

$$\times\exp[\mathrm{i}(\boldsymbol{k}\cdot\boldsymbol{R}_j+\boldsymbol{k}'\cdot\boldsymbol{R}_{j'})](\boldsymbol{q}\cdot\boldsymbol{e}_{\boldsymbol{k}s})(\boldsymbol{q}\cdot\boldsymbol{e}_{\boldsymbol{k}'s'}) \qquad (2.7.42)$$

$$\times[\delta_{\boldsymbol{k},-\boldsymbol{k}'}\delta_{ss'}\exp(-\mathrm{i}\omega_{\boldsymbol{k}s}t)-\delta_{\boldsymbol{k}',-\boldsymbol{k}}\delta_{ss'}\exp(\mathrm{i}\omega_{\boldsymbol{k}s}t)]$$

$$=\frac{1}{N}\sum_{\boldsymbol{k}s}(\hbar/2m\omega_{\boldsymbol{k}s})\exp(\mathrm{i}\boldsymbol{k}\cdot(\boldsymbol{R}_j-\boldsymbol{R}_{j'}))|\boldsymbol{q}\cdot\boldsymbol{e}_{\boldsymbol{k}s}|^2$$

$$\times[\exp(-\mathrm{i}\omega_{\boldsymbol{k}s}t)-\exp(\mathrm{i}\omega_{\boldsymbol{k}s}t)]$$

(we have employed the relation $\boldsymbol{e}_{-\boldsymbol{k},s}=\boldsymbol{e}^*_{\boldsymbol{k},s}$, which may be obtained from an exploration of the properties of the matrix $\mathscr{G}$). The commutator of the operators with which we are concerned is a c number. In this case,

$$\exp A\exp B=\exp(A+B)\exp\tfrac{1}{2}[A,B]_-\ . \qquad (2.7.43)$$

This equation may be proved as follows. Introduce the operator

$$S(\lambda) = \exp(-\lambda A)\exp\lambda(A+B) \; . \tag{2.7.44}$$

Then

$$\begin{aligned} dS(\lambda)/d\lambda &= -\exp(-\lambda A)A\exp\lambda(A+B) + \exp(-\lambda A)(A+B)\exp\lambda(A+B) \\ &= \exp(-\lambda A)B\exp\lambda(A+B) \\ &= \exp(-\lambda A)B\exp(\lambda A)S(\lambda) \; . \end{aligned} \tag{2.7.45}$$

In its turn,

$$\frac{d}{d\lambda}\exp(-\lambda A)B\exp(\lambda A) = -[A,B]_- \; ,$$

i.e.,

$$\exp(-\lambda A)B\exp(\lambda A) = B - \lambda[A,B]_- \; . \tag{2.7.46}$$

Substitution of (2.7.46) into (2.7.46) yields

$$dS(\lambda)/d\lambda = \{B - \lambda[A,B]_-\}S(\lambda) \; . \tag{2.7.47}$$

The solution of (2.7.47) becomes

$$S(1) = \exp\left[\int_0^1 d\lambda(B-\lambda[A,B]_-)\right] = \exp B\exp(\tfrac{1}{2}[A,B]_-) \; . \tag{2.7.48}$$

The possibility of applying the conventional method of solving differential equations to (2.7.47) is due to the $[A,B]_-$ being a c number and may be proved in about the same way as (2.7.28).
Thus, it remains for us to find

$$\begin{aligned} \langle\exp\mathrm{i}\boldsymbol{q}\cdot[\hat{\boldsymbol{u}}_{j'} - \hat{\boldsymbol{u}}_j(t)]\rangle = \langle\exp\Bigg(&\sum_{ks}(\hbar/2Nm\omega_{ks})^{1/2}(\boldsymbol{q}\cdot\boldsymbol{e}_{k,s}) \\ &\times\{[\exp(\mathrm{i}\boldsymbol{k}\cdot\boldsymbol{R}_{j'}) - \exp(\mathrm{i}\boldsymbol{k}\cdot\boldsymbol{R}_j - \mathrm{i}\omega_{ks}t)]\hat{b}_{k,s} \\ &-[\exp(\mathrm{i}\boldsymbol{k}\cdot\boldsymbol{R}_{j'}) - \exp(\mathrm{i}\boldsymbol{k}\cdot\boldsymbol{R}_j + \mathrm{i}\omega_{ks}t)]\hat{b}^+_{-k,s}\}\Bigg) \\ \equiv \langle\exp C\rangle \; . \end{aligned} \tag{2.7.49}$$

Since the operator involved in the exponent is a linear combination of the operators $\hat{b}, \hat{b}^+$, Wick's theorem is applicable to it:

$$\langle \exp C\rangle = \sum_{n=0}^{+\infty} \frac{1}{n!}\langle C^n\rangle = \sum_{n=0}^{+\infty} \frac{1}{(2n)!}\underbrace{\langle CC\dots C\rangle}_{2n}$$

$$= \sum_{n=0}^{+\infty} \frac{1}{(2n)!}\frac{(2n)!}{n!\,2^n}\underbrace{\langle C^2\rangle\langle C^2\rangle\dots\langle C^2\rangle}_{n} \tag{2.7.50}$$

$$= \sum_{n=0}^{+\infty} \frac{1}{n!}(\tfrac{1}{2}\langle C^2\rangle)^n = \exp(\tfrac{1}{2}\langle C^2\rangle)\ .$$

Here we have allowed for $\langle C^{2n+1}\rangle = 0$. The number of the various ways in which $2n$ similar objects may be broken up into pairs is obtainable from the total number, the permutation $(2n)!$ being obtained by dividing the number of permutations of pairs $n!$ by the number of permutations of objects in each pair 2^n.

Substituting (2.7.50) into (2.7.49) gives

$$\langle \exp \mathrm{i}\boldsymbol{q}\cdot[\hat{\boldsymbol{u}}_{j'} - \hat{\boldsymbol{u}}_j(t)]\rangle = \exp\Bigg(\frac{1}{2}\sum_{\substack{\boldsymbol{kk}'\\ ss'}}[\hbar/2mN(\omega_{\boldsymbol{k}s}\omega_{\boldsymbol{k}'s'})^{1/2}]$$
$$\times (\boldsymbol{q}\cdot\boldsymbol{e}_{\boldsymbol{k}s})(\boldsymbol{q}\cdot\boldsymbol{e}_{\boldsymbol{k}'s'})\langle\{[\exp(\mathrm{i}\boldsymbol{k}\cdot\boldsymbol{R}_{j'}) - \exp(\mathrm{i}\boldsymbol{k}\cdot\boldsymbol{R}_j - \mathrm{i}\omega_{\boldsymbol{k}s}t)]$$
$$\times \hat{b}_{\boldsymbol{k}s} - [\exp(\mathrm{i}\boldsymbol{k}\cdot\boldsymbol{R}_{j'}) - \exp(\mathrm{i}\boldsymbol{k}\cdot\boldsymbol{R}_j + \mathrm{i}\omega_{\boldsymbol{k}s}t)]\hat{b}^+_{-\boldsymbol{k}s}\}$$
$$\times\{[\exp(\mathrm{i}\boldsymbol{k}'\cdot\boldsymbol{R}_{j'}) - \exp(\mathrm{i}\boldsymbol{k}'\cdot\boldsymbol{R}_j - \mathrm{i}\omega_{\boldsymbol{k}'s'})]\hat{b}_{\boldsymbol{k}'s'}$$
$$- [\exp(\mathrm{i}\boldsymbol{k}'\cdot\boldsymbol{R}_{j'}) - \exp(\mathrm{i}\boldsymbol{k}'\cdot\boldsymbol{R}_j + \mathrm{i}\omega_{\boldsymbol{k}'s'}t)]\hat{b}^+_{-\boldsymbol{k}'s'}\}\rangle\Bigg) \tag{2.7.51}$$

$$= \exp\Bigg(-\sum_{\boldsymbol{k}s}[\hbar/2mN\omega_{\boldsymbol{k}s}]|\boldsymbol{q}\cdot\boldsymbol{e}_{\boldsymbol{k}s}|^2(2N_{\boldsymbol{k}s}+1)$$
$$\times\{1 - \cos[\boldsymbol{k}\cdot(\boldsymbol{R}_j - \boldsymbol{R}_{j'}) + \omega_{\boldsymbol{k}s}t]\}\Bigg)\ .$$

Here we have taken into account the equations

$$\begin{aligned}&\langle \hat{b}^+_\nu \hat{b}_{\nu'}\rangle = (\exp(\beta\hbar\omega_\nu) - 1)^{-1}\delta_{\nu\nu'} \equiv N_\nu\delta_{\nu\nu'}\ ,\\ &\langle \hat{b}_\nu \hat{b}^+_\nu\rangle = (1+N_\nu)\delta_{\nu\nu'}\ , \qquad \langle \hat{b}_\nu\hat{b}_{\nu'}\rangle = \langle \hat{b}^+_\nu \hat{b}^+_{\nu'}\rangle = 0\ ,\end{aligned} \tag{2.7.52}$$

and also the fact that the functions $\omega_{\boldsymbol{k}s}$, $|\boldsymbol{q}\cdot\boldsymbol{e}_{\boldsymbol{k}s}|^2$ are even, having replaced $\boldsymbol{k}$ on the left-hand side of the equation by $-\boldsymbol{k}$. Finally, allowing for (2.7.51, 42, 43), we find

$$\langle \exp(-\mathrm{i}\boldsymbol{q}\cdot\hat{\boldsymbol{u}}_j(t))\exp(\mathrm{i}\boldsymbol{q}\cdot\hat{\boldsymbol{u}}_{j'})\rangle = \exp\left\{-\sum_{\boldsymbol{k}s}(\hbar)|\boldsymbol{q}\cdot\boldsymbol{e}_{\boldsymbol{k}s}|^2/2mN\omega_{\boldsymbol{k}s}) \times [(2N_{\boldsymbol{k}s}+1)(1-\cos\Omega_{\boldsymbol{k}s}^{jj'})-\mathrm{i}\sin\Omega_{\boldsymbol{k}s}^{jj'}]\right\} . \tag{2.7.53}$$

where

$$\Omega_{\boldsymbol{k}s}^{jj'} = \boldsymbol{k}\cdot(\boldsymbol{R}_{j'}-\boldsymbol{R}_j)+\omega_{\boldsymbol{k}s}t . \tag{2.7.54}$$

Formula (2.7.53) also may be represented as

$$\langle \exp(-\mathrm{i}\boldsymbol{q}\cdot\hat{\boldsymbol{u}}_j(t))\exp(\mathrm{i}\boldsymbol{q}\cdot\hat{\boldsymbol{u}}_{j'})\rangle = \exp(-2w_{\boldsymbol{q}})\exp\left\{\sum_{\boldsymbol{k}s}\frac{\hbar|\boldsymbol{q}\boldsymbol{e}_{\boldsymbol{k}s}|^2}{2mN\omega_{\boldsymbol{k}s}} \times [(N_{\boldsymbol{k}s}+1)\exp(\mathrm{i}\Omega_{\boldsymbol{k}s}^{jj'}) + N_{\boldsymbol{k}s}\exp(-\mathrm{i}\Omega_{\boldsymbol{k}s}^{jj'})]\right\}. \tag{2.7.55}$$

The multiplier $(-2w_{\boldsymbol{q}})$, where

$$w_{\boldsymbol{q}} = \sum_{\boldsymbol{k}s}(\hbar|\boldsymbol{q}\boldsymbol{e}_{\boldsymbol{k}s}|^2/2mN\omega_{\boldsymbol{k}s})(N_{\boldsymbol{k}s}+\tfrac{1}{2}) , \tag{2.7.56}$$

is called the Debye–Waller factor. It may be shown by direct calculation (an exercise for the reader) that

$$2w_{\boldsymbol{q}} = \langle(\boldsymbol{q}\cdot\hat{\boldsymbol{u}}_j)^2\rangle . \tag{2.7.57}$$

Calculate $w_{\boldsymbol{q}}$ in the Debye model (for $\sigma = 1$). Then

$$\sum_s|\boldsymbol{q}\cdot\boldsymbol{e}_{\boldsymbol{k}s}|^2 \to \boldsymbol{q}^2 , \quad \frac{1}{3N}\sum_{\boldsymbol{k}s} \to 3\int_0^{\omega_{\mathrm{D}}}\frac{d\omega\,\omega^2}{\omega_{\mathrm{D}}^3}$$

(Sect. 2.2) and, according to (2.7.56),

$$\begin{aligned} w_{\boldsymbol{q}} &= \frac{3\hbar q^2}{2m\omega_{\mathrm{D}}^3}\int_0^{\omega_{\mathrm{D}}} d\omega\,\omega\left(\frac{1}{2}+\frac{1}{\exp(\hbar\omega/k_{\mathrm{B}}T)-1}\right) \\ &= \frac{3\hbar^2q^2}{2mk_{\mathrm{B}}\vartheta_{\mathrm{D}}}\left[\frac{1}{4}+\left(\frac{T}{\vartheta_{\mathrm{D}}}\right)^2\int_0^{\vartheta_{\mathrm{D}}/T}\frac{x\,dx}{\mathrm{e}^x-1}\right], \end{aligned} \tag{2.7.58}$$

where we have introduced a new variable $x = \hbar\omega/k_{\mathrm{B}}T$. When $T \ll \vartheta_{\mathrm{D}}$,

$$w_{\boldsymbol{q}} = \frac{3\hbar^2q^2}{8mk_{\mathrm{B}}\vartheta_{\mathrm{D}}}\left[1+\frac{2\pi^2}{3}\left(\frac{T}{\vartheta_{\mathrm{D}}}\right)^2\right]; \tag{2.7.59}$$

when $T \gg \vartheta_D$,

$$w_q = 3\hbar^2 q^2 T / 2 k_B \vartheta_D^2$$

(here the formula $\int_0^\infty x dx/(e^x - 1) = \pi^2/6$ is used).

Next we discuss the physical meaning of the results obtained.

2.7.4 Elastic Scattering

Represent the second cofactor in (2.7.55) as a series

$$\begin{aligned}
&\langle \exp[-i\boldsymbol{q}\cdot\boldsymbol{u}_j(t)]\exp(i\boldsymbol{q}\cdot\boldsymbol{u}_{j'})\rangle = \exp(-2\omega_q)\\
&\times \sum_{n=0}^{+\infty}\frac{1}{n!}\sum_{k_1 s_1}\dots\sum_{k_n s_n}(\hbar|\boldsymbol{q}\cdot\boldsymbol{e}_{k_1 s_1}|^2/2mN\omega_{k_1 s_1})\dots\\
&\dots(\hbar|\boldsymbol{q}\cdot\boldsymbol{e}_{k_n s_n}|^2/2mN\omega_{k_n s_n})[(N_{k_1 s_1}+1)\\
&\times\exp(i\boldsymbol{k}_i(\boldsymbol{R}_{j'}-\boldsymbol{R}_j)+i\omega_{k_1 s_1}t)+N_{k_1 s_1}\exp(i\boldsymbol{k}_i(\boldsymbol{R}_j-\boldsymbol{R}_{j'})+i\omega_{k_n s_n}t)]\times\dots\\
&\dots[(N_{k_n s_n}+1)\exp(i\boldsymbol{k}_n\cdot(\boldsymbol{R}_j-\boldsymbol{R}_{j'})+i\omega_{k_n s_n}t)\\
&+(N_{k_n s_n}[\exp(-i\boldsymbol{k}_n(\boldsymbol{R}_j-\boldsymbol{R}_{j'})-i\omega_{k_n s_n}t)]\ .
\end{aligned}\tag{2.7.60}$$

According to (2.7.41), it is necessary to carry out integration with respect to t and summation over j, j'. Here

$$\int_{-\infty}^{+\infty} dt \exp(i\Omega t) = 2\pi\delta(\Omega)\ , \tag{2.7.61}$$

$$\sigma_k = \sum_{jj'}\exp(i\boldsymbol{k}\cdot(\boldsymbol{R}_j - \boldsymbol{R}_{j'})) = N^2\sum_g \delta_{k,b_g^*}\ , \tag{2.7.62}$$

where $\boldsymbol{b}_g^*$ stands for reciprocal-lattice vectors (1.5.12, 1.3.3). Equation (2.7.62) may be proved as follows:

$$\sigma_k = \sum_{jj'}\exp(i\boldsymbol{k}\cdot(\boldsymbol{R}_j - \boldsymbol{R}_{j'} + \boldsymbol{R}_l)) = \exp(i\boldsymbol{k}\cdot\boldsymbol{R}_l)\sigma_k\ ,$$

where $\boldsymbol{R}_l$ is an arbitrary direct-lattice vector. Therefore, $\sigma_k = 0$ if $\exp(i\boldsymbol{k}\cdot\boldsymbol{R}_l) \neq 1$. But if $\exp(i\boldsymbol{k}\cdot\boldsymbol{R}_l) = 1$, i.e., $\boldsymbol{k}$ is equal to some reciprocal-lattice vector, then $\sigma_k = N^2$. Substituting (2.7.60) into (2.7.41) with allowance for (2.7.61), we arrive at the result that the term with $n = 0$ contributes to $\boldsymbol{S}(\boldsymbol{q}, \omega)$, the contribution being proportional to $\delta(\omega) = \delta((\varepsilon_{k'} - \varepsilon_k)/\hbar)$. Thus, this term describes elastic scattering (i.e., a scattering without a change in energy). The corresponding contribution to $S(q, \omega)$ is

$$S_{el}(q, \omega) = [N/(2\pi)^3]^2\, 2\pi\delta(\omega)\sum_g \delta_{q,b_g^*} \exp(-2w_q) \; . \tag{2.7.63}$$

The last cofactor, viz., the Debye–Waller factor, determines, according to (2.7.58, 59), the temperature dependence of the intensity of elastic scattering. The latter, according to (2.7.63), occurs only in certain directions defined by the equation $q = b_g^*$, i.e.,

$$k' = b_g^* + k \;, \quad \varepsilon_{k'} = \varepsilon_k \;, \tag{2.7.64}$$

which, as has been shown in Chap. 1, is equivalent to the Wulf–Bragg condition (1.5.13).

Here we discuss an important problem relating to the difference between coherent and incoherent lattice scattering. Suppose that we consider an alloy in which the probability of an atom of species A being located on the jth site is x and that for an atom of species B is $1-x$. Instead of (2.7.3) for the interaction potential between the particle scattered and the crystal we then will have

$$V(r) = \sum_j [\eta_j v_A(r - r_j) + (1 - \eta_j) v_B(r - r_j)] \;, \tag{2.7.65}$$

where $\eta_j = 1$ if an atom of species A is located in the jth site, and $\eta_j = 0$ if the site accommodates a B atom. Repeating the entire derivation (Sect. 2.7.1), we obtain, in place of (2.7.16),

$$\begin{aligned} \frac{dw_{k' \to k}}{dv_{k'}} &= \frac{1}{(2\pi)^6} \frac{1}{\hbar^2} \int_{-\infty}^{+\infty} dt \exp(i\omega t) \\ &\times \sum_{jj'} \langle \exp(-iq \cdot r_j(t)) \exp(iq \cdot r_{j'}) \rangle [|v_A(q)|^2 \\ &\times \eta_j \eta_{j'} + v_A^*(q) v_B(q)(1 - \eta_j)\eta_{j'} + v_A(q) \\ &\times v_B^*(q) \eta_j (1 - \eta_{j'}) + |v_B(q)|^2 (1 - \eta_j)(1 - \eta_{j'})] \end{aligned} \tag{2.7.66}$$

(for neutrons, ordinary isotopes may act as alloy constituents since, as distinct from X rays, neutrons interact chiefly with nuclei).

Formula (2.7.66) will now be averaged with respect to the distribution of atoms in the lattice. On the assumption that the atoms are distributed randomly and independently, we obtain

$$\overline{\eta_j \eta_{j'}} = (\bar{\eta}_{j'})^2 = x^2 \quad \text{when} \quad j \neq j', \overline{\eta_j^2} = \bar{\eta}_j = x \;,$$

i.e.,

$$\overline{\eta_j\eta_{j'}} = x[\delta_{jj'} + (1-\delta_{jj'})x] = x^2 + \delta_{jj'}x(1-x) \ , \tag{2.7.67}$$

$$\overline{(1-\eta_j)(1-\eta_{j'})} = (1-x)^2 + \delta_{jj'}x(1-x) \ ,$$

$$\overline{\eta_j(1-\eta_{j'})} = (1-\delta_{jj'})x(1-x) \ .$$

Now we average the terms enclosed in the curly brackets in (2.7.66):

$$\begin{aligned} &|v_A(\boldsymbol{q})|^2\overline{\eta_j\eta_{j'}} + v_A^*(\boldsymbol{q})v_B(\boldsymbol{q})\overline{(1-\eta_j)\eta_{j'}} + v_A(\boldsymbol{q})v_B^*(\boldsymbol{q})\overline{\eta_j(1-\eta_{j'})} + |v_B(\boldsymbol{q})|^2 \\ &\quad \times \overline{(1-\eta_j)(1-\eta_{j'})} = |xv_A(\boldsymbol{q}) + (1-x)v_B(\boldsymbol{q})|^2 \\ &\quad + \delta_{jj'}x(1-x)|v_A(\boldsymbol{q}) - v_B(\boldsymbol{q})|^2 \ . \end{aligned} \tag{2.7.68}$$

The first term describes the same scattering processes as those in a perfect crystal, with the replacement $v(\boldsymbol{q}) \to xv_A(\boldsymbol{q}) + (1-x)v_B(\boldsymbol{q})$; the second term contains $\delta_{jj'}$. Therefore, the expression for the scattering cross section will involve $\sum_{jj'}\delta_{jj'}\exp[i\boldsymbol{k}\cdot(\boldsymbol{R}_j - \boldsymbol{R}_{j'})] = N$ rather than $\sum_{jj'}\exp[i\boldsymbol{k}\cdot(\boldsymbol{R}_j - \boldsymbol{R}_{j'})]$. Instead of (2.7.16, 63), the elastic scattering probability is given by the expression

$$\begin{aligned} \left.\frac{dw_{\boldsymbol{k}\to\boldsymbol{k}'}}{dv_{\boldsymbol{k}'}}\right|_{\mathrm{el}} = &\frac{2\pi\delta(\omega)}{(2\pi)^6\hbar^2}\exp(-2W_{\boldsymbol{q}})\Bigg(|xv_A(\boldsymbol{q}) + (1-x)v_B(\boldsymbol{q})|^2N^2\sum_{\boldsymbol{g}}\delta_{\boldsymbol{q},\boldsymbol{b}_{\boldsymbol{g}}^*} \\ &+ |v_A(\boldsymbol{q}) - v_B(\boldsymbol{q})|^2N\Bigg) \ . \end{aligned} \tag{2.7.69}$$

The second term in the square brackets in (2.7.69), specific to a partially disordered crystal, is nonzero for all scattering angles, and not only for those defined by (2.7.64). This term describes what is known as the incoherent scattering processes, producing a background against which sharp coherent scattering peaks are discriminated [in the scattering directions defined by (2.7.64), the intensity of the peaks are a factor of N higher than that of the background]. The actual ratio of background intensity to peak intensity (as well as the peak width) is determined by the resolving power of the measuring equipment.

2.7.5 Inelastic Scattering

Consider now the term with $n = 1$ in (2.7.60). The corresponding contribution to $S(\boldsymbol{q}, \omega)$, according to (2.7.41, 60–62), is equal to

$$S_1(\boldsymbol{q},\omega)=N(2\pi)^{-5}\exp(-2w_{\boldsymbol{q}})\sum_{\boldsymbol{k}_1 s_1}\sum_{\boldsymbol{g}}(\hbar|\boldsymbol{e}_{\boldsymbol{k}_1 s_1}\cdot\boldsymbol{q}|^2/2m\omega_{\boldsymbol{k}_1 s_1})$$
$$\times[(N_{\boldsymbol{k}_1 s_1}+1)\delta_{\boldsymbol{k}_1,\boldsymbol{b}_g^*}\delta(\omega+\omega_{\boldsymbol{k}_1 s_1})+N_{\boldsymbol{k}_1 s_1}\delta_{\boldsymbol{k}_1\boldsymbol{b}_g^*}\delta(\omega-\omega_{\boldsymbol{k}_1 s_1})]\ . \quad (2.7.70)$$

The first term in the square brackets describes inelastic scattering processes with $\omega=-\omega_{\boldsymbol{k}_1 s_1}$, i.e.,

$$\varepsilon_{\boldsymbol{k}'}-\varepsilon_{\boldsymbol{k}}=-\hbar\omega_{\boldsymbol{k}_1 s_1}\ , \quad (2.7.71)$$

which evidently corresponds to the creation (emission) of a phonon in the state $\boldsymbol{k}_1 s_1$. The law of conservation of momentum holds here to within an arbitrary reciprocal-lattice vector

$$\boldsymbol{k}'=\boldsymbol{k}-\boldsymbol{k}_1+\boldsymbol{b}_g^* \quad (2.7.72)$$

(cf. the qualitative discussion of the phonon–phonon collisions in Sect. 2.3.3). The second term enclosed in the square brackets in (2.7.70) describes the absorption of a phonon in the state $\boldsymbol{k}_1$, s_1, the emission probability–absorption probability ratio (derived by Einstein in 1917 for photons) being equal to

$$(N_{\boldsymbol{k}_1 s_1}+1)/N_{\boldsymbol{k}_1 s_1}\ .$$

Similarly, the terms with $n=2, 3, \ldots$ in (2.7.60) describe processes with the participation of two, three, ..., n phonons. The probability of many-phonon processes may be shown to be small subject to the condition that

$$q^2 r_0^2 \ll 1\ , \quad (2.7.73)$$

with r_0 being the characteristic amplitude of lattice vibrations. Figure 2.18 shows that

$$|\boldsymbol{k}-\boldsymbol{k}'|=(k^2+k'^2-2kk'\cos\vartheta)^{1/2}=[(k-k')^2+4kk'\sin^2(\vartheta/2)]^{1/2}\ . \quad (2.7.74)$$

For light scattering,

$$\varepsilon_k=\hbar kc\equiv\hbar\omega\ ,\qquad \varepsilon_{k'}=\hbar k'c\equiv\hbar\omega'$$

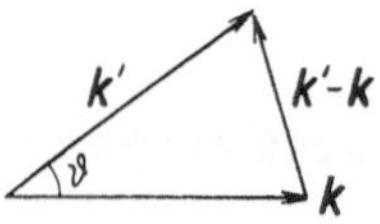

Fig. 2.18. Variation of particle momentum during scattering through an angle ϑ

(with c being the velocity of light), and for the creation (annihilation) process of a phonon with a frequency ω_1 and wave vector $\boldsymbol{q}_1$, we obtain, in keeping with (2.7.71, 72, 74),

$$\begin{aligned} &|\omega-\omega'|=\omega_1(\boldsymbol{q}_1)\\ &|\boldsymbol{b}_g^*\pm\boldsymbol{q}_1|=c^{-1}[\omega_1^2+4\omega(\omega\pm\omega_1)\sin^2(\vartheta/2)]^{1/2}\\ &\qquad\approx 2\omega c^{-1}|\sin(\vartheta/2)| \ . \end{aligned} \tag{2.7.75}$$

Here we have taken into account that $\omega_1 \ll \omega$. For light the wavelength is

$$\lambda \sim 10^3 \div 10^4\,\text{Å} \ , \quad \text{and} \quad \omega/c=2\pi/\lambda \ll d^{-1} \ ,$$

where d is the lattice period. Therefore, practically, $\omega_1(\boldsymbol{q}_1)\approx\omega_1(\boldsymbol{b}_g^*)=\omega_1(0)$; i.e., the variation of the frequency of scattered light in processes involving the creation and annihilation of one phonon is equal to the threshold frequency of some phonon branch (certainly an optical one, since for acoustic phonons $\omega_1(0)=0$ and the corresponding processes make only small contributions to the "truly elastic" scattering).

The phenomenon of the variation of the frequency of light when scattered from molecules and crystals is called combination scattering (in English literature it is often referred to as the Raman effect). It was discovered by *Mandelstamm* and *Landsberg* [2.22] and, independently, by *Raman* and *Krishnan* [2.23]. This effect admits of a purely classical interpretation: the permittivity ε of a medium whose particles perform oscillations of frequency ω_1 will be an oscillating function of time t and vary obeying the law $\varepsilon(t)=\varepsilon_0(1+\alpha\cos\omega_1 t)(\alpha\ll 1)$. Since the permittivity determines the intensity of a medium-scattered electromagnetic field, the latter will be modulated by the form of the function $\varepsilon(t)$, and this will lead to the occurrence of harmonics with an altered frequency:

$$\cos\omega t\cos\omega_1 t=\tfrac{1}{2}[\cos(\omega+\omega_1)t+\cos(\omega-\omega_1)t] \ .$$

As far as the intensity of scattered light or of neutrons is concerned, the classical theory describes it correctly only for sufficiently high temperatures $T\gg\vartheta_{\text{D}}$, where the phonons may be treated classically.

In highest orders with respect to the parameter (2.7.73) (many-phonon processes), the scattering probability involves terms with $\omega'=\omega\pm 2\omega_1$, $\omega\pm\omega_1\pm\omega_2, \omega\pm 2\omega_1\pm\omega_2, \ldots$, where $\omega_1, \omega_2, \ldots$ are the threshold frequencies of optical phonons.

In contrast to light scattering, neutron scattering enables a determination not only of the threshold frequencies but also of the entire phonon spectrum $\omega_1(\boldsymbol{q}_1)$. For this purpose, neutrons that have a velocity on the order of the velocity of sound must be used, and then (2.7.71, 72) may be satisfied for any value of $\boldsymbol{q}_1$, and not only in the region (2.7.75), as with light scattering. The neutron scattering method is now a major tool in determining lattice vibration spectra.

2.7.6 The Mössbauer Effect

The formalism outlined above may be applied, practically without alteration, to the theory of one more very important and convenient method of investigating the properties of a solid, that is, the *Mössbauer* effect [2.24], also called recoil-free nuclear gamma resonance (NGR). In English literature it is sometimes referred to as recoil-free gamma-ray fluorescence.

Let us consider two like nuclei, one of which is excited while the other is in the ground state (Fig. 2.19). The first nucleus emits a gamma energy quantum E_0 which, it would seem, should be capable of absorption by the second nucleus which, in turn, should pass to the excited state $|1\rangle$. However, the situation is not so simple as that: when the first nucleus emits a gamma quantum, this photon carries away not only the energy E_0 but also the momentum E_0/c, and accordingly, the first nucleus recoils and receives some kinetic energy. That is to say, the energy carried away by the gamma quantum is not E_0 but $E < E_0$. The latter may be determined from the laws of conservation of momentum and energy

$$E_0 = E + p^2/2M \; , \qquad E/c = p \; , \tag{2.7.76}$$

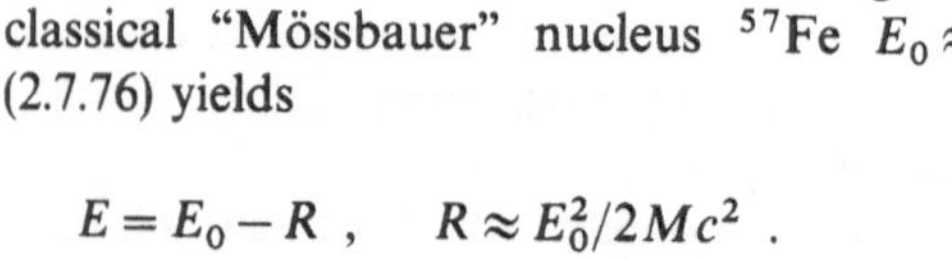

Fig. 2.19. Resonance absorption without nuclear recoil.

with M being the nuclear mass. Using the condition $E_0 \ll Mc^2$ (say for the classical "Mössbauer" nucleus ^{57}Fe $E_0 \approx 1.4 \times 10^4$ eV, $Mc^2 \approx 6 \times 10^{10}$ eV), (2.7.76) yields

$$E = E_0 - R \; , \qquad R \approx E_0^2/2Mc^2 \; . \tag{2.7.77}$$

Similarly, for the second nucleus to absorb the gamma quantum, it must possess the energy $E_0 + R$. The absorption (emission) probability, as is known from quantum mechanics, has the form depicted in Fig. 2.20a, where Γ is the natural width of the level $|1\rangle$ ($\Gamma = \hbar|\tau$, with τ being the lifetime of the nucleus in the state $|1\rangle$, i.e., the inverse probability of the $|1\rangle \rightarrow |0\rangle$ transition per unit time). Then, as is seen from Fig. 2.20b, when $\Gamma \ll R$ the probability of the

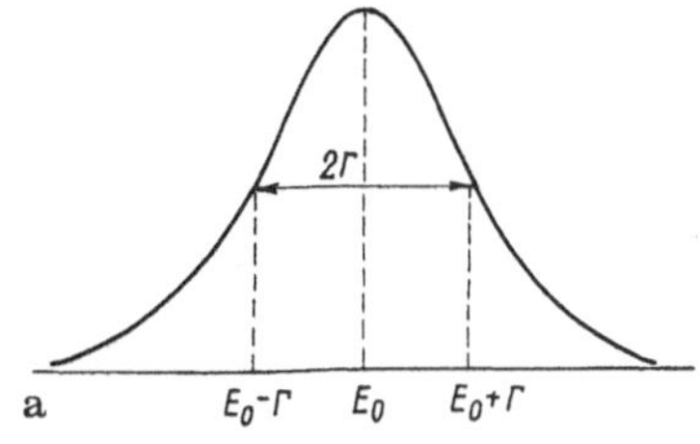

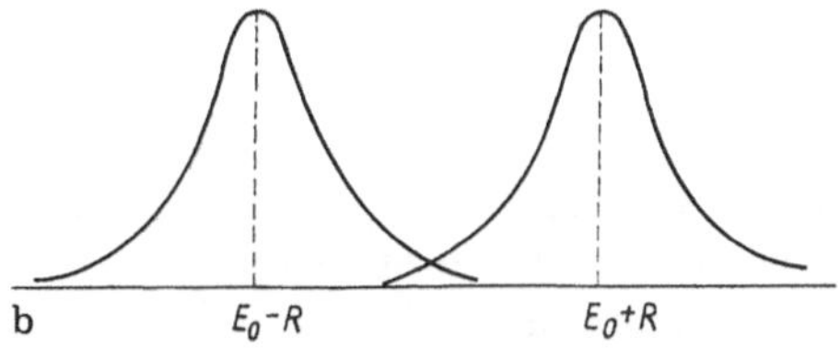

Fig. 2.20. Natural absorption or emission line width without allowance for nucleus recoil (**a**), effect of recoil at $R \gtrsim \Gamma$ on resonance absorption probability (**b**)

resonance absorption of the emitted gamma quantum by the second nucleus is very small. Normally $\Gamma \lesssim 10^{-5}$ eV, $R \sim 10^{-2}$ eV, i.e., $\Gamma \ll R$. Hence it is practically impossible to observe NGR in experiments with free nuclei (say, in a gas).

The nucleus in a solid is quite a different thing. If it is rigidly bound to the crystal, the latter, taken as a whole, will recoil; it stands to reason that the recoil in this case is infinitely small (a factor of $N \sim 10^{23}$ lower than that for a free nucleus, where N is the number of atoms in the crystal). Insofar as the motion with respect to the crystal goes, not any energy but only the one equal to the sum of the energies of some number of phonons may be imparted to the crystal, there existing the probability that a gamma quantum will be emitted from the crystal without generating a single phonon (this corresponds to elastic neutron scattering). That gamma quantum will have an energy precisely equal to E_0 and may be absorbed by a similar nucleus bound rigidly to the crystal. Thus, we need to determine the probability of recoil-free gamma quantum emission in the crystal.

Assume that the crystal is first in the quantum state $|l\rangle$. The probability for the nucleus j to have the value of the momentum $\hbar \boldsymbol{k}$ here is equal to

$$w_{\boldsymbol{k}} = |\langle \boldsymbol{k}|l\rangle|^2 \equiv |\int d\boldsymbol{r}_j \exp(-i\boldsymbol{k}\cdot\boldsymbol{r}_j)/(2\pi)^{3/2} \times \Psi_l(\boldsymbol{r}_1, \ldots, \boldsymbol{r}_j, \ldots, \boldsymbol{r}_N)|^2 \,, \quad (2.7.78)$$

where $\langle \boldsymbol{k}|l\rangle$ stands for the expansion coefficients of the wave function $|l\rangle$ in plane waves $|\boldsymbol{k}\rangle$ with $\boldsymbol{r} = \boldsymbol{r}_j$ (2.6.16):

$$|l\rangle = \sum_{\boldsymbol{k}} |\boldsymbol{k}\rangle\langle \boldsymbol{k}|l\rangle \,. \quad (2.7.79)$$

Upon emission of a gamma quantum with momentum

$$\hbar q = E_0/c \,, \quad (2.7.80)$$

we should, by the law of conservation of momentum, have $w_k \to w_{k-q}$. Thus the final state $|f\rangle$ should be

$$|f\rangle = \sum_k |k-q\rangle\langle k|l\rangle \equiv \exp(-\mathrm{i}q\cdot r_j) \times \sum_k |k\rangle\langle k|l\rangle = \exp(-\mathrm{i}q\cdot r_j)|l\rangle \ , \tag{2.7.81}$$

where we have used the property

$$|k-q\rangle = \exp(-\mathrm{i}q\cdot r_j)|k\rangle$$

and the completeness condition $\sum_k |k\rangle\langle k| = 1$. The final state $|f\rangle$ must be expanded in the eigenstates of the nucleus $|n\rangle$, and the probability of the $l\to n$ transition during the emission of a gamma quantum is

$$P_{l\to n} = |\langle n|f\rangle|^2 = |\langle n|\exp(-\mathrm{i}q\cdot r_j)|l\rangle|^2 \ . \tag{2.7.82}$$

If, as a function of its energy E, the intensity of the gamma quantum emitted is equal to $I_0(E)$ (Fig. 2.20a), then the intensity of the gamma quantum emitted during the $l\to n$ transition will be $I_0(E+E_n-E_l)$ (so that, in keeping with the law of conservation of energy, the maximum shifts by $\Delta E = E_n - E_l$). The total intensity, upon averaging with respect to the initial states and upon summation over the final ones, is

$$I(E) = \sum_{nl} I_0(E+E_n-E_l)|\langle n|\exp(-\mathrm{i}q\cdot r_j)|l\rangle^2 \varrho_l \ . \tag{2.7.83}$$

Expand $I_0(E+E_n-E_l)$ as a Fourier integral:

$$I_0(E+E_n-E_l) = \int_{-\infty}^{\infty} dt(2\pi\hbar)^{-1}\exp\left[-\frac{\mathrm{i}t}{\hbar}(E+E_n-E_l)\right]I_0(t) \ . \tag{2.7.84}$$

Substituting (2.7.84) into (2.7.83) and carrying out the same transformations as those used in going from (2.7.2) to (2.7.10), we find

$$\begin{aligned} I(E) &= \sum_{nl}\int_{-\infty}^{+\infty} dt(2\pi\hbar)^{-1}\exp(-\mathrm{i}Et/\hbar)I_0(t)\varrho_l \\ &\quad\times\langle l|\exp(\mathrm{i}E_l t/\hbar)\exp(-\mathrm{i}q\cdot r_j)\exp(-\mathrm{i}E_n t/\hbar)|n\rangle \\ &\quad\times\langle n|\exp(\mathrm{i}q\cdot r_j)|l\rangle = \int_{-\infty}^{+\infty} dt(2\pi\hbar)^{-1}\exp(-\mathrm{i}Et/\hbar) \\ &\quad\times I_0(t)\langle\exp(-\mathrm{i}q\cdot r_j(t))\exp(\mathrm{i}q\cdot r_j)\rangle \\ &= \int_{-\infty}^{+\infty} dt(2\pi\hbar)^{-1}\exp(-\mathrm{i}Et/\hbar)I_0(t)\langle\exp(-\mathrm{i}q\cdot u_j(t))\exp(\mathrm{i}q\cdot u_j)\rangle \ . \end{aligned} \tag{2.7.85}$$

The average involved in (2.7.85) has already been calculated, see (2.7.60). Substituting (2.7.60) into (2.7.85) and allowing for (2.7.84), we see that the term with $n=0$ in (2.7.60) yields a term with $I_0(E)$ in (2.7.85) (i.e., an unaltered line), and the terms with $n=1, 2, \ldots$ yield lines displaced by ω_1, $2\omega_1$, $\omega_1+\omega_2$, etc. Thus, the intensity of an undisplaced (Mössbauer) line is equal to

$$I_{\mathrm{M}}(E)=\exp(-2w_{\boldsymbol{q}})I_0(E) \; ; \tag{2.7.86}$$

i.e., it is defined by the Debye–Waller factor. According to (2.7.58, 59), as the temperature is lowered, the quantity $I_{\mathrm{M}}(E)$ increases (since $w_{\boldsymbol{q}}$ decreases).

It has already been noted that the natural width of nuclear levels is very small, and therefore the possibility of observing an unbroadened line allows very accurate measurements of the various quantities. Suppose, for example, that a transmitter crystal and a receiver crystal differ somewhat in electronic structure (say, in the character of chemical bonds or in the magnetic state of electrons). Owing to the interaction of electrons with the nucleus (responsible for the hyperfine structure of atomic spectra) the nuclear energy levels are shifted, very insignificantly in terms of nuclear scales, but nevertheless this displacement may be much larger than the width Γ, which also is very small. Then the resonance will be detuned, and the intensity of the Mössbauer signals will be very low. The resonance may be restored thanks to the Doppler effect: if the source moves relative to the absorber with a velocity v, the gamma quantum frequency will be shifted by

$$\Delta E=E_0 v/c \tag{2.7.87}$$

and this shift may compensate for the displacement of the nuclear levels as a result of the interaction with electrons already at small v. By varying the intensity of absorption, we can measure fields induced by electrons on the nucleus, extracting very important information on electronic structure [2.25–27]. The method is so accurate that it has even allowed the effect of the gravitational shift of electromagnetic radiation frequency, predicted by general relativity theory, to be observed under terrestrial conditions.

2.8 Conclusion

Crystal lattice theory is in some respects the prototype of solid-state theory in general. The following major points may be noted:

1) *Introduction of phonons.* Strikingly, it is possible to proceed almost rigorously from a consideration of a system of strongly interacting atoms in the crystal to a consideration of an ideal phonon gas. The only necessary condition

(in addition to periodicity) is that the amplitude of (quantum and thermal) oscillations should be small compared to the lattice constant.

When this condition is violated, the crystal simply melts; that is to say, far from the melting point the interaction between phonons (anharmonicity) may always be taken into account as a small perturbation (which, incidentally, sometimes gives rise to qualitatively new effects—Sect 2.3).

2) *Special character of the energy spectrum.* The energy spectrum consists of bands of allowed energy values $\varepsilon = \hbar\omega$ separated by energy gaps; energy-band splitting results from the doubling of the lattice period (2.1.2).

3) *Possibility of the formation of a "local" level.* A local level splits off from the phonon energy band in an ideal crystal, near a defect (Sect. 2.4).

4) *Possibility of measuring the energy spectrum* by neutron, electron, and photon scattering experiments. The scattering probability relates to the density correlation function (Sect. 2.7).

All these points, as will be seen, play an important, and sometimes decisive, role in the theory of the electrical, magnetic, optical, and other properties of solids. Emphasis will be given to the introduction of weakly interacting quasi-particles, or elementary excitations, instead of a system of particles that interact strongly. This is actually the major concept of the whole quantum theory of the condensed state (the tremendous importance of this concept was dramatically demonstrated by Landau in his theory of the superfluidity of helium [2.28]).

In connection with the general approach of the book, we do not dwell upon the properties specific to the lattice. These include primarily mechanical properties (elastic properties, plastic deformation), electrical properties (including ferroelectricity), and the theory of crystal lattice defects (vacancies, dislocations, etc.). These problems are better treated phenomenologically, in fact, independently of quantum solid-state theory (except for some very interesting problems such as quantum diffusion, the interaction of electrons with dislocations, etc.). For a detailed treatment of these properties, see [2.13, 29–34].

3. Simple Metals: The Free Electron–Gas Model

This chapter elaborates on the simplest model of the electronic subsystem of metals, viz., the free-electron Fermi gas model. We trace the evolution of the electron theory of metals from the discovery of the electron to the creation of quantum mechanics and quantum statistics. We calculate and discuss in detail a very large number of properties of the electron Fermi gas—thermodynamic, magnetic, transport, high-frequency, and others. It is shown how discrepancies between theory and fundamental experimental facts ("disasters" of electron theory) have led us to an increasingly sophisticated understanding of metals.

3.1 Types of Metals

Metallic-bonded solids are divided into pure metals, that is, chemical elements, and their alloys and compounds (with metals and nonmetals). In this chapter we consider the theory of pure metals. Pure metals, depending on the structure of the electronic shell of atoms, are classified into two classes, viz., normal and transition metals. The normal metals (on which we focus our attention here) are subdivided into groups according to the valence of their atoms:

a. *Univalent metals*—alkali metals Li, Na, K, Rb, Cs, Fr; noble metals Cu, Ag, Au.

b. *Bivalent metals*—alkaline-earth metals Be, Mg, Ca, Sr, Ba, Ra; analogs of the noble metals Zn, Cd, Hg.

c. *Trivalent metals*—Al, Ga, In, Tl.

d. *Quadrivalent metals*—Sn, Pb.

e. *Quantivalent metals* (semimetals?)—As, Sb, Bi, Po.

The transition metals are divided into three large groups—*d* metals, *f* metals, and mixed *d–f* metals—which are further divided as follows:

a. 3*d metals* Sc, Ti, V, Cr, Mn, Fe, Co, Ni (iron group).

b. 4*d metals* Y, Zr, Nb, Mo, Tc, Ru, Rh, Pd (palladium group).

c. 5*d metals* La, Hf, Ta, W, Re, Os, Ir, Pt (platinum group).

d. 4*f metals* Ce, Pr, Nd, Pm, Sm, Eu, Gd, Tb, Dy, Ho, Er, Tm, Yb, Lu (rare-earth metals, i.e., REM or lanthanides).

e. 6*d*–5*f metals* Ac, Th, Pa, U (actinides); to these we also may add the transuranium elements Np, Pu, Am, Cm, Bk, Cf, Es, Fm, Md, No, Lr.

3.2 Physical Properties of the Metallic State. Conduction Electrons

To begin with, we address the question of what property should be singled out as the most remarkable characteristic of metal. Since the time of Lomonosov metals have been recognized by their malleability and characteristic luster. However, more distinct criteria are their electrical and thermal properties, viz., high specific electrical and thermal conductivities, σ and κ. In typical metals at room temperature $\sigma \sim 10^6 - 10^4\ \Omega^{-1}\,\mathrm{cm}^{-1}$, and for the typical metalloid—sulfur—$\sigma \sim 10^{-17}\ \Omega^{-1}\,\mathrm{cm}^{-1}$.

After identification of the microscopic nature of electrical conductivity associated with the transport of matter in electrolytes (Faraday laws) and following the discovery of cathode rays (electrons), the problem arose of determining experimentally the microscopic nature of current carriers in metals.

Riecke was the first to show that current carriers in metals are not connected with the main mass of atoms (by contrast with electrolytes). In his experiments a current was continuously passed through three cylinders (copper, aluminium, copper) closely fitted together. On completion of that experiment, exceptional in its duration (1898–1909), Riecke did not find any traces of electrolysis. He thus proved that the current in metals is not associated with ion transfer, and it seemed likely that electrons were the current carriers in metals. However, experiments designed to prove this were also negative.

Stewart and Tolman [3.1] finally proved that the current in metals is transferred by electrons. In Strasbourg, L.I. Mandelstamm and N.D. Papaleksi had first attempted such an experiment. They discontinued their work because of the outbreak of the 1914–1918 war. They had, however, pointed out that, if Riecke's experiments were given an electronic interpretation and the electrons participating in the current were assumed to be loosely bound to the ion lattice, an elecric inertia effect could be observed. In the first experiments it was observed in the following fashion: A cylindrical solenoid made up of a metallic wire was set into rapid rotation about its axis and then stopped abruptly. A noticeable current pulse then passed through the leads that were connected to the ends of the solenoid (through the axis of revolution) to a sensitive galvanometer, so that the pulse charge Q could be measured. As the solenoid was slowed down, a negative acceleration $\dot{\boldsymbol{v}}$ ($\boldsymbol{v}$ is the linear velocity of the solenoid winding) arose which built up an effective field $\boldsymbol{E}_{\mathrm{eff}} = -m\dot{\boldsymbol{v}}/e$, with m being the mass and e the charge of the current carrier. Using Ohm's law $\boldsymbol{j} = \sigma \boldsymbol{E}_{\mathrm{eff}}$ and integrating over the solenoid winding of length l and ohmic resistance R, and also over current pulse duration, we readily find

$$e/m = v_0 l/QR \ , \tag{3.2.1}$$

where v_0 is the velocity prior to braking. Since l, v_0, and R are known parameters of the setup and Q is easy to measure, the sign and magnitude of the specific charge e/m of the current carrier can be determined. The experiments unambiguously gave a negative sign for the charge carriers and showed with high accuracy ($\lesssim 10\%$) that e/m coincided with similar measurements for cathode rays. In later experiments the coincidence was brought up to 0.2% [3.2]. Thus metals may be regarded as electronic conductors.

After the type of current carriers has been ascertained, the question arises of whether a physical criterion exists which distinguishes metals from other electronic conductors (for example, semiconductors). Only if this question is answered in the affirmative may the theory of the metallic state be spoken of as a special domain of theoretical physics.

To establish this criterion, we again turn to an analysis of the electrical properties of electronic conductors. As stated in the foregoing, metals exhibit large values of specific electrical conductivity. However, its magnitude may not be an unambiguous indication, since there are metals and nonmetals, i.e., electronic conductors, whose σ values can coincide under certain conditions. It is therefore more correct to choose the criterion not to be a random value of σ or specific resistivity $\varrho = \sigma^{-1}$ under some random external conditions (T, p, etc.) but the temperature trend of the function $\varrho(T)$ in the region of low temperatures. Figure 3.1 schematically presents the run of the $\varrho(T)$ curve for a metal and a semiconductor. Comparison of the curves shows that when the temperature approaches 0 K the metals exhibit a sharp decrease in ϱ. For instance, in the case of pure cadmium, when T is lowered from 273 to 1.61 K, the $\varrho_{273\,\mathrm{K}}/\varrho_{1.61\,\mathrm{K}}$ ratio is on the order of 10^6. The purer the metal and the less the ideality of its crystal lattice is perturbed, the smaller the value of ϱ resulting from extrapolation of the curve to the so-called residual resistivity $\varrho_R = \varrho(0)$. Recently, a detailed study has been made of the increase of $\varrho(T)$ at ultralow temperatures (Fig. 3.1b), which was detected as far back as the thirties. This increase is due to the influence of magnetic impurities (Kondo effect). An ideally pure metal crystal at $T = 0$ K may be expected to have $\varrho(0) = 0$ (ideal conductor). This state should not be confused with the phenomenon of superconductivity, which occurs at $T \neq 0$ K and the nature of which is altogether different (Chap. 5). Conversely, in nonmetallic

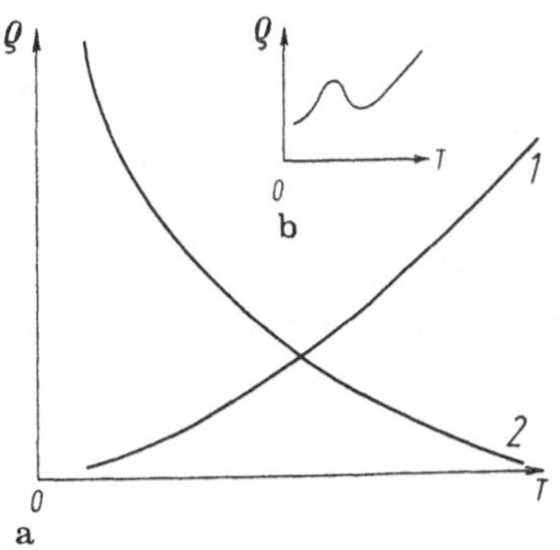

Fig. 3.1. (**a**) Temperature dependence of specific electrical resistivity for metals (*1*) and semiconductors (*2*); (**b**) the same dependence in metals with magnetic impurities (scaled up)

electronic conductors at $T \to 0$ K, the quantity ϱ tends to infinity (ideal insulator).

This analysis allows us to choose the temperature trend of electrical resistivity at low temperatures as the physical criterion of the metallic state: metals are solids in which the $\varrho(T)$ with $T \to 0$ K falls sharply reaching a minimum value, and semiconductors or insulators are solids in which $\varrho(T) \to \infty$.

If we combine all that has been said in Chap. 1 with regard to metals and itinerant electrons in them, with the experimental criterion established here, the following conclusions may be arrived at. Small values of ϱ in the vicinity of 0 K indicate that near the ground state in the metal there are already many itinerant electrons capable of acceleration in weak external fields. Later we will discuss what a weak or strong electric field in a metal signifies (Chap. 4). These electrons will be called conduction electrons. Using this concept, we may paraphrase somewhat the criterion formulated above: In metals at low temperatures near the ground state there are conduction electrons. In nonmetals there are practically no conduction electrons and they occur in inconsiderable amounts when the temperature is raised or when subject to some other excitation, for example, when exposed to light, etc.

From what follows it will be clear that this criterion is a fundamental one—indeed, all physical properties of electronic conductors are due to the presence of conduction electrons in them.

Another natural question is, what is the density of conduction electrons, n? It certainly may not be answered definitely without solving the problem of the motion of an electron in a crystal. But it is qualitatively reasonable to assume that in a metal the n is of the order of the density of crystal sites, N, and in nonmetals at low temperatures $n \ll N$. This conjecture is supported in part by the fact that all elements of the first three periodic table columns easily lose their valence electrons and, as is known, in the solid state all of them (except boron) are metals. Hence the conclusion is forced upon us that n is equal to the density of valence electrons. However, care should be exercised here since a large value of σ does not indicate unambiguously a large number of conduction electrons. This is clearly seen from the Lorentz formula $\boldsymbol{j} = ne\boldsymbol{v}$; i.e., the same current-density value may be obtained using two quantities: the density n of current carriers and their velocity v.

To gain a better understanding of conduction electrons, we consider the effect of an externally applied constant electric field of strength E on electrons that are in two opposite situations, viz., in the free state in vacuum and tightly bound in an atom.

The first case is realized in a low-density electron beam in a vacuum tube. The field, viewed as a small perturbation, leads to a continuous linear increase with time t of the electronic momentum component along the field E because of the increase of the kinetic energy of the electron. For a field aligned with the x axis we have

$$p_x(t) = p_{0x} + eE_x t \ . \tag{3.2.2}$$

In the second case, in an atom, the electron is subject to a strong nuclear field ($E_{at} \approx 10^8$ V/cm). Therefore, the effect of an externally applied field depends largely on the magnitude of the field. When $E \ll E_{at}$ no changes with time occur in the atom. The electron remains in the stationary state, which differs little from the unperturbed one. This is known from the observation of the Stark effect in "weak" fields. The cause of this influence of the field on the electron in the atom is easy to understand. A constant field is the limit of a variable field with $\omega \to 0$. Therefore, it transfers energy to an electron in a classical way, i.e., by infinitely small portions ($\hbar\omega \to 0$). By the action of a constant field, transitions between states with infinitely close energy levels may occur. Hence a clearly indispensable condition for the existence of an accelerating effect of a weak constant field is the presence of a continuous energy spectrum. This condition is not sufficient. In any solid there exist, for example, acoustic phonons of frequency $\omega \to 0$; however, those states are currentless and do not directly affect the metallic-state criterion. For atoms with a discrete spectrum the accelerating effect is absent subject to the condition that $E \ll E_{at}$. If this condition is violated, the probability of cold atomic ionization arises.

To which of the two cases is the behavior of the conduction electron in the metal the closer? Letting a current pass through a metallic wire, we place the wire into a thermostat so that, in spite of heat liberation, the temperature remains constant. Then the wire will be in a state of equilibrium and, according to Ohm's law, $\boldsymbol{j} \sim \boldsymbol{E}$. Thus, the behavior of conduction electrons under the accelerating effect of the field is something in between the behavior of a free electron and that of an atomic electron. The behavior of conduction electrons differs from that of an electron beam in vacuum in that, with the field turned on, the conduction electrons remain in a stationary state (after a short transient period during which this state sets in) rather than undergoing unlimited acceleration (3.2.2). Acceleration occurs only during a finite time span τ, and we obtain $\Delta p_x = p_x(\bar{\tau}) - p_{0x} \sim E\bar{\tau}$. Then the acceleration ceases and the electron transfers the accumulated energy to the ions of the crystal in which Joule–Lenz heat is released uniformly over the entire volume. In a vacuum tube accelerated electrons lose energy at the anode only. We come to the conclusion that although there is no current when $E = 0$, there are, infinitely close to it, excited states with $j \neq 0$ even at infinitely weak E. It may be thought that when $E = 0$ conduction electrons are not localized near lattice sites either; otherwise they would be incapable of participating in the current with the field being arbitrarily weak. To draw particular conclusions about the properties of conduction electrons, we need to proceed to a construction of a quantitative theory using reasonable approximations.

3.3 Classical Conduction-Electron Theory (Drude–Lorentz Theory)

The Drude–Lorentz theory is obviously crude, but owing to its "physical insight" it has retained significance to this day. A major difference between the behavior of a metallic conduction electron in a field and that of an electron in vacuum boils down to the fact that the extra velocity gained in the field turns out to be finite, as is the acceleration time (mean free time) $\bar{\tau}$, which depends on temperature. The acceleration process starts and terminates with collisions of very short duration ($\bar{\tau}_{imp} \ll \bar{\tau}$), when the electron transfers to the lattice all of the accumulated energy and the momentum; simultaneously, an equilibrium thermal distribution of electrons with respect to velocities sets in. The details of the collision mechanism are not important to us, except for the fact itself that they exist, just as in the case of the kinetic theory of ideal gases.

When $E=0$, the thermal motion is isotropic and $j=0$. However, as seen from Fig. 3.2, states with an anisotropic velocity distribution (with $j \neq 0$) lie in the immediate vicinity of this state. The statistical character of the equilibrium of conduction electrons leads to current fluctuations (shot effect) whose existence is direct evidence for the presence of excited states with $j \neq 0$ in the immediate proximity to the state with $j=0$. This provides indirect justification for the concepts of the nature of metallic conductivity which have been adopted here. What is implied here is the inevitably present noise in electronic devices which proves the "atomicity" of electricity in the same way as the existence of Brownian motion proves the atomicity of matter. In terms of the model considered, one more stringent approximation is introduced in which electrons are viewed as particles that do not interact with each other. This one-electron model disregards the electrostatic (Coulomb) repulsion between electrons.

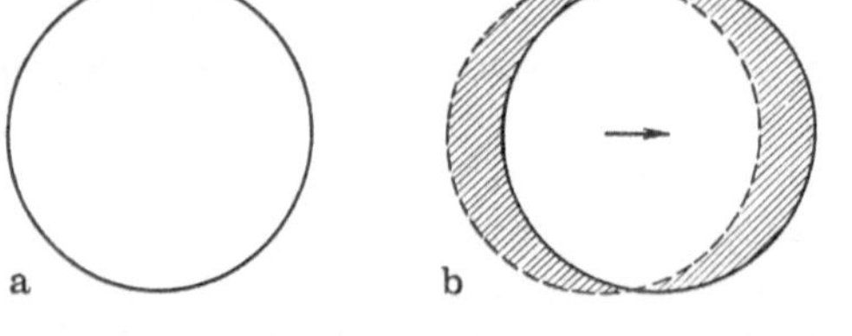

Fig. 3.2. Electron velocity distribution in metal: $E = 0$ (**a**), $E \neq 0$ (**b**)

All these assumptions are natural and, to some extent, have been maintained in quantum models of the metallic state. Taking into account the tight coupling of electrons to the lattice and between each other, one may talk of their effective individualization in terms of the aforementioned quasiparticles.

Let us now proceed to quantitative computations in terms of the Drude model. We start by estimating the time $\bar{\tau}$. To this end, we will determine the specific electrical conductivity σ. Assume that a field $\boldsymbol{E}$ is applied along a metallic wire. The field acts on the electron with a force $\boldsymbol{eE}$, which imparts to it an acceleration $\boldsymbol{a}$ equal to

$$\boldsymbol{a}=e\boldsymbol{E}/m \ .$$

During the time $\bar{\tau}$ between two successive collisions the electrons receive an additional mean velocity $\Delta \boldsymbol{v}_E=\boldsymbol{a}\bar{\tau}/2=e\boldsymbol{E}\bar{\tau}/2m$. Substituting it into the Lorentz formula, we find

$$\boldsymbol{j}=ne\overline{\Delta \boldsymbol{v}_E}=ne^2\bar{\tau}\boldsymbol{E}/2m=\sigma\boldsymbol{E} \ ; \tag{3.3.1}$$

Thus, the specific electrical conductivity is

$$\sigma=ne^2\bar{\tau}/2m \ . \tag{3.3.2}$$

Expression (3.3.2) can be obtained in another way. Consider the equation of motion of an electron which is subject, in addition to the force eE_x, to a frictional force, proportional to its velocity, $2mfdx/dt$, with f being the frictional coefficient

$$m\frac{d^2x}{dt^2}=eE_x-2mf\frac{dx}{dt} \ .$$

The solution of this equation for the initial conditions $(dx/dt)_{t=0}=0$ will have the form

$$v=(eE_x\tau_0)/2m[1-\exp(-t/\tau_0)] \ ,$$

where $\tau_0=f^{-1}$. The stationary state is reached when the values of t are such that the inequality $\exp(-t/\tau_0)\ll 1$ holds, and the corresponding velocity will be

$$\bar{v}=eE_x\tau_0/2m \ .$$

Hence, by virtue of (3.3.4), we again find (3.3.2).

In this fashion Ohm's law and the expression for σ in terms of atomic quantities were derived for the first time theoretically. As subsequent theoretical developments have shown, (3.3.2) is closer to reality than may seem at first sight. From dimensional considerations, every theory that exploits the concept of the mean free time should lead to (3.3.2) for σ. The right-hand side of (3.3.2) involves three unknown quantities: n, $\bar{\tau}$, and m (although in the derivation m was considered to be the mass of a free electron; but this is not obligatory, as will be discussed later). There is, however, only one equation for determining them from the experimental data for σ. Therefore they cannot be determined separately. It merely follows from (3.3.2) that in the region of applicability of Ohm's law the density n and time $\bar{\tau}$ are independent of E. If we assume, in addition, that n in a metal is practically independent of T (by contrast with semiconductors) and is commensurate with the number of valence electrons in unit volume of the metal, then (3.3.2) may be employed to estimate the time $\bar{\tau}$.

Let us make such an estimate for pure gold, assuming it to be univalent. Using the values $n = 5.95 \cdot 10^{22}\,\mathrm{cm}^{-3}$, $e = 4.80 \cdot 10^{-10}$ CGSE, $m = 9.10 \cdot 10^{-28}$ g, and the experimental data for σ_{Au}, we find the values of $\bar{\tau}$ for different T as summarized in Table 3.1. As is seen from the tabulated values, $\bar{\tau}$ increases by five orders of magnitude when the specimen is cooled from room temperature to liquid-helium temperature. To get a better idea of what this time is, we will compare these values with the characteristic electronic times in atoms. The period of orbital revolution of an electron in an atom, $\bar{\tau}_{\mathrm{at}}$, is known to be $\sim 10^{-15}$ s. Hence the times $\bar{\tau}$ are of an altogether different order of magnitude. At normal temperatures, $\bar{\tau}$ is a factor of 100 higher than $\bar{\tau}_{\mathrm{at}}$, whereas at low temperatures $\bar{\tau}$ approaches the mean lifetime of excited atomic states (10^{-8} s). This difference of $\bar{\tau}$ from $\bar{\tau}_{\mathrm{at}}$ results from the fact that a macroscopic current is flowing in the metal. The current is absent in a metallic vapor, where the accelerating effect on an atomic electron lasts no longer than 10^{-15} s, which is due to the localization of the atomic electron ($\sim 10^{-8}$ cm).

Table 3.1. Mean free time $\bar{\tau}$ and mean free path $\bar{l}$. The quantity $\bar{\tau}$ has been determined according to (3.3.2) from measurements of the specific electrical conductivity of gold, and $\bar{l}$ comes from (3.3.3)

T [K]	273.2	90.1	20.4	11.1	4.2
$\bar{\tau}$ [s]	$5.9 \cdot 10^{-11}$	$2.2 \cdot 10^{-13}$	$9.8 \cdot 10^{-12}$	$1.4 \cdot 10^{-10}$	$6 \cdot 10^{-9}$
$\bar{l}$ [cm]	$5.9 \cdot 10^{-6}$	$2.2 \cdot 10^{-5}$	$9.8 \cdot 10^{-4}$	$1.4 \cdot 10^{-2}$	$6 \cdot 10^{-1}$

The inequality $\bar{\tau} \gg \bar{\tau}_{\mathrm{at}}$ indicates once again that conduction electrons are delocalized. We specify the freedom of their motion by the mean free path $\bar{l}$, that is, the distance which they cover during the time $\bar{\tau}$. This is certainly true in a quasi-classical approximation. Thus, $\bar{l} = \bar{\tau}\bar{v}$, $\bar{v}$ being the mean velocity of an electron. It seems at first sight that for this velocity we should take the values obtained from the formulas of classical Maxwell–Boltzmann statistics. However (see below), conduction electrons obey Fermi–Dirac quantum statistics; their mean velocity $\bar{v}$ depends weakly on T and is close to the velocity of an electron in an atom. The latter velocity way, according to Bohr's atomic theory, be defined as the orbital velocity. If the period of revolution is $\bar{\tau}_{\mathrm{at}} \sim 10^{-15}$ s and the length of the orbit is $2\pi r_{\mathrm{at}} \sim 10^{-7}$ cm, the result for $\bar{v}$ will be $\sim 10^{8}\,\mathrm{cm\,s}^{-1}$. In keeping with the classical kinetic theory, to this mean velocity there corresponds the temperature $T = m\bar{v}^2/3k_{\mathrm{B}} \approx 9 \cdot 10^{4}$ K, i.e., about a hundred thousand degrees. Using the formula

$$\bar{l} = \bar{\tau} v_{\mathrm{at}} , \qquad (3.3.3)$$

we obtain the values of l in the second row of Table 3.1. As may be deduced, at normal temperatures $\bar{l} \sim 10^{-6}$ cm; i.e., it is hundreds of times larger than the lattice parameter of the metal. Again, this indicates that conduction electrons are delocalized.

Following the Drude model, we now pass on to a derivation of the Joule–Lenz law according to which the power Q, released in unit volume of a conductor per unit time, when a current passes through it by the action of a field E, is

$$Q = \sigma E^2 \; , \tag{3.3.4}$$

where σ is the same quantity as that involved in Ohm's law (3.3.1). As already stated, the extra energy $\Delta\mathcal{E}_E$ acquired by an electron in a field E during a time $\bar{\tau}$ is supplied to the lattice. Aligning the field E with the x axis, we obtain

$$\Delta\mathcal{E}_E = \sum_i \frac{m}{2}\left[\left(v_{0x}^{(i)} + \frac{eE}{m}\bar{\tau}\right)^2 - v_{0x}^{(i)2}\right] ,$$

where $v_0^{(i)} = (v_{0x}^{(i)}, v_{0y}^{(i)}, v_{0z}^{(i)})$ is the equilibrium velocity of an i electron for $E=0$; the summation is taken over all electrons in unit volume. After elementary transformations, we find

$$\Delta\mathcal{E}_E = eE\bar{\tau} \sum_i v_{0x}^{(i)} + ne^2\bar{\tau}^2 E^2/2m \; ,$$

and, since $\sum_i v_{0x}^{(i)} = 0$ because the thermal motion is random, the result for the power per unit time will be

$$Q = \Delta\mathcal{E}_E/\bar{\tau} = ne^2\bar{\tau}E^2/2m = \sigma E^2 \; . \tag{3.3.5}$$

The electrical conductivity σ in the Joule–Lenz law (3.3.4) derived here tallies exactly with the value (3.3.2) obtained earlier. Thus, the Drude model is consistent with the derivation of Ohm's law (3.3.1) and the assumption of complete transfer of energy accumulated by electrons in a field to the metallic lattice. The Drude model achieved complete success in the derivation of the Wiedemann–Franz law (1853), according to which the ratio of thermal conductivity to electrical conductivity is a universal linear function of temperature for many metals:

$$\kappa/\sigma = LT \; , \tag{3.3.6}$$

with L being the Lorentz number. Here we quote some numerical values of the Lorentz number:

Element	Li	As	Pt	Zn	Au	Cu	Mo
L [10^8 WΩ/K^2]	2.21	2.31	2.51	2.31	2.35	2.23	2.61

To derive (3.3.6), we need to calculate the coefficient κ involved in the equation of heat transfer W in the presence of a temperature gradient ∇T. We restrict our attention to the case when T depends on one coordinate x. Then

$$W=\kappa\partial T/\partial x \ . \tag{3.3.7}$$

If we assume that heat carriers in metals are chiefly conduction electrons, we need to calculate the number of particles passing through unit cross section of the conductor S_A with the coordinate x (Fig. 3.3) per unit time. This is equal to $n|v_x|/2$, where $|v_x|$ is the modulus of the mean velocity of an electron along the x axis. The cross sections spaced out at equal distances $\pm l$ along the x axis from the cross section S_A will be labeled S_B and S_C, respectively. On these cross sections the electrons have suffered, on the average, the last collision with the lattice, before reaching the cross section S_A. Again, a rigorous statistical calculation is replaced by the use of mean-free-path values. From kinetic theory it is known that $l=2\bar{\tau}\bar{v}/3$. Thus, the total energy transferred by electrons through S_A (from left to right) is equal to

$$W=\tfrac{1}{2}n|v_x|[(mv^2/2)_{x-l}-(mv^2/2)_{x+l}] \ . \tag{3.3.8}$$

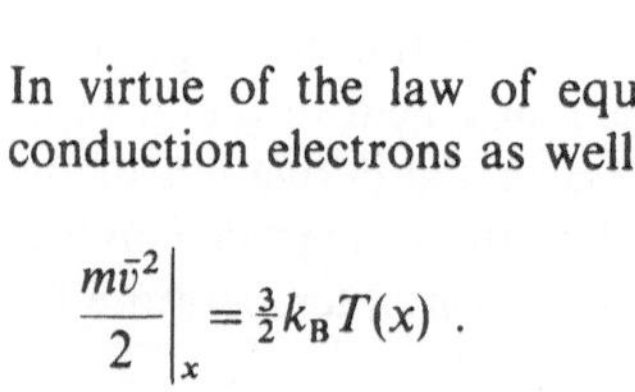

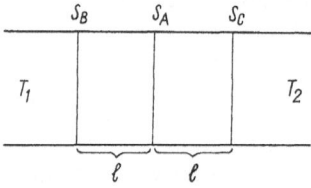

Fig. 3.3. Thermal conductivity calculation according to Drude's theory

In virtue of the law of equipartition of energy, which we believe holds for conduction electrons as well, we may set

$$\left.\frac{m\bar{v}^2}{2}\right|_x=\tfrac{3}{2}k_B T(x) \ . \tag{3.3.9}$$

Using (3.3.8, 9) yields

$$W\approx\tfrac{1}{2}nk_B\bar{v}^2\bar{\tau}\frac{\partial T}{\partial x} \tag{3.3.10}$$

to within small quantities on the order of l. Here we have obtained a theoretical derivation of (3.3.7) and the expression for the specific thermal conductivity

$$\kappa = nk_B\bar{v}^2\bar{\tau}/2 = 3k_B^2 n\bar{\tau}T/2m \ , \tag{3.3.11}$$

where, again, $\bar{v}^2$ is taken to be equal to $3k_BT/m$ according to (3.3.9). From (3.3.11) we see that κ, just as σ in (3.3.2), depends on the quantities m, n, and $\bar{\tau}$ in one and the same combination. In the κ/σ ratio these quantities therefore cancel out and

$$\kappa/\sigma = 3(k_B/e)^2 T \approx 2.24 \cdot 10^{-8}\ \mathrm{W}\Omega\mathrm{K}^{-2}\mathrm{T} \ . \tag{3.3.12}$$

Thus, Drude's theory provided a substantiation for the experimental Wiedemann–Franz relation not only from a qualitative but also from a quantitative standpoint: the values of the Lorentz numbers obtained from (3.3.6, 12) turned out to be very close to each other. The derivation (3.3.12) was the triumph of Drude's theory. However, more thorough experiments on a large number of metals and over a wider T range have shown (3.3.6) to be approximate in character and to fail in a number of cases. This indicates that the theory is approximate.

Let us scrutinize the major difficulties encountered in Drude's theory. To start with, we dwell upon the problem concerning the $\sigma(T)$ relation. According to (3.3.3), the σ involved in (3.3.2) will take the form

$$\sigma = (e^2/2m)(n\bar{l}/\bar{v}) \ . \tag{3.3.13}$$

From the kinetic theory of gases we know that $\bar{v} \sim T^{1/2}$. On the other hand, experiment yielded the result that at room temperature $\sigma \sim T^{-1}$. One therefore had to assume that in (3.3.13) $n\bar{l} \sim T^{-1/2}$ (since the factor $e^2/2m$ is independent of T). Such a dependence is difficult to understand in terms of classical theory. In the kinetic theory of gases, $\bar{l}$ is defined as the inverse of the cross section of scattering from N obstacles with which a moving particle meets in unit volume, i.e., $\bar{l} \sim (\pi NR^2)^{-1}$ (R is the effective scattering cross-section radius). Thus, if we neglect the temperature dependence of R, then $\bar{l}$ does not depend on T at all. Insofar as the electron density n goes, it is hard to imagine, as has been noted above, that it varies with T and, what is more, decreases. Still more difficulties were encountered in attempts to account for the $\sigma(T)$ relation at low temperatures (where $\sigma \sim T^5$). Similar difficulties arise in attempts to explain the $\kappa(T)$ dependence.

Also incomprehensible were large values of $\bar{\tau}$ and $\bar{l}$ compared with the atomic times and the lattice parameter, respectively. It appeared more natural to expect the value of $\bar{l}$ in the crystal to be on the order of the lattice constant $d \sim 10^{-8}$ cm.

The theory received its greatest blow in connection with the so-called heat capacity catastrophe. According to classical statistics, conduction electrons should participate in the thermal motion on a par with ions, immediately affecting the magnitude of the heat capacity. Since the energy per degree of freedom is $k_B T/2$, the mean energy of a free conduction electron is equal to $3k_B T/2$. Taking the density of the electron gas to be equal to the density of lattice sites, we obtain for the heat capacity of the metal

$$C_{met} = C_{lat} + C_{el} = 3R + 3R/2 \approx 9 \cdot 10^3 \text{ cal/mol K} , \tag{3.3.14}$$

i.e., a value which is 50% larger than that given by the Dulong and Petit law (2.2.2). The measurement accuracy here assures a maximum error of not over 1%. Therefore, if we assume that the contribution of conduction electrons to the heat capacity is very small because they are small in number, it is necessary to take the conduction-electron density to constitute not more than 1% of the density of lattice sites. But then it would be necessary to increase correspondingly the values of $\bar{\tau}$ and $\bar{l}$ in (3.3.2, 11) which, even without this increase, turns out to be anomalously large from the viewpoint of classical concepts.

Historically, the heat capacity catastrophe was a test for the entire electronic theory of metals. Significantly, this disaster arose not from any particular model simplifications but from the fundamental postulates of classical statistics. Lorentz and others tried to "mend" Drude's model by a more accurate statistical derivation of the formulas quoted. But that did not help. What was needed was not the mending of individual "holes" of the theory but a radical change in its fundamentals.

3.4 Itinerant Electron Theory According to Frenkel

The first paper in which a step was made to rehabilitate Drude's theory is due to *Frenkel* [3.3]. He proceeded from Bohr's atomic model. A metallic vapor in the normal state is a totality of neutral atoms that suffer relatively rare collisions; there are no conduction electrons in a vapor. However, the outer (valence) electrons in metallic atoms are bound to the ion cores more loosely than in the atoms of dielectrics. According to Bohr, the orbit of the valence electron of a metallic atom has the shape of an extended ellipse. As the atoms come closer together in the course of condensation, the mean distances between adjacent atoms become on the order of or even less than the diameter of the "orbits" of the valence electrons. Therefore they do not remain firmly bound to a particular core. A large probability of a transition from one core to an adjacent one, etc. arises.

Thus, when the vapor condenses, the valence electrons become free to move over the entire metal. *Frenkel* identified these itinerant electrons with conduction electrons. He assumed that each of them, freeing itself from its individual core, would be bound more tightly to the entire lattice (here he was the first to approach correctly the explanation of the nature of the metallic bond). Then the problem concerning the velocity of itinerant electrons was solved. In isolated atoms the velocity of valance electrons in an orbit is as high as 10^8 cm/s. When electrons become itinerant, this velocity does not decrease but even increases by approximately 10–15%. According to Frenkel, this follows from the virial theorem of classical mechanics, in keeping with which the following relation holds for a system of particles with Coulomb bonding forces [3.4]:

$$2\mathscr{E}_{\mathrm{kin}} = -\mathscr{E}_{\mathrm{pot}} , \tag{3.4.1}$$

with $\mathscr{E}_{\mathrm{kin}}$ and $\mathscr{E}_{\mathrm{pot}}$ being the mean values of the kinetic and the potential energy of the system respectively. The total energy is

$$\mathscr{E}_{\mathrm{tot}} = \mathscr{E}_{\mathrm{kin}} + \mathscr{E}_{\mathrm{pot}} = -\mathscr{E}_{\mathrm{kin}} . \tag{3.4.2}$$

When the vapor condenses, the total negative bonding energy increases in absolute magnitude by the latent heat of sublimation $|\eta|$:

$$|\mathscr{E}_{\mathrm{tot}}|_{\mathrm{crystal}} = |\mathscr{E}_{\mathrm{tot}}|_{\mathrm{vapor}} + |\eta| . \tag{3.4.3}$$

According to (3.4.3), the quantity $|\mathscr{E}_{\mathrm{kin}}|_{\mathrm{crystal}}$ increases by the same amount. The order of magnitude of this increase per atom is estimated in the following fashion: The value of $|\eta|$ is about 20–40 kcal/mol K; per atom we thus have $|\eta|/N_A \sim 1$ eV. The kinetic energy of a valence electron in an atom is equal to 7 eV. Hence the velocity of itinerant electrons is $\sim 15\%$ higher than that of a valence electron of an atom; i.e., it also is $\sim 10^8$ cm/s, as was already adopted in the calculation of $\bar{l}$ in Table 3.1.

The most important result of Frenkel's theory is the elimination of the catastrophe in the heat capacity. Already when created and at 0 K, itinerant electrons turn out to be effectively heated to 10^4–10^5 K, and therefore heating the metal up to ~ 300 K alters their energy by not more than a few percent. These electrons do not participate actively in the thermal motion within the metal, nor do they substantially affect its heat capacity. With $T \gg \vartheta_D$, metals, just as dielectrics, obey the Dulong and Petit law, and when $T \ll \vartheta_D$ they obey the Debye law, except for the lowest temperatures (see below).

Proceeding from the picture of itinerant electrons, Frenkel derived all the formulas of Drude's theory. Because of interactions, the motion of itinerant electrons is still constrained; that is, the site vacated by one of them is replaced by another so that their density remains constant. However, Frenkel did not consider the many-electron problem but replaced it by a one-electron approximation, considering the mechanical trajectory of an electron along a giant molecule, viz., considering the crystal as a quantized orbit. This treatment holds good only at zero temperature. Lattice perturbations that arise, for example, from thermal oscillations at $T > 0$ K "break" the orbit of an itinerant electron up into segments which give, on the average, the free path $\bar{l}$. Perturbations of the ideality of the crystal are responsible for the finiteness of σ and κ. Frenkel succeeded in obtaining a correct $\sigma(T)$ dependence at high T and a correct estimate of the quantity σ ($10^5\,\Omega^{-1}\,\mathrm{cm}^{-1}$). By contrast with Drude's theory, the effect of the field boils down chiefly to the variation of the probability of elementary displacements along different directions rather than the occurrence of an additional velocity on the mean free path.

Frenkel's basic idea that the velocities of electrons in metals are on the order of intratomic velocities is still valid. A more accurate and complete quantitative description of the properties of metals is given in the Fermi–Dirac quantum gas model.

3.5 Application of Fermi–Dirac Quantum Statistics to the Conduction-Electron Gas

We consider a system of N conduction electrons as a free gas. In this system the electrons move within the metal, i.e., a potential box (Fig. 3.4) with a rectangular potential barrier (the exact shape of the barrier is unimportant to us at the moment). The motion of an electron is then defined by the Schrödinger equation

$$\Delta\psi(\boldsymbol{r})+\frac{2m}{\hbar^2}(\varepsilon+U)\psi(\boldsymbol{r})=0 \tag{3.5.1}$$

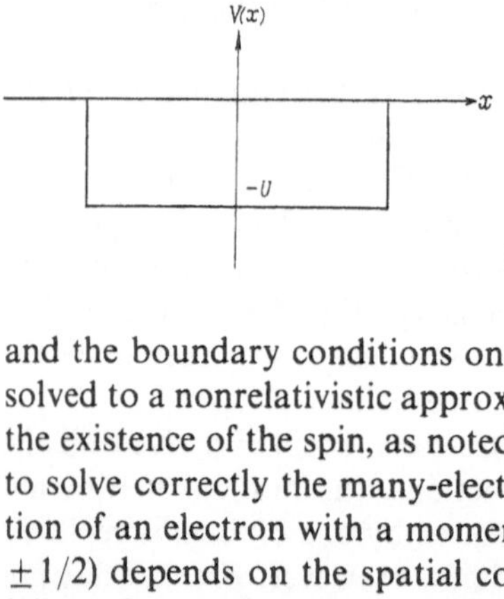

Fig. 3.4. Rectangular potential box for electrons in metal

and the boundary conditions on the surface of the metal. The problem will be solved to a nonrelativistic approximation. Therefore (3.5.1) involves no spin, but the existence of the spin, as noted above, should be taken into account in order to solve correctly the many-electron problem (discussed later). The wave function of an electron with a momentum $\boldsymbol{p}$ and spin quantum number σ (equal to $\pm 1/2$) depends on the spatial coordinates $\boldsymbol{r}$ (x, y, z) and spin projection s. As follows from (3.5.1) (for $U=\mathrm{const}$), the dependence of the wave function on the coordinates has the form of a plane wave $\exp(\mathrm{i}\boldsymbol{k}\cdot\boldsymbol{r})$, where $\boldsymbol{k}$ is a wave vector which is related to $\boldsymbol{p}$ by the equation

$$\hbar\boldsymbol{k}=\boldsymbol{p}\ . \tag{3.5.2}$$

The coordinate function is multiplied by the spin function $\varphi_\sigma(s)$ (its form is immaterial here), and the total wave function has the form

$$\psi_{k\sigma}(\boldsymbol{r}, s)=\alpha \exp(\mathrm{i}\boldsymbol{k}\cdot\boldsymbol{r})\varphi_\sigma(s) \ , \tag{3.5.3}$$

with α being a normalization factor. The energy spectrum in this case has the form of a quadratic dispersion relation

$$\varepsilon=p^2/2m^*=m^*v^2/2 \ ; \tag{3.5.4}$$

the asterisk for the mass indicates that the mass of a conduction electron may differ from that of a free electron (Fermi quasiparticle). Our problem possesses a quasi-continuous energy spectrum. The momentum $\boldsymbol{p}$ varies quasi-continuously since $k_x=\pi\kappa_x/L_x$, $k_y=\pi\kappa_y/L_y$, $k_z=\pi\kappa_z/L_z$, where L_x, L_y, and L_z are the potential box edges, which are equal to the product $N^{1/3}d$, and κ_x, κ_y, $\kappa_z=0, \pm 1, \pm 2, \dots$ [see the analogous definition of the vector $\boldsymbol{q}$ in (2.1.7, 8)]. In a finite crystal the spectrum (3.5.4) is discrete. However, when the dimensions of the solid are large compared to the lattice parameter ($L \gg d$), the distances between adjacent levels are very small and the spectrum is practically continuous. Deviations will be observed for very small particles [3.5], when $L \gtrsim d$.

The problem under consideration also possesses a high degree of degeneracy. To each value of the energy ε which, according to (3.5.4), depends on the modulus of the vector $\boldsymbol{p}$, there corresponds a set of wave functions (3.5.3) that depend also on the direction of the vector $\boldsymbol{p}$. Therefore at the outset we must determine the degree of degeneracy $g(\varepsilon)$. To this end, we calculate the number of states in which the electrons possess the value of the momentum within the limits from p to $p+dp$ for all possible directions of the vector $\boldsymbol{p}$. To these states in $\boldsymbol{p}$ space there corresponds a set of phase points that fill up a spherical layer of volume $4\pi p^2 dp$ (Fig. 3.5), which, from (3.5.4), is equal to

$$2^{5/2}\pi m^{*\,3/2}\varepsilon^{1/2}d\varepsilon \ . \tag{3.5.5}$$

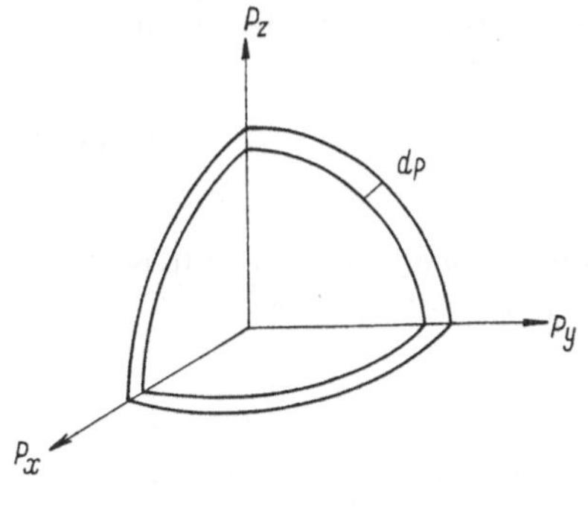

Fig. 3.5. Calculation of the degree of degeneracy of the Fermi gas, $g(\varepsilon)$

To find $g(\varepsilon)$ we have to divide by this unit phase cell volume $\Delta p_x \Delta p_y \Delta p_z$, which is determinable from the Heisenberg uncertainty principle

$$\Delta x \Delta y \Delta z \Delta p_x \Delta p_y \Delta p_z \sim h^3 \; . \tag{3.5.6}$$

The quantity $\Delta x \Delta y \Delta z$ involved in (3.5.6) gives the accuracy of localization of an electron in normal space. In the present case it is equal to the volume of the metal, V, since what counts here is that the electron is in the metal. Then, according to (3.5.6), the momentum determination accuracy, i.e., the size of the unit phase cell in $\boldsymbol{p}$ space, is

$$\Delta p_x \Delta p_y \Delta p_z \approx h^3/V \; . \tag{3.5.7}$$

Such a derivation of the unit volume in phase space does not give the correct numerical coefficients. However, a straightforward calculation of the number of states such as that performed in Chap. 2 will lead precisely to (3.5.8). To allow for the spin degeneracy (in the absence of a magnetic field and spin forces), the quotient of the left-hand side of (3.5.6) by the right-hand side of (3.5.7) is multiplied by 2. This yields

$$g(\varepsilon) = \frac{1}{h^3} 4\pi (2m^*)^{3/2} V \varepsilon^{1/2} \equiv C\varepsilon^{1/2} \; . \tag{3.5.8}$$

To find the distribution function $f(\varepsilon)$ of electrons in states with different ε, the degree of degeneracy (density of states) should be multiplied by a statistical quantity, i.e., the mean with respect to the occupation number of states with different $\varepsilon - \overline{n(\varepsilon)}$.

The mean occupation numbers for noninteracting particles obeying the Pauli exclusion principle are determined by the Fermi–Dirac distribution function. The latter may be derived in a fashion analogous to the treatment of two-level systems in Sect. 2.5.

Let $\{\nu\}$ be a one-particle quantum state, and ε_ν the corresponding energy. The total energy of any quantum state of a system of many noninteracting particles then may be determined by specifying occupied and free states

$$\mathscr{E} = \sum_\nu \varepsilon_\nu n_\nu \; , \tag{3.5.9}$$

where $n_\nu = 0$ for free states and $n_\nu = 1$ for occupied ones. In contrast to Sect. 2.5, we explore here a system with a prescribed number of particles

$$N = \sum_\nu n_\nu \; . \tag{3.5.10}$$

For this reason we should use a large canonical ensemble, introducing a chemical potential ζ. Since we consider the electrons to be noninteracting and each state ν to be filled up independently, we have

$$\bar{n}_\nu = \frac{\sum\limits_{n=0,1} n \exp\left[\left(\frac{\zeta-\varepsilon_\nu}{k_B T}\right)n\right]}{\sum\limits_{n=0,1} \exp\left[\left(\frac{\zeta-\varepsilon_\nu}{k_B T}\right)n\right]} = \left[\exp\left(\frac{\varepsilon_\nu-\zeta}{k_B T}\right)+1\right]^{-1},$$

that is,

$$\overline{n(\varepsilon)} = \left[\exp\left(\frac{\varepsilon-\zeta}{k_B T}\right)+1\right]^{-1}. \tag{3.5.11}$$

Thus the function $f(\varepsilon)$ for the quadratic dispersion relation (3.5.4) has the form

$$f(\varepsilon) = g(\varepsilon)\overline{n(\varepsilon)} = 4\pi(2m^*)^{3/2}\, Vh^{-3}/\{\exp[(\varepsilon-\zeta)/k_B T]+1\} \tag{3.5.12}$$

The chemical potential $\zeta(T)$ is determined from the requirement that the total number of particles remains constant:

$$N = \int\limits_0^\infty f(\varepsilon)d\varepsilon\ . \tag{3.5.13}$$

As distinct from classical statistics, (3.5.13) is an integral equation and, in the general case, is incapable of solution. In what follows we therefore consider limiting cases.

3.5.1 The Case of $T = 0$ K

At $T = 0$ the system is in its lowest-energy (ground) state. In classical statistics all the electrons would be in a $\boldsymbol{p}$-space cell with $\varepsilon = 0$. However, in Fermi–Dirac statistics this is forbidden by the Pauli exclusion principle, and the electrons will fill up most densely only the $\boldsymbol{p}$-space cells around the point $\varepsilon = 0$ so that their total energy $\mathscr{E}$ is minimized. With the quadratic dispersion relation (3.5.4), the volume thus filled with electrons has the shape of a sphere with an isoenergetic spherical surface, called a Fermi surface (in the case of an arbitrary dispersion relation the shape of the surface may differ arbitrarily from a sphere), and with maximum boundary energy $\varepsilon_{0\,\max}$, a Fermi energy ε_F. The quantity ε_F is defined by the electron density

$$n = N/V\ . \tag{3.5.14}$$

Indeed, the volume of the Fermi sphere occupied by N electrons with two in each cell (3.5.7), because of spin degeneracy, is equal to $(4\pi/3)p_{0\,\max}^3$, where $p_{0\,\max}$ is the largest magnitude of the momentum of an electron at $T = 0$ K, i.e., the Fermi momentum p_F. The quotient of this volume by the cell volume (3.5.7) is

$$4\pi V p_F^3/3h^3 = N/2 \ . \tag{3.5.15}$$

From (3.5.14, 15), we obtain

$$\begin{aligned} p_F &= h(3n/8\pi)^{1/3} = \hbar(3\pi^2 n)^{1/3} \ , \\ k_F &= p_F/\hbar = (3\pi^2 n)^{1/3} \ , \\ \varepsilon_F &= p_F^2/2m^* = (h^2/2m^*)(3n/8\pi)^{2/3} = (\hbar^2/2m^*)(3\pi^2 n)^{2/3} \ , \end{aligned} \tag{3.5.16}$$

or

$$n = (8\pi/3)(p_F/h)^3 = (8\pi/3)[(2m^*)^{3/2}/h^3]\varepsilon_F^{3/2} \ , \tag{3.5.17}$$

and for the Fermi velocity

$$v_F = p_F/m^* = (h/m^*)(3n/8\pi)^{1/3} = (\hbar/m^*)(3\pi^2 n)^{1/3} \ . \tag{3.5.18}$$

If we choose the value of n to be $\sim 10^{22}$ cm^{-3}, the result will be $\sim$ 1–10 eV (or 10^{-12}–10^{-13} erg) for the quantity ε_F and 10^8 cm/s for the quantity v_F. Table 3.2 summarizes the values of v_F, p_F, and ε_F calculated for a number of metals. As is seen from the table, the Fermi gas contains particles that possess enormous velocities and energies, even at $T = 0$.

Table 3.2. Numerical values of the velocity v_F, momentum ϱ_F, and energy $\varepsilon_F = \zeta_0$ on the Fermi surfaces of normal univalent metals

Elements	Li	Na	K	Cu	Ag	Au
v_F [cm s$^{-1}\cdot 10^{-8}$]	1.28	1.04	0.84	1.58	1.31	1.39
ϱ_F [g cm s$^{-1}\cdot 10^{19}$]	1.17	0.95	0.77	1.44	1.20	1.27
$\varepsilon_F = \zeta_0$ [erg $\cdot 10^{12}$]	7.6	5.0	3.3	11.37	8.84	8.9
$\varepsilon_F = \zeta_0$ [eV]	4.74	3.16	2.06	7.10	5.52	5.56

Our objective now is to identify the functions $\overline{n(\varepsilon)}$ and $f(\varepsilon)$ for 0 K. As can be deduced from (3.5.11), the function $\overline{n(\varepsilon)}$ will depend substantially on the sign of the difference $\varepsilon - \zeta_0$, with ζ_0 being the value of $\zeta(T)$ for 0 K.

For $\varepsilon < \zeta_0 \lim_{T\to 0} \exp[(\varepsilon - \zeta_0)/k_B T] = 0$ and $\overline{n_0(\varepsilon)} = 1$;

for $\varepsilon > \zeta_0 \lim_{T\to 0} \exp[(\varepsilon - \zeta_0)/k_B T] = \infty$ and $\overline{n_0(\varepsilon)} = 0$.

Thus, when $\varepsilon = \zeta_0$, the exponent $[(\varepsilon - \zeta_0)/k_B T]$ changes sign and $\overline{n_0(\varepsilon)}$ varies discontinuously from 1 to 0 (Fig. 3.6). The discontinuous $\overline{n_0(\varepsilon)}$ curve is called the Fermi step. Hence the chemical potential at 0 K clearly coincides with the Fermi energy

$$\zeta_0 = \varepsilon_F \equiv \varepsilon_{0\,max} \; . \tag{3.5.19}$$

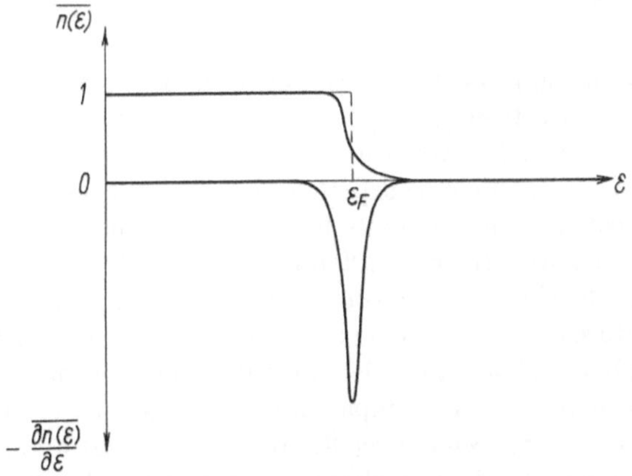

Fig. 3.6. Fermi–Dirac function $\overline{n(\varepsilon)}$ and its derivative $\overline{\partial n(\varepsilon)}/\partial\varepsilon$ (dashed line refers to the case $T = 0$)

The distribution function $f(\varepsilon)$ for $T = 0$, in view of (3.5.12), is equal to (Fig. 3.7)

$$f_0(\varepsilon) = \begin{cases} 0 \; , & \varepsilon > \zeta_0 \\ 4\pi(2m^*)^{3/2} V h^{-3} \varepsilon^{1/2} \; , & \varepsilon < \zeta_0 \; . \end{cases} \tag{3.5.20}$$

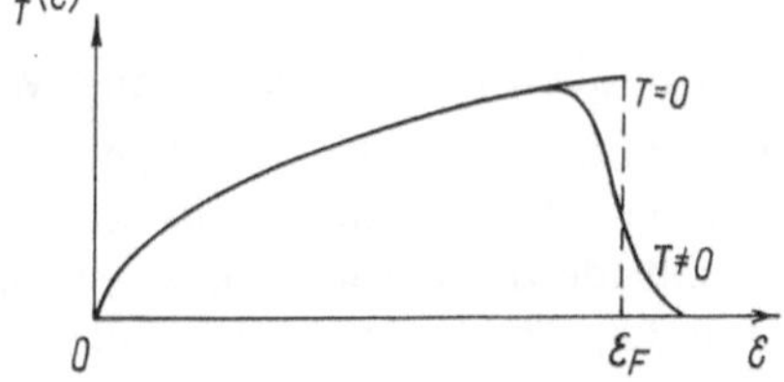

Fig. 3.7. Energy distribution function $f(\varepsilon)$ of electrons in a metal (dashed line refers to the case $T = 0$)

The mean energy of an electron at 0 K is

$$\bar{\varepsilon}_0 = \int_0^{+\infty} d\varepsilon\varepsilon f_0(\varepsilon) \Big/ \int_0^{+\infty} d\varepsilon f_0(\varepsilon) = 3\zeta_0/5 \equiv 3\varepsilon_F/5 \; . \tag{3.5.21}$$

According to classical theory, this quantity is equal to $\bar{\varepsilon}_{cl} = 3k_B T/2$ and for 0 K $\bar{\varepsilon}_{cl} = 0$. In the classical theory, (3.5.21) could be obtained only at very high T, viz., at the effective gas degeneracy temperature ϑ_{el} at 0 K, which is determined from the formula

$$\varepsilon_0 = 3\varepsilon_F/5 = 3k_B\vartheta_{el}/2 \ . \tag{3.5.22}$$

Taking the values of ζ_0 (Table 3.2) to be 10^{-11}–10^{-12} erg, we obtain

$$\vartheta_{el} = 2\zeta_0/5k_B \approx 10^4 \div 10^5\,\mathrm{K} \ . \tag{3.5.23}$$

Thus, due to the Pauli exclusion principle, the energy of the ground state is high, and therefore the quantum electron gas appears to be heated to very high degeneracy temperatures ϑ_{el}. For real temperatures $T \ll \vartheta_{el}$ the statistical behavior of conduction electrons should then be expected to deviate substantially from the predictions of the classical theory. Only for $T \gg \vartheta_{el}$ should they be expected to exhibit classical behavior. However, it follows from (3.5.37–42) that, if the density n is close to the density of lattice sites ($\sim 10^{22}\,\mathrm{cm}^{-3}$), ϑ_{el} exceeds not only the melting temperature of the metal but also the temperature at which it vaporizes. In all real metals the conduction electron gas is therefore highly degenerate. Only with small densities (for example, in semiconductors, where $n \ll 10^{22}\,\mathrm{cm}^{-3}$) is the value of ϑ_{el} appreciably smaller than 10^4 K, and we are sometimes in a position to describe conduction electrons using classical theory.

3.5.2 The Low-Temperature Case ($T > 0$ K, but $T \ll \vartheta_{el}$)

In the low-temperature case it is useful to introduce the small dimensionless parameter

$$T/\vartheta_{el} \sim k_B T/\zeta_0 \ll 1 \ , \tag{3.5.24}$$

the power series expansion of which is used in quantitative calculations. We begin by qualitatively ascertaining how the form of the functions $\overline{n(\varepsilon)}$ and $f(\varepsilon)$ will change in comparison with the 0 K case (Figs. 3.6, 7). With increasing T the electrons begin to occupy the $\boldsymbol{p}$-space cells outside the Fermi surface, vacant electron sites (holes) remaining beneath it (Fig. 3.8). The Fermi surface will start

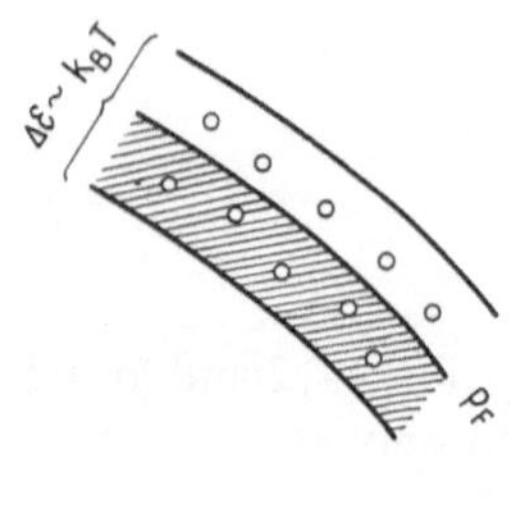

Fig. 3.8. Thermal smearing of the Fermi surface (the clear area shows electrons, the shaded area shows holes)

to "diffuse" (will be smeared). The width of the smeared layer (Fig. 3.8) will be related to the Fermi sphere radius in the same way that the temperature T responsible for the smearing is related to the quantity ϑ_{el}. Thus, when $T > 0\,\mathrm{K}$, the $\overline{n(\varepsilon)}$ and $f(\varepsilon)$ curves (Figs. 3.6, 7) will exhibit very sharp but continuous fall-offs, called "Maxwell tails", in the region of the discontinuity at $\varepsilon = \zeta_0$. For $T > 0\,\mathrm{K}$ it follows from a simple analysis of (3.5.11, 12) that with $\varepsilon \to \infty$ the $\overline{n(\varepsilon)}$ and $f(\varepsilon)$ always tend to zero, and when $\varepsilon = 0$ the quantity $\overline{n(\varepsilon)} \approx 1$ and $f(\varepsilon) \to 0$ (if $\zeta \gg k_B T$). For $\varepsilon = \zeta$ we have $\exp[(\varepsilon - \zeta)/k_B T] = 1$ and, consequently, at the inflection point of the Maxwell tail $\overline{n(\zeta)} = 1/2$ and $f(\zeta) = g(\zeta)/2$. The quantity $\zeta(T)$ turns out to be somewhat smaller than ζ_0, as will be shown later (3.5.51).

As T approaches ϑ_{el} the smearing should increase, and in the limit $T \gtrsim \vartheta_{el}$ the $\overline{n(\varepsilon)}$ and $f(\varepsilon)$ curves turn into classical statistics functions. We confine ourselves to the strong degeneracy case.

To solve particular problems, we follow Frenkel and introduce the density of thermally excited electrons present in the smeared band near the Fermi surface. Since the width of this band is related to the radius of the Fermi sphere as T/ϑ_{el},

$$n_T \approx nT/\vartheta_{el} \; . \tag{3.5.25}$$

The distribution function of excited electrons has the form of a Maxwell tail and therefore we can apply classical theory to it. We illustrate this with the example of the heat capacity of a degenerate electron gas. To do this, we insert into the Dulong and Petit formula $C_v^{(cl)} = 3nk_B/2$, using the density n according to (3.5.25) in place of the density n_T. This yields

$$C_v^{(quan)} = 3n_T k_B = 3nk_B T/2\vartheta_{el} \; . \tag{3.5.26}$$

As seen from the above equation, in real metals at room temperature (where the Dulong and Petit law holds) the ratio $T/\vartheta_{el} \sim 0.01$ to 0.001 and therefore the contribution of electrons to the heat capacity constitutes 1.0 to 0.1% of the ionic contribution. In this fashion, the disaster with the heat capacity for a degenerate Fermi gas is completely resolved. Only electrons of density n_T contribute to the heat capacity. Electrons with $\varepsilon \lesssim \zeta - k_B T$ cannot accept energy from the heater, since they would enter a state already occupied, a situation which is forbidden by the Pauli exclusion principle.

Returning to the more rigorous calculations, we consider the expression for determining the chemical potential $\zeta(T)$ and the energy $\mathscr{E}(T)$ of the electron gas, and specify how the calculation can be carried out under the condition (3.5.24). The quantity $\zeta(T)$ is determined from (3.5.13):

$$N = C \int_0^{+\infty} d\varepsilon \overline{n(\varepsilon)}\, \varepsilon^{1/2} \; . \tag{3.5.27}$$

Integration by parts yields

$$N = -(2C/3)\int_0^{+\infty} d\varepsilon\varepsilon^{3/2}\partial\overline{n(\varepsilon)}/\partial\varepsilon \ . \tag{3.5.28}$$

For the total energy of the gas we also have

$$\mathscr{E} = C\int_0^{+\infty} d\varepsilon\varepsilon^{3/2}\overline{n(\varepsilon)} = -(2C/5)\int_0^{+\infty} d\varepsilon\varepsilon^{5/2}\partial\overline{n(\varepsilon)}/d\varepsilon \ . \tag{3.5.29}$$

Thus we need to calculate a typical integral

$$I = C\int_0^{+\infty} d\varepsilon\alpha(\varepsilon)\partial\overline{n(\varepsilon)}/\partial\varepsilon \ , \tag{3.5.30}$$

where $\alpha(\varepsilon)$ is a continuous function ε. For convenience, we introduce

$$x = (\varepsilon - \zeta)/k_B T \ . \tag{3.5.31}$$

Then

$$\alpha(\varepsilon) \rightarrow \tilde{\alpha}(x) \ ,$$

and (3.5.30) becomes

$$I = C\int_{-\zeta/k_B T}^{+\infty} dx\tilde{\alpha}(x)\partial n(x)/dx \ . \tag{3.5.32}$$

In the case of strong degeneracy, when (3.5.24) holds, the falloff of the $\overline{n(x)}$ curve in the vicinity of $x = 0$ is very abrupt, and the derivative $\partial n(x)/\partial x$ is close to a δ function with respect to x (Fig. 3.6) and is exactly equal to it at 0 K. This allows a very simple method of approximate calculation (3.5.30).

Using (3.5.24), the lower limit in (3.5.32) may be replaced by $-\infty$:

$$I \approx C\int_{-\infty}^{\infty} dx\tilde{\alpha}(x)\partial\overline{n(x)}/\partial x \ . \tag{3.5.33}$$

Assuming that the function $\tilde{\alpha}(x)$ near $x = 0$ may be expanded in a power series of

$$\tilde{\alpha}(x) = \tilde{\alpha}(0) + \left(\frac{\partial\tilde{\alpha}}{\partial x}\right)_0 x + \frac{1}{2}\left(\frac{\partial^2\tilde{\alpha}}{\partial x^2}\right)_0 x^2 + \dots \tag{3.5.34}$$

and substituting (3.5.34) into (3.5.33) yields

$$I \approx C \int_{-\infty}^{\infty} dx\, \partial\overline{n(x)}/\partial x \left[\tilde{\alpha}(0) + \left(\frac{\partial \tilde{\alpha}}{\partial x}\right)_0 x + \right.$$

$$\left. + \frac{1}{2}\left(\frac{\partial^2 \tilde{\alpha}}{\partial x^2}\right) x^2 + \dots \right] = I_0 + I_1 + I_2 + \dots \, . \tag{3.5.35}$$

The first term I_0 in (3.5.35) can be calculated if the lower limit of the integral is exact. This term is equal to

$$I_0 = C\tilde{\alpha}(0) \int_{-\zeta/k_B T}^{+\infty} dx\, \partial\overline{n(x)}/\partial x = -C\tilde{\alpha}(0) \, . \tag{3.5.36}$$

The second term in (3.5.35) is equal to zero because of the fúnction $\partial\overline{n(x)}/\partial x$ is even. Indeed, by virtue of (3.5.11, 31),

$$\partial\overline{n(x)}/\partial x = -e^x/(e^x+1)^2 = -e^{-x}/(e^{-x}+1)^2 =$$

$$= \partial\overline{n(-x)}/\partial(-x) \, . \tag{3.5.37}$$

The third term in (3.5.35) has the form

$$I_2 = -C(\partial^2\tilde{\alpha}/\partial x^2)_0 \int_0^{+\infty} dx x^2 e^{-x}/(e^{-x}+1)^2 \, . \tag{3.5.38}$$

The integral involved in (3.5.38) has already been calculated in Sect. 2.5 (2.5.6). Thus we find

$$I_2 = -(\pi^2/6)C(\partial^2\tilde{\alpha}/\partial x^2)_0 \, . \tag{3.5.39}$$

Adding (3.5.36) to (3.5.39) and proceeding from $\tilde{\alpha}(x)$ to $\alpha(\varepsilon)$, we have

$$I \approx -C\left(\alpha(\zeta) + \frac{\pi^2}{6} \left.\frac{\partial^2 \alpha(\varepsilon)}{\partial \varepsilon^2}\right|_{\varepsilon=\zeta} (k_B T)^2\right) \, . \tag{3.5.40}$$

The rapidity of convergence of (3.5.40) is determined by the closeness of the derivative $-\partial\overline{n(\varepsilon)}/\partial\varepsilon$ to the δ function. The zero expansion term (3.5.35) is obtained if the quantity $-\partial\overline{n(\varepsilon)}/\partial\varepsilon$ is set equal to $\delta(\varepsilon-\zeta)$ and the following terms take into account the correction for the finite "width" of the function $-\partial\overline{n(\varepsilon)}/\partial\varepsilon$. Subject to the strong degeneracy condition (3.5.24), all statistical parameters of the Fermi gas may be calculated to zero approximation if the distribution function $\overline{n(\varepsilon)}$ is replaced by a Fermi step and $-\partial\overline{n(\varepsilon)}/\partial\varepsilon$ is assumed to be a δ function.

Using (3.5.40), we calculate $\zeta(T)$ and $\mathscr{E}(T)$ according to (3.5.18, 29). In these cases we have $\alpha(\varepsilon)=\varepsilon^{3/2}$ and $\alpha(\varepsilon)=\varepsilon^{5/2}$, respectively. Substituting these expressions into (3.5.40), we find

$$N \approx \tfrac{2}{3}C\zeta^{3/2}[1+(\pi^2/8)(k_B T/\zeta)^2] \ , \tag{3.5.41}$$

$$\mathscr{E} \approx \tfrac{2}{5}C\zeta^{5/2}[1+(5\pi^2/8)(k_B T/\zeta)^2] \ . \tag{3.5.42}$$

Since $k_B T \ll \zeta$, the second terms in (3.5.41, 42) are small, which justifies the approximation (3.5.35). Comparing (3.5.41) with (3.5.42), we see that to within the terms $\sim (k_B T/\zeta)^4$

$$\mathscr{E} \approx \tfrac{3}{5}N\zeta[1+(\pi^2/2)(k_B T/\zeta)^2] \ . \tag{3.5.43}$$

To calculate $\zeta(T)$, we use (3.5.19, 16) which, in combination with (3.5.41), yields

$$N = \tfrac{2}{3}C\zeta_0^{3/2} \approx \tfrac{2}{3}C\zeta^{3/2}[1+(\pi^2/8)(k_B T/\zeta)^2] \tag{3.5.44}$$

or

$$\zeta = \zeta_0[1-(\pi^2/12)(k_B T/\zeta_0)^2] \ . \tag{3.5.45}$$

It is seen from (3.5.45) that in the case of strong degeneracy the point at which the function $\overline{n(\varepsilon)}$ (with $\overline{n(\varepsilon)}=1/2$) "falls off" most abruptly with increasing T shifts weakly towards the coordinate origin from the Fermi energy ζ_0, i.e., $\zeta(T)$ differs little from ζ_0, as has been assumed above. From (3.5.45, 43), the mean energy of a conduction electron for $T>0\,\mathrm{K}$ (but for $k_B T \ll \zeta_0$) is

$$\begin{aligned}\bar{\varepsilon} &= \tfrac{3}{5}\zeta_0[1+(5\pi^2/12)(k_B T/\zeta_0)^2]= \\ &= \bar{\varepsilon}_0[1+(5\pi^2/12)(k_B T/\zeta_0)^2] = \varepsilon_0+\gamma' T^2/2 \ ,\end{aligned} \tag{3.5.46}$$

where

$$\gamma' = \pi^2 k_B^2/2\zeta_0 \ . \tag{3.5.47}$$

The results thus obtained may be generalized for the case of an arbitrary dispersion relation $\varepsilon(\boldsymbol{p})$. The function $g(\varepsilon)$ will then take an arbitrary form; in place of (3.5.27–29) we will have

$$N = \int_0^{+\infty} g(\varepsilon)\overline{n(\varepsilon)}d\varepsilon = -\int_0^{+\infty}\left[\int_0^{\varepsilon} g(x)dx\right]\left(\frac{\partial\overline{n(\varepsilon)}}{\partial\varepsilon}\right)d\varepsilon \ , \tag{3.5.48}$$

$$\mathscr{E} = \int_0^{+\infty} \varepsilon g(\varepsilon)\overline{n(\varepsilon)}\, d\varepsilon = -\int_0^{+\infty} \left[\int_0^{\varepsilon} g(x) x dx\right]\left(\frac{\partial \overline{n(\varepsilon)}}{\partial \varepsilon}\right) d\varepsilon \ . \tag{3.5.49}$$

In view of (3.5.32, 40), we obtain approximately, instead of (3.5.41, 42),

$$N = \int_0^{\zeta} d\varepsilon g(\varepsilon) + (\pi^2/6)(k_B T)^2 g'(\varepsilon) \ , \tag{3.5.50}$$

$$\mathscr{E} = \int_0^{\zeta} d\varepsilon \varepsilon g(\varepsilon) + (\pi^2/6)(k_B T)^2 [\zeta g'(\zeta) + g(\zeta)] \ , \tag{3.5.51}$$

with $g'(\zeta)$ being a derivative $\partial g(\varepsilon)/\partial \varepsilon$ for $\varepsilon = \zeta$.

Expressing ζ in the form $\zeta = \zeta_0 + \delta\zeta (|\delta\zeta| \ll \zeta_0)$ and expanding (3.5.50, 51) in $\delta\zeta$, we find, similar to (3.5.45, 46),

$$\zeta = \zeta_0 - (\pi^2/6)(k_B T)^2 g'(\zeta_0)/g(\zeta_0) \ , \tag{3.5.52}$$

$$\mathscr{E} = \mathscr{E}_{T=0} + \gamma \frac{T^2}{2} \ , \tag{3.5.53}$$

where

$$\gamma = \pi^2 k_B^2 g(\zeta_0)/3 \ . \tag{3.5.54}$$

The electronic heat capacity at constant volume is

$$C_v = \partial \mathscr{E}/\partial T = \gamma T \ , \tag{3.5.55}$$

which is a more exact result than (3.5.26). Thus low-temperature measurements of the heat capacity of metals permit a straightforward determination of the density of states at the Fermi level, $g(\zeta_0)$, which is a very important characteristic.

With $T = 0$, the energy for the free-electron mode is equal to (3.5.43, 11)

$$\mathscr{E} = 3N\zeta_0/5 = (3h^2 N/10m^*)(3N/8\pi V)^{2/3} \ . \tag{3.5.56}$$

The pressure of the Fermi gas at $T = 0$ is defined by the formula

$$p_0 = -\partial \mathscr{E}/\partial V = 2\mathscr{E}/3V = \tfrac{2}{5} n\zeta_0 \ . \tag{3.5.57}$$

A numerical estimate with $n \sim 10^{22}\,\mathrm{cm}^{-3}$ and $\zeta_0 \sim 10^{-11}\,\mathrm{erg}$ yields $p_0 \sim 10^{11}\,\mathrm{dyn/cm^2} \sim 10^5\,\mathrm{atm}$. This enormous pressure arises from the high electron density. It is easy to understand why the electrons do not disperse through the crystal surface, the "strength" of which is, as first pointed out by

Frenkel, of an electronic nature. A negatively charged electron, as it leaves the metal, is subject to the attraction of the positive charge of the ionic lattice. This attractive force is approximately estimated by the Coulomb attraction of an electron to its positive image charge, see (3.1.28). At small distances from the surface this force is very large, and near the surface a high potential barrier arises which is capable of withstanding substantial electron pressure.

3.5.3 Atomic Volume, Compressibility, and Strength of Metals

We wish to enlarge upon the nature of this barrier, which assures the stability of the metal. The interaction of an electron and its image produces a negative pressure; when they are in equilibrium a balancing (equalizing) positive pressure (3.5.57) occurs. The potential energy of an electron with respect to an ion is equal to

$$\varepsilon_{\text{pot}} = -\alpha' e^2/r, \tag{3.5.58}$$

where α' is a numerical constant ~ 1, and r is the mean distance between the electron and the nearest ion. The quantity r is calculated in terms of the electron density

$$r \approx \beta n^{-1/3}, \tag{3.5.59}$$

with the numerical coefficient $\beta \sim 1$ being determined by the type of crystal. Substitution of (3.5.59) into (3.5.58) yields

$$\varepsilon_{\text{pot}} = -\alpha e^2 n^{1/3} \quad (\alpha = \alpha'/\beta) .$$

Inserting n from (3.5.14), we obtain for the potential energy

$$\mathscr{E}_{\text{pot}} = N\varepsilon_{\text{pot}} = -\alpha e^2 N^{4/3} V^{-1/3} . \tag{3.5.60}$$

The negative pressure p_- is equal to

$$p_- = -\partial \mathscr{E}_{\text{pot}}/\partial V = -\alpha N^{4/3} e^2/3 V^{1/3} = -\alpha e^2 \pi^{1/3}/3 . \tag{3.5.61}$$

At equilibrium, the sum of p_- and p_0 from (3.5.57) is equal to zero:

$$p = p_- + p_0 = 0 , \tag{3.5.62}$$

whence, from (3.5.10, 57, 61), we obtain

$$r_{\text{at}} \approx n^{1/3} \approx (3/10)(3/8\pi)^{2/3}(h^2/me^2) \sim 10^{-8}\ \text{cm} . \tag{3.5.63}$$

The above value is very close to that of the radius of the first hydrogen orbit $h^2/4\pi^2 me^2$. This elementary calculation has shown that the zero-point kinetic energy of a degenerate electron gas and the positive pressure (3.5.57) due to this energy have a reasonable dynamic meaning and give a correct estimate of the atomic volume in the metal. Now we calculate the compressibility of the metal. Note that (3.5.62), because of (3.5.57, 61) is the condition for the minimum of the total energy of the crystal:

$$\frac{d\mathscr{E}}{dr}=\frac{d}{dr}(\mathscr{E}_{\mathrm{pot}}+\mathscr{E}_{\mathrm{kin}})=0 \ . \tag{3.5.64}$$

According to (3.5.57, 59, 60),

$$\mathscr{E}=-A/r+B/r^2 \ , \tag{3.5.65}$$

where $A=\alpha e^2 N$, $B=\beta h^2 N/2m$.

If the metal is compressed or extended by an external pressure, its energy $\mathscr{E}$ increases in comparison with $\mathscr{E}_0$ in the absence of a pressure. For small deformations

$$\mathscr{E}-\mathscr{E}_0\approx\tfrac{1}{2}(\partial^2\mathscr{E}/\partial r^2)_0(r-r_0)^2 \ , \tag{3.5.66}$$

where r and r_0 stand for the atomic radii of a distorted and an undistorted crystal. The linear expansion term with respect to $(r-r_0)$ is absent due to (3.5.64), and the terms higher than second-order will be neglected.

Express the atomic radii r and r_0 in terms of the volume of the metal:

$$V_0=\beta^{-3}Nr_0^3 \ , \qquad V=\beta^{-3}Nr^3 \ , \qquad \beta^{-3}\sim 1 \ . \tag{3.5.67}$$

This gives, instead of (3.5.66),

$$\mathscr{E}-\mathscr{E}_0\approx\frac{1}{2}\left(\frac{\partial^2\mathscr{E}}{\partial r^2}\right)_0\frac{\beta^6(V-V_0)^2}{9N^2r_0^4} \ .$$

From (3.5.64, 65) we have

$$(\partial^2\mathscr{E}/\partial r^2)_0=-2A/r_0^3+6B/r_0^4=Ar_0^3 \ ,$$

and, consequently,

$$\mathscr{E}-\mathscr{E}_0=(A\beta^6/2)(V-V_0)^2/9N^2r_0^7 \ . \tag{3.5.68}$$

The energy of an elastically deformed solid is normally represented as

$$\mathscr{E}-\mathscr{E}_0=(K/2)(V-V_0)^2/V_0 \ , \tag{3.5.69}$$

with K being the bulk modulus (compressibility). Comparing (3.5.68) with (3.5.69) and taking into consideration (3.5.64, 65), we find the statistical definition of the bulk modulus

$$K = A\beta^6 V_0/9N^2 r_0^7 = \alpha e^2 N/9V_0 r_0 = 2\mathscr{E}/9V_0 \ , \tag{3.5.70}$$

where we have made use of the virial theorem (3.4.2). From (3.5.70) we can readily estimate the order of magnitude of the modulus K. Substituting $N/V_0 \sim (1 \div 3) \cdot 10^{22}\,\text{cm}^{-3}$, $e \sim 4.8 \cdot 10^{-10}$ CGSE, $\alpha \sim (1 \div 3)$, we obtain $K_{\text{theor}} \sim 10^{10} \div 10^{11}\,\text{erg/cm}^3 \approx 10^4 \div 10^5$ atm. For alkali metals experiment gives $K_{\text{Na}} \approx 0.6 \cdot 10^5$ atm, $K_{\text{K}} \approx 0.3 \cdot 10^5$ atm, and $K_{\text{Li}} \approx 1.1 \cdot 10^5$ atm. Thus, the agreement between theory and experiment is quite good. Further, we estimate the strength of the metal according to Frenkel, i.e., determine the maximum pressure p_- which the metal can withstand without undergoing fracture. The condition for the maximum has the form $dp/dV = 0$ or, because $p = -d\mathscr{E}/dV$, $d^2\mathscr{E}/dV^2 = 0$.

According to (3.5.65, 67),

$$\frac{d^2\mathscr{E}}{dV^2} = \frac{d}{dV}\left(\frac{d\mathscr{E}}{dr}\frac{dr}{dV}\right) = \frac{d}{dV}\left[\left(\frac{A}{r^2} - \frac{2B}{r^3}\right)\frac{\beta^3}{3r^2 N}\right] = 0 \ ,$$

whence we find

$$B = 2Ar_{\text{max}}/5 \ . \tag{3.5.71}$$

In this equation r_{max} is the largest attainable radius for the atomic volume and corresponds to the maximum possible negative pressure beyond which the metal starts to fracture ($p_{-\text{max}}$). By definition,

$$p = -d\mathscr{E}/dV = -(\beta^3/3r^2 N)(d\mathscr{E}/dr) \ ,$$

or, using (3.5.65),

$$p = -(A/r^2 - 2B/r^3)(\beta^3/3r^2 N) \ . \tag{3.5.72}$$

Substituting B from (3.5.71) and A from (3.5.65) into (3.5.72), we find

$$p_{-\text{max}} = -A\beta^3/15Nr_{\text{max}}^4 = -\alpha\beta^3 e^2/15r_{\text{max}}^4 \ . \tag{3.5.73}$$

Setting $\alpha\beta^3/15 \sim 1$, $e \approx 4.8 \cdot 10^{-10}$ CGSE, and $r_{\text{max}} \approx 10^{-8}$ cm, we obtain the estimated value $p_{-\text{max}} \sim 10^{11}\,\text{dyn/cm}^2 \sim 10^5$ atm. We know from experiment that the actual strength of metals is hundreds of times smaller than this theoretical limit. This happens because uniform extension is not feasible and overstressed regions arise inevitably in the sample. Also, as first pointed out by

Ioffe, the presence of defects on the surface, due to the concentration of stresses near them, leads to an effective lowering of the strength of the sample. Therefore, the estimate (3.5.73) may be regarded as the upper limit of the strength of metals.

3.5.4 Paramagnetism of a Degenerate Electron Gas

All metals and alloys possess magnetic properties. Two classes are distinguished: (1) metals and alloys in which no atomic magnetic ordering (ferro-, antiferro- or ferrimagnetic) occurs in the absence of an external magnetic field, whatever the conditions may be, and (2) metals and alloys in which this ordering is observed in a certain temperature interval. The second case usually requires that at least one of the constituents be a transition element. Considering, at this point, normal metals, we restrict our attention to a study of the first type of metallic magnet, viz., weakly magnetic diamagnets or paramagnets.

Table 3.3 lists the magnetic susceptibilities of normal metals. As can be seen from the table, alkali and alkaline-earth metals (except for Be) are paramagnets.

Table 3.3. Atomic (χ_A) and specific (χ) magnetic susceptibilities of weakly magnetic (diamagnetic and paramagnetic) normal metals at $T \approx 300$ K

Element	Density [$g\,cm^{-3}$]	$\chi_A \cdot 10^6$	$\chi \cdot 10^6$
Li	0.534	+24.6	+1.89
Na	0.9725	+16.1	+0.68
K	0.862	+21.35	+0.47
Rb	1.532	+18.2	+0.33
Cs	1.90	+29.9	+0.42
Be	1.8477	−9.02	−1.83
Mg	1.74	+13.25	+0.95
Ca	1.54	+44.0	+1.7
Sr	2.63	+91.2	+2.65
Ba	3.65	+20.4	+0.56
Cu	8.96	−5.41	−0.76
Ag	10.5034	−21.56	−2.1
Au	19.32	−29.59	−2.9
Zn	7.131	−11.4	−1.24
Cd	8.65	−19.7	−1.52
Hg	13.6902	−33.3	−2.25
Al	2.70	+16.7	+1.67
Ga	5.91	−21.7	−1.84
In	7.31	−12.6	−0.8
α-Te	11.85	−58.0	−3.37
Sn white	7.2984	+4.5	+0.276
Sn gray	5.8466	−3.7	−0.184
Pb	11.3415	−24.86	−1.36
Bi	9.80	−284.0	−13.0

Their susceptibility is low ($\chi_{pm} \sim 10^{-6}$) and is practically independent of temperature. Of the group-III and group-IV elements, only aluminium and white tin are paramagnets. The other normal metals are diamagnets which (except for Bi) exhibit low susceptibility ($|\chi_{dm}| \sim 10^{-6}$) and $\partial\chi_{dm}/dT \approx 0$.

Several questions arise here: Why are some of the metals paramagnets, and some of them diamagnets, and why is their susceptibility practically independent of temperature?

A free electron, possessing an intrinsic mechanical moment (i.e., a spin) and an electric charge, should also possess a magnetic moment μ_{el}. Relativistic mechanics tells us that the spin-magnetic moment of a free electron is equal to the Bohr magneton

$$\mu_B = |e|\hbar/2mc \ . \tag{3.5.74}$$

If conduction electrons obeyed the laws of classical statistics, their paramagnetism would be similar to that of normal gases. Specifically, the temperature dependence of the susceptibility of a gas of density n would obey the Curie law

$$\chi_{pm}(T) = n\mu_B^2/k_B T \ . \tag{3.5.75}$$

The factor of 3 is absent in the denominator of (3.5.75) because the spin-magnetic moment may have only two projections relative to the field rather than any projection, as is assumed in the classical theory.

In explaining the magnetic properties of normal metals, we must take into account that the "source" of magnetism is the ion core lattice and itinerant electrons (the contribution of nuclear magnetic moments may be neglected because of their smallness). The simplest case is that of the alkali metals, in which the electronic shells of ion cores are identical to the closed shells of the atoms of inert gases possessing weak diamagnetism. Therefore, the paramagnetism that occurs in these metals is due to the paramagnetism of the conduction electrons, i.e., $\chi_{pm}^{obs} = \chi_{pm}^{el} - |\chi_{dm}^{ion}|$. The same is true of the alkali-earth metals (except for Be), Al, and β-Sn. In all other cases $|\chi_{dm}^{ion}| > \chi_{pm}^{el}$, and therefore these metals are diamagnetic (in reality, the diamagnetism may sometimes arise also from conduction electrons; for Bi, e.g., see Sect. 3.5.5). The first to pay attention to this circumstance was Dorfmann [3.6] who also pointed out another fact: if one compares the observed atomic susceptibility of normal diamagnetic metals with the susceptibility of their ions (obtained, for example, from salts or salt solutions containing these ions), the quantity $|\chi_{dm}^{met}|$ is always smaller than $|\chi_{dm}^{ion}|$. Thus, for copper, silver, and gold we have

Element	Cu	Ag	Au
χ_{dm}^{met}, 10^6	−5.41	−21.56	−29.59
χ_{dm}^{ion}, 10^6	−18.0	−31.0	−45.8

Proceeding from this fact, Dorfmann inferred that conduction electrons possessed paramagnetism and it was this paramagnetism that decreased the diamagnetic susceptibility in going from ions to metals. Since $\chi_{\rm dm}^{\rm met}$, as a rule, is independent of T, the paramagnetism of conduction electrons, according to Dorfmann, must also be independent of T. Thus the absence of a dependence such as (3.5.75) for all normal paramagnetic and diamagnetic metals turned out to be one of the most conclusive proofs of the inability of classical physics to explain the properties of electrons in a metal. Thus the papers concerned with the explanation of the temperature-independent paramagnetism and diamagnetism of normal metals [3.6–9] may have marked the beginning of not only the quantum theory of the magnetic properties of these materials, but also of quantum (electronic) solid-state theory as a whole.

We start with the simplest case of the paramagnetism of a Fermi electron gas. This paramagnetism occurs if $\chi_{\rm pm}^{\rm el} \gg |\chi_{\rm dm}^{\rm ion}|$ and the diamagnetism of the electrons themselves may be neglected. These requirements are met best in the alkali metals.

In the absence of an externally applied magnetic field the total magnetic moment of the Fermi gas at 0 K is equal to zero because of the complete compensation of electronic spins. This shows that the Pauli exclusion principle in a system of fermions leads to a substantial dependence of their energy on the magnetic moment, even if the magnetic forces are disregarded. As will be seen later, this dependence appears even when we take into account the electrostatic interaction of electrons, thus permitting an explanation of the ferro- and antiferromagnetism (Chap. 5).

In magnetizing a highly degenerate electron gas, the magnetic field has to move the electrons from states with energy $\varepsilon < \zeta_0$ to states with $\varepsilon > \zeta_0$. The energetic influence of the field is determined by the quantity $\mu_{\rm B} H$; with fields $H \approx 10^4$ Oe, it will amount to $\sim 10^{-16}$ erg, which is much less than $\zeta_0 \sim 10^{-12}$ erg.

Therefore, in the main, the redistribution of electrons by the action of a field occurs in a thermal-smearing band of width $\sim k_{\rm B} T$ near the Fermi level. To estimate the paramagnetic susceptibility, one may employ, according to Frenkel, the classical formula (3.5.74). However, for thermally excited electrons, (3.5.25) implies

$$\chi_{\rm pm}^{\rm el} = n_T \mu_{\rm B}^2 / k_{\rm B} T = n \mu_{\rm B}^2 / k_{\rm B} \vartheta_{\rm el} \ . \tag{3.5.76}$$

In this approximation the temperature independence of the paramagnetism of an electron gas is explained immediately. In addition, we obtain from (3.5.76) a numerical estimate of $\chi_{\rm pm}^{\rm el} \sim 10^{-6} \div 10^{-7}$ which tallies with the data summarized in Table 3.3. Only at very high temperatures ($T \gtrsim \vartheta_{\rm el}$) will the electron gas behave as a classical one, obeying (3.5.76). Since $\vartheta_{\rm el}$ is higher than the vaporization temperature of metals, the classical paramagnetism is virtually unobservable in normal metals. Exceptions occur only for a few poor metals which have a low conduction electron density and a low ϑ_{el}. Let it be emphasized that what is meant here by normal metals is solely nontransition metals.

Let us now consider the changes which the Fermi distribution undergoes by the action of a field. With $H=0$ and with the dispersion relation being quadratic, the density of states $g(\varepsilon)$ has the form of a parabola. The density of states may be introduced separately for electrons with "right-hand" and "left-hand" spin projections (sometimes one also talks of spins directed "upward" or "downward") $g_+(\varepsilon)$ and $g_-(\varepsilon)$; these functions also will have the shape of parabolas (Fig. 3.9a)

$$g_+(\varepsilon)=g_-(\varepsilon)=\tfrac{1}{2}g(\varepsilon)\ . \tag{3.5.77}$$

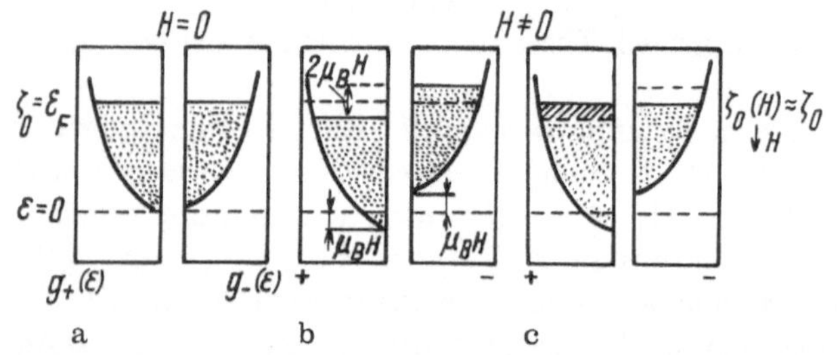

Fig. 3.9. Density of states for electrons in the metal with spins up and down: without a magnetic field (**a**), in a magnetic field upon shifting of subbands (**b**), upon establishment of a new equilibrium in the field (**c**)

With the field on, the subband with right-hand spins will be displaced downward along the ε axis by quantity $\mu_B H$, and the subband with left-hand spins will be displaced upward by the same quantity, since the energy of the magnetic moment μ_B with respect to the field H is

$$-\mu_B H\ . \tag{3.5.78}$$

Thus the subbands will be displaced relative to each other by $2\mu_B H$ (Fig. 3.9b). The equal occupancy of the subbands does not correspond any longer to the energy minimum of the system. At equilibrium, some of the electrons will pass from the subband of left-hand spins to the subband of right-hand spins (Fig. 3.9c). This corresponds to the magnetized state of conduction electrons known as Pauli paramagnetism. We neglect the variation of the function $g(\varepsilon)$ in the $\mu_B H$ interval near $\varepsilon=\zeta_0$ (because even when $H\sim 10^4$ Oe, $\sim 8\cdot 10^5$ A/m $\mu_B H\sim 10^{-16}$ erg $\ll \zeta_0 \sim 10^{-12}$ erg $=10^{-19}$ J). The variation of the concentration of electrons with right-hand spins then is equal to $\delta n_+ \approx +g_+(\zeta_0)\mu_B H V^{-1}$, and for left-hand spin electrons $\delta n_- \approx -g_-(\zeta_0)\,\mu_B H V^{-1}$. If we assume in the same approximation that (3.5.77) holds also for $H \neq 0$, the result for the magnetization in a field will be

$$I = (\delta n_+ - \delta n_-)\mu_B = [g_+(\zeta_0) + g_-(\zeta_0)]\mu_B^2 H V^{-1} = \\ = g(\zeta_0)\mu_B^2 H V^{-1} . \tag{3.5.79}$$

Using (3.5.8, 19), we find the susceptibility [3.7]

$$\chi_{pm}^{el} = g(\zeta_0)\mu_B^2/V^{-1} = (12m^* \mu_B^2/h^2) \times \\ \times (\pi/3)^{2/3} n^{1/3} = 3n\mu_B^2/2k_B \vartheta_{el} . \tag{3.5.80}$$

Expression (3.5.80) differs from the approximate formula (3.5.76) only in that the former contains the factor 3/2. What is significant is that, according to (3.5.80), the quantity χ_{pm}^{el} to a first approximation is determined by the density of electronic states at the Fermi surface. The heat-capacity formula (3.5.55) shows that the coefficient γ, like χ_{pm}^{el}, is determined by the quantity $g(\zeta_0)$. From an analysis of the experimental values of the quantities χ_{pm}^{el} and γ one therefore may also obtain values for $g(\zeta_0)$. The values obtained are fairly close in magnitude but are still at variance, with a discrepancy that cannot be attributed to measurement error. The point is that the observed χ_{pm}^{el} is not equal to the theoretical quantity (3.5.80) because of the diamagnetic contribution of the ionic lattice and the electron gas itself (Sect. 3.5.5). But even if they are taken into account, the discrepancy in the determination of $g(\zeta_0)$ from thermal and magnetic measurements persists. This is so because we have assumed, without justification, that the magnetic moment of a conduction electron in a metal is equal to μ_B. However, because of the interactions that take place in a metal, a renormalization of the magnetic moment of an electron occurs:

$$\mu_{el} = \mu_B(1 + \alpha_M) , \tag{3.5.81}$$

where α_M is the relative variation of the moment due to internal interactions (Chap. 5).

In a more rigorous calculation of χ_{pm}^{el} we should write, in place of (3.5.79),

$$I = \mu_{el} \int d\varepsilon (g(\varepsilon)/2V)[\overline{n(\varepsilon - \mu_{el} H)} - \overline{n(\varepsilon + \mu_{el} H)}] . \tag{3.5.82}$$

In this case (3.5.11) for the $\overline{n(\varepsilon)}$ allows for the subband displacement (Fig. 3.9), and the density of states $g(\varepsilon)$ is assumed to be constant (in what follows we will justify this approximation), the factor 1/2 taking into account the separation into subbands. In (3.5.82) we need to take into account that the quantity ζ is also a function of H. According to (3.5.27), ζ is found from the condition that the total number of particles be constant:

$$N = \int d\varepsilon [g(\varepsilon)/2][\overline{n(\varepsilon - \mu_{el} H)} + \overline{n(\varepsilon + \mu_{el} H)}] . \tag{3.5.83}$$

As can be seen from this equation, the quantity ζ may depend not only on T, but also on H. If we now allow for the fact that normally $\mu_{el}H \ll \zeta_0$, the integrand functions in (3.5.83) may be expanded in a power series of the dimensionless parameter $\mu_{el}H/\zeta_0$. It follows from this expansion that the H dependence of N is defined by the second-order terms $(\mu_{el}H/\zeta_0)^2$ only, i.e.,

$$N = \int d\varepsilon g(\varepsilon)\overline{n(\varepsilon)} + O(H^2) \ . \tag{3.5.84}$$

Consequently, ζ only depends on H starting with quadratic terms. This weak dependence may be neglected, since what interests us is the linear dependence of magnetization on H. Taking the quantity $\partial\zeta/\partial H$ to be ≈ 0, we expand $\overline{n(\varepsilon)}$ in (3.5.82) in a power series of the field and confine ourselves to terms that are not higher than first order. Then we obtain, instead of (3.5.79),

$$I = -(\mu_{el}^2/V)H\int d\varepsilon g(\varepsilon)[\partial\overline{n(\varepsilon)}/\partial\varepsilon] + O(H^3) \ , \tag{3.5.85}$$

and, using (3.5.40), the result for the susceptibility will be

$$\chi_{pm}^{el} = \frac{\mu_{el}^2}{V}\left[g(\zeta) + \frac{\pi^2}{6}(k_B T)^2\frac{\partial^2 g(\zeta)}{\partial\zeta^2}\right] . \tag{3.5.86}$$

Allowing for (3.5.52) for the ζ and carrying out the expansion

$$g(\zeta) \approx g(\zeta_0) + (\zeta - \zeta_0)\left.\frac{\partial g(\zeta)}{\partial\zeta}\right|_{\zeta=\zeta_0} \tag{3.5.87}$$

up to terms of order T^2, we find, with the help of (3.5.87), for (3.5.86)

$$\chi_{pm}^{el} = \frac{\mu_{el}^2}{V}g(\zeta_0)\left\{1 + \frac{\pi^2}{6}(k_B T)^2\left[\frac{g''(\zeta_0)}{g(\zeta_0)} - \left(\frac{g'(\zeta_0)}{g(\zeta_0)}\right)^2\right]\right\} . \tag{3.5.88}$$

If we take, instead of the arbitrary function $g(\zeta_0)$, its value for the quadratic dispersion relation, (3.5.88) yields, by virtue of (3.5.8, 19)

$$\begin{aligned}\chi_{pm}^{el} &= (3\mu_{el}^2 N/2V\zeta_0)[1 - (\pi^2/12)(k_B T/\zeta_0)^2] \\ &= (12m^* \mu_{el}^2/h^2)(\pi/3)^{2/3} n^{1/3}[1 - (\pi^2/12)(k_B T/\zeta_0)^2] \ ,\end{aligned} \tag{3.5.89}$$

whence it is seen that, when $T = 0$ K, (3.5.89) goes into (3.5.80). Allowing for the fact that $\partial\overline{n(\varepsilon)}/\partial\varepsilon = -\partial\overline{n(\varepsilon)}/\partial\zeta$, we may write, instead of (3.5.85),

$$I = \frac{\mu_{el}^2}{V}\left(\frac{\partial}{\partial\zeta}\int d\varepsilon g(\varepsilon)\overline{n(\varepsilon)}\,H = \frac{\mu_{el}^2}{V}\frac{\partial N}{\partial\zeta}H = \mu_{el}^2\frac{\partial n}{\partial\zeta}H \ .$$

The result for the susceptibility will then be

$$\chi_{\rm pm}^{\rm el} = (n\mu_{\rm el}^2/k_{\rm B}T)[F'_{1/2}(\tilde{\zeta})/F(\tilde{\zeta})] \;, \qquad \tilde{\zeta} = \zeta/k_{\rm B}T \;,$$

with

$$F_\alpha(y) = \int_0^{+\infty} x^\alpha dx/(\exp(x-y)+1)^{-1} \;,$$

$$x = \varepsilon/k_{\rm B}T \;, \qquad y = \zeta/k_{\rm B}T \;, \qquad F'_\alpha(y) = \partial F_\alpha(y)\partial y \;,$$

and

$$n = C\int_0^{+\infty} \varepsilon^{1/2} d\varepsilon(\exp[(\varepsilon-\zeta)/k_{\rm B}T]+1)^{-1}$$
$$= C(k_{\rm B}T)^{3/2}F_{1/2}(\tilde{\zeta}) = C\int_0^{\zeta_0} d\varepsilon\varepsilon^{1/2} = 2C\zeta_0^{3/2}/3 \;.$$

In the classical limit $k_{\rm B}T \gg \zeta_0$ we have $F_{1/2}(\tilde{\zeta}) = \Gamma(3/2)\exp\tilde{\zeta}$, where $\Gamma(y)$ is a gamma function, and, consequently, we obtain for $\chi_{\rm pm}^{\rm el}$ the Curie law

$$\chi_{\rm pm}^{\rm el} \approx n\mu_{\rm el}^2/k_{\rm B}T \;. \tag{3.5.90}$$

Because of the quantum laws for the spin, there is no factor 1/3 here, either.

3.5.5 Diamagnetism of a Degenerate Electron Gas According to Landau

As stated in the foregoing, the treatment of the magnetic properties in Sect. 3.5.4 is incomplete because conduction electrons possess not only paramagnetism but also diamagnetism. (In normal measurements of the magnetic susceptibility of a metal one actually measures the sum of the paramagnetic and the diamagnetic susceptibility of electrons and ion cores. However, with magnetic resonance methods the paramagnetic susceptibility component can be separated).

According to the general theorem of classical statistics [3.10, 11], the diamagnetic susceptibility of electrons should be equal to zero. Indeed, all magnetic properties are determined from the partition function

$$Z = \int d^{3N}pd^{3N}x(N!h^{3N})^{-1}\exp(-\mathscr{H}(p,x)/k_{\rm B}T) \;, \tag{3.5.91}$$

where $\mathscr{H}$ is the Hamiltonian of a system of N particles, which depends on all coordinates x_j and momenta p_i [in (3.5.91) all of them are labeled by x and p]. Classical electrodynamics shows that the magnetic field is included in the

Hamiltonian by the replacement $p_j \to p_j - e_j A(x_j)/c$ (with e_j being the charge of the j particle, and A the vector potential) without altering the form of the functional dependence $\mathscr{H}(p, x)$. With this replacement of variables, the integral (3.5.91) remains invariant and, in consequence, all derivatives of the thermodynamic potential $\phi = -k_B T \ln Z$ with respect to magnetic field (magnetization, susceptibility, etc.) are equal to zero. The impossibility of explaining the magnetic properties of a substance in classical terms is one of the "catastrophes" of classical physics. Langevin's theory of paramagnetism and diamagnetism was not consistent in this respect, for it actually postulated the existence of stationary electron orbits as well as magnetic moments of atoms and molecules. In fact, these assumptions cannot be substantiated classically.

According to *Bohr* [3.12], the disappearance of diamagnetism in a metallic sample of finite dimensions is clearly explained by the fact that the external-field-induced diamagnetic moment within a metal is exactly compensated for by the back moment of the random path of electrons reflected from the boundaries of the sample (Fig. 3.10). However, Landau [3.9] discovered the remarkable fact that a quantum free-electron gas possesses a nonzero diamagnetism. According to the classical theory, the projection of the motion of free electrons onto the plane perpendicular to the magnetic field that acts on them has the form of a closed path (cyclotron orbits) and the motion is of a periodic charater. In passing to quantum theory, the periodic motion is quantized and therefore, when the magnetic field is turned on, the electronic energy changes and a diamagnetic effect arises.

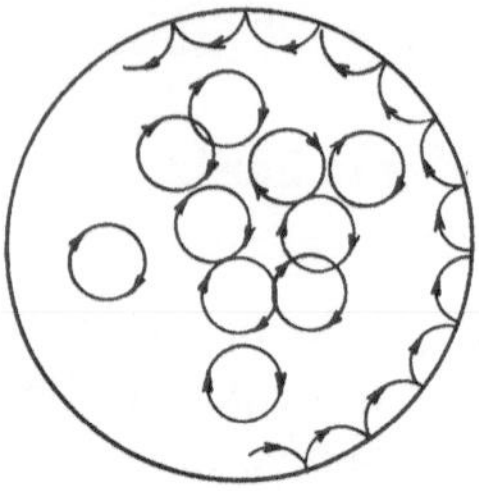

Fig. 3.10. Proving the absence of the diamagnetism of the classical electron gas according to Bohr

To begin with, we consider the motion of an electron in a uniform constant magnetic field parallel to the z axis in a quasi-classical approximation. If the electron is in a bound state in the atom, its motion in a circular orbit of radius $\boldsymbol{r}$ in a Coulomb nuclear field of charge $|e|$ is described by the equation of classical dynamics $m\omega_0^2 r = e^2/r^2$, with $\omega_0 = (e^2/mr^3)^{1/2}$ being the natural cyclic frequency of an electron in an atom. If the atom is placed in a field $\boldsymbol{H}$, it will be subject—in addition to the Coulomb force e^2/r^2—to a Lorentz force (e/c) $(\boldsymbol{v} \times \boldsymbol{H})$, where $\boldsymbol{v} = \boldsymbol{\omega} \times \boldsymbol{r}$ is the orbital velocity of the electron. If the vector $\boldsymbol{H}$ is perpendicular to the orbital plane, the equation of motion takes the form

$$m\omega^2 r = e^2/r^2 + er\omega H/c \ ,$$

and, consequently,

$$\omega = -(|e|H/2mc) \pm [\omega_0^2 + (eH/2mc)^2]^{1/2} .$$

In the case of weak fields the quantity $\omega_0 \sim 10^{16}\,\mathrm{s}^{-1}$, and for $H \sim 10^4$ Oe $\sim 8 \cdot 10^5$ A/m and $|e|H/2mc \sim 10^{11}\,\mathrm{s}^{-1}(|e|H/2mc \ll \omega_0)$ we have $|\omega| = \omega_0 \pm |e|H/2mc \equiv \omega_0 \pm \omega_L$, where $\omega_L = |e|H/2mc$ is the Larmor frequency of an electron in an atom. For a free particle the orbital radius $r \to \infty$, i.e., $\omega_0 \to 0$, and therefore $\omega = -\omega_L \pm \omega_L$, whence it follows immediately that the cyclotron frequency of a free electron in a uniform magnetic field is equal to

$$\omega_H = 2\omega_L = |e|H/mc \; ; \tag{3.5.92}$$

i.e., it is twice as large as the Larmor frequency of an electron in an atom with the magnitude of the field being the same.

In quantum mechanics the motion of an electron in an electromagnetic field is described by the Hamiltonian operator

$$\hat{\mathscr{H}} = (1/2\,m)\left(\boldsymbol{p} - \frac{e}{c}\boldsymbol{A}(\boldsymbol{r})\right)^2 + eV(\boldsymbol{r}) \; ,$$

where $\boldsymbol{p}$ is the momentum operator ($\boldsymbol{p} = -\mathrm{i}\hbar\nabla$), $\boldsymbol{A}(\boldsymbol{r}) \equiv \boldsymbol{A}(x, y, z)$ the operator of the vector potential of the field $\boldsymbol{H} = \operatorname{curl} \boldsymbol{A}$, $\boldsymbol{E} = -\nabla V - \dfrac{1}{c}\dfrac{\partial \boldsymbol{A}}{\partial t}$, and V the operator of its scalar potential. When only the magnetic field ($V = 0$) is present, we have

$$\hat{\mathscr{H}} = (1/2\,m)\left(\boldsymbol{p} - \frac{e}{c}\boldsymbol{A}\right)^2 . \tag{3.5.93}$$

In the case of a uniform magnetic field aligned with the z axis, $\boldsymbol{H} = (0, 0, H)$, the vector potential is equal to

$$A_x = -Hy \; , \quad A_y = A_z = 0 \; . \tag{3.5.94}$$

Substituting (3.5.94) into (3.5.93) yields for the Schrödinger equation (3.5.1)

$$-\frac{\hbar^2}{2m}\Delta\psi - \frac{\mathrm{i}e\hbar}{mc}Hy\frac{\partial\psi}{\partial x} + \frac{e^2H^2}{2mc^2}y^2\psi = \mathscr{E}\psi \; . \tag{3.5.95}$$

In (3.5.95) we separate variables and try a solution in the form

$$\psi(\boldsymbol{r}) = \exp(\mathrm{i}\alpha x + \mathrm{i}\beta z)\varphi(y) \; , \tag{3.5.96}$$

where α and β are some constants that have the dimensions of cm^{-1}. Inserting (3.5.96) into (3.5.95), we find the equation for $\varphi(y)$:

$$-\frac{\hbar^2}{2m}\frac{d^2\varphi(y)}{dy^2}+\frac{e\hbar\alpha}{mc}Hy\varphi(y)+\frac{e^2H^2}{2mc^2}y^2\varphi(y)=\left(\mathscr{E}-\frac{\hbar^2\alpha^2}{2m}-\frac{\hbar^2\beta^2}{2m}\right)\varphi(y)\ . \tag{3.5.97}$$

Introducing the new variable

$$y=y'-\hbar\alpha c/eH\ , \tag{3.5.98}$$

and using the notation (3.5.92), and

$$\varepsilon=\mathscr{E}-\hbar^2\beta^2/2m\ , \tag{3.5.99}$$

equation (3.5.97) then assumes the form of the equation for an oscillator of mass m and frequency ω_H:

$$-\frac{\hbar^2}{2m}\frac{d^2\varphi}{dy'^2}+\frac{m\omega_H^2y'^2}{2}\varphi=\varepsilon\varphi\ . \tag{3.5.100}$$

From the solution of the quantum oscillator problem [1.12] we know that $\varepsilon=\hbar\omega_H(l+1/2)$, $l=0,1,2\ldots$ Thus, from (3.5.96, 99), the energy of a particle in a uniform constant magnetic field H will be

$$\mathscr{E}(l,\beta)=\hbar\omega_H(l+1/2)+\hbar^2\beta^2/2m\ . \tag{3.5.101}$$

The second term in (3.5.101) gives the kinetic energy of motion of an electron along a field, the generalized momentum along the z axis being equal to

$$p_z=\hbar\beta\ ; \tag{3.5.102}$$

i.e., just as in the classical case, the energy spectrum remains continuous for the momentum component along the field. The first term gives a discrete effective-oscillator spectrum (Landau levels) rather than a continuous spectrum $(p_x^2+p_y^2)/2m$ for $H=0$. Using (3.5.74, 92, 102), (3.5.101) may be rearranged as

$$\mathscr{E}(l,p_z)=2\mu_B H(l+1/2)+p_z^2/2m\ . \tag{3.5.103}$$

The partial quantization of the energy of electrons in a magnetic field may be represented in the following fashion. With $H=0$, the energy spectrum corresponding to the motion of an electron in the x, y plane is quasi-continuous, viz. $(p_x^2+p_y^2)/2m$, as portrayed by a continuous band in Fig. 3.11, from $\varepsilon=0$ to

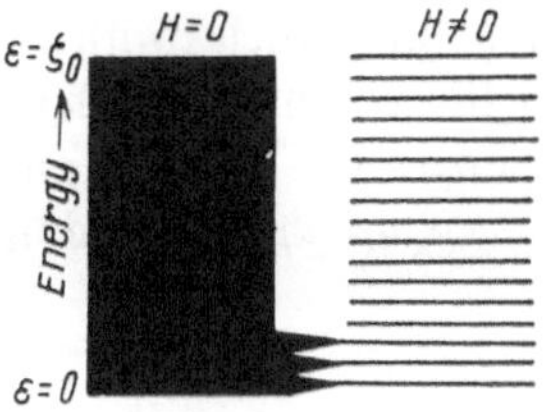

Fig. 3.11. Landau levels arising from the continuous spectrum

$\varepsilon = \zeta_0$. When $H \neq 0$, the band splits into strips of width

$$\Delta\varepsilon = 2\mu_B H[(l+1) + 1/2 - l - 1/2] = 2\mu_B H \ , \tag{3.5.104}$$

each of which converts to one discrete Landau level that lies exactly in the middle of the strip (Fig. 3.11). The degree of degeneracy of each discrete level will be proportional to $2\mu_B H$. The emergence of Landau levels for conduction electrons in a magnetic field may also be represented in p space. Figure 3.12 depicts a Fermi sphere octant ACB. The occurrence of Landau levels boils down to replacement of the continuous sphere by a set of inscribed concentric cylinders that have a common axis p_z and are spaced a distance of $2\mu_B H$ apart.

To determine the diamagnetic susceptibility of an electron gas, we have to calculate the partition function $Z(T, H)$ and then determine the thermodynamic potential $\phi(T, H) = -k_B T \ln Z$ and the magnetization

$$I = -\partial\phi/\partial H = k_B T \partial \ln Z(T, H)/\partial H \ . \tag{3.5.105}$$

For simplicity, we start by considering the Maxwell statistics case. Then

$$Z(T, H) = \sum_l g_l \exp(-\varepsilon_l/k_B T) \ . \tag{3.5.106}$$

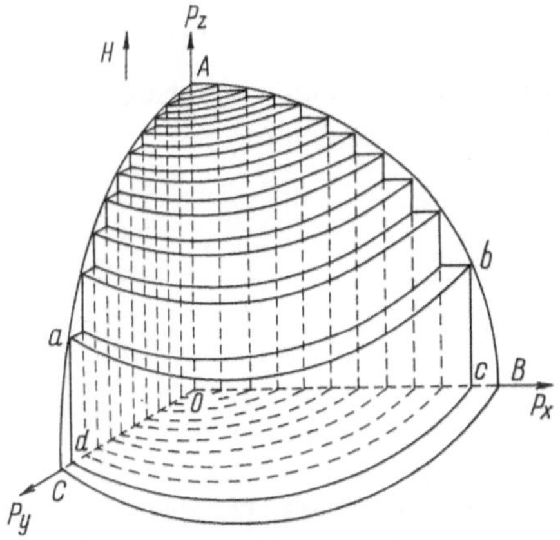

Fig. 3.12. Manifestation of Landau levels in momentum space

At the outset, we find the values of the statistical weight g_l for particles with the energy spectrum (3.5.103), i.e., the number of states in the volume of a phase space as a cylindrical ring of height dp_z, internal radius $p_\perp = (p_x^2 + p_y^2)^{1/2}$, and width $dp_\perp$ (Fig. 3.13). The volume of the ring is equal to $2\pi p_\perp dp_\perp dp_z$. According to (3.5.103), in quantum mechanics

$$p_\perp^2/2m \to \varepsilon - p_z^2/2m = 2\mu_B H(l + 1/2) \ ,$$

$$p_\perp dp_\perp \to 2m\mu_B H \Delta l = 2m\mu_B H \ ,$$

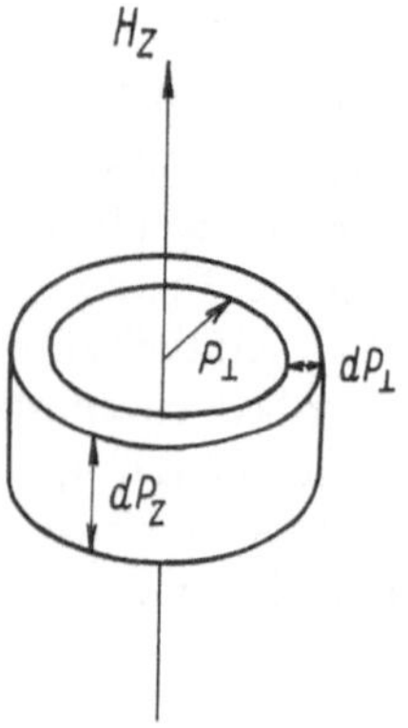

Fig. 3.13. Calculation of the statistical weight for electrons in a magnetic field

since $\Delta l = 1$. If we recall that the size of a unit cell according to (3.5.7) is equal to h^3/V and if we take account of the spin degeneracy, the result for g_l, obtained by replacing the μ_B according to (3.5.74), will be

$$g_l = 2\pi p_\perp dp_\perp dp_z \, 2V/h^3 = 2\pi \cdot 2m\mu_B H(2V/h^3)dp_z = 2|e| VH dp_z/h^2 c \ . \tag{3.5.107}$$

Substitution of (3.5.103, 107) into (3.5.106) yields the following [here we introduce a new variable $x^2 = p_z^2/2mk_B T$ and use the conditions

$$\int_{-\infty}^{+\infty} dx \exp(-x^2) = \pi^{1/2}$$

and

$$\sum_{l=0}^{+\infty} \exp[-(2l+1)\mu_B H/k_B T] = e^{-y}(1 + e^{-2y} + e^{-4y} + \ldots)$$
$$= e^{-y}(1 - e^{-2y})^{-1} = (2\operatorname{sh} y)^{-1} \ ,$$

where $y = \mu_B H/k_B T$]:

$$Z(T,H) = \sum_{l=0}^{+\infty}\int_{-\infty}^{+\infty} dp_z 2|e|VH/h^2c \times \exp\{-[\mu_B H(2l+1) + p_z^2/2m](k_B T)^{-1}\} \tag{3.5.108}$$

$$= (|e|VH/h^2c)(2\pi m k_B T)^{1/2}[\mathrm{sh}(\mu_B H/k_B T)]^{-1}$$

$$= (2\pi m k_B T/h^2)^{1/2}(\mu_B H/k_B T)[\mathrm{sh}(\mu_B H/k_B T)]^{-1} .$$

Using (3.5.105), we obtain for the magnetization

$$I = Nk_B T\partial \ln Z/\partial H = -N\mu_B[\mathrm{cth}(\mu_B H/k_B T) - k_B T/\mu_B H] \equiv -N\mu_B L(\mu_B H/k_B T) , \tag{3.5.109}$$

with $L(x) = \mathrm{cth} x - x^{-1}$ being a Langevin function. For weak magnetic fields and not very low temperatures—i.e., subject to the condition that $\mu_B H \ll k_B T$—(3.5.109) simplifies to $L(x) \approx x^{-1} + x/3 - x^3/45 + \ldots - x^{-1}$, and we obtain for the susceptibility the Curie law

$$\chi_{dm}^{el} = -n\mu_B^2/3k_B T . \tag{3.5.110}$$

It follows from (3.5.110) that in the "classical" case the diamagnetic susceptibility of an ideal electron gas is equal in absolute magnitude to one third the paramagnetic susceptibility: see (3.5.75) or (3.5.90).

In the derivation of (3.5.110), we assumed that, although possessing a quantum spectrum, the electron gas obeys classical statistics, whereas actually it obeys Fermi–Dirac statistics. However, we can find from (3.5.110) the correct expression for χ_{dm}^{el} without performing, again, a statistical calculation but, using the already familiar Frenkel method, by replacing the total density n in (3.5.110) by its thermally excited part (3.5.25). If, in addition, we allow for the factor 3/2 [transformation from (3.5.76 to 80)], the result for the diamagnetic susceptibility will be [3.9]

$$\chi_{dm}^{el} = -n\mu_B^2/2\zeta_0 = -(4m\mu_B^2/h^2)(\pi/3)^{2/3}n^{1/3} . \tag{3.5.111}$$

Let us now look at a more rigorous derivation of (3.5.111). To this end, we must find the thermodynamic potential $\Omega(T,\zeta)$ of an ideal Fermi gas with a spectrum ε_ν. This potential can be readily found using the distribution function (3.5.11) and the thermodynamic relation

$$N \equiv \sum_\nu \overline{n(\varepsilon_\nu)} \equiv \sum_\nu \{\exp[(\varepsilon_\nu - \zeta)/k_B T] + 1\}^{-1} = -\partial\Omega/\partial\zeta . \tag{3.5.112}$$

Integration with respect to ζ yields

$$\Omega = -k_B T \sum_\nu \ln\{1 + \exp[(\zeta - \varepsilon_\nu)/k_B T]\} + C\ , \tag{3.5.113}$$

where C is independent of ζ. For $\zeta/k_B T \to \infty$ we should have $\Omega = \mathscr{E} - \zeta N = \sum_{\nu, \varepsilon < \zeta_0}(\varepsilon_\nu - \zeta_0)$, whence we find $C = 0$. Inserting into (3.5.113) the Landau spectrum (3.5.101) in place of ε_ν and setting $V = 1$, i.e., $N = n$, we obtain

$$\begin{aligned} \Omega = &-(8\pi m/h^3)\mu_B H k_B T \sum_{l=0}^{+\infty} \int_{-\infty}^{+\infty} dp_z \ln(1 \\ &+ \exp\{[\zeta - 2\mu_B H(l + 1/2) - p_z^2/2m](k_B T)^{-1}\})\ . \end{aligned} \tag{3.5.114}$$

Here we have employed (3.5.107) for g_l per unit volume. The summation in (3.5.114) may be carried out with the help of the Euler formula [3.13], which, to within the approximations adopted, has the form

$$\sum_{l=a}^{b-1} f(l + 1/2) = \int_a^b dx\, f(x) - \frac{1}{24} f'(b) + \frac{1}{24} f'(a)\ . \tag{3.5.115}$$

Formula (3.5.115) is applicable provided that the function $f(x)$ is close to a linear one in the interval (a, b) between two values of the argument x that differ by unity, i.e., under the condition

$$|f(x + 1/2) - f(x - 1/2) - f'(x)| \ll |f(x)|\ .$$

The integrand in (3.5.114) varies appreciably in the interval for l, which is equal to $k_B T/\mu_B H$ at some points of the entire interval of l, i.e., $0 < l < \infty$. Therefore, the Euler formula (3.5.115) can only be used for calculating (3.5.114) provided that

$$\mu_B H/k_B T \ll 1\ , \tag{3.5.116}$$

i.e., in the region of weak fields and not very low temperatures. Under these conditions, according to (3.5.116), (3.5.115) for Ω becomes

$$\begin{aligned} \Omega = &-8\pi m \mu_B H k_B T h^{-3} \int_{-\infty}^{\infty} dp_z \int_0^{+\infty} dx \ln\{1 + \exp[(\zeta - 2\mu_B H_x \\ &- p_z^2/2m)(k_B T)^{-1}]\} + (2\pi m/3h^3)(\mu_B H)^2 \\ &\times \int_{-\infty}^{+\infty} dp_z \{1 - \exp[(p_z^2/2m - \zeta)(k_B T)^{-1}]\}^{-1}\ . \end{aligned} \tag{3.5.117}$$

Proceeding to the new variables (first $2\mu_B Hx = y$, then $p_z^2/2m = z$, and, finally, $y + z = \eta$) and changing the order of integration with respect to z and η, we obtain, instead of (3.5.117),

$$\begin{aligned} \Omega = &-[4\pi(2m)^{3/2}/h^3]k_B T \int_0^\infty d\eta\eta^{1/2} \ln\{1 \\ &+ \exp[(\zeta - \eta)/k_B T]\} + [\pi(2m)^{3/2}/3h^3](\mu_B H)^2 \\ &\times \int_0^{+\infty} dz z^{1/2}/\{1 + \exp[(z - \zeta)/k_B T]\}^{-1} . \end{aligned} \tag{3.5.118}$$

Applying to the integrals in (3.5.118) the procedure employed to calculate the integral (3.5.32), we find for the zero approximation

$$\Omega = -16\pi(2m)^{3/2}\zeta^{5/2}/15h^3 + [2\pi(2m)^{3/2}\zeta^{1/2}/3h^3](\mu_B H)^2 . \tag{3.5.119}$$

For the magnetization we have $I = -(\partial\Omega/\partial H)_\zeta$. Using (3.5.119) with $\zeta = \zeta_0$, we obtain

$$I = -[4\pi(2m)^{3/2}/3h^3]\zeta_0^{1/2}\mu_B^2 H .$$

Replacing ζ_0 according to (3.5.19), we find for the diamagnetic susceptibility according to Landau

$$\chi_{dm}^{el} = -(4\pi m/3h^2)(3n/\pi)^{1/3}\mu_B^2 , \tag{3.5.120}$$

which coincides exactly with (3.5.111). Using in (3.5.119) expansions in higher powers of $\mu_B H_{el}$, it is possible to calculate the dependence of χ_{dm}^{el} on T and H for strong degeneracy [3.14]. However, the result obtained is valid only for free electrons with a quadratic dispersion relation and when (3.5.115) holds.

One important remark must be made. When we wrote the quadratic dispersion relation (3.5.4), we labeled the conduction-electron mass m^* because it may differ from the mass of a free electron in vacuum. On the other hand, the Bohr magneton μ_B involved in the formulas for χ_{dm}^{el} has appeared as the replacement of a definite group of constants entering into the expression of ω_H according to (3.5.92) [transformation from (3.5.101) to (3.5.103)]. Therefore, strictly speaking, (3.5.120) should be rewritten as

$$\chi_{dm}^{el} = -(4m^*\mu_B^2/h^2)(\pi/3)^{2/3}(m/m^*)^2 \tag{3.5.121}$$

[one factor m^* arises from (3.5.19) for ζ_0]. Thus the total susceptibility χ_{dm}^{el} of an electron gas, when the dispersion relation is quadratic but the mass is only the effective mass, will have the form

$$\chi^{\mathrm{el}}_{\mathrm{tot}} = \chi^{\mathrm{el}}_{\mathrm{pm}} + \chi^{\mathrm{el}}_{\mathrm{dm}} = \frac{12m^*\mu_{\mathrm{B}}^2}{h^2}\left(\frac{\pi}{3}\right)^{2/3}\left[1 - \frac{1}{3}\left(\frac{m}{m^*}\right)^2\right] . \qquad (3.5.122)$$

Equation (3.5.122) demonstrates that $\chi^{\mathrm{el}}_{\mathrm{tot}} = 2\chi^{\mathrm{el}}_{\mathrm{pm}}/3$ only when $m^* = m$. But if $m^* > m/\sqrt{3}$, the electron gas is always paramagnetic overall. Conversely, for $m^* < m/\sqrt{3}$ it is always diamagnetic. Thus, Landau diamagnetism is fundamentally observable in a gas of electrons with small effective masses ($m^* < m/\sqrt{3}$).

3.5.6 Oscillatory Effects in the Fermi Gas

In the general case the energy of an electron in a magnetic field is the sum of the energies (3.5.78, 103):

$$\varepsilon(l, p_z, H) = p_z^2/2m^* + \hbar\omega_H(2l + 1) - \mu_{\mathrm{el}} H \ . \qquad (3.5.123)$$

Substituting (3.5.123) into (3.5.113), we can, in principle, calculate the magnetic susceptibility of an electron gas without dividing it into a diamagnetic and a paramagnetic part, as was done in the foregoing, and also take more rigorously into account all the features peculiar to Fermi–Dirac statistics. These calculations are very cumbersome and for this reason we restrict our attention to an elementary method of ascertaining some peculiarities of the magnetic properties of the Fermi gas.

As far back as 1931 De Haas and van Alphen discovered experimentally that the magnetic susceptibility in bismuth varied in a periodic fashion when the magnetic field was varied in the region of low temperatures. This phenomenon may be explained if we give up the restriction imposed by the condition $\mu_{\mathrm{B}} H \ll k_{\mathrm{B}} T$, which is at the foundation of all the preceding calculations, and perform a rigorous calculation using the thermodynamic potential (3.5.113). Qualitatively, this can be understood if we turn to Fig. 3.11, as well as to the elementary conclusion drawn in Sect. 3.5.5, showing that the degree of degeneracy of Landau levels is determined by the magnitude of the field according to (3.5.107). If this number is larger than the number of electrons N, all of them will be "accommodated" at the level with $l = 0$; with decreasing field the number of sites diminishes and the electrons will then start to "migrate" on to the next level with $l = 1$, etc. Therefore, the magnetic and, generally speaking, all properties of the electrons will alter periodically with varying magnetic field magnitude, and the susceptibility will not only vary in magnitude but also change sign, as is observed experimentally.

Figure 3.11 allows the period of these variations to be readily evaluated from the inverse of the magnetic field $\Delta(1/H)$. We consider two field values $H_1 > H_2$

for which the number of Landau levels with energies smaller than or equal to ζ_0 is equal to N' or $N'+1$, respectively. Then

$$\zeta_0/2\mu_B H_1 = N' \ , \quad \zeta_0/2\mu_B H_2 = N'+1 \ ,$$

whence we find for the oscillation period

$$\Delta(1/H) = 1/H_2 - 1/H_1 = 2\,\mu_B/\zeta_0 \ . \tag{3.5.124}$$

Peierls [3.15] investigated the magnetism of electrons subject to the condition

$$\mu_B H > k_B T \ . \tag{3.5.125}$$

He considered a two-dimensional electron gas at $T = 0$ K in a magnetic field perpendicular to the plane of the field. In this situation the energy levels are, according to (3.5.123), equal to

$$\varepsilon_l = \mu_B H(2l+1) \ .$$

In the two-dimensional case the degree of degeneracy, according to (3.5.107), is

$$g_l = 2|e|HS/ch \equiv \beta H \ , \quad \beta = 2|e|S/ch \ , \tag{3.5.126}$$

with S being the area of the system. If g_l is larger than the total number of electrons N, all of them occupy the state with $l = 0$ and the total energy $\mathscr{E}$ is $-N\mu_B H$. According to (3.5.107), the magnetization $I = n\mu_B$ and zero susceptibility correspond to this energy. As the magnetic field H is decreased, the total energy $\mathscr{E}$ decreases until the g_l become smaller than N. According to the Pauli exclusion principle, some of the electrons will then pass to levels with $l = 1$ and, in consequence, with decreasing field the energy increases; i.e., the system becomes paramagnetic. It is easy to find the general expression for the total energy $\mathscr{E}$, when N electrons completely fill r lowest-lying Landau levels and partially fill the $r+1$ level, according to the inequalities

$$r\beta H < N < (r+1)\beta H \tag{3.5.127}$$

or

$$N/(r+1) < \beta H < N/r \ .$$

Here we assume that $\beta H/N < 1$.

Recall that $l = 0, 1, 2, \ldots$. Therefore, to level number r in the formula for the energy corresponds the quantum number $l = r-1$, etc. The energy of the electrons that wholly fill r first levels is equal to

$$\beta H \sum_{l=0}^{r-1} \mu_B H(2l+1) = \beta\mu_B H^2 r^2 \ .$$

The energy $N - r\beta H$ of the electrons that partially fill the $r+1$ level is

$$(N - r\beta H)\mu_B H(2r+1) \ .$$

The total energy is equal to the sum of these two expressions

$$\mathscr{E}(H, r) = N\mu_B(2r+1)H - r(r+1)\mu_B \beta H^2 \ .$$

Consequently, in the interval of field values defined by (3.5.127) the magnetic moment of the system of electrons is

$$M \equiv IV = -\partial\mathscr{E}(H, r)/\partial H = -N\mu_B(2r+1-2r(r+1)\beta H/N) \ . \qquad (3.5.128)$$

The magnetic moment as a function of the integral parameter r is a discrete function of H. As noted above, when $\beta H/N > 1$, the magnetic moment $M = -N\mu_B$ and is independent of H [this is also seen from (3.5.128) for $r = 0$]. At $1/2 \leqslant \beta H/N \leqslant 1$ the electrons pass to the level with the quantum number 1. When $\beta H/N = 1$ and $r = 1$, the magnetic moment varies stepwise from $M = -N\mu_B$ to $M = N\mu_B$ and then, with $r = 1$, varies according to (3.5.127) in a linear fashion up to $-N\mu_B$ (with the field $\beta H/N = 1/2$). As the field is decreased further, the electrons start to pass to the level with quantum number 2. Again, with $\beta H/N = 1/2$ and $r = 2$, the magnetic moment M here jumps, according to (3.5.128), to the value of $N\mu_B$ and then varies linearly up to $M = -N\mu_B$ for $\beta H/N = 1/3$, etc. Thus, as the field is decreased, the moment reverses sign at certain intervals and undergoes discontinuities at field values for which $\beta H/N = 1/r$ (Fig. 3.14).

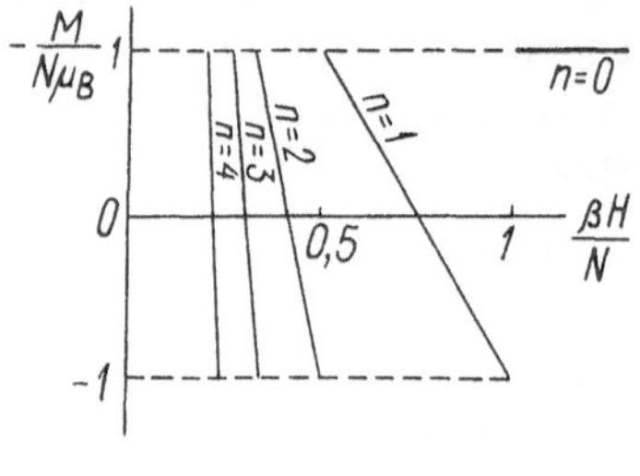

Fig. 3.14. De Haas–van Alphen effect in a plane model

From (3.5.103) it follows that a real three-dimensional energy spectrum is continuous and should therefore exhibit no discontinuities in the magnetic moment. Having calculated the thermodynamic potential (3.5.113) for the spectrum (3.5.103) more exactly, Peierls showed that here the function $M(H)$ indeed possesses no discontinuities, as has since been confirmed by numerous experiments. We will return to this problem later on (Chap. 4).

3.5.7 Thermionic Emission (the Richardson Effect)

Up to this point we have not been particularly concerned with the surface of the metal or the shape of the potential near it. In treating the withdrawal of electrons through the surface, the potential barrier shape becomes substantial. As a rule, electrons do not evaporate from a metal spontaneously. This is evident from the fact that under normal conditions a metal does not change its charge appreciably. In fact, electrons only escape through the surface of a metal when acted upon by some external factor, e.g., temperature, applied electric field, etc. As the electron leaves the metal, some work is done on it. To determine this work, we need to ascertain the behavior of the electron at the surface.

Figure 3.15 is a diagram of the electron energy. In the free-electron model the potential energy at the surface inside and outside the metal is constant; that is, the electron is free. But outside the metal the potential energy is larger. The potential jump $\Delta V(x) = W$ at the boundary of the metal specifies the work function of the electron. (We assume that the surface of the metal coincides with the yz plane and the normal to the metal is directed along the x axis.) With this potential shape, the metal can be considered as a potential box.

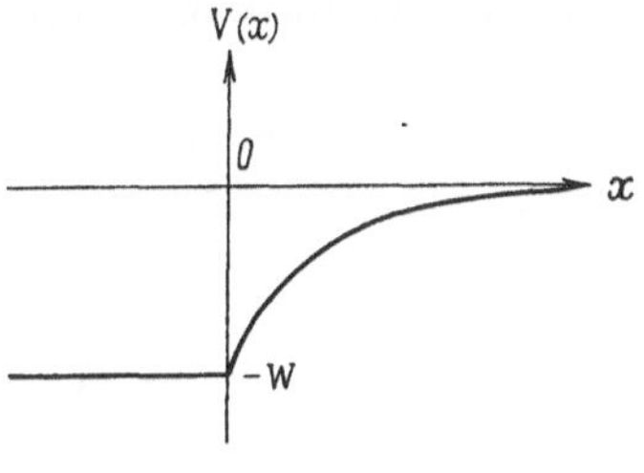

Fig. 3.15. Potential energy of an electron near the surface of the metal

The model is very crude. Actually, the $V(\boldsymbol{r})$ within the metal is a periodic function and therefore the electron in the metal may not be regarded as free. The potential jump at the surface is by no means sharp because the electron, when beyond the surface but still near it, is not free but is attracted by the metal. Nor is the surface of the metal ever ideal, in either the geometric or the chemical sense. It always possesses chemical inhomogeneities and other crystal structure defects. Nevertheless, even in terms of this model, we can obtain a qualitative description of the thermionic emission, i.e., the evaporation of electrons from a metal as a function of temperature and in the absence of an externally applied electric field.

First we specify more exactly the form of the potential $V(x)$ outside the metal. The electron, as it leaves the metal and moves farther away from the uncharged surface of the metal, is, as has been stated in Sects. 3.5.2, 3, subject to the attractive force of the mirror image at the surface. Suppose that the electron is separated from the surface by a distance that considerably exceeds the atomic size ($x \gg r_{at}$) but is smaller than the size of the plane sections of the surface of the metal. Then the attractive force of the mirror image is

$$F_i = -e^2/4x^2 \ ,$$

and the potential energy of this "image field" is

$$V_i = -e^2/4x \ . \tag{3.5.129}$$

From (3.5.129) we see that when $x \to \infty$ $V_i \to 0$ the potential energy V_i on the surface of the metal tends to minus infinity for $x = 0$. On the other hand, in the potential box model we assume that inside the metal ($x \leqslant 0$) $V = -W$. Since the details of the exact shape of the potential in the immediate vicinity of the surface ($x = 0$) are irrelevant to us, we assume the "image field" potential outside the surface to have the form (Fig. 3.15)

$$V_i(x) = -e^2/(4x + e^2/W) \ . \tag{3.5.130}$$

We see that for $x = 0$ the potential $V_i = -W$, and for large $x \gg e^2/4W$ (3.5.130) practically does not differ from (3.5.129). Also, because of the fact that inside the metal electrons obey the Fermi–Dirac distribution (3.5.11), their maximum energy at 0 K is equal to ζ_0.

Therefore, the work function w of the electrons escaping from the Fermi surface is defined not by the entire height of the potential barrier W but by the difference between it and the Fermi energy ζ_0, i.e.,

$$w = W - \zeta_0 \ . \tag{3.5.131}$$

This definition of the work function differs substantially from the classical one. With strong degeneracy and with $T > 0$ K, the determination of the quantity w according to (3.5.131) is quite legitimate since the number of electrons with $\varepsilon > \zeta_0$ is vanishingly small.

Our object now is to calculate the thermionic current, which is equal to the number of electrons that evaporate from unit surface of the metal per unit time. The dispersion relation is assumed to be quadratic. Since the potential (3.5.130) depends only on x, only the momentum component p_x is of interest to us. The energy of the electron which can escape from the metal through the surface (y, z) should exceed the sum of the potential barrier energy and the quantity $(p_y^2 + p_z^2)/2m^*$, i.e.,

$$\varepsilon \geqslant W + (p_y^2 + p_z^2)/2m^* = \zeta_0 + w + (p_y^2 + p_z^2)/2m^* \equiv \varepsilon_1 \ . \tag{3.5.132}$$

The total number of electrons per square centimeter per second with a momentum between p_x and $p_x + dp_x$ that penetrate the metal is equal to

$$\frac{2}{h^3}\bar{n}(p_x, p_y, p_z)v_x dp_x dp_y dp_z = \frac{2}{h^3}\bar{n}(p_x, p_y, p_z)\frac{\partial\varepsilon}{\partial p_x} dp_x dp_y dp_z \ . \tag{3.5.133}$$

The flux of electrons can then be found by integrating (3.5.133) over all values of p_x, p_y, and p_z that satisfy (3.5.132). As we do so, we take an integral over $d\varepsilon$ ($d\varepsilon = (\partial\varepsilon/\partial p_x)\,dp_x$). This gives

$$\begin{aligned}\frac{2}{h^3}\iint\limits_{-\infty}^{+\infty} dp_y dp_z \int\limits_{\varepsilon_1}^{+\infty} d\varepsilon\{1+\exp[(\varepsilon-\zeta)/k_B T]\}^{-1} \\ = \frac{2k_B T}{h^3}\iint\limits_{-\infty}^{+\infty} dp_y dp_z \ln(1+\exp\{-(k_B T)^{-1} \\ \times [w+(p_y^2+p_z^2)/2m^*]\}) \ .\end{aligned} \tag{3.5.134}$$

Here do not use the distribution function that gives the number of electrons in volume V with energies ranging from ε to $\varepsilon + d\varepsilon$—i.e., the function from (3.5.12)—but rather the function $\varphi(p_x, p_y, p_z)$, which gives the number of electrons per unit volume with momentum components from p_x to $p_x + dp_x$, etc., and which is equal to $(2/h^3)\times\bar{n}(p_x, p_y, p_z)$. (In taking an integral over $d\varepsilon$ we replace the variables $\exp[(\varepsilon-\zeta)/k_B T] \equiv u$. The replacement yields $k_B T\int_{u_1}^{+\infty}[u(u+1)]^{-1}\cdot du$, this integral being easy to take.) Under normal conditions $w \gg k_B T$ (it is known from experiment that $w \sim 1 \div 10$ eV, and at normal temperatures $k_B T \sim 0.01$ eV); therefore,

$$\begin{aligned}\ln([1+\exp\{-(k_B T)^{-1}[w+(p_y^2+p_z^2)/2m^*]\}) \\ \approx \exp\{-(k_B T)^{-1}[w+(p_y^2+p_z^2)/2m^*]\} \ .\end{aligned}$$

Upon integration over dp_y and dp_z, we find for the thermionic current

$$j_{th} = (4\pi e m^* k_B/h^3)T^2\exp(-w/k_B T) \equiv AT^2\exp(-w/k_B T) \ . \tag{3.5.135}$$

The factor $\exp(-w/k_B T)$ plays the main role in (3.5.135); in comparison with this factor, the term T^2 is practically unobservable experimentally. In the classical theory (Richardson) the result obtained instead of (3.5.135) was

$$j_{th}^{cl} = en(k_B T/2\pi m)^{1/2}\exp(-W/k_B T) \ , \tag{3.5.136}$$

derived in the same way as (3.5.135) if in place of the Fermi distribution one substitutes into (3.5.135) the classical function $\exp[(\zeta_{cl}-\varepsilon)/k_B T]$. It is nearly impossible to perceive the difference between the factors T^2 and $T^{1/2}$ in (3.5.135, 136) in the presence of the factors $\exp(-w/k_B T)$ or $\exp(-W/k_B T)$. However, both formulas differ substantially in that in the quantum formula the exponent involves the quantity $-w/k_B T$, and in the classical formula it involves the quantity $-W/k_B T$; i.e., they involve different work functions.

The decrease of the work function in the quantum case may be qualitatively explained as follows: the Fermi energy ζ_0 entails an internal electron pressure (3.5.57) $p \sim n\zeta_0$ which facilitates the penetration of electrons through the potential barrier ("squeezes out" the electrons). Let it be also noted that in the quantum formula (3.5.135) the factor A has a universal value ($A \approx 120\ \mathrm{A/cm^2\,K^2}$) for all metals, whereas in the classical formula (3.5.136) it involves n—a quantity that is individual for each particular metal. Table 3.4 shows that the experimental values of A coincide with the theoretical ones only in order of magnitude. This is due to the crudeness of the model used in the derivation of (3.5.135) [Ref. 3.6.6, Sect. 30].

Table 3.4. Thermoelectric constants A of some metals [3.16]

Metal	A [A/cm² K²]	Metal	A [A/cm² K²]
Ca	60	Pt	10^4
Cs	160	Ta	50
Mo	60	Th	60
Ni	27	W	60

3.6 Transport Phenomena

3.6.1 The Boltzmann Kinetic Equation

Conduction electrons in a metal may be subject to an electric fields, a magnetic field, and a temperature gradient. In addition, they suffer collisions with each other, the ionic lattice, its various defects, etc., and a dynamic equilibrium results. In this situation, the electrons lose the energy and momentum they have received from the field in scattering processes. To study a set of transport phenomena (electrical conductivity, thermal conductivity, Hall effect, etc.) we use the Boltzmann kinetic equation. Now we concentrate on its derivation.

The state of the system is statistically described with the help of a distribution function in the phase space of coordinates and velocities, which may also depend on time $t: \bar{n}(\boldsymbol{r}, \boldsymbol{v}, t)$. The number of electrons in a volume element of the six-dimensional phase space is equal to $d\tau_r\, d\tau_v$.

$$\bar{n}(\boldsymbol{r}, \boldsymbol{v}, t)\, d\tau_r d\tau_v \ .$$

The normalization conditions for the function n are chosen as

$$\int d\tau_r\, d\tau_v \bar{n}(\boldsymbol{r}, \boldsymbol{v}, t) = N \ , \tag{3.6.1}$$

where $\bar{n}$ is integrated over the entire phase space.

To find a kinetic equation satisfied by the functions $\bar{n}(\boldsymbol{r}, \boldsymbol{v}, t)$ in the presence of externally applied fields and scattering processes, we single out a volume element in the phase space and consider all possible variations of these functions in the volume element. The quantity $\bar{n}$ can vary with time for a number of reasons: first, there is the explicit $\bar{n}$ dependence on t, determined by the partial derivative of $\bar{n}$ with respect to time: $\partial\bar{n}(\boldsymbol{r}, \boldsymbol{v}, t)/\partial t$; second, the diffusive variation of $\bar{n}$ owing to the transport of particles from one segment of $\boldsymbol{r}$ space to another and a field-induced variation of $\bar{n}$ owing to the acceleration $\boldsymbol{a} = (a_x, a_y, a_z)$ in external fields. Thus, those particles which at a time t were in a cell with coordinates $x - v_x dt$, $y - v_y dt$, $z - v_z dt$; $v_x - a_x dt$, $v_y - a_y dt$, $v_z - a_z dt$ will emerge at a time $t + dt$ in a given phase-space cell with the mean coordinates $x, y, z; v_x, v_y, v_z$. This will be sufficiently exact if the time interval dt is so small ($dt \ll \bar{\tau}$) that collisions do not have enough time to alter the statistical distribution itself (i.e., $\boldsymbol{v}$ and $\boldsymbol{a}$). The magnitudes of the unit-cell volumes may then be thought not to vary during the time dt either; i.e., $d\tau_{\boldsymbol{r}-\boldsymbol{v}dt}\, d\tau_{\boldsymbol{v}-\boldsymbol{a}dt} = d\tau_r\, d\tau_v$. The diffusive and field-induced variation of $\bar{n}$ therefore will be

$$(\delta\bar{n})_{\mathrm{dif}} + (\delta\bar{n})_{\mathrm{field}} = \bar{n}(\boldsymbol{r} - \boldsymbol{v}\, dt, \boldsymbol{v} - \boldsymbol{a}dt, t + dt) - \bar{n}(\boldsymbol{r}, \boldsymbol{v}, t) \ ,$$

or, in virtue of the smallness of dt,

$$\left(\frac{\partial\bar{n}}{\partial t}\right)_{\mathrm{dif}} + \left(\frac{\partial\bar{n}}{\partial t}\right)_{\mathrm{field}} = -\boldsymbol{v}\cdot \nabla_r\bar{n} - \boldsymbol{a}\cdot \nabla_v\bar{n} \ , \tag{3.6.2}$$

where ∇_r and ∇_v are the gradient operators in $\boldsymbol{r}$ and $\boldsymbol{v}$ space, respectively.

Third, the quantity $\bar{n}(\boldsymbol{r}, \boldsymbol{v}, t)$ may also vary because of discrete variations in the velocities of particles at the time of collision. For computational purposes, we will introduce the concept of the mean number of collisions (the so-called molecular disorder hypothesis). Denote the probability that in unit time a particle with a velocity $\boldsymbol{v}'$ will cause $\boldsymbol{v}$ to vary and will get into the interval $\boldsymbol{v} + d\boldsymbol{v}$ by $\nu(\boldsymbol{v}, \boldsymbol{v}')[1 - \bar{n}(\boldsymbol{r}, \boldsymbol{v}', t)]$. Then, using the molecular disorder hypothesis, the density of the number of particles whose velocity $\boldsymbol{v}$ assumes in unit time any other value $\boldsymbol{v}'$ is

$$b_- = \bar{n}(\boldsymbol{r}, \boldsymbol{v}, t) \int d\boldsymbol{v}'\, \nu(\boldsymbol{v}, \boldsymbol{v}')[1 - \bar{n}(\boldsymbol{r}, \boldsymbol{v}', t)] \ . \tag{3.6.3}$$

The factor $1 - \bar{n}$ in $b_\pm$ is associated with the Pauli principle, which forbids a scattering into an occupied state ($\bar{n} = 1$). The density of the number of particles which in unit time acquire a velocity $\boldsymbol{v}$, having prior to this any other value $\boldsymbol{v}'$, will be

$$b_+ = [1 - \bar{n}(\boldsymbol{r}, \boldsymbol{v}, t)] \int d\boldsymbol{v}' \, \nu(\boldsymbol{v}', \boldsymbol{v}) \bar{n}(\boldsymbol{r}, \boldsymbol{v}', t) \ . \tag{3.6.4}$$

Because of collisions, for the total variation of $\bar{n}$ in unit time,

$$\left(\frac{\partial n}{\partial t}\right)_{\text{col}} = b_+ - b_- \ . \tag{3.6.5}$$

The sum of (3.6.2, 5) and $\partial \bar{n}/\partial t$ gives the total variation of the distribution function in unit time. From continuity considerations this variation is equal to zero; i.e.,

$$\frac{d\bar{n}}{dt} = \frac{\partial \bar{n}}{\partial t} + \left(\frac{\partial \bar{n}}{\partial t}\right)_{\text{dif}} + \left(\frac{\partial \bar{n}}{\partial t}\right)_{\text{field}} + b_+ - b_- = 0 \ . \tag{3.6.6}$$

With steady-state processes the distribution function $\bar{n}$ is explicitly independent of t and therefore

$$\partial \bar{n}/\partial t = 0 \ . \tag{3.6.7}$$

For such processes, on performing the replacement according to (3.6.3), (3.6.6, 7) then yield

$$\boldsymbol{v} \cdot \nabla_r \bar{n} + \boldsymbol{a} \cdot \nabla_v \bar{n} = b_+ - b_- \ . \tag{3.6.8}$$

This is the Boltzmann kinetic equation. It shows that in the stationary case the diffusive and field-induced variations are completely compensated for by scattering processes and that in a homogeneous metallic sample without a temperature gradient or external fields (for $\nabla_r \bar{n} = 0$)

$$b_+ = b_- \ ; \tag{3.6.9}$$

i.e., the numbers of particles that, when scattered, lose and acquire a given velocity are equal to each other. Hence we can find, in particular, the equilibrium value of the distribution function $\bar{n}_0$. In the presence of fields, $b_+ \neq b_-$ and $\bar{n}$ differs from $\bar{n}_0$. If the fields are weak, we may assume that at each point of the phase space the function $\bar{n}$ differs little from its equilibrium value; i.e.,

$$\bar{n} = \bar{n}_0 + \bar{n}_1 \ , \quad \bar{n}_1 \ll \bar{n}_0 \ . \tag{3.6.10}$$

In solving the differential equation (3.6.8), the following simplifying assumptions are normally introduced. First, electrons suffer, in the main, only elastic collisions with metallic ions. Light electrons (compared to ions) transfer to ions only an insignificant part of their total kinetic energy (only that accumulated during a time τ in a field). Apart from this, since the electron scattering during collisions is assumed to be isotropic, the collision probabilities $\nu(\boldsymbol{v}, \boldsymbol{v}')$ in (3.5.13, 14) do not depend on the direction of the vectors $\boldsymbol{v}$ and $\boldsymbol{v}'$ but only on the angle between them. Thus, according to the first assumption, when electrons collide the electronic energy hardly changes, i.e., $|\boldsymbol{v}| \approx |\boldsymbol{v}'|$, and therefore the function $\nu(\boldsymbol{v}, \boldsymbol{v}')$ depends on the angle between $\boldsymbol{v}$ and $\boldsymbol{v}'$, and on $|\boldsymbol{v}|$.

We now proceed to calculate the $b_+ - b_-$ in (3.6.8). Using the same assumptions, the function $\nu(\boldsymbol{v}, \boldsymbol{v}')$ is nonzero subject only to the condition that $|\boldsymbol{v}| = |\boldsymbol{v}'|$. It therefore follows immediately from the requirement of the finiteness of the total collision probability $\int d\boldsymbol{v}' \nu(\boldsymbol{v}, \boldsymbol{v}')$ that when $\boldsymbol{v} = \boldsymbol{v}'$ the function $\nu(\boldsymbol{v}, \boldsymbol{v}')$ should transform to a δ function $\delta(v - v')$. To avoid dealing with discontinuous functions, we introduce a collision function

$$\eta(v; \vartheta, \varphi; \vartheta', \varphi') \sin\vartheta' \, d\vartheta' \, d\varphi' \ , \tag{3.6.11}$$

which defines the probability that a particle moving with a velocity $\boldsymbol{v}$ in the direction ϑ, φ will deviate when scattered and will move with the same velocity in the direction ϑ', φ' (to within an element of the solid angle $\sin\vartheta' \, d\vartheta' \, d\varphi'$). An evident relationship exists between the functions ν and η:

$$\eta(v; \vartheta, \varphi; \vartheta', \varphi') = \int_0^{+\infty} dv' \, v'^2 \, \nu(\boldsymbol{v}, \boldsymbol{v}') \ . \tag{3.6.12}$$

Using (3.6.12), we rearrange (3.6.3, 4) in the forms

$$b_- = \bar{n}(\boldsymbol{r}, \boldsymbol{v}; t) \int d\varphi' \, d\vartheta' \sin\vartheta' [1 - \bar{n}(\boldsymbol{r}, \boldsymbol{v}'; t)] \eta \, (v; \vartheta, \varphi; \vartheta', \varphi') \tag{3.6.13}$$

and

$$b_+ = [1 - \bar{n}(\boldsymbol{r}, \boldsymbol{v}; t)] \int d\varphi' \, d\vartheta' \sin\vartheta' \, \bar{n}(\boldsymbol{r}, \boldsymbol{v}'; t) \eta(v; \vartheta', \varphi'; \vartheta, \varphi) \ . \tag{3.6.14}$$

If we strengthen the second assumption by supposing that the collision function does not depend on angles at all but depends only on v, then replacing the $\bar{n}$ according to (3.6.10) yields

$$b_+ - b_- = \{[(4\pi)^{-1} \int d\vartheta \sin\vartheta \, d\varphi \, \bar{n}_1] - \bar{n}_1\} 4\pi\eta(v) \ . \tag{3.6.15}$$

Equation (3.6.15) shows that $b_+ - b_-$ in this approximation depends on the perturbation of the distribution function $\bar{n}_1$ in a linear fashion. The factor $4\pi\eta(v)$ in (3.6.15) is equal to the total number of collisions which an electron suffers in unit time. Its inverse has the dimensions of time and is a familiar quantity, viz., the mean free time q (Sect. 3.3):

$$\tau(v) = [4\pi\eta(v)]^{-1} . \tag{3.6.16}$$

In the approximation adopted, this time is equal to the time during which a statistical equilibrium sets in when external fields are turned on or off. That is, it is equal to the relaxation time for a nonequilibrium distribution function. (If the quantity η is largely dependent on angles, the energy relaxation time and the momentum relaxation time may differ from each other—see Chap. 5.) Indeed, for $\partial\bar{n}/\partial t \neq 0$ and $(\partial\bar{n}/\partial t)_{\text{dif}} = (\partial\bar{n}/\partial t)_{\text{field}} = 0$, (3.6.6) yields, using (3.6.15, 16),

$$\frac{\partial\bar{n}}{\partial t} = \frac{\partial\bar{n}_1}{\partial t} = b_+ - b_- = -[\bar{\tau}(v)]^{-1}[\bar{n}_1 - (4\pi)^{-1}\int d\varphi\, d\vartheta \sin\vartheta\, \bar{n}_1] . \tag{3.6.17}$$

The quantity $\bar{n}_1$ may normally be represented as

$$\bar{n}_1 = v_i\chi(v) , \tag{3.6.18}$$

where $i = x, y, z$, and therefore $\int d\varphi\, d\vartheta \sin\vartheta \bar{n}_1 = \chi(v)\int d\varphi\, d\vartheta \sin\vartheta v_i = 0$. From (3.6.17) we then have

$$\bar{n}_1(t) = \bar{n}_1(0)\exp(-t/\bar{\tau}) . \tag{3.6.19}$$

Thus, we have adopted here the approximation of one relaxation time, which often works well when comparing with experiment. However, in some cases it is necessary to introduce several relaxation times for the various phenomena studied. Aside from this, we must take into account the kind of equilibrium distribution function $\bar{n}_0$ involved in the collision term $b_+ - b_-$. It may be a function which depends on the local density of particles, $\varrho(\boldsymbol{r})$ (with spatially inhomogeneous systems), i.e., $\bar{n}_0[\varrho(\boldsymbol{r})]$, or it may depend on the mean electron density $\bar{\varrho}$ (independent of coordinates). Thus, in (3.6.10) we have set

$$\bar{n} = \bar{n}_0(\varrho) + \bar{n}_1 .$$

Therefore, (3.6.17) is more correctly written as

$$\begin{aligned} b_+ - b_- &= -[\bar{n} - \bar{n}_0(\varrho)]/\bar{\tau} = -[\bar{n} - \bar{n}_0(\varrho)]/\bar{\tau} \\ &\quad + [\bar{n}_0(\varrho(r)) - \bar{n}_0(\varrho)]/\bar{\tau} = -\bar{n}_1/\bar{\tau} + \delta_n \bar{n}_0/\bar{\tau} , \end{aligned}$$

with

$$\delta_n \bar{n}_0 = \bar{n}_0[\varrho(r)] - \bar{n}_0(\bar{\varrho}) .$$

The term with $\delta_n \bar{n}_0$ in the Boltzmann equation is often omitted, but when we deal with spatial inhomogeneity [for $\varrho = \varrho(\boldsymbol{r})$], it must be taken into account.

Proceeding to particular transport effects, we must recall how the acceleration $\boldsymbol{a}$ in (3.6.2) in the presence of an electric $\boldsymbol{E}$ and magnetic $\boldsymbol{H}$ field is determined in the quasi-classical approximation. Here Newton's equation of motion reads

$$m\boldsymbol{a} = e\boldsymbol{E} + \frac{e}{c}(\boldsymbol{v} \times \boldsymbol{H}) \ , \tag{3.6.20}$$

where the first term gives the magnitude of the Coulomb electric force, and the second term the magnitude of the Lorentz magnetic force.

3.6.2 Electrical Conductivity

Let us consider a spatially homogeneous case in which $T = \text{const}$, the magnetic field is absent, and the electric field is aligned with the x axis: $\boldsymbol{E} = (E_x, 0, 0)$. From (3.6.20) we then have

$$a_x = \frac{e}{m} E_x \ , \qquad a_y = a_z = 0 \ . \tag{3.6.21}$$

The kinetic equation under these conditions takes the form

$$\frac{e}{m} E_x \frac{\partial \bar{n}}{\partial v_x} = b_+ - b_- \ . \tag{3.6.22}$$

Replace $\bar{n}$ according to (3.6.10), and the $b_+ - b_-$ according to (3.6.17). Because of the approximations adopted, we assume that the addition $\bar{n}_1$, destroying the isotropy of $\bar{n}_0$, in a first approximation may be expressed in the form of (3.6.18), i.e.,

$$\bar{n}_1 = v_x \chi(v) \ . \tag{3.6.23}$$

The quantity $\chi(v)$ here is an unknown small function which depends only on the absolute value of the velocity v (or energy ε). This choice of the form of $\bar{n}_1$ is natural from symmetry considerations, and also because it satisfies the conditions $\int d\tau_r d\tau_v \bar{n}_1 = 0$ and $\int d\tau_r d\tau_v \bar{n}_1 v^2 = 0$, which follow from the normalization (3.6.11) and the definition of the mean quadratic velocity $\overline{v^2}$ with the help of the equilibrium function $\bar{n}_0$. Formula (3.6.23) may be regarded as a first-order term in the expansion of the function $\bar{n}$ in powers of E_x (linearized Boltzmann equation). As a result, we get, instead of (3.6.22),

$$\frac{e}{m} E_x \frac{\partial n_0}{\partial v_x} = -\frac{v_x}{\bar{\tau}(v)} \chi(v) . \tag{3.6.24}$$

In (3.6.24) we proceed from differentiation with respect to v_x to differentiation with respect to ε, using the dispersion relation (3.5.4). From (3.6.24) we then find

$$\chi(v) = -\bar{\tau}(\varepsilon) e E_x \frac{\partial \bar{n}_0}{\partial \varepsilon} . \tag{3.6.25}$$

To integrate over $\boldsymbol{p}$, we determine the number of states in the intervals from p_x, p_y, p_z to $p_x + dp_x, p_y + dp_y, p_z + dp_z$ in the volume V. Quantum mechanics shows that one state is in the phase-space volume of size h^3 (3.5.6). The phase volume occupied by electrons is equal to the product of the volume in $\boldsymbol{p}$ and $\boldsymbol{r}$ space. Thus, to the above momentum interval corresponds the phase volume $V dp_x dp_y dp_z$, and the number of quantum states in it (with allowance for spin degeneracy) is equal to

$$d\tau_p = \frac{2V}{h^3} dp_x dp_y dp_z = m^3 d\tau_v \equiv 2V\left(\frac{m}{h}\right)^3 dv_x dv_y dv_z . \tag{3.6.26}$$

Here m^3 has appeared because of the replacement $\boldsymbol{p} = m\boldsymbol{v}$. The expression for the current density along the x axis will be

$$j_x = \frac{e}{V} \int v_x \bar{n}_1 \, d\tau_p = \frac{e}{V} \int v_x^2 \chi(v) \, d\tau_p = \frac{e^2 E_x}{V} \int v_x^2 \bar{\tau}(\varepsilon) \frac{\partial \bar{n}_0}{\partial \varepsilon} d\tau_p . \tag{3.6.27}$$

In (3.6.27) we replace v_x^2 by $1/3\, v^2$ and proceed to spherical coordinates in $\boldsymbol{v}$ space. As a result of the spherical symmetry, $\overline{v_x^2} = \overline{v_y^2} = \overline{v_z^2} = 1/3\, v^2$, where the dash from above denotes averaging over angles. Upon integration over the angular variables, we have

$$j_x = -(8\pi e^2 m^3/3h^3) E_x \int_0^{+\infty} dv v^4 \bar{\tau}(\varepsilon) \frac{\partial \bar{n}_0}{\partial \varepsilon} .$$

From the integration over dv we now move on to the integration over $d\varepsilon\,[mv\, dv = d\varepsilon$ and $v^3 = (2/m)^{3/2} \varepsilon^{3/2}]$. This gives

$$j_x = -(16\sqrt{2}\pi e^2 m^{1/2}/3h^3) E_x \int d\varepsilon \varepsilon^{3/2} \bar{\tau}(\varepsilon) \frac{\partial \bar{n}_0}{\partial \varepsilon} . \tag{3.6.28}$$

For $k_B T \ll \zeta_0$ the quantity $\partial \bar{n}_0/\partial \varepsilon$ may be taken to be approximately equal to $-\delta(\varepsilon - \zeta_0)$. Therefore,

$$j_x = (16\sqrt{2}\pi e^2 m^{1/2}/3h^3)\zeta_0^{3/2}\bar{\tau}(\zeta_0)E_x \ . \tag{3.6.29}$$

Using (3.5.17), the final result will be

$$j_x = (ne^2\bar{\tau}(\zeta_0)/m)E_x$$

and, consequently,

$$\sigma = ne^2\bar{\tau}(\zeta_0)/m \ . \tag{3.6.30}$$

Outwardly, (3.6.30) coincides with the classical formula (3.3.2) or (3.3.5). However, (3.6.30) involves the mean free time of the electrons on the Fermi surface rather than the mean free time of the classical theory. This derivation indicates that only the electrons present in the narrow smearing band near the Fermi energy play an active part in electrical conductivity. Like Drude's theory, the degenerate gas theory is incapable of yielding more detailed information on $\bar{\tau}(\zeta_0)$ since it disregards the interactions of electrons with thermal vibrations of the ionic lattice.

Here, just as in (3.3.3), we introduce the mean free path length $\bar{l}(\zeta_0)$:

$$\bar{l}(\zeta_0) = v(\zeta_0)\bar{\tau}(\zeta_0) \ . \tag{3.6.31}$$

The velocity is taken for electrons with energies ζ_0. According to (3.5.18), its value is approximately equal to 10^8 cm/s. By contrast with Drude's theory, the degenerate gas theory provides a natural substantiation for the constancy of the thermal velocity of a conduction electron and for the fact that the large value of $\bar{l}(\zeta_0)$ is determined straightforwardly by the large value of $\tau(\zeta_0)$. The estimates of $\bar{l}$ presented in Table 3.6. become quite reasonable. However, as before, we still must ascertain why these values are so large despite the "crowding" in the crystal.

3.6.3 Thermal Conductivity and the Wiedemann–Franz Relation

With thermal conductivity calculations we may no longer assume that $\nabla_r\bar{n} = 0$, and therefore we must take the term $(\partial n/\partial t)_{\text{dif}}$ in (3.6.2) into account. If we assume that the temperature varies only along x and only $\partial T/\partial x$ is nonzero, then we should take only the summand $\partial\bar{n}/\partial x$ into account in the quantity $\nabla_r\bar{n}$. Thus, (3.6.24) in the same linear approximation will take a more general form

$$v_x\frac{\partial\bar{n}_0}{\partial x} + \frac{eE_x}{m}\frac{\partial n_0}{\partial v_x} = -\frac{v_x}{\bar{\tau}(x)}\chi(v) \ , \tag{3.6.32}$$

whence, upon proceeding from $\partial/\partial v_x$ to $\partial/\partial\varepsilon$, we find, in place of (3.6.25),

$$\chi(v) = -\bar{\tau}(v)\left(\frac{\partial \bar{n}_0}{\partial x} + eE_x \frac{\partial \bar{n}_0}{\partial \varepsilon}\right) . \tag{3.6.33}$$

To calculate the thermal conductivity, we find the thermal flux W_x in the presence of E_x and $\partial T/\partial x$. According to Lorentz, the expression for the thermal flux has the form

$$W_x = \int d\tau_p \bar{n} v_x (mv^2/2) = 1/2 \int d\tau_p v_x^2 v^2 \chi(v) . \tag{3.6.34}$$

Prior to substituting (3.6.33) into (3.6.34), we note that $\partial\bar{n}_0/\partial x$ here is dependent on the temperature gradient. The quantity $\bar{n}_0$, according to (3.5.11), is a function of $\xi \equiv (\varepsilon - \zeta)/k_B T$. Therefore,

$$\begin{aligned} \frac{\partial \bar{n}_0}{\partial x} &= \frac{\partial n_0}{\partial T}\frac{\partial T}{\partial x} = \frac{\partial \bar{n}_0}{\partial \xi}\frac{\partial \xi}{\partial T}\frac{\partial T}{\partial x} \\ &= \frac{\partial n_0}{\partial \varepsilon}\frac{\partial \varepsilon}{\partial \xi}\frac{\partial \xi}{\partial T}\frac{\partial T}{\partial x} = -\frac{\partial \bar{n}_0}{\partial \varepsilon}\left[\frac{\varepsilon}{T} + T\frac{\partial}{\partial T}\left(\frac{\zeta}{T}\right)\right]\frac{\partial T}{\partial x} . \end{aligned} \tag{3.6.35}$$

Substituting (3.6.35) into (3.6.33), and then into (3.6.28) for j_x and into (3.6.34), we find, after using (3.6.26) and integrating over angles,

$$j_x = \frac{e}{m}\left\{K_1\left[eE_x - T\frac{\partial}{\partial T}\left(\frac{\zeta}{T}\right)\frac{\partial T}{\partial x}\right] + K_2 \frac{1}{T}\frac{\partial T}{\partial x}\right\} , \tag{3.6.36}$$

$$W_x = \frac{1}{m}\left\{K_2\left[eE_x - T\frac{\partial}{\partial T}\left(\frac{\zeta}{T}\right)\frac{\partial T}{\partial x}\right] + K_3 \frac{1}{T}\frac{\partial T}{\partial x}\right\} . \tag{3.6.37}$$

The contracted notation introduced here for I-type integrals from (3.5.30) is as follows:

$$K_j = -\frac{N}{\zeta_0^{3/2}} \int_0^{+\infty} d\varepsilon \varepsilon^{j+1/2} \bar{\tau}(\varepsilon) \frac{\partial \bar{n}_0}{\partial \varepsilon} . \tag{3.6.38}$$

According to (3.5.40), these integrals are equal to

$$K_j = \frac{N}{\zeta_0^{3/2}}\left\{\zeta^{j+1/2}\bar{\tau}(\zeta) + \frac{\pi^2}{6}(k_B T)^2 \frac{\partial^2}{\partial \varepsilon^2}[\bar{\tau}(\varepsilon)\varepsilon^{j+1/2}]|_{\varepsilon=\zeta} + \dots\right\} . \tag{3.6.39}$$

To determine the thermal conductivity coefficient κ, we must set $j_x = 0$ in (3.6.37) for $\partial T/\partial x \neq 0$. From this condition we find the electric thermodiffuse field E_x, and substitute it into (3.6.37). As a result, we find

$$\kappa = -\frac{W_x}{\partial T/\partial x} = \frac{1}{m}\frac{K_1 K_3 - K_2^2}{K_1 T} \ . \tag{3.6.40}$$

If we take for K_j the first approximation from (3.6.39), then the numerator in (3.6.40) will go to zero. Therefore the thermal conductivity due to conduction electrons is a second-order phenomenon in $k_B T/\zeta_0$. To calculate it we must allow for the second-order expansion terms in (3.6.40), since the zero approximation in (3.6.40) corresponds to an equilibrium distribution at $T = 0\,\mathrm{K}$. In calculating the electrical conductivity at $\partial T/\partial x = 0$—i.e., for a purely electrical effect—we can confine ourselves to this approximation, but in calculations of thermal phenomena we must take into account the influence of temperature on the state of the system and, consequently, on $\bar{n}(\varepsilon)$. Calculating the numerator on the right-hand side of (3.6.40) in the second approximation, we find

$$K_1 K_3 - K_2^2 \approx (1/3)\pi^2 (k_B T)^2 n^2 [\bar{\tau}(\zeta_0)]^2 \ .$$

Obviously all the terms containing $\partial^2\bar{\tau}/\partial\varepsilon^2$ and $\partial\bar{\tau}/\partial\varepsilon$ cancel out in this expression. The denominator (3.6.39) can be calculated in the zero approximation

$$K_1 T \approx n\bar{\tau}(\zeta_0) T \ .$$

Therefore, using (3.6.31), the final result is

$$\kappa = \frac{\pi^2}{3}\frac{k_B^2}{m} n\bar{\tau}(\zeta_0) T = \frac{\pi^2}{3}\frac{k_B^2}{m}\frac{n\bar{l}(\zeta_0)}{v(\zeta_0)} T \ . \tag{3.6.41}$$

Formula (3.6.41) coincides outwardly with the classical formula (3.3.11) for κ with the replacement of $\bar{\tau}$ by $\bar{\tau}(\zeta_0)^2$. As first noted by *Frenkel* [3.3], (3.6.41) satisfies the familiar ratio of the kinetic theory of gases $\kappa = (1/3)\, l(\zeta_0)\, v(\zeta_0) C_v^{\mathrm{el}}$.

From a comparison of (3.6.30) for σ and (3.6.41) for κ we obtain the Wiedemann–Franz ratio, already familiar from (3.3.12):

$$\kappa/\sigma = (\pi^2/3)(k_B/e)^2 T \approx 2.43 \cdot 10^{-8}\, T \ \mathrm{W}\,\Omega/\mathrm{K}^2 \ . \tag{3.6.42}$$

The theoretical expression for the Lorentz number $L = \kappa/\sigma T = 2.43 \cdot 10^{-8}$ $\mathrm{W}\,\Omega/\mathrm{K}^2$ is in somewhat better agreement with the mean experimental value of L than the classical formula (3.3.12).

3.6.4 Thermoelectric Phenomena

We now investigate the release of heat in a metal when an electric current passes through it and in the presence of a temperature gradient arising from the work of the electric field on conduction electrons. We may question whether these phenomena can be properly described without allowance for the interaction of electrons with the ionic lattice, but in fact, just as in the derivation of the Joule–Lenz law (3.3.5), the mechanism of this interaction is quite unimportant to a qualitative study of the question posed, since the energy piled up by the electrons in the field is transferred to the lattice as a consequence of the law of conservation of energy. However, there does exist a specific thermoelectric effect, in which the ionic-lattice vibrations (phonons) play a decisive role. This is the effect of electron entrainment by phonons, predicted by *Gurevitch* [3.18] and discovered experimentally by *Frederikse* in 1953 and independently by *Geballe* [3.16].

Let us consider thermoelectric effects for a chemically and structurally homogeneous wire. Suppose that a current j_x and a temperature gradient $\partial T/\partial x$ are directed along its axis. The rate of heat release dQ/dt in unit volume is then equal to

$$dQ/dt = j_x E_x - \partial W_x/\partial x \ , \tag{3.6.43}$$

where $j_x E_x$ is the specific power released by the electric current, and $\partial W_x/\partial x$ is the thermal current divergence which, by the continuity equation, is equal to the rate of heat flow inside the unit volume considered. Express E_x and W_x in terms of j_x with the help of (3.6.37, 38). Then, using (3.6.28, 38, 40), we find

$$\begin{aligned} E_x &= \frac{m}{e^2}\frac{1}{K_1} j_x + \frac{T}{e}\frac{d}{dT}\left(\frac{\partial \zeta}{\partial T}\right)\frac{\partial T}{\partial x} + \frac{K_2}{K_1}\frac{1}{eT}\frac{\partial T}{\partial x} \\ &= \frac{j_x}{\sigma} + \frac{T}{e}\frac{d}{dT}\left(\frac{\partial \zeta}{\partial T}\right)\frac{\partial T}{\partial x} + \frac{K_2}{K_1}\frac{1}{eT}\frac{\partial T}{\partial x} \ , \end{aligned} \tag{3.6.44}$$

$$W_x = \frac{K_2}{K_1}\frac{j_x}{e} + \frac{1}{m}\frac{K_2^2 - K_1 K_3}{K_1 T}\frac{\partial T}{\partial x} = \frac{K_2}{K_1}\frac{j_x}{e} - \kappa \frac{\partial T}{\partial x} \ . \tag{3.6.45}$$

Substitution of (3.6.44, 45) into (3.6.43) yields

$$\frac{dQ}{dT} = \frac{1}{\sigma} j_x^2 - \frac{j_x}{e} T \frac{\partial}{\partial x}\left(\frac{K_2}{TK_1} - \frac{\zeta}{T}\right) + \frac{\partial}{\partial x}\left(\kappa \frac{\partial T}{\partial x}\right) \ . \tag{3.6.46}$$

The first term on the right-hand side of (3.3.4) gives the magnitude of the Joule–Lenz heat (3.3.4), which is of a purely electric origin. The third term, independent of j_x, gives the magnitude of the heat supplied to a given volume element owing to thermal conductivity. The two effects do not depend on the mutual direction of the electric and thermal current and are even. The second term, conversely, pertains to the class of odd effects; i.e., it reverses sign when the direction of j_x or $\partial T/\partial x$ is reversed. This term describes all the thermoelectric phenomena governed by the interrelation of the thermal and electric processes occurring in electronic conductors. We deal here with three of these fundamental phenomena: Thomson heat, Peltier heat, and the thermoelectromotive force (thermal emf).

Thomson heat. Thomson heat is a reversible heat which is released in a homogeneous but nonuniformly heated conductor ($\partial T/dx \neq 0$) when a current is passed through it. Qualitatively, the effect may be explained as follows. If, with a given current direction, the electrons move from the more heated to the less heated end of the conductor, they transfer the extra energy to the ionic lattice (Thomson heat release). With the current flowing in the opposite direction, the electrons available in the colder sections of the conductor arrive at the more heated ones and increase their energy due to the more heated ions (Thomson heat absorption). Of course, this explanation is superficial, since it disregards the thermal emf, the entrainment by phonons, etc., which may largely influence the effect and even entail the reversal of its sign.

The magnitude of Thomson heat is immediately obtainable from the second term on the right-hand side of (3.6.46). In this term we merely replace the quantity $\partial/\partial x$ by $(\partial T/\partial x)\partial/\partial T$:

$$\frac{dQ_T}{dt} = -\frac{T}{e}\frac{\partial}{\partial T}\left(\frac{K_2}{TK_1} - \frac{\zeta}{T}\right) j_x \frac{\partial T}{\partial x} \equiv -\kappa_T j_x \frac{\partial T}{\partial x} . \qquad (3.6.47)$$

Examination of (3.6.47) shows that the Thomson heat is proportional to the current density and temperature gradient. The κ_T is termed the Thomson coefficient. Like thermal conductivity, in the zero approximation for the integrals K_j it is equal to zero; i.e., it is a second-order effect with respect to $k_B T/\zeta_0$. Therefore, calculating K_2/K_1 in the second approximation according to (3.6.39), we find

$$\kappa_T = \frac{\pi^2}{3}\frac{k_B^2}{e} T\left(\frac{3}{2\zeta_0} + \frac{1}{\bar{\tau}(\zeta_0)}\frac{\partial \bar{\tau}(\varepsilon)}{\partial \varepsilon}\bigg|_{\varepsilon=\zeta_0}\right) . \qquad (3.6.48)$$

Expression (3.6.48) may also be rearranged as

$$\kappa_T = \frac{\pi^2}{2}\frac{k_B^2}{e}\frac{T}{\zeta_0}\left(1 + \frac{2}{3}\frac{\partial \ln \bar{\tau}(\varepsilon)}{\partial \varepsilon}\bigg|_{\varepsilon=\zeta_0}\right) , \qquad (3.6.49)$$

whereas the classical Lorentz theory yielded the expression

$$\kappa_T^{el} = 3k_B/2e . \qquad (3.6.50)$$

Comparing (3.6.49) with (3.6.50), we see that for the κ_T quantum theory yields a value which is hundreds of times smaller ($k_B T/\zeta_0 \lesssim 10^{-2}$) than that given by classical theory, and this is in qualitative agreement with experiment. For temperatures that are not too low, the linear dependence of κ_T on T (3.6.49) is also well borne out by experiment. Insofar as the sign of κ_T goes, it is positive for some metals (Cu, Ag, Au, Hg, . . .) and negative for others (Pb, Pt, Fe, . . .). It is only the sign of the derivative $\partial \ln \bar{\tau}(\varepsilon)/\partial\varepsilon|_{\varepsilon=\zeta_0}$ in (3.6.49) that is responsible for the sign of the effect (recall that $e = -|e|$).

Peltier heat. Suppose now that the temperature gradient is absent ($\partial T/\partial x = 0$) but the wire consists of two pieces of different metals, I and II (Fig. 3.16). Then, with $j_x \neq 0$, the second term on the right-hand side of (3.6.46) will be different from zero at the junction a, b of the two metals, where the quantity standing under the derivative sign changes its value abruptly. Thus, it should be expected that when a current is passed through the junction of two homogeneous and uniformly heated metals heat release or heat absorption occurs, in addition to Joule–Lenz heat. This additional heat is called Peltier heat. It may be explained by the fact that the mean energy of the electrons that participate in the current depends on the dispersion relation, electron concentration, and scattering mechanism; therefore, the mean electron energy is different in different metals. And, consequently, when passing from one metal to another, the electrons either donate the excess energy or receive it from the lattice (depending on the direction of the current). In the first case the Peltier heat is released; in the second case it is absorbed.

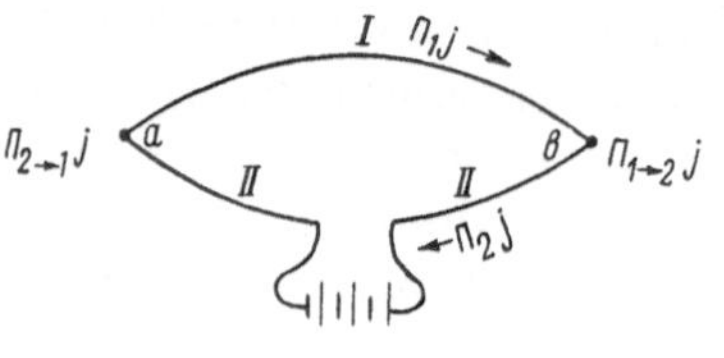

Fig. 3.16. Peltier effect

To find the Peltier heat, we integrate the second term on the right-hand side of (3.6.46) over dx between any two points that lie on either side of the junction of the metals, I and II:

$$\frac{dQ_{1\to 2}}{dt} = \frac{T}{e} j_x \int_1^2 dx \frac{\partial}{\partial x}\left(\frac{K_2}{TK_1} - \frac{\zeta}{T}\right).$$

The factor standing before the j_x is called the Peltier coefficient. According to (3.6.48), it is equal to

$$\Pi_{1\to 2} = \frac{T}{e}\left[\left(\frac{K_2}{TK_1} - \frac{\zeta}{T}\right)_2 - \left(\frac{K_2}{TK_1} - \frac{\zeta}{T}\right)_1\right]$$

$$= \frac{\pi^2}{2}\frac{k_B T}{e}\left[\frac{k_B T}{\zeta_{02}}\left(1 + \frac{2}{3}\left.\frac{d\ln\bar{\tau}_2(\varepsilon)}{d\ln\varepsilon}\right|_{\varepsilon=\zeta_{02}}\right)\right. \tag{3.6.51}$$

$$\left. - \frac{k_B T}{\zeta_{01}}\left(1 + \frac{2}{3}\left.\frac{d\ln\bar{\tau}_1(\varepsilon)}{d\ln\varepsilon}\right|_{\varepsilon=\zeta_{01}}\right)\right] .$$

From (3.6.49, 51) we find the relation, known from thermodynamics, between the Peltier and the Thomson coefficient:

$$\frac{d}{dT}\left(\frac{\Pi_{1\to 2}}{T}\right) = \frac{\kappa_{T_2} - \kappa_{T_1}}{T} . \tag{3.6.52}$$

Equation (3.6.51) is difficult to compare with experiment because the quantity $d\ln\bar{\tau}(\varepsilon)/d\ln\varepsilon$ is unknown in our model. However, quantum theory, just as in the case of the Thomson effect, yields a correct estimate of the quantity $\Pi_{1\to 2}$ as compared with the overstated value ($\sim k_B T/e$) in the classical theory.

Thermoelectromotive force (*thermal emf*) or Seebeck effect. In a circuit made up of different metals whose junctions are maintained at different temperatures a thermal emf arises. Consider a circuit composed of two metals, I and II, with two junctions: b at a temperature T_1 and c at T_2 (Fig. 3.17), with $T_1 > T_2$. The points a and d are maintained at equal temperature $T (T_1 > T > T_2)$. The thermal emf $\mathscr{E}_T$ sought for is equal to the potential difference $\varphi_a - \varphi_d$ arising in such an open circuit:

$$\mathscr{E}_T = \varphi_a - \varphi_d = -\int_a^d E_x dx = -\oint E_x dx . \tag{3.6.53}$$

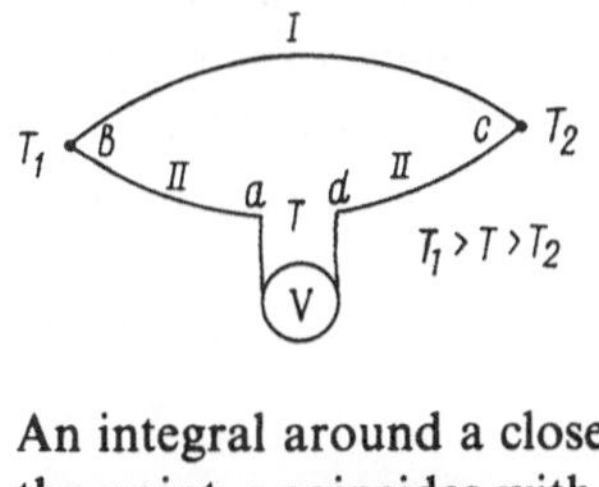

Fig. 3.17. Thermal emf

An integral around a closed circuit may also be taken, since in a closed circuit the point a coincides with the point d.

The occurrence of a thermal emf may be explained as follows. If there is a temperature gradient in the conductor, the electrons at the hot end acquire higher energies (velocities) and start flowing from the hot to the cold end where the negative charge begins to increase, a noncompensated positive charge persisting at the hot end. This process will continue until a dynamic equilibrium sets in as a result of the potential difference that creates a reverse electron flow. The algebraic sum of the potential differences in the circuit generates what is known as the bulk thermal emf component. The second component, i.e., the contact component, arises from the temperature dependence of the contact potential difference. With the temperature of the junctions b and c being different, a potential difference arises which makes a contribution to the potential difference.

With the circuit open there is no electric current ($j_x = 0$). Therefore, according to (3.6.36), the electric field intensity produced by the thermal diffusion of electrons is equal to

$$E_x = \frac{T}{e}\frac{\partial}{\partial x}\frac{\zeta}{T} + \frac{K_2}{K_1}\frac{1}{eT}\frac{\partial T}{\partial x} .$$

Substituting this value of E_x into (3.6.53), we find (in the approximation $\partial\zeta/\partial T \approx 0$)

$$\mathscr{E}_T = -\frac{1}{e}\oint dx\left(\frac{K_2}{K_1 T}\frac{\partial T}{\partial x} - \frac{\zeta}{T}\frac{\partial T}{\partial x}\right) .$$

We now pass on to a new integration variable (from dx to dT) and break the integral up into two terms with respect to the metals. This gives

$$\mathscr{E}_T = \frac{1}{e}\int_{T_1}^{T_2} dT\left[\left(\frac{K_2}{K_1 T} - \frac{\zeta}{T}\right)_{II} - \left(\frac{K_2}{K_1 T} - \frac{\zeta}{T}\right)_I\right] . \tag{3.6.54}$$

Comparing (3.6.54) with (3.6.51), we find the relation between the thermal emf and the Peltier heat

$$\mathscr{E}_T = -\int dT\,\Pi_{I\to II}/T \tag{3.6.55}$$

in full agreement with the conclusions of thermodynamics [3.19, 20].

Again, (3.6.54) shows that in the zeroth approximation the quantity $\mathscr{E}_T$ is equal to zero. In the second-order approximation, according to (3.6.51) we obtain (if it is assumed that $\partial \ln \bar{\tau}/\partial\varepsilon|_{\varepsilon=\zeta_0}$ is independent of temperature)

$$\mathscr{E}_T = -\frac{\pi^2 k_B^2}{6e}\left\{\left[\frac{1}{\zeta_0}\left(1 + \frac{2}{3}\frac{\partial \ln \bar{\tau}}{\partial \ln \varepsilon}\bigg|_{\varepsilon=\zeta_0}\right)_I \right.\right. \tag{3.6.56}$$

$$\left.\left. - \left[\frac{1}{\zeta_0}\left(1 + \frac{2}{3}\frac{\partial \ln \bar{\tau}}{\partial \ln \varepsilon}\bigg|_{\varepsilon=\zeta_0}\right)_{II}\right]\right\}(T_1^2 - T_2^2) .$$

The classical electron theory of metals predicts, to a first approximation, a nonzero thermal emf

$$\mathscr{E}_T^{\mathrm{cl}} = -(k_B/e)\ln(n_I/n_{II})(T_1 - T_2) .$$

Thus, different temperature dependences of $\mathscr{E}_T$ are obtained. Also, the quantum theory predicts a smaller absolute value of $\mathscr{E}_T$ because of the factor $k_B T/\zeta_0$. Apart from this, the differential thermal emf (for $T_1 - T_2 = dT$), according to Nernst's quantum theory for $T \to 0$, yields $\mathscr{E}_T \to 0$, whereas the classical theory leads to a finite value of $\mathscr{E}_T$.

3.6.5 Galvanomagnetic Phenomena

We consider two major galvanomagnetic phenomena: the Hall effect and magnetoresistance. These effects arise in a current-carrying conductor subject to an externally applied magnetic field transverse to the current direction. As will be seen, the Fermi gas model does not provide a complete explanation of these effects. Nevertheless, we will treat them in terms of this model, which will enable us to reveal some of the difficulties encountered.

Assume that our current-carrying conductor (aligned with the x axis) is placed in a constant uniform magnetic field $\boldsymbol{H}$ oriented along the z axis (Fig. 3.18). An electron travelling with a velocity $\boldsymbol{v} = (v_x, v_y, v_z)$ is then subject to the sum of Coulomb and Lorentz forces (3.6.20). In keeping with the geometry of the fields (Fig. 3.18) $\boldsymbol{E} = (E_x, E_y, 0)$, $\boldsymbol{H} = (0, 0, H_z)$, (3.6.20) will take the form

$$m^* a_x = eE_x + \frac{e}{c} v_y H_z \ ,$$

$$m^* a_y = eE_y - \frac{e}{c} v_x H_z \ , \qquad (3.6.57)$$

$$m^* a_z = 0 \ .$$

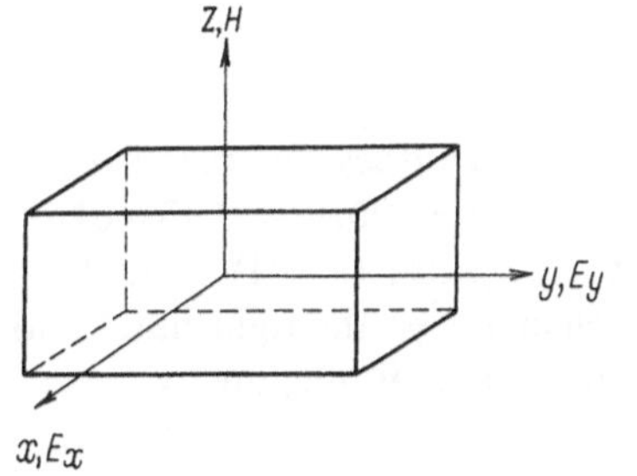

Fig. 3.18. Hall effect

The magnetic field H_z, as it deviates the electron along the y axis, produces a transverse electric field E_y, in addition to the external longitudinal field E_x along the current j_x. Thus, according to (3.6.2),

$$\left(\frac{\partial\bar{n}}{\partial t}\right)_{\text{dif}}+\left(\frac{\partial\bar{n}}{\partial t}\right)_{\text{field}}=\frac{\partial\bar{n}}{\partial x}v_x+\frac{\partial\bar{n}}{\partial y}v_y-\left(eE_x+\frac{e}{c}v_yH_z\right)\frac{\partial\bar{n}}{\partial v_x}$$
$$-\left(eE_y-\frac{e}{c}v_xH_z\right)\frac{\partial\bar{n}}{\partial v_y}\ . \tag{3.6.58}$$

The distribution function perturbed by the fields and $\partial T/\partial x$ will be sought in the form of (3.6.10), where the perturbation $\bar{n}_1$ is represented as [3.21]

$$\bar{n}_1=v_x\chi_x(v)+v_y\chi_y(v)\ . \tag{3.6.59}$$

Instead of (3.6.24) we then have for the collision number

$$b_+-b_-=-\frac{1}{\tau}(v_x\chi_x+v_y\chi_y)\ . \tag{3.6.60}$$

In (3.6.58) the terms $E_{x,y}(\partial\bar{n}_1/\partial v)_{x,y}$ should be neglected, in view of their smallness. When acting on an arbitrary function that depends only on the velocity modulus, the operator $\frac{e}{c}H_z\left(v_y\frac{\partial}{\partial v_x}-v_x\frac{\partial}{\partial v_y}\right)$ involved in (3.6.58) gives a zero:

$$v_y\frac{\partial f(v^2)}{\partial v_x}-v_x\frac{\partial f(v^2)}{\partial v_y}=2(v_yv_x-v_xv_y)\frac{\partial f(v^2)}{\partial v^2}=0\ .$$

Substitution of (3.6.59, 60) into (3.6.58) with allowance for what has been stated above yields

$$v_x\frac{\partial\bar{n}_0}{\partial x}+v_y\frac{\partial\bar{n}_0}{\partial y}-m^*\frac{\partial\bar{n}_0}{\partial\varepsilon}(eE_xv_x+eE_yv_y)$$
$$-\frac{eH_z}{m^*c}(v_y\chi_x-v_x\chi_y)=\frac{1}{\bar{\tau}}(v_x\chi_x+v_y\chi_y)\ . \tag{3.6.61}$$

Equation (3.6.61) shows that the magnetic field per se does not perturb the statistical equilibrium, since at $E_x=E_y=0$ and $\partial\bar{n}/\partial x=\partial\bar{n}/\partial y=0$, (3.6.61) is satisfied for $\chi_x=\chi_y=0$ and for an arbitrary H_z. Since (3.6.61) should be satisfied whatever the values of v_x and v_y, the coefficients on the right-hand and left-hand sides of (3.6.61) with v_x and v_y are equal in magnitude, giving

$$\frac{\partial\bar{n}_0}{\partial x}-eE_x\frac{\partial\bar{n}_0}{\partial\varepsilon}=\bar{\tau}^{-1}\chi_x-\frac{eH_z}{m^*c}\chi_y\ ,$$
$$\frac{\partial\bar{n}_0}{\partial y}-eE_y\frac{\partial\bar{n}_0}{\partial\varepsilon}=\frac{eH_z}{m^*c}\chi_x+\bar{\tau}^{-1}\chi_y\ . \tag{3.6.62}$$

Before solving these two equations for the required functions χ_x and χ_y, we introduce the contracted notation

$$L_x = \frac{\partial \bar{n}_0}{\partial x} - eE_x \frac{\partial n_0}{\partial \varepsilon} , \quad L_y = \frac{\partial \bar{n}_0}{\partial y} - eE_y \frac{\partial \bar{n}_0}{\partial \varepsilon} , \tag{3.6.63}$$

$$s = -eH_z\bar{\tau}/m^*c .$$

Observed that the dimensionless quantity s is equal to the ratio of the mean free path $\bar{l} = v\bar{\tau}$ to the cyclotron orbital radius (3.5.91)

$$r_H = v/\omega_H = m^*cv/|e|\,H_z . \tag{3.6.64}$$

Indeed,

$$s = |e|\,\bar{\tau}H_z/m^*c = |e|\,\bar{l}H_z/m^*vc = \bar{l}/r_H . \tag{3.6.65}$$

Thus, substituting (3.6.63) into (3.6.62), we find

$$\chi_x = \bar{\tau}(L_x - sL_y)/(1+s^2) , \quad \chi_y = \bar{\tau}(L_y + sL_x)/(1+s^2) . \tag{3.6.66}$$

Owing to the presence of two perturbation functions, χ_x and χ_y, we obtain, in place of (3.6.36, 37), two electric current terms and two thermal current terms in which we replace $d\tau_p$, according to (3.6.26), by $dv_x dv_y dv_z$. The term with $\bar{n}_0$ and the term with χ_y in j_x and W_x cancel out by symmetry, and the term with χ_x in j_y and W_y cancels out because the integrand is odd:

$$j_x = (e/3V)\int d\tau_p v^2 \chi_x \tag{3.6.67}$$
$$= (8\pi e m^{*3}/3h^3) \int_0^{+\infty} dv v^4 \bar{\tau}(L_x - sL_y)/(1+s^2) ,$$

$$j_y = (e/3V)\int d\tau_p v^2 \chi_y$$
$$= (8\pi e m^{*3}/3h^3) \int_0^{+\infty} dv v^4 \bar{\tau}(L_y + sL_x)/(1+s^2) , \tag{3.6.68}$$

$$W_x = (m^*/6V)\int d\tau_p v^4 \chi_x$$
$$= (8\pi m^{*4}/6h^3) \int_0^{+\infty} dv v^6 \bar{\tau}(L_x - sL_y)/(1+s^2) , \tag{3.6.69}$$

$$W_y = (m^*/6V)\int d\tau_p v^4 \chi_y$$
$$= (8\pi m^{*4}/6h^3) \int_0^{+\infty} dv v^6 \bar{\tau}(L_y + sL_x)/(1+s^2) , \tag{3.6.70}$$

Formulas (3.6.67–70) allow all transverse galvanomagnetic and thermomagnetic phenomena to be described quantitatively. Here we restrict our attention to the Hall effect and magnetoresistance (the Thomson effect).

Isothermal Hall effect. The gist of the isothermal Hall effect is that an electric field E_y occurs which is transverse to the current j_x induced by a magnetic field H_z. The procedure for measuring this phenomenon is depicted in Fig. 3.18. Under isothermal conditions we have: $\partial T/\partial x = \partial T/\partial y = 0$, i.e., $T = \text{const}$ (and, consequently, $\partial \bar{n}_0/\partial x = \partial \bar{n}_0/\partial y = 0$). If the electric circuit along the y axis is open, then $j_y = 0$. Hence we find the intensity of the Hall field E_y. Then (3.6.67, 68), with the L_x and L_y replaced according to (3.6.63), will take the form

$$j_x = -8\pi e^2 m^{*3}/3h^3 \left[E_x \int_0^{+\infty} dv v^4 \frac{\partial \bar{n}_0}{\partial \varepsilon} \frac{\bar{\tau}}{1+s^2} - E_y \int_0^{+\infty} dv v^4 \frac{\partial \bar{n}_0}{\partial \varepsilon} \frac{\bar{\tau} s}{1+s^2} \right] , \tag{3.6.71}$$

$$0 = E_y \int_0^{+\infty} dv v^4 \frac{\partial \bar{n}_0}{\partial \varepsilon} \frac{\bar{\tau}}{1+s^2} + E_x \int_0^{+\infty} dv v^4 \frac{\partial \bar{n}_0}{\partial \varepsilon} \frac{\bar{\tau} s}{1+s^2} . \tag{3.6.72}$$

In these formulas we proceed to the integration over the energy $d\varepsilon$ and find $v^4 dv = (2\varepsilon)^{3/2}(m^*)^{-5/2} d\varepsilon$. For the quadratic dispersion relation (3.5.4)

$$j_x = L_1 E_x - L_2 E_y , \tag{3.6.73}$$

$$0 = L_2 E_x + L_1 E_y , \tag{3.6.74}$$

where we have introduced the contracted notation

$$L_1 = -\frac{16\pi e^2 (2m^*)^{1/2}}{3h^3} \int_0^{+\infty} d\varepsilon \frac{\partial \bar{n}_0}{\partial \varepsilon} \frac{\varepsilon^{3/2} \tau(\varepsilon)}{1+s^2(\varepsilon)} , \tag{3.6.75}$$

$$L_2 = -\frac{16\pi e^2 (2m^*)^{1/2}}{3h^3} \int_0^{+\infty} d\varepsilon \frac{\partial \bar{n}_0}{\partial \varepsilon} \frac{\varepsilon^{3/2} \bar{\tau}(\varepsilon) s(\varepsilon)}{1+s^2(\varepsilon)} . \tag{3.6.76}$$

Solving (3.6.73, 74) for E_x and E_y yields

$$E_x = L_1 j_x/(L_1^2 + L_2^2) , \tag{3.6.77}$$

$$E_y = -L_2 j_x/(L_1^2 + L_2^2) . \tag{3.6.78}$$

From (3.6.77) we find the specific electrical conductivity in a magnetic field, $\sigma(H_z)$, or the magnetoresistance $\varrho(H_z) = 1/\sigma(H_z)$:

$$\varrho(H_z) = L_1/(L_1^2 + L_2^2) \tag{3.6.79}$$

and from (3.6.78) the transverse electric Hall field. If, in calculating the integrals

(3.6.75, 76), we confine ourselves to the zero approximation (3.6.39), then, from (3.6.63), we have

$$L_2 = s(\zeta_0)L_1 \equiv \omega_H \bar{\tau}(\zeta_0)L_1 \ . \tag{3.6.80}$$

So, substituting (3.6.80) into (3.6.79) and then into (3.6.78), we obtain

$$E_y = e\bar{\tau}(\zeta_0)H_z j_x/m^* c\sigma(H_z) \ . \tag{3.6.81}$$

Equation (3.6.81) shows that the Hall field E_y, with neglect of the H_z dependence of electrical conductivity (see below), depends on the magnetic field H_z and the longitudinal current j_x in a linear fashion. Thus we have derived a formula for the odd transverse galvanomagnetic Hall effect (E_y changes sign with sign reversal of H_z or j_x). The factor before $H_z j_x$ in Eq. (3.6.81) is called the Hall constant

$$R = e\bar{\tau}(\zeta_0)/m^* c\sigma(H_z) \ . \tag{3.6.82}$$

We will see (3.6.86) that, in the first approximation,

$$\sigma(H_z) = \sigma(0) \ , \tag{3.6.83}$$

where $\sigma(0)$ is defined according to (3.6.30). Therefore, using (3.6.83), the result for the Hall constant found from (3.6.80) is

$$R = 1/nec \ . \tag{3.6.84}$$

From (3.6.84) we see that the sign of R is determined by the sign of the current carrier (electronic) charge e, and the value of R by the electron density n (as in classical theory).

Table 3.5 presents R values obtained from experiment and calculated according to (3.6.84) by inserting, instead of n, the number of valence electrons per unit volume. For a number of metals the agreement between theory and experiment is good, especially for liquid alkali metals where the assumption that n coincides with the number of atoms in unit volume apparently works best. However, there are also substantial differences between theory and experiment. First, for a number of metals (Be, Zn, Cd, etc.) experiment yields a sign of the Hall constant other than that which theory predicts according to (3.6.84). Second, in a number of metals (ferromagnetic metals, Bi, and others) the observed value of the constant is several orders of magnitude larger than the theoretical values.

This "catastrophe" in the Hall constant in both classical and quantum free-electron gas models arises from the crudeness of these models, and is eliminated only in the more exact band model (Chap. 4).

Table 3.5. Experimental and theoretical values of the Hall constant R for different metals

Metals	$R \cdot 10^{12}$ [V cm/A G] $R \cdot 10^{14}$ [V m/A T]	
	experiment [3.22]	theory [3.19]
Cu	−0.54	−0.74
Ag	−0.9	−1.04
Au	−0.7	−1.05
Li	−1.7	−1.31
Na	−2.1	−2.4
Be	+7.7	−0.25
Zn	+1.0	−0.46
Cd	+0.53	−0.65
Al	−0.3	−0.34

Using the approximate formula (3.6.84), the product of the Hall constant and the specific electrical conductivity $R\sigma$ is directly proportional to the mobility of a conduction electron, μ, i.e., the velocity it acquires in a field of unit intensity ($E_x = 1$). Indeed, according to (3.3.1), we have

$$\mu = v/E_x = \sigma/ne \ ,$$

and, from (3.6.84), we find

$$\mu/c = \sigma R \ . \tag{3.6.85}$$

It follows from experimental data for the values of σR for metals and electrolytes that the mobility μ varies comparatively little from metal to metal (except for metals exhibiting anomalously large values of $|R|$) and is hundreds of times larger than the mobility of ions in liquid electrolytes.

Variation of electrical resistivity in a transverse magnetic field (*magnetoresistance*). From (3.6.79) we may expect that $\sigma(H_z)$ differs from $\sigma(0)$. As is clear from symmetry considerations, this effect should not depend on the sign of field. Therefore, at least with weak fields, the magnetoresistance $\Delta\varrho_H = \varrho(H_z) - \varrho(0)$ should depend quadratically on field: $\Delta\varrho_H \sim H_z^2$.

The cause of this phenomenon may be explained as follows. By the action of the Lorentz force (3.6.57), the trajectories of the conduction electrons are distorted, and their mean free path along the accelerating field E_x decreases (Fig. 3.19), entailing an increase in resistivity: $\Delta\varrho_H > 0$. However, in the zero approximation for the integrals (3.6.75, 76), we arrive at the result (3.6.83). Indeed, from (3.6.79) we find in this approximation, allowing for (3.6.76, 80)

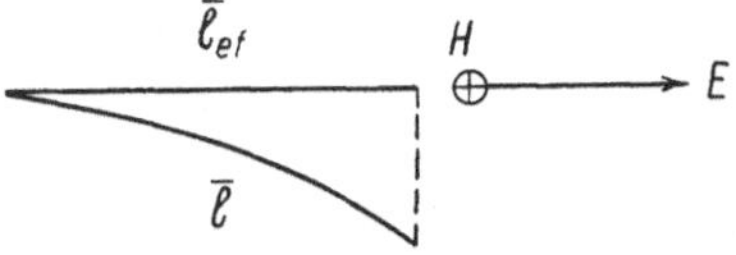

Fig. 3.19. Effective decrease in the mean free path of an electron in a magnetic field

$$\begin{aligned}\sigma(H_z) &\approx L_1[1+s^2(\zeta_0)] \approx [16\pi e^2(2m^*)^{1/2}/3h^3] \\ &\quad \times \zeta_0^{3/2}\bar{\tau}(\zeta_0)[1+s^2(\zeta_0)]^{-1}[1+s^2(\zeta_0)] \\ &= ne^2\bar{\tau}(\zeta_0)/m^* = \sigma(0) \ .\end{aligned} \tag{3.6.86}$$

Thus, just as for thermoelectric phenomena, the magnetoresistance in the zero approximation is equal to zero. Therefore, the integrals L_1 and L_2 from (3.6.75, 76) have to be calculated up to second order (3.6.39). A little manipulation yields

$$\Delta\varrho_H/\varrho_0 = BH_z^2/(1+CH_z^2) \ , \tag{3.6.87}$$

where

$$\begin{aligned}B &= \frac{\pi^2}{12}\{[|e|\bar{\tau}(\zeta_0)/m^*c]\,(k_B T/\zeta_0)\}^2 \ , \\ C &= [|e|\bar{\tau}(\zeta_0)/m^*c]^2 = [R\sigma(0)]^2 \ .\end{aligned} \tag{3.6.88}$$

Formula (3.6.87) confirms that to second order the magnetoresistance depends on field in a quadratic fashion. For weak fields ($CH_z^2 \ll 1$) the quantity $\Delta\varrho_H/\varrho_0$ varies approximately according to the parabolic law in proportion to BH_z^2 (Fig. 3.20). When the fields are strong ($CH_z^2 \gg 1$), the quantity $\Delta\varrho_H/\varrho_0$ tends to saturation $\approx B/C$.

Let us ascertain the meaning of weak and strong fields in this case and the physical cause of saturation. The boundary between these fields may con-

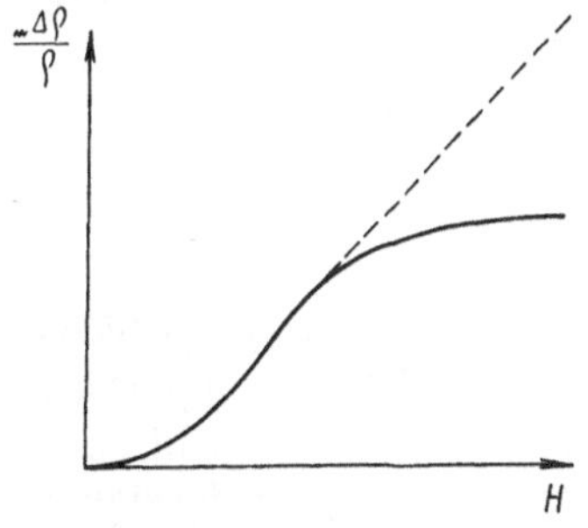

Fig. 3.20. Dependence of magnetoresistance on external magnetic field

ventionally be determined from the equality $CH_z^2 = 1$. According to (3.6.63), this signifies that the radius of the electron orbit is equal to the mean free path: $r_H = \bar{l}$. Thus, in a weak magnetic field $\bar{l} \ll r_H$, the electron trajectories are only slightly distorted. Conversely, in strong fields $\bar{l} \gg r_H$, the electron may make several turns in the cyclotron orbit in the magnetic field before it suffers a collision. In strong fields the resistivity therefore does not build up but reaches saturation.

For some time, the most thorough magnetoresistance measurements had been carried out by *Kapitsa* in 1929 [3.23]. His experiments showed that in low fields (3.6.87) gives a good fit to experimental data. In high fields the situation is much more complicated. First of all, normal metals possessing good conductivity exhibited no saturation up to the highest fields attained by Kapitsa ($\sim 3 \cdot 10^5$ Oe $\approx 2.4 \cdot 10^7$ A/m). Saturation was detected only in the case of poor metals (bismuth) and semiconductors (germanium, silicon). Secondly, the numerical value of the coefficient B, calculated according to (3.6.88), turned out to be approximately a factor of 10^4 lower than the observed value. In addition, most of the metals were found to display a large region of linear dependence of magnetoresistance on external field—this is the Kapitsa law.

In some cases the linear trend turned out to be intermediate between two quadratic dependences, or occurring before the transition to saturation, but sometimes deviations from the linear trend could not be observed even in very high fields of several hundred kilooersteds (dotted line in Fig. 3.20).

Finally, it was found experimentally that the electrical resistivity varies in a longitudinal magnetic field ($H_x \neq 0$). According to the electron gas model, in both the classical and the quantum version, such an effect should not take place. As stated above, a magnetic field is incapable of introducing asymmetry into the distribution function along the field direction, since it does not affect the parallel component of the electron velocity.

All these difficulties are associated with the major simplifying assumptions of the Fermi free-electron gas model, which employs an isotropic quadratic dispersion relation. It may be expected that these difficulties will be resolved only in a more rigorous theory (Chap. 4).

3.7 High-Frequency Properties

3.7.1 Basic Equations

One of the characteristic properties of metals is their luster. In every day life we often distinguish metals from nonmetals by metallic luster. Of great importance in electrical engineering is the so-called skin effect (nonpenetration of an ac electromagnetic field into a metal). Finally, a vast and rapidly developing domain of solid-state physics is the study of the various types of electromagnetic waves in metals and semiconductors. Many of these phenomena, relating to the propagation of an ac electromagnetic field in semiconductors, may be understood satisfactorily in terms of the free-electron gas model.

We should start from a system of Maxwell equations

$$\operatorname{div} \boldsymbol{b} = 0 \ , \tag{3.7.1}$$

$$\operatorname{div} \boldsymbol{D} = 4\pi \varrho_0 \ , \tag{3.7.2}$$

$$\operatorname{curl} \boldsymbol{E} = -\frac{1}{c}\frac{\partial \boldsymbol{b}}{\partial t} \ , \tag{3.7.3}$$

$$\operatorname{curl} \boldsymbol{h} = \frac{1}{c}\frac{\partial \boldsymbol{D}}{\partial t} + \frac{4\pi}{c}\boldsymbol{j}_0 \ . \tag{3.7.4}$$

The quantities $\boldsymbol{b}$ and $\boldsymbol{D}$ here stand for the magnetic and electric induction; $\boldsymbol{h}$ and $\boldsymbol{E}$ are the strengths of an ac magnetic field and an electric field, respectively; ϱ_0 and $\boldsymbol{j}_0$ are external-charge and current densities, which will be assumed to be equal to zero. For the system of equations (3.7.1–4) to be of closed form, we need also to set down the so-called material equations relating $\boldsymbol{D}$ and $\boldsymbol{h}$ to $\boldsymbol{E}$ and $\boldsymbol{b}$:

$$\boldsymbol{D} = \boldsymbol{E} + 4\pi \boldsymbol{P} \ , \tag{3.7.5}$$

$$\boldsymbol{h} = \boldsymbol{b} - 4\pi \boldsymbol{M} \ . \tag{3.7.6}$$

The polarization $\boldsymbol{P}$ is the field-induced mean dipole moment of unit volume

$$\boldsymbol{P} = \frac{1}{V}\left\langle \sum_i e_i \boldsymbol{x}_i \right\rangle \ ,$$

where the summation is taken over all i particles with x_i coordinates and e_i charges; the angular brackets denote averaging over a nonequilibrium distribution in a field. The quantity $\boldsymbol{P}$ is related to the induced current density

$$\boldsymbol{j} = \frac{1}{V}\left\langle \sum_i e_i \frac{d\boldsymbol{x}_i}{dt} \right\rangle$$

by the equation

$$\partial \boldsymbol{P}/\partial t = \boldsymbol{j} \ . \tag{3.7.7}$$

The magnetization $\boldsymbol{M}$ is usually small in normal metals because of the smallness of the susceptibility χ. In spite of this, a number of high-frequency phenomena exist in which it plays a decisive role (e.g., electron-spin resonance). We will not consider such phenomena here, setting $\boldsymbol{b} = \boldsymbol{h}$.

We will be concerned with a monochromatic electromagnetic field of frequency ω with $\boldsymbol{h}, \boldsymbol{E} \sim \exp(-\mathrm{i}\omega t)$, $\partial/\partial t \rightarrow -\mathrm{i}\omega$. Taking account of what has been stated above, we transcribe (3.7.1–4) into

$$\operatorname{div} \boldsymbol{h} = 0 \ , \tag{3.7.8}$$

$$\operatorname{div} \boldsymbol{D} = 0 \ , \tag{3.7.9}$$

$$\operatorname{curl} \boldsymbol{E} = \frac{\mathrm{i}\omega}{c} \boldsymbol{h} \ , \tag{3.7.10}$$

$$\operatorname{curl} \boldsymbol{h} = -\frac{\mathrm{i}\omega}{c} \boldsymbol{D} \ . \tag{3.7.11}$$

Equations (3.7.8, 9) follow from (3.7.10. 11) due to the identity div curl = 0 and will henceforth not be considered:

$$\operatorname{curl} \operatorname{curl} \boldsymbol{E} = \frac{\mathrm{i}\omega}{c} \operatorname{curl} \boldsymbol{H} = \frac{\omega^2}{c^2} \boldsymbol{D} \ . \tag{3.7.12}$$

According to (3.7.5, 7),

$$\boldsymbol{D} = \boldsymbol{E} + \frac{4\pi\mathrm{i}}{\omega} \boldsymbol{j} \ . \tag{3.7.13}$$

Substituting (3.7.13) into (3.7.12) and using the identity curl curl $\equiv \nabla \operatorname{div} - \Delta$, we obtain the equation

$$\nabla \operatorname{div} \boldsymbol{E} - \Delta \boldsymbol{E} = \frac{\omega^2}{c^2} \left(\boldsymbol{E} + \frac{4\pi\mathrm{i}}{\omega} \boldsymbol{j} \right) \ . \tag{3.7.14}$$

The current density $\boldsymbol{j}$ is expressed in terms of $\boldsymbol{E}$ from the solution of the kinetic equation. The question arises, What alterations must be made here in comparison with the situation outlined in Sect. 3.6?

First of all, the nonequilibrium addition to the distribution function will depend explicitly on time as $\exp(-\mathrm{i}\omega t)$ if the latter is the time dependence of the externally applied field involved in the kinetic equation and if we solve this equation to a linear approximation with respect to $\boldsymbol{E}$, the term $\partial \bar{n}_1/\partial t = -\mathrm{i}\omega \bar{n}_1$ may then be grouped together with the term $b_+ - b_- = -\bar{n}_1/\bar{\tau}$ in (3.6.6). This will lead to the replacement

$$\bar{\tau}^{-1} \rightarrow \bar{\tau}^{-1} - \mathrm{i}\omega \ , \qquad \tau \rightarrow \tau(1 - \mathrm{i}\omega\bar{\tau})^{-1} \tag{3.7.15}$$

in all the formulas written in Sect. 3.6. Furthermore, an ac electric field $\boldsymbol{E}$ is inevitably accompanied by an ac magnetic field $\boldsymbol{h}$, and it becomes necessary in, for example, (3.6.22) to perform the replacement $\boldsymbol{E}\cdot\partial\bar{n}_0/\partial\boldsymbol{v}\to(\boldsymbol{E}+c^{-1}(\boldsymbol{v}\times\boldsymbol{h}))\cdot\partial\bar{n}_0/\partial\boldsymbol{v}$, in addition to the replacement of (3.7.14). Nothing. however, is altered since

$$(\boldsymbol{v}\times\boldsymbol{h})\cdot\frac{\partial\bar{n}_0}{\partial\boldsymbol{v}}=m^*(\boldsymbol{v}\times\boldsymbol{h})\cdot\boldsymbol{v}\,\frac{\partial\bar{n}_0}{\partial\varepsilon}=0\ .$$

The wave field $\boldsymbol{h}$ should not be confused with the external magnetic field $\boldsymbol{H}$, which, in keeping with Sect. 3.6.5, enters into the kinetic equation as the terms $\sim Hn_1$; allowance for the terms $\sim hn_1$ would be an excess of accuracy since $h\sim E$ and $\bar{n}_1\sim E$. Thus, this point needs no alteration in comparison with Sect. 3.6.

Finally—what is most important—an ac electric field is necessarily spatially inhomogeneous. That is to say, $\bar{n}_1$ will also be inhomogeneous, and in the kinetic equation we have to allow for the diffusion term $(v\,\nabla_r)\bar{n}_1$, which was not taken into account in Sect. 3.6. Allowance for this term leads to what is known as nonlocality (spatial dispersion) effects, which we will consider in Chap. 5 when we discuss the plasma model of a metal.

Here the diffusion term will be neglected, as it is small compared with $\partial\bar{n}_1/\partial t$ or b_+-b_-, provided

$$v/\delta\ll\max(\omega,\bar{\tau}^{-1})\ , \qquad (3.7.16)$$

where v is the characteristic electron velocity (in metals this is the Fermi velocity), and δ is the reference length over which the field (and, consequently, $\bar{n}_1$) varies substantially. Assuming (3.7.16) to be fulfilled (local regime), the current $\boldsymbol{j}$ is expressed in terms of the field $\boldsymbol{E}$ by the formulas of Sects 3.6.2 and 3.6.5 with the replacement of (3.7.15).

3.7.2 Skin Effect

To start with, we consider the penetration of an electromagnetic field into a metal in the absence of a dc magnetic field. Then, according to (3.7.15), (3.6.30)

$$\boldsymbol{j}=\frac{ne^2}{m}\bar{\tau}(1-\mathrm{i}\omega\bar{\tau})^{-1}\,\boldsymbol{E}\ . \qquad (3.7.17)$$

Substitution of (3.7.17) into (3.7.14) yields

$$\nabla\mathrm{div}\,\boldsymbol{E}-\Delta\boldsymbol{E}=\omega^2c^{-2}\varepsilon(\omega)\boldsymbol{E}\ , \qquad (3.7.18)$$

where the notation introduced is

$$\varepsilon(\omega) = 1 + 4\pi i n e^2 \bar{\tau}/m\omega(1 - i\omega\bar{\tau}) = 1 + i\bar{\tau}\omega_p^2/\omega(1 - i\omega\bar{\tau}) \tag{3.7.19}$$

and $\varepsilon(\omega)$ is the dielectric constant of the metal at a frequency ω. The quantity

$$\omega_p = (4\pi n e^2/m)^{1/2} \tag{3.7.20}$$

of dimension s^{-1} is called the plasma frequency; its meaning will be clarified later on. For good metals $\omega_p \sim 10^{16}\,s^{-1}$.

Let the metal occupy a semispace $x > 0$ and let the electromagnetic wave be normally incident at its surface (a generalization for the case of inclined incidence presents no difficulties). Then $\boldsymbol{E}$ depends only on x. The electric field vector should be normal to the propagation direction, i.e., $E_x = 0$. Then

$$\operatorname{div} \boldsymbol{E} = \partial E_y/\partial y + \partial E_z/\partial z = 0 \ , \quad \Delta \boldsymbol{E} = \partial^2 \boldsymbol{E}/\partial x^2 \ ,$$

and (3.7.18) takes the form

$$\partial^2 \boldsymbol{E}/\partial x^2 - (\omega^2/c^2)\varepsilon(\omega)\boldsymbol{E} = 0 \ . \tag{3.7.21}$$

The solutions to (3.7.21) behave differently for different ω. At the outset we explore the low-frequency region

$$\omega\bar{\tau} \ll 1 \tag{3.7.22}$$

(for good metals this signifies $\omega \ll 10^9\,s^{-1}$ at low temperatures and $\omega \ll 10^{12}$–$10^{13}\,s^{-1}$ at high temperatures). The quantity $\omega\bar{\tau}$ may be neglected in comparison with the unity in (3.7.19). In addition, $\omega \ll \bar{\tau}\omega_p^2$ [this follows from (3.7.22) since $\omega_p\bar{\tau}$ is always much larger than unity]. The quantity $\varepsilon(\omega)$ then becomes

$$\varepsilon(\omega) = i\bar{\tau}\omega_p^2/\omega = 4\pi i\sigma/\omega \ , \tag{3.7.23}$$

with σ being the static conductivity of the metal. Insert (3.7.23) into (3.7.22). The general solution of the differential equation (3.7.21) has the form

$$\begin{aligned} E(x) &= A\exp[(4\pi i\sigma\omega/c^2)^{1/2}x] \\ &\quad + B\exp[-(4\pi i\sigma\omega/c^2)^{1/2}x] \\ &\equiv A\exp[(1+i)x/\delta] + B\exp[-(1+i)x/\delta] \ , \end{aligned} \tag{3.7.24}$$

where

$$\delta = (c^2/2\pi\omega\sigma)^{1/2} \ . \tag{3.7.25}$$

The first term in (3.7.24) increases exponentially as the wave penetrates into the metal. Therefore it should be eliminated. Then we have

$$E(x) = E(0)\exp[-(1+\mathrm{i})x/\delta] \ . \tag{3.7.26}$$

Thus, the field practically vanishes in the bulk of the metal at distances of $\sim \delta$ from the surface (and oscillates with the same characteristic period). For good metals (such as copper) at frequencies $\omega \sim 10^{10}$–$10^{11}\,\mathrm{s}^{-1}$, we find $\delta \sim 10^{-5}$ cm. This phenomenon, characterized by the nonpenetration of a high-frequency field into a metal, is called the skin effect, and δ is said to be the skin depth.

Let us verify the condition for the applicability of the local approximation (3.7.16) which, allowing for (3.7.22), assumes the form

$$\bar{l} \ll \delta \ , \tag{3.7.27}$$

with $\bar{l} = v(\zeta_0)\bar{\tau}$ being the mean free path length. At low temperatures in pure samples the $\bar{l}$ reaches 0.1 cm, (3.7.27) being violated. The skin effect in the nonlocal regime ($\bar{l} \gtrsim \delta$) is said to be anomalous. The depth of penetration of a field into a metal, δ^*, for a hyperanomalous skin effect $\bar{l} \gg \delta$ is easy to evaluate from the qualitative effective-electron consideration proposed by Pippard [3.24]. If a field penetrates into a metal to a depth δ^*, which is small compared to the free path length, only the electrons that move at small angles relative to the surface $\varphi \lesssim \delta^*/\bar{l}$ will interact with the electromagnetic field (Fig. 3.21), since only during a small fraction of free path time do the other electrons "sense" the field and receive energy from it. The number of such effective electrons, $n_{\mathrm{eff}} = \delta^* n/\bar{l}$, should, according to Pippard, be substituted into the conductivity σ, which will be replaced by $\sigma_{\mathrm{eff}} = \sim \delta^*\sigma/\bar{l}$. The expression for δ^* should involve σ_{eff} rather than σ:

$$\delta^* \sim (c^2/2\pi\omega\sigma_{\mathrm{eff}})^{1/2} \sim \delta(\bar{l}/\delta^*)^{1/2} \tag{3.7.28a}$$

or

$$\delta^* \sim \delta(\bar{l}/\delta)^{1/3} \gg \delta \ , \tag{3.7.28b}$$

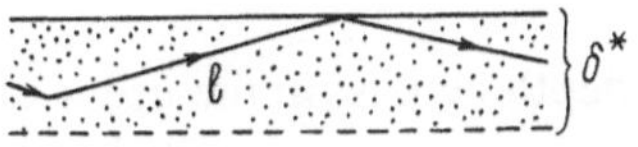

Fig. 3.21. Effective electrons in an anomalous skin effect

where δ depends on frequency ω as $\omega^{-1/2}$, and δ^* as $\omega^{-1/3}$. To construct a detailed theory of the anomalous skin effect, the kinetic equation must be solved with allowance for not only the diffusion term but also the electron scattering on the surface of a sample [3.25].

We now calculate the reflection coefficient for an electromagnetic wave reflected from the surface of a metal with a normal skin effect. From optics we know that the expression for the reflection coefficient of a normally incident electromagnetic wave is [3.26]

$$R = |[\sqrt{\varepsilon(\omega)} - 1]/[\sqrt{\varepsilon(\omega)} + 1]|^2 \ , \tag{3.7.29}$$

where the branch of the root is chosen in such a way that $\mathrm{Im}\{\sqrt{\varepsilon(\omega)}\} > 0$. For $\varepsilon(\omega)$ we substitute expression (3.7.23) and find, allowing for the inequality $\omega \ll \omega_p^2 \bar{\tau} \sim \sigma$,

$$R = \left|\frac{1 + \mathrm{i} - (\omega/2\pi\sigma)^{1/2}}{1 + \mathrm{i} + (\omega/2\pi\sigma)^{1/2}}\right|^2 \approx \frac{[1 - (\omega/2\pi\sigma)^{1/2}]^2 + 1}{[1 + (\omega/2\pi\sigma)^{1/2}]^2 + 1} \approx 1 - 2\left(\frac{\omega}{2\pi\sigma}\right)^{1/2} . \tag{3.7.30}$$

This is the Hagen–Rubens formula. The quantity R is close to unity, which means that the metal reflects electromagnetic waves well.

Let us proceed to a consideration of the high-frequency region

$$\omega\bar{\tau} \gg 1 \ . \tag{3.7.31}$$

Here, according to (3.7.19),

$$\varepsilon(\omega) = 1 - \frac{\omega_p^2}{\omega^2}\left(1 - \frac{\mathrm{i}}{\omega\bar{\tau}}\right) . \tag{3.7.32}$$

For $\omega < \omega_p$ we have $\mathrm{Re}\{\varepsilon(\omega)\} < 0$, and the field does not penetrate deep into the metal but decays according to the law

$$\begin{aligned} E(x) &= E(0)\exp[-\omega c^{-1}\sqrt{-\varepsilon(\omega)}x] \\ &\approx E(0)\exp\{-[(\omega_p^2 - \omega^2)/c^2]^{1/2} x\} \end{aligned} \tag{3.7.33}$$

rather than obeying (3.7.26). The field penetration depth here is on the order of $c/\omega_p \sim 10^{-5}$–10^{-6} cm (if $|\omega - \omega_p| \sim \omega_p$).

For $\omega > \omega_p$ we have $\mathrm{Re}\{\varepsilon(\omega)\} > 0$, and the field penetrates into the metal practically without decreasing [with neglect of the small parameter $(\omega\bar{\tau})^{-1} < (\omega_p\bar{\tau})^{-1} \sim 10^{-3}$–$10^{-6}$]. The transparency of thin sodium films exposed to ultraviolet radiation was discovered experimentally by *Wood* [3.27] and explained theoretically by *Kronig* [3.28].

3.7.3 Cyclotron Resonance

Now we proceed to a discussion of the high-frequency properties of metals in a dc uniform magnetic field $\boldsymbol{H}$. Let us align it with the z axis and examine an electromagnetic wave polarized in such a way that $E_z = 0$. The formulas for the induced current density can be obtained similar to (3.6.73, 74):

$$j_x = L_1 E_x - L_2 E_y \ , \tag{3.7.34}$$

$$j_y = L_2 E_x + L_1 E_y \ , \tag{3.7.35}$$

where, allowing for the replacement of (3.7.15), and to the zero order in $k_B T/\zeta_0$,

$$L_1 = (ne^2 \bar{\tau}/m)(1 - \mathrm{i}\omega\bar{\tau})^{-1} \{1 + [\omega_H \bar{\tau}/(1 - \mathrm{i}\omega\bar{\tau})]^2\}^{-1} \ , \tag{3.7.36}$$

$$L_2 = \omega_H \bar{\tau} L_1/(1 - \mathrm{i}\omega\bar{\tau}) \ . \tag{3.7.37}$$

It is convenient to go over to what are known as circular components:

$$j_\pm = j_x \pm \mathrm{i} j_y \ , \qquad E_\pm = E_x \pm \mathrm{i} E_y \ . \tag{3.7.38}$$

Multiplying (3.7.35) by $\pm\mathrm{i}$ and combining it with (3.7.34) yields

$$\begin{aligned} j_\pm &= (L_1 \pm \mathrm{i} L_2) E_\pm = (ne^2 \bar{\tau}/m)(1 - \mathrm{i}\omega\bar{\tau})^{-1} \\ &\quad \times [1 \pm \mathrm{i}\omega_H \bar{\tau}/(1 - \mathrm{i}\omega\bar{\tau})] \{1 + [\bar{\tau}\omega_H/(1 - \mathrm{i}\omega\bar{\tau})]^2\}^{-1} E_\pm \\ &= (ne^2 \bar{\tau}/m) \{[1 - \mathrm{i}(\omega \mp \omega_H)\bar{\tau}]/[(1 - \mathrm{i}\omega\bar{\tau})^2 + (\omega_H \bar{\tau})^2]\} E_\pm \\ &= \sigma_\pm(\omega) E_\pm \ , \end{aligned} \tag{3.7.39}$$

where

$$\sigma_\pm(\omega) = (ne^2 \bar{\tau}/m)\,(1 - \mathrm{i}(\omega \pm \omega_H)\bar{\tau})^{-1} \ . \tag{3.7.40}$$

For the circular induction vector components $D_\pm$ we find

$$D_\pm = E_\pm + 4\pi\mathrm{i}\omega^{-1} j_\pm = \varepsilon_\pm(\omega) E_\pm \ , \tag{3.7.41}$$

with

$$\varepsilon_\pm(\omega) = 1 + \mathrm{i}\omega_p^2 \bar{\tau}/\omega[1 - \mathrm{i}(\omega \pm \omega_H)\bar{\tau}] \ . \tag{3.7.42}$$

Assuming the field to be approximately uniform, we calculate the mean power component in unit voluem Q. According to the Joule–Lenz law, using a complex representation for the field and current,

$$\boldsymbol{j} \to (1/2)[\boldsymbol{j}\exp(-\mathrm{i}\omega t) + \boldsymbol{j}^*\exp(\mathrm{i}\omega t)] \ ,$$
$$\boldsymbol{E} \to (1/2)[\boldsymbol{E}\exp(-\mathrm{i}\omega t) + \boldsymbol{E}^*\exp(\mathrm{i}\omega t)] \ ,$$
$$Q = \boldsymbol{jE} \to (1/4)[\boldsymbol{j}\exp(-\mathrm{i}\omega t) + \boldsymbol{j}^*\exp(\mathrm{i}\omega t)]$$
$$\times [\boldsymbol{E}\exp(-\mathrm{i}\omega t) + \boldsymbol{E}^*\exp(\mathrm{i}\omega t)] \to (1/2)\,\mathrm{Re}\{\boldsymbol{j}\cdot\boldsymbol{E}^*\} \ .$$

Here we have eliminated the rapidly oscillating terms proportional to $\exp(\pm 2\mathrm{i}\omega t)$—this corresponds to a calculation of the mean absorbed power during a time interval that is much larger than the field variation period. Thus,

$$Q = \tfrac{1}{2}\,\mathrm{Re}\{\boldsymbol{j}\cdot\boldsymbol{E}^*\} \ . \tag{3.7.43}$$

Represent (3.7.43) in terms of circular components

$$\begin{aligned} Q &= \tfrac{1}{2}\mathrm{Re}\{j_x E_x^* + j_y E_y^*\} = \tfrac{1}{2}\mathrm{Re}\{\tfrac{1}{2}(j_+ + j_-)\tfrac{1}{2}(E_+^* + E_-^*) \\ &\quad + \tfrac{1}{2}(j_+ - j_-)i^{-1}\tfrac{1}{2}(E_-^* - E_+^*)i^{-1}\} \\ &= \tfrac{1}{2}\mathrm{Re}\{j_+ E_+^* + j_- E_-^*\} = \tfrac{1}{2}\mathrm{Re}\{\sigma_+(\omega)E_+E_+^* + \sigma_-(\omega)E_-E_-^*\} \\ &= \tfrac{1}{2}\mathrm{Re}\{\sigma_+(\omega)|E_+|^2\} + \tfrac{1}{2}\mathrm{Re}\{\sigma_-(\omega)|E_-|^2\} \ . \end{aligned} \tag{3.7.44}$$

For the real part of $\sigma_\pm(\omega)$ we have, according to (3.7.40),

$$\begin{aligned} \mathrm{Re}\{\sigma_\pm(\omega)\} &= (ne^2\bar{\tau}/m)[1 + (\omega \pm \omega_H)^2\bar{\tau}^2]^{-1} \\ &= (ne^2/m\bar{\tau})[(\omega \pm \omega_H)^2 + \bar{\tau}^{-2}]^{-1} \ . \end{aligned} \tag{3.7.45}$$

Subject to the condition that

$$|\omega_H|\bar{\tau} \gg 1, \tag{3.7.46}$$

the quantity (3.7.45) and, consequently, the absorbed power (3.7.46) exhibit, as a function of frequency (or field), a sharp maximum for a positive circular component when

$$\omega = |\omega_H| = |e|\,H/mc \tag{3.7.47}$$

(recall that $e < 0$ and therefore $\omega_H < 0$). The phenomenon of the resonance absorption of electromagnetic energy with the field frequency coinciding with the rotation frequency of electrons in a magnetic field is called cyclotron resonance.

However, one very important circumstance is not taken into account here. At frequencies that satisfy conditions (3.7.46, 47) the field will obviously be localized in a narrow skin (Fig. 3.22a). Here it follows from (3.7.46) that

$$r_H \ll \bar{l} \ ,$$

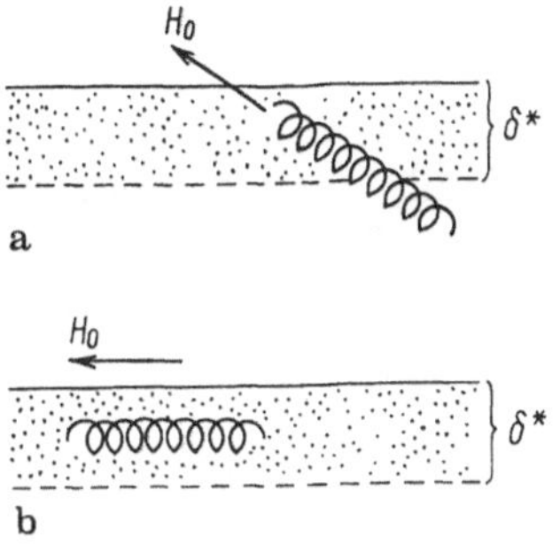

Fig. 3.22. Cyclotron resonance in metals: (**a**) inclined orientation of the magnetic field with the surface of the sample; (**b**) parallel orientation

where $r_H = v_F/|\omega_H|$ is the radius of the Larmor orbit, and $\bar{l} = v_F\bar{\tau}$ is the mean free path length. For pure metals at low temperatures in fields $H \approx 10^4$ Oe, $r_H \approx 10^{-3}$ cm the mean free path length is approximately equal to 0.1 cm. The skin δ is about 10^{-5} cm deep; i.e., we have the inequality

$$\delta \ll r_H \ll \bar{l} \tag{3.7.48}$$

which are the conditions for an anomalous skin effect. That is, spiraling in the magnetic field, the electron rapidly leaves the skin and ceases to participate in the absorption of electromagnetic energy.

An exception is a magnetic field oriented strictly parallel to the surface of the metal (Fig. 3.22b), or at least oriented so that the angle of inclination of the field is small compared to δ/τ_H. Only then will a resonance really be observed in metals [3.29]. Its mechanism is entirely similar to the mechanism of the acceleration of charged particles in a cyclotron (which serves as justification for the name of the effect itself). The electron returns into the skin after a time period $2\pi/|\omega_H|$. If the condition holds

$$\omega = n|\omega_H| \tag{3.7.49}$$

($n = 1, 2, 3, \ldots$), then the phase of the electric field will be the same when the electron returns into the skin, and the power $e\boldsymbol{E}\boldsymbol{v}$ (with $\boldsymbol{v}$ being the velocity of an electron) received from the field will always be positive. But if condition (3.7.49) is violated, the electron will give the acquired energy back to the field since $\boldsymbol{E}\boldsymbol{v}$ will change sign in a random fashion. After a time $\bar{\tau}$ all of the energy acquired by the electron is then transferred to the lattice and released as heat. Our simple treatment gives the resonance condition (3.7.49) only for $n = 1$, although, generally speaking, resonance occurs with any n.

Cyclotron resonance is more conveniently observed with the frequency fixed by varying the magnetic field intensity.

In semiconductors the skin effect is absent, and resonance is observed for an arbitrary direction of the field relative to the surface (to distinguish it from resonance in metals, it is sometimes said to be diamagnetic). This resonance was first predicted theoretically by *Dorfmann* [3.30] and independently by *Dingle* [3.31].

As will be seen in Chap 4, the free-electron model is inadequate for a description of cyclotron resonance in metals. Specifically, the resonance frequencies may differ drastically from the value given by (3.7.47) and the sign of ω_H (and, consequently, the field polarization at which the resonance occurs) may also be opposite to that predicted by the free-electron model.

3.7.4 Electromagnetic Waves in Metals

As already stated in Chap. 2, one of the primary objectives of quantum solid-state theory is to study the various elementary excitations, i.e., weakly interacting collective modes that may exist in a system of strongly interacting particles. The electromagnetic waves in metals should also be included in these elementary excitations. Such excitations are coupled oscillations of an electromagnetic field and the electronic subsystem, together with the charge densities induced by that field. Electromagnetic waves in metals actually arise from the electron–electron interaction. This problem will be treated in more detail in Chap. 5, but some of the fundamental concepts will be considered here.

We will assume that the field depends on coordinates according to the law

$$\boldsymbol{E}(\boldsymbol{r}) = \boldsymbol{E} \exp \mathrm{i} \boldsymbol{k} \cdot \boldsymbol{r} \tag{3.7.50}$$

(plane wave). A similar dependence holds for the current density

$$\boldsymbol{j} = \hat{\sigma}(\omega) \boldsymbol{E}(\boldsymbol{r}) \ , \tag{3.7.51}$$

where $\sigma(\omega)$ is, generally speaking, a tensor quantity (3.7.34, 35). Substituting (3.7.50, 51) into (3.7.14), we obtain

$$\sum_j \left[k^2 \delta_{ij} - k_i k_j - \frac{\omega^2}{c^2} \varepsilon_{ij}(\omega)\right] E_j = 0 \ , \tag{3.7.52}$$

with $i, j = x, y, z$,

$$\varepsilon_{ij}(\omega) = \delta_{ij} + \frac{4\pi \mathrm{i}}{\omega} \sigma_{ij}(\omega) \ . \tag{3.7.53}$$

Allowance for the spatial dispersion would result in ε_{ij} depending not only on ω but also on k. This would lead to the occurrence of a large number of new types of waves and entail an appreciable alteration of some of the properties of the earlier wave types. Here we restrict our attention to the simplest type of wave under local conditions.

In the high-frequency region (3.7.31), the condition for the neglect of spatial dispersion (3.7.16) takes the form

$$k v_{\mathrm{F}} \ll \omega \ . \tag{3.7.54}$$

We start by considering an isotropic medium in the absence of a magnetic field. Then

$$\varepsilon_{ij}(\omega) = \varepsilon(\omega)\,\delta_{ij} \ , \tag{3.7.55}$$

where $\varepsilon(\omega)$ in the metal is given by (3.7.32) (the imaginary part will be neglected).

Substitution of (3.7.55) into (3.7.52) yields

$$\left[k^2 - \frac{\omega^2}{c^2}\varepsilon(\omega)\right]\boldsymbol{E} = \boldsymbol{k}(\boldsymbol{k}\cdot\boldsymbol{E}) \ . \tag{3.7.56}$$

If $\boldsymbol{k}\boldsymbol{E} = 0$ (transverse field), we find

$$k^2 = \frac{\omega^2}{c^2}\varepsilon(\omega_t) \tag{3.7.57}$$

or, allowing for (3.7.32),

$$\omega_{\mathrm{t}}^2 = \omega_{\mathrm{p}}^2 + k^2 c^2 \ . \tag{3.7.58}$$

Thus, ω_{p} is the threshold frequency of transverse electromagnetic waves in metals in the absence of a magnetic field. When $\omega < \omega_{\mathrm{p}}$, in keeping with the results obtained in Sect. 3.7.1, such waves do not propagate in metals.

Now let $\boldsymbol{k}\cdot\boldsymbol{E}$ be other than zero (longitudinal field). Scalar multiplication of both sides of (3.7.56) by $\boldsymbol{k}$ gives

$$\varepsilon(\omega_{\mathrm{l}}) = 0 \ , \tag{3.7.59}$$

i.e.,

$$\omega_{\mathrm{l}} = \omega_{\mathrm{p}} \ . \tag{3.7.60}$$

Thus, ω_{p} is also the frequency of longitudinal electromagnetic waves in metals. With neglect of the spatial dispersion, the quantity ω_{l} is independent of $\boldsymbol{k}$ (as with the ionic lattice, Chap. 2).

Longitudinal oscillations (also called plasma oscillations or Langmuir oscillations in honor of *Langmuir*, who considered similar oscillations in a gaseous plasma) interact strongly with an electric field produced by a beam of fast electrons. Electrons lose energy to generate plasmons (quanta of plasma oscillations). Similar to the procedure in Chap. 2 for neutrons and phonons, the formula for the corresponding energy losses (said to be characteristic) can be derived from the law of conservation of energy

$$\Delta\varepsilon = -\hbar\omega_p \ . \tag{3.7.61}$$

The theory of these losses is treated in greater detail in Chap. 5.

In the absence of a magnetic field one more wave type exists in metals—surface plasma waves (surface plasmons).

Let the metal occupy a semispace $x > 0$, and let the wave propagate along the y axis:

$$\boldsymbol{E} = \boldsymbol{E}(x)\exp(\mathrm{i}ky) \ , \quad \boldsymbol{h} = \boldsymbol{h}(x)\exp(\mathrm{i}ky) \ . \tag{3.7.62}$$

The system of Maxwell equations (3.7.8–11) for an isotropic medium breaks apart into uncoupled systems of equations for E_x, E_y, h_z (the so-called p polarization) and E_z, h_x, h_y(s polarization). The waves of interest are p polarized. Substituting (3.7.62) into (3.7.8–11) and allowing for the fact that in an isotropic medium $\boldsymbol{D} \parallel \boldsymbol{E}$, we obtain

$$\frac{\partial D_x}{\partial x} + \mathrm{i}kD_y = 0 \ , \tag{3.7.63}$$

$$\frac{\partial E_y}{\partial x} - \mathrm{i}kE_x = \frac{\mathrm{i}\omega}{c}h_z \ , \tag{3.7.64}$$

$$\mathrm{i}kh_z = -\frac{\mathrm{i}\omega}{c}D_x \ , \tag{3.7.65}$$

$$\frac{\partial h_z}{\partial x} = \frac{\mathrm{i}\omega}{c}D_y \ . \tag{3.7.66}$$

Equation (3.7.66) follows from (3.7.63, 65) and will not be considered henceforth. Substitution of (3.7.65) into (3.7.64) yields

$$\frac{\partial E_y}{\partial x} - \mathrm{i}kE_x = -\frac{\mathrm{i}\omega^2}{kc^2}D_x \ . \tag{3.7.67}$$

We need to solve the system (3.7.63, 67) with the addition of the material equations

$$\boldsymbol{D} = (1 - \omega_p^2/\omega^2)\boldsymbol{E} \ , \quad x > 0 \ , \quad \boldsymbol{D} = \boldsymbol{E} \ , \quad x < 0 \ , \tag{3.7.68}$$

and the boundary conditions

$$D_x(+0) = D_x(-0) \ , \tag{3.7.69}$$

$$E_y(+0) = E_y(-0) \tag{3.7.70}$$

(continuity of the components $\boldsymbol{D}$ and $\boldsymbol{E}$ which are, respectively, normal and tangential to the surface).

For $x > 0$ we obtain, differentiating (3.7.67) with respect to x and allowing (3.7.63, 68),

$$\partial^2 E_y/\partial x^2 - [k^2 - c^{-2}(\omega^2 - \omega_p^2)]\, E_y = 0 \ . \tag{3.7.71}$$

When

$$\omega^2 < c^2 k^2 + \omega_p^2 \ , \tag{3.7.72}$$

the $E_y(x)$ decays exponentially:

$$E_y(x) = E_y(0) \exp\{-[k^2 - c^{-2}(\omega^2 - \omega_p^2)]^{1/2}\, x\} \ . \tag{3.7.73}$$

Similarly, we find for $x > 0$

$$\partial^2 E_y/\partial x^2 - (k^2 - \omega^2/c^2)\, E_y = 0 \ . \tag{3.7.74}$$

For

$$\omega^2 < c^2 k^2 \ , \tag{3.7.75}$$

the quantity $E_y(x)$ also falls off exponentially on the other side of the surface

$$E_y(x) = E_y(0) \exp\,[(k^2 - \omega^2/c^2)^{1/2} x] \ . \tag{3.7.76}$$

Waves where all of the energy is localized near the surface $x = 0$ are called surface waves. On obtaining a result, we must check (3.7.75) [(3.7.72) follows from (3.7.75)]. It follows from (3.7.70) that $E_y(+0) = E_y(-0) \equiv E_y(0)$.

Equations (3.7.67, 68) yield

$$\left.\frac{\partial E_y}{\partial x}\right|_{x=+0} = \left(\frac{ik}{1 - \omega_p^2/\omega^2} - \frac{i\omega^2}{kc^2}\right) D_x(+0)$$
$$= -\frac{ik}{1 - \omega_p^2/\omega^2}\, \frac{k^2c^2 - \omega^2 + \omega_p^2}{kc^2}\, D_x(+0) \ , \tag{3.7.77}$$

$$\left.\frac{\partial E_y}{\partial x}\right|_{x=-0} = i\,\frac{k^2 c^2 - \omega^2}{kc^2} D_x(-0) \ .$$

From (3.7.69, 71, 76, 77) we find, finally, an equation for determining the $\omega(k)$ relation:

$$(\omega_p^2/\omega^2 - 1)(k^2c^2 - \omega^2 + \omega_p^2)^{-1/2} = (k^2c^2 - \omega^2)^{-1/2} . \tag{3.7.78}$$

Solving (3.7.78) we obtain

$$\omega^4 - 2\omega^2(k^2c^2 + \omega_p^2/2) + \omega_p^2 k^2 c^2 = 0 \tag{3.7.79}$$

or

$$\omega^2 = k^2c^2 + \omega_p^2/2 - (\omega_p^4/4 + k^4c^4)^{1/2} , \tag{3.7.80}$$

where one of the roots of (3.7.79) is selected in keeping with (3.7.72). In the limiting cases of small and large k we obtain (Fig. 3.23)

$$\omega \approx \begin{cases} ck & k \ll \omega_p/c \\ \omega_p/\sqrt{2} & k \gg \omega_p/c . \end{cases} \tag{3.7.81}$$

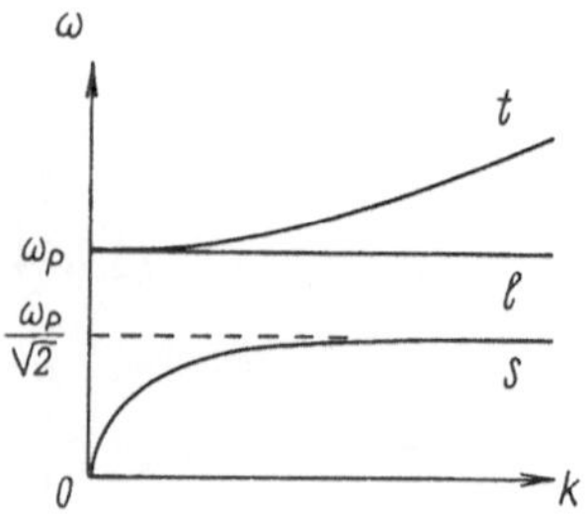

Fig. 3.23. Dispersion curves for electromagnetic waves in the metal in the absence of a magnetic field: t, l, s stand, respectively, for transverse, longitudinal and surface waves

Finally, we briefly examine simple waves that are capable of propagating in a metal in a magnetic field—helicons. We confine ourselves to the case of wave propagation along the magnetic field and set $\boldsymbol{k} \parallel \boldsymbol{H}$, $\boldsymbol{E} \perp \boldsymbol{H}$. Rather than use the system of equations (3.7.52), we straightforwardly employ (3.7.14)—for plane waves (3.7.50)—with $\boldsymbol{E} \perp \boldsymbol{k}$, $\operatorname{div} \boldsymbol{E} = 0$, $\Delta \boldsymbol{E} = -k^2 \boldsymbol{E}$, and (3.7.41)

$$[k^2 - \omega^2 c^{-2} \varepsilon_\pm(\omega)] E_\pm = 0 . \tag{3.7.82}$$

Consider the case

$$\omega \bar{\tau}^{-1} \ll |\omega_H| . \tag{3.7.83}$$

Here we always have $\omega_p^2/\omega|\omega_H| \gg 1$, and the unity in (3.7.42) may be neglected in comparison with the second term. We then obtain

$$\varepsilon_\pm(\omega) \approx \mp \omega_p^2/\omega_H\omega = \pm \omega_p^2/|\omega_H|\omega . \tag{3.7.84}$$

Set $E_- = 0$, $E_+ \neq 0$, i.e.,

$$E_x = \mathrm{i} E_y \ . \tag{3.7.85}$$

According to (3.7.82, 84),

$$k^2 = \frac{\omega^2}{c^2} \frac{\omega_p^2}{|\omega_H|\omega} = \frac{\omega}{c^2|e|H/mc} \frac{4\pi ne^2}{m} = \frac{4\pi n|e|}{Hc}\omega$$

or

$$\omega = cHk^2/4\pi n|e| \ . \tag{3.7.86}$$

The corresponding low-frequency waves have, according to (3.7.82), circular polarization and are called helicons. There are no such waves with opposite polarization. The condition for the existence of helicons at small k coincides with the condition for the occurrence of cyclotron resonance (3.7.46), and the condition for the neglect of spatial dispersion (3.7.83) has the form

$$k v_\phi \ll \omega_H \quad \text{or} \quad k r_H \ll 1 \tag{3.7.87}$$

instead of (3.7.54). The properties of the metal at small k enter into the helicon dispersion relation (3.7.86) only in terms of the total electron density n.

The possible existence of helicons in metals was first pointed out by *Konstantinov* and *Perel* [3.32] and, independently, by *Aigrain* [3.33].

Concluding our treatment of the high-frequency properties of metals, we may note that, on the whole, the free-electron model successfully explains many of these properties at a qualitative level, although needing refinement for a more quantitative approach.

3.8 Conclusions

Summing up the results of the electron Fermi gas theory considered above, we see that its major achievement, just as in the success of Frenkel's itinerant electrons theory, is in the resolution of the disaster with the electronic heat capacity and thus in the rehabilitation of the electron theory of metals as a whole. Furthermore, it has been shown that only thermally active electrons, small in number, with an energy close to the Fermi energy, are essential to the conductivity and other electronic properties. The Fermi gas model also gives information about the magnitude of the heat capacity and its T dependence, a more accurate understanding of the metallic bond and work function, an explanation of T-independent paramagnetism and of diamagnetism due to Landau and oscillation effects, and an explanation of the smallness of thermoelectric effects. Weak points have been revealed in the model, for example, the impossibility of accounting for the temperature trend of the kinetic coefficients at low temperatures, the disaster with the sign of thc Hall constant, and difficulties in explaining the phenomenon of magnetoresistance.

The greatest difficulty with both the Fermi gas theory and the classical theory is the conclusion about the magnitude of the mean free time of a conduction electron, which is enormous compared to typical atomic times ($\sim 10^{-16}$ s), or, put another way, about very large values of the mean free path, which at room temperature reach a factor of 100 higher than the lattice parameters, and at low temperatures ($T \lesssim 10$ K) reach macroscopic values (~ 0.01 to 0.1 cm).

To resolve these difficulties, one should construct a consistent quantum-mechanical theory of electron motion in the periodic potential field of the crystal lattice of the metal and try to understand why, despite the strong interaction between electrons and ion cores, the conduction electron enjoys so great a freedom and does not suffer collisions with each lattice site.

Such a theory should enable us to solve, along with the multitudinous partial problems, one of the problems of paramount importance: why some solids or liquids are metals and others are semiconductors or insulators.

Although not in completely quantitative form, it is possible to solve these problems in terms of the band model, a further stage of development of solid-state theory, on which we concentrate in the next chapter.

4. Band Theory

This chapter treats the quantum-mechanical theory of electron motion in a crystal lattice in a one-particle approximation. We focus on conceptual problems of band theory—the idea of quasi-particles, the effect of external electric and magnetic fields on electron motion, the metal–nonmetal criterion, effects of disorder. Simple model problems (for example, the one-dimensional array) are considered in detail. A general idea is given of methods for computing the electron energy spectrum of crystals. Consideration is given to topical problems of solid state physics such as metal–insulator transitions, Anderson localization, topological electronic transitions, etc. In connection with the theory of disordered systems, the Green's function (resolvent) method is outlined.

4.1 Preliminary Observations and the One-Dimensional Model

4.1.1 Electron Waves in a Crystal

A major result of the classical and quantum electron gas model was the conclusion that conduction electrons have a very large mean free time $\bar{\tau}$ and path length $\bar{l}$. Frenkel was the first to point out that these come from the wave properties of electrons [4.1]. The orbital angular momenta of thermally excited electrons near the Fermi surface are equal to $p_F \approx hn^{1/3}$ (3.5.11). The electron density n is related to the lattice parameter as

$$n \approx d^{-3} , \tag{4.1.1}$$

and, consequently,

$$p_F \approx h/d . \tag{4.1.2}$$

Using the expression for the de Broglie wavelength,

$$\lambda = h/p , \tag{4.1.3}$$

and (4.1.2), we arrive at the result that the electrons near the Fermi surface have $\lambda \approx d$; i.e., the wavelength falls short of 10^{-7} cm.

Let s_1 be the intensity of electron waves scattered by single atoms, and N_a the number of atoms in a small volume V ($V^{1/3} \ll \lambda_{detr}$). All the atoms of the volume then oscillate in phase, and the net amplitude of the waves scattered in this volume is proportional to N_a. The total intensity is equal to $s_1 N_a^2$. Its part due to the scattering on fluctuations of the number of atoms, determined by the difference between the actual number N_a and its mean value $\bar{N}_a$, is equal to

$$s_1 \overline{(n_a - \bar{n}_a)^2}\, V^2 = s_1\, \overline{(\Delta n_a)^2}\, V^2 \ ,$$

with $n_a = N_a/V$ being the density. The unit-volume scattering coefficient is equal to

$$s = s_1 \overline{(\Delta n_a)^2}\, V \ . \tag{4.1.4}$$

For very short waves $\lambda \lesssim V^{1/3}$ scattered waves are incoherent and therefore it is not their net amplitude but their total intensity that is proportional to the number of scatterers. Instead of (4.1.4) we then obtain

$$s = s_1 \overline{n_a} \ . \tag{4.1.5}$$

If the scattering medium is a gas, then (4.1.4, 5) just coincide, since it is known from the gas–kinetic theory that

$$\overline{(\Delta N_a)^2} = N_a \ , \qquad \overline{(\Delta n_a)^2} = \overline{n_a}/V \ .$$

In a gas the mean free path length is

$$\bar{l} = (\pi a^2 \overline{n_a})^{-1} = s^{-1} \ ,$$

where a is the radius of the cross section of an atom for an electron. Hence it is also seen from (4.1.4) that the scattering coefficient s_1 is equal to the effective cross section of an isolated atom: πa^2. Formula (4.1.4) for an arbitrary medium may be represented as outwardly coincident with (4.1.5) for a gas

$$s = \pi a_1^2 \bar{n}_a \ ,$$

where a_1 is the effective atomic radius, which, according to (4.1.4), has the form

$$a_1^2 = a^2 \overline{(\Delta n_a)^2}\, V/\overline{n_a} \ . \tag{4.1.6}$$

Thus, the scattering characteristics for a gas may also be used for a solid phase, if we replace the true atomic radius by the effective radius defined in (4.1.6). It is readily seen that $a_1 < a$ depends on T through $\overline{(\Delta n_a)^2}$ and $\overline{n_a}$. This dependence is easy to determine. We will consider not a fixed volume V but a given number of atoms, N_a. At equilibrium, these atoms occupy an equilibrium volume $\bar{V}$.

Because of the thermal motion, volume fluctuations $V - \bar{V}$ arise involving an energy fluctuation which, according to (3.5.69), is equal to

$$K(V-\bar{V})^2/2V \, . \tag{4.1.7}$$

Since the probability of the deviation $\Delta V = V - \bar{V}$ is equal to $\exp(-K(\Delta V)^2/2Vk_BT)$, the result for the mean-square fluctuation will be

$$\overline{(\Delta V)^2} = \int_{-\infty}^{\infty} d\Delta V (\Delta V)^2 \exp(-K(\Delta V)^2/2Vk_BT) \left\{ \int_{-\infty}^{\infty} d\Delta V \exp(-K(\Delta V)^2/2Vk_BT) \right\}^{-1} = Vk_BTK^{-1} \, .$$

From the condition $N_a = \text{const}$ we find $\Delta V/V = -\Delta n_a/n_a$. Therefore,

$$\overline{(\Delta n_a)^2}\, \bar{V} = \bar{n}_a^2 k_B T K^{-1} \, . \tag{4.1.8}$$

Substitution of (4.1.8) into (4.1.6) yields

$$a_1^2/a^2 = \bar{n}_a k_B T/K \, . \tag{4.1.9}$$

With Na at $T = 300$ K and with $\overline{n_a} = 10^{22}$ cm^{-3}, $K = 3 \cdot 10^{10}$ erg/cm^3, and so we find from (4.1.9)

$$a_1^2/a^2 \approx 10^{-2} \, ;$$

i.e., the effective atomic radius is tens of times smaller than the true one. The effective radius may be identified with the mean amplitude of thermal vibrations of an atom in a crystal. By the definition of the bulk modulus, the variation of the potential energy of an atom in the case of its displacement a_1 from the equilibrium position is, according to (4.1.7), equal to $Ka_1^2/\bar{n}_a a^2$. At the temperature T this quantity will be on the order of k_BT. Hence we immediately obtain (4.1.9). Substituting (4.1.8) into (4.1.4) and replacing s in terms of l^{-1}, we find for electrons in a metal

$$\bar{l} = K/s_1 \bar{n}_a^2 k_B T \, . \tag{4.1.10}$$

Formulas (4.1.10) and (3.3.2) or (3.3.3) give the correct dependence $\sigma(T) \propto T^{-1}$ (at room temperatures) and the correct value for $\bar{l} \approx 10^{-5}$ cm.

On the whole, this calculation of $\bar{l}$ resolves the difficulties of the gas model correctly. However, it is inadequate from the quantitative point of view, since it disregards the influence of the periodic crystalline field and employs, as before, a "free" electron described by a plane wave. A consistent quantum (band) theory for a periodic crystalline field was first proposed by Bloch [4.2].

4.1.2 The Array of Rectangular Potential Barriers

Before expounding the band theory of a three-dimensional crystal, we will explore two examples of a one-dimensional periodic field.

We start by considering the problem of rectangular barriers aligned with the x axis (Fig. 4.1) [4.3, 4], of height V_0 and width a; the potential wells between them have the width b. Thus, the period is equal to

$$c = a + b \ . \tag{4.1.11}$$

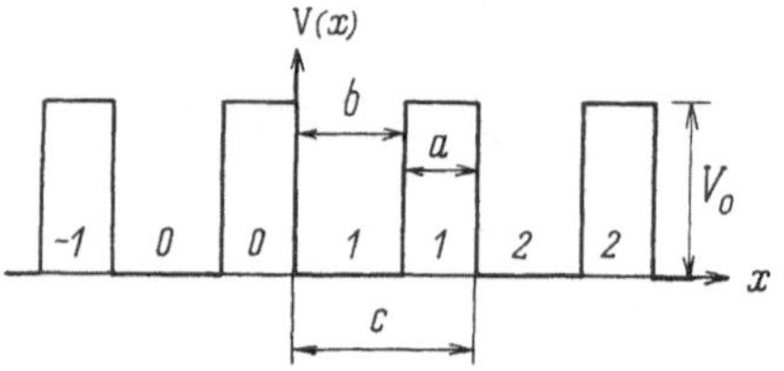

Fig. 4.1. Array of rectangular potential barriers

We choose the origin of coordinates to be at the right-hand end of one of the barriers. A barrier whose right-hand side has the coordinate $x = nc$ ($n = 0, \pm 1, \ldots$) will be called an nth barrier, and a well with the right-hand side at the point $x = nc - a$ an nth well.

The Schrödinger equation has the form

$$\frac{d^2\psi(x)}{dx^2} + \frac{2m}{\hbar^2}[E - V(x)]\psi(x) = 0 \ . \tag{4.1.12}$$

We try a solution in the form

$$\psi(x) = \exp(ikx)u(x) \ , \tag{4.1.13}$$

where $u(x)$ is a periodic function x with a period c

$$u(x + nc) = u(x) \ . \tag{4.1.14}$$

This form of solution is natural from physical considerations. That such solutions exist will be seen from what follows, and Sect. 4.1.4 will show that all the solutions to (4.1.12) have the form of (4.1.13, 14).

Suppose at the outset that the energy E lies in an interval below the potential barrier height

$$0 < E < V_0 \ . \tag{4.1.15}$$

Introduce the notation

$$\alpha^2 = 2mE/\hbar^2 \ , \qquad \beta^2 = 2m(V_0 - E)/\hbar^2 \ . \tag{4.1.16}$$

Substitution of (4.1.13, 16) into (4.1.12) yields for the situation inside a barrier $(-a \leqslant x < 0)$

$$\frac{d^2u}{dx^2} + 2ik\frac{du}{dx} + (\alpha^2 - k^2)u = 0 \ ,$$

and for the situation inside a well $(0 \leqslant x < b)$

$$\frac{d^2u}{dx^2} + 2ik\frac{du}{dx} - (\beta^2 + k^2)u = 0 \ . \tag{4.1.17}$$

The respective solutions of (4.1.17) for the situation inside an nth well and inside an nth barrier will be

$$u_n(x) = A_n \exp[i(\alpha - k)x] + B_n \exp[-i(\alpha + k)x]$$

and

$$u_n(x) = C_n \exp[(\beta - ik)x] + D_n \exp[-(\beta + ik)x] \ , \tag{4.1.18}$$

with A_n, B_n, C_n, and D_n being arbitrary constants. Accordingly, inside an nth well

$$\psi(x) = A_n \exp(i\alpha x) + B_n \exp(-i\alpha x) \ ,$$

and inside an nth barrier

$$\psi(x) = C_n \exp(\beta x) + D_n \exp(-\beta x) \ . \tag{4.1.19}$$

Using (4.1.14), we find for the wells $(0 \leqslant x \leqslant b)$

$$\begin{aligned} &A_0 \exp[i(\alpha - k)x] + B_0 \exp[-i(\alpha + k)x] \\ &\quad = A_n \exp[i(\alpha - k)(x + nc)] + B_n \exp[-i(\alpha + k)(x + nc)] \ , \end{aligned} \tag{4.1.20}$$

and for the barriers $(0 \geqslant x \geqslant -a)$

$$\begin{aligned} &C_0 \exp[(\beta - ik)x] + D_0 \exp[-(\beta + ik)x] \\ &\quad = C_n \exp[(\beta - ik)(x + nc)] + D_n \times \exp[-(\beta + ik)(x + nc)] \ . \end{aligned} \tag{4.1.21}$$

Equations (4.1.20, 21) yield

$$\begin{aligned}A_n &= A_0 \exp[-\mathrm{i}(\alpha-k)nc]\\ B_n &= B_0 \exp[\mathrm{i}(\alpha+k)nc]\\ C_n &= C_0 \exp[-(\beta-\mathrm{i}k)nc]\\ D_n &= D_0 \exp[(\beta+\mathrm{i}k)nc]\ .\end{aligned} \tag{4.1.22}$$

Thus, the periodicity requirements decrease the number of arbitrary constants in (4.1.19) to four: A_0, B_0, C_0, and D_0. These should be appropriately chosen so that the wave function and its derivative are continuous at the barrier boundaries (with $x = 0$, b). Using (4.1.19, 22), this gives

$$\begin{aligned}&A_0 + B_0 = C_0 + D_0\ ,\\ &\mathrm{i}(\alpha-k)A_0 - \mathrm{i}(\alpha+k)B_0 = (\beta-\mathrm{i}k)C_0 + (\beta+\mathrm{i}k)D_0\ ,\\ &\exp[\mathrm{i}(\alpha-k)b]A_0 + \exp[-\mathrm{i}(\alpha+k)b]B_0\\ &\quad = \exp[-(\beta-\mathrm{i}k)a]C_0 + \exp[(\beta+\mathrm{i}k)a]D_0\ ,\\ &\mathrm{i}(\alpha-k)\exp[\mathrm{i}(\alpha-k)b]A_0 - \mathrm{i}(\alpha+k)\exp[-\mathrm{i}(\alpha+k)b]B_0\\ &\quad = (\beta-\mathrm{i}k)\exp[-(\beta-\mathrm{i}k)a]C_0 - (\beta+\mathrm{i}k)\exp[(\beta+\mathrm{i}k)a]D_0\ .\end{aligned} \tag{4.1.23}$$

The system (4.1.23) has nonzero solutions if its determinant is equal to zero, i.e.,

$$\cosh\beta a\cos\alpha b + \frac{\beta^2-\alpha^2}{2\alpha\beta}\sinh\beta a\sin\alpha b = \cos kc\ . \tag{4.1.24}$$

Hence, allowing for (4.1.16), we find the energy as a function of k. Since k is real valued, the quantity $\cos kc$ is real valued, too, and therefore the left-hand side of (4.1.24) lies within the limits

$$-1 \leqslant \cosh\beta a\cos\alpha b + \frac{\beta^2-\alpha^2}{2\beta\alpha}\sinh\beta a\sin\alpha b \leqslant 1\ . \tag{4.1.25}$$

Inequalities (4.1.25) are consistent with (4.1.12). We consider a numerical example: $2mV_0/b^2 = 144$, $a = b/24$, $\varepsilon = E/V_0$. Equation (4.1.25) then becomes

$$\begin{aligned}-1 \leqslant \Bigg[&\cosh\frac{(1-\varepsilon)^{1/2}}{2}\cos 12\,\varepsilon^{1/2} + \frac{1-2\varepsilon}{2[\varepsilon(1-\varepsilon)]^{1/2}}\\ &\times\sinh\frac{(1-\varepsilon)^{1/2}}{2}\sin 12\,\varepsilon^{1/2}\Bigg] \leqslant 1\ .\end{aligned} \tag{4.1.26}$$

The oscillating function $F(\varepsilon)$, enclosed in square brackets, is presented in Fig. 4.2a, from which it is seen that for $\varepsilon < 0.03\,(E < 0.03\,V_0)$ the $F(\varepsilon)$ curve lies higher than the ordinate $+1$. The energies $E < 0.03\,V_0$ therefore do not satisfy (4.1.26) and are not allowed. For the ε that lie in the interval $0.03 \leqslant \varepsilon \leqslant 0.06$ indicated by a bold line on the abscissa axis, all the values of E $(0.03\,V_0 \leqslant E \leqslant 0.06\,V_0)$ are allowed since $|F(\varepsilon)| \leqslant 1$. For $\varepsilon > 0.06$ the function $F(\varepsilon) < -1$, and therefore the energies $E > 0.06\,V_0$ are not allowed, etc. In the energy interval explored, $0 \leqslant \varepsilon \leqslant 1$, there are four allowed energy bands:

$$
\begin{aligned}
&0.03\,V_0 \leqslant E \leqslant 0.06\,V_0, \text{ first band;}\\
&0.13\,V_0 \leqslant E \leqslant 0.25\,V_0, \text{ second band;}\\
&0.33\,V_0 \leqslant E \leqslant 0.51\,V_0, \text{ third band;}\\
&0.64\,V_0 \leqslant E \leqslant 0.99\,V_0, \text{ fourth band.}
\end{aligned}
\tag{4.1.27}
$$

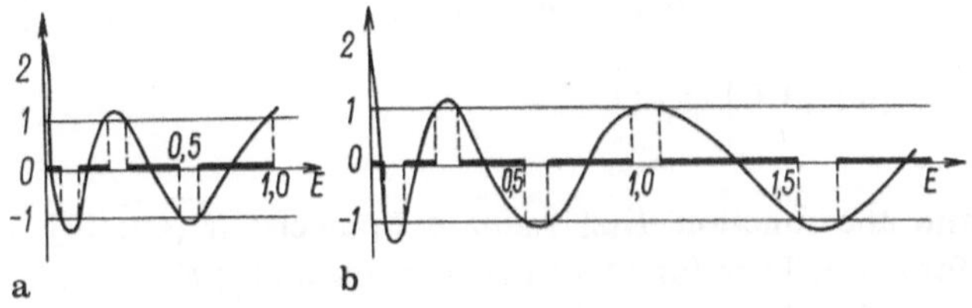

Fig. 4.2. Function $F(\varepsilon)$: $\varepsilon < 1$ (4.1.26) (**a**); $\varepsilon > 1$ (4.1.30) (**b**)

Thus, the energy spectrum of an electron in the periodic field of an array of rectangular barriers has the form of allowed energy bands that are separated by forbidden energy gaps. In Fig. 4.2 this is shown by thick and thin areas, and in Fig. 4.3 by shadowed and empty strips. The horizontal lines in Fig. 4.3a give, with $E < V_0$, a discrete electron spectrum for one potential well (when $E > V_0$ the spectrum is continuous). Meanwhile, passing from one well to a periodic sequence of wells (with barriers of atomic sizes $\approx 10^{-8}$ cm) splits each discrete level up into a band (Fig. 4.3b).

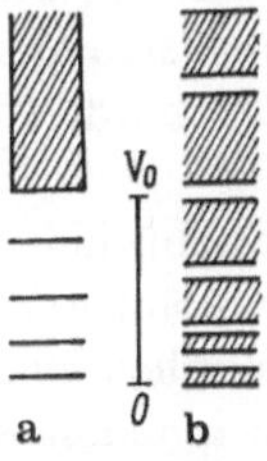

Fig. 4.3. Energy spectrum in the model of rectangular potential barriers of height V_0: one potential well (**a**); an array of barriers (**b**)

Next, we consider the case $E > V_0$. According to (4.1.16), $\beta = i\gamma$ is an imaginary quantity in which

$$\gamma = \left[\frac{2m}{\hbar^2}(E - V_0)\right]^{1/2} \tag{4.1.28}$$

is real valued. Equations (4.1.17–24) remain the same as they were for $E < V_0$, so in (4.1.24) we may replace β by $i\gamma$. Using the equations $\cosh ix = \cos x$, $\operatorname{sh} ix = i \sin x$, we obtain, in place of (4.1.25),

$$-1 \leqslant \cos \gamma a \cos \alpha b - \frac{\alpha^2 + \gamma^2}{2\alpha\gamma} \sin \gamma a \sin \alpha b \leqslant 1 \ . \tag{4.1.29}$$

For the same numerical example we have

$$-1 \leqslant \left[\cos \frac{(\varepsilon - 1)^{1/2}}{2} \cos 12\,\varepsilon^{1/2} + \frac{1 + 2\varepsilon}{2(\varepsilon(\varepsilon - 1))^{1/2}} \times \sin \frac{(\varepsilon - 1)^{1/2}}{2} \sin 12\varepsilon^{1/2}\right] \leqslant 1 \ . \tag{4.1.30}$$

Here $\varepsilon = E/V_0 > 1$, and the function $F(\varepsilon)$ [square brackets in (4.1.30)] is portrayed in Fig. 4.2b for $\varepsilon > 1$. Therefore, the energy spectrum for $E > V_0$ also has the shape of continuous bands separated by forbidden gaps. This part of the spectrum is depicted as shaded bands in Fig. 4.3b. A comparison with Fig. 4.3a shows that the periodicity (with $E > V_0$) splits up the continuous spectrum of a single well into bands separated by gaps, and for $E < V_0$ it splits up the isolated levels into continuous bands that are also separated by gaps.

Determine the position and width of the gaps for $E \gg V_0$. From (4.1.30) we find, up to the quantities $\sim \xi \equiv V_0/E$,

$$-1 \leqslant \cos \alpha c + \sin \alpha c\,(\alpha a \xi/2) \leqslant 1 \ . \tag{4.1.31}$$

As seen from (4.1.31), the expression for $F(\varepsilon)$, enclosed in square brackets, is equal to $(-1)^n$ whenever the argument $\alpha c = (2mE/\hbar^2)^{1/2} c = n\pi$ holds ($n > 0$ is an integer). For αc values somewhat smaller than $n\pi$, $\sin \alpha c$ and $\cos \alpha c$ are of different sign and $|F(\varepsilon)| < 1$; the corresponding energy values therefore are allowed. Conversely, when $\alpha c > n\pi$, $\sin \alpha c$ and $\cos \alpha c$ are of the same sign and $|F(\varepsilon)| > 1$. From the right (i.e., for $\alpha c > n\pi$) each point $\alpha c = n\pi$ is thus bordered by a region of forbidden energies.

We wish to ascertain the physical meaning of the boundary of a forbidden region. As follows from (4.1.16, 19), the quantity $\hbar\alpha$ is similar to the momentum of a free particle, with the only difference that, instead of being multiplied by a constant factor, the phase term $\exp(\pm i\alpha x)$ involved in the wave function is multiplied by A_n or B_n, which, from (4.1.22, 23), are energy dependent. This quantity is called the quasimomentum (with α being the wave number). The gap boundary condition will then be $\alpha = \pi n/c$, i.e., the Wulf–Bragg formula familiar from (1.5.11). If we introduce the electron wavelength $\lambda = 2\pi/\alpha$, this condition then assumes the standard form of (1.5.11)

$$2c = n\lambda \ , \tag{4.1.32}$$

since, for a one-dimensional array, $\vartheta = \pi/2$ and $\sin\vartheta = 1$. Thus, the energy gap arises when the electron wave suffers a complete Wulf–Bragg internal reflection which hinders the passage of an electron current across the array.

To evaluate the gap width we introduce into (4.1.31) the notation $\alpha c\xi/2 = \tan\varphi$. It is readily inferred that $F(\varepsilon)$ will reach the values $(-1)^n$ not only when $\alpha c = n\pi$ but also when $\cos(\alpha c - \varphi) = (-1)^n \cos\varphi$, provided that $\alpha c = n\pi + 2\varphi$. The width of the gap therefore is equal to 2φ. For large n (small ξ) we may set $\varphi = \tan^{-1}(\alpha c\xi/2) \approx \alpha c\xi/2$ and $2\varphi \approx c(2m)^{1/2}\, V_0/\hbar E^{1/2}$. Consequently, the gap width diminishes in proportion to $E^{-1/2}$ or n^{-1}, since the boundaries of the gaps are determined by the condition $\alpha c = (2mE/\hbar^2)^{1/2} c = n\pi$. As the barrier height V_0, specifying the coupling of an electron to the lattice, increases, the width of the allowed bands decreases. In the limit $V_0 \to \infty$ we obtain a discrete spectrum of an isolated well. Conversely, for $V_0 \to 0$ we have the case of a free electron with a continuous spectrum. We will show that the band spectrum thus obtained is typical of electrons in any periodic field.

4.1.3 Linear Atomic Array

Now we consider the motion of an electron in a one-dimensional array of N ions. To avoid the difficulties associated with the boundary conditions, we form the array into a loop (Sect. 2.1.1 and Fig. 4.4). We assign to the ions the numbers $j = 1, 2, \ldots N$. The $N + 1$ number will coincide with 1. The energy $V(x)$ is equal to the sum of the potential energies of the ions—(this assumption is possible if the energy of the first excited state, E_1, is sufficiently far from the ground-state

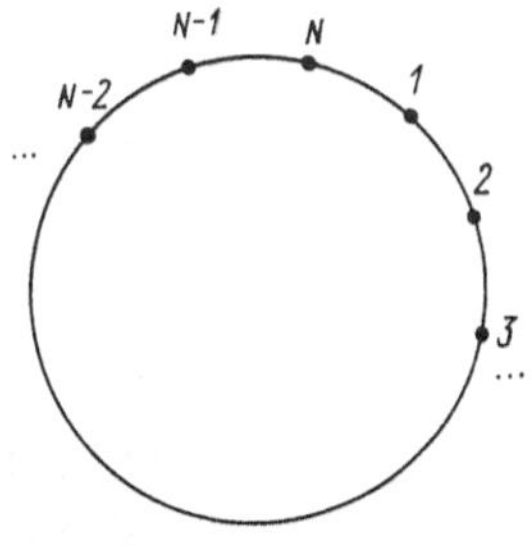

Fig. 4.4. Closed atomic array

energy E_0 so that the energy difference $E_1 - E_0$ is much larger than the splitting of these energies in the field of the array—

$$V(x) = \sum_{j=1}^{N} v(x - jd) \tag{4.1.33}$$

(with d being the distance between neighboring ions). The energy also possesses the periodicity property

$$V(x) = V(x + nd) \ , \tag{4.1.34}$$

The quantity $V(x)$ is actually not equal to (4.1.33), owing to the "distortion" of the potential in the interatomic regions which arises from the interaction of atoms. However, this discrepancy will be neglected here. The electron wave function $\psi(x)$ satisfies the Schrödinger equation (4.1.12) with the potential (4.1.33). We seek this function as a sum of ground-state wave functions $\varphi_j(x)$ for the electron of an isolated atom

$$\psi(x) = \sum_{j=1}^{N} a_j \varphi_j(x) = \sum_{j=1}^{N} a_j \varphi(x - jd) \ . \tag{4.1.35}$$

This form of solution would be exact if the $\varphi_j(x)$ formed a complete system of orthonormalized functions. But in reality this is not the case, since even in a single atom there is an infinite set of such functions. In spite of this, the expansion (4.1.35) will be viewed as exact. A solution in the form of (4.1.35) signifies that the electron has a finite probability of being located at any array site. The coefficients a_j from (4.1.35) are determined by substituting the latter into (4.1.13), allowing for the fact that the $\varphi_j(x)$ satisfy the Schrödinger equation of a single atom,

$$-\frac{\hbar^2}{2m}\frac{d^2\varphi_j}{dx^2} + v_j(x)\varphi_j = E_0 \varphi_j \ . \tag{4.1.36}$$

As a result,

$$\sum_{j=1}^{N} a_j \{E - E_0 - [V(x) - v_j(x)]\} \varphi_j(x) = 0 \ . \tag{4.1.37}$$

The normalization condition for the functions

$$\int_0^{Nd} dx \, |\varphi_j(x)|^2 = 1 \tag{4.1.38}$$

and the approximate condition of their orthogonality

$$\int_0^{Nd} dx\, \varphi_j^*(x)\, \varphi_{j'}(x) \approx 0 \quad (j \neq j) \tag{4.1.39}$$

are assumed to be fulfilled. The acceptability of (4.1.39) is seen from Fig. 4.5: where $\varphi_j(x)$ differs appreciably from zero ($x \approx jd$), the quantity $\varphi_{j'}(x)$ ($j' \neq j$) is close to zero; conversely, where $\varphi_{j'}(x)$ has a maximum ($x \approx j'd$), $\varphi_j(x)$ is close to zero. Over the entire interval from 0 to Nd, the product $\varphi_j^*(x) \cdot \varphi_{j'}(x)$ therefore is close to zero. Premultiplying (4.1.37) by $\varphi_{j'}^*(x)$ and integrating over the entire array, allowing for (4.1.38, 39), and replacing j' by j and replacing the summation subscript by i, we obtain

$$(\varepsilon - \alpha)\, a_j - \sum_{i=1}^{(N-1)/2} \beta_i (a_{j+i} + a_{j-i}) = 0 \ , \tag{4.1.40}$$

$$\varepsilon = E - E_0 \ . \tag{4.1.41}$$

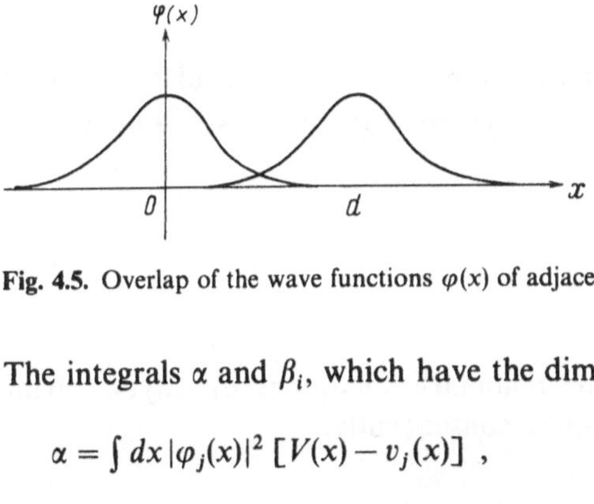

Fig. 4.5. Overlap of the wave functions $\varphi(x)$ of adjacent atoms

The integrals α and β_i, which have the dimensions of energy, are as follows:

$$\alpha = \int dx\, |\varphi_j(x)|^2\, [V(x) - v_j(x)] \ , \tag{4.1.42}$$

that is, the energy of the interaction of the electron with "foreign" array ions [the subscript $i(j)$ is omitted here since, as may be shown, α is independent of the atomic number]:

$$\beta_i = \int dx\, \varphi_j^*(x)\, [V(x) - v_{j\pm i}(x)]\, \varphi_{j\pm i}(x) \ , \tag{4.1.43}$$

that is, the "transfer integral", or, put another way, the amplitude of the probability that the electron will move from site j and $j \pm i$ under the influence of the potential $V(x) - v_{j\pm i}(x)$. Since $\varphi_j(x)$ decreases rapidly as x moves away from the site j, in (4.1.43) we confine ourselves to the integral for $i = 1$, i.e., the nearest-neighbor approximation. Equation (4.1.40) then takes the form

$$(\varepsilon - \alpha)\, a_j - \beta_j (a_{j+1} + a_{j-1}) = 0 \ ; \quad j = 1, 2, \ldots N \tag{4.1.44}$$

i.e., we obtain a system of N homogeneous first-degree equations in finite differences with $N+1$ unknowns ($a_1, a_2, \ldots a_N$ and ε). We then try solutions of (4.1.44) in the form

$$a_j = a_0 \exp(ikjd) \ , \quad j = 1, 2, \ldots N \ , \tag{4.1.45}$$

where a_0 is determined from the normalization conditions of (4.1.35), and k is a parameter that has the dimensions of cm^{-1}. For (4.1.35) we thus find

$$\psi(x) = \sum_{j=1}^{N} a_0 \exp(ikjd)\,\varphi(x-jd) = \exp(ikx)\,u(x) \ , \tag{4.1.46}$$

where

$$\begin{aligned} u(x) &= a_0 \sum_{j=1}^{N} \exp[-ik(x-jd)]\,\varphi(x-jd) \\ &= u(x+nd) \end{aligned} \tag{4.1.47}$$

is periodic with the period of the array. As in the previous problem, the wave function is a plane wave modulated with the array period. Substituting (4.1.44, 45), we find the energy

$$\begin{aligned} \varepsilon &= \alpha + 2\beta_1 \cos kd \ , \\ E &= E_0 + \alpha + 2\beta_1 \cos kd \ . \end{aligned} \tag{4.1.48}$$

The parameter k is found from the boundary conditions $a_j = a_{j+N}$, giving $\exp(ikNd) = 1$, whence $kNd = 2\pi n$, and, consequently,

$$k = \frac{2\pi n}{Nd} \ , \quad n = 0, \pm 1, \pm 2, \ldots \tag{4.1.49}$$

Equation (4.1.48) shows that ε is a periodic function of k. The N different solutions of (4.1.45) and the values of the energies $\varepsilon(k)$ correspond to the energy interval

$$-\pi/d \leqslant k < \pi/d \ , \tag{4.1.50}$$

known as the Brillouin zone (2.1.8), and, for the numbers n, (4.1.49) therefore yields

$$-N/2 < n \leqslant N/2 \ . \tag{4.1.51}$$

In conformity with the general rule of quantum mechanics, the quantity $|a_j|^2$ defines the probability that the electron will be located at the jth atom of the array. According to (4.1.45), it is equal to $|a_0|^2$ and is independent of the site number; the electron is located with equal probability at any site and is free to move along the array, in spite of the potential barriers between the atoms. From the normalization condition for ψ and from (4.1.38, 39) we obtain the result that $\sum_{j=1}^{N}|a_j|^2 = 1$ and, using (4.1.45),

$$|a_0|^2 = 1/N \ . \tag{4.1.52}$$

From (4.1.48) we see that the coupling of N atoms into an array results in the atomic level E_0 splitting into a band with width determined by the integral β_1. If we assume that α and β_1 are negative, then (4.1.48) may be transcribed into

$$E(k) = E_0 - |\alpha| - 2|\beta_1|\cos kd \ . \tag{4.1.53}$$

The period of the function $E(k)$, according to (4.1.50), is equal to $2\pi/d$ (Fig. 4.6) with an energy minimum at $k = 0$ ($E_{\min} = E_0 - |\alpha| - 2|\beta_1|$), and an energy maximum at $k = \pm\pi/d$ ($E_{\max} = E_0 - |\alpha| + 2|\beta_1|$). The difference between these energies gives the energy bandwidth

$$\Delta E = E_{\max} - E_{\min} = 4|\beta_1| \ . \tag{4.1.54}$$

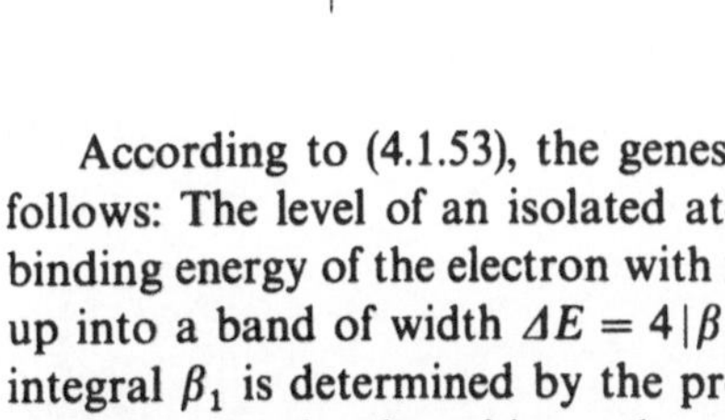

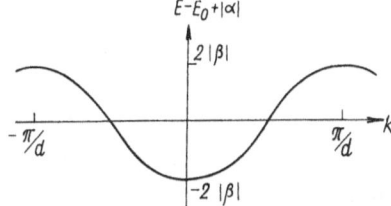

Fig. 4.6. Function $E(k)$ from (4.1.53)

According to (4.1.53), the genesis of the electron energy in the array is as follows: The level of an isolated atom, E_0, lowers by a value $|\alpha|$ equal to the binding energy of the electron with foreign ions and then this shifted level splits up into a band of width $\Delta E = 4|\beta_1|$ (Fig. 4.7). The order of magnitude of the integral β_1 is determined by the product $\varphi_j(x)\,\varphi^*_{j\pm 1}(x)$. If we move the atoms away from each other, this product will decrease. It also depends on which state φ_j describes. For outer orbits the product is larger than for inner orbits, since $\varphi_j(x)$ and $\varphi_{j\pm 1}(x)$ for the former overlap in a large volume. An estimate for outer orbits (say, 3s in sodium) yields

$$\beta_1 \approx 10^{-12}\,\text{erg} \ , \tag{4.1.55}$$

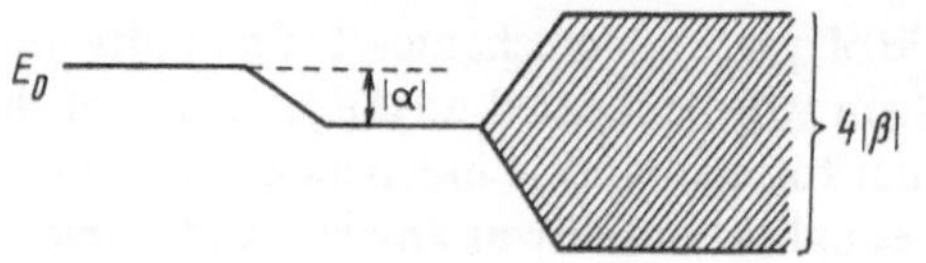

Fig. 4.7. Displacement and splitting of the discrete atomic level E_0 in an atomic array

and for inner ones (say, 1s)

$$\beta_1 \approx 10^{-20}\,\text{erg}\ ; \tag{4.1.56}$$

i.e., the bandwidth of valence electrons is many orders of magnitude larger than that of inner-shell electrons. An indirect indication of a remaining quasi discreteness of low-lying energy levels is the characteristic X-ray spectra, which practically maintain their line character in all states of aggregation. We may say then that the atomic array model is closer to a real crystal in the case of narrow bands.

The quantity $\hbar/|\beta_1|$, having the dimensions of time, specifies the transition time τ, during which the electron passes from site to site. It follows from (4.1.55) that for outer-shell electrons $\tau \approx \hbar/|\beta_1| \approx 10^{-15}$ s; i.e., on the order of the times of intratomic processes. For inner-shell electrons we have, from (4.1.56), the transition time $\tau \approx 10^{-7}$ s, which, on an atomic scale, is a very large time: the electron manages to "orbit" the nucleus many times before passing on to an adjacent atomic nucleus.

We now ascertain the dependence of the energy E on k in the energy band (Fig. 4.6). To clear up the physical significance of k, we consider the mean value of the current operator in the array. From quantum mechanics we have for the mean current

$$\langle J \rangle = \frac{e\hbar}{2mi} \int dx \left(\psi^* \frac{\partial \psi}{\partial x} - \psi \frac{\partial \psi^*}{\partial x} \right) . \tag{4.1.57}$$

Substituting ψ^* and ψ from (4.1.46, 52), we find

$$\langle J \rangle = -\frac{2e\hbar}{m} \gamma \sin kd\ , \tag{4.1.58}$$

where

$$\gamma = \pm \int_0^{Nd} dx\ \varphi(x) \frac{\partial \varphi(x \mp d)}{\partial x} . \tag{4.1.59}$$

For the mean electron velocity we thus have

$$\langle v \rangle = \langle J \rangle / e = \frac{2\hbar\gamma}{m} \sin kd\ . \tag{4.1.60}$$

The quantity γ, just as β_1, depends on the distance between the sites and on the (valence or inner-shell) electrons to which the function $\varphi(x)$ corresponds. Namely, γ is vanishingly small for inner-shell electrons, whereas an estimate for valence electrons yields $\gamma \approx 10^8\ \mathrm{cm}^{-1}$. For $|\sin kd| \approx 1$ we find from (4.1.60) $\langle v \rangle \approx 10^8$ cm/s; i.e., in states that are not very close to the "bottom" of the band ($|\sin kd| \approx 1$), the mean velocity of valence electrons is close to the Fermi velocity of free electrons. But there is a substantial difference—the energy of the electron in the array and its mean velocity are related to each other differently to a free electron and its mean velocity: $\varepsilon = m^* \langle v \rangle^2/2$. This is seen from a comparison of the $E(k)$ and $\langle v \rangle (k)$ relations presented in Fig. 4.6, from which it follows that in the states of the lower portion of the band $\langle v \rangle$ increases with E, but in the states of the upper half of the band the mean velocity diminishes with increasing energy, which is altogether incomprehensible from the standpoint of the free gas theory. This behavioral feature of the electron in the array results from its interaction with the periodic field. By $E(k)$ we should understand not the energy of an isolated electron but the energy of the array as a whole, for which it is not absolutely necessary to maintain the relation that holds for a free electron. Only for the states that are near the bottom (low energies) or top (highest energies) of the band may the energy of the electron in the array be brought into the classical form. From (4.1.53) we have

$$E - E_{\min} = 2|\beta_1|(1 - \cos kd) \ .$$

The condition $k = 0$ corresponds to the minimum band level; expanding $\cos kd$ in the vicinity of $k = 0$ up to second-order terms in kd, we obtain

$$\varepsilon' = E - E_{\min} \approx |\beta_1| k^2 d^2 \ . \tag{4.1.61}$$

Similarly, according to (4.1.53), we get

$$\varepsilon'' = E - E_{\max} = -2|\beta_1|(1 + \cos kd) \ . \tag{4.1.62}$$

The condition $k = \pm\pi/d$ corresponds to the maximum band level; performing expansion to the same order, we find

$$\varepsilon'' \approx -|\beta_1| k'^2 d^2 \ , \qquad k' = k - \pi/d \tag{4.1.63}$$

We now introduce the quantities $p = \hbar k$ and $p' = \hbar k'$, which are called quasimomenta and have the dimensions of momentum; then

$$\varepsilon' = |\beta_1| d^2 p^2/\hbar^2 \ , \qquad \varepsilon'' = -|\beta_1| d^2 p'^2/\hbar^2 \ . \tag{4.1.64}$$

A correlation may be established between (4.1.64) and the corresponding formula for a free electron, if we call the quantity $\hbar^2/2|\beta_1|d^2$, having the dimensions of mass, the effective mass $|m^*|$ of the electron in the array

$$|m^*| = \hbar^2/2|\beta_1|d^2 . \tag{4.1.65}$$

It follows from (4.1.63) that for the levels near the bottom of the band $m^* > 0$, and near the top $m^* < 0$. The latter inequality expresses the unusual relation found between the electron energy and the mean electron velocity in the array. We will show that the existence of $m^* < 0$ has the result that rather than accelerate the electron in the lattice the externally applied field decelerates it, leading, under certain conditions, to an apparent occurrence of positive charges that participate in the current of the metal (Sect. 3.7.5).

4.1.4 Rigorous Theory of Electron Motion in a One-Dimensional Array

By contrast with the problems treated in Sects. 4.1.2, 3, no assumptions will be made as to the form of the potential $V(x)$, except for allowance for periodicity (4.1.34) and other assumptions of a general character. Thereby we will ascertain the properties of the electron in the array, which follow directly from the periodicity of the array. We will proceed from (4.1.12) which is a Hill-type differential equation. Without solving it explicitly, we will do an analysis that will help elucidate the essential features of the electron's behavior in a one-dimensional array with a period d.

It follows from the potential periodicity (4.1.34) that if $\psi(x)$ is some solution of (4.1.12) for energy E, then $\psi(x+d)$ is also a solution of (4.1.12) with the same energy. This conclusion, a central point of the theory, is a mathematical expression of the fact that the electron in a periodic array has no steady states with a localization near one atom. If we were to assume that a steady state of this type exists, then, according to the above, with E being the same, an infinite set of other steady states with a localization at other atoms should exist. But this is mathematically impossible, since (4.1.12) for this E has two linearly independent solutions at the most.

Thus the wave nature of the electron excludes its localization near array sites, and the freedom of movement is not the property of special potentials (Sects. 4.1.2, 3), but a general consequence of periodicity. Since (4.1.12) is a linear differential second-order equation, any one of its solutions may be represented as a linear combination of two basis solutions, $\psi_1(x)$ and $\psi_2(x)$, for a given E. They are linearly independent; i.e., a relation such as $\gamma_1\psi_1(x)+\gamma_2\psi_2(x)\equiv 0$, with γ_1 and γ_2 being nonzero constants, does not exist between them. This one-dimensional problem is two-fold degenerate.

Compare this general theorem with the basic property of the solutions of (4.1.12) that $\psi_1(x+d)$ and $\psi_2(x+d)$ are also its solutions for the same E; i.e., they may be represented in the form

$$\psi_1(x+d) = \alpha_{11}\psi_1(x) + \alpha_{12}\psi_2(x)$$

and

$$\psi_2(x+d) = \alpha_{21}\psi_1(x) + \alpha_{22}\psi_2(x) \; . \tag{4.1.66}$$

The values of the coefficients α_{ij}, generally speaking, depend on the choice of the basis $\psi_1(x)$ and $\psi_2(x)$. It may always be chosen in such a fashion that the matrix α_{ij} is diagonal. In the replacement of x by $x+d$ the basis functions are then multiplied by some factor

$$\psi(x+d) = \lambda\psi(x) \; , \tag{4.1.67}$$

where λ is one of the two eigenvalues of the matrix α_{ij}. Proceeding from the basic property (4.1.66), we may write

$$\psi(x) = \beta\psi_1(x) + \gamma\psi_2(x) \tag{4.1.68}$$

and thereby reduce the search for ψ to a search for the β and γ that correspond to it. Substituting (4.1.68) into (4.1.67) and making use of (4.1.66), we obtain for β and γ a system of two linear homogeneous equations

$$\begin{aligned} &(\alpha_{11} - \lambda)\beta + \alpha_{12}\gamma = 0 \\ &\alpha_{21}\beta + (\alpha_{22} - \lambda)\gamma = 0 \; . \end{aligned} \tag{4.1.69}$$

They have nonzero solutions if the determinant of the system is equal to zero:

$$\lambda^2 - (\alpha_{11} + \alpha_{22})\lambda + \alpha_{11}\alpha_{22} - \alpha_{12}\alpha_{21} = 0 \; . \tag{4.1.70}$$

Since all the coefficients in (4.1.12) are real, the functions ψ_1, ψ_2, and the numbers α_{ij} in (4.1.66) may be chosen to be real. Owing to the fact that the trace of the matrix α_{ij} and its determinant do not alter when we proceed to a new (possibly complex) basis, $\alpha_{11} + \alpha_{22}$ and $\alpha_{11}\alpha_{22} - \alpha_{12}\alpha_{21}$ must be real. Equation (4.1.70) then is a quadratic equation with real coefficients for λ. We consider three cases.

In the first case, (4.1.70) has two different real roots λ_1 and λ_2. Substituting them successively into (4.1.69), we find two systems of coefficients β and γ and two real functions $\psi(x)$ that satisfy (4.1.67). When $\lambda_1 \neq \lambda_2$, these functions are linearly independent and therefore may be the basis of (4.1.12).

In the second case, (4.1.70) has two complex conjugate roots

$$\lambda_1^* = \lambda_2 \; . \tag{4.1.71}$$

The corresponding functions are also complex conjugate ones

$$\psi_1^*(x) = \psi_2(x) \; . \tag{4.1.72}$$

For these functions we may repeat the same lines of argument as those in the first case.

In the third case, (4.1.70) has one real root λ. Here one real solution exists which satisfies (4.1.67). Therefore, if (4.1.70) has a multiple root, the properties of all the solutions of (4.1.12), as a rule, may not be described by functions (4.1.67), although at least one such function most certainly exists among these solutions.

The above properties of the solutions of (4.1.12) are wholly determined by the value of the parameter E. Find the form of the roots of (4.1.70) and differentiate (4.1.66) with respect to

$$\begin{aligned}\psi_1'(x+d) &= \alpha_{11}\psi_1'(x) + \alpha_{12}\psi_2'(x)\\ \psi_2'(x+d) &= \alpha_{21}\psi_1'(x) + \alpha_{22}\psi_2'(x)\ .\end{aligned} \tag{4.1.73}$$

From (4.1.66, 73) we obtain

$$\begin{vmatrix}\psi_1(x+d) & \psi_2(x+d)\\ \psi_1'(x+d) & \psi_2'(x+d)\end{vmatrix} = \begin{vmatrix}\psi_1(x) & \psi_2(x)\\ \psi_1'(x) & \psi_2'(x)\end{vmatrix}\begin{vmatrix}\alpha_{11} & \alpha_{12}\\ \alpha_{21} & \alpha_{22}\end{vmatrix}\ .$$

The derivative of the Wronskian $\begin{vmatrix}\psi_1 & \psi_2\\ \psi_1' & \psi_2'\end{vmatrix}$ with respect to x is, according to (4.1.12), equal to $\psi_1\psi_2'' - \psi_2\psi_1'' = 0$, whence it follows that

$$\begin{vmatrix}\alpha_{11} & \alpha_{12}\\ \alpha_{21} & \alpha_{22}\end{vmatrix} = \alpha_{11}\alpha_{22} - \alpha_{12}\alpha_{21} = 1 \tag{4.1.74}$$

(from the invariance of the Wronskian). From (4.1.70, 74) we get the result that the product of the roots is equal to

$$\lambda_1\lambda_2 = 1\ . \tag{4.1.75}$$

If the roots are complex conjugates, then, as follows from (4.1.75), $|\lambda_1|^2 = |\lambda_2|^2 = 1$, and we may set

$$\lambda_1 = e^{i\xi}\ , \quad \lambda_2 = e^{-i\xi}\ , \tag{4.1.76}$$

with ξ being real. Without a loss in generality, it may be assumed that $0 \leqslant \xi \leqslant \pi$. Formula (4.1.67) now yields

$$\psi_1(x+d) = e^{i\xi}\psi_1(x)\ , \quad \psi_2(x+d) = e^{-i\xi}\psi_2(x)\ . \tag{4.1.77}$$

Introduce the functions $u_{1,2}(x) = \psi_{1,2}(x)\exp(\mp i\xi x/d)$; then $u(x)$ is periodic with period d and

$$\psi_{1,2}(x) = \exp(\pm i\xi x/d)\,u_{1,2}(x)\ , \tag{4.1.78}$$

where

$$u_{1,2}(x+nd)=u_{1,2}(x)\ , \qquad u_2(x)=u_1(-x)=u_1^*(x)\ . \tag{4.1.79}$$

Thus, if to the energy E there correspond complex conjugate λ_1 and λ_2, the solutions of (4.1.78) are physically permissible: these are plane waves that are modulated in phase with the period of the array. But if λ_1 and λ_2 are real, (4.1.12) has no allowed solutions, for, as can be readily shown, they diverge when $x \to \pm\infty$. Indeed,

$$\lambda_{1,2}=\pm e^{\pm\xi}\ , \qquad \psi_{1,2}(x)=\pm e^{\pm\xi x}u_{1,2}(x)\ , \tag{4.1.80}$$

$$u_{1,2}(x+nd)=u_{1,2}(x)\ .$$

With multiple roots $\lambda_1=\lambda_2$, we have from (4.1.75)

$$\lambda_2=1\ , \qquad \lambda=\pm 1\ , \tag{4.1.81}$$

and to each E there corresponds one solution of the type (4.1.67) (with $\lambda=+1$ or -1), which satisfies the condition

$$\psi(x)=\pm\psi(-x)\ . \tag{4.1.82}$$

To visualize this better, we introduce an auxiliary quantity

$$f(E)=\lambda_1+\lambda_2\ . \tag{4.1.83}$$

For the first type of solution we obtain from (4.1.76) the result that

$$f(E)=2\cos\xi\ , \tag{4.1.84}$$

i.e., $|f(E)|\leqslant 2$. For the second type

$$f(E)=\pm 2\cosh\xi\ , \tag{4.1.85}$$

so that $|f(E)|>2$. Finally, the third type of solutions results from the second type with $\xi=0,\pi$; i.e., it corresponds to $|f(E)|=2$. Therefore, if E is such that $|f(E)|<2$, there are two linearly independent complex conjugate solutions of (4.1.12), which are given by (4.76–79). Any other solution may be represented as a linear combination of these two. If the energy is such that $|f(E)|>2$, there are two linearly independent real solutions of (4.1.12), determinable from (4.1.80), which, however, should be rejected as unnormalizable. Finally, if $|f(E)|=2$, one solution exists, having the properties of (4.1.82).

Thus the structure of the energy spectrum of the array is wholly determined by the form of the function $f(E)$, plotted in Fig. 4.8 (see Fig. 4.2a, b). The domain enclosed between the straight lines parallel to the abscissa axis at $f(E) = \pm 2$ is called the internal region, and the rest the external region. Every E for which $f(E)$ lies in the external region is forbidden, and the E for which $f(E)$ lies in the internal region (intervals $E_2 - E_1, E_4 - E_3, E_6 - E_5$, and $E_8 - E_7$) is an allowed and two-fold degenerate energy value; finally, the E for which $f(E)$ lies at the boundary between the regions [at $f(E) = \pm 2$] corresponds to a special case (see below). Figure 4.8 shows that this case occurs for discrete E values.

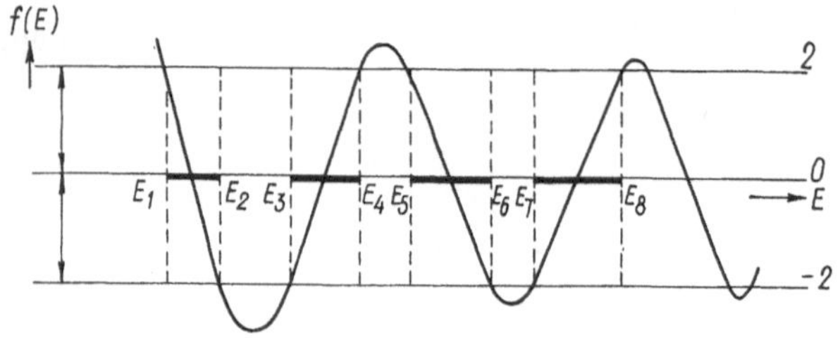

Fig. 4.8. Function $f(E)$ from (4.1.83)

To find the energy spectrum, we consider two limiting cases. One of them is when negative energies are very large in magnitude; i.e., when (4.1.12) at finite $V(x)$ takes the form

$$\frac{d^2\psi}{dx^2} - k^2\psi = 0 , \quad k^2 = -2mE/\hbar^2 . \tag{4.1.86}$$

Solutions to (4.1.86) will be $\psi_{1,2}(x) = e^{\pm kx}$. Comparing them with (4.1.81), we see that starting from some point, a large negative E is bound to become unallowable. Equation (4.1.85) shows that in this situation $f(E) = 2\cosh kd \to +\infty$, for $k \to \infty$ when $E \to -\infty$. Thus the extreme left-hand branch of $f(E)$ has a shape such as that portrayed in Fig. 4.8.

If $E > 0$ and is very large, then $2mE/\hbar^2 = k^2$ and $k \to \infty$, and the wave functions have the form of plane waves: $\psi_1(x) = \exp(ikx)$, $\psi_2(x) = \exp(-ikx)$. Therefore, with $E \to +\infty$, we are certain, sooner or later, to get into the region of allowed solutions, which in the limit transform to free-electron plane waves. This is clear, for when $E \gg |V(x)|$, the inhomogeneities of $V(x)$ cease to affect the electron motion. Thus, as it goes over to the right-hand extremity of Fig. 4.8, the $f(E) = 2\cos kd$ curve becomes periodic, being confined within the internal region and touching the boundary straight lines at the extreme points only. In the intermediate region of finite E the $f(E)$ curve is always finite for, as (4.1.85) shows, with $f(E) \to \infty$ and $\xi \to \infty$ both functions (4.1.80) would go to infinity in the entire semispace, which is impossible. In all cases except for $E \to -\infty$, the

$f(E)$ curve is therefore finite and regular. Figure 4.8 reproduces schematically the oscillatory trend of the $f(E)$ curve (see Fig. 4.2). As is seen from the figure, the spectrum of an electron in a one-dimensional array consists of continuous allowed energy bands (heavy-line segments on the E axis) separated by forbidden energy gaps (thin-line segments on the E axis). This is similar to the results of model problems (Sects. 4.1.2, 3).

Making use of the $f(E)$ curve, we will determine the numeration of the wave functions (4.1.78). In the internal region we break this curve up into branches between adjacent maxima and minima; if these extrema lie in the external region, what we will understand by branches will be segments of $f(E)$ between the point of its "entry" into the internal region and the point of "exit" closest to it. To each branch corresponds a determinate set of wave functions, called a band. Thus, the steady states of the electron in a one-dimensional array are subdivided into classes that form a discrete sequence of bands.

To each interior point $f(E)$ corresponds a pair of functions (4.1.78) defined by the band number ζ and the value of the function $f(E)$. To specify each of the functions, we make use of the quantity ξ from (4.1.76) or (4.1.86): $\xi = -\, i\ln\lambda_1 = i\ln\lambda_2 = \cos^{-1}[f(E)/2]$. For instance, to the first of the functions (4.1.78) we ascribe the value $\xi = -i\ln\lambda_1$, and the second $-\xi = -i\ln\lambda_2$, differing in sign. Each wave function then is defined by the band number ζ and quantity ξ. However, as is manifest from (4.1.76), ξ is ambiguous and is determined to within $2\pi n$ (with n an arbitrary integer). To lift this ambiguity, we henceforth consider a ξ that satisfies the inequality

$$-\pi < \xi \leqslant \pi \ . \tag{4.1.87}$$

This choice of ξ (reduced zone) is not unique. What is essential is that not a single physical result should alter when ξ is replaced by $\xi + 2\pi n$. An exception is the value $\xi = -\pi$ in (4.1.87), associated with special conditions at the band boundaries (treated later). Instead of ξ we sometimes will use, just as in Sects. 4.1.2, 3 and Chap. 2, the quantity $k = \xi/d$.

A complete system of electron wave functions in an infinite one-dimensional lattice with a constant d thus has the form

$$\psi(k\zeta; x) = \exp(ikx)\, u(k\zeta; x) \ , \tag{4.1.88}$$

$$u(k\zeta; x + nd) = u(k\zeta; x) \ . \tag{4.1.89}$$

Apart from this, the following relations hold:

$$u(-k\zeta; x) = u^*(k\zeta; x) = u(k\zeta; -x) \ . \tag{4.1.90}$$

The band number runs over an infinite discrete series of integers. At each ζ the variable k runs over a continuous series of values, for example, those lying in the region (4.1.87). When $k = 0, \pi/d$, the functions (4.1.78) become real, and we will have, in place of (4.1.90),

$$\psi(k\zeta;x)=\psi^*(k\zeta;x)=\pm\psi(k\zeta;-x)\ ,\quad k=0,\pi/d\ . \tag{4.1.91}$$

The physical meaning of (4.1.88, 89) is the same as that of (4.1.78, 79). In each steady state the electron is "smeared" along the array; that is, the position probability density for the electron is

$$\varrho(k\zeta;x)=|\psi(k\zeta;x)|^2=|u(k\zeta;x)|^2\ . \tag{4.1.92}$$

This density has a period d and, according to (4.1.90), is an even function x.

The wave functions (4.1.88) are complex, except for the case (4.1.91). We may therefore expect both the mean electron velocity and the allied current density in the array to be other than zero. From the periodicity of (4.1.89) it follows that these two quantities are the same in all cells of the array. Therefore, we are in a position to say that the electron propagates through the entire ideal array without being scattered. An exception to this rule may be expected for states (4.1.91) with $k=0,\pi/d$, when the function is real and the average current is equal to zero. In all the other cases the band electron in many ways resembles a free electron. Thus a reversal of the sign of k changes the sign of the velocity and current. The degeneracy in an array therefore signifies, just as with a free particle, that an electron moving in a given direction possesses the same energy as that of an electron moving with the same velocity in the inverse direction.

We now find the relation between the mean electron velocity and the mean electron energy. To this end, we have to calculate the matrix element of the coordinate operator. We will start by ascertaining the orthogonality and normalization conditions for the functions (4.1.88). For this purpose we substitute these functions into the integral

$$S=\int_{-\infty}^{\infty}dx\,\psi^*(k'\zeta';x)\psi(k''\zeta'';x)\ .$$

Using (4.1.89), we obtain

$$S=\lim_{G\to\infty}\sum_{l=-G}^{G}\exp[\mathrm{i}(\xi''-\xi')l]\int_0^d dx\exp[\mathrm{i}(\xi''-\xi')x/d]\,u^*(\xi'\zeta';x)u(\xi''\zeta'';x)\ .$$

The integral over the array is here represented as the sum over its cells. On account of (4.1.89), these integrals depend on l only in terms of $\exp[\mathrm{i}(\xi''-\zeta')l]$. The summation over l reduces to a geometric progression, yielding

$$\begin{aligned}\sum_{l=-G}^{G}\exp[\mathrm{i}(\xi''-\xi')l]&=\frac{-2\mathrm{i}\sin(\xi'-\xi'')G}{\exp\mathrm{i}(\xi''-\xi')-1}\\&\equiv\frac{2\sin(\xi'-\xi'')G}{\xi'-\xi''}\,\frac{\mathrm{i}(\xi''-\xi')}{\exp\mathrm{i}(\xi''-\xi')-1}\ .\end{aligned}$$

Using the identity

$$\lim_{G\to\infty} \frac{\sin G\xi}{\xi} = \pi\delta(\xi) \ , \tag{4.1.93}$$

we obtain the result

$$S = 2\pi\delta(\xi' - \xi'') \int_0^d dx\, u^*(\xi'\zeta';\, x) u(\xi''\zeta'';\, x) \ . \tag{4.1.94}$$

From the orthogonality condition for a continuous series of eigenvalues we see that the factor $\delta(\xi' - \xi'')$ assures that these values are satisfied for all $\xi' \neq \xi''$. It remains for us to explore the case $\xi' = \xi''$ and $\zeta' \neq \zeta''$. Figure 4.8 then shows that the first and the second states undoubtedly correspond to different energies. Proceeding from the general rules of quantum mechanics, we thus may assert that for $\zeta' \neq \zeta''$ the second term involved in (4.1.94) will go to zero, too:

$$\int_0^d dx u^*(\xi\zeta';\, x) u(\xi\zeta'';\, x) = \delta_{\zeta'\zeta''}/2\pi \ , \tag{4.1.95}$$

with $\delta_{\zeta'\zeta''} = 1$ for $\zeta' = \zeta''$ and with $\delta_{\zeta'\zeta''} = 0$ for $\zeta' \neq \zeta''$. Equation (4.1.94) then becomes

$$\int_{-\infty}^{\infty} dx \psi^*(\xi'\zeta';\, x) \psi(\xi''\zeta'';\, x) = \delta(\xi' - \xi'')\delta_{\zeta'\zeta''} \ . \tag{4.1.96}$$

The general expression for the matrix element of the coordinate operator is equal to

$$(\xi'\zeta'|x|\xi''\zeta'') = \int_{-\infty}^{\infty} dx \psi^*(\xi'\zeta';\, x) x \psi(\xi''\zeta'';\, x) \ .$$

According to (4.1.88), we may write

$$\begin{aligned} x\psi(\xi''\zeta'';\, x) &= -\mathrm{i}d \frac{\partial}{\partial \xi''} \psi(\xi''\zeta'';\, x) + \mathrm{i}d \exp(\mathrm{i}\xi'' x/d) \frac{\partial}{\partial \xi''} u(\xi''\zeta'';\, x) \\ &\equiv (\hat{x}_1 + \hat{x}_2)\psi(\xi''\zeta'';\, x) \ . \end{aligned} \tag{4.1.97}$$

In what follows we will demonstrate that, with (4.1.97) being multiplied by $\psi^*(\xi'\zeta';\, x)$ and integrated over dx, both terms on the right-hand side of the equation yield expressions that are nonzero in two different cases, and therefore it is convenient to consider these expressions separately.

For the first term we have

$$(\xi'\zeta'|\hat{x}_1|\xi''\zeta'') = id \int_{-\infty}^{\infty} dx\psi^*(\xi'\zeta'; x)\frac{\partial}{\partial\xi''}\psi(\xi''\zeta''; x) .$$

By differentiating (4.1.96) with respect to ξ'', this expression is brought into the form

$$(\xi'\zeta'|\hat{x}_1|\xi''\zeta'') = id\delta'(\xi' - \xi'')\delta_{\zeta'\zeta''} , \tag{4.1.98}$$

where $\delta'(\xi)$ is the derivative of $\delta(\xi)$. Therefore (4.1.98) is different from zero only provided that both states lie in the same band; if, besides, they are infinitely close to each other (i.e., ξ' is infinitely close to ξ''), (4.1.98) has a singularity such as the derivative of the δ function.

We wish to calculate the contribution that the second term (4.1.97) makes to the matrix element $\hat{x}$. Since the function $(\partial/\partial\xi'')u(\xi''\zeta''; x)$ is periodic with respect to x with period d, we may use all the transformations that were carried out in the derivation of (4.1.96). This gives

$$(\xi'\zeta'|\hat{x}_2|\xi''\zeta'') = 2\pi i\delta(\xi' - \xi'')\int_0^d dxu^*(\xi'\zeta'; x)\frac{\partial}{\partial\xi'}u(\xi'\zeta''; x) . \tag{4.1.99}$$

Expression (4.1.99) is nonzero only when $\xi' = \xi''$, i.e., for states with the same ξ. We will show that if $V(x)$ is even, then, even subject to this condition, (4.1.99) is nonzero only for $\zeta' \neq \zeta''$, i.e., for states of different bands. With this aim in view, we differentiate (4.1.95) with respect to ξ:

$$\int_0^d dx\left[u^*(\xi\zeta'; x)\frac{\partial}{\partial\xi}u(\xi\zeta''; x) + u(\xi\zeta''; x)\frac{\partial}{\partial\xi}u^*(\xi\zeta'; x)\right] = 0 .$$

Using (4.1.90), we transform the second integral. Then

$$\int_0^d dx\, u(\xi\zeta''; x)\frac{\partial}{\partial\xi}u^*(\xi\zeta'; x) = \int_0^d dx\, u^*(\xi\zeta''; x)\frac{\partial}{\partial x}u(\xi\zeta'; x) .$$

Substitution of this expression into the input equation yields

$$\int_0^d dx\left[u^*(\xi\zeta''; x)\frac{\partial}{\partial\xi}u(\xi\zeta'; x) + u^*(\xi\zeta'; x)\frac{\partial}{\partial\xi}u(\xi\zeta''; x)\right] = 0 . \tag{4.1.100}$$

With $\zeta' = \zeta''$, (4.1.100) takes the form

$$\int_0^d dxu^*(\xi\zeta; x)\frac{\partial}{\partial\xi}u(\xi\zeta; x) = 0 . \tag{4.1.101}$$

It is exactly this expression which is involved in (4.1.99) at $\zeta' = \zeta''$.

Thus rigid selection rules exist for the matrix element of $\hat{x}$ between two states in the array. Two cases may be mentioned in which this matrix element is not equal to zero. In one of them we have the explicit formula (4.1.98), irrespective of the particular shape of the potential $V(x)$. In the other we deal with elements between states with equal quasimoment. When the potential energy $V(x)$ is even they are nonzero if the states belong to different bands (4.1.99). By contrast with the previous case, the explicit form of the elements in (4.1.99) depends on the function $u(\xi\zeta;\, x)$, i.e., the shape of the potential $V(x)$.

The result obtained is important when considering the effect of a constant or low-frequency electric field on a lattice electron, for the matrix element of the coordinate is related to the probability of the electron passing from the state $\xi''\zeta''$ to the field-induced state $\xi'\zeta'$.

We now find the mean velocity of the electron in the states (4.1.88), making use of the general equation of motion for the operators

$$\dot{\hat{x}} = -\frac{\mathrm{i}}{\hbar}(\hat{x}\hat{\mathscr{H}} - \hat{\mathscr{H}}\hat{x}) \,. \tag{4.1.102}$$

This may be transcribed in matrix form in the representation of eigenfunctions of $\hat{\mathscr{H}}$. We label the number of these states by κ. Using the matrix multiplication rules, we can represent (4.1.102) as

$$(\kappa'|\dot{\hat{x}}|\kappa'') = -\frac{\mathrm{i}}{\hbar}\int d\kappa[(\kappa'|\hat{x}|\kappa)(\kappa|\hat{\mathscr{H}}|\kappa'') - (\kappa'|\hat{\mathscr{H}}|\kappa)(\kappa|\hat{x}|\kappa'')] \,. \tag{4.1.103}$$

In the diagonal representation adopted for $\hat{\mathscr{H}}$, we find

$$(\kappa'|\dot{\hat{x}}|\kappa'') = \frac{\mathrm{i}}{\hbar}(\kappa'|\hat{x}|\kappa'')[E(\kappa'') - E(\kappa')] \,. \tag{4.1.104}$$

Here $E(\kappa)$ stands for the energy eigenvalues of the operator $\hat{\mathscr{H}}$ (Hamiltonian). Equation (4.1.104) shows that the selection rules that hold for the velocity operator are the same as those for the coordinate operator. The diagonal velocity matrix element defining the mean value of the velocity is nonzero when the matrix element $(\kappa'|\hat{x}|\kappa'')$ goes to infinity. To determine the mean velocity, we substitute into (4.1.104) the coordinate matrix element (4.1.98):

$$(\xi'\zeta|\dot{\hat{x}}|\xi''\zeta) = -\frac{d}{\hbar}\delta'(\xi' - \xi'')[E(\xi'\zeta) - E(\xi''\zeta)] \,. \tag{4.1.105}$$

The square bracket in (4.1.105) may be written as

$$E(\xi'\zeta) - E(\xi''\zeta) = (\xi' - \xi'')\frac{\partial E(\xi'\zeta)}{\partial \xi'} + \tfrac{1}{2}(\xi' - \xi'')^2\frac{\partial^2 E(\xi'\zeta)}{\partial \xi'^2} + \dots \,.$$

Using the familiar properties of the δ function—$x\delta'(x) = -\delta(x)$, $x^2\delta'(x) = 0, \ldots,$—we obtain

$$(\xi'\zeta|\hat{\dot{x}}|\xi''\zeta) = \frac{d}{\hbar}\frac{\partial E(\xi'\zeta)}{\partial \xi'}\delta(\xi' - \xi'') .$$

The factor $\delta(\xi' - \xi'')$ indicates that within the limits of one band the velocity operator matrix is diagonal and the mean velocity value sought after is equal to

$$\dot{x}(\xi, \zeta) = \frac{d}{\hbar}\frac{\partial E(\xi, \zeta)}{\partial \xi} = \frac{1}{\hbar}\frac{\partial E(k, \zeta)}{\partial k} . \tag{4.1.106}$$

Expression (4.1.106) defines the general relation of the mean velocity of the electron in the array to the energy of the electron, and is a generalization of the de Broglie formula (4.1.3) for a free electron.

Let us now return to the energy spectrum. With the help of (4.1.106) we may prove that the extrema $f(E)$—i.e., the points at which $\partial f(E)/\partial E = 0$—never lie in the inner domain, shown in Fig. 4.8. But, according to (4.1.84), in the region of allowed solutions,

$$\frac{\partial f(E)}{\partial E} = -\frac{2\sin\xi}{\partial E/\partial \xi} . \tag{4.1.107}$$

Expression (4.1.107) is equal to zero only in two cases.

First, $\sin\xi = 0$ and $\partial E/\partial\xi \neq 0$, the extremum will be at $\xi = 0, \pi$; i.e., they will lie on one of the straight line boundaries shown in Fig. 4.8. This is the case for a free particle. For the electron in an array this occurs only when the potential $V(x)$ has a special form. As will be shown later for the three-dimensional case (Sect. 4.2), this possibility is virtually never realized.

Second, if $\sin\xi \neq 0$ and $\partial E/\partial\xi = \infty$, this is the only case in which the extremum of the function $f(E)$ can lie in the internal region (Fig. 4.8). But this is not feasible, for, according to (4.1.106), $\partial E/\partial\xi$ is proportional to the mean electron velocity, which is always finite. Since (4.1.107) offers no other possibilities, the postulate that the extrema of $f(E)$ lie in the external region of Fig. 4.8 has been proved. From this theorem we again have the rigorous corollary that the electron energy spectrum in the array has the shape of continuous bands that are separated by forbidden energy bands (gaps). Each allowed energy band corresponds to a band of states. Inside each band the energy and other dynamic characteristics of the electron vary continuously, and at the boundaries of the bands they undergo a finite discontinuity.

Since a real crystal is finite, it is worthwhile to find the number of states in a band. We proceed in the same way as in Sect. 4.1.3 (4.1.49), imposing periodicity conditions on the wave functions of the electron

$$\psi(x) = \psi(x + Nd) \ . \tag{4.1.108}$$

Then, from (4.1.88), we arrive at the condition

$$\xi = \frac{2\pi n}{N} \ , \quad n = 0, \pm 1, \pm 2, \ldots \ . \tag{4.1.109}$$

Since within the band the ξ's satisfy (4.1.87), the possible values of n are bounded by the inequalities

$$-N/2 < n \leqslant N/2 \ ; \tag{4.1.110}$$

i.e., the number of states in the band is equal to the total number of atoms N (4.1.51).

We can ascertain in what fashion $E(\xi)$ varies inside the band. Since the energy $E(\xi)$ is an even function of ξ, i.e.,

$$E(\xi) = E(-\xi) \ , \tag{4.1.111}$$

it suffices to clear up the problem concerning the ξ dependence of E, for example, for the interval $0 \leqslant \xi \leqslant \pi$, within which E should vary monotonically, for $f(E)$ is a single-valued function of E. [Formula (4.1.111) follows simply from the fact that $\psi(\xi\zeta; x)$ and $\psi(-\xi\zeta; x)$ are complex conjugate, regardless of the condition $V(x) = V(-x)$.] In the first band, which corresponds to the smallest energies, the dependence increases with E, whereas in the successive band it diminishes, etc., by turns. At the boundaries of the bands the derivative $\partial E/\partial\xi$ is equal to zero. This is seen from (4.1.107), since for $\xi = 0$ or π the quantity $\partial f(E)/\partial E \neq 0$, and then

$$\frac{\partial E}{\partial \xi} = 0 \ , \quad \xi = 0, \pi \ . \tag{4.1.112}$$

Furthermore, only one wave function corresponds to the values $\xi = 0, \pi$. This can be demonstrated proceeding from (4.1.12). Differentiating (4.1.12) with respect to ξ yields

$$\frac{\partial^2}{\partial x^2}\frac{\partial \psi}{\partial \xi} + \frac{2m}{\hbar^2}(E - V)\frac{\partial \psi}{\partial \xi} = -\frac{2m}{\hbar^2}\frac{\partial E}{\partial \xi}\psi \ .$$

Thus we see that if $\partial E/\partial\xi = 0$, then $\partial\psi/\partial\xi$, along with ψ, is also a solution of (4.1.12). Formula (4.1.88) shows that

$$\frac{\partial \psi}{\partial \xi} = \frac{\mathrm{i}x}{d}\psi + \exp\left(\frac{\mathrm{i}\xi x}{d}\right)\frac{\partial u}{\partial \xi} \ , \tag{4.1.113}$$

and hence $\partial\psi/\partial\xi$ and ψ are linearly independent. Thus, with $f(E) = \pm 2$, the functions ψ and $\partial\psi/\partial\xi$ form the basis of solutions to (4.1.12). As is seen from (4.1.113), when $x \to \pm\infty$ the function $\partial\psi/\partial\xi$ increases indefinitely and is therefore not allowed. Consequently, for $\xi = 0$ or $\xi = \pi$, the function ψ from (4.1.88) is a unique allowed solution of (4.1.12).

Let us now ascertain the general shape of the electron energy spectrum in an ideal one-dimensional array. We will plot the quasimomentum ξ as abscissa (by marking the points 0, $\pm\pi$, $\pm 2\pi, \ldots$) and the energy $E(\xi, \zeta)$ as the ordinate (Fig. 4.9). To begin with, we consider the region of values of ξ from (4.1.108, 109); i.e., we confine ourselves to one period of the function $E(\xi, \zeta)$. Figure 4.9 shows that for the various bands (here we deal with two bands, ζ_1 and ζ_2) the energy spectrum consists of a number of segments of curves that overlie each other, provided that the variation of ξ is bounded by the conditions $-\pi < \xi \leqslant \pi$. This method of depicting the energy spectrum is, as stated earlier, called the reduced-zone method. It is, however, possible to continue the curve of the periodic function $E(\xi)$ along the entire abscissa axis, relinquishing this condition. This is the extended-zone method.

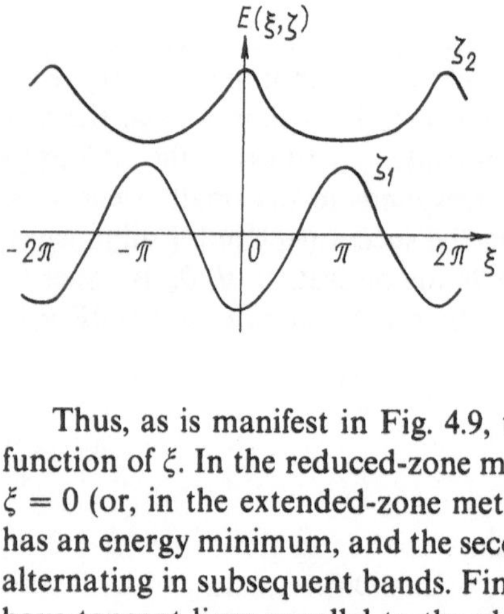

Fig. 4.9. Energy band spectrum in the one-dimensional model in the case of two bands (ζ_1 and ζ_2)

Thus, as is manifest in Fig. 4.9, the energy E is a monotonic even periodic function of ξ. In the reduced-zone method the first energy band (with $\zeta = 1$) for $\xi = 0$ (or, in the extended-zone method, for $\xi = 2\pi n$, where $n = \pm 1, \pm 2, \ldots$) has an energy minimum, and the second band (with $\zeta = 2$) an energy maximum, alternating in subsequent bands. Finally, when $\xi = 0, \pm\pi, \pm 2\pi, \ldots$ the curves have tangent lines parallel to the abscissa axis ($\partial E/\partial\xi = 0$).

Let us compare the plot in Fig. 4.9 with the plot of the energy versus the momentum of a free electron. In the extended-zone method we assume that in the first band $|\xi| = (1/\hbar)\,|p|d$ varies from 0 to π, and in the second band from π to 2π, etc. From a comparison of the parabola in Fig. 4.17, depicting the energy spectrum of a free electron, with the energy curves of the electron in the array for different bands (i.e., different values of ζ) in different zones, we see that the presence of the potential field of the array makes the parabola discontinuous and bends the broken ends in such a fashion that they approach the zone boundaries with the tangent line parallel to the abscissa axis.

Formula (4.1.106) enables us to get an idea of the mean velocity of the electron in the states (4.1.88) from the shape of the energy spectrum. The mean velocity of the electron in the array is normally different from zero. This follows from (4.1.107), since, if inside a band $\partial E/\partial\xi = 0$, then $\partial f(E)/\partial E = \infty$, which is inadmissible with the potential $V(x)$ being regular. In each of the states (4.1.88) lying inside a band a nonzero current is associated with the electron. The current goes to zero only at the band edges, where $\partial E/\partial\xi = 0$. The velocity is an odd function of ξ, and therefore states with $+\xi$ and $-\xi$, having the same energy (4.1.111), possess opposite velocities. Figure 4.10 shows a plot of the mean velocity and energy for the first band. In the right-hand half of the band $(0 \leqslant \xi \leqslant \pi)$ the velocity is positive, and in the left-hand half it is negative. Starting from $\xi = 0$, the velocity increases in absolute magnitude in both band halves; at $\xi = \pm\pi/2$ it reaches its highest value and then decreases, reaching zero at the band edges. By exploiting group theory, we can prove that the $\dot{x}(\xi)$ curve can have no more than one maximum inside the band [4.5].

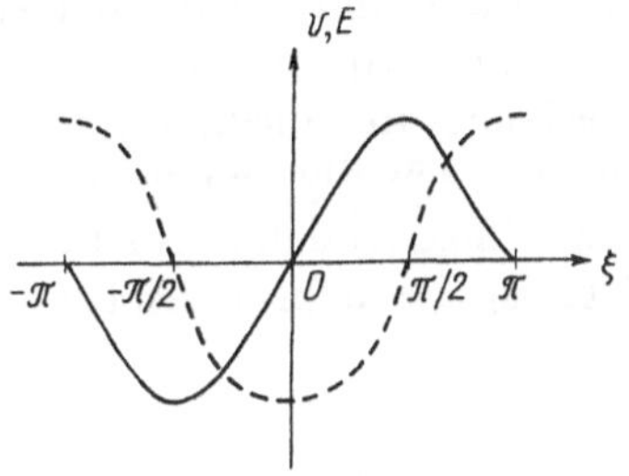

Fig. 4.10. Dependence of the energy E (dashed line) and mean velocity of a band electron v (solid line) on quasi-momentum ξ (wave number) in the one-dimensional case for the first band

The most important result of band theory (apart from the "free smearing" of the electron along the array) is that, as ξ approaches the band edges, the absolute magnitude of the current tends to zero. Comparison of Figs. 4.9, 10 shows that at $\partial^2E/\partial\xi^2 > 0$ the relation of the energy to the mean velocity is qualitatively similar to the free electron case—to an increase in energy there corresponds an increase in velocity. The situation is different for $\partial^2E/\partial\xi^2 < 0$. Here the velocity decreases with increase in E. This anomaly, as already noted, may formally be described by introducing a negative effective electron mass. Let us compare this result with the model problem (Sect. 4.1.3) for narrow bands, for which the segment of the $f(E)$ curve between the boundary straight lines in the allowed region may be approximated by a straight line (Fig. 4.8):

$$E = E_0 - |\alpha| + 2\beta\cos\xi \quad . \tag{4.1.114}$$

In our method of numbering states in bands, β should reverse sign from band to band: (4.1.114) coincides with the model formula (4.1.53). From (4.1.114) we have for the mean velocity (4.1.106)

$$\langle \dot{x} \rangle = -2\beta \frac{d}{\hbar} \sin \xi \ , \tag{4.1.115}$$

which is similar to the model formula (4.1.60).

Our conclusions practically exhaust the general properties of an electron in a one-dimensional array of atoms [4.6]. The unfortunate fact remains that they cannot be applied wholly to a real three-dimensional case. However, results for the latter can be obtained with the help of group theory, using methods that allow a simple generalization to three dimensions.

In conclusion we note one more point. The Schrödinger equation with a periodic potential is formally equivalent to the equation of a classical harmonic oscillator with a frequency that depends periodically on time, e.g.,

$$\frac{d^2 x(t)}{dt^2} + \omega_0^2 (1 + h \cos \Omega t) x(t) = 0 \ .$$

Here the $\psi(x)$ act as $x(t)$, and the quantity $k^2 = 2mE/\hbar^2$ as ω_0^2; Ω corresponds to $2\pi/d$. It is known from mechanics [3.4] that in the vicinity of the points $\omega_0 = n\Omega/2$ ($n = \pm 1, \pm 2, \pm 3, \ldots$), i.e., for $k = \pm n\pi/d$, $x(t)$ increases exponentially; that is, $|\psi(x)| \to \infty$ when $x \to \infty$. In mechanics this phenomenon is known as parametric resonance, and the quantity $x(t)$ actually remains finite owing to nonlinear terms (anharmonicity), the Schrödinger equation is strictly linear, and so such solutions have to be discarded, giving rise to forbidden bands in the vicinity of $k = \pm n\pi/d$.

4.2 General Theory of the Electron Motion in a Three-Dimensional Crystal

4.2.1 Bloch's Theorem

The potential energy of an electron in an ideal three-dimensional infinite lattice possesses the following basic periodicity property:

$$V(\boldsymbol{r} + \boldsymbol{R}_m) = V(\boldsymbol{r}) \ , \tag{4.2.1}$$

with $\boldsymbol{R}_m$ being translation vectors (1.2.9). The property (4.2.1) does not embrace all of the symmetry properties (e.g., point symmetry, Sect. 4.2.4); however, the relevant properties will be introduced when required. Using (4.2.1), $V(\boldsymbol{r})$ may be expanded in a Fourier series

$$V(\boldsymbol{r}) = \sum_{g=(g_1, g_2, g_3)} V_{\boldsymbol{b}_g^*} \exp(\mathrm{i}\boldsymbol{b}_g^* \cdot \boldsymbol{r}) \ , \tag{4.2.2}$$

where $\boldsymbol{b}_g^*$ are reciprocal lattice vectors. It follows from the real-valuedness of $V(\boldsymbol{r})$ that

$$V_{\boldsymbol{b}_g^*} = V_{-\boldsymbol{b}_g^*} \ . \tag{4.2.3}$$

In the three-dimensional case, (4.1.12) will take the form

$$\hat{\mathscr{H}}_{\text{crys}}\psi(\boldsymbol{r}) \equiv \left[-\frac{\hbar^2}{2m}\Delta + V(\boldsymbol{r})\right]\psi(\boldsymbol{r}) = E\psi(\boldsymbol{r}) \ . \tag{4.2.4}$$

This is a differential equation with partial derivatives, and therefore we cannot do an analysis similar to that in a one-dimensional problem. We may, however, show with group theory that the major results of Sect. 4.1.4 hold here as well.

To start with, we will demonstrate that the wave functions of an electron in a three-dimensional crystal also are plane waves modulated by the periodicity of the lattice [4.2]. We introduce translation operators $\hat{T}(\boldsymbol{R}_m)$, which are defined by the equations

$$\hat{T}(\boldsymbol{R}_m)f(\boldsymbol{r}) = f(\boldsymbol{r}+\boldsymbol{R}_m) \ , \tag{4.2.5}$$

with $f(\boldsymbol{r})$ being an arbitrary function of the coordinates. These operators form an Abelian translation group Γ, for they commute with each other: $\hat{T}(\boldsymbol{R}_m)\hat{T}(\boldsymbol{R}_n) = \hat{T}(\boldsymbol{R}_n)\hat{T}(\boldsymbol{R}_m)$. It follows from (4.2.1) and the form of the Hamiltonian (4.2.4) that the translation operators commute with the Hamiltonian

$$[\hat{T}(\boldsymbol{R}_n), \hat{\mathscr{H}}_{\text{crys}}]_- = 0 \ . \tag{4.2.6}$$

As is seen from (4.2.6), $\psi(\boldsymbol{r})$, which is an eigenfunction of $\hat{\mathscr{H}}_{\text{crys}}$, may be simultaneously the eigenfunction of all group-Γ operators (4.1.67):

$$\hat{T}(\boldsymbol{R}_m)\psi(\boldsymbol{r}) = \psi(\boldsymbol{r}+\boldsymbol{R}_m) = \lambda(\boldsymbol{R}_m)\psi(\boldsymbol{R}_m) \ , \tag{4.2.7}$$

where $\lambda(\boldsymbol{R}_m)$ stands for the eigenvalues of the operator $\hat{T}(\boldsymbol{R}_m)$. By the property

$$\hat{T}(\boldsymbol{R}_m)\hat{T}(\boldsymbol{R}_n) = \hat{T}(\boldsymbol{R}_m+\boldsymbol{R}_n) \tag{4.2.8}$$

we have

$$\lambda(\boldsymbol{R}_m)\lambda(\boldsymbol{R}_n) = \lambda(\boldsymbol{R}_m+\boldsymbol{R}_n) \tag{4.2.9}$$

or

$$\ln\lambda(\boldsymbol{R}_m) + \ln\lambda(\boldsymbol{R}_n) = \ln\lambda(\boldsymbol{R}_m+\boldsymbol{R}_n) \ .$$

A unique solution of (4.2.9) is

$$\ln \lambda(\boldsymbol{R}_m) = \mathrm{i}\boldsymbol{k} \cdot \boldsymbol{R}_m ,$$

where $\boldsymbol{k}$ is an arbitrary (generally speaking, complex) vector that does not depend on m. The second equality in (4.2.8) may therefore be written as

$$\psi(\boldsymbol{r} + \boldsymbol{R}_m) = \exp(\mathrm{i}\boldsymbol{k} \cdot \boldsymbol{R}_m)\psi(\boldsymbol{r}) . \tag{4.2.10}$$

Since $|\psi(\boldsymbol{r} + \boldsymbol{R}_m)|^2$ should be limited for $|\boldsymbol{R}_m| \to \infty$, the components of the vector $\boldsymbol{k}$ are purely real. Equation (4.2.10) expresses Bloch's theorem [4.2], which defines the most general properties of electronic states in a crystal.

Similar to the transformation from (4.1.77) to (4.1.78), the electron wave function may, using (4.2.10), be represented in the form

$$\psi_{\boldsymbol{k}}(\boldsymbol{r}) = \exp(\mathrm{i}\boldsymbol{k} \cdot \boldsymbol{r})\, u_{\boldsymbol{k}}(\boldsymbol{r}) . \tag{4.2.11}$$

According to Bloch's theorem (4.2.10),

$$\psi_{\boldsymbol{k}}(\boldsymbol{r} + \boldsymbol{R}_m) = \exp[\mathrm{i}\boldsymbol{k} \cdot (\boldsymbol{r} + \boldsymbol{R}_m)]\, u_{\boldsymbol{k}}(\boldsymbol{r} + \boldsymbol{R}_m) = \exp(\mathrm{i}\boldsymbol{k} \cdot \boldsymbol{R}_m)\exp(\mathrm{i}\boldsymbol{k} \cdot \boldsymbol{r})\, u_{\boldsymbol{k}}(\boldsymbol{r}) ;$$

i.e., the function $u_{\boldsymbol{k}}(\boldsymbol{r})$ satisfies the periodicity conditions

$$u_{\boldsymbol{k}}(\boldsymbol{r} + \boldsymbol{R}_m) = u_{\boldsymbol{k}}(\boldsymbol{r}) , \tag{4.2.12}$$

as well as the condition for the potential in (4.2.11). From the latter equation we can see that, just as in Sect. 4.1.4, the electron wave function is a plane wave modulated by the periodicity of the lattice. Also, it follows from (4.2.11) that electrons cannot be localized near lattice sites; this is a general conclusion not associated with any particular assumption.

We consider one more representation of the function $\psi_{\boldsymbol{k}}(\boldsymbol{r})$. Using the vectors $\boldsymbol{b}^*_g$, the modulating function $u_{\boldsymbol{k}}(\boldsymbol{r})$ may, by (1.3.2), be expanded in a Fourier series. Then for $\psi_{\boldsymbol{k}}(\boldsymbol{r})$ we have, according to (4.2.11),

$$\psi_{\boldsymbol{k}}(\boldsymbol{r}) = \sum_g a(\boldsymbol{k}, \boldsymbol{b}^*_g)\exp[\mathrm{i}(\boldsymbol{k} + \boldsymbol{b}^*_g) \cdot \boldsymbol{r}] . \tag{4.2.13}$$

We choose the system of coordinates in such a way that its unit vectors are reciprocal-lattice vectors (1.3.3), i.e.,

$$\boldsymbol{k} = k_1\boldsymbol{b}_1 + k_2\boldsymbol{b}_2 + k_3\boldsymbol{b}_3 , \qquad k_i = \boldsymbol{k} \cdot \boldsymbol{a}_i . \tag{4.2.14}$$

If the components of the vector $\boldsymbol{k}$ satisfy the inequalities

$$-\pi \leqslant k_i d_i < \pi , \qquad i = 1, 2, 3 \tag{4.2.15}$$

any other vector $\boldsymbol{k}$ may be represented as $\boldsymbol{k}+\boldsymbol{b}_g^*$, for, when the k_i run over all the values of the interval (4.2.15), and the g_i over all the integers, the components of the vector $(\boldsymbol{k}+\boldsymbol{b}_g^*)_i = k_i + 2\pi g_i/d_i$ run over all possible values.

To solve the problem completely we must determine the coefficients $a(\boldsymbol{k}, \boldsymbol{b}_g^*)$ in (4.2.13) so that (4.2.14) is satisfied. Specific techniques for computing these coefficients are outlined in what follows. Substituting (4.2.12, 13) into (4.2.4), premultiplying the latter by $\exp(-i\boldsymbol{k}\cdot\boldsymbol{r})$, performing integration over dr_i between $-\infty$ and ∞, and using the properties of the δ functions, we arrive at the result that the $a(\boldsymbol{k}, \boldsymbol{b}_g^*)$ satisfy the system of linear homogeneous equations

$$\left[E-\frac{\hbar^2}{2m}(\boldsymbol{k}+\boldsymbol{b}_g^*)^2\right]a(\boldsymbol{k}, \boldsymbol{b}_g^*) = \sum_{g'} V_{\boldsymbol{b}_{g'}^*}a(\boldsymbol{k}, \boldsymbol{b}_{g-g'}^*) \ . \tag{4.2.16}$$

Substituting (4.2.11) into (4.2.4), we find the equation for the modulating function $u_{\boldsymbol{k}}(\boldsymbol{r})$:

$$\Delta u_{\boldsymbol{k}}(\boldsymbol{r}) + 2i\boldsymbol{k}\cdot\nabla u_{\boldsymbol{k}}(\boldsymbol{r}) + \frac{2m}{\hbar^2}\left[E-\frac{\hbar^2 k^2}{2m} - V(\boldsymbol{r})\right]u_{\boldsymbol{k}}(\boldsymbol{r}) = 0 \ . \tag{4.2.17}$$

Equations (4.2.4, 16, 17) are equivalent and henceforth we will make use of all three notations. In virtue of the homogeneity of the system (4.2.16), the value of one of the coefficients—for example, $a(\boldsymbol{k}, 0)$—may be assigned arbitrarily. Then the values of the other $a(\boldsymbol{k}, \boldsymbol{b}_g^*)$ (if E and $\boldsymbol{k}$ are given) can be determined unambiguously. Put another way, for those $\boldsymbol{k}$ which lie in the interval (4.2.15), the values of $a(\boldsymbol{k}, 0)$ may be prescribed independently of each other; whilst the coefficients of the form $a(\boldsymbol{k}, \boldsymbol{b}_g^*)(\boldsymbol{b}_g^* \neq 0)$ are determined unambiguously through $a(\boldsymbol{k}, 0)$ and E. The system (4.2.17) has nonzero solutions if its determinant Δ is equal to zero; it depends on E, $\boldsymbol{k}$, and all $V_{\boldsymbol{b}_g^*}$.

The secular equation $\Delta(E, \ \boldsymbol{k}, V_{\boldsymbol{b}_g^*}) = 0$ has, generally speaking, an infinite number of roots which are functions of the vector $\boldsymbol{k}$. In a periodic field the electron energy spectrum therefore is an assembly of individual allowed energy bands

$$E = E_\zeta(\boldsymbol{k}) \ , \quad \zeta = 1, 2, 3 \ldots . \tag{4.2.18}$$

Thus assigning the quasimomentum $\boldsymbol{k}$ does not suffice for the energy to be determined unambiguously. In addition, the band number ζ should be assigned. The set of numbers ζ is characterized by the fact that the energy spectrum $E_\zeta(\boldsymbol{k})$ (with a given $\boldsymbol{k}$) of (4.2.16) is discrete, and it is possible to determine the smallest value of E_ζ (for a given $\boldsymbol{k}$); i.e., the spectrum (4.2.18) is bounded from below and there are no multiples among the values of $E_\zeta(\boldsymbol{k})$.

The first two postulates are quite rigorous and are introduced as assumptions in order not to go into complicated mathematical proofs. For the system $\Delta = 0$ to have multiple roots, the coefficients $V_{\boldsymbol{b}_g^*}$ specifying the system must satisfy some extra symmetry conditions. Whatever $V_{b_g^*}$ may be, there will certainly be no degeneracy. In other words, the fact that several wave functions correspond to the same quasimomentum $\boldsymbol{k}$ and the same energy E does not follow straightforwardly from (4.2.1), but may be a consequence of some additional (point) symmetry. A rigorous theory shows that such "superdegeneracy" does not occur often, and when it does it is only for an infinitesimal share of $\boldsymbol{k}$ values.

For each band number ζ the energy is a periodic function of the wave vector $\boldsymbol{k}$ with the period of the reciprocal lattice. This may be proved from the form of the system (4.2.16) and its determinant. It is seen from (4.2.16) that the replacement of $\boldsymbol{k}$ by $\boldsymbol{k} + \boldsymbol{b}_g^*$ merely denotes a different sequence for the equations of the system—i.e., the system remains invariant—and the roots of the equation $\Delta = 0$ therefore also will remain unchanged. Thus,

$$E_\zeta(\boldsymbol{k} + \boldsymbol{b}_g^*) = E_\zeta(\boldsymbol{k}) \; ; \tag{4.2.19}$$

i.e., the energy is periodic with the reciprocal-lattice period. Then

$$E_\zeta(\boldsymbol{k}) = \sum_m E_{\boldsymbol{R}_m \zeta} \exp(\mathrm{i}\boldsymbol{k} \cdot \boldsymbol{R}_m) \; . \tag{4.2.20}$$

Let us now determine the numbers of the energy levels in (4.2.18), i.e., renumber all the possible states of the electron with a given $\boldsymbol{k}$. Consider the renumbered sequence of wave functions for a given value of $\boldsymbol{k}$ infinitely close to that selected initially. Then each of the functions of this sequence is infinitely close to the corresponding function in the initial sequence. In this fashion we can embrace the entire region of $\boldsymbol{k}$ values, determinable according to (4.2.15), since the system (4.2.16) is capable of solution for any $\boldsymbol{k}$. Now we are in a position to classify the functions $\psi_{\boldsymbol{k}\zeta}(\boldsymbol{r})$ from (4.2.11 or 13) by uniting all the functions of the same number ζ into one class. Within the latter, the $\psi_{\boldsymbol{k}\zeta}(\boldsymbol{r})$ varies continuously and unambiguously with $\boldsymbol{k}$ in the region (4.2.15). Thus two functions for infinitely close $\boldsymbol{k}$ and $\boldsymbol{k}'$ but with $\zeta \neq \zeta'$, are not infinitely close. These classes of functions are called bands. The basic conclusion of band theory then says that each steady state of an electron in an ideal lattice is unambiguously determined by a given quasimomentum $\boldsymbol{k}$ and band number ζ: $\psi_{\boldsymbol{k}\zeta}(\boldsymbol{r})$ and $E_\zeta(\boldsymbol{k})$.

The counting of the states in a band is done in the same way as in the one-dimensional case (4.1.108–110). We introduce periodicity conditions according to which the $\psi_{\boldsymbol{k}\zeta}(\boldsymbol{r})$ recur when any one of the lattice unit cells is carried over to any one of the vectors of the form $G_1 a_1$, $G_2 a_2$, $G_3 a_3$, where G_1, G_2, and G_3 are large integers. The product $G_1 G_2 G_3 = N$ gives the total number of unit cells in the crystal. This periodicity requirement when applied to (4.1.11, 12) results in

$$k_i = \frac{2\pi n_i}{d_i G_i} \, , \quad i = 1, 2, 3 \tag{4.2.21}$$

with n_1, n_2, and n_3 being integers. Since inside a band the possible values of k_i are limited, in addition to this requirement, by the inequalities (4.2.15), each n_i varies within the limits

$$-\frac{G_i}{2} \leqslant n_i < \frac{G_i}{2} \tag{4.2.22}$$

and hence assumes, inside the band, G_i values. Therefore, the total number of functions in the band is equal to N.

4.2.2 Brillouin Zones

In addition to the band construction considered above, there is another method, that proposed by *Brillouin* [4.7]. It follows from (4.2.19) that the region of possible $\boldsymbol{k}$ values should meet the sole requirement that their values should be sufficient in number to allow any vector k' in reciprocal space to be represented as

$$\boldsymbol{k}' = \boldsymbol{k} + \boldsymbol{b}_g^* \ , \tag{4.2.23}$$

where $\boldsymbol{b}_g^*$ is any vector of the reciprocal lattice (1.5.12). Following Brillouin, we construct this region as follows. We take a definite $\boldsymbol{b}_g^*$ and construct two planes, normal to $\boldsymbol{b}_g^*$ and a distance $\frac{1}{2}|\boldsymbol{b}_g^*|$ from the origin. The equations for these planes will be

$$\boldsymbol{k}\cdot\boldsymbol{b}_g^* = \pm\tfrac{1}{2}|\boldsymbol{b}_g^*|^2 \ . \tag{4.2.24}$$

The points of $\boldsymbol{k}$ space that lie within the layer between the planes (4.2.24) will be called interior ones, and the rest of the points exterior ones. In addition, the points that lie on one of the planes will be ranked with the interior ones, and those that lie on the other plane will be ranked with the exterior ones (Fig. 4.11).

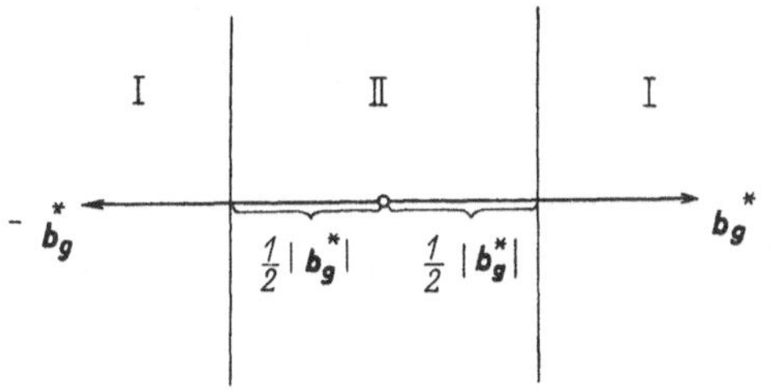

Fig. 4.11. Construction of Brillouin zone boundaries: region of *exterior* points of $\boldsymbol{k}$ space (I); region of *interior* points of $\boldsymbol{k}$ space (II)

None of the interior points may be translated to any one of the other interior points by means of the transformation (4.2.23). Conversely, any exterior point may be obtained from some interior point by successive application of the transformation (4.2.23).

From the origin we draw vectors $\boldsymbol{b}_g^*$ to all the other sites and construct planes (4.2.24) for them. The intersection of these planes will give us a series of closed volumes whose boundaries will be Brillouin boundaries. The innermost volume will be some complicated polyhedron. Its interior points are not interconnected by the transformation (4.2.23), which connects them only to the exterior points. This polyhedron is the first Brillouin zone. It does not coincide with the zone constructed using the previous method (except for lattices whose primitive vectors $\boldsymbol{a}_1, \boldsymbol{a}_2, \boldsymbol{a}_3$ and $\boldsymbol{b}_1, \boldsymbol{b}_2, \boldsymbol{b}_3$ generate regular parallelepipeds), and its boundaries may not be expressed by inequalities such as (4.2.15). But the set of wave functions (number of states) in the zone or its volume is the same in both constructions; they are merely arranged in a different sequence. We have repeatedly mentioned Brillouin zones. It is readily surmised that the equations for their boundaries (4.2.24) coincide exactly with the Wulf–Bragg formula (1.5.11).

As an example, we examine a plane square lattice with a parameter a. The vector of the reciprocal lattice here is equal to $\boldsymbol{b}_g^* = (2\pi/a)(g_1\boldsymbol{e}_1 + g_2\boldsymbol{e}_2)$, and (4.2.24) takes the form

$$k_x g_1 + k_y g_2 = \frac{\pi}{a}(g_1^2 + g_2^2) \ . \tag{4.2.25}$$

Exhaustive search for all possible combinations of the integers g_1 and g_2 allows us to find the equations of the lines—the boundaries of the zones (Fig. 4.12) $g_1 = \pm 1, g_2 = 0$; $k_x = \pm \pi/a$; $g_1 = 0, g_2 = \pm 1$; $k_y = \pm \pi/a$. These four straight lines bound the square (1 in Fig. 4.12) with the side $2\pi/a$ and with the center at $k_x = k_y = 0$. This square is the first Brillouin zone of the plane square lattice. The sides of the square are also parts of the boundaries of the second

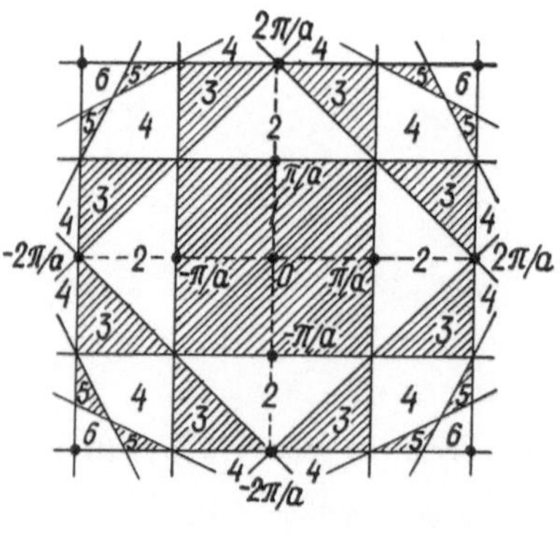

Fig. 4.12. Construction of Brillouin zones for a simple square lattice of atoms

zone. The successive zone boundaries are found from the conditions $g_1 = 1$, $g_2 = \pm 1$; $k_x \pm k_y = 2\pi/a$; $g_1 = -1$, $g_2 = \pm 1$; $k_x \pm k_y = 2\pi/a$. Together with the four sides of the square these four straight lines delineate the second zone. The latter, just as in the one-dimensional case, consists not of one region of $\boldsymbol{k}$ space but of four similar rectangular isosceles triangles whose hypotenuses are the sides of the square of the first zone. The eight legs of these four triangles serve again as part of the boundaries of the third zone. Its other sixteen sides are found from the equations $g_1 = \pm 2$, $g_2 = 0$; $\pm 2k_x = 4\pi/a$; $g_1 = 0$, $g_2 = \pm 2$; $\pm 2k_y = 4\pi/a$, and also from the equations of the straight lines whose intersection has given the square of the first zone.

Figure 4.12 shows that the third zone consists already of eight separate rectangular triangles whose hypotenuses adjoin the eight legs of the triangles of the second zone. Employing this procedure, we can readily construct the successive zones (Fig. 4.12). Two important conclusions may be drawn from this construction: (1) the overall area of each zone is the same and equal to the area of the square of the first zone; (2) translation by a reciprocal-lattice vector parallel to the k_x, k_y axes, enables sections of the second, third, and other zones to be mapped into, and to fill completely, the area of the square of the first zone (Fig. 4.12). As already said, the portrayal of all zones in the entire $\boldsymbol{k}$ space is the extended-zone representation, and the projection to the first zone the reduced-zone representation. Thus all Brillouin zones are bounded by Wulf–Bragg surfaces. As for the interior points of the first zone, we can say that they can be reached from the fiducial point without crossing any of the Wulf–Bragg surfaces at all. The interior points of the second zone are likewise reachable from the points of the first zone (and vice versa) if one of the points of the Wulf–Bragg surface is crossed, etc. Also, the first Brillouin zone is a Wigner–Seitz primitive cell in the reciprocal lattice (Sect. 1.2).

The generalization for the three-dimensional case is evident. For a simple cubic lattice, (4.2.24) will take the form

$$k_x g_1 + k_y g_2 + k_z g_3 = \frac{\pi}{a}(g_1^2 + g_2^2 + g_3^2) \ . \tag{4.2.26}$$

The boundaries of the first zone are obtainable from the conditions $g_i = \pm 1$; $g_j + g_k = 0$ ($i, j, k = 1, 2, 3$); and $k_x = k_y = k_z = \pm \pi/a$. As a result, for the first zone in the simple cubic (sc) lattice we obtain a cube with the center at $\boldsymbol{k} = 0$ and with edges of length $2\pi/a$ parallel to the coordinate axes (Fig. 4.13a—also shown are some points that possess symmetry). The second zone consists of four straight pyramids with square bases—faces of the cube of the first zone, whose height is equal to π/a (Fig. 4.13b). The lateral triangular faces of these pyramids are given by the cross sections of the planes: $k_x \pm k_y = \pm 2\pi/a$; $k_x \pm k_z = \pm 2\pi/a$; $k_y \pm k_z = \pm 2\pi/a$. Again, the total volume of the six pyramids of the second zone is equal to the volume of the cube of the first zone. By its outer surface, the second zone has the shape of a dodecahedron. Figure 4.13c depicts the third

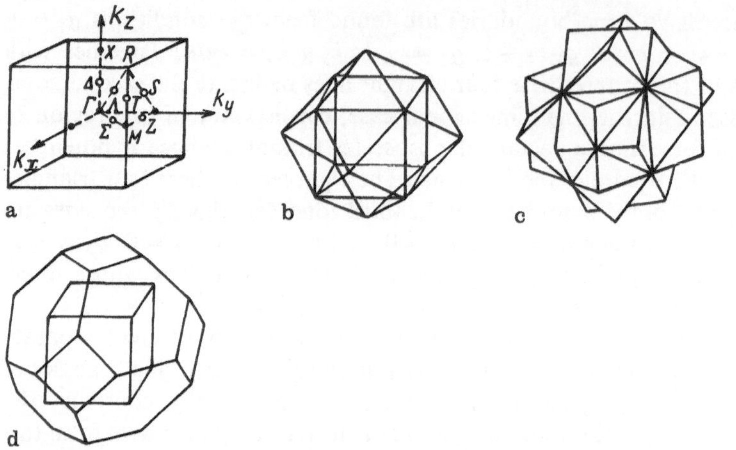

Fig. 4.13. Brillouin zones in the three-dimensional case for an s.c. lattice: first (**a**), second (**b**), third (**c**), fourth (**d**). With the first zone some symmetry points (Γ, X, R, M) and lines (Δ, Σ, T, Λ, S, Z) connecting these are shown

zone, whose sections touch the surface of the second zone. The fourth zone is portrayed in Fig. 4.14d.

Our objective now is to find the first zone of a bcc lattice. Using the matrix A_{bcc} and the rules of determining reciprocal-lattice vectors $\boldsymbol{b}_g^*$ according to (1.3.4, 1.5.12), we arrive at the result

$$\boldsymbol{b}_1 = \frac{1}{a}(\boldsymbol{e}_1 + \boldsymbol{e}_2)\,, \quad \boldsymbol{b}_2 = \frac{1}{a}(-\boldsymbol{e}_1 + \boldsymbol{e}_2)\,, \quad \boldsymbol{b}_3 = \frac{1}{a}(-\boldsymbol{e}_2 + \boldsymbol{e}_3)\,.$$

The reciprocal lattice for a body-centered-cubic crystal is an fcc lattice. For the vectors $\boldsymbol{b}_g^*$ we have

$$\boldsymbol{b}_g^* = 2\pi(g_1\boldsymbol{b}_1 + g_2\boldsymbol{b}_2 + g_3\boldsymbol{b}_3) = \frac{2\pi}{a}[(g_1 - g_2)\boldsymbol{e}_1 + {} \\ + (g_2 - g_3)\boldsymbol{e}_2 + (g_1 + g_3)\boldsymbol{e}_3]\,.$$

We readily infer that there are twelve nonzero minimum-length vectors $\boldsymbol{b}_g^*$

$$\frac{2\pi}{a}(\pm\boldsymbol{e}_1 \pm \boldsymbol{e}_2)\,, \quad \frac{2\pi}{a}(\pm\boldsymbol{e}_1 \pm \boldsymbol{e}_3)\,, \quad \frac{2\pi}{a}(\pm\boldsymbol{e}_2 \pm \boldsymbol{e}_3)\,.$$

In the system of coordinates k_x, k_y, k_z these vectors are diagonals that go from the origin, four lines on each of the coordinate planes. Thus the boundaries of the first Brillouin zone of a bcc lattice are generated by twelve planes that are perpendicular to these diagonals and intersect them at a distance of π/a from the origin. Accordingly, we obtain the first zone of a bcc lattice in the form of a rhombic dodecahedron such as that portrayed in Fig. 4.14, in which the most important symmetry points and lines are also indicated.

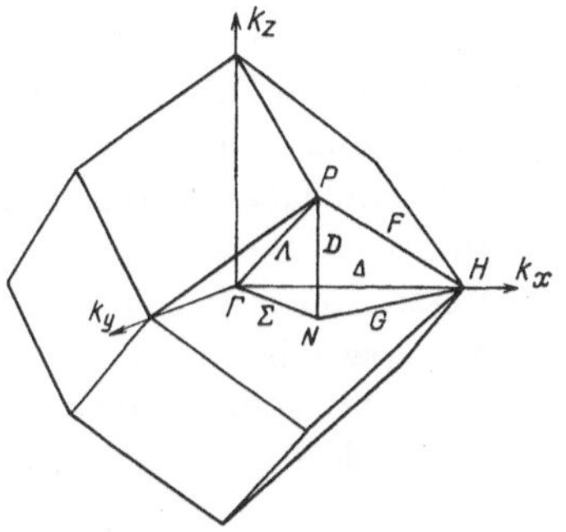

Fig. 4.14. First Brillouin zone in the three-dimensional case for the bcc lattice, with some symmetry points (Γ, H, P, N) and lines (Δ, Σ, Λ, F, D, G) connecting these

With an fcc lattice, using the matrix A_{fcc} and the reciprocal-lattice vectors, just as in the previous case, we obtain the result that the reciprocal lattice here will be bcc. For the reciprocal lattice we find the fourteen shortest vectors $2\pi/a\,(\pm \boldsymbol{e}_1 \pm \boldsymbol{e}_2 \pm \boldsymbol{e}_3)$ and $\pm 2\pi/a \cdot 2\boldsymbol{e}_i$ ($i = 1, 2, 3$), which, by (4.2.26), define the octahedron with truncated vertices. This is shown in Fig. 4.15, again indicating some of the symmetry points and lines.

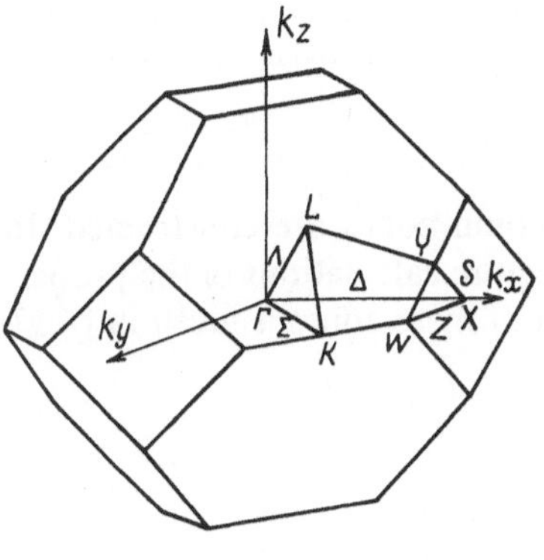

Fig. 4.15. First Brillouin zone in the three-dimensional case for the fcc lattice, with some symmetry points (Γ, X, K, L, U, W) and lines (Δ, Σ, Λ, S, Z) connecting these

For an hcp lattice, we use the matrix A_{hcp} and the reciprocal-lattice vectors, $\boldsymbol{b}_1 = a^{-1}(\boldsymbol{e}_1 + \boldsymbol{e}_2/\sqrt{3})$; $\boldsymbol{b}_2 = a^{-1}(-\boldsymbol{e}_1 + \boldsymbol{e}_2/\sqrt{3})$; $\boldsymbol{b}_3 = c^{-1}\boldsymbol{e}_3$. Here the axes of the direct and reciprocal lattices coincide. For the vectors $\boldsymbol{b}_g^*$ we have

$$\boldsymbol{b}_g^* = 2\pi\left[\frac{1}{a}(g_1 - g_2)\boldsymbol{e}_1 + \frac{1}{\sqrt{3}a}(g_1 + g_2)\boldsymbol{e}_2 + \frac{1}{c}g_3\boldsymbol{e}_3\right].$$

Hence there are eight shortest vectors b_g^*. In virtue of (4.2.26), the first zone for an hcp lattice is of the form shown in Fig. 4.16a. The situation here is complicated by the fact that there are no energy discontinuities on the lower and upper hexagonal surfaces. This can be shown if we calculate the structure factor in (1.5.25), which vanishes for the aforementioned two planes (we leave the proof as an exercise for the reader). For the hcp lattice, with two atoms in the unit cell with cartesian coordinates (0, 0, 0; 1/2, 1/6, 1/2), we must consider the second zone, which is depicted in Fig. 4.16b. As a result, we will obtain, instead of the first zone, the so-called composite zone or the first Jones zone, which owes its name to *Jones* [4.5]. This zone is bounded by the lateral faces of the first Brillouin zone (Fig. 4.16a) and incorporates all the surfaces of the second zone (Fig. 4.16b) that lie above and below the point of intersection of the first and second zones. The Jones zone is the smallest region of $\boldsymbol{k}$ space, bounded by the planes on which the energy undergoes a discontinuity. This zone, as distinct from the Brillouin zone, contains a fractional number of electronic states per atom $n = 2-(3/4)(a/c)^2 \cdot [1-(a/c)^2/4]$.

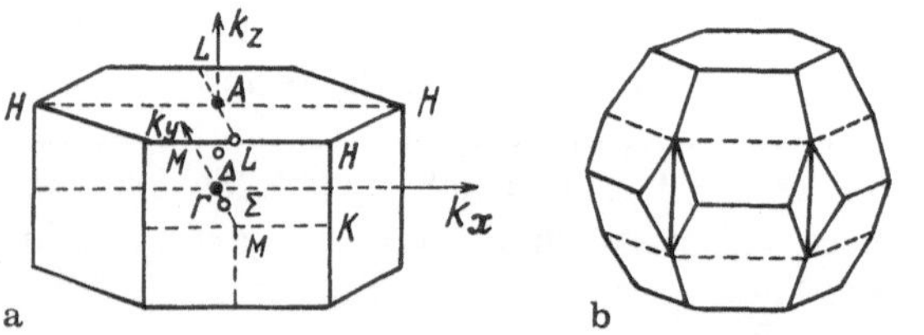

Fig. 4.16. First and second Brillouin zones in the three-dimensional case for the hcp lattice [4.5]: first zone, some symmetry points (A, H, L, M, Γ, K) and lines (Δ, Σ) connecting these are indicated (**a**); second zone (**b**)

The examples given above show how Brillouin zones are constructed. In what follows these zones are used for band-structure calculations of the properties of solids. A more comprehensive treatment of these zones is given in [4.5].

4.2.3 Electron Energy Spectrum

Our objective now is to consider the energy spectrum of an electron in a crystal. To all states of a zone there corresponds a quasi-continuous energy band of finite width (for a finite crystal), since all the parameters specifying the state inside a zone (band) are continuous. By contrast with the one-dimensional case, this exhausts all the cases yielded by a rigorous theory. In the one-dimensional case, the separation of states into zones (bands) also corresponds to an unambiguous separation of the energy levels, but here there is always a gap between two bands of adjacent zones. In the three-dimensional case, this is possible but not obligatory.

We will examine energy curves, $E(\boldsymbol{k})$, for $\boldsymbol{k}$ values along three straight lines that pass through the origin of coordinates in $\boldsymbol{k}$ space in some three directions a, b, and c for two adjacent bands ζ_1 and ζ_2. Let these lines intersect the boundaries of the zones at points $\boldsymbol{k}_a$, $\boldsymbol{k}_b$, and $\boldsymbol{k}_c$, respectively. For each direction a, b, and c, energy gaps are always present on these boundaries: $E(\zeta_1, \boldsymbol{k}_a) < E(\zeta_2, \boldsymbol{k}_a)$, etc. But for different directions a, b, and c different energy relations occur. Thus, the highest point $1(b)$ of the energy $E(\zeta_1, \boldsymbol{k}_b)$ of the first band lies higher than the lowest point $2(a)$ of the energy $E(\zeta_2, \boldsymbol{k}_a)$, which belongs to the second band. The same is true of the points $\boldsymbol{k}_c$ of the first band $1(c)$ and the second band $2(b)$, etc. This is the energy spectrum of a metal with overlapping bands of adjacent zones (Fig. 4.17a).

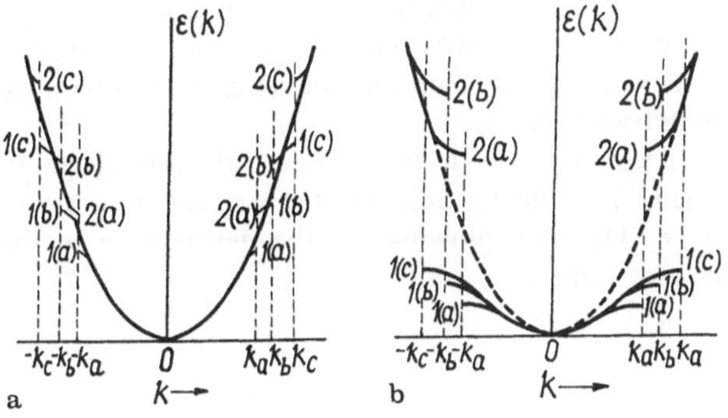

Fig. 4.17. Relative position of overlapping energy bands in the three-dimensional crystal of the metal (**a**), the same for an insulator or semiconductor with energy gaps (**b**)

Of course, cases are liable to occur in which the energy gaps are present for all directions in $\boldsymbol{k}$ space. This situation is portrayed in Fig. 4.17b, which shows that adjacent energy bands do not overlap for all three directions chosen, a, b, and c (as well as for other directions). This is the case of a semiconductor or insulator, and also of a metal with partially filled electronic states in the lowest-lying energy band.

In the literature the term "band" is frequently used to imply a set of energy levels corresponding to electronic states rather than a set of electronic states as such. Therefore, one talks of an "overlap of bands" rather than an overlap of the energy bands of different zones. Such usage of the term "band" results in some vagueness. Actually, a classification of states in terms of zones may always be carried out exactly and unambiguously. True, when the energy E pertains to two states of different zones ψ_1 and ψ_2, we may treat any one of their linear combinations $\alpha\psi_1 + \beta\psi_2$ as an electron wave function and thereby alter the classification of states by zones. But this is not reasonable, for the electron transitions that occur under the influence of externally applied fields obey different rules, depending on whether the initial and final states belong to one and the same or different zones. The difference persists, irrespective of whether or not the corresponding energy bands overlap. This is why the states of the electron should be classified by zones.

In a one-dimensional array the energy in the Brillouin zone always depends on $|\boldsymbol{k}|$ in a monotonic fashion, a consequence of the two-fold degeneracy of each E. We will show further that the $E(\boldsymbol{k})$ function may be highly varied in form, and not necessarily monotonic. To each E, within the confines of even one zone, there corresponds a whole continuum of wave functions. The character of the continuum—i.e., the form of the energy surface $E(\boldsymbol{k}) = \text{const}$—cannot be judged in terms of a rigorous theory unless recourse is made to group theory. Otherwise, we may only assert that, if ψ is one of the solutions corresponding to the particular E given, ψ^* is also its solution. Hence it follows that, if the surface $E(\boldsymbol{k}) = \text{const}$ passes through point $\boldsymbol{k}$, then, from (4.2.11), it must pass through point $-\boldsymbol{k}$ as well. Other statements of this kind may be made, if we take into account the additional symmetry properties.

To ascertain the features peculiar to the electron states at the boundaries of the zones, we must first generalize the formula for the current in a three-dimensional lattice. Using (4.2.11), and allowing for the dependence of the electron wave function one zone number,

$$\psi_{\boldsymbol{k}\zeta}(\boldsymbol{r}) = \exp(\mathrm{i}\boldsymbol{k}\cdot\boldsymbol{r})u_{\boldsymbol{k}\zeta}(\boldsymbol{r}) \ . \tag{4.2.27}$$

We define the normalization conditions for (4.2.27) by introducing a system of coordinates with unit vectors—principal lattice vectors. Then

$$\boldsymbol{r} = x^{(1)}\boldsymbol{a}_1 + x^{(2)}\boldsymbol{a}_2 + x^{(3)}\boldsymbol{a}_3 \ , \tag{4.2.28}$$

where $x^{(j)}$ stands for numbers that vary by an integer during transition from one lattice site to another. By analogy with Sect. 4.1.4, the normalization conditions are

$$\begin{aligned}\int d\boldsymbol{r}\psi^*_{\boldsymbol{k}'\zeta'}(\boldsymbol{r})\psi_{\boldsymbol{k}''\zeta''}(\boldsymbol{r}) &= C\lim_{G\to\infty}\sum_{l_1,l_2,l_3=-G}^{G}\int_{l_1}^{l_1+1}dx^{(1)}\int_{l_2}^{l_2+1}dx^{(2)}\int_{l_3}^{l_3+1}dx^{(3)}\\ &\times\exp[\mathrm{i}(k_1''-k_1')x^{(1)}+\mathrm{i}(k_2''-k_2')x^{(2)}+(k_3''-k_3')x^{(3)}]\\ &\times u^*_{\boldsymbol{k}'\zeta'}(\boldsymbol{r})u_{\boldsymbol{k}''\zeta''}(\boldsymbol{r}) = (2\pi)^3\delta(k_1'-k_1'')\delta(k_2'-k_2'')\\ &\times\delta(k_3'-k_3'')\int_{V_0}d\boldsymbol{r}u^*_{\boldsymbol{k}'\zeta'}(\boldsymbol{r})u_{\boldsymbol{k}''\zeta''}(\boldsymbol{r}) \ ,\end{aligned}$$

where C is the Jacobian of the transformation from normal rectangular coordinates x, y, z to $x^{(1)}$, $x^{(2)}$, $x^{(3)}$; and V_0 is the volume of an elementary parallelepiped. As follows from the expression obtained above,

$$\int_{V_0} d\boldsymbol{r} u^*_{\boldsymbol{k}'\zeta'}(\boldsymbol{r}) u_{\boldsymbol{k}''\zeta''}(\boldsymbol{r}) = \delta_{\zeta'\zeta''}/(2\pi)^3 \tag{4.2.29}$$

$$\int d\boldsymbol{r} \psi^*_{\boldsymbol{k}'\zeta'}(\boldsymbol{r}) \psi_{\boldsymbol{k}''\zeta''}(\boldsymbol{r}) = \prod_{i=1}^{3} \delta(k'_i - k''_i) \ . \tag{4.2.30}$$

We now determine the matrix elements $\boldsymbol{r}$, by transcribing (4.2.27) for $x^{(j)}$:

$$\psi_{\boldsymbol{k}\zeta}(\boldsymbol{r}) = \exp[\mathrm{i}(k_1 x^{(1)} + k_2 x^{(2)} + k_3 x^{(3)})] u_{\boldsymbol{k}\zeta}(\boldsymbol{r}) \ .$$

Therefore,

$$x^{(i)} \psi_{\boldsymbol{k}\zeta}(\boldsymbol{r}) = -\mathrm{i} \frac{\partial \psi}{\partial k_i} + \mathrm{i} \exp(\mathrm{i}\boldsymbol{k} \cdot \boldsymbol{r}) \frac{\partial u}{\partial k_i} \ . \tag{4.2.31}$$

Just as in the one-dimensional case, we divide the operator $\hat{\boldsymbol{r}}$ into two terms, $\hat{\boldsymbol{r}}_1$ and $\hat{\boldsymbol{r}}_2$. Similar to (4.1.97–99), we obtain

$$(k'\zeta'|x_1^{(j)}|k''\zeta'') = \mathrm{i}\delta'(k'_j - k''_j)\delta(k'_l - k''_l)\delta(k'_m - k''_m)\delta_{\zeta'\zeta''} \ , \quad j \neq l \neq m = 1, 2, 3 \tag{4.2.32}$$

$$(k'\zeta'|x_2^{(j)}|k''\zeta'') = \mathrm{i} \int d\boldsymbol{r} \exp[\mathrm{i}(\boldsymbol{k}'' - \boldsymbol{k}') \cdot \boldsymbol{r}] u^*_{\boldsymbol{k}'\zeta'}(\boldsymbol{r}) \frac{\partial}{\partial \boldsymbol{k}''_j} u_{\boldsymbol{k}''\zeta''}(\boldsymbol{r}) \ . \tag{4.2.33}$$

Applying to the right-hand side of (4.2.33) the same procedure as in the derivation of (4.2.30), we find

$$\begin{aligned}(k'\zeta'|x_2^{(j)}|\boldsymbol{k}''\zeta'') = {} & (2\pi)^3 \delta(k'_1 - k''_1)\delta(k'_2 - k''_2) \\ & \times \delta(k'_3 - k''_3) \mathrm{i} \int_{V_0} d\boldsymbol{r} u^*_{\boldsymbol{k}'\zeta'}(\boldsymbol{r}) \frac{\partial}{\partial k'_j} u_{\boldsymbol{k}'\zeta''}(\boldsymbol{r}) \ .\end{aligned} \tag{4.2.34}$$

In contrast to the one-dimensional case, (4.2.34), as a rule, does not go to zero when $\zeta' = \zeta''$, if the $u_{\boldsymbol{k}\zeta}(\boldsymbol{r})$ do not possess any special symmetry properties. Also note that the diagonal matrix elements of the coordinates at $\zeta' = \zeta''$—i.e., the mean values of the coordinates of the electron in the eigenstate $\boldsymbol{k}\zeta$—do not play any role in the physically interesting phenomena involved. But when $\zeta' \neq \zeta''$ the matrix elements indicate interband transitions under the influence of a constant and slowly varying electric field and are therefore physically interesting. Formula (4.2.34) shows that in the three-dimensional case we deal here with the same rigid selection rules as those in the one-dimensional problem; that is, transitions are possible when the quasimomenta in the final and initial states are equal. This has important implications for the optical properties of crystals.

In electron current calculations the operator $\hat{r}_1$ is essential. Its matrix element (4.2.32) has the general form, regardless of the form of $u_{k\zeta}(r)$. In vector form

$$\hat{r}_1 = \mathrm{i}\,\nabla_k \ , \tag{4.2.35}$$

and (4.2.33) may be transcribed in vector form as

$$(k'\zeta'|\hat{r}_2|k''\zeta'') = \mathrm{i}(2\pi)^3\delta(k'_1 - k''_1)\delta(k'_2 - k''_2) \times \delta(k'_3 - k''_3)\int d\boldsymbol{r}\, u^*_{k'\zeta'}(\boldsymbol{r})\,\nabla_{k'} u_{k'\zeta''}(\boldsymbol{r}) \ . \tag{4.2.36}$$

The mean electron velocity value can be found from the equation of motion

$$\hat{\dot{r}} = -\frac{\mathrm{i}}{\hbar}(\hat{r}\hat{\mathscr{H}} - \hat{\mathscr{H}}\hat{r}) = -\frac{\mathrm{i}}{\hbar}[(\hat{r}_1\hat{\mathscr{H}} - \hat{\mathscr{H}}\hat{r}_1) + (\hat{r}_2\hat{\mathscr{H}} - \hat{\mathscr{H}}\hat{r}_2)] \ . \tag{4.2.37}$$

From the foregoing we clearly see that the second term on the right-hand side of (4.2.37) does not contribute to the mean value of the operator $\hat{\dot{r}}$, because the diagonal matrix elements $\hat{r}_2$ are finite, and therefore

$$\langle k\zeta|[\hat{\mathscr{H}}, \hat{r}_2]_-|k\zeta\rangle = [E(k,\zeta) - E(k,\zeta)]\langle k\zeta|\hat{r}_2|k\zeta\rangle = 0 \ .$$

Thus, for the mean value of the operator $\hat{\dot{r}}$ it suffices to take into account the first term of (4.2.37). Using (4.2.35), we find

$$\langle v(k,\zeta)\rangle \equiv \langle\hat{\dot{r}}(k,\zeta)\rangle = \frac{1}{\hbar}\nabla_k E(k,\zeta) \ . \tag{4.2.38}$$

Formula (4.2.38), just as (4.1.106), is a generalization of the de Broglie formula. From (4.2.38) we see that within the eigenstates the electron in the crystal normally possesses a nonzero mean velocity; i.e., it is free to move over the crystal like a free electron. The magnitude and direction of $\langle v(k,\zeta)\rangle$ are determined by the form of the constant energy surfaces $E(k,\zeta) = \text{const}$.

For the velocity operator in the crystal we have the same rigid selection rules as those in an array. The matrix elements of this operator are nonzero only for $k' = k''$. The diagonal elements are determined from (4.2.38), and the off-diagonal elements, corresponding to interband transitions ($\zeta' \neq \zeta''$), are related to the off-diagonal elements of the operator $\hat{r}_2$. From (4.2.36) we can readily obtain the formula

$$(\boldsymbol{k}'\zeta'|\hat{\boldsymbol{r}}|\boldsymbol{k}''\zeta'') = -\frac{(2\pi)^3}{\hbar}\delta(k_1'-k_1'')\delta(k_2'-k_2'')\delta(k_3'-k_3'')$$
$$\times [E(\boldsymbol{k}'\zeta') - E(\boldsymbol{k}'\zeta'')] \int_{V_0} d\boldsymbol{r} u^*_{\boldsymbol{k}'\zeta'}(\boldsymbol{r}) \tag{4.2.39}$$
$$\times \nabla_{\boldsymbol{k}'} u_{\boldsymbol{k}'\zeta''}(\boldsymbol{r}) , \quad \zeta' \neq \zeta'' .$$

We conclude this section by clearing up the question of the electron current at the boundaries of a zone. In the interior of a zone the wave functions can be arranged in different ways and, depending on the mode of arrangement, the states ascribed to a zone boundary will be different. Having no opportunity to enlarge upon this problem, we restrict ourselves to the following postulate: For a number of crystals, by virtue of their specific symmetry, the current vector in eigenstates is normal (orthogonal) to certain planes of $\boldsymbol{k}$ space. Whatever the zone construction method, these planes normally turn out to be boundary planes [Ref. 4.5, Sect. 17].

4.2.4 The Properties of Constant Energy Surfaces

Thus far we have used only translational crystal symmetry. However, we can also consider other symmetry elements—proper and improper rotations and reflections that form the point group of the crystal, Q (Sect. 1.2). Then, just as in the case of (4.2.1), we have

$$QV(\boldsymbol{r}) = V(\boldsymbol{r}) . \tag{4.2.40}$$

Each of these operations can be compared with the coordinate transformation

$$x_i = \sum_j a_{ij}x_j , \quad i,j = 1,2,3 , \tag{4.2.41}$$

where the 3×3 matrix elements of the transformations $\|a_{ij}\|$ satisfy the conditions

$$\sum_i a_{ij}a_{ik} = \delta_{jk} . \tag{4.2.42}$$

The Schrödinger equation (4.2.17) in this case will not be invariant under the transformation (4.2.41) because of the term $2\mathrm{i}(\boldsymbol{k}\cdot\nabla)u_{\boldsymbol{k}\zeta}(\boldsymbol{r})$. To assure invariance, we need simultaneously to transform the wave vector $\boldsymbol{k}$ as well. (We leave the proof for the reader; see Jones [4.5].) Thus

$$k_i' = \sum_j a_{ij}k_j . \tag{4.2.43}$$

It then follows from the invariance (4.2.17) that in each band ζ $E(\boldsymbol{k}, \zeta) = E(\boldsymbol{k}', \zeta)$ or

$$QE(\boldsymbol{k}, \zeta) = E(\boldsymbol{k}, \zeta) \; ; \tag{4.2.44}$$

i.e., in each band the energy $E(\boldsymbol{k}, \zeta)$ as a function of $\boldsymbol{k}$ possesses the complete point symmetry of the crystal.

Let us now ascertain some general properties of the constant energy surfaces (including Fermi surfaces) associated with crystal symmetry. In particular, just as for the function $E(\boldsymbol{k}, \zeta)$, the surfaces $E(\boldsymbol{k}, \zeta) = \text{const}$ should possess the complete symmetry of the point group of the crystal. Also note that there is a direct correlation between the transformation in $\boldsymbol{r}$ space (direct lattice) and the transformation in $\boldsymbol{k}$ space (reciprocal lattice); namely, if the direct lattice possesses symmetry axes and planes, we may talk of corresponding parallel axes and planes in the reciprocal lattice. This may be proved using the definition of the reciprocal lattice as a set of vectors $\boldsymbol{b}_g$, for which the number $\boldsymbol{b}_g \cdot \boldsymbol{R}_n$ is an integer for any direct-lattice vector $\boldsymbol{R}_n$. But if $\boldsymbol{R}'_n$, related to $\boldsymbol{R}_n$ by transformation (4.2.41), also belongs to the direct lattice, and $\boldsymbol{b}'_g$ is related to $\boldsymbol{b}_g$ by the same transformation, then, from condition (4.2.42), $\boldsymbol{b}'_g \cdot \boldsymbol{R}'_n = \boldsymbol{b}_g \cdot \boldsymbol{R}_n$ and, consequently, $\boldsymbol{b}'_g$ belongs to the reciprocal lattice, proving our assertion.

If the direct lattice possesses a symmetry plane $y = z$, then the plane in $\boldsymbol{k}$ space will be $k_y = k_z$. Therefore, the continuous function $E(\boldsymbol{k}, \zeta)$ is symmetric with respect to any symmetry plane in $\boldsymbol{k}$ space. Hence on any such plane

$$\boldsymbol{n} \cdot \nabla_{\boldsymbol{k}} E(\boldsymbol{k}, \zeta) = 0 \; , \tag{4.2.45}$$

with $\boldsymbol{n}$ being the unit vector of the normal to the symmetry plane. If (4.2.45) is not met, then, by symmetry conditions, the $\nabla_{\boldsymbol{k}} E(\boldsymbol{k}, \zeta)$ would undergo a discontinuity within the Brillouin zone, which contradicts the requirement that $E(\boldsymbol{k}, \zeta)$ with its derivatives be continuous in the interior of the zone. The vector $\nabla_{\boldsymbol{k}} E$ is normal to the constant energy surface. By (4.2.45), it is normal also to the normal to the symmetry plane. Therefore, the surfaces $E(\boldsymbol{k}, \zeta) = \text{const}$ should intersect the symmetry plane at right angles, enabling us to judge the topology of the isoenergetic surfaces (including Fermi surfaces) inside the Brillouin zone. Also, we may ascertain the general behavior of these surfaces at the boundaries of the zones.

In this context, the concepts of contiguous and intersecting energy bands must be recalled. Figure 4.18a presents the trends in the $E(\boldsymbol{k}, \zeta)$ functions along some line in $\boldsymbol{k}$ space: a degeneracy occurs at point $\boldsymbol{k}_0$; i.e., the bands touch. Figure 4.18b depicts the case of overlapping bands. In some regions of the zone we have $E(\boldsymbol{k}', \zeta_1) > E(\boldsymbol{k}'', \zeta_2)$, but for each particular $\boldsymbol{k}$ given the inequality $E(\boldsymbol{k}, \zeta_2) > E(\boldsymbol{k}, \zeta_1)$ always holds. Cases are possible in which the bands both touch and cross each other.

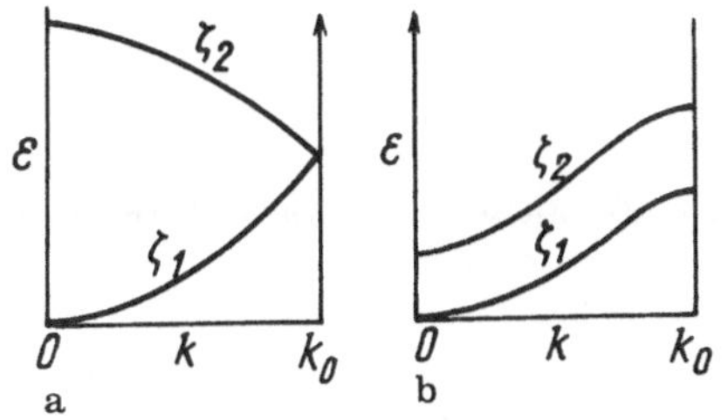

Fig. 4.18. Energy bands in the three-dimensional crystal for two adjacent Brillouin zones ζ_1 and ζ_2: touching (**a**) and crossing (**b**)

As an example of point-symmetry applications, we examine the shape of the constant energy surfaces when they cross the boundaries of the Brillouin zones, and also the symmetry planes inside the zones. We consider two opposing boundary surfaces of a zone, which are normal to the vectors $\boldsymbol{lB}$ and $-\boldsymbol{lB}$ and bisect them. If the point group of the crystal has a symmetry plane (l_1, l_2, l_3), the Brillouin zone also has one; it passes through the origin and is normal to the vectors $\pm\boldsymbol{lB}$. Therefore, by symmetry,

$$(\nabla E\cdot\boldsymbol{lB})_a = -(\nabla E\cdot\boldsymbol{lB})_b \ , \qquad (4.2.46)$$

where a and b (Fig. 4.19) are equivalent points on the opposite faces of the zone. In turn, these points represent one and the same electron state; therefore $(\nabla E)_a = (\nabla E)_b$. Then it follows from (4.2.46) that the normal derivative of the energies on both faces is zero; i.e., the constant energy surface crosses these faces at right angles. Figures 4.19a, b show isoenergetic lines on plane $k_z = 0$ in the Brillouin zone of a bcc lattice. All these lines intersect the straight lines portrayed in Figs. 4.19a, b (the inner lines, which are the intersections of the symmetry planes inside a zone with plane $k_z = 0$, and the outer lines, which are the intersections of the zone boundaries with plane $k_z = 0$) at right angles; i.e., at the point of intersection $\nabla E = 0$. However, ∇E does not equal zero on all the faces of the Brillouin zone in an fcc lattice [Ref. 4.5, Sect. 17].

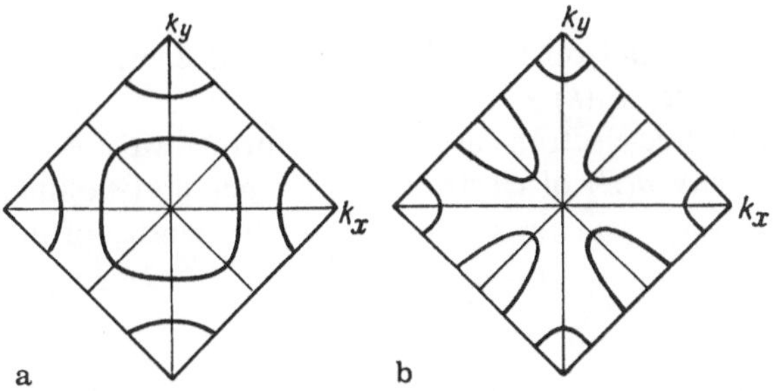

Fig. 4.19. Intersection of isoenergetic surfaces with zone boundaries and symmetry planes in zones

4.2.5 Density of Electron States in Energy Bands. Topological Electronic Transitions

In the band theory of solids we often have to calculate the various physical quantities F, which are the weighted sums of the corresponding one-electron characteristics $F_\zeta(\boldsymbol{k})$ over the quantum states (defined by the wave vectors $\boldsymbol{k}$) of the electron energy bands of the crystal. We did so on more than one occasion using the free-electron gas model (3.5.27, 47, 48, 81, 82). The general form of the corresponding band theory formulas turns out to be as follows:

$$F = 2\sum_{\zeta,\boldsymbol{k}} F_\zeta(\boldsymbol{k}) \ . \tag{4.2.47}$$

The factor 2 in this equation arises from spin degeneracy. For the sum over $\boldsymbol{k}$ to be calculated, we need to recall the definition of the electron wave vector $\boldsymbol{k}$ (Sect. 3.5). As follows from the definition, to each possible value of $\boldsymbol{k}$ in $\boldsymbol{k}$ space there corresponds the volume $\Delta\boldsymbol{k} = (2\pi)^3/L_xL_yL_z = 8\pi^3/V$ (with V being the volume of the usual r space). The region of volume $V_{\boldsymbol{k}}$ in $\boldsymbol{k}$ space contains $V_{\boldsymbol{k}}{:}(8\pi^3/V)$ possible values of $\boldsymbol{k}$ and the unit volume of $\boldsymbol{k}$ space ($V_{\boldsymbol{k}} = 1$); i.e., the density of states in it is equal to $V/8\pi^3$. Hence we may immediately write the sum in (4.2.47) over $\boldsymbol{k}$ in the following form:

$$2\sum_{\boldsymbol{k}} F_\zeta(\boldsymbol{k}) = \frac{V}{4\pi^3}\sum_{\boldsymbol{k}} \Delta\boldsymbol{k}\, F_\zeta(\boldsymbol{k}) \ . \tag{4.2.48}$$

In the limit of infinitely large systems ($V \to \infty$ and $\Delta\boldsymbol{k} \to 0$) and if the function $F_\zeta(\boldsymbol{k})$ varies insignificantly over the interval $\Delta\boldsymbol{k}$ in $\boldsymbol{k}$ space, the sum in (4.2.48) may be replaced by the integral

$$\lim_{V\to\infty} \frac{2}{V}\sum_{\boldsymbol{k}} F_\zeta(\boldsymbol{k}) = \int \frac{d\boldsymbol{k}}{4\pi^3} F_\zeta(\boldsymbol{k}) \ . \tag{4.2.49}$$

The quantities $F_\zeta(\boldsymbol{k})$ normally depend on the wave vector and energy-band number through the electron energy $E_\zeta(\boldsymbol{k})$.

Pushing the analogy with the free-electron case (Chap. 3) further, we are in a position to determine the density $g(E)$ of electronic levels (per unit volume $V=1$) and in band theory. In this event the expression for the density of the quantity F itself will be

$$f = \lim_{V\to\infty} \frac{F}{V} = \sum_{\zeta}\int \frac{d\boldsymbol{k}}{4\pi^3} F_\zeta(\boldsymbol{k}) \tag{4.2.50}$$

or

$$f = \int dEg(E)F(E) \ . \tag{4.2.51}$$

Comparing (4.2.50) and (4.2.51) we obtain

$$g(E) = \sum_{\zeta} g_{\zeta}(E) \ , \tag{4.2.52}$$

where the density of electron energy levels in the ζ band is equal to

$$g_{\zeta}(E) = \int \frac{d\boldsymbol{k}}{4\pi^3} \delta[E - E_{\zeta}(\boldsymbol{k})] \ , \tag{4.2.53}$$

with the integral taken over any unit cell of $\boldsymbol{k}$ space.

It is actually more convenient to determine $g_{\zeta}(E)$ in a slightly different fashion. Just as in the case of a free-electron gas (3.5.5, 8), (but subject now to the restriction that $V = 1$), we have

$$g_{\zeta}(E)dE = dN_{\zeta}(E) \ , \tag{4.2.54}$$

where $dN_{\zeta}(E)$ is the number of allowed one-particle levels, i.e., $\boldsymbol{k}$ values, in the energy interval between E and $E + dE$.

As we have already seen, the number of admissible vectors $\boldsymbol{k}$ in the ζ band in the above energy interval dE is equal to the ratio of the volume of that portion of the unit cell in $\boldsymbol{k}$ space in which the energy lies within the limits $E \leqslant E_{\zeta}(\boldsymbol{k}) \leqslant E + dE$, i.e., $\int_{E \leqslant E_{\zeta}(\boldsymbol{k}) \leqslant E + dE} d\boldsymbol{k}$, to the volume $\Delta \boldsymbol{k} = (2\pi)^3/V$ corresponding to one allowed $\boldsymbol{k}$ value. Thus, (4.2.54) becomes

$$g_{\zeta}(E)dE = \int_{E \leqslant E_{\zeta}(\boldsymbol{k}) \leqslant E + dE} d\boldsymbol{k}/4\pi^3 \ . \tag{4.2.55}$$

Since dE is an infinitesimally small quantity, the volume integral on the right-hand side of (4.2.55) can readily be transformed to a surface integral; in the free-electron gas case that was a spherical layer (Sect. 3.5). Labeling the distance between the constant energy surfaces $S_{\zeta}(E)$ and $S_{\zeta}(E + dE)$ at point $\boldsymbol{k}$ by $\delta\mathscr{K}(\boldsymbol{k})$, we then obtain, instead of (4.2.55),

$$g_{\zeta}(E)dE = \int_{S_{\zeta}(E)} \frac{dS}{4\pi^3} \delta\mathscr{K}(\boldsymbol{k}) \ . \tag{4.2.56}$$

Our purpose now is to find the explicit expressions of the quantity $\delta\mathscr{K}(\boldsymbol{k})$ in terms of the energy $E_{\zeta}(\boldsymbol{k})$. The surface $S_{\zeta}(E)$ is isoenergetic, so the energy gradient $\nabla_{\boldsymbol{k}} E_{\zeta}(\boldsymbol{k})$ will be a normal to it and is equal in magnitude to the rate of change of

$E_\zeta(\boldsymbol{k})$ along this normal. Hence we have

$$E + dE = E + |\nabla_{\boldsymbol{k}} E_\zeta(\boldsymbol{k})| \delta\mathscr{K}(\boldsymbol{k})$$

or

$$\delta\mathscr{K}(\boldsymbol{k}) = dE / |\nabla_{\boldsymbol{k}} E_\zeta(\boldsymbol{k})| \ . \tag{4.2.57}$$

Equations (4.2.56, 57) yield

$$g_\zeta(E) = \int\limits_{S_\zeta(E)} \frac{dS}{4\pi^3} \frac{1}{|\nabla_{\boldsymbol{k}} E_\zeta(\boldsymbol{k})|} \ . \tag{4.2.58}$$

From (4.2.58) it can be readily shown (an exercise for the reader) that the result for $g(E)$ with a free-electron gas will be (3.5.8), again only for $V = 1$.

In the electron energy bands, the $E_\zeta(\boldsymbol{k})$ functions display at least one minimum and one maximum, the gradient $\nabla_{\boldsymbol{k}} E_\zeta(\boldsymbol{k})$ being equal to zero. At these points the integrand on the right-hand side of (4.2.58) exhibits a singularity. It may be shown that these singularities are integrable and the densities of states, $g_\zeta(E)$, remain finite. However, singularities also arise on the $\partial g_\zeta(E)/\partial E$ curve—i.e., kinks occur in the slope of the $g_\zeta(E)$ curve—these are the Van Hove singularities which have already been discussed (Chap. 2, Fig. 2.15). As we pointed out there, Van Hove singularities occur also when the values of the energy E_c correspond to saddle points in $\boldsymbol{k}$ space. When $E = E_c$, the topology of the constant energy surfaces alters (for instance, the heretofore separated parts of the isoenergetic surfaces can merge).

The shape of the $g_\zeta(E)$ curve is schematically presented in Fig. 4.20. The character of Van Hove singularities can be determined in a fashion analogous to the treatment for phonons (Sect. 2.2). In the vicinity of the critical point $\boldsymbol{k}_c$, where $\nabla_{\boldsymbol{k}} E_\zeta(\boldsymbol{k}_c) = 0$, the expansion of the energy in wave vector differences $\boldsymbol{y} = \boldsymbol{k} - \boldsymbol{k}_c$ begins with quadratic terms

$$Z \equiv E_\zeta(\boldsymbol{k}) - E_\zeta(\boldsymbol{k}_c) = \alpha_1 y_1^2 + \alpha_2 y_2^2 + \alpha_3 y_3^2 \tag{4.2.59}$$

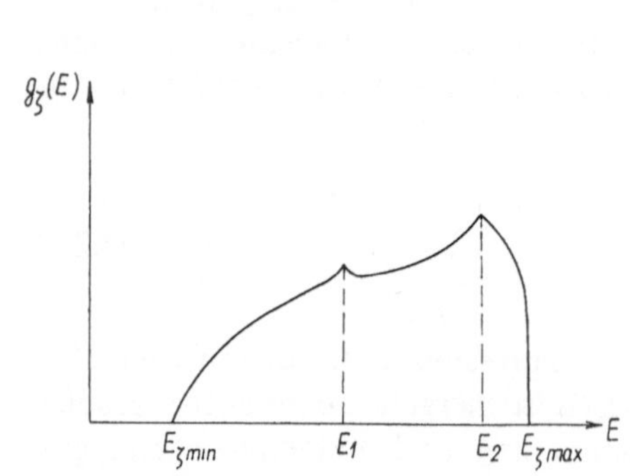

Fig. 4.20. The density of state curve with Van Hove singularities E_1, E_2, $E_{\min}$ and $E_{\max}$

[cf. (2.2.22)]. If all $\alpha_i < 0$ ($i = 1, 2, 3$), the critical point corresponds to the top of the band $E_\zeta(\boldsymbol{k})$. This signifies that the full constant energy surface, defined by the equation

$$E = E_\zeta(\boldsymbol{k}) \quad \text{for some } \zeta , \tag{4.2.60}$$

contains, when $z < 0$, an ellipsoid with the principal axes

$$y_i = (z/\alpha_i)^{1/2} , \tag{4.2.61}$$

which contracts for $z \to -0$. Similarly, if all $\alpha_i > 0$, we deal with the bottom of the band, and the ellipsoid contracts to a point for $Z \to +0$. In both cases, when $z = 0$ the topology of the isoenergetic surface changes; i.e., new sheets emerge and disappear. The corresponding contribution to the density of states has a unilateral singularity of the form $z^{1/2}$ [cf. (2.2.23)]. If one of the α_i has a sign opposite to that of the other (for example, $\alpha_1 > 0$, $\alpha_2 > 0$, $\alpha_3 < 0$), we deal with a saddle point. Then, as is known from analytic geometry, with $z > 0$, (4.2.59) describes a hyperboloid of one sheet, and for $z < 0$ it describes a hyperboloid of two sheets. Finally, when $z = 0$ the topology of the isoenergetic surface also changes—its two uncoupled parts coalesce. It can be shown that the behavior of Van Hove singularities here is the same as when a new sheet appears or disappears [4.8].

The Fermi surface plays the most important role in metals, corresponding in energy to the highest occupied level, $E = E_F = \zeta_0$ at $T = 0$ K. As shown in Chap. 3, nearly all properties of metals (except for ultrahigh-frequency properties) are determined by electrons with $E = \zeta_0$ so that, if the topology of the Fermi surface changes, this strongly affects all the properties of the metal and, in fact, a phase transition, called a topological electronic transition [4.9] will occur. It may be shown that this is a 2.5-order transition, according to Ehrenfest's classification; i.e., the singularity in the thermodynamic potential Ω has the form

$$\delta\Omega \propto |z|^{5/2}\vartheta(\pm z) , \tag{4.2.62}$$

with $z = E_F - E_c$, $\vartheta(x > 0) = 1$, and $\vartheta(x < 0) = 0$. The result (4.2.62) follows immediately from the character of the Van Hove singularity in the density of states

$$\delta g(E) = \alpha(E - E_c)^{1/2}\vartheta(E - E_c) \tag{4.2.63}$$

and from the formula for the thermodynamic potential Ω at $T = 0$ K (Sect. 3.5.5):

$$\Omega = \sum_{\nu, \varepsilon_\nu < \zeta_0} (\varepsilon_\nu - \zeta_0) = \int_{-\infty}^{\zeta_0} dE g(E)(E - \zeta_0) . \tag{4.2.64}$$

Indeed, substituting (4.2.63) into (4.2.64), we find the result for the part of the potential Ω which is due to the Van Hove singularity:

$$\delta\Omega = 0\ , \quad \zeta_0 < E_c\ , \tag{4.2.65}$$

$$\begin{aligned}\delta\Omega &= \alpha\int_{E_c}^{\zeta_0} dE(E-\zeta_0)(E-E_c)^{1/2} = \alpha\int_{E_c}^{\zeta_0} dE[(E-E_c)^{3/2} \\ &\quad -(\zeta_0-E_c)(E-E_c)^{1/2}] = \alpha[\tfrac{2}{5}(\zeta_0-E_c)^{5/2}-\tfrac{2}{3}(\zeta_0-E_c)^{5/2}] \\ &= -\frac{4\alpha}{15}(\zeta_0-E_c)^{5/2}\ , \quad \zeta_0 > E_c\end{aligned}$$

which proves (4.2.62). The singularity in the electronic heat capacity C_e in the case of a topological electronic transition is on the order of $|z^{1/2}|\vartheta(\pm z)$, since it is determined directly by the density of states at the Fermi level (3.5.53, 54). It may be shown [4.10] that the singularity has the same relaxation time as that at the Fermi level, $\tau(\zeta_0)$, and, accordingly, the same electrical resistivity (Sect. 3.6). Recently, these anomalies have been observed experimentally [4.11] on Al whiskers (the topology of the Fermi surface in these experiments altered owing to external deformations applied to the crystal). Note that, since the thermal emf $\mathscr{E}_T$ is defined by the quantity $\partial \ln\tau(\zeta_0)/\partial\zeta_0$ (Sect. 3.6.4), the singularity in it is more pronounced than in electrical resistivity. In fact, if we assume that

$$\tau(\zeta_0) = \tau_0 + A(\zeta_0 - E_c)^{1/2}\vartheta(\zeta_0 - E_c)\ , \tag{4.2.66}$$

where τ_0 is a smooth contribution to $\tau(\zeta_0)$, then

$$\delta\mathscr{E}_T \propto \frac{\partial}{\partial\zeta_0}\ln\tau(\zeta_0) \approx \frac{A}{2\tau_0}(\zeta_0 - E_c)^{-1/2}\vartheta(\zeta_0 - E_c)\ . \tag{4.2.67}$$

Therefore, measurement of thermal emf is a very sensitive tool for detecting and studying topological electronic transitions [4.10].

4.3 Nearly-Free-Electron Approximation

4.3.1 Statement of the Problem

To make the general theory more concrete we examine the limiting case of nearly free electrons. The first to solve this problem was *Peierls* [4.12]. We have stated that many of the phenomena which occur in the metal can be qualitatively understood on the basis of the model of completely free electrons. It may turn out that in a real crystal the interactions of the electron with ions and other itinerant electrons may be well compensated and so the net periodic potential $V(\boldsymbol{r})$ is small. In this case perturbation theory can be used and the free-electron model counts as the zeroth approximation, and the $V(\boldsymbol{r})/E$ ratio as the small perturbation. A difficulty may arise here due to the fact that in the zero approximation the spectrum is continuous. However, using an auxiliary technique, we can reduce this problem to one with a discrete spectrum.

The electron in an arbitrary periodic field has a wave function (4.2.13) whose expansion coefficients satisfy (4.2.16). An innumerable multitude of systems (4.2.16) exist, since to each $\boldsymbol{k}$ there corresponds a particular system and the components of $\boldsymbol{k}$ assume a whole continuum of values (4.2.22). Each of these systems of equations may be considered independently of the other, and each $\boldsymbol{k}$ assigned a set of states forming an infinite but discrete sequence to which conventional perturbation theory may be applied. The system with $V(\boldsymbol{r}) = \text{const}$ or $V(\boldsymbol{r}) = 0$ plays the role of an unperturbed one. Here the system (4.2.16) takes the form

$$\left[E - \frac{\hbar^2}{2m}(\boldsymbol{k} + \boldsymbol{b}_g^*)^2\right] a(\boldsymbol{k}, \boldsymbol{b}_g^*) = 0 \ ,$$

and its solutions are

$$a(\boldsymbol{k}, \boldsymbol{b}_g^*) = \alpha \delta_{\boldsymbol{b}_g^* \boldsymbol{q}} \ , \tag{4.3.1}$$

where $\boldsymbol{q}$ is a particular vector b_g^* for each solution; the energy is equal to

$$E(\boldsymbol{k} + \boldsymbol{q}) = \frac{\hbar^2}{2m}(\boldsymbol{k} + \boldsymbol{q})^2 \ . \tag{4.3.2}$$

4.3.2 Empty-Lattice Model

At the outset we formulate the problem of a free electron as a particular case of its motion in a periodic field. To each of the solutions (4.3.1) there corresponds a plane wave

$$\psi(\boldsymbol{r}) = \alpha \exp(\mathrm{i}\xi \cdot \boldsymbol{r}) \tag{4.3.3}$$

with a wave vector

$$\xi = \boldsymbol{k} + \boldsymbol{q} \ . \tag{4.3.4}$$

These vectors differ by the reciprocal-lattice vectors. According to Sect. 4.2.4, (4.2.16) and, consequently, (4.3.3) satisfy three conditions. (1) The spectrum

(4.3.2) is always discrete, since the vector $\boldsymbol{q}$ is defined by three integral numbers (q_1, q_2, q_3) and, therefore, should undergo only a finite increment. (2) Among the levels (4.3.2) there is one which is the lowest. (3) Among these energies there are, as a rule, no equal energies when $\boldsymbol{k}$ is arbitrary. Therefore, ignoring the possible complication of multiple roots, the classification of states by bands can be used for a free electron as well. Rather than define the (4.3.3) by the wave vector ξ, we assign to it the quasimomentum $\hbar\boldsymbol{k}$ and band number ζ. To distribute the states of a free electron, we may follow Sect. 4.2.4 and arrange all the energies (4.3.2) for a given $\boldsymbol{k}$ in ascending order by renumbering them with the numbers ζ. On performing this manipulation for all $\boldsymbol{k}$, the states with the same number ζ are grouped into a band (zone). However, one vector $\boldsymbol{q}$ may not necessarily correspond to a single band. This becomes clear if we write the difference between two energies for the same $\boldsymbol{k}$ and different $\boldsymbol{q}$:

$$E(\boldsymbol{k}+\boldsymbol{q}') - E(\boldsymbol{k}+\boldsymbol{q}'') = \frac{\hbar^2}{2m}[2\boldsymbol{k}(\boldsymbol{q}'-\boldsymbol{q}'') + \boldsymbol{q}'^2 - \boldsymbol{q}''^2] \; . \tag{4.3.5}$$

The sign of (4.3.5) gives the order in which the states $\boldsymbol{k}$, $\boldsymbol{q}'$ and $\boldsymbol{k}$, $\boldsymbol{q}''$ are arranged by bands. If it is positive, the states with $\boldsymbol{q}'$ belong to the band whose number ζ is larger than that of a band with $\boldsymbol{q}''$, and vice versa. Equation (4.3.5) shows that for the same $\boldsymbol{q}'$ and $\boldsymbol{q}''$ but different $\boldsymbol{k}$ the sign of this expression may be different. This certainly does not mean that a muddle may arise in the classification of the states of a free electron, for, as long as (4.3.5) is nonzero, everything is clear. Simply and solely, the numbers of the bands may not be unambiguously related to some $\boldsymbol{q}$ from (4.3.2) (see below). A special case occurs when

$$2\boldsymbol{k}(\boldsymbol{q}'-\boldsymbol{q}'') + \boldsymbol{q}'^2 - \boldsymbol{q}''^2 = 0 \; . \tag{4.3.6}$$

For the particular $\boldsymbol{k}$ given, the system (4.2.16) has multiple energies. According to (4.3.6), those $\boldsymbol{k}$ which are special lie on the segments of the planes in $\boldsymbol{k}$.space (i.e., they are relatively small in number). If $\boldsymbol{k}$ pertains to one of these planes, the energy

$$E = \frac{\hbar^2}{2m}(\boldsymbol{k}+\boldsymbol{q}')^2 = \frac{\hbar^2}{2m}(\boldsymbol{k}+\boldsymbol{q}'')^2$$

has two states that cannot be unambiguously assigned to a band, since the bands touch one of the planes (4.3.6). The states that are infinitely close to those for which this "superdegeneracy" occurs are unambiguously in the bands ζ and $\zeta+1$. As for the "superdegenerate" states themselves, they may be arbitrarily assigned to both bands. The possibility that three or four bands may touch simultaneously cannot be dismissed, but this will occur only on straight lines or at special points in $\boldsymbol{k}$ space.

To illustrate, we begin by considering the case of a one-dimensional array having a period d and a reciprocal lattice with a period $2\pi/d$; the role of the vector $\boldsymbol{q}$ will be assumed by the quantity $2\pi\tilde{q}/d$ (with the $\tilde{q}$ being integers), and the role of $\boldsymbol{k}$ by the $\tilde{k}/d$ ratio, where $\tilde{k}$ satisfies (4.1.87). Then (4.3.2) will take the form

$$E(\tilde{k}+2\pi\tilde{q})=\frac{\hbar^2}{2md^2}(\tilde{k}+2\pi\tilde{q})^2\ ,\qquad \tilde{q}=0,\ \pm 1,\ \pm 2\ldots . \tag{4.3.7}$$

The lowest band in terms of E values corresponds to $\tilde{q}=0$. The second band, according to (4.3.7), consists of states with $-\pi<\tilde{k}\leqslant 0$ and $\tilde{q}=1$; $0<\tilde{k}\leqslant\pi$ and $\tilde{q}=-1$; etc. (Fig. 4.21). The energy dispersion curves that correspond to these bands are segments of a parabola.

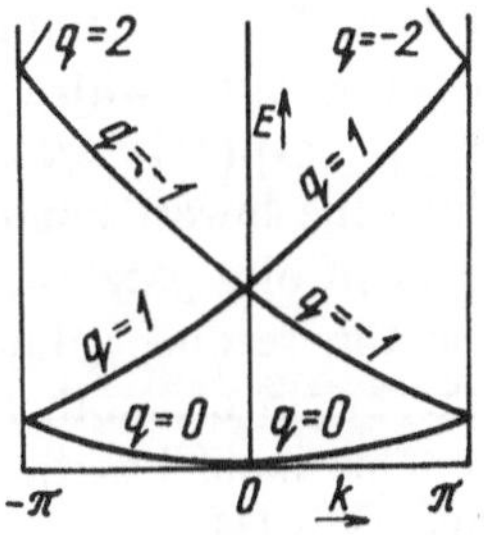

Fig. 4.21. Energy bands in the one-dimensional case in the empty-lattice model

This method is called the "empty-lattice" model [4.13, 14]. In the plot we clearly see the special significance of the points with $\tilde{k}=0$ (except for the lowest) and with $\tilde{k}=\pi$ (as already mentioned, the points with $\tilde{k}=-\pi$ need not be taken into account, and the entire drawing should be conceived of as rolled onto a cylinder of unit radius). With $\tilde{k}=0$ each even band touches a successive odd band, and when $\tilde{k}=\pi$, conversely, an odd band touches an even one. We arrive at the same result if we employ (4.3.6), which here has the form $\tilde{k}=-\pi(\tilde{q}'+\tilde{q}'')$. Because of the inequalities (4.1.87), we need only to take into account those $\tilde{k}$ for which $\tilde{q}'+\tilde{q}''=0$ or $\tilde{q}'+\tilde{q}''=-1$, as is evident from Fig. 4.21.

Now we generalize this for the case of a three-dimensional crystal, using the Brillouin zones. With an sc lattice the role of the vector $\boldsymbol{q}$ is played by the quantity $(2\pi/d)$ (q_1, q_2, q_3), where the q_i are integers, and the role of $\boldsymbol{k}$ by the quantity $(2\pi/d)$ $(\kappa_x, \kappa_y, \kappa_z)$, where κ_j $(j=x, y, z)$, are, by virtue of (4.2.22), bounded by the conditions

$$-\tfrac{1}{2}<\kappa_j\leqslant\tfrac{1}{2}\ . \tag{4.3.8}$$

Introducing the notation

$$\varepsilon=md^2E/2\pi^2\hbar^2\ , \tag{4.3.9}$$

we have, instead of (4.3.7),

$$\varepsilon_{kq} = (\kappa_x + q_1)^2 + (\kappa_y + q_2)^2 + (\kappa_z + q_3)^2 \ , \tag{4.3.10}$$

and the expression for the wave function is

$$\psi_{kq}(\boldsymbol{r}) = \exp\left\{\frac{2\pi \mathrm{i}}{d}[(\kappa_x + q_1)x + (\kappa_y + q_2)y + (\kappa_z + q_3)z]\right\} . \tag{4.3.11}$$

Now we wish to consider the variation of ε when k varies along the $\Delta\,[0, 0, 1]$ axis (Fig. 4.13a) between the symmetry points $\Gamma(0, 0, 0)$ and $X(0, 0, 1/2)$ and the line $\Delta(0, 0, \kappa_z)$. Therefore, we have on this axis $\varepsilon_\Gamma = q_1^2 + q_2^2 + q_3^2$, $\varepsilon_\Delta = q_1^2 + q_2^2 + (q_3 + \kappa_z)^2$ $(0 \leqslant \kappa_z < 1/2)$, $\varepsilon_X = q_1^2 + q_2^2 + (1/2 + q_3)^2$. Similar expressions are easy to obtain for any other points and axes. The minimum value of ε corresponds to the point Γ with $\boldsymbol{q} = (0, 0, 0)$. The energy curve with κ_z varying along the Δ axis has the form $\varepsilon_\Delta^{(1)} = \kappa_z^2$ and $\psi_{\kappa_z, 0} = \exp(2\pi \mathrm{i} \kappa_z z/d)$; the limiting energy values are $\varepsilon_\Gamma^{(1)} = 0$ and $\varepsilon_X^{(1)} = 1/4$. Thus, the lowest band of width $\Delta\varepsilon^{(1)} = 1/4$ in an sc lattice has the shape of a segment of a parabola (Fig. 4.22). For $\varepsilon_X^{(1)} = 1/4$ the vector $\boldsymbol{q}$ may also assume the second value $\boldsymbol{q} = (0, 0, -1)$. Therefore, the point $\varepsilon_X = 1/4$, $\kappa_z = 1/2$ is the origin of another band $\varepsilon_\Delta^{(2)} = (1 - \kappa_z)^2$ with $\psi_{\kappa_z, -1} = \exp[(2\pi \mathrm{i}(\kappa_z - 1)z/d)]$. This band has the shape of the next segment of parabola between the points $\varepsilon_X^{(2)} = 1/4$, $\kappa_z = 1/2$ and $\varepsilon_\Gamma^{(2)} = 1$, $\kappa_z = 0$ with the bandwidth $\Delta\varepsilon^{(2)} = 3/4$ (Fig. 4.22). To the upper end of this parabola correspond five more values of the vectors $\boldsymbol{q}$: $(0, 0, 1)$, $(0, \pm 1, 0)$, and $(\pm 1, 0, 0)$. This gives five extra energy bands (segments of parabola)

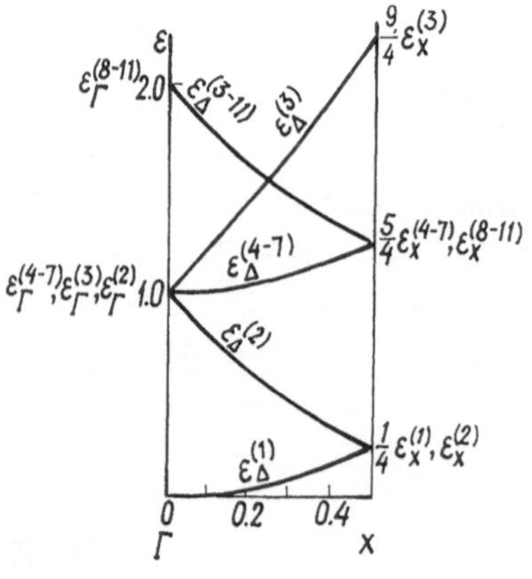

Fig. 4.22. Energy bands in the three-dimensional case in the sc lattice for the direction in $\boldsymbol{k}$ space along the Δ axis from point Γ to point X (see Fig. 4.13a); the superscript of ε_Δ in parentheses denotes simultaneously band number and band degeneracy

$$\varepsilon_{\Delta}^{(3)} = (1+\kappa_z)^2 \;, \quad \psi_{\kappa_z, q_3=1} = \exp\left[\frac{2\pi i}{d}(\kappa_z+1)z\right] \;,$$

$$\varepsilon_{\Delta}^{(4-7)} = 1+\kappa_z^2 \;, \quad \psi_{\kappa_z, q_1=\pm 1} = \exp\left[\frac{2\pi i}{d}(\pm x+\kappa_z z)\right] \;,$$

$$\psi_{\kappa_z, q_2=\pm 1} = \exp\left[\frac{2\pi i}{d}(\pm y+\kappa_z z)\right] \;.$$

We actually have not five but two bands, one, $\varepsilon_{\Delta}^{(3)}$ (just as $\varepsilon_{\Delta}^{(1)}$ and $\varepsilon_{\Delta}^{(2)}$), is nondegenerate, and the other, $\varepsilon_{\Delta}^{(4-7)}$, is fourfold degenerate (q_1, $q_2 = \pm 1$) (Fig. 4.22). When $\kappa_z = 1/2$ the upper point of the band $\varepsilon_{\Delta}^{(3)}$ has the energy $\varepsilon_X^{(3)} = 9/4$, and therefore its width $\Delta\varepsilon^{(3)} = \varepsilon_X^{(3)} - \varepsilon_\Gamma^{(3)} = 5/4$. With $\kappa_z = 1/2$ the respective quantities for the band $\varepsilon_{\Delta}^{(4-7)}$ are $\varepsilon_X^{(4-7)} = 5/4$ and $\Delta\varepsilon_{\Delta}^{(4-7)} = \varepsilon_X^{(4-7)} - \varepsilon_\Gamma^{(4-7)} = 1/4$. At the point $\varepsilon_X^{(4-7)}$, $\kappa_z = 1/2$, four more values of $\boldsymbol{q}$ are possible: $(\pm 1, 0, -1)$ and $(0, \pm 1, -1)$. The corresponding fourfold degenerate band will be $\varepsilon_{\Delta}^{(8-11)} = 1+(1-\kappa_z)^2$ (Fig. 4.22), and

$$\psi_{\kappa_z, q_1=\pm 1} = \exp\left\{\frac{2\pi i}{d}[\pm x+(\kappa_z-1)z]\right\} \;,$$

$$\psi_{\kappa_z, q_2=\pm 1} = \exp\left\{\frac{2\pi i}{d}[\pm y+(\kappa_z-1)z]\right\} \;.$$

This process of constructing bands in the free-electron energy spectrum may be carried on by exhaustively searching for all the possible values of $\boldsymbol{q}$. By contrast with the one-dimensional case, Fig. 4.22 shows that only the first two bands do not overlap; further we deal with an overlap that becomes complicated if we take into account the other senses of variation of $\boldsymbol{k}$.

Similar constructions may also be performed with other lattice types, for example, as illustrated in Fig. 4.23, with a bcc lattice for the direction Δ between symmetry points Γ and H. Note that using point symmetry we are in a position to classify the wave functions at certain points of the Brillouin zone [4.5].

The results obtained enable us to draw conclusions about the properties of those electrons which are qualitatively similar to free ones. This applies primarily to the width of their energy bands. According to (4.3.9), the order of magnitude is determined by the universal $\hbar^2/md^2$ relation; amounting to $\approx 10^{-12}$ erg or $\approx$1–10 eV, if we consider m to be the mass of a free electron ($\approx 10^{-28}$ g). Since, in the lattice, m is replaced by the effective mass m^*, which may be an order of magnitude larger or smaller than m, the $\Delta\varepsilon$, too, may differ from 10^{-12} erg by an order of magnitude in both senses.

Let us return, once again, to the problem of the dependence of the free-electron energy on the quasimomentum components inside a band for the three-dimensional case. As long as we are concerned with the first band, where,

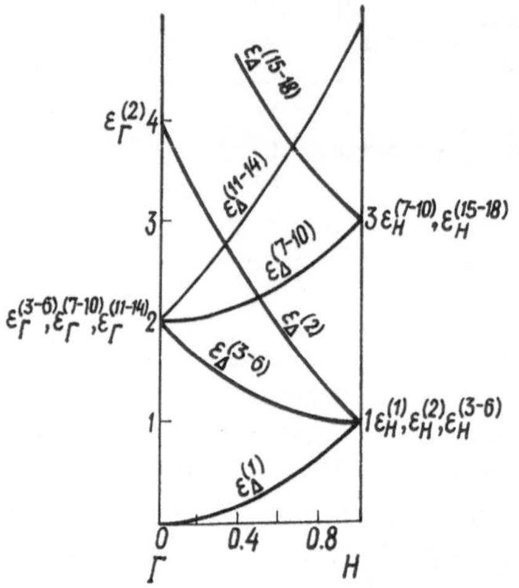

Fig. 4.23. Energy bands in the three-dimensional case in the bcc lattice in the empty-lattice model for the direction in $\boldsymbol{k}$ space along the Δ axis from point Γ to point H (see Fig. 4.14); the superscript of ε_Δ in parentheses denotes band number and band degeneracy

according to (4.3.2), $\boldsymbol{q}=0$ and the vectors $\boldsymbol{k}$ and ξ coincide, the energy is a monotonic function of $|\boldsymbol{k}|$. In the subsequent zones, as is seen from the examples in Figs. 4.22, 23, the situation becomes more complicated. It stands to reason that the constant energy surfaces, because of the quadratic dispersion relation (4.3.2), are made up of portions of spherical surfaces also in the bands, but for different $\boldsymbol{k}$ we have to take surfaces with different centers, leading to a complicated picture.

We wish to illustrate this with reference to the example of a plane square lattice. Figure 4.24 presents constant energy curves for the first three bands in a reduced scheme. In the first zone they are concentric circles (Fig. 4.24a). In the second zone (Fig. 4.24b) they are more complicated, pillowlike curves. The electron energy here may not be thought of as unambiguously determined by the modulus of the vector $\boldsymbol{k}$, for it depends also on the direction of this vector. True, here one may also talk of the monotonicity of $E(\boldsymbol{k})$ for states with the same direction of $\boldsymbol{k}$; the energy always decreases with increasing $|\boldsymbol{k}|$. In the third zone (Fig. 4.24c) this behavior persists no longer. For some directions of $\boldsymbol{k}$ the energy increases with increasing $|\boldsymbol{k}|$, whereas for other directions it decreases. Finally, there are directions of $\boldsymbol{k}$ for which the energy varies in a nonmonotonic fashion (for example, those which occur in the area enclosed between curves 4). Apart from this, if we take into account the effect of even a weak periodic field, the constant energy lines near the boundaries of the Brillouin zones depicted in Fig. 4.24 will undergo a distortion and cease to be segments of the circumference, as is shown by the curves in Figs. 4.24d, e.

This example shows that all assumptions as to the shape of the function $E(\boldsymbol{k}, \zeta)$ should be handled circumspectly. All these conclusions may be drawn merely on the basis of rigorous symmetry considerations. We will return to this subject in Sect. 4.6.

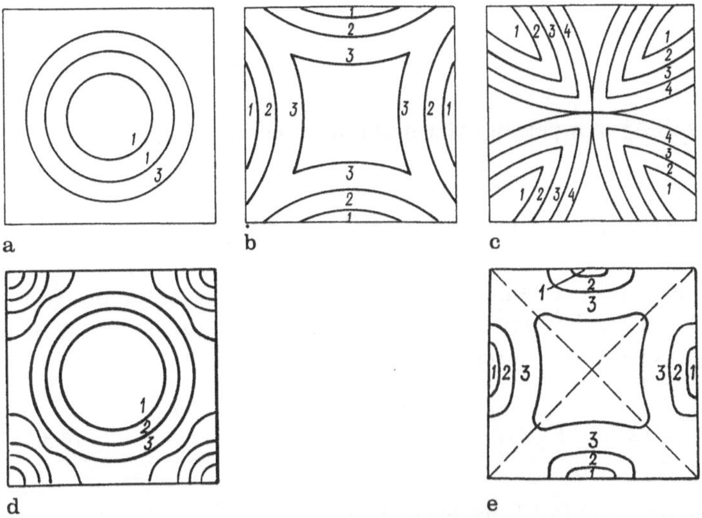

Fig. 4.24. Isoenergetic lines in the square plane of the lattice for the first three reduced Brillouin zones: first (**a**), second (**b**), third (**c**). The digits indicate the number of the isoenergetic surface in the order of increasing energy; (**d**), (**e**) demonstrate modifications of the isoenergetic curves due to the periodic potential in comparison with the empty-lattice model, cf. (**a**), (**b**)

4.3.3 Allowance for a Weak Periodic Field

We transcribe the fundamental equation (4.2.16) into a somewhat different form,

$$\left[E - \frac{\hbar^2}{2m}(\boldsymbol{k} + \boldsymbol{b}_g^*)^2\right] a(\boldsymbol{k}, \boldsymbol{b}_g^*) = \sum_{g'} V_{\boldsymbol{b}_g^* - \boldsymbol{b}_{g'}^*} a(\boldsymbol{k}, \boldsymbol{b}_{g'}^*) \ , \tag{4.3.12}$$

on the assumption that the $V_{\boldsymbol{b}_g^*} \neq 0$ but are small. Two cases will be considered in which the vectors $\xi = \boldsymbol{k} + \boldsymbol{b}_g^*$ are firstly well away from, and secondly close to the Brillouin planes. In the first case, (4.3.12) has no multiple eigenvalues in the zero approximation. We may assume that the energy (4.3.2) does not lie close to the other value, either. We try a solution of (4.3.12) in the form

$$\begin{aligned} a(\boldsymbol{k}, \boldsymbol{b}_g^*) &= \alpha\delta_{\boldsymbol{b}_g^* \boldsymbol{q}} + \alpha'_{\boldsymbol{b}_g^*} + \alpha''_{\boldsymbol{b}_g^*} + \dots \\ E &= \frac{\hbar^2}{2m}(\boldsymbol{k} + \boldsymbol{q})^2 + \varepsilon' + \varepsilon'' + \dots , \end{aligned} \tag{4.3.13}$$

where $\alpha', \alpha'', \varepsilon', \varepsilon''$, etc. are small. Substituting (4.3.13) into (4.3.12), we have in the first approximation

$$\frac{\hbar^2}{2m}[(\boldsymbol{k} + \boldsymbol{q})^2 - (\boldsymbol{k} + \boldsymbol{b}_g^*)^2]\alpha'_{\boldsymbol{b}_g^*} = (V_{\boldsymbol{b}_g^* - \boldsymbol{q}} - \varepsilon'\delta_{\boldsymbol{b}_g^* \boldsymbol{q}})\alpha \ .$$

The equation with $\boldsymbol{b}_g^* = \boldsymbol{q}$ yields

$$\varepsilon' = V_{000} \, . \tag{4.3.14}$$

The equations with $\boldsymbol{b}_g^* \neq \boldsymbol{q}$ permit determination of all $\alpha'_{\boldsymbol{b}_g^*}$:

$$\alpha'_{\boldsymbol{b}_g^*} = \frac{2m}{\hbar^2} V_{\boldsymbol{b}_g^* - \boldsymbol{q}} \frac{\alpha}{(\boldsymbol{k}+\boldsymbol{q})^2 - (\boldsymbol{k}+\boldsymbol{b}_g^*)^2} \, . \tag{4.3.15}$$

The undefined quantity α is found from the normalization condition. Comparison of (4.2.13 and 31) shows that to satisfy the normalization condition with respect to k_1, k_2, k_3, we need to set

$$\sum_g {}^* a(\boldsymbol{k}, \boldsymbol{b}_g^*)|^2 = (2\pi)^{-3} V_0^{-1} \, . \tag{4.3.16}$$

In the zero approximation (4.3.1) we hence have

$$\alpha = (2\pi)^{-3/2} V_0^{-1/2} \, . \tag{4.3.17}$$

This normalization persists also in the first approximation. Just as in the general theory, we have, according to (4.3.13),

$$\alpha'_{\boldsymbol{q}} = 0 \, . \tag{4.3.18}$$

Thus, the solution in the first approximation has been found. Equations (4.3.15) and (4.2.15) give

$$\psi_{\boldsymbol{k}}(\boldsymbol{r}) = (2\pi)^{-3/2} V_0^{-1/2} \exp[\mathrm{i}(\boldsymbol{k}+\boldsymbol{q})\cdot\boldsymbol{r}] \times \left\{ 1 + \frac{2m}{\hbar^2} \sum_g {}^* V_{\boldsymbol{b}_g^*} \exp(\mathrm{i}\boldsymbol{b}_g^*\cdot\boldsymbol{r}) \Big/ [(\boldsymbol{k}+\boldsymbol{q})^2 - (\boldsymbol{k}+\boldsymbol{q}+\boldsymbol{b}_g^*)^2] \right\} , \tag{4.3.19}$$

where the asterisk next to $\sum$ denotes that the term 0, 0, 0 is eliminated from the sum.

Formula (4.3.14), in which $V_{000} = \text{const}$ or 0, shows that a weak periodic field in the first approximation exerts no influence on the energy if $\boldsymbol{k}$ is not close to the boundary of the zone. Even in the first approximation, a modulating factor arises for (4.3.19). This enables us to ascertain the exact meaning of the concept of a nearly-free electron: the concept is applicable as long as

$$|V_{\boldsymbol{b}_g^*}| \ll |E(\boldsymbol{\xi} + \boldsymbol{b}_g^*) - E(\boldsymbol{\xi})| \tag{4.3.20}$$

for the Fourier transforms of the potential with all $\boldsymbol{b}_g^*$. Equation (4.3.20) is not satisfied even for small $V_{b_g^*}$, if at least for one $\boldsymbol{b}_g^*$, $E(\boldsymbol{\xi}+\boldsymbol{b}_g^*)$ lies close to $E(\boldsymbol{\xi})$; i.e., the end of the vector $\boldsymbol{\xi}$ lies close to the zone boundary. If this is not the case, the right-hand side of (4.3.20), as pointed out above, $\hbar^2/md^2$, i.e., ≈ 1–10 eV. Therefore, if $|V_{b_g^*}|$ does not exceed 0.1–0.01 eV, the electron behaves almost as a free one. This estimate of $|V_{b_g^*}|$ appears implausible at first sight; $V(\boldsymbol{r})$ is the potential of ions with a deep and narrow well at the lattice sites. For many $\boldsymbol{b}_g^*$ (i.e., those which largely exceed the size of the first Brillouin zone), the Fourier transforms of $V(\boldsymbol{r})$ will therefore be not small. But the approximation works well in practice, for it is actually not the potential of the ions but a much smaller potential that acts on the electron (Sect. 4.6).

Now we consider the second case, in which the end of the vector $\boldsymbol{k}$ is close to the Brillouin plane, i.e.,

$$\boldsymbol{k}(\boldsymbol{q}'-\boldsymbol{q}'')+\tfrac{1}{2}(\boldsymbol{q}'^2-\boldsymbol{q}''^2)=0 \tag{4.3.21}$$

for the particular values of $\boldsymbol{q}'$ and $\boldsymbol{q}''$ given. Then the unperturbed energies $E(\boldsymbol{k}+\boldsymbol{q}')$ and $E(\boldsymbol{k}+\boldsymbol{q}'')$ are close, and if the end of $\boldsymbol{k}$ lies at the boundary of the zone, equal in magnitude. We wish to take into account the perturbation of these "nearly degenerate" states $(\boldsymbol{k},\boldsymbol{q}')$ and $(\boldsymbol{k},\boldsymbol{q}'')$.

Try the solution of (4.3.12) in the form

$$a(\boldsymbol{k},\boldsymbol{b}_g^*)=\alpha\delta_{\boldsymbol{b}_g^*\boldsymbol{q}'}+\beta\delta_{\boldsymbol{b}_g^*\boldsymbol{q}''}+\alpha'_{\boldsymbol{b}_g^*}+\alpha''_{\boldsymbol{b}_g^*}+\dots, \tag{4.3.22}$$

where the corrections $\alpha', \alpha'', \dots$ may be regarded as small compared to α and β, which are of the same order of magnitude. Next, we substitute (4.3.22) into (4.3.12) and pay attention to those equations in which $\boldsymbol{b}_g^*=\boldsymbol{q}'$ and $\boldsymbol{b}_g^*=\boldsymbol{q}''$. If we throw away all the terms of second and higher orders, then

$$\begin{aligned}&[E-E(\boldsymbol{k}+\boldsymbol{q}')]\alpha-V_{\boldsymbol{q}'-\boldsymbol{q}''}\beta=0\ ,\\ &V_{\boldsymbol{q}''-\boldsymbol{q}'}\alpha-[E-E(\boldsymbol{k}+\boldsymbol{q}'')]\beta=0\end{aligned} \tag{4.3.23}$$

(with $V_{000}=0$). The solutions of the homogeneous system (4.3.23) are nonzero if their determinant is equal to zero. Allowing for (4.2.3), the result for the electron energy in the crystal will be

$$\begin{aligned}E_\pm(\boldsymbol{k})=&\frac{E(\boldsymbol{k}+\boldsymbol{q}')+E(\boldsymbol{k}+\boldsymbol{q}'')}{2}\\ &\pm\left[\frac{[E(\boldsymbol{k}+\boldsymbol{q}')-E(\boldsymbol{k}+\boldsymbol{q}'')]^2}{4}+|V_{\boldsymbol{q}'-\boldsymbol{q}''}|^2\right]^{1/2}.\end{aligned} \tag{4.3.24}$$

Thus, the difference between the two perturbed energy levels, corresponding to the unperturbed $E(\boldsymbol{k}+\boldsymbol{q}')$ and $E(\boldsymbol{k}+\boldsymbol{q}'')$, remains finite, however small the difference between the latter may be. Even if the end of the vector $\boldsymbol{k}$

lies at the boundary of the zone and the unperturbed level is degenerate—$E(\boldsymbol{k}+\boldsymbol{q}') = E(\boldsymbol{k}+\boldsymbol{q}'') = E_0$—(4.3.24) yields two perturbed levels

$$E_{\pm} = E_0 \pm |V_{q'-q''}| , \tag{4.3.25}$$

which for $|V_{q'-q''}| \neq 0$ are always different from each other. Therefore, no matter how weak the periodic potential may be, it removes the "superdegeneracy" when the bands touch (see above), and in the first approximation we have an energy gap (4.3.25).

Consider (4.3.23, 24), i.e., the situation in which the end of $\boldsymbol{k}$ does not lie exactly on the plane of the zone boundary (4.3.21). Imagine that a perpendicular is dropped from the end of $\boldsymbol{k}$ to this plane. Designate the new vector (small under the statement of the problem) by $\boldsymbol{\eta}$. Further, draw a vector $\boldsymbol{K}_0$ from the origin to the end of $\boldsymbol{\eta}$ and denote the relevant energy $E(\boldsymbol{K}_0+\boldsymbol{q}') = E(\boldsymbol{K}_0+\boldsymbol{q}'') = E_0$. It is clear that

$$\boldsymbol{k} = \boldsymbol{K}_0 - \boldsymbol{\eta} . \tag{4.3.26}$$

Hence

$$E(\boldsymbol{k}+\boldsymbol{q}') = E_0 - \frac{\hbar^2}{m}(\boldsymbol{K}_0+\boldsymbol{q}')\cdot\boldsymbol{\eta} + \frac{\hbar^2}{2m}\eta^2 , \tag{4.3.27}$$

$$E(\boldsymbol{k}+\boldsymbol{q}'') = E_0 - \frac{\hbar^2}{m}(\boldsymbol{K}_0+\boldsymbol{q}'')\cdot\boldsymbol{\eta} + \frac{\hbar^2}{2m}\eta^2 , \tag{4.3.28}$$

and, consequently,

$$\tfrac{1}{2}[E(\boldsymbol{k}+\boldsymbol{q}') + E(\boldsymbol{k}+\boldsymbol{q}'')] = E_0 - \frac{\hbar^2}{2m}(2\boldsymbol{K}_0+\boldsymbol{q}'+\boldsymbol{q}'')\cdot\boldsymbol{\eta} + \frac{\hbar^2}{2m}\eta^2 ,$$

$$E(\boldsymbol{k}+\boldsymbol{q}') - E(\boldsymbol{k}+\boldsymbol{q}'') = -\frac{\hbar^2}{m}(\boldsymbol{q}'-\boldsymbol{q}'')\cdot\boldsymbol{\eta} .$$

The vector $\boldsymbol{\eta}$ is normal to the plane (4.3.23) and parallel to the vector $\boldsymbol{q}'-\boldsymbol{q}''$, and the vector $2\boldsymbol{K}_0+\boldsymbol{q}'+\boldsymbol{q}''$ is normal to $\boldsymbol{q}'-\boldsymbol{q}''$ [since (4.3.23), which $\boldsymbol{K}_0$ should satisfy, may be written as $(2\boldsymbol{K}_0+\boldsymbol{q}'+\boldsymbol{q}'')\cdot(\boldsymbol{q}'-\boldsymbol{q}'') = 0$]. Then we obtain

$$\tfrac{1}{2}[E(\boldsymbol{k}+\boldsymbol{q}') + E(\boldsymbol{k}+\boldsymbol{q}'')] = E_0 + \frac{\hbar^2}{2m}\eta^2 , \tag{4.3.29}$$

$$E(\boldsymbol{k}+\boldsymbol{q}') - E(\boldsymbol{k}+\boldsymbol{q}'') = \pm\frac{\hbar^2}{m}|\boldsymbol{q}'-\boldsymbol{q}''||\boldsymbol{\eta}| . \tag{4.3.30}$$

Substituting this into (4.3.24), we obtain the net result

$$E_\pm = E_0 + \frac{\hbar^2 \eta^2}{2m} \pm \left[\frac{\hbar^4}{4m^2} (\boldsymbol{q}' - \boldsymbol{q}'')^2 \eta^2 + |V_{\boldsymbol{q}' - \boldsymbol{q}''}|^2 \right]^{1/2} . \tag{4.3.31}$$

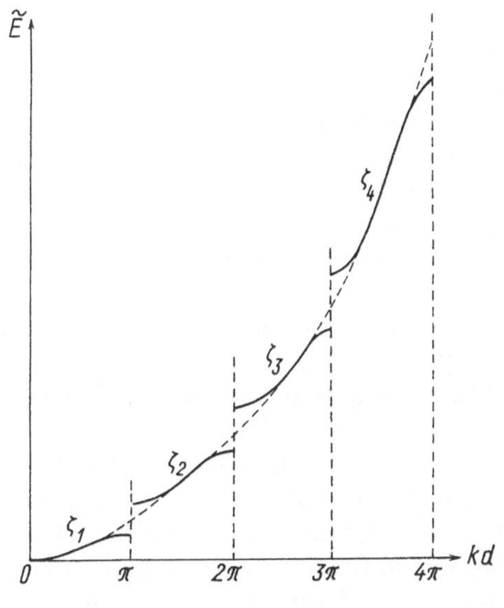

Fig. 4.25. Dependence of energy $\tilde{E} = E(8md^2/\hbar^2)$ on the quasimomentum $\boldsymbol{k}$ (wave vector) of the band electron near the Brillouin zone boundary. The first four zones $\zeta_1, \zeta_2, \zeta_3, \zeta_4$ and the positive values of the quasimomentum are indicated

This dependence is portrayed schematically in Fig. 4.25. For $\eta = 0$, (4.3.29) goes into (4.3.25). An essential fact, not immediately seen from (4.3.24), is manifest in (4.3.31). The vector $\boldsymbol{K}_0$ and the energy E_0 do not vary if we displace the end of $\boldsymbol{k}$ along the normal to the boundary of the zone; i.e., the quantity E_0 is determined only by the projection of $\boldsymbol{k}$ to the boundary plane. It is only through η that the perturbed energy depends on the normal component $\boldsymbol{k}$ to this plane. However, the quantity η involved in (4.3.31) is quadratic [here lies the fundamental difference of (4.3.31) from (4.3.27, 28) for an unperturbed problem]. Therefore, the normal component of the energy gradient is, by (4.3.31), equal to

$$\begin{aligned} \boldsymbol{n}\,\nabla_k E_\pm = \frac{\partial E_\pm}{\partial |\eta|} = \frac{\hbar^2}{m} |\eta| \Bigg\{ 1 + \frac{\hbar^2}{m^2} |\boldsymbol{q}' - \boldsymbol{q}''|^2 \\ \times \left[\frac{\hbar^2}{m^2} (\boldsymbol{q}' - \boldsymbol{q}'')^2 \eta^2 + 4 |V_{\boldsymbol{q}' - \boldsymbol{q}''}|^2 \right]^{-1/2} \Bigg\} \end{aligned} \tag{4.3.32}$$

also for $\eta = 0$ and $\boldsymbol{n} \cdot \nabla_k E_\pm = 0$; i.e., at the boundary (4.3.21) $\nabla_k E$ lies in its plane, and the energy surface $E = \text{const}$, constructed according to (4.3.30), is normal to (4.3.21).

This derivation vividly demonstrates the effect of a periodic field on the energy spectrum. Far from the planes (4.3.21), the energy surfaces do not alter in the first approximation. Before the field is turned on, the constant energy surfaces of the zone for the planes (4.3.21) touch their counterparts from the neighboring zones. With the field on, a finite discontinuity arises between these surfaces of equal energy; they fold in such a way that they approach the plane on which they touched earlier and have a normal tangent to that plane. This is particularly evident in the one-dimensional representation (for some one direction in $\boldsymbol{k}$ space), as is clearly indicated by the dashed segments of the curve (parabola) in Fig. 4.25. For the values of $k = \pi n/d$ ($n = 1, 2, 3, \ldots$), where we earlier dealt with the touching of adjacent bands (Fig. 4.18), we now have the divergence of the energy curves for adjacent zones in different directions, which leads to energy gaps. At the points of discontinuity, the energy curve approaches the straight lines parallel to the ordinate axis with the tangent parallel to the abscissa axis.

According to (4.2.44), expression (4.3.32) defines the component of the mean electron velocity with respect to the normal to the zone boundary plane: $\langle \boldsymbol{v}(\boldsymbol{k}, \zeta)\rangle_n$. Thus, the mean velocity component goes to zero on these planes. This comes from the "vanishing-of-current theorem" of the rigorous theory (Sect. 4.2.3). For the one-dimensional case this is an exact relation (4.1.112). In the three-dimensional problem, the condition $\langle \boldsymbol{v}(\boldsymbol{k}, \zeta)\rangle_n = 0$ on all the planes (4.3.21) is a much more severe statement than that made at the end of Sect. 4.2.3. In this drastic form, however, it is valid only in the first approximation, since, per se, the planes (4.3.21) appear in the theory if what we take as the zero approximation is the free electron. These planes may always be drawn, but there are no grounds for asserting that $\langle \boldsymbol{v}(\boldsymbol{k}, \zeta)\rangle_n$ on them is always equal to zero in exact theory, too. The vanishing of $\langle \boldsymbol{v}(\boldsymbol{k}, \zeta)\rangle_n$ stems from the fact that in the vicinity of the planes (4.3.21) the wave function in the potential field undergoes substantial changes, which can be traced with the aid of (4.3.23).

Let us assume, for illustration, that $E(\boldsymbol{k} + \boldsymbol{q}') > E(\boldsymbol{k} + \boldsymbol{q}'')$. Therefore the sign in (4.3.30) should be positive, and we find from (4.3.23) with (4.3.24, 29, 30)

$$\beta = V_{\boldsymbol{q}'' - \boldsymbol{q}'} \left\{ \frac{\hbar^2}{2m} |\boldsymbol{q}' - \boldsymbol{q}''||\boldsymbol{\eta}| \pm \left[\frac{\hbar^2}{4m^2} (\boldsymbol{q}' - \boldsymbol{q}'')^2 \boldsymbol{\eta}^2 + |V_{\boldsymbol{q}' - \boldsymbol{q}''}|^2 \right]^{1/2} \right\}^{-1} \alpha \,. \tag{4.3.33}$$

We see that as long as

$$|E(\boldsymbol{k} + \boldsymbol{q}') - E(\boldsymbol{k} + \boldsymbol{q}'')| = \frac{\hbar^2}{m} |\boldsymbol{q}' - \boldsymbol{q}''||\boldsymbol{\eta}| \gg 2 |V_{\boldsymbol{q}' - \boldsymbol{q}''}| \,, \tag{4.3.34}$$

either $|\beta/\alpha| \gg 1$ or $|\alpha/\beta| \gg 1$; i.e., the linear combination $\alpha\delta_{\boldsymbol{b}_g^*\boldsymbol{q}'} + \beta\delta_{\boldsymbol{b}_g^*\boldsymbol{q}''}$ is respectively close to one of the unperturbed values $\alpha\delta_{\boldsymbol{b}_g^*\boldsymbol{q}'}$ or $\beta\delta_{\boldsymbol{b}_g^*\boldsymbol{q}''}$. This is what corresponds to the assumption that there is no degeneracy (4.3.20, 34); Equation (4.3.24) also yields the unperturbed energy values. In the opposite case

$$\frac{\hbar^2}{m}|\boldsymbol{q}'-\boldsymbol{q}''||\boldsymbol{\eta}| \ll 2|V_{\boldsymbol{q}'-\boldsymbol{q}''}| \,, \tag{4.3.35}$$

the amplitudes α and β, by (4.3.33), are of the same order of magnitude; i.e., the wave function turns out to be substantially different from that of a travelling wave, (4.3.33), and is a superposition of two travelling waves whose amplitudes are of the same order of magnitude. This is particularly true for $\eta = 0$, when (4.3.33) has the form

$$\beta = \pm\alpha V_{\boldsymbol{q}''-\boldsymbol{q}'}/|V_{\boldsymbol{q}''-\boldsymbol{q}'}| \,, \tag{4.3.36}$$

i.e., $|\alpha| = |\beta|$. Introduce the notation

$$V_{\boldsymbol{b}_g^*} = |V_{\boldsymbol{b}_g^*}|\exp(\mathrm{i}\varphi_{\boldsymbol{b}_g^*}) \,, \tag{4.3.37}$$

so that, according to (4.3.22) (with neglect of all the small corrections α', α'', etc.), we then have in the zero approximation

$$a(\boldsymbol{k},\boldsymbol{b}_g^*) = \alpha[\delta_{\boldsymbol{b}_g^*\boldsymbol{q}'} \pm \exp(-\mathrm{i}\varphi_{\boldsymbol{q}'-\boldsymbol{q}''})\delta_{\boldsymbol{b}_g^*\boldsymbol{q}''}] \,, \tag{4.3.38}$$

or, proceeding according to (4.2.13) to the usual expression for the wave function,

$$\psi_{\boldsymbol{k}}(\boldsymbol{r}) = \alpha\{\exp[\mathrm{i}(\boldsymbol{K}_0+\boldsymbol{q}')\cdot\boldsymbol{r}] \pm \exp[-\mathrm{i}\varphi_{\boldsymbol{q}'-\boldsymbol{q}''} + \mathrm{i}(\boldsymbol{K}_0+\boldsymbol{q}'')\cdot\boldsymbol{r}]\} \,. \tag{4.3.39}$$

The subscript 0 of $\boldsymbol{K}_0$ here emphasizes that (4.3.39) holds for the exact equality $E(\boldsymbol{K}_0+\boldsymbol{q}') = E(\boldsymbol{K}_0+\boldsymbol{q}'')$. From (4.3.39) it follows that in the state described by this function the velocity component, normal to the plane (4.3.21), is equal to zero. In addition, as an exercise, the reader can obtain from (4.3.39) a number of derivations of the exact one-dimensional problem treated in Sect. 4.1.4. In (4.3.39) the vectors $\boldsymbol{K}_0+\boldsymbol{q}'$ and $\boldsymbol{K}_0+\boldsymbol{q}''$ are equal in magnitude, and the vector of their difference $\boldsymbol{q}'-\boldsymbol{q}''$ is normal to (4.3.21). Therefore the unit vectors $(\boldsymbol{K}_0+\boldsymbol{q}')/|\boldsymbol{K}_0+\boldsymbol{q}'|$ and $(\boldsymbol{K}_0+\boldsymbol{q}'')/|\boldsymbol{K}_0+\boldsymbol{q}''|$ are interrelated as the unit vectors of an incident beam and a beam reflected from the plane (4.3.39). Thus the state described by one of the functions (4.3.39) may be considered as a reflection of the travelling wave from this plane with a phase equal to the difference of the incident and reflected waves. To some extent, this result may be viewed as the vindication of the classical "wave" treatment in Sect. 4.1.1. We should, however, bear in mind that the form of ψ (4.3.39) has been obtained in the approximation of the nearly-free-electron problem and this purely wave form is by no means obligatory for the exact theory.

We could go further and find successive approximations for ψ and E. But this would be pointless, for no qualitatively new result would be obtained.

It would be interesting now to consider the opposite limiting case of strong-coupled electrons, but we wish to defer this discussion until Sect. 4.6.3.

In Sect. 4.3 we noted the possibility of some compensation for the effects which the ion cores and other band electrons have on a given band electron. As a result of this compensation, the total effective periodic potential has a small amplitude. The fairly good agreement between the conclusions of the band model in the nearly-free-electron approximation and the experimental results provides evidence that this actually occurs, at least in normal metals. The above approximation may therefore be regarded not as an abstract illustration of the laws of the quantum motion of electrons in a periodic potential field but as a good description of the real situation in crystalline conductors. In all appearance, two chief physical causes may be pointed out to explain why strong interactions between band electrons, on the one hand, and ion cores and other band electrons, on the other, actually reduce the potential to a weak effective periodic potential (now often referred to as the pseudopotential, Sect. 4.6.4). First, the Pauli principle comes into force here, preventing band electrons from being located close to ion cores, where core shell electrons are available. Second, other band electrons, owing to their high mobility, may cause the positive potential acting on a particular band electron to decrease appreciably due to a substantial screening effect. These two important circumstances, occurring in the ion–electron system of solids, will be discussed in greater detail in Sect. 4.6 and Chap. 5.

4.4 Effect of an Electric Field on Electronic States

4.4.1 Acceleration and Effective Electron Mass

Computing the various characteristics of solids requires a knowledge of the influence which an externally applied electric and magnetic field exert on an electron. Specifically, in order to account for the existence of metals and nonmetals and to formulate exactly the concept of the conduction electron, we must consider its acceleration in an electric field. Let us assume that at $t < 0$ the electron is in the band eigenstate

$$\psi_{\boldsymbol{k}\zeta}(\boldsymbol{r}, t) = \psi_{\boldsymbol{k}\zeta}(\boldsymbol{r}) \exp\left[-\frac{\mathrm{i}}{\hbar} E(\boldsymbol{k}, \zeta) t\right] , \tag{4.4.1}$$

we now turn on a constant uniform electric field $\boldsymbol{F}$ with a scalar potential $-\boldsymbol{F} \cdot \boldsymbol{r}$. The change of state of the electron is described by the time-dependent Schrödinger equation

$$i\hbar \frac{\partial \psi(\boldsymbol{r}, t)}{\partial t} = (\hat{\mathscr{H}}_{\text{crys}} - e\boldsymbol{F}\cdot\boldsymbol{r})\psi(\boldsymbol{r}, t) , \tag{4.4.2}$$

where $\hat{\mathscr{H}}_{\text{crys}}$ has been defined according to (4.2.4). We seek $\psi(\boldsymbol{r}, t)$ as an expansion in the stationary states of the unperturbed problem (4.4.1)

$$\psi(\boldsymbol{r}, t) = \sum_{\boldsymbol{k}'\zeta'} \psi_{\boldsymbol{k}'\zeta'}(\boldsymbol{r}, t)\alpha_{\boldsymbol{k}'\zeta'}(t) . \tag{4.4.3}$$

Substituting (4.4.3) into (4.4.2), we obtain

$$\begin{aligned} &\sum_{\boldsymbol{k}'\zeta'} \psi_{\boldsymbol{k}'\zeta'}(\boldsymbol{r})\exp\left[-\frac{i}{\hbar}E(\boldsymbol{k}', \zeta')t\right] \\ &\quad\times\left[E(\boldsymbol{k}', \zeta') + i\hbar\frac{\partial}{\partial t}\right]\alpha_{\boldsymbol{k}'\zeta'}(t) = \sum_{\boldsymbol{k}'\zeta'} \alpha_{\boldsymbol{k}'\zeta'}(t)[E(\boldsymbol{k}', \zeta') \\ &\quad - e\boldsymbol{F}\cdot\boldsymbol{r}]\psi_{\boldsymbol{k}'\zeta'}(\boldsymbol{r})\exp\left[-\frac{i}{\hbar}E(\boldsymbol{k}', \zeta')t\right] . \end{aligned} \tag{4.4.4}$$

Multiplying (4.4.4) by $\psi^*_{\boldsymbol{k}\zeta}(\boldsymbol{r})$ and performing the integration over $d\boldsymbol{r}$, we obtain, with allowance for the orthogonality property (4.2.30),

$$\begin{aligned} \frac{\partial \alpha_{\boldsymbol{k}\zeta}(t)}{\partial t} &= \frac{ie\boldsymbol{F}}{\hbar}\cdot \sum_{\boldsymbol{k}'\zeta'} \exp\left\{\frac{i}{\hbar}[E(\boldsymbol{k}, \zeta) - E(\boldsymbol{k}', \zeta')]t\right\} \\ &\quad\times (\boldsymbol{k}\zeta|\hat{\boldsymbol{r}}|\boldsymbol{k}'\zeta')\alpha_{\boldsymbol{k}'\zeta'}(t) . \end{aligned} \tag{4.4.5}$$

Substituting the coordinate matrix elements (4.2.35, 36) into (4.4.5) and transposing the term with $\hat{\boldsymbol{r}}_1$ into the left-hand side, we find [4.15]

$$\begin{aligned} \frac{\partial \alpha_{\boldsymbol{k}\zeta}(t)}{\partial t} + \frac{e\boldsymbol{F}}{\hbar}\cdot\frac{\partial \alpha_{\boldsymbol{k}\zeta}(t)}{\partial \boldsymbol{k}} &= -\frac{e}{\hbar}\boldsymbol{F}\sum_{\zeta'} \exp\left\{\frac{i}{\hbar}[E(\boldsymbol{k}, \zeta) - E(\boldsymbol{k}, \zeta')]t\right\}\alpha_{\boldsymbol{k}\zeta'}(t) \\ &\quad\times \int d\boldsymbol{r}\, u^*_{\boldsymbol{k}\zeta}(\boldsymbol{r})\, \nabla_{\boldsymbol{k}} u_{\boldsymbol{k}\zeta}(\boldsymbol{r}) . \end{aligned} \tag{4.4.6}$$

The term with $\zeta' = \zeta$ may be excluded by the replacement

$$\frac{1}{\hbar}E(\boldsymbol{k}, \zeta) \to \frac{1}{\hbar}E(\boldsymbol{k}, \zeta) - \frac{ie\boldsymbol{F}}{\hbar}\cdot\int d\boldsymbol{r}\, u^*_{\boldsymbol{k}\zeta}(\boldsymbol{r})\, \nabla_{\boldsymbol{k}} u_{\boldsymbol{k}\zeta}(\boldsymbol{r}) \tag{4.4.7}$$

or

$$\alpha_{\boldsymbol{k}\zeta} \to \alpha_{\boldsymbol{k}\zeta}\exp\left[-\frac{e\boldsymbol{F}t}{\hbar}\cdot\int d\boldsymbol{r}\, u^*_{\boldsymbol{k}\zeta}(\boldsymbol{r})\, \nabla_{\boldsymbol{k}} u_{\boldsymbol{k}\zeta}(\boldsymbol{r})\right] .$$

For all attainable fields $\boldsymbol{F}$ the second term involved in (4.4.7) is small and may be thrown away, for it does not lead to substantial effects. This approximation boils down to the necessity of excluding the term with $\zeta' = \zeta$ in the sum over ζ' in (4.4.6). At the outset we neglect the interband transitions, replacing the right-hand side of (4.4.6) by zero (it will be proved in Sect. 4.4.2 that their contribution is negligibly small). A solution to the Cauchy problem for the equation

$$\left(\frac{\partial}{\partial t} + \frac{e\boldsymbol{F}}{\hbar}\cdot\frac{\partial}{\partial \boldsymbol{k}}\right)\alpha_{\boldsymbol{k}\zeta}(t) = 0 \tag{4.4.8}$$

is the function

$$\alpha_{\boldsymbol{k}\zeta}(t) = \alpha_{\boldsymbol{k} - e\boldsymbol{F}t/\hbar, \zeta}(0) \ . \tag{4.4.9}$$

If, with $t = 0$, the system was in the state (4.4.1), then, according to (4.4.3, 9),

$$\psi(\boldsymbol{r}, t) = \sum_{\boldsymbol{k}'} \psi_{\boldsymbol{k}'\zeta}(\boldsymbol{r}, t)\delta_{\boldsymbol{k}, \boldsymbol{k}' - e\boldsymbol{F}t/\hbar} = \psi_{\boldsymbol{k}(t)\zeta}(\boldsymbol{r}, 0) \ , \quad \boldsymbol{k}(t) = \boldsymbol{k} + \frac{e\boldsymbol{F}t}{\hbar} \ . \tag{4.4.10}$$

We calculate the mean value of the velocity operator in the state (4.4.10). Allowing for (4.2.38), we find

$$\langle \boldsymbol{v}(\boldsymbol{k}\zeta, t)\rangle = (\boldsymbol{k}(t)\zeta|\hat{\boldsymbol{v}}|\boldsymbol{k}(t)\zeta) = \frac{1}{\hbar}\frac{\partial E(\boldsymbol{k}(t), \zeta)}{\partial \boldsymbol{k}} \ . \tag{4.4.11}$$

From (4.4.10, 11) we determine the mean acceleration

$$a_i(\boldsymbol{k}\zeta, t) = \frac{dv_i(\boldsymbol{k}\zeta, t)}{dt} = \sum_j m_{ij}^{-1}(\boldsymbol{k}(t), \zeta)e\boldsymbol{F}_j \ , \tag{4.4.12}$$

where $i, j = x, y, z$ and we have introduced the tensor for effective inverse masses

$$m_{ij}^{-1}(\boldsymbol{k}, \zeta) = \frac{1}{\hbar^2}\frac{\partial^2 E(\boldsymbol{k}, \zeta)}{\partial k_i \partial k_j} \ . \tag{4.4.13}$$

For a free electron $E(\boldsymbol{k}) = \hbar^2\boldsymbol{k}^2/2m$, $m_{ij}^{-1} = \delta_{ij}/m$ [which justifies the name of (4.4.13)]. From (4.4.12) we obtain Newton's law

$$m\boldsymbol{a} = e\boldsymbol{F} \ . \tag{4.4.14}$$

Generally speaking, the tensor (4.4.13) is not a multiple of the unit tensor, and therefore the acceleration of the electron in the crystal is not necessarily aligned with the field. It becomes aligned when the field is oriented along one of the principal axes of the tensor (4.4.13). However, the factor of proportionality between acceleration and force [eigenvalue of the tensor (4.4.13) m_α^{-1}] is not necessarily positive. Near $E(\boldsymbol{k}, \zeta)_{\min}$ all $m_\alpha > 0$, and near $E(\boldsymbol{k}, \zeta)_{\max}$ all $m_\alpha < 0$. In the latter case the electron is accelerated not against but along the field; i.e., it will behave like a positively charged particle. As noted above, these states are referred to as hole states, and band theory has thereby resolved the Hall effect "disaster" (Chap. 3).

Let us assume that the electron at $t=0$ has a quasimomentum $\hbar\boldsymbol{k}$ near the bottom of the band and is accelerated against the field. The quasimomentum of the electron varies according to (4.4.10) and, after a certain time span, emerges in the region of the top of the band; as this takes place, the sign of the acceleration is reversed. As a result, by contrast with a free electron, an electron in a crystal, when subject to a constant electric field, performs oscillation, so that its velocity and, consequently, the current, oscillate [4.16]. This is a result of (4.2.5) and the periodicity of the $E(\boldsymbol{k}, \zeta)$ function. As an example, we examine an sc lattice of period d with the field aligned along one of the translation vectors (x axis). The period t_0 of the oscillations in field F will then be determined by

$$k_x(t_0) = k_x + |e| F t_0/\hbar = k_x + 2\pi/d \ ,$$

i.e.,

$$t_0 = \frac{2\pi\hbar}{|e| F d} \ , \tag{4.4.15}$$

since the variation of k_x by $2\pi/d$ denotes a return to the previous state. The typical field value for metals being $F \approx 10^{-6}$ V/cm, (4.4.15) yields $t_0 \approx 1$ s, which is many orders of magnitude larger than the mean free time in the purest samples. This is because the periodicity of electron motion in an electric field in metals will be completely distorted by collisions. As a result we can assume quite accurately that the acceleration of an electron in a metal is constant, although current-density fluctuations could probably be observed in semiconductors.

4.4.2 Zener Breakdown

To consider the role of the interband transitions on the right-hand side of (4.4.6), which were disregarded in the foregoing, we now proceed to a somewhat different basis for the wave-function expansion (4.4.3), namely,

$$\psi(\boldsymbol{r}, t) = \sum_{\boldsymbol{k}'\zeta'} \tilde{\alpha}_{\boldsymbol{k}'\zeta'}(t) \psi_{\boldsymbol{k}'\zeta'}(\boldsymbol{r}) \exp\left[-\frac{\mathrm{i}}{\hbar} \int_0^t dt' \cdot E_{\zeta'}(\boldsymbol{k}'(t')) \right] . \tag{4.4.16}$$

The input equation now takes into account the variation of the electron wave vector $\boldsymbol{k} \to \boldsymbol{k}(t)$ in a single-band approximation. In calculating the derivative $\partial\psi(\boldsymbol{r}, t)/\partial t$, allowance must be made also for the time dependence of the function $\psi_{\boldsymbol{k}'(t)\zeta'}(\boldsymbol{r})$. As is clear from the foregoing, the resulting terms will cancel the contribution made by the matrix elements of the operator $-e\boldsymbol{F}\cdot\hat{\boldsymbol{r}}_1 = -\mathrm{i}e\boldsymbol{F}\cdot\nabla_{\boldsymbol{k}}$. As a result, we obtain the following equation (4.4.6):

$$\frac{\partial}{\partial t}\tilde{\alpha}_{\boldsymbol{k}\zeta}(t) = -\frac{e\boldsymbol{F}}{\hbar}\cdot\sum_{\zeta'\neq\zeta}\int d\boldsymbol{r}u^*_{\boldsymbol{k}(t)\zeta}(\boldsymbol{r})\nabla_{\boldsymbol{k}}u_{\boldsymbol{k}(t)\zeta'}(\boldsymbol{r})$$
$$\times\exp\left\{\frac{\mathrm{i}}{\hbar}\int_0^t dt'[E(\boldsymbol{k}(t'),\zeta)-E(\boldsymbol{k}(t'),\zeta')]\right\}\tilde{\alpha}_{\boldsymbol{k}\zeta'}(t)\ , \tag{4.4.17}$$

where the term with $\zeta' = \zeta$ is discarded as a result of the replacement (4.3.7). Introducing the notation

$$\lambda_{\zeta\zeta'}(\boldsymbol{k}) = -\frac{e\boldsymbol{F}}{\hbar}\cdot\int d\boldsymbol{r}u^*_{\boldsymbol{k}\zeta}(\boldsymbol{r})\,\nabla_{\boldsymbol{k}}u_{\boldsymbol{k}\zeta'}(\boldsymbol{r})$$

and integrating (4.4.17) with respect to t, we modify this equation to

$$\tilde{\alpha}_{\boldsymbol{k}\zeta}(t) = \tilde{\alpha}_{\boldsymbol{k}\zeta}(0) + \sum_{\zeta'\neq\zeta}\int_0^t d\tau\,\lambda_{\zeta\zeta'}(\boldsymbol{k}(\tau))$$
$$\times\exp\left\{\frac{\mathrm{i}}{\hbar}\int_0^\tau d\tau'[E(\boldsymbol{k}(\tau'),\zeta)-E(\boldsymbol{k}(\tau'),\zeta')\right\}\tilde{\alpha}_{\boldsymbol{k}\zeta'}(t)\ . \tag{4.4.18}$$

Neglecting interband transitions, we now substitute $\tilde{\alpha}_{\boldsymbol{k}\zeta'}$ into (4.4.18) and set $\tilde{\alpha}_{\boldsymbol{k}\zeta}(0) = \delta_{\zeta\eta}$ (i.e., at $t = 0$ the electron is in the state $|\boldsymbol{k}\eta\rangle$). The breakdown probability $w_{\eta\to\zeta}$ is equal to $|\alpha_{\boldsymbol{k}\zeta}(\infty)|^2$, i.e.,

$$w_{\eta\to\zeta} = \left|\int_0^\infty d\tau\exp\left\{\frac{\mathrm{i}}{\hbar}\int_0^t d\tau'[E(\boldsymbol{k}(\tau'),\zeta)-E(\boldsymbol{k}(\tau'),\eta)]\right\}\lambda_{\zeta\eta}(\boldsymbol{k}(\tau))\right|^2\ . \tag{4.4.19}$$

Further, for simplicity, we consider a one-dimensional array. We make the substitution of the variables $\tau \to k-(eF/\hbar)\tau$, and replace the lower limit of the integrals (4.4.19) by $-\infty$. We need to exclude the effects of the instantaneous switching-on of the field since, in reality, it always comes into play in an infinitesimally slow fashion in comparison with the atomic time scale. Then nothing depends on the instant at which the field is turned on, and we may choose this instant in an indefinite past. The substitution results in

$$w_{\eta\to\zeta} = \left(\frac{\hbar}{eF}\right)^2\left|\int_{-\infty}^\infty dk\,\lambda_{\zeta\eta}(k)\exp\left\{\frac{\mathrm{i}}{eF}\int_{-\infty}^k dk'[E(k',\zeta)-E(k',\eta)]\right\}\right|^2\ . \tag{4.4.20}$$

The integrand in (4.4.20) contains a rapidly oscillating term as $F \to 0$ and, consequently, the integral can be calculated using the saddle-point method [4.17] (the first to use this for computing transition probabilities in quantum mechanics was Landau [4.18]). The fundamental contribution to the integral is made by point k_0, where the phase derivative of the oscillatory function with respect to k is equal to zero:

$$E(k_0, \zeta) = E(k_0, \eta) \ . \tag{4.4.21}$$

In the vicinity of k_0 the phase varies smoothly, and far from k_0 it varies very rapidly. The integrand thus reverses sign in a random fashion and is, on the average, equal to zero. Furthermore, k_0 does not have real values (except for the degenerate case, with which we will not be concerned here) and for $F \to 0$ the quantity $w_{\eta \to \zeta}$ may be shown to tend to zero more rapidly than any power of F [4.17]. If the point k_0 lies in a complex plane and the functions $E(k, \zeta)$, $\lambda_{\zeta\eta}(k)$ are analytic, the path of integration (real axis) may be distorted so that it passes through k_0. In this case, at k_0 the path of integration will have a maximum for the integrand. Therefore, up to the preexponential factors,

$$\begin{aligned} w_{\eta \to \zeta} &\propto \left| \exp\left\{ \frac{\mathrm{i}}{|e|F} \int_C dk [E(k, \zeta) - E(k, \eta)] \right\} \right|^2 \\ &\approx \exp\left\{ -\frac{2}{|e|F} \int_0^{\kappa_0} d\kappa [E(\tilde{k} + \mathrm{i}\kappa, \zeta) - E(\tilde{k} + \mathrm{i}\kappa, \eta)] \right\} \ . \end{aligned} \tag{4.4.22}$$

The path C is depicted in Fig. 4.26; we have taken account of the fact that

$$\left| \exp\left\{ \frac{\mathrm{i}}{|e|F} \int_{C_1} dk [E(k, \zeta) - E(k, \eta)] \right\} \right|^2 = 1 \ ,$$

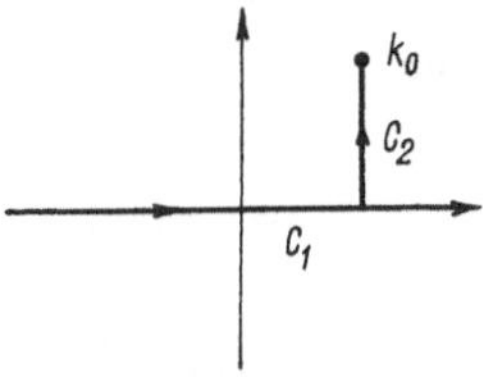

Fig. 4.26. Path of integration for calculating the integral (4.4.22)

and introduced the notation $\kappa_0 = \mathrm{Im}\, k_0$, $\tilde{k} = \mathrm{Re}\, k_0$. The probability of a breakdown is particularly large near the Brillouin-zone boundary. In the weak-binding approximation (4.3.24)

$$\begin{aligned} E(k, \zeta) - E(k, \eta) &= \{[E(k+q) - E(k+q')]^2 + 4|V_{q'-q}|^2\}^{1/2} \\ &= \left[\left(\frac{2\pi\hbar^2}{md^2} \right)^2 \left(k + \frac{\pi}{d} \right)^2 + \Delta^2 \right]^{1/2} \ , \end{aligned} \tag{4.4.23}$$

where we have chosen $q=0$, $q'=2\pi/d$, $k\approx -\pi/d$ and introduced the notation $2|V_{q-q'}|=\Delta$ for the energy gap between the bands. It follows from (4.4.23) that

$$\begin{aligned} k_0 &= -\frac{\pi}{d}+\frac{imd\Delta}{2\pi\hbar^2}\,, \\ w &\approx \exp\left\{-\frac{2}{|e|F}\int_0^{md\Delta/2\pi\hbar^2} d\kappa\left[\Delta^2-\left(\frac{2\pi^2\hbar^2}{md^2}\cdot\kappa\right)^2\right]^{1/2}\right\} \\ &= \exp\left(-\frac{md\Delta^2}{4|e|F\hbar^2}\right)\,. \end{aligned} \tag{4.4.24}$$

Thus the interband transition probability is, in fact, negligibly small when

$$F \lesssim \frac{md\Delta^2}{4|e|\hbar^2}\,. \tag{4.4.25}$$

4.4.3 Quantum Theory of the Electric Inertia Effect

In connection with the effective-mass concept, we may ask why the measured values of e/m in electric inertia effect experiments such as those staged by Stewart and Tolman (Sect. 3.2) coincide with those for free electrons, and whether or not this contradicts band theory. Such doubts, we will see, prove to be unfounded (although they are expressed in the literature from time to time [3.2]).

The problem of the electric inertia effect may be treated in the most general form, with allowance for the electron–electron interaction. Let us consider the motion of a metal, as a whole, with a varying velocity $\boldsymbol{v}(t)$. The coordinates of all ion cores will then depend on time according to the formula

$$\boldsymbol{R}_i(t)=\boldsymbol{R}_i+\int_0^t dt'\,\boldsymbol{v}(t')\,. \tag{4.4.26}$$

In the adiabatic approximation (Sect. 1.9) the time-dependent Schrödinger equation for the electron wave function for this nuclear arrangement has the form

$$\begin{aligned} \hat{\mathscr{H}}\psi(\boldsymbol{r},t) &\equiv \left[-\frac{\hbar^2}{2m}\sum_j \Delta_j+\sum_{j'<j} W(\boldsymbol{r}_{j'}-\boldsymbol{r}_j)+\sum_{ij} G(\boldsymbol{r}_j-\boldsymbol{R}_i(t))\right]\psi(\boldsymbol{r}_j,t) \\ &= i\hbar\frac{\partial\psi(\boldsymbol{r}_j,t)}{\partial t}\,, \end{aligned} \tag{4.4.27}$$

where W is the potential energy of the interaction of conduction electrons with each other, and G with ion cores.

We pass on to new electronic coordinates

$$r'_j = r_j - \int_0^t dt' \, v(t') \tag{4.4.28}$$

and represent the electron wave function as

$$\psi(r_j, t) = \tilde{\psi}(r'_j, t) \exp\left[\frac{i}{\hbar}\lambda(t) + \frac{i}{\hbar}\mu(t) \cdot \sum_j r'_j\right], \tag{4.4.29}$$

where $\lambda(t)$ and $\mu(t)$ are, for the time being, arbitrary real-valued time functions. The transformation (4.4.29) maintains the normalization of the wave function, and the current density of the number of particles will take the form

$$\begin{aligned} j &= \frac{\hbar}{2mi} \sum_j [\psi^*(r_j, t) \nabla_j \psi(r_j, t) - \psi(r_j, t) \nabla_j \psi^*(r_j, t)] \\ &= \frac{\hbar}{2mi} \sum_j [\psi^*(r_j, t)]^2 \nabla_j \left(\frac{\psi(r_j, t)}{\psi^*(r_j, t)}\right) \\ &= \frac{\hbar}{2mi} \sum_j [\tilde{\psi}^*(r'_j, t) \nabla'_j \tilde{\psi}(r'_j, t) - \tilde{\psi}(r'_j, t) \nabla'_j \tilde{\psi}^*(r'_j, t)] \\ &\quad + \frac{\mu(t)}{m} = j' + \frac{\mu(t)}{m}, \end{aligned} \tag{4.4.30}$$

with $\nabla'_j = \partial / \partial r'_j$. As will be seen, the transformation (4.4.29, 30) corresponds, with a specific choice of the functions $\lambda(t)$ and $\mu(t)$, to the transition to a new noninertial system of coordinates. In the transformation to new coordinates the Hamiltonian of (4.4.27) becomes

$$\hat{\mathscr{H}}' = -\frac{\hbar^2}{2m} \sum_j \Delta'_j + \sum_{j<j'} W(r'_j - r'_{j'}) + \sum_{ij} G(r'_j - R_i(0)) . \tag{4.4.31}$$

Taking account of (4.4.29),

$$\begin{aligned} \Delta'_j \psi(r_j, t) &= \exp\left[\frac{i}{\hbar}\lambda(t) + \frac{i}{\hbar}\mu(t) \cdot \sum_j r'_j\right] \\ &\quad \times \left[\Delta'_j \tilde{\psi}(r'_j, t) + \frac{2i}{\hbar}\mu(t) \cdot \nabla'_j \tilde{\psi}(r'_j, t) - \frac{\mu^2(t)}{\hbar^2} \tilde{\psi}(r'_j, t)\right], \end{aligned} \tag{4.4.32}$$

$$\frac{\partial\psi(\boldsymbol{r}_j,t)}{\partial t}=\exp\left[\frac{\mathrm{i}}{\hbar}\lambda(t)+\frac{\mathrm{i}}{\hbar}\boldsymbol{\mu}(t)\cdot\sum_j\boldsymbol{r}'_j\right]\left\{\frac{\partial\tilde{\psi}(\boldsymbol{r}'_j,t)}{\partial t}\right.$$

$$+\frac{\mathrm{i}}{\hbar}\left(\frac{\partial\lambda}{\partial t}+\frac{\partial\boldsymbol{\mu}}{\partial t}\cdot\sum_j\boldsymbol{r}'_j\right)\tilde{\psi}(\boldsymbol{r}'_j,t)+\boldsymbol{v}(t) \tag{4.4.33}$$

$$\left.\cdot\left[\sum_j\nabla'_j\tilde{\psi}(\boldsymbol{r}'_j,t)+\frac{\mathrm{i}}{\hbar}\boldsymbol{\mu}\tilde{\psi}(\boldsymbol{r}'_j,t)\right]\right\}.$$

Substituting (4.4.31–33) into the Schrödinger equation (4.4.27), we see that the latter simplifies drastically if we set

$$\boldsymbol{\mu}(t)=m\boldsymbol{v}(t) \tag{4.4.34}$$

$$-\frac{\partial\lambda(t)}{\partial t}+\boldsymbol{\mu}(t)\cdot\boldsymbol{v}(t)=\frac{\boldsymbol{\mu}^2(t)}{2m}. \tag{4.4.35}$$

The terms containing $\nabla'_j\tilde{\psi}(\boldsymbol{r}'_j,t)$ here cancel out because of (4.4.34), and most of the extra terms proportional to $\tilde{\psi}$ cancel out due to (4.4.35). Only one extra term, equal to $-\partial\boldsymbol{\mu}/\partial t\cdot\sum_j\boldsymbol{r}'_j\tilde{\psi}(\boldsymbol{r}'_j,t)$, remains in the Schrödinger equation. As a result, the equation for $\tilde{\psi}$ has the form

$$\left[-\frac{\hbar^2}{2m}\sum_j\Delta'_j+\sum_{ij}G(\boldsymbol{r}'_j-\boldsymbol{R}_i)+\sum_{j<j'}W(\boldsymbol{r}'_{j'}-\boldsymbol{r}'_j)\right. \tag{4.4.36}$$

$$\left.+m\frac{d\boldsymbol{v}}{dt}\cdot\sum_j\boldsymbol{r}'_j\right]\tilde{\psi}(\boldsymbol{r}'_j,t)=\mathrm{i}\hbar\frac{\partial\tilde{\psi}(\boldsymbol{r}'_j,t)}{\partial t}.$$

Besides, in view of (4.4.34, 35),

$$\lambda(t)=\int_0^t dt'\frac{m\boldsymbol{v}^2(t')}{2}. \tag{4.4.37}$$

As follows from (4.4.34, 30), upon transformation of the wave function, $\boldsymbol{j}=\boldsymbol{j}'+\boldsymbol{v}(t)$; i.e., we are merely dealing with a transition to an inertial frame of reference. If $\boldsymbol{v}(t)=\text{const}$, the Schrödinger equation for the function $\tilde{\psi}$ (4.4.36) simply coincides with the input equation (4.4.27) for the function ψ. This is a manifestation of the Galilean invariance of nonrelativistic quantum mechanics. As is seen from a comparison of (4.4.2, 36), the acceleration leads to an "inertial force", alternatively known as an electric inertia force.

$$\boldsymbol{F}'=-\frac{m}{e}\frac{d\boldsymbol{v}}{dt}. \tag{4.4.38}$$

This expression was used in the classical treatment of the Stewart and Tolman experiments in Sect. 3.2. The formula is seen to include the free-electron mass. However, the character of the motion of an electron subject to this fictitious field may be substantially different from the counterpart for a free electron. On the other hand, this is not essential to the interpretation of the experiment by Stewart and Tolman, since they utilize an experimental value for the conductivity that is certainly determined by the effective mass rather than the free-electron mass.

4.5 The Metal–Semiconductor Criterion

4.5.1 The Metal–Nonmetal Criterion in Band Theory

So far in this chapter we have been considering the properties of one-electron states. However, describing the properties of solids even in terms of the band model necessitates allowance for many-particle effects. In the simplest approximation one completely neglects the dynamic electron–electron interaction and takes into account only the statistical correlation, which is due to Pauli's exclusion principle and is described by the Fermi–Dirac function (3.5.9) (the scope of applicability of this approach is discussed below; this is actually one of the central problems of the entire theory of condensed media). The heat capacity, paramagnetism, etc. are determined in the same way as in electron gas theory (Chap. 3), but with the "band" density of states (4.2.58). Of supreme importance to the metal–nonmetal criterion are the electrical properties (Chap. 3). In the ground state ($T = 0$ K) metals possess free electrons, whereas no free electrons are available in nonmetals. Free electrons are accelerated by a weak electric field. In such a field we may neglect the interband transitions, the probability of which, according to (4.4.25), is exponentially small, and replace $\boldsymbol{k}(t)$ by $\boldsymbol{k}$ in (4.4.12). If the electrons do not interact, the result for the acceleration of all of them (equal to the time derivative of the total current) will be

$$W_i = \sum_{\boldsymbol{k}\zeta} a_i(\boldsymbol{k}\zeta)\overline{n(\boldsymbol{k}\zeta)} = \sum_j eF_j \sum_{\boldsymbol{k}\zeta} m_{ij}^{-1}(\boldsymbol{k}\zeta)\overline{n(\boldsymbol{k}\zeta)} \ , \tag{4.5.1}$$

with the summation over $\boldsymbol{k}$ being carried out over the first Brillouin zone. For $T = 0$ K the $\overline{n(\boldsymbol{k}\zeta)}$ satisfies the plot given in Fig. 3.6. If, for some ζ, the quantity $E(\boldsymbol{k}, \zeta)$ is less than or equal to E_{F} for all $\boldsymbol{k}$, this band does not contribute to (4.5.1), since

$$\sum_{\boldsymbol{k}} m_{ij}^{-1}(\boldsymbol{k}\zeta) = 0 \ . \tag{4.5.2}$$

This can be proved by substituting (4.2.20) into (4.4.13); as we do so, the term with $\boldsymbol{R}_m = 0$ cancels out, and for $\boldsymbol{R}_m \neq 0$ we have

$$\sum_{\boldsymbol{k}} \exp(\mathrm{i}\boldsymbol{k} \cdot \boldsymbol{R}_m) = 0 \ . \tag{4.5.3}$$

Formula (4.5.3) may be proved as follows:

$$\begin{aligned}
\sum_{\boldsymbol{k}} \exp(\mathrm{i}\boldsymbol{k} \cdot \boldsymbol{R}_m) &= -R_m^{-2} \sum_{\boldsymbol{k}} \Delta_{\boldsymbol{k}} \exp(\mathrm{i}\boldsymbol{k} \cdot \boldsymbol{R}_m) \\
&= -R_m^{-2} V_0 \int_{\mathrm{BZ}} \frac{d\boldsymbol{k}}{(2\pi)^3} \Delta_{\boldsymbol{k}} \exp(\mathrm{i}\boldsymbol{k} \cdot \boldsymbol{R}_m) \\
&= -R_m^{-2} V_0 \int \frac{d\boldsymbol{S}_{\boldsymbol{k}}}{(2\pi)^3} \nabla_{\boldsymbol{k}} \exp(\mathrm{i}\boldsymbol{k} \cdot \boldsymbol{R}_m) \\
&= -\mathrm{i}\boldsymbol{R}_m R_m^{-2} V_0 \int \frac{d\boldsymbol{S}_{\boldsymbol{k}}}{(2\pi)^3} \exp(\mathrm{i}\boldsymbol{k} \cdot \boldsymbol{R}_m) = 0 \ .
\end{aligned}$$

Here BZ denotes the first Brillouin zone, $\int d\boldsymbol{S}_{\boldsymbol{k}}$ is taken over its boundary, and $d\boldsymbol{S}_{\boldsymbol{k}}$ is an element of area in space, multiplied by the unit normal. In the transformation from the volume integral to the surface integral we have exploited Gauss' theorem; the latter integral is equal to zero, since at the opposite zone edges 1, 2

$$d\boldsymbol{S}_{\boldsymbol{k}_1} = -d\boldsymbol{S}_{\boldsymbol{k}_2}; \ \exp(\mathrm{i}\boldsymbol{k}_2 \cdot \boldsymbol{R}_m) = \exp[\mathrm{i}(\boldsymbol{k}_1 + \boldsymbol{b}_g^*) \cdot \boldsymbol{R}_m] = \exp(\mathrm{i}\boldsymbol{k}_1 \cdot \boldsymbol{R}_m) \ .$$

Thus the electrons of completely filled or empty bands are not accelerated by an electric field. This is readily interpretable. With neglect of the zener breakdown, the total current of ζ-band electrons is equal to

$$j_\zeta = e \sum_{\boldsymbol{k}} \boldsymbol{v}\left(\boldsymbol{k} + \frac{e\boldsymbol{F}}{\hbar} t, \zeta\right) \overline{n(\boldsymbol{k}\zeta)} = e \sum_{\boldsymbol{k}} \boldsymbol{v}(\boldsymbol{k}, \zeta) \overline{n\left(\boldsymbol{k} - \frac{e\boldsymbol{F}}{\hbar} t, \zeta\right)} \tag{4.5.4}$$

[see (4.4.11)]. An electric field causes quasi-rotation of the distribution function in $\boldsymbol{k}$ space—$\boldsymbol{k} \to \boldsymbol{k} - e\boldsymbol{F}t/\hbar$ followed by reduction to the first zone. If $\overline{n(\boldsymbol{k}\zeta)} = 1$, j_ζ will be independent of t and at time zero is equal to zero by symmetry considerations: $E(\boldsymbol{k}, \zeta) = E(-\boldsymbol{k}, \zeta)$; whence $\boldsymbol{v}(\boldsymbol{k}, \zeta) = -\boldsymbol{v}(-\boldsymbol{k}, \zeta)$; consequently, $j_\zeta(t) = 0$.

Thus, in band theory, a solid is a metal only if some of the energy bands are partially occupied or overlap. For example, any crystal that contains an odd number of electrons per unit cell may be a metal.

As it is, the total number of states in a band is even, since a twofold spin degeneracy occurs [for electrons that do not interact dynamically, their energy is independent of spin (4.2.4)]. For a metal though, there is an odd number of electrons, and therefore one or several bands will be partially filled. For example, if we assume that the 3*s* states of Na form an energy band (in the spirit of the problem posed in Sect. 4.1.3), the latter is exactly half full. On the other hand, solid hydrogen (at least at pressures that are not too high) is not a metal. There is one electron per hydrogen atom, but it is energetically more favorable for the atoms to combine in H_2 molecules. As the hydrogen atoms do this, they occupy sites that are obviously nonequivalent (the interatomic spacing in a molecule is much smaller than the intermolecular spacing), and the number of electrons per unit cell turns out to be even. By contrast, the number of electrons in, say, Ca is even; nevertheless, solid Ca is a metal. Thus, some energy bands overlap and therefore are only partly filled. Figures 4.27a–d provide a schematic picture of the energy bands and the filling of the energy levels with electrons for an insulator, a semiconductor, and a metal.

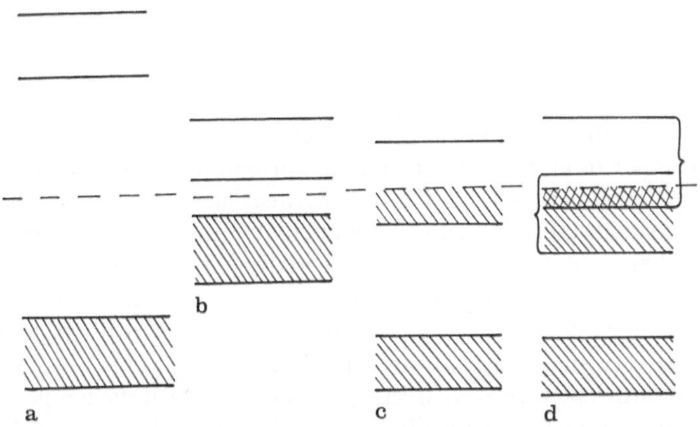

Fig. 4.27. Energy band filling (shadowed) in the ground state of an insulator (**a**), a semiconductor (**b**), and a metal (**c**, **d**)

The above metal–nonmetal criterion considered by Wilson [4.19] in band theory accounts for a large body of experimental evidence and is the basis for considering nearly all of the properties of solids. However, a number of problems arise with the generalization of this criterion for noncrystalline condensed states (liquid mercury is a metal, whereas water is not) and with allowance for the electron–electron interaction. For example, experiment provides conclusive evidence for the fact that in most rare-earth metals 4*f* electrons do not participate in conductivity, although the 4*f* band is only partly filled. This is due to their strong Coulomb repulsion and the very narrow 4*f* band. Although formally these problems lie outside the scope of band theory, they will be allotted some space here in view of their supreme importance.

Nonmetals are divided into insulators (dielectrics) and semiconductors (although this has no profound physical meaning and is only based on convention; see Sect. 1.8). Solids in which, in the ground state, the energy gap G between the last occupied (valence) band and the first empty (conduction) band is not larger than 1 to 2 eV are called semiconductors. Solids with $G \gtrsim 2$ eV are said to be insulators. Special mention must be made of gapless (zero-gap) semiconductors (gray tin, HgTe, HgSe, etc.), where the conduction band and the valence band touch at one point, determined by the symmetry properties of the crystal (Fig. 4.28). When in the ground state, these materials are nonmetals, since they have no partially filled bands. However, near the ground state the energy spectrum of these zero-gap semiconductors is continuous, just as in metals.

Fig. 4.28. Energy spectrum of a gapless semiconductor

This formulation of the criterion for differentiating between the state of a semiconductor and that of an insulator is only a convention. A more distinct and physically clear formulation to distinguish between insulators and semiconductors is to consider their electrical conductivity. Insulators are then viewed as crystals that exhibit chiefly ionic conductivity (a small admixture of electronic conductivity at very high temperatures is immaterial). Semiconductors are regarded as crystals that exhibit, long before the onset of observable ionic conductivity at very high temperatures, a pronounced electronic conductivity whose temperature dependence is directly opposite (in sign of the temperature derivative) to that of metals. This difference in the physical nature of the electrical conductivities of an insulator and a semiconductor arises from the difference in the size of the energy gap between the occupied valence band and the empty conduction band at $T = 0$ K.

We now evaluate the temperature dependence of the number of current carriers in pure (intrinsic) semiconductors. To do this, we determine their chemical potential ζ. The energies will be counted from the top of the valence band. The number of thermally excited band electrons at temperature T is equal to

$$N_e = \int_0^{D_e} dE g_e(E) \left[\exp\left(\frac{E - \zeta + G}{k_B T} \right) + 1 \right]^{-1} , \tag{4.5.5}$$

where D_e is the bandwidth, and $g_e(E)$ the density of states in it. Assuming that

$$G - \zeta \gg k_B T , \tag{4.5.6}$$

we find

$$N_e \approx \exp\left(\frac{\zeta - G}{k_B T}\right) \int_0^{D_e} dE g_e(E) \exp(-E/k_B T) \ . \tag{4.5.7}$$

Similarly, the number of the unoccupied states (holes) that have arisen in the band is equal to

$$N_h \approx \exp(-\zeta/k_B T) \int_0^{D_h} dE g_h(E) \exp(-E/k_B T) \ , \tag{4.5.8}$$

where D_h is the width, $g_h(E)$ is the valence-band density of states, and it is assumed that

$$\zeta \gg k_B T \ . \tag{4.5.9}$$

Evidently, in an intrinsic semiconductor

$$N_e = N_h \ . \tag{4.5.10}$$

As has already been noted for phonons in Sect. 2.3 (see Sect. 3.5.8), we may assert that near the band edge in the three-dimensional case

$$g_e(E) = A_e E^{1/2} \ , \quad g_h(E) = A_h E^{1/2} \ , \tag{4.5.11}$$

with A_e and A_h being some constant quantities. For $k_B T \ll D_e$, D_h a small E makes a fundamental contribution to the integrals (4.5.7, 8). Substituting (4.5.11) into (4.5.7, 8), we find from (4.5.10)

$$\zeta \approx \frac{G}{2} + \frac{k_B T}{2} \ln \frac{A_h}{A_e} \ . \tag{4.5.12}$$

Conditions (4.5.6, 9) evidently hold for $\zeta \approx 1$ eV and $G \approx 2$ eV and at a reasonable temperature. Now we substitute (4.5.11) into (4.5.7):

$$\begin{aligned} N_e = N_h &= \exp(-G/2k_B T)(A_h A_e)^{1/2} \int_0^{D_e} dE E^{1/2} \exp(-E/k_B T) \\ &= \exp(-G/2k_B T)(A_h A_e)^{1/2}(k_B T)^{3/2} \\ &\quad \cdot \int_0^{D_e/k_B T} dx x^{1/2} e^{-x} \approx (\pi A_h A_e/4)^{1/2} (k_B T)^{3/2} \exp(-G/2k_B T) \ , \end{aligned} \tag{4.5.13}$$

$$\left(D_e \gg k_B T \ , \quad \int_0^{\infty} dx x^{1/2} e^{-x} = \sqrt{\pi}/2\right) .$$

Formula (4.5.13) gives the temperature dependence of the number of current carriers for intrinsic semiconductors. The electrical properties of real semiconductors, however, are determined chiefly by impurities.

If we implant, say, an As atom into a Ge crystal, the following will occur: The Ge atom is quadrivalent, so each site in its crystal (diamond-type lattice, Fig. 1.22) has four nearest neighbors. The As atom has five valence electrons, which are bound to the nucleus comparatively weakly. If the Ge atom in the lattice is replaced by the As atom, four electrons serve to form valence bonds (Fig. 1.34), and the fifth may become practically free. This is so because of the following: The radius of the bound state of an electron with positive charge $|e|$ in vacuum is on the order of $\hbar^2/me^2$, and the binding energy is known to be equal to $\Delta E = me^4/2\hbar^2 \approx 13.5$ eV. In a medium the interaction of charges becomes ε times weaker: $e^2 \to e^2/\varepsilon$, where ε is the static dielectric constant, and the mass m should be replaced by the effective mass m^*. The radius of the bound state is then

$$a \approx \frac{\hbar^2 \varepsilon}{m^* e^2} = \frac{\varepsilon m}{m^*} \frac{\hbar^2}{me^2} \approx 0.5 \frac{\varepsilon m}{m^*} \text{Å} , \tag{4.5.14}$$

and the binding energy is

$$\Delta E = \frac{m^*}{m\varepsilon^2} \frac{me^4}{2\hbar^2} \approx 13.5 \frac{m^*}{m\varepsilon^2} \text{eV} . \tag{4.5.15}$$

For semiconductors the constant ε is rather large: $\varepsilon \gtrsim 10$. This is due to the small band gap of G; when $G = 0$ the system becomes a metal and $\varepsilon = \infty$. As will be seen in Sect. 4.6.6, the smallness of G leads also to a small m^*/m. As a result, a may reach a magnitude on the order of dozens of interatomic distances (this is what allows us to describe the influence of the medium of the macroscopic, i.e., averaged, quantity ε). The binding energy, for example, for an As impurity in Ge, diminishes in comparison with the ionization potential of As in vacuum from 9.8 eV to 0.013 eV. Such "shallow" bound states are easily destroyed by thermal motion already at $T \approx 100$ K.

Impurities that donate electrons are said to be donors. We may consider in an analogous fashion some trivalent impurity in Ge that supplies holes to the valence band (Fig. 1.30). Such impurities are called acceptors.

Figure 4.29 is a sketch of the energy spectrum for an extrinsic semiconductor. The donor levels and all electron-occupied levels at $T = 0$ K lie beneath the conduction band and are separated from it by a small distance (their ionization energy $\approx \Delta E$), whereas the acceptor levels, not occupied by electrons at $T = 0$ K, overlie the top of the valence band and are separated from it by a small distance (their ionization energy E_a counted from the top of the valence band, E_v).

Later we will return to some of the fundamental points of the theory of semiconductors. More detailed information on this area of solid-state theory is available in the literature [2.34; 3.19; 4.20, 21].

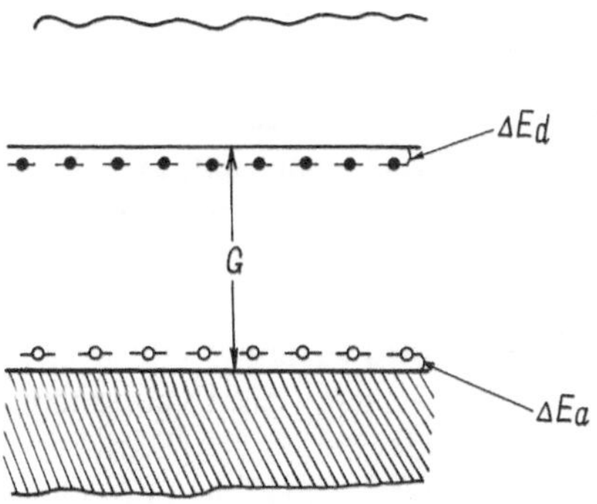

Fig. 4.29. Energy spectrum of a doped semiconductor with specification of acceptor ΔE_a and donor ΔE_d levels

4.5.2 The Peierls Transition

Up to this point we have assumed that the lattice is rigid and the electrons do not interact. The fundamental problem of the metal–nonmetal criterion is that these two assumptions underlying the band model are liable to incur gross errors. Here we concentrate on a purely qualitative treatment of the most important points. To start with, we relinquish the assumption of the lattice being rigid (as before, the electrons are assumed to be noninteracting). We examine a linear array with one half-full energy band (Fig. 4.30a). In the one-dimensional case there are $(Nd/2\pi)\Delta k$ states (with d being the lattice period) for the interval of values of the quasiwave vector Δk (Chap. 2). Therefore, the occupied states in the half-filled case are those with $k \leqslant \pi/2d$ (Fig. 4.30a). We consider a minor

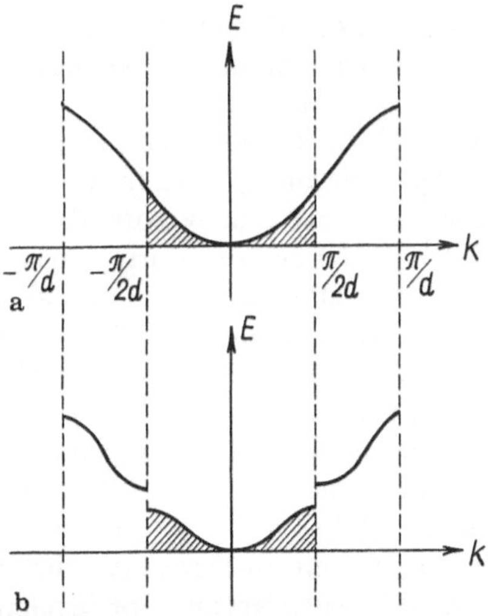

Fig. 4.30. Lowering of the total energy of band electrons when the lattice period is doubled (Peierls transition): prior to doubling (**a**), subsequent to doubling (**b**)

distortion of the array, where every other atom shifts slightly from its equilibrium position. The lattice period then is not d any longer, but $2d$ (Fig. 4.30b), and a discontinuity occurs at the new Brillouin zone boundaries (Sects. 4.1, 3). As this takes place, the energy of all the occupied states lowers somewhat, and that of the unoccupied states rises (Fig. 4.30b), leading to a gain in the total band electron energy in the array and, consequently, to a decrease in the total energy of the one-dimensional crystal [Ref. 2.7, Chap. 5]. Broadly speaking, this distortion may also entail an increase in core electron energy. However, in metals this is small compared with the metallic bond energy due to conduction electrons.

Thus distortion has led to the occurrence of an energy gap at the Fermi level, and a metal has turned into a nonmetal. Clearly, another distortion will not lead to an appreciable energy gain, since the energies of the states above the gap increase by approximately the same amount as that by which the energies of the states below the gap decrease (4.3.24). Obviously, for any band occupancy it is possible in the one-dimensional case to choose a distortion such that a gap opens at $k = k_F$. The corresponding period of the resulting "superlattice" will be

$$D = 2\pi/2k_F \ . \tag{4.5.16}$$

If D is commensurate with d, i.e., $D = pd/q$, where p and q are integers that have no common divisor ($p > q$), the resulting "superlattice" has a period pd. Here Bloch's theorem is applicable and we can exploit the theory outlined in this chapter. But if D and d are incommensurate, we are confronted with the complicated and little-explored problem of the electron motion in a nearly periodic potential [4.22].

The previous treatment (although not totally rigorous) shows that in the ground state a one-dimensional array cannot be a metal. This conclusion is supported experimentally for some quasi-one-dimensional systems (i.e., those consisting of arrays with a weak interaction between them), which are semiconductors [with a superstructure (4.5.16)] at low temperatures and change into metals as the temperature is raised (Peierls transition).

In the three-dimensional case, generally speaking, it is impossible by distorting the lattice to attain a substantial lowering of the energy of all the occupied states (this requires that the Fermi surface coincide exactly with the new Brillouin-zone boundary). Nonetheless, if the Fermi surface nearly coincides with the zone boundary, the total energy gain attained in the crystal by decreasing the band energy at the points of contact may be appreciable. In this situation, only a local spectrum readjustment at determinate points of $\boldsymbol{k}$ space may occur and so there is no metal–nonmetal transition.

These qualitative considerations enable us to understand the empirical Hume–Rothery rule [1.32, 3.16] relating to the structure of some alloys that arrange themselves in an ordered fashion (for example, Cu-Zn). A typical feature of such alloys is that at certain concentrations of the constituents, superstructure phases are produced in which the ions of different elements are regularly arranged. According to this rule, the stability limits of the various phases (bcc, fcc, etc.) correspond to quite definite mean electron concentrations that are very close to the values for which the Fermi sphere touches the bands for the corresponding structures on the inside of the boundary surfaces. [The spherical Fermi surface approximation, which corresponds to quasi-free electrons (Sect. 4.3), gives a fairly good fit to experimental data for many real metals.] The energy gain due to the local gap arising in the electron spectrum at the points of contact stabilizes the relevant phases.

The considerations set forth above show the importance of the contribution which noncentral forces can make to the lattice energy of the metal (with a fixed lattice structure the band energy of electrons certainly cannot be represented as a sum of pairwise ion–ion interaction energies). For reasons close to those outlined above, some metallic-bonded compounds undergo structural transitions accompanied by distortion of a highly symmetric initial lattice. The character of these distortions is associated with the shape of the electron energy spectrum.

4.5.3 The Mott Transition

Here we will try to ascertain how seriously the neglect of the electron–electron interaction in band theory may affect the metal–nonmetal criterion. Part of the electron–electron interaction may be taken into account by including it into the crystalline potential (self-consistent field, Sect. 4.6.1). The many-particle effects that cannot be included in this way are said to be correlational; we concentrate on these effects in this section.

We examine a crystal with the number of electrons equal to the number of sites. According to band theory, the ground state of this many-electron system is metallic and has a half-full energy band, there being two electrons with antiparallel spins in each orbital (Bloch) state (for simplicity, we exclude the band-overlap case). The probability of finding on some crystal site an electron with spin up is equal to 1/2, that for an electron with spin down is the same, and, consequently, the probability of finding on some site two electrons with opposite spins is equal to 1/4. This state is undoubtedly advantageous in terms of a gain in band (kinetic) energy, since the lowest one-electron states are the ones occupied. However, the large share of doubly occupied sites leads to an increase in electron–electron repulsion energy in comparison with the so-called homopolar state in which one electron sits on each site. On the other hand, the electron in this state is localized in a small volume, which, according to the Heisenberg uncertainty relation, leads to large fluctuation of momentum and to an increase in kinetic energy.

We label the characteristic Coulomb repulsive interaction energy by U, and the bandwidth (i.e., the characteristic kinetic energy of electrons) by W. The above lines of argument show that with $U \ll W$ the interaction will lead only to some small corrections to band theory, but, on the other hand, when $U \gtrsim W$, the ground state changes radically and each electron is localized on its site. Clearly, this state will be nonconducting. (Incidentally, a rigorous proof does not exist as yet. Moreover, the metallic state in the one-dimensional case turns out to be unstable for an arbitrary U [4.23].) Somewhere at $U \approx W$ a metal–nonmetal transition should occur [4.24], either smoothly or sharply. All this remains to be confirmed experimentally [4.25, 4.26].

Thus the electron motion in narrow energy bands differs drastically from that predicted by band theory. Realistically, such narrow, partially filled bands only originate from d or f states. Classical examples of materials that should be metals in keeping with band theory but actually are not, because of the correlation, are some transition-metal oxides such as NiO (for more details see Mott [4.27]). Such materials are said to be Mott insulators, and the metal–semiconductor transition due to correlation is called the Mott transition. Thus far, the theory of this transition has not been constructed in a form that is in any way complete; some of the related problems are discussed below.

4.5.4 Disordered Systems

Here we consider the result of abandoning another fundamental assumption of band theory, namely, strict spatial periodicity and, as a consequence, Bloch's theorem. Experiment shows that, in their electronic properties, liquid metals differ little from crystalline metals. This phenomenon was qualitatively explained by *Shubin* [1.41]. According to Shubin, for the usual concepts of the quantum theory of solids to be applicable to systems devoid of spatial periodicity, two conditions must be fulfilled: First, there should exist steady current-carrying states. Second, the potential fluctuation scattering in the system should not be very strong (in today's more refined form this condition says that the mean free path length should be large compared to the electron wavelength) so that the lifetime of an electron in the eigenstate is much larger than the collision time. The first requirement holds in the three-dimensional case if the potential itself is sufficiently small. To prove this, it suffices to apply perturbation theory for a continuous spectrum. The wave function in a weak potential has the shape of a plane wave with small additions and, consequently, the corresponding state can carry current. Let us emphasize, however, that in the one- and two-dimensional cases this correction, found from perturbation theory, diverges and a radical readjustment of states may be expected even for a weak potential.

The problem as to why the potential in liquid metals (and in metals in general) may be regarded as small was touched on in Sect. 4.3 and will be discussed in some detail in Sect. 4.6.4. The smallness of the potential also assures the fulfillment of the second condition.

Now we proceed to a case which violates the second condition; i.e., the free mean path length is comparable with the electron wavelength (the latter in metals corresponds to the interatomic spacing). The wave functions $\psi(x)$ of electrons on neighboring sites will be totally uncorrelated (Fig. 4.31). *Anderson* [4.28] has shown that with a large amount of disorder, the electrons turn out to be localized in some region of space and their states, therefore, do not carry current.

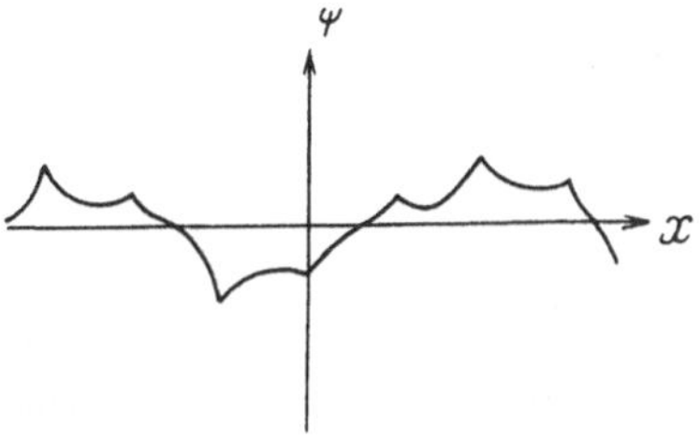

Fig. 4.31. Shape of the wave function $\psi(x)$ with strong disorder

Consider a liquid or amorphous semiconductor. In a crystalline semiconductor the density of states has the shape portrayed in Fig. 4.32a, showing the presence of an energy gap. In a disordered system the energy of an electron on a site fluctuates as a function of environment, which results in the band edges being smeared. A so-called pseudogap forms (Fig. 4.32b). The states at the center of the pseudogap are formed by a small number of sites, because these states require large and therefore improbable energy fluctuations. The distance between the centers is on the average large and the wave functions of the electrons sitting on them do not overlap. Therefore, the states with the energy lying at the center of the pseudogap are localized, whereas the states at the edges of the pseudogap are, in some way, more similar to the usual band states. Critical energy values exist separating localized states from current-carrying states (such values are called mobility thresholds). If, due to a change in electron concentration or under the effect of high pressure, the Fermi level passes through the mobility threshold, the insulator (semiconductor) will turn into a metal or vice versa. This situation is called an Anderson transition (Sect. 4.9).

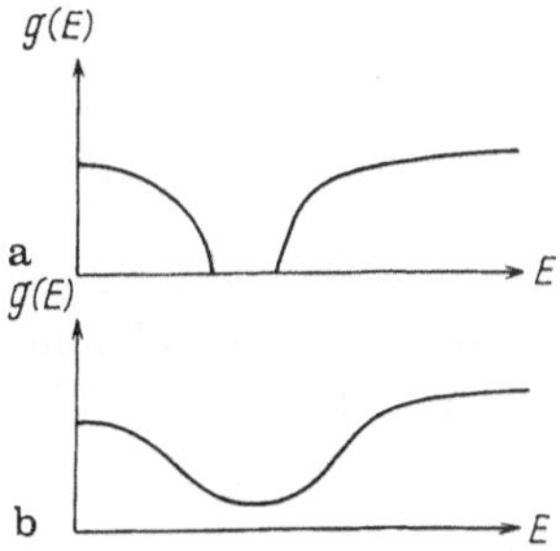

Fig. 4.32. Formation of a "pseudogap": density of states in energy positions as a function of energy, in the presence of a gap (**a**), in the presence of a pseudogap (**b**)

This concludes our discussion of "nonband" effects in connection with the fundamental problem of the metal–insulator criterion. Now that we have reviewed the range of applicability of band theory we proceed to a consideration of some of its specific problems. Although modern solid-state theory goes far beyond the scope of the band model, the latter has a sufficiently broad spectrum of applications, primarily for normal metals and crystalline semiconductors.

4.6 Computing the Electron Energy Spectrum of Crystals

4.6.1 Self-Consistent Field Approximation

As stated previously, the interaction between electrons in solids may not, generally speaking, be viewed as weak (at least it is not weak in atoms that constitute a solid). Therefore, any energy spectrum calculation that has pretensions at least to be qualitatively in agreement with experiment has to allow for the electron–electron interaction in a real solid. On the other hand, because of the complexity of the many-particle problem, some results can only be obtained if one formally preserves the scheme of the band model, i.e., by setting up a one-electron, wave-function Schrödinger equation, which only partly takes into account the effects of Coulomb repulsion [4.29].

We consider the many-electron Schrödinger equation for a crystal (1.9.6). In this equation we first make the replacements $G(\boldsymbol{R}_i, \boldsymbol{r}_j) \rightarrow V_n(\boldsymbol{r}_j)$ and $W(\boldsymbol{r}_j, \boldsymbol{r}_{j'}) \rightarrow e^2/|\boldsymbol{r}_j - \boldsymbol{r}_{j'}|$. Intuitively, the possibility of treating the nuclei as "frozen" seems obvious (Sect. 1.9). However, this (adiabatic) approximation is very difficult to substantiate rigorously. (The problem is considered in greater detail in Chap. 5.) Equation (1.9.6) is equivalent to the extremum condition for the functional $\langle \Psi | \mathscr{H} | \Psi \rangle$ for a particular wave-function normalization given

$$\begin{gathered}
\delta \sum_{\substack{\sigma_1 = \pm 1/2 \ldots \\ \sigma_N = \pm 1/2}} \int d\boldsymbol{r}_1 \ldots d\boldsymbol{r}_N \, \Psi^*(\boldsymbol{r}_1\sigma_1, \ldots, \boldsymbol{r}_N\sigma_N) \\
\cdot (\hat{\mathscr{H}} - E)\, \Psi(\boldsymbol{r}_1\sigma_1, \ldots, \boldsymbol{r}_N\sigma_N) = 0 \ , \\
\sum_{\sigma_1 \ldots \sigma_N} \int d\boldsymbol{r}_1 \ldots d\boldsymbol{r}_N \, |\Psi(\boldsymbol{r}_1\sigma_1, \ldots, \boldsymbol{r}_N\sigma_N)|^2 = 1 \ ,
\end{gathered} \tag{4.6.1}$$

where δ is the variation symbol. By varying $\langle \Psi | \hat{\mathscr{H}} | \Psi \rangle - E \langle \Psi | \Psi \rangle$ with respect to Ψ^*, we obtain a Schrödinger equation. The E plays the role of an indefinite Lagrange factor. According to the Pauli exclusion principle, the function Ψ is antisymmetric with respect to permutations of the coordinates and spins of any two electrons

$$\Psi(q_1 \ldots q_i \ldots q_j \ldots q_N) = -\Psi(q_1 \ldots q_j \ldots q_i \ldots q_N) , \tag{4.6.2}$$

where the notation introduced is $q_i = (\boldsymbol{r}_i \sigma_i)$. For noninteracting particles $\Psi(q_1, q_2, \ldots, q_N)$ has the form of the determinant involved in the one-electron functions $\Psi_\nu(q)$:

$$\Psi(q_1 \ldots q_N) = (N!)^{-1/2} \begin{vmatrix} \psi_{\nu_1}(q_1) \ldots \psi_{\nu_1}(q_N) \\ \psi_{\nu_2}(q_1) \ldots \psi_{\nu_2}(q_N) \\ \ldots\ldots \\ \psi_{\nu_N}(q_1) \ldots \psi_{\nu_N}(q_N) \end{vmatrix}$$

$$= (N!)^{-1/2} \sum_{\hat{\mathscr{P}}} \varepsilon_{\hat{\mathscr{P}}} \hat{\mathscr{P}} \psi_{\nu_1}(q_1) \ldots \psi_{\nu_N}(q_N) , \tag{4.6.3}$$

with ν_i being the quantum numbers of occupied electronic states, and $\hat{\mathscr{P}}$ the permutation of the indices $\nu_1, \ldots, \nu_N$, the quantity $\varepsilon_{\hat{\mathscr{P}}} = \pm 1$ depending on the parity of permutation.

We make use of the direct variational method; i.e., we seek the extremum of the functional under the variation sign in (4.6.1) on a class of functions of the form (4.6.3), where $\psi_\nu(q)$ are some trial functions. This will give the best quasi-one-electron Hartree–Fock (or self-consistent field) approximation in terms of the variational principle. First we calculate the normalization integral

$$\begin{aligned} \langle \Psi | \Psi \rangle &= \int dq_1 \ldots dq_N |\Psi(q_1 \ldots q_N)|^2 \\ &= (N!)^{-1} \sum_{\hat{\mathscr{P}}\hat{\mathscr{P}}'} \varepsilon_{\hat{\mathscr{P}}} \varepsilon_{\hat{\mathscr{P}}'} \int dq_1 \ldots dq_N [\hat{\mathscr{P}}' \psi^*_{\nu_1}(q_1) \ldots \\ &\quad \ldots \psi^*_{\nu_N}(q_N)][\hat{\mathscr{P}} \psi_{\nu_1}(q_1) \ldots \psi_{\nu_N}(q_N)] , \end{aligned} \tag{4.6.4}$$

where $\int dq_1 = \sum_{\sigma_1} \int d\boldsymbol{r}_1$. We now require that the additional conditions

$$\int dq \psi^*_\nu(q) \psi_{\nu'}(q) = \delta_{\nu\nu'} \tag{4.6.5}$$

be fulfilled. Only the terms with $\hat{\mathscr{P}} = \hat{\mathscr{P}}'$ then contribute to (4.6.3), all of them being equal to unity. The number of the various permutations $\hat{\mathscr{P}}$ is equal to $N!$, so (4.6.4) is automatically equal to unity.

We now calculate the functional $\langle \Psi | \mathscr{H} | \Psi \rangle$. Consider, as an example, the most complicated term

$$\begin{aligned} \langle \Psi | \tfrac{1}{2} \sum_{ij}{}' |\boldsymbol{r}_i - \boldsymbol{r}_j|^{-1} | \Psi \rangle &= (N!)^{-1} \sum_{\hat{\mathscr{P}}\hat{\mathscr{P}}'} \varepsilon_{\hat{\mathscr{P}}} \varepsilon_{\hat{\mathscr{P}}'} \\ &\cdot \int dq_1 \ldots dq_N [\hat{\mathscr{P}}' \psi^*_{\nu_1}(q_1) \ldots \psi^*_{\nu_N}(q_N)][\hat{\mathscr{P}} \psi_{\nu_1}(q_1) \ldots \\ &\ldots \psi_{\nu_N}(q_N)] \tfrac{1}{2} \sum_{ij}{}' |\boldsymbol{r}_i - \boldsymbol{r}_j|^{-1} . \end{aligned} \tag{4.6.6}$$

The terms with q_i, q_j in (4.6.6) will be separated out and integrated over the other variables with allowance for (4.6.5). The contribution to (4.6.6) is made by two sets of permutations $\hat{\mathscr{P}}' = \hat{\mathscr{P}}$ and $\hat{\mathscr{P}}' = \hat{\bar{\mathscr{P}}}$, where $\hat{\bar{\mathscr{P}}}$ differs from $\hat{\mathscr{P}}$ in that it has the permutation $v_i \rightleftarrows v_j$. Here $\varepsilon_{\mathscr{P}}^2 = 1$, $\varepsilon_{\mathscr{P}}\varepsilon_{\bar{\mathscr{P}}} = -1$, and we have

$$\langle \Psi | \tfrac{1}{2} \sum_{ij}' |\boldsymbol{r}_i - \boldsymbol{r}_j|^{-1} | \Psi \rangle = \tfrac{1}{2} \sum_{vv'} \int dq\, dq' \psi_v^*(q) \cdot \psi_{v'}^*(q') |\boldsymbol{r} - \boldsymbol{r}'|^{-1} [\psi_v(q)\psi_{v'}(q') - \psi_v(q')\psi_{v'}(q)] \ . \tag{4.6.7}$$

Again, the factor $(N!)^{-1}$ cancels out by combinatorial considerations. Analogously,

$$\begin{aligned} \langle \Psi | \hat{\mathscr{H}} | \Psi \rangle = & \sum_v \int dq \psi_v^*(q) \hat{\mathscr{H}}_0 \psi_v(q) \\ & + \frac{e^2}{2} \sum_{vv'} dq\, dq' \psi_v^*(q) \psi_{v'}^*(q') |\boldsymbol{r} - \boldsymbol{r}'|^{-1} \\ & \cdot [\psi_v(q)\psi_{v'}(q') - \psi_v(q')\psi_{v'}(q)] \ , \end{aligned} \tag{4.6.8}$$

$$\hat{\mathscr{H}}_0 = -\frac{\hbar^2}{2m} \Delta + V_n(\boldsymbol{r}) \ .$$

Taking into account (4.6.5), we solve the variational problem

$$\frac{\delta}{\delta \psi_v^*(q)} \left[\langle \Psi | \hat{\mathscr{H}} | \Psi \rangle - \sum_{v'v} \lambda_{vv'} \int dq \psi_v^*(q) \psi_{v'}(q) \right] = 0 \ . \tag{4.6.9}$$

With allowance for (4.6.8), and carrying out the variation, we obtain

$$\begin{aligned} & \left[-\frac{\hbar^2}{2m} \Delta + V_n(\boldsymbol{r}) + e^2 \int d\boldsymbol{r}' \frac{\varrho(\boldsymbol{r}')}{|\boldsymbol{r} - \boldsymbol{r}'|} \right] \psi_{is}(\boldsymbol{r}\sigma) \\ & - e^2 \sum_{js'\sigma'} \int d\boldsymbol{r}' \psi_{js'}^*(\boldsymbol{r}'\sigma') \psi_{is}(\boldsymbol{r}'\sigma') |\boldsymbol{r} - \boldsymbol{r}'|^{-1} \\ & \cdot \psi_{is}(\boldsymbol{r}\sigma) = \sum_{js'} \lambda_{is,\, js'} \psi_{js'}(\boldsymbol{r}\sigma) \ . \end{aligned} \tag{4.6.10}$$

Here $is = v$, $js' = v'$, where i, j are orbital quantum numbers, and s, s' spin quantum numbers; and

$$\varrho(\boldsymbol{r}) = \sum_{js\sigma} |\psi_{js}(\boldsymbol{r}\sigma)|^2 \tag{4.6.11}$$

is the total electron density at point $\boldsymbol{r}$. At the moment we are taking into account the possible dependence of the orbital part of the wave function on spin projection (spin-polarized self-consistent field, or spin-unrestricted Hartree–Fock approximation). Perform now the unitary transformation of the functions $\psi_{is}(\boldsymbol{r}\sigma)$:

$$\tilde{\psi}_\nu(q) = \sum_\mu c^*_{\nu\mu}\psi_\mu(q) \ , \qquad \psi_\mu(q) = \sum_\nu c_{\nu\mu}\tilde{\psi}_\nu(q) \ . \tag{4.6.12}$$

Here

$$\sum_\mu \psi^*_\mu(q)\psi_\mu(q') = \sum_\nu \tilde{\psi}^*_\nu(q)\tilde{\psi}_\nu(q') \ . \tag{4.6.13}$$

Using the unitary properties of the matrix $\|c_{\nu\mu}\|$, we have

$$\sum_\mu c^*_{\nu\mu}c_{\nu'\mu} = \delta_{\nu\nu'} \ , \tag{4.6.14}$$

which allows us to obtain (4.6.12, 13). Substituting (4.6.12, 13) into (4.6.10), we arrive at the result that (4.6.10) maintains its form if we make the replacement $\psi \to \tilde{\psi}$, $\lambda_{\nu\nu'} \to \sum_{\mu\mu'} c^*_{\mu\nu}\lambda_{\mu\mu'}c_{\mu'\nu'}$. By virtue of the hermiticity of $\|\lambda_{\mu\mu'}\|$, evident from (4.6.10), a unitary matrix $\|c_{\mu\nu}\|$ exists that diagonalizes $\|\lambda_{\mu\nu}\|$. Only this representation will be used. Denoting the eigenvalues of the matrix $\|\lambda_{\mu\nu}\|$ in terms of ε_ν and omitting the tilde, we find

$$\begin{aligned}&\left[-\frac{\hbar^2}{2m}\Delta + V_n(\boldsymbol{r}) + e^2\int\frac{d\boldsymbol{r}'\varrho(\boldsymbol{r}')}{|\boldsymbol{r}-\boldsymbol{r}'|}\right]\psi_{is}(\boldsymbol{r}\sigma)\\ &\quad - e^2\sum_{js'\sigma'}\int d\boldsymbol{r}'\psi^*_{js'}(\boldsymbol{r}'\sigma')\psi_{is}(\boldsymbol{r}'\sigma')|\boldsymbol{r}-\boldsymbol{r}'|^{-1}\psi_{js'}(\boldsymbol{r}\sigma) = \varepsilon_{is}\psi_{is}(\boldsymbol{r}\sigma) \ . \end{aligned} \tag{4.6.15}$$

Equations (4.6.15) are said to be Hartree–Fock equations. The third term enclosed in square brackets is a Coulomb potential, which is produced by all the other electrons acting on a given electron [4.30]. The last term on the left-hand side of (4.6.15), called the exchange interaction, is due to the Pauli exclusion principle and is of a purely quantum origin [4.31]. This interaction is nonlocal in character. Therefore, the relevant equations are very complicated, and currently are not often used to compute the energy spectrum of solids.

Before passing over to further simplifications, we wish to discuss the physical meaning of the parameters ε_{is}. Multiplying (4.6.15) by $\psi^*_{is}(\boldsymbol{r}\sigma)$, taking an integral over $\boldsymbol{r}$, and carrying out summation over σ, we find, with allowance for (4.6.5),

$$\varepsilon_{is} = [is|is] + \sum_{js'}\{[is, js'|is, js'] - [is, js'|js', is]\} \ , \tag{4.6.16}$$

where the notation introduced is

$$[\alpha|\beta] = \int dq \psi_\alpha^*(q) \hat{\mathscr{H}}_0 \psi_\beta(q) \; ,$$
$$[\alpha\beta|\gamma\delta] = e^2 \int dq dq' \psi_\alpha^*(q) \psi_\beta^*(q') |\boldsymbol{r}-\boldsymbol{r}'|^{-1} \psi_\gamma(q) \psi_\delta(q') \; . \tag{4.6.17}$$

The quantity (4.6.17) is equal to the contribution that the terms depending on the coordinates of an electron in the state $|is\rangle$ make to the total energy $\langle \Psi|\hat{\mathscr{H}}|\Psi\rangle$. Therefore, this quantity is equal, with the sign reversed, to the energy required to remove an electron from the state $|is\rangle$, if we neglect any resulting variation of the wave functions of the other electrons (Koopmans' theorem).

Let us discuss the meaning of the restriction made. The removal of one electron leads to a relative variation of the wave functions of all the other electrons by an amount $\propto N^{-1}$. Since the first variation of the total energy with respect to the wave functions is equal to zero, the change is energy due to the variations of ψ_{is} by an amount $\propto N^{-1}$ constitutes $\propto N^{-2}$. When N functions are varied, this change in energy is on the order of N^{-1}; i.e., it is negligibly small [4.32].

We consider N states and N electrons, and for this reason all the states are occupied. Correlation effects may lead, however, to partial filling of some states (as with d and f metals and many of their compounds). Here the entire scheme of our treatment becomes dubious. For example, the $4f$ electrons in a Eu atom in a EuO crystal, hardly sense the presence of electrons on other atoms (because of the small falloff radius of the wave function), while interacting very strongly with other $4f$ electrons in the same atom. Therefore, the variation of the wave function of the other "essential" electrons when this electron is removed is large. The conditions under which Koopmans' theorem and the Hartree–Fock method in general may be applied to compounds of d and f elements are not clear. On the other hand, computational practice shows that some energy characteristics of these materials, calculated using some version of the self-consistent field approximation, give a fairly good fit to experiment [4.33, 34].

As mentioned previously, the Hartree–Fock equations are complicated for band calculations. A certain inconvenience in purely theoretical terms is caused by the fact that the effective Hamiltonian itself depends on the number of the state for which the calculation is performed. Therefore, the last term on the left-hand side of (4.6.15) is replaced by some potential that depends only on the density of particles with a certain spin projection

$$\left\{-\frac{\hbar^2}{2m}\Delta + V_n(\boldsymbol{r}) + e^2 \int d\boldsymbol{r}' \frac{\varrho(\boldsymbol{r}')}{|\boldsymbol{r}-\boldsymbol{r}'|} \right. \tag{4.6.18}$$
$$\left. + v_{xc}[\varrho_\uparrow(\boldsymbol{r}), \varrho_\downarrow(\boldsymbol{r})] \right\} \psi_{is}(\boldsymbol{r}\sigma) = \varepsilon_{is} \psi_{is}(\boldsymbol{r}\sigma) \; .$$

The quantity v_{xc} is called the exchange-correlation potential, and the method based on the replacement of the nonlocal exchange term in (4.6.15) by some averaged potential is referred to as the Hartree–Fock–Slater method.

The nature of the exchange-correlation interaction may be understood from the requirement that the wave function be antisymmetric (4.6.2), from which it follows that for electrons with parallel spins the probability of their being located at the same point is equal to zero. Therefore, by including the potential produced by spin-up electrons and the potential produced by spin-down electrons into the self-consistent field acting on a spin-up electron, we reassess the Coulomb repulsive interaction energy of electrons with parallel spins, since, on the average, these electrons are separated by large distances. The radius of what is known as the Fermi hole around an electron with the spin oriented upward—i.e., the radius of the region in which there are no other such electrons— is on the order of $[\varrho_\uparrow(r)]^{-1/3}$, the corresponding gain in Coulomb energy being about $e^2[\varrho_\uparrow(r)]^{1/3}$. The so-called Slater $X\alpha$ method employs the equation

$$\left[-\frac{\hbar^2}{2m}\Delta + V_n(r) + e^2\int\frac{dr'\varrho(r')}{|r-r'|} - 3\alpha e^2 \cdot\left(\frac{3\varrho_\uparrow(r)}{4\pi}\right)^{1/3}\right]\psi_{i\uparrow}(r) = \varepsilon_{i\uparrow}\psi_{i\uparrow}(r)\ , \quad (4.6.19)$$

which is similar for a spin directed downward. The α in this equation is a numerical parameter, which is somewhat less than unity. For the rules of selecting α values, other problems of the $X\alpha$ method, and its relation to the Hartree–Fock method, we refer the reader to Slater [4.33].

Upon solving (4.6.19) for the initial choice of $\varrho_\uparrow$, $\varrho_\downarrow$, the functions ψ_{is} thus found are used to construct new $\varrho_\uparrow$, $\varrho_\downarrow$, etc., until a self-consistency of the required accuracy is attained. In a self-consistent solution for d and f elements we sometimes have $\varepsilon_{i\uparrow} \neq \varepsilon_{i\downarrow}$, $\varrho_\uparrow \neq \varrho_\downarrow$. This indicates the existence of magnetic moments and some magnetic ordering, which are not considered here. We confine ourselves to a system with $\varrho_\uparrow = \varrho_\downarrow = \varrho/2$; the spin subscripts of ψ_i and ε_i will be omitted. In this case (said to be spin restricted)

$$v_{xc}[\varrho(r)] = -\frac{3}{2}\alpha e^2\left(\frac{3\varrho(r)}{\pi}\right)^{1/3}\ . \quad (4.6.20)$$

A major drawback to the $X\alpha$ method is that the correlation of electrons with antiparallel spins is neglected. A great deal more complicated and exact expressions have been proposed for the exchange-correlation potential; we do not consider them, for they require the use of many-body theory [4.35–37].

4.6.2 Solving the Schrödinger Equation. Formulation of the Problem and the Cellular Method

We now investigate practical methods of solving the Schrödinger equation (4.2.4) for a particular given potential. According to Bloch's theorem (4.2.10), for the form of the wave function $\psi_{k\zeta}(\boldsymbol{r})$ to be known everywhere, it suffices to know its form in any one of the unit cells of the crystal. This function is chosen in its most symmetric form—as a Wigner–Seitz cell—which in the direct lattice is constructed in the same way as the first Brillouin zone in the reciprocal lattice (Sects. 1.2, 4.2.2). The boundaries of this cell $\boldsymbol{r}_s$ are given, similar to (4.2.24), by

$$\boldsymbol{r}_s \cdot \boldsymbol{R}_i = \pm \tfrac{1}{2} |\boldsymbol{R}_i|^2 \ , \tag{4.6.21}$$

where $\boldsymbol{R}_i$ is the translation vector connecting a Bravais lattice site with all the neighboring sites. The wave function satisfies Bloch's theorem, so the following boundary conditions are imposed on the values of its first derivative at the boundary of the cell,

$$\begin{aligned} &\psi_{k\zeta}(\boldsymbol{r}+\boldsymbol{R}) = \mathrm{e}^{\mathrm{i}\boldsymbol{k}\cdot\boldsymbol{R}}\, \psi_{k\zeta}(\boldsymbol{r}) \ , \\ &\frac{\partial \psi_{k\zeta}(\boldsymbol{r}+\boldsymbol{R})}{\partial \boldsymbol{n}(\boldsymbol{r}+\boldsymbol{R})} = -\,\mathrm{e}^{\mathrm{i}\boldsymbol{k}\cdot\boldsymbol{R}} \frac{\partial \psi_{k\zeta}(\boldsymbol{r})}{\partial \boldsymbol{n}(\boldsymbol{r})} \ , \end{aligned} \tag{4.6.22}$$

where $\boldsymbol{r}$ and $\boldsymbol{r}+\boldsymbol{R}$ belong to opposite cell faces, $\boldsymbol{R}$ is the translation vector that connects them (Fig. 4.33), and $\partial/\partial\boldsymbol{n}$ is the derivative with respect to the normal to the cell surface, with $\boldsymbol{n}(\boldsymbol{r}+\boldsymbol{R}) = -\,\boldsymbol{n}(\boldsymbol{r})$. The Schrödinger equation is of second order, and therefore the boundary conditions for $\psi_{k\zeta}$ and $\partial\psi_{k\zeta}/\partial\boldsymbol{n}$ suffice; for higher-order derivatives they will be fulfilled automatically.

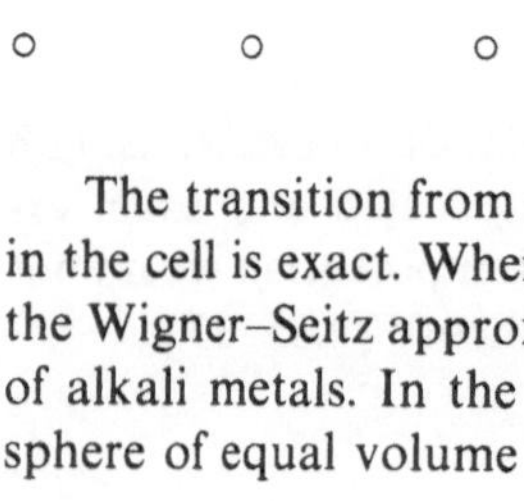

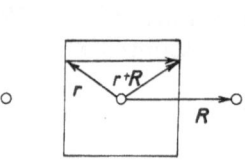

Fig. 4.33. Derivation of the boundary conditions on the surfaces of the Wigner–Seitz cell

The transition from the Schrödinger equation in the crystal to the equation in the cell is exact. When band theory was in its infancy, an important tool was the Wigner–Seitz approximation, which enable estimation of the binding energy of alkali metals. In the Wigner–Seitz approximation the cell is replaced by a sphere of equal volume (of radius r_s). It is postulated that the potential in this sphere is spherically symmetric, and the case of $\boldsymbol{k} = 0$ (bottom of conduction band) is considered. Then (4.6.22) holds if ψ depends only on $|\boldsymbol{r}|$ and satisfies the condition

$$\frac{\partial\psi(|\boldsymbol{r}|)}{\partial|\boldsymbol{r}|} = 0 \ , \quad |\boldsymbol{r}| = r_s \ . \tag{4.6.23}$$

This problem differs from the definition of the energy of the s state of the atom only in that it has the boundary condition (4.6.23) instead of the condition $\psi(|\boldsymbol{r}|) \to 0$ for $|\boldsymbol{r}| \to \infty$. This change results in a shift of eigenenergies. By lowering the energy of, say, the 3s level of Na (bottom of the 3s band) in comparison with the energy of a free atom, we can estimate the binding energy per atom, and, by varying r_s—i.e., actually the density—and exploring the energy variation, we can evaluate the compressibility, in fairly good agreement with experiment. The Wigner–Seitz method, however, has a very narrow scope of applicability and, with the advent of powerful computers, has been superseded by more exact methods, although they are more cumbersome computationally.

If we expand the required function in some orthonormalized set of functions satisfying (4.6.22), the Schrödinger equation will transform into an infinite system of linear algebraic equations for the expansion coefficients. The ordinary approximation is to "truncate" this expansion. We wish to enlarge upon this problem.

As discussed in Sect. 4.6.1, the Schrödinger equation is equivalent to the variational principle

$$\delta \int_{\Omega} d\boldsymbol{r}\psi^*(\boldsymbol{r})(\hat{\mathscr{H}} - E)\psi(\boldsymbol{r}) = 0 \ , \tag{4.6.24}$$

where $\hat{\mathscr{H}}$ is a one-particle Hamiltonian (4.2.4), and the integration is taken over the Wigner–Seitz cell Ω. As in Sect. 4.6.1, we assume the wave function and energy to be spin independent. The potential $V(\boldsymbol{r})$ is assumed to be an independent function and not to vary. The $\psi_{\boldsymbol{k}}(\boldsymbol{r})$ may be sought as an expansion in some finite set of functions $\varphi_\mu(\boldsymbol{k}, \boldsymbol{r})$ (not necessarily orthonormalized) satisfying the conditions (4.6.22)

$$\psi_{\boldsymbol{k}}(\boldsymbol{r}) = \sum_{\mu} c_\mu(\boldsymbol{k})\varphi_\mu(\boldsymbol{k}, \boldsymbol{r}) \ . \tag{4.6.25}$$

Here the coefficients $c_\mu^*(\boldsymbol{k})$ and $c_\mu(\boldsymbol{k})$ are treated as trial parameters. This is what we call the approximate variational Ritz–Galerkin method. It stands to reason that the more functions $\varphi_\mu(\boldsymbol{k}, \boldsymbol{r})$ we take and the happier the choice made, the more exact the method. The various band calculation methods differ from each other in the choice of the set of $\{\varphi_\mu\}$. We substitute (4.6.25) into (4.6.24) and carry out the variation with respect to $c_\mu^*(\boldsymbol{k})$ (the calculation is similar to that presented in Sect. 4.6.1, but is much simpler). This yields a system of equations

$$\sum_\nu [\mathscr{H}_{\mu\nu}(\boldsymbol{k}) - ES_{\mu\nu}(\boldsymbol{k})]c_\nu(\boldsymbol{k}) = 0 \ , \tag{4.6.26}$$

where

$$\begin{aligned} \mathscr{H}_{\mu\nu}(\boldsymbol{k}) &= \int_\Omega d\boldsymbol{r}\,\varphi_\mu^*(\boldsymbol{k},\boldsymbol{r})\hat{\mathscr{H}}\varphi_\nu(\boldsymbol{k},\boldsymbol{r}) \ , \\ S_{\mu\nu}(\boldsymbol{k}) &= \int_\Omega d\boldsymbol{r}\,\varphi_\mu^*(\boldsymbol{k},\boldsymbol{r})\varphi_\nu(\boldsymbol{k},\boldsymbol{r}) \ . \end{aligned} \tag{4.6.27}$$

Variation with respect to $c_\mu(\boldsymbol{k})$ yields a system of equations which differ from (4.6.26) only in complex conjugation. The solvability condition for (4.6.26) has the form

$$\det \| \mathscr{H}_{\mu\nu}(\boldsymbol{k}) - ES_{\mu\nu}(\boldsymbol{k}) \| = 0 \ . \tag{4.6.28}$$

Numerical solution of (4.6.28) enables us to determine the spectrum $E(\boldsymbol{k}\zeta)$.

4.6.3 The LCAO Method and Tight-Binding Approximation

Let us try to choose the wave functions of electrons in an atom, $\chi_\mu(\boldsymbol{r})$, as the basis functions, similar to the treatment in the one-dimensional model problem considered in Sect. 4.1.3. Admittedly, they do not satisfy Bloch's boundary conditions, but the solution of (4.1.46) obtained in the model problem points to a way out of this difficulty. We introduce the functions

$$\varphi_\mu(\boldsymbol{k},\boldsymbol{r}) = N^{-1/2}\sum_p \chi_\mu(\boldsymbol{r}-\boldsymbol{R}_p)\exp(\mathrm{i}\boldsymbol{k}\cdot\boldsymbol{R}_p) \ , \tag{4.6.29}$$

which, as can be readily verified, satisfy Bloch's theorem and, consequently, the boundary conditions (4.6.22). The functions φ_μ are linear combinations of atomic functions, to which the LCAO (linear combinations of atomic orbitals) method owes its name. Substitute (4.6.29) into (4.6.27):

$$\begin{aligned} \mathscr{H}_{\mu\nu}(\boldsymbol{k}) &= N^{-1}\sum_{pq}\exp[\mathrm{i}\boldsymbol{k}\cdot(\boldsymbol{R}_q-\boldsymbol{R}_p)]\int_\Omega d\boldsymbol{r}\,\chi_\mu^*(\boldsymbol{r}-\boldsymbol{R}_p)\hat{\mathscr{H}}\chi_\nu(\boldsymbol{r}-\boldsymbol{R}_q) \\ &= \sum_n \exp(-\mathrm{i}\boldsymbol{k}\cdot\boldsymbol{R}_n)\int_\Omega d\boldsymbol{r}\,\chi_\mu^*(\boldsymbol{r}-\boldsymbol{R}_n)\hat{\mathscr{H}}\chi_\nu(\boldsymbol{r}) \ , \\ S_{\mu\nu}(\boldsymbol{k}) &= \sum_n \int_\Omega d\boldsymbol{r}\exp(-\mathrm{i}\boldsymbol{k}\cdot\boldsymbol{R}_n)\chi_\mu^*(\boldsymbol{r}-\boldsymbol{R}_n)\chi_\nu(\boldsymbol{r}) \ . \end{aligned} \tag{4.6.30}$$

Here we have made the replacement of variables in the integral $\boldsymbol{r} \to \boldsymbol{r} - \boldsymbol{R}_q$ and have introduced the notation $\boldsymbol{R}_n = \boldsymbol{R}_p - \boldsymbol{R}_q$.

In principle, if we take a sufficient number of basis functions (4.6.29), we can obtain a fairly good description of the energy spectrum of the crystal by substituting (4.6.30) into (4.6.28). The wave functions of the continuous spectrum of the atom are, as a matter of fact, not included in the set of the expansion so that the system of functions (4.6.29) is not complete and the accuracy of the method is restricted. Clearly, the LCAO method reproduces the core states—at least the *d* band—better than, say, the states of conduction electrons in alkali metals. This is so because the states of electrons in narrow bands, with a weak overlap of the wave functions on different sites, are largely determined by the atomic states, whereas an ordinary plane wave that would provide a fairly good description of the state of conduction electrons in metals is very difficult to construct by use of (4.6.29) one has to take the complete set of atomic states including the functions of the continuous spectrum of the atom.

On the other hand, as is obvious from Sect. 4.5.3, in narrow energy bands the correlation effects are substantial. Therefore, the wave functions and one-electron energies thus found should be substituted further into some many-electron model; i.e., they are related to the observables only indirectly. Nevertheless, the approximation concerned has still been used in band calculations, although not as frequently as some of the more sophisticated methods (see below).

In the various model calculations the tight-binding approximation [4.2] is frequently used, in which the basis of the LCAO expansion contains one function: one atomic level, one band. This approximation, strictly speaking, is applicable when the bandwidth is much smaller than the distance to the neighboring atomic level (Fig. 4.34). As a model approximation, it may, however, be employed in a more general case, since it frequently reproduces the character of the spectrum qualitatively correctly. From (4.6.26) we obtain

$$E_\mu(\boldsymbol{k}) = \mathscr{H}_{\mu\mu}(\boldsymbol{k})/S_{\mu\mu}(\boldsymbol{k}) \ . \tag{4.6.31}$$

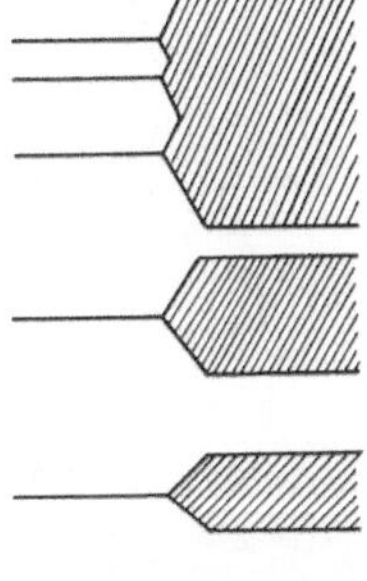

Fig. 4.34. Derivation of the applicability conditions for the tight-binding approximation

Normally we assume that $S_{\mu\mu}(\boldsymbol{k}) \approx 1$, i.e.; the overlap integral of the atomic functions on different sites is neglected:

$$\int_{\Omega} d\boldsymbol{r}\chi_{\mu}^{*}(\boldsymbol{r}-\boldsymbol{R}_n)\chi_{\mu}(\boldsymbol{r}) \approx \delta_{n0} \ . \tag{4.6.32}$$

Substitution of (4.6.30) into (4.6.31) yields

$$E_{\mu}(\boldsymbol{k}) = \varepsilon_{\mu} + \Delta\varepsilon_{\mu} + \sum_{n \neq 0} \beta_{\mu n} \exp(-\mathrm{i}\boldsymbol{k}\cdot\boldsymbol{R}_n) \ . \tag{4.6.33}$$

Here

$$\begin{aligned} \beta_{\mu n} &= \int_{\Omega} d\boldsymbol{r}\, \chi_{\mu}^{*}(\boldsymbol{r}-\boldsymbol{R}_n)\hat{\mathscr{H}}\chi_{\mu}(\boldsymbol{r}) \ , \\ \varepsilon_{\mu} + \Delta\varepsilon_{\mu} &= \int_{\Omega} d\boldsymbol{r}\, \chi_{\mu}^{*}(\boldsymbol{r})\hat{\mathscr{H}}\chi_{\mu}(\boldsymbol{r}) \ , \end{aligned} \tag{4.6.34}$$

with ε_{μ} denoting the energy of the atomic level, and $\Delta\varepsilon_{\mu}$ the shift of this level owing to the variation of the potential in comparison with the atomic potential. The fact that the integration (4.6.34) is not over the entire space but only over the cell is immaterial, since the falloff radius of $\chi_{\mu}(\boldsymbol{r})$ is small compared with r_s; $\beta_{\mu n}$ are called transfer integrals, which decrease rapidly with increasing n. The nearest-neighbor approximation consists of only the integral $\beta_{\mu 1}$ being retained in (4.6.33). By symmetry, this integral for s states (only) depends on $|\boldsymbol{R}_n|$ alone—not on its direction

$$E_{\mu}(\boldsymbol{k}) = \varepsilon_{\mu} + \Delta\varepsilon_{\mu} + \beta_{\mu 1}\sum_{\delta} \exp(-\mathrm{i}\boldsymbol{k}\cdot\boldsymbol{R}_{\delta}) \ , \tag{4.6.35}$$

where the integral is summed over the nearest neighbors.

Using the expression for the translation vectors in simple lattices (sc, bcc, fcc, see Chap. 1), we obtain (the calculus is left as an exercise)

$$E_{\mu}(\boldsymbol{k}) = \varepsilon_{\mu} + \Delta\varepsilon_{\mu} + 2\beta_{\mu 1}(\cos k_x a + \cos k_y a + \cos k_z a) \tag{4.6.36}$$

for sc lattices,

$$\begin{aligned} E_{\mu}(\boldsymbol{k}) = \varepsilon_{\mu} + \Delta\varepsilon_{\mu} + 4\beta_{\mu 1}\bigg(& \cos\frac{k_x a}{2}\cos\frac{k_y a}{2} + \cos\frac{k_x a}{2} \\ & \times \cos\frac{k_z a}{2} + \cos\frac{k_y a}{2}\cos\frac{k_z a}{2}\bigg) \end{aligned} \tag{4.6.37}$$

for fcc lattices, and

$$E_\mu(\boldsymbol{k}) = \varepsilon_\mu + \Delta\varepsilon_\mu + 8\beta_{\mu 1}\cos\frac{k_x a}{2}\cos\frac{k_y a}{2}\cos\frac{k_z a}{2} \tag{4.6.38}$$

for bcc lattices, a being the edge of the elementary cube. It follows from (4.6.38) that the bandwidth in the approximation of nearest neighbors with the coordination number z is equal to

$$\begin{aligned} \Delta E_\mu &= 2z|\beta_{\mu 1}| \ , \\ \Delta E_\mu &= 16|\beta_{\mu 1}| \ , \end{aligned} \tag{4.6.39}$$

for sc and bcc lattices, and for fcc lattices, respectively.

As follows from the exact formula (4.2.20), the band energy is exactly represented in the form of (4.6.33) if β is taken to imply some undefined parameters.

We can also introduce relevant wave functions, similar to $\varphi_\mu(\boldsymbol{r})$, such that the condition holds

$$\psi_{k\zeta}(\boldsymbol{r}) = N^{-1/2}\sum_n \varphi_{n\zeta}(\boldsymbol{r})\exp(\mathrm{i}\boldsymbol{k}\cdot\boldsymbol{R}_n) \ , \tag{4.6.40}$$

where $\psi_{k\zeta}(\boldsymbol{r})$ is an exact Bloch function. The functions $\varphi_{n\zeta}(\boldsymbol{r})$ are said to be Wannier functions. For narrow bands they resemble atomic functions and, at any rate, exhibit a maximum on the nth site. To find the explicit form of $\varphi_{n\zeta}(\boldsymbol{r})$, we multiply (4.6.40) by $N^{-1/2}\exp(-\mathrm{i}\boldsymbol{k}\cdot\boldsymbol{R}_m)$ and perform a summation over k with respect to the Brillouin zone. Making use of (4.5.3), we obtain

$$\begin{aligned} N^{-1/2}\sum_k \psi_{k\zeta}(\boldsymbol{r})\exp(-\mathrm{i}\boldsymbol{k}\cdot\boldsymbol{R}_m) &= N^{-1}\sum_n \varphi_{n\zeta}(\boldsymbol{r}) \\ \cdot\sum_k \exp[\mathrm{i}\boldsymbol{k}\cdot(\boldsymbol{R}_n-\boldsymbol{R}_m)] &= \sum_n \varphi_{n\zeta}(\boldsymbol{r})\delta_{nm} = \varphi_{m\zeta}(\boldsymbol{r}) \ . \end{aligned} \tag{4.6.41}$$

In contrast to the atomic functions $\chi_\mu(\boldsymbol{r}-\boldsymbol{R}_m)$, the Wannier functions are strictly orthonormalized:

$$\begin{aligned} \int_\Omega d\boldsymbol{r}\,\varphi^*_{m\zeta}(\boldsymbol{r})\varphi_{m'\zeta'}(\boldsymbol{r}) &= N^{-1}\sum_{kk'}\int_\Omega d\boldsymbol{r}\,\psi^*_{k\zeta}(\boldsymbol{r})\psi_{k'\zeta'}(\boldsymbol{r}) \\ &\quad\cdot\exp(\mathrm{i}\boldsymbol{k}\cdot\boldsymbol{R}_m-\mathrm{i}\boldsymbol{k}'\cdot\boldsymbol{R}_{m'}) = N^{-1}\sum_{kk'}\exp(\mathrm{i}\boldsymbol{k}\cdot\boldsymbol{R}_m \\ &\quad-\mathrm{i}\boldsymbol{k}'\cdot\boldsymbol{R}_{m'})\delta_{kk'}\delta_{\zeta\zeta'} = \delta_{\zeta\zeta'}N^{-1}\sum_k \exp[\mathrm{i}\boldsymbol{k}\cdot(\boldsymbol{R}_m-\boldsymbol{R}_{m'})] \\ &= \delta_{\zeta\zeta'}\delta_{mm'} \ . \end{aligned} \tag{4.6.42}$$

Here and henceforth we assume the Bloch function to be normalized to a single cell (Sect. 4.2.3)

$$\int_{\Omega} d\boldsymbol{r}\,\psi^*_{\boldsymbol{k}\zeta}(\boldsymbol{r})\psi_{\boldsymbol{k}'\zeta'}(\boldsymbol{r}) = \delta_{\boldsymbol{k}\boldsymbol{k}'}\delta_{\zeta\zeta'} \ . \tag{4.6.43}$$

Wannier functions are utilized in considering the quasi-classical dynamics of an electron, localized states, some magnetic phenomena, etc.

4.6.4 The Orthogonalized Plane Waves (OPW) Method. Pseudopotential

As stated previously, the LCAO method is of little use for describing the states of conduction electrons in metals. The plane wave (PW) method suits this purpose better. The basis functions chosen in the PW method are as follows:

$$\chi_\mu(\boldsymbol{k}, \boldsymbol{r}) = \exp[\mathrm{i}(\boldsymbol{k} + \boldsymbol{b}^*_g)\cdot\boldsymbol{r}]/V_0^{1/2} \ . \tag{4.6.44}$$

The functions $\chi_\mu(\boldsymbol{k}, \boldsymbol{r})$ satisfy (4.6.22). The expansion of $\psi_{\boldsymbol{k}\zeta}(\boldsymbol{r})$ in (4.6.44) was employed in (4.2.13). The above expansion is convenient in proving the various properties of the Bloch functions, but is practically unacceptable for real computations because the convergence of the method is very poor; for reasonable results to be obtained, a large number of functions have to be included in the expansion of $\psi_{\boldsymbol{k}\zeta}(\boldsymbol{r})$. The point is that, before arriving at the states of conduction electrons, we need to reproduce the ground states $\psi_c(\boldsymbol{k}, \boldsymbol{r})$, which lie lower in energy and are completely dissimilar to plane waves. It is far more convenient to seek the conduction electron states by the LCAO method. Herring [4.38] suggested that orthogonalized plane waves (OPW) be used as the expansion basis

$$\varphi_\mu(\boldsymbol{k}, \boldsymbol{r}) = \chi_\mu(\boldsymbol{k}, \boldsymbol{r}) - \sum_c \psi_c(\boldsymbol{k}, \boldsymbol{r}) \int_{\Omega} d\boldsymbol{r}'\,\psi^*_c(\boldsymbol{k}, \boldsymbol{r}')\chi_\mu(\boldsymbol{k}, \boldsymbol{r}') \ . \tag{4.6.45}$$

These waves satisfy (4.6.22) and, besides, are orthogonal to all the core states (it is, in a way, a matter of convenience to decide which states exactly should be included in the core states):

$$\int_{\Omega} d\boldsymbol{r}\,\varphi^*_\mu(\boldsymbol{k}, \boldsymbol{r})\psi_c(\boldsymbol{k}, \boldsymbol{r}) = 0 \ . \tag{4.6.46}$$

Equation (4.6.46) follows immediately from (4.6.45) and the orthonormalization of the core states.

In the expansion in the functions (4.6.45), the core states cancel out, and we immediately seek the conduction-electron states, which are already well described by a linear combination of a small number of OPW. The convergence improves drastically.

The OPW method is rather more than merely a lucky computational subterfuge. It has given birth to the concept of the pseudopotential, which has exerted a tremendous influence on the entire solid-state theory [1.31, 4.39, 40].

The main idea of the pseudopotential method is as follows: In many cases we are interested in a small segment of the energy spectrum near the Fermi energy; only these states determine the thermodynamics of the metal, its kinetic properties at frequencies that are not too high, phonon spectra, etc. (but not optical or X-ray spectra). An effective Hamiltonian can be selected in which this area of the spectrum coincides with the exact region (for example, the core states are thrown away, as was done in the OPW method). The throwing away of a large number of bound states corresponds to an abrupt decrease in the depth of the potential well on each site. One may expect that the effective potential (pseudopotential) that has arisen will already be so weak that it can be taken into account using perturbation theory. As mentioned in Sect. 4.2.5, the possibility of constructing this pseudopotential accounts for the surprising success of the free-electron approximation and, especially, the nearly-free-electron approximation in describing conduction electrons in metals. The pseudopotential is defined ambiguously and may be introduced in many ways.

One of the possible methods of constructing a pseudopotential is as follows: Let the Schrödinger equation have a spectrum $E_1, E_2, \ldots$ and relevant eigenfunctions $\psi_1, \psi_2, \ldots$. The energy levels of interest are those beginning from $(n+1)$. Try ψ_m for $m \geq n+1$ in the form

$$\psi = \varphi - \sum_{m=1}^{n} \psi_m(\psi_m, \varphi) \,, \tag{4.6.47}$$

with (ψ, φ) being a scalar product. Here

$$(\psi, \psi_m) = 0 \,, \quad m = 1, 2, \ldots, n \tag{4.6.48}$$

since $(\psi_m, \psi_{m'}) = 0$ for $m \neq m' \leqslant n$. Substituting (4.6.47) into (4.2.4) gives

$$\hat{\mathscr{H}}\varphi - \sum_{m=1}^{n} (\hat{\mathscr{H}}\psi_m)(\psi_m, \varphi) = E\varphi - \sum_{m=1}^{n} E\psi_m(\psi_m, \varphi)$$

or

$$\hat{\mathscr{H}}_{\text{ef}}\varphi = E\varphi \,, \tag{4.6.49}$$

where

$$\hat{\mathscr{H}}_{\text{ef}} = \hat{\mathscr{H}} + \sum_{m=1}^{n} (E - E_m)\psi_m(\psi_m, \ldots) \equiv -\frac{\hbar^2}{2m}\Delta + V(\boldsymbol{r}) + \hat{V}_{\text{p}} \,, \tag{4.6.50}$$

$V + \hat{V}_{\text{p}}$ is the pseudopotential.

The functions $\psi_{n+1}, \psi_{n+2}, \ldots$ and the corresponding eigenvalues $E_{n+1}, E_{n+2}, \ldots$ can be obtained from (4.6.49), since for $\psi_m (m > n)$ the second term in (4.6.47, 50) is identically equal to zero. We exclude the eigenvalues $E_1, E_2, \ldots, E_n$. The pseudoHamiltonian (4.6.50) is nonlocal and, besides, depends on energy, although the latter is immaterial, since we can always add to $\hat{\mathscr{H}}_{\text{ef}}$ an arbitrary term of the form

$$\hat{\mathscr{H}}' = \sum_{m=1}^{n} \psi_m(f_m, \ldots) , \tag{4.6.51}$$

where f_m are arbitrary functions of coordinates and energy. This comes from the fact that the operator $\hat{\mathscr{H}}'$ has zero matrix elements between any two functions $\psi_m, \psi_{m'} (m, m' > n)$ and, therefore, does not affect the energy spectrum.

The nonlocal nature of the pseudopotential means that its matrix elements between plane waves $(\boldsymbol{k}| V + \hat{V}_{\text{p}}|\boldsymbol{k}')$ depending not only on the difference $\boldsymbol{k} - \boldsymbol{k}'$ but also on each $\boldsymbol{k}$, $\boldsymbol{k}'$. This fact should be borne in mind in pseudopotential calculations involving the use of perturbation theory. In practice, the pseudopotential is constructed either analytically or by selecting a model with several fitting parameters, which are determined from experiment and then used to interpret a great many other experiments.

As follows from the definition of the pseudopotential in (4.6.50), it is a sum of the periodic potential of a real crystal, $V(\boldsymbol{r})$ in (4.2.4), and some operator $\hat{V}_{\text{p}}$ which is involved in (4.6.50). The diagonal matrix elements of these summands are of different sign, negative for $V(\boldsymbol{r})$ (in the ion core region) and positive for the operator V_{p}: $(\psi|\hat{V}_{\text{p}}|\psi) = \sum_{m=1}^{n} (E - E_m)|(\varphi, \psi_m)|^2$ (since $E > E_m$ for $m \leqslant n$). This difference in sign may lead, as stated in the foregoing, to an appreciable decrease of the effective potential. Thus we have ascertained why conduction electrons in metals may, to a good approximation, be viewed as free.

In the free-electron model the Fermi surface is a sphere of radius k_{F}. The magnitude of this radius is determined by the electron density n according to (3.33). If the density is such that the diameter $2k_{\text{F}}$ is larger than the minimum "diameter" of the first Brillouin zone (for example, $2\pi/a$, as depicted in Fig. 4.12), then the Fermi sphere will enter partially also into the second zone and, with further increase in density, into the third zone, etc. If we then go from the extended-zone scheme (Fig. 4.12) to the reduced-zone scheme, we will get the picture of multiply connected Fermi surfaces, as already shown in Fig. 4.23. Experimental determination of Fermi surfaces with the help of the de Haas–van Alphen effect etc. provides in many cases a good confirmation of the theoretical prediction of their shape on the basis of Harrison's method [4.32].

4.6.5 The Augmented Plane Waves (APW) Method

In 1937 Slater proposed a very fruitful approach to the calculation of the energy spectrum of crystals [4.41]. His method is based on the following idea: At the center of the Wigner–Seitz cell the potential $V(\boldsymbol{r})$ is chiefly of an atomic character; near the boundaries of the cell it is approximately constant (Fig. 4.35). For simplicity, we adopt the approximation

$$V(\boldsymbol{r}) = \begin{cases} V(|\boldsymbol{r}|) , & |\boldsymbol{r}| < r_0 \\ V_0^{\mathrm{MT}} , & |\boldsymbol{r}| > r_0 , \end{cases} \tag{4.6.52}$$

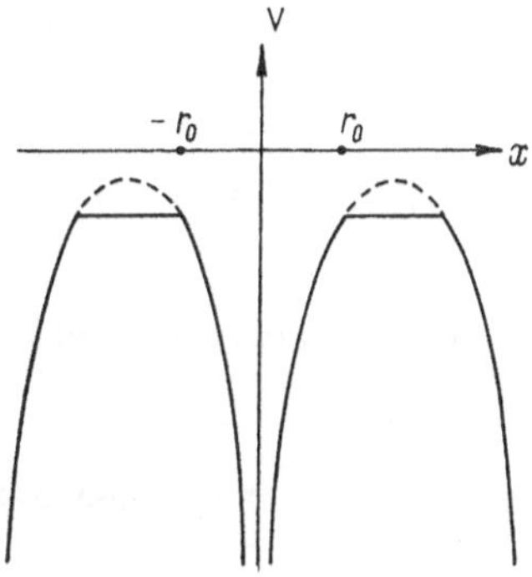

Fig. 4.35. Definition of the muffin-tin (MT) approximation

where r_0 is the radius of some sphere inscribed into the Wigner–Seitz cell. This approximation is called the muffintin (MT) approximation, and the sphere is called the MT sphere, or the Slater sphere. There is certainly some arbitrariness in the choice of the potential, but it is assumed that the MT approximation does not affect results very much.

The zero of energy is chosen in such a way that $V_0^{\mathrm{MT}} \equiv 0$. Due to the spherical symmetry of the potential, the wave functions within the MT sphere may be sought as an expansion in spherical harmonics $Y_{lm}(\vartheta, \varphi)$

$$\chi(\boldsymbol{r}) = \sum_{lm} \alpha_{lm} R_l(r, E)\, Y_{lm}(\vartheta, \varphi) , \quad r = |\boldsymbol{r}| , \tag{4.6.53}$$

where ϑ and φ are the polar angles of the vector $\boldsymbol{r}$, and $R_l(r, E)$ satisfies the equation

$$\left(\frac{\partial^2}{\partial r^2} + \frac{2}{r}\frac{\partial}{\partial r} - \frac{l(l+1)}{r^2}\right) R_l(r, E) + \frac{2m}{\hbar^2}[E - V(r)] R_l(r, E) = 0 , \tag{4.6.54}$$

$$r < r_0$$

and the condition of finiteness at $r = 0$. A unique solution of this problem exists, whatever E may be. In the Slater method the wave functions in which the expansion is carried out are chosen in the form (we recall that V_0 is the unit-cell volume)

$$\varphi_\mu(\boldsymbol{k},\boldsymbol{r}) = \begin{cases} V_0^{-1/2}\exp[\mathrm{i}(\boldsymbol{k}+\boldsymbol{b}^*_{g_\mu})\cdot\boldsymbol{r}], & r > r_0 \\ V_0^{-1/2}\sum_{lm} \mathrm{i}^l a_{lm}^{(\mu)} R_l(r,E)\, Y_{lm}(\vartheta,\varphi), & r < r_0\,. \end{cases} \tag{4.6.55}$$

These are augmented plane waves, APW. Evidently, they satisfy the Bloch-boundary conditions. The coefficients $a_{lm}^{(\mu)}$ may be determined from the requirement that the functions φ_μ be continuous at $r = r_0$. Using the familiar expansion of a plane wave in spherical functions

$$\exp[\mathrm{i}(\boldsymbol{k}+\boldsymbol{b}^*_{g_\mu})\cdot\boldsymbol{r}] = 4\pi\sum_{lm}\mathrm{i}^l j_l(|\boldsymbol{k}+\boldsymbol{b}^*_{g_\mu}|r)\, Y_{lm}(\vartheta,\varphi)\, Y^*_{lm}(\vartheta_\mu,\varphi_\mu)\,, \tag{4.6.56}$$

where j_l stands for spherical Bessel functions, and ϑ_μ and φ_μ are the polar angles of the vector $\boldsymbol{k}+\boldsymbol{b}^*_{g_\mu}$, we find

$$a_{lm}^{(\mu)} = 4\pi j_l(|\boldsymbol{k}+\boldsymbol{b}^*_{g_\mu}|r)_0\, Y^*_{lm}(\vartheta_\mu,\varphi_\mu)/R_l(r_0,E)\,. \tag{4.6.57}$$

Unfortunately, the basis functions $\varphi_\mu(\boldsymbol{k},\boldsymbol{r})$ have no continuous derivative. Slater avoided this difficulty by variational means. The variational principle (4.6.24) in the Wigner–Seitz cell is equivalent to the following:

$$\delta I = 0\,,$$

$$I = \frac{\hbar^2}{2m}\int_\Omega d\boldsymbol{r}|\nabla\psi|^2 + \int_\Omega d\boldsymbol{r}\psi^*(\boldsymbol{r})[V(\boldsymbol{r})-E]\psi(\boldsymbol{r})\,. \tag{4.6.58}$$

To prove this, we make use of the Green formula

$$\begin{aligned}\int_\Omega d\boldsymbol{r}[\psi^*(\boldsymbol{r})\Delta\psi(\boldsymbol{r}) + |\nabla\psi(\boldsymbol{r})|^2] &= \int_\Omega d\boldsymbol{r}\,\nabla(\psi^*(\boldsymbol{r})\nabla\psi(\boldsymbol{r})) \\ &= \int d\boldsymbol{S}\cdot\psi^*(\boldsymbol{r})\frac{\partial\psi(\boldsymbol{r})}{\partial\boldsymbol{n}}\,,\end{aligned} \tag{4.6.59}$$

where the integration is carried out over the cell boundaries, and $d\boldsymbol{S}$ is an element of area. By virtue of (4.6.22), as can be readily deduced, the integral on the right-hand side of (4.6.59) is identically equal to zero and the functional (4.6.58) coincides with the variable in (4.6.24). The latter variable functional, however, is meaningless for the class of functions that have no continuous derivative, since $\Delta\psi(\boldsymbol{r})$ does not exist for them.

The functional (4.6.58), retains its meaning also for trial functions with a derivative that is discontinuous when $r = r_0$. We seek $\psi_{\boldsymbol{k}}(\boldsymbol{r})$ as an expansion of

(4.6.25) in functions (4.6.55) so as to minimize $I[\psi]$. We divide the integration in (4.6.58) into two regions: $r < r_0$ and $r > r_0$. Using (4.6.59) for each of the regions, we obtain

$$\begin{aligned} I = & \int_{r<r_0} d\boldsymbol{r}\psi^*(\boldsymbol{r})\left[-\frac{\hbar^2}{2m}\Delta + V(\boldsymbol{r}) - E\right]\psi(\boldsymbol{r}) \\ & + \int_{r>r_0} d\boldsymbol{r}\psi^*(\boldsymbol{r})\left[-\frac{\hbar^2}{2m}\Delta + V(\boldsymbol{r}) - E\right]\psi(\boldsymbol{r}) \\ & + r_0^2\int d\omega\left[\psi^*(r_0,\vartheta,\phi)\frac{\partial\psi(r,\vartheta,\varphi)}{\partial r}\bigg|_{r=r_0-0} - \psi^*(r_0,\vartheta,\varphi)\right. \\ & \left.\cdot\frac{\partial\psi(r,\vartheta,\varphi)}{\partial r}\bigg|_{r=r_0+0}\right], \end{aligned} \tag{4.6.60}$$

with $d\omega$ an element of the solid angle. The first term in (4.6.60) on the class of functions (4.6.55) goes to zero in virtue of (4.6.54); in the second term we may set $V(\boldsymbol{r}) = 0$. Substituting $\psi_{\boldsymbol{k}}(\boldsymbol{r})$ as a series (4.6.25) into (4.6.60) and varying it with respect to $c^*_\mu(\boldsymbol{k})$, we obtain a system of linear equations (4.6.26) and the spectrum equation in the form of (4.6.28). Since the basis of the expansion here depends on energy E, (4.6.28) does not have the form of an ordinary eigenvalue problem. This shortcoming is compensated for by the rapid convergence of the method, i.e., the comparatively small order of magnitude of the determinant. Good results have already been obtained with this method.

The APW method is an ideological basis for other methods employing the MT approximation, which differ only in the way solutions join at $r = r_0$. A technique that has recently been enjoying great popularity is the Green's-function method, or the Korringa–Kohn–Rostoker (KKR) method, expounded in a review by *Ziman* [4.40]. A detailed bibliography of band calculations for specific compounds is contained in *Slater*'s monograph [4.33]. The methods of band model calculations have also been considered in great detail in [4.42, 43].

4.6.6 $\boldsymbol{k}\cdot\boldsymbol{p}$ Perturbation Theory

Except for the model pseudopotential method, the above techniques of calculating the energy spectrum of a solid employ only the most general information about its properties—i.e., its chemical composition and crystal structure. This leads to some rather gross approximations in the energy-spectrum calculations. Furthermore, all these methods are cumbersome to compute. Thus, for a simple interpretation of experiments, we very frequently and successfully employ empirical methods that allow the energy spectrum to be expressed in terms of some parameters that are determined from experiment. One of the most important of these methods is the $\boldsymbol{k}\cdot\boldsymbol{p}$ perturbation theory in the theory of semiconductors.

The properties of semiconductors are determined by the electron states near the bottom of the conduction band and by the hole states near the top of the valence band, because the number of carriers is small. Thus only a small region of $\boldsymbol{k}$ space is important.

As an example, we consider the states near the bottom of the conduction band, the bottom itself corresponding to $\boldsymbol{k} = 0$. We make use of a Schrödinger equation taking into account the Bloch form of the functions (4.2.11):

$$\left[\frac{\hat{\boldsymbol{p}}^2}{2m} + V(\boldsymbol{r})\right] u(\boldsymbol{k}, \boldsymbol{r}) \exp(\mathrm{i}\boldsymbol{k}\cdot\boldsymbol{r}) = E u(\boldsymbol{k}, \boldsymbol{r}) \exp(\mathrm{i}\boldsymbol{k}\cdot\boldsymbol{r}) , \tag{4.6.61}$$

where $\hat{\boldsymbol{p}} = -\mathrm{i}\hbar\nabla$. Using the relation

$$\hat{\boldsymbol{p}}^2 u(\boldsymbol{k}, \boldsymbol{r}) \exp(\mathrm{i}\boldsymbol{k}\cdot\boldsymbol{r}) = \exp(\mathrm{i}\boldsymbol{k}\cdot\boldsymbol{r})(\hat{\boldsymbol{p}}^2 + \hbar^2\boldsymbol{k}^2 + 2\hbar\boldsymbol{k}\hat{\boldsymbol{p}}) u(\boldsymbol{k}, \boldsymbol{r}) , \tag{4.6.62}$$

(4.6.61, 62) yield

$$\begin{gathered}\left[\frac{\hat{\boldsymbol{p}}^2}{2m} + V(\boldsymbol{r}) + W(\hat{\boldsymbol{p}}, \boldsymbol{k})\right] u(\boldsymbol{k}, \boldsymbol{r}) = E u(\boldsymbol{k}, \boldsymbol{r}) , \\ W(\hat{\boldsymbol{p}}, \boldsymbol{k}) = \frac{\hbar^2\boldsymbol{k}^2}{2m} + \frac{\hbar\boldsymbol{k}\cdot\hat{\boldsymbol{p}}}{m} .\end{gathered} \tag{4.6.63}$$

Let us assume that we know the solution of (4.6.63) for $\boldsymbol{k} = 0$, i.e., the eigenfunctions $u_\zeta(0, \boldsymbol{r}) \equiv \psi_{0\zeta}(\boldsymbol{r})$ and the spectrum $E_\zeta(0)$. The operator $W(\boldsymbol{k}, \hat{\boldsymbol{p}})$ will be regarded as a perturbation. On the assumption that $E_\eta(0)$ is non-degenerate [the first-order terms in $\boldsymbol{k}$ are equal to zero since at the point of minimum $E(\boldsymbol{k})$ we have $\partial E/\partial \boldsymbol{k} = 0$], the result for the state $E_\eta(\boldsymbol{k})$ with which we are concerned will be, up to second-order terms in $\boldsymbol{k}_0$ [Ref. 1.12, Sect. 38],

$$E_\eta(\boldsymbol{k}) \approx E_\eta(0) + \frac{\hbar^2\boldsymbol{k}^2}{2m} + \sum_{\zeta\neq\eta} \frac{\left|\left\langle \eta \left| \dfrac{\hbar\boldsymbol{k}\cdot\boldsymbol{p}}{m} \right| \zeta \right\rangle\right|^2}{E_\eta(0) - E_\zeta(0)} , \tag{4.6.64}$$

where we have introduced the notation

$$\langle \eta | \hat{A} | \zeta \rangle = \int d\boldsymbol{r}\, \psi_{0\eta}^*(\boldsymbol{r}) \hat{A} \psi_{0\zeta}(\boldsymbol{r}) \tag{4.6.65}$$

and allowed for the fact that $\langle \eta | \hat{\boldsymbol{p}} | \zeta \rangle = \langle \zeta | \hat{\boldsymbol{p}} | \eta \rangle^*$. Formula (4.6.64) may be represented in the form

$$E_\eta(\boldsymbol{k}) = E_\eta(0) + \frac{\hbar^2}{2}\sum_{i,j} m_{ij}^{*\,-1} k_i k_j \ , \tag{4.6.66}$$

where

$$m_{ij}^{*\,-1} = m^{-1}\left[\delta_{ij} + m^{-1}\sum_{\zeta\neq\eta} \times\left(\frac{\langle\eta|\hat{p}_i|\zeta\rangle\langle\zeta|\hat{p}_j|\eta\rangle + \langle\eta|\hat{p}_j|\zeta\rangle\langle\zeta|\hat{p}_i|\eta\rangle}{E_\eta(0) - E_\zeta(0)}\right)\right] . \tag{4.6.67}$$

Comparison of (4.6.67) with (4.4.13) shows that $m_{ij}^{*\,-1}$ is the tensor of inverse effective masses, which defines the accelerating effect of an electric field. As is seen, for a nondegenerate band, (4.6.67) is exact; (4.6.67) was first derived by Seitz [3.16].

In semiconductors the energy gap G between the bottom of the conduction band and the top of the valence band is comparatively small (for simplicity, we assume that the top is also at $\boldsymbol{k} = 0$, the so-called direct gap). Therefore, the corresponding contribution to, say, $m_{xx}^{*\,-1}$ is large and positive

$$\Delta m_{xx}^{*\,-1} = \frac{2|\langle h|\hat{p}_x|e\rangle|^2}{m^2 G} \approx \frac{\hbar^2}{m^2 d^2 G} \gg m^{-1} \ , \tag{4.6.68}$$

where $\langle h|\hat{p}_x|e\rangle$ is used to label the matrix element of $\hat{p}_x$ between the top of the valence band and the bottom of the conduction band. If, according to the selection rule, this contribution is not identically equal to zero (one of the functions is an s-type function, and the other a p-type function), it can be evaluated as $\hbar/d$, with d being the lattice period. Equation (4.6.68) explains why the effective mass of a conduction electron in semiconductors is often much smaller than the mass of a free electron, a circumstance that was turned to account in Sect. 4.5.1.

For a degenerate band the energy does not have the simple parabolic form of (4.6.66). The energy is found by solving the secular equation, i.e., from the perturbation theory for a degenerate level [4.21].

4.6.7 Fermi Surfaces in Real Metals

A variety of experimental techniques together with the methods of band-structure calculation that we have just considered allow us to determine the most important characteristic of a metal: the shape of its Fermi surface [4.44]. We would expect a large variety of shapes, depending on the particular metal (Sect. 3.1). A more complicated picture would certainly be expected for intermetallics and ordered metallic alloys, let alone highly defective crystals and disordered alloys.

The simplest case is that of normal univalent metals, which, in turn, are divided into alkali and noble metals. For alkali metals the ion core has the clectronic configuration of the atoms of inert gases, the electrons of the configuration being tightly bound to the corresponding atomic nucleus. Therefore, the band structure of the electronic shell of these cores is an assembly of fully occupied and very narrow energy bands, which lie appreciably lower than the conduction band and may, in principle, be described with high accuracy by the LCAO method (Sect. 4.6.3). If we recall the shape of the first Brillouin zone for the bcc lattice, in which all alkali metals crystallize (for Li and Na this is only true at temperatures that are not too low; otherwise these metals have hcp lattices), the minimum distance from the center of the zone (in quasimomentum space) to the face closest to the center is determined by the spacing between the points Γ and N (Fig. 4.14). We can readily show that this distance (in terms of the reciprocal value of the edge a of an elementary cube, Fig. 1.13b) is equal to

$$\Gamma N = \frac{1}{\sqrt{2}}\frac{2\pi}{a} \approx 0.707\frac{2\pi}{a} \ . \tag{4.6.69}$$

On the other hand, the Fermi sphere of a univalent metal in the free-electron approximation in the same terms has a radius (3.5.16)

$$k_{\mathrm{F}} = \left(\frac{3}{4\pi}\right)^{1/3}\frac{2\pi}{a} \approx 0.620\,\frac{2\pi}{a} \ . \tag{4.6.70}$$

Thus, $\Gamma N > k_{\mathrm{F}}$ and, consequently, the Fermi sphere lies entirely inside the first Brillouin zone. Experiments concerned with the measurement of the de Haas–van Alphen effect [4.45] have shown that deviations of the real Fermi surface in alkali metals from the ideal sphere in the places of least distances ($\Gamma N - k_{\mathrm{F}}$) do not exceed hundredths and tenths of a percent in the quantity ($k_{\mathrm{F}}^{\mathrm{exp}}/k_{\mathrm{F}}^{\mathrm{sphere}}$) in sodium and potassium and three percent in rubidium and cesium. (Thus far, because of its complicated structure due to the martensitic transformation at 77 K, experiments aimed at determining the shape of the Fermi surface in lithium have not been possible; nor has the shape of the Fermi surface been determined experimentally for francium, because of its radioactivity.)

Thus alkali metals illustrate very well possible applications of the free-electron approximation. But of course we have to bear in mind that it is necessary to use a weak pseudopotential and that the effective mass of a conduction electron in these metals may differ somewhat from the mass of a free electron.

At first sight the picture observed for the univalent noble metals—copper, silver, and gold—will seem to be the same as that for alkali metals. However, this is not the case since, along with the ion core similar to the atom of inert gases, these noble metals have either a $3d$ shell that has just filled (in going from nickel to copper) or a filled $4d$ shell (as in silver). In gold there is, in addition, a filled $4f$ shell, but it lies deep and is therefore close to the tightly bound electronic shell of the atom of the inert gas. Since the d bands have energies that lie just below the Fermi energy and since the width of the corresponding bands is not very narrow, they have an appreciable effect on the character of the constant energy surfaces of conduction electrons in noble metals. This influence, as calculations have shown [4.46], results in the following: When the Fermi sphere of conduction electrons "is inscribed" into the first Brillouin zone of the fcc crystals of copper, silver, and gold, the spherical shape of the Fermi surface undergoes a sharper distortion than is the case with alkali metals. Specifically, around the center L of the hexagonal face of the first Brillouin zone of the fcc crystal the Fermi surface touches the boundary of the zone (Fig. 4.15). Figure 4.36a–d [4.44] presents the general shape of the Fermi surface in the first Brillouin zone, and also the

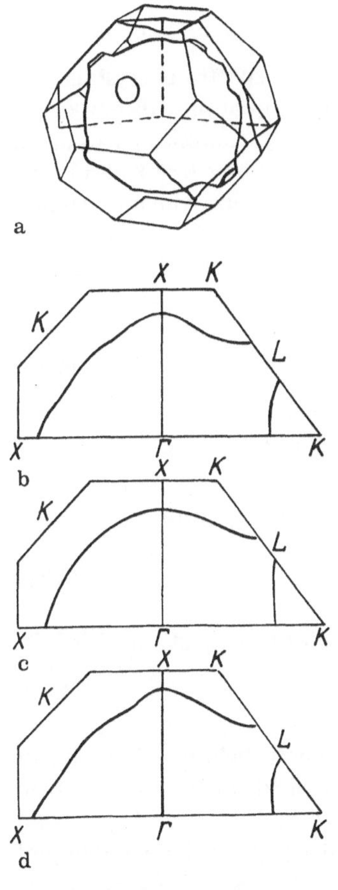

Fig. 4.36. Fermi surface of noble metals: general view (**a**); cross sections for Cu (**b**), Ag (**c**), and Au (**d**)

detailed trend of the Fermi boundary for a cross section of the zone for all the three noble metals. A point to be noted is that this picture has also been fully confirmed in measurements of the de Haas–van Alphen effect (Sect. 4.7).

A most interesting feature is the presence of "necks" on the Fermi surfaces, which, on the whole, are close to a sphere. The necks change the topology of the Fermi surface drastically (for example, they are responsible for the occurrence of open surfaces and cross sections which are absent in the case of "separate" spheres in alkali metals).

Still more complicated is the picture of the Fermi surfaces for multivalent, transition, and RE metals. For lack of space we do not enlarge on this problem but refer the reader to the literature [2.34, 4.8, 44].

4.7 Band Electrons in a Magnetic Field

4.7.1 The Effective Hamiltonian

One of the most important ways of studying the energy spectra of crystals is by investigating their magnetic properties. The rigorous quantum theory of the effect of a magnetic field $\boldsymbol{H}$ on band states is far more complicated than the electric field theory (Sect. 4.4) and is not treated in detail here (see Kohn [4.47] and Blount [4.48]). On the other hand, a quasi-classical approximation suffices in most cases, and we focus on this approximation here. The Hamiltonian of an electron in a potential field $V(\boldsymbol{r})$ and a constant magnetic field $\boldsymbol{H}$ has the form

$$\hat{\mathscr{H}} = \frac{\hat{\mathscr{P}}^2}{2m} + V(\boldsymbol{r}) \ ,$$
$$\hat{\mathscr{P}} = -\mathrm{i}\hbar\nabla - \frac{e}{c}\boldsymbol{A}(\boldsymbol{r}) \ , \qquad (4.7.1)$$

where $\boldsymbol{A}(\boldsymbol{r})$ is a vector potential, which is defined by the equation

$$\boldsymbol{H} = \operatorname{curl} \boldsymbol{A}(\boldsymbol{r}) \ . \qquad (4.7.2)$$

One of the possible choices of $\boldsymbol{A}$ for a constant uniform magnetic field is

$$\boldsymbol{A} = \tfrac{1}{2}(\boldsymbol{H} \times \boldsymbol{r}) \ . \qquad (4.7.3)$$

Since the Hamiltonian of the interaction with a magnetic field is unbounded (it contains terms proportional to r and r^2), even a weak field H causes the electron spectrum to readjust drastically (just as with an electric field, Sect. 4.4). For an electric field, however, collisions impose a rigid restriction on the possibility of observing readjustment effects [the inequality $t_0 \ll \bar{\tau}$, with t_0 being the oscillation period in an electric field (4.4.15), and $\bar{\tau}$ the time between collisions, is never fulfilled]. With a magnetic field the same condition for free electrons is of the form

$$\omega_H \bar{\tau} \gg 1 \, , \tag{4.7.4}$$

where ω_H is the cyclotron frequency (3.5.90). When $H \approx 10^4$ Oe $\approx 8 \cdot 10^5$ A/m, $\bar{\tau} \approx 10^{-9}$ s, $\omega_H \bar{\tau} \approx 10^2$; i.e., the inequality (4.7.4) holds good and we are in a position to observe effects of the dynamics of electrons in magnetic fields. The best conditions for this will be high magnetic fields H—i.e., large values of ω_H—and low temperatures—i.e., large values of $\bar{\tau}$. The fundamental difference between an electric field and a magnetic field lies in the fact that the former increases the kinetic energy of an electron, whereas the latter does not do any work on it and does not cause its energy to change. A very strong electric field therefore cannot be built up in solids, because of the heating and destruction of the sample, whereas no similar limitations exist for a magnetic field.

The problem of strictly determining the eigenfunctions and eigenstates of the Hamiltonian (4.7.1) with a periodic potential $V(\boldsymbol{r})$ is extremely complicated. However, it simplifies drastically with sufficiently weak fields, since the variation of the energy of an electron subject to a weak magnetic field is small compared to typical band energies. An exact criterion will be obtained somewhat later. In the majority of cases, real magnetic fields in metals may be regarded as weak.

As we saw in Sect. 4.4, the interband transition probability is exponentially small in a weak electric field. From the very derivation of this result we see that this is not necessarily related to an electric field but holds for any sufficiently smooth and weak perturbation. Therefore we start by considering the one-band approximation, then estimate the interband transition ("magnetic breakdown") probability and thereby justify the assumption made. Thus, the solution of the Schrödinger equation with the Hamiltonian (4.7.1) [4.49] will be sought as an expansion in Wannier functions (4.6.41) in a single band, according to (4.6.40). These functions are orthonormal and are as good an expansion basis as Bloch's functions

$$\psi_\zeta(\boldsymbol{r}) = \sum_m a_m \varphi_{m\zeta}(\boldsymbol{r}) \, . \tag{4.7.5}$$

Wannier functions may be represented as

$$\begin{aligned} \varphi_{m\zeta}(\boldsymbol{r}) &= N^{-1/2} \sum_{\boldsymbol{k}} \psi_\zeta(\boldsymbol{k}, \boldsymbol{r}) \exp(-\mathrm{i}\boldsymbol{k} \cdot \boldsymbol{R}_m) \\ &= N^{-1/2} \sum_{\boldsymbol{k}} \psi_\zeta(\boldsymbol{k}, \boldsymbol{r} - \boldsymbol{R}_m) = \exp\left(-\frac{\mathrm{i}}{\hbar} \hat{\boldsymbol{p}} \cdot \boldsymbol{R}_m\right) \varphi_{0\zeta}(\boldsymbol{r}) \, , \end{aligned} \tag{4.7.6}$$

where $\hat{\boldsymbol{p}} = -\mathrm{i}\hbar\nabla$, $\exp(\mathrm{i}\hat{\boldsymbol{p}}\cdot\boldsymbol{a}/\hbar) = \exp(\boldsymbol{a}\nabla)$ is the translation operator for the vector $\boldsymbol{a}$,

$$\exp(\boldsymbol{a}\nabla)f(\boldsymbol{r}) = \sum_{n=0}^{\infty}\frac{1}{n!}(\boldsymbol{a}\nabla)^n f(\boldsymbol{r}) = f(\boldsymbol{r}+\boldsymbol{a}) \ , \tag{4.7.7}$$

where a Taylor-series expansion is used.

Instead of (4.7.6), we can also use other functions as the expansion basis

$$\exp\left(-\frac{\mathrm{i}}{\hbar}\hat{\boldsymbol{\Pi}}\cdot\boldsymbol{R}_m\right)\varphi_{0\zeta}(\boldsymbol{r}) \ ,$$

where

$$\hat{\boldsymbol{\Pi}} = \hat{\boldsymbol{p}} + \frac{e}{c}\boldsymbol{A} = \hat{\boldsymbol{p}} + \frac{e}{2c}(\boldsymbol{H}\times\boldsymbol{r}) \ . \tag{4.7.8}$$

An advantage of the operator $\hat{\boldsymbol{\Pi}}$ is that it commutes with the operator $\hat{\mathscr{P}}$ involved in the Hamiltonian (4.7.1)

$$\begin{aligned}[\hat{\Pi}_i, \hat{\mathscr{P}}_j]_- &= \left[\hat{p}_i - \frac{e}{2c}\varepsilon_{ilk}H_l\hat{r}_k, \quad \hat{p}_j + \frac{e}{2c}\varepsilon_{ipq}H_p\hat{r}_q\right]_- \\ &= -\frac{\mathrm{i}e\hbar}{2c}\varepsilon_{ijp}H_p - \frac{\mathrm{i}e\hbar}{2c}\varepsilon_{ipj}H_p = 0 \ ,\end{aligned} \tag{4.7.9}$$

where ε_{ijk} is an absolutely antisymmetric unit tensor and a canonical commutation relation

$$[\hat{p}_i, \hat{r}_k]_- = -\mathrm{i}\hbar\delta_{ik} \tag{4.7.10}$$

is used. Thus we seek the wave function in the form

$$\psi_\zeta(\boldsymbol{r}) = \sum_m \alpha_m \exp\left(-\frac{\mathrm{i}}{\hbar}\hat{\boldsymbol{\Pi}}\cdot\boldsymbol{R}_m\right)\varphi_{0\zeta}(\boldsymbol{r}) \ . \tag{4.7.11}$$

Substituting (4.7.11) into the Schrödinger equation, multiplying it by

$$\varphi_{0\zeta}^*(\boldsymbol{r})\exp\left(\frac{\mathrm{i}}{\hbar}\hat{\boldsymbol{\Pi}}\cdot\boldsymbol{R}_n\right) ,$$

and carrying out the integration over the unit cell, we obtain

$$\sum_m(\mathscr{H}_{nm} - ES_{nm})\alpha_m = 0 \ , \tag{4.7.12}$$

where

$$\mathscr{H}_{nm} = \int d\mathbf{r}\varphi^*_{0\zeta}(\mathbf{r})\exp\left(\frac{\mathrm{i}}{\hbar}\hat{\mathbf{\Pi}}\cdot\mathbf{R}_n\right)\hat{\mathscr{H}}\exp\left(-\frac{\mathrm{i}}{\hbar}\hat{\mathbf{\Pi}}\cdot\mathbf{R}_m\right)\varphi_{0\zeta}(\mathbf{r}) \tag{4.7.13}$$

$$S_{nm} = \int d\mathbf{r}\varphi^*_{0\zeta}(\mathbf{r})\exp\left(\frac{\mathrm{i}}{\hbar}\hat{\mathbf{\Pi}}\cdot\mathbf{R}_n\right)\exp\left(-\frac{\mathrm{i}}{\hbar}\hat{\mathbf{\Pi}}\cdot\mathbf{R}_m\right)\varphi_{0\zeta}(\mathbf{r}) \ . \tag{4.7.14}$$

The commutator of the operators $\hat{\mathbf{\Pi}}\cdot\mathbf{R}_n$ and $\hat{\mathbf{\Pi}}\cdot\mathbf{R}_m$ is a c number. Therefore we may use the identity (2.7.42)

$$\exp\left(\frac{\mathrm{i}}{\hbar}\hat{\mathbf{\Pi}}\cdot\mathbf{R}_n\right)\exp\left(-\frac{\mathrm{i}}{\hbar}\hat{\mathbf{\Pi}}\cdot\mathbf{R}_m\right) = \exp\left[\frac{\mathrm{i}}{\hbar}\hat{\mathbf{\Pi}}\cdot(\mathbf{R}_n - \mathbf{R}_m)\right] \cdot\exp\left\{-\frac{1}{2\hbar^2}[\hat{\Pi}^i, \hat{\Pi}^j]_- R^i_n R^j_m\right\} , \tag{4.7.15}$$

where $i, j = x, y, z$, and the repeating vector superscripts here and henceforth implying summation. We calculate the commutator involved in (4.7.15). Analogous to (4.7.9), we have

$$[\hat{\Pi}^{(i)}, \hat{\Pi}^{(j)}]_- = -\frac{\mathrm{i}e\hbar}{c}\varepsilon_{ijl}H_l \tag{4.7.16}$$

$$\exp\left(\frac{\mathrm{i}}{\hbar}\hat{\mathbf{\Pi}}\cdot\mathbf{R}_n\right)\exp\left(-\frac{\mathrm{i}}{\hbar}\hat{\mathbf{\Pi}}\cdot\mathbf{R}_m\right) = \exp\left[\frac{\mathrm{i}}{\hbar}\hat{\mathbf{\Pi}}\cdot(\mathbf{R}_n - \mathbf{R}_m)\right]\exp\left\{-\frac{\mathrm{i}e}{2\hbar c}(\mathbf{R}_n\times\mathbf{R}_m)\cdot\mathbf{H}\right\} . \tag{4.7.17}$$

Similarly,

$$\exp\left(\frac{\mathrm{i}}{\hbar}\hat{\mathbf{\Pi}}\cdot\mathbf{R}_m\right) = \exp\left(\frac{\mathrm{i}}{\hbar}\hat{\mathbf{p}}\cdot\mathbf{R}_m + \frac{\mathrm{i}e}{2\hbar c}(\mathbf{R}_m\times\mathbf{H})\cdot\mathbf{r}\right) = \exp\left(\frac{\mathrm{i}e}{2\hbar c}(\mathbf{R}_m\times\mathbf{H})\cdot\mathbf{r}\right)\exp\left(\frac{\mathrm{i}}{\hbar}\hat{\mathbf{p}}\cdot\mathbf{R}_m\right) , \tag{4.7.18}$$

since $\mathbf{R}_m\cdot(\mathbf{R}_m\times\mathbf{H}) = 0$. It follows from (4.7.18) that

$$\left[\exp\left(\frac{\mathrm{i}}{\hbar}\hat{\mathbf{\Pi}}\cdot\mathbf{R}_m\right), \hat{\mathscr{H}}\right]_- = 0 \ . \tag{4.7.19}$$

The first term in the Hamiltonian $\hat{\mathscr{P}}^2/2m$ evidently commutes, by virtue of (4.7.9), with the exponent, and so

$$\exp\left(\frac{i}{\hbar}\hat{\boldsymbol{\Pi}}\cdot\boldsymbol{R}_m\right)V(\boldsymbol{r}) = \exp\left(\frac{ie}{2\hbar c}(\boldsymbol{R}_m\times\boldsymbol{H})\cdot\boldsymbol{r}\right)$$

$$\cdot\exp\left(\frac{i}{\hbar}\hat{\boldsymbol{p}}\cdot\boldsymbol{R}_m\right)V(\boldsymbol{r}) = \exp\left(\frac{ie}{2\hbar c}(\boldsymbol{R}_m\times\boldsymbol{H})\cdot\boldsymbol{r}\right)$$

$$\cdot V(\boldsymbol{r}+\boldsymbol{R}_m)\exp\left(\frac{i}{\hbar}\hat{\boldsymbol{p}}\cdot\boldsymbol{R}_m\right)$$

$$= V(\boldsymbol{r})\exp\left(\frac{i}{\hbar}\hat{\boldsymbol{\Pi}}\cdot\boldsymbol{R}_m\right),$$

in view of the condition $V(\boldsymbol{r}+\boldsymbol{R}_m) = V(\boldsymbol{r})$. Substitution of (4.7.18, 19) into (4.7.13) yields

$$\mathscr{H}_{nm} = \exp\left\{\frac{ie}{2\hbar c}(\boldsymbol{R}_m\times\boldsymbol{R}_n)\cdot\boldsymbol{H}\right\}\int d\boldsymbol{r}\,\varphi^*_{0\zeta}(\boldsymbol{r})\exp\left[\frac{i}{\hbar}\hat{\boldsymbol{\Pi}}\cdot(\boldsymbol{R}_n-\boldsymbol{R}_m)\right]\varphi_{0\zeta}(\boldsymbol{r})\ . \tag{4.7.20}$$

Only small r are important in (4.7.20), because of the rapid decrease of the function $\varphi_{0\zeta}(\boldsymbol{r})$. Therefore, the second term in $\hat{\boldsymbol{\Pi}}$ from (4.7.9) may be neglected. The second term is on the order of eHd/c, with d being the lattice period, whereas the first term, as it acts on the function $\varphi_{0\zeta}$, gives a value on the order of $\hbar/d$. Thus, we require that the following inequality be fulfilled:

$$l_H = \left(\frac{c\hbar}{|e|H}\right)^{1/2} \gg d\ . \tag{4.7.21}$$

In metals, d is the only characteristic length that determines band structure, since $k_F \approx d^{-1}$. In the general case, (4.7.21) should involve, instead of d, some combination of the lattice period, forbidden-energy-band width, Fermi momentum, etc. In this inequality the quantity l_H is called the magnetic length; when $H \approx 10^4$ Oe, $l_H \approx 10^{-5}$–10^{-6} cm. In the integral (4.7.20) we set $\hat{\boldsymbol{\Pi}} \approx \hat{\boldsymbol{p}}$; this yields

$$\int d\boldsymbol{r}\,\varphi^*_{0\zeta}(\boldsymbol{r})\exp\left[\frac{i}{\hbar}\hat{\boldsymbol{p}}\cdot(\boldsymbol{R}_n-\boldsymbol{R}_m)\right]\hat{\mathscr{H}}\varphi_{0\zeta}(\boldsymbol{r}) \tag{4.7.22}$$

$$= \int d\boldsymbol{r}\,\varphi^*_{n\zeta}(\boldsymbol{r})\hat{\mathscr{H}}\varphi_{m\zeta}(\boldsymbol{r}) = \sum_{\boldsymbol{k}} E_\zeta(\boldsymbol{k})\exp[i\boldsymbol{k}\cdot(\boldsymbol{R}_m-\boldsymbol{R}_n)]\ .$$

Equation (4.7.22) follows directly from the definition of the Wannier functions (4.6.41). Similarly, we find

$$S_{nm} \approx \delta_{nm} \ . \tag{4.7.23}$$

Taking account of (4.7.20, 22), we transform $\mathscr{H}_{nm}$:

$$\begin{aligned}
\mathscr{H}_{nm} &= \exp\left\{\frac{\mathrm{i}e}{2\hbar c}(\boldsymbol{R}_n \times \boldsymbol{R}_m)\cdot \boldsymbol{H}\right\} \sum_{\boldsymbol{k}} E_\zeta(\boldsymbol{k}) \\
&\quad \times \exp[\mathrm{i}\boldsymbol{k}\cdot(\boldsymbol{R}_m - \boldsymbol{R}_n)] = \exp\left\{-\frac{\mathrm{i}e}{2\hbar c}[(\boldsymbol{R}_m - \boldsymbol{R}_n)\times \boldsymbol{R}_n]\cdot \boldsymbol{H}\right\} \\
&\quad \times \sum_{\boldsymbol{k}} E_\zeta(\boldsymbol{k}) \exp[\mathrm{i}\boldsymbol{k}\cdot(\boldsymbol{R}_m - \boldsymbol{R}_n)] = \sum_{\boldsymbol{k}} E_\zeta(\boldsymbol{k}) \\
&\quad \times \exp\left\{\mathrm{i}\left[\boldsymbol{k} + \frac{e}{\hbar c}\boldsymbol{A}(\boldsymbol{R}_n)\right]\cdot(\boldsymbol{R}_m - \boldsymbol{R}_n)\right\} \\
&= \sum_{\boldsymbol{k}} E_\zeta\left[\boldsymbol{k} - \frac{e}{\hbar c}\boldsymbol{A}(\boldsymbol{R}_n)\right] \exp[\mathrm{i}\boldsymbol{k}\cdot(\boldsymbol{R}_m - \boldsymbol{R}_n)] \\
&= \sum_{\boldsymbol{k}} E_\zeta\left[\boldsymbol{k} - \frac{e}{\hbar c}\boldsymbol{A}\left(\mathrm{i}\frac{\partial}{\partial \boldsymbol{k}}\right)\right] \exp[\mathrm{i}\boldsymbol{k}\cdot(\boldsymbol{R}_m - \boldsymbol{R}_n)] \ .
\end{aligned} \tag{4.7.24}$$

The order in which the operators $\boldsymbol{k}$ and $\partial/\partial \boldsymbol{k}$ act is unimportant here because, when $\boldsymbol{k}$ and $(e/\hbar c)\boldsymbol{A}(\mathrm{i}\partial/\partial \boldsymbol{k})$ commute, a magnetic field arises that is not multiplied by the large factor $\boldsymbol{R}_n$; such terms in weak fields may be neglected in comparison with the results of the operation of $\mathrm{i}\partial/\partial \boldsymbol{k}$ on $\exp(\mathrm{i}\boldsymbol{k}\cdot\boldsymbol{R}_n)$.

With allowance for (4.7.23, 24), the Schrödinger equation assumes the form

$$E\alpha_n = \sum_{\boldsymbol{k}} \alpha(\boldsymbol{k}) E_\zeta\left[\boldsymbol{k} - \frac{e}{\hbar c}\boldsymbol{A}\left(\mathrm{i}\frac{\partial}{\partial \boldsymbol{k}}\right)\right] \exp(-\mathrm{i}\boldsymbol{k}\cdot\boldsymbol{R}_n) \ , \tag{4.7.25}$$

where

$$\alpha(\boldsymbol{k}) = \sum_m \alpha_m \exp(\mathrm{i}\boldsymbol{k}\cdot\boldsymbol{R}_m) \ . \tag{4.7.26}$$

Performing integration by parts and neglecting the noncommutativity of the operators $\boldsymbol{k}$ and $(e/\hbar c)\boldsymbol{A}(\mathrm{i}\partial/\partial \boldsymbol{k})$ for the reasons already stated, we obtain

$$E\alpha_n = \sum_{\boldsymbol{k}} \exp(-\mathrm{i}\boldsymbol{k}\cdot\boldsymbol{R}_n) E_\zeta\left[\boldsymbol{k} + \frac{e}{\hbar c}\boldsymbol{A}\left(\mathrm{i}\frac{\partial}{\partial \boldsymbol{k}}\right)\right] \alpha(\boldsymbol{k}) \ . \tag{4.7.27}$$

Formula (4.7.27) may be proved by straightforward series expansion, allowing for $\boldsymbol{A}(-\mathrm{i}\partial/\partial \boldsymbol{k}) = -\boldsymbol{A}(\mathrm{i}\partial/\partial \boldsymbol{k})$.

Multiplying (4.7.27) by $\exp(\mathrm{i}\boldsymbol{k}\cdot\boldsymbol{R}_n)$ and carrying out the summation over n, we obtain

$$\begin{aligned} &\hat{\mathscr{H}}_{\mathrm{ef}}\alpha(\boldsymbol{k}) = E\alpha(\boldsymbol{k}) \;, \\ &\hat{\mathscr{H}}_{\mathrm{ef}} = E_\zeta\left[\boldsymbol{k}+\frac{e}{\hbar c}\boldsymbol{A}\left(\mathrm{i}\frac{\partial}{\partial \boldsymbol{k}}\right)\right] \;. \end{aligned} \tag{4.7.28}$$

Thus we have obtained an effective Hamiltonian that describes the motion of an electron in a magnetic field. This Hamiltonian is expressed in terms of canonical variables $\hat{\boldsymbol{p}} = \hbar\hat{\boldsymbol{K}}$ and $\hat{\boldsymbol{r}} = \mathrm{i}\partial/\partial\boldsymbol{K}$, which satisfy the commutation relations (4.7.10). The equations of motion for these operators have the standard form

$$\begin{aligned} &\hbar\dot{\boldsymbol{K}} = -\frac{\partial}{\partial \boldsymbol{r}}E_\zeta\left[\boldsymbol{K}-\frac{e}{\hbar c}\boldsymbol{A}(\boldsymbol{r})\right] \\ &\dot{\boldsymbol{r}} = \frac{\partial}{\partial \boldsymbol{K}}E_\zeta\left[\boldsymbol{K}-\frac{e}{\hbar c}\boldsymbol{A}(\boldsymbol{r})\right] \;. \end{aligned} \tag{4.7.29}$$

The characteristic size of the orbit of an electron in a magnetic field is on the order of r_H (3.6.68). Assuming $k_{\mathrm{F}} \approx d^{-1}$, we obtain

$$r_H = v_{\mathrm{F}}/\omega_H \approx \frac{\hbar mc}{md|e|H} = \frac{\hbar c}{d|e|H} = l_H^2/d \;.$$

On the other hand, the amplitude of quantum oscillations of an electron near the classical orbit (uncertainty of coordinate) is on the order of $(\hbar/m\omega_H)^{1/2} = l_H$. Consequently, when the inequality (4.7.21) holds, the electron motion is also described by a narrow wave packet ($r_H \gg l_H$), too. Then we may proceed from the operators $\hat{\boldsymbol{K}}$ and $\hat{\boldsymbol{r}}$ to the momentum and coordinate of the wave-packet center and understand (4.7.29) as an equation for averages.

Let us introduce

$$\boldsymbol{k} = \boldsymbol{K}-\frac{e}{\hbar c}A(\boldsymbol{r}) \;. \tag{4.7.30}$$

The velocity of an electron then is equal to

$$\dot{\boldsymbol{r}} = \frac{1}{\hbar}\frac{\partial E_\zeta(\boldsymbol{k})}{\partial \boldsymbol{k}} = \boldsymbol{v} \;. \tag{4.7.31}$$

We calculate the rate of change of the kinematic wave vector $\boldsymbol{k}$. Differentiating (4.7.30) with respect to time and taking into account (4.7.29, 31), we find

$$\dot{k}_j = -\frac{1}{\hbar}\frac{\partial}{\partial r_j} E_\zeta\left[K - \frac{e}{\hbar c} A(r)\right] - \frac{e}{\hbar c}\dot{A}_j(r)$$
$$= \frac{e}{\hbar c}\left(v_i \frac{\partial A_i}{\partial r_j} - v_j \frac{\partial A_j}{\partial r_i}\right), \tag{4.7.32}$$

or, in vector form,

$$\dot{k} = \frac{e}{\hbar c}(v \times H) \tag{4.7.33}$$

(Lorentz force). The equivalence of (4.7.32, 33) can be readily proved by writing out the components with allowance for (4.7.2). Equations (4.7.31, 33) serve as a basis for describing the classical dynamics of an electron with an arbitrary dispersion relation [i.e., exhibiting the $E_\zeta(k)$ dependence].

Finally, let us estimate at what field the magnetic breakdown sets in—i.e., the field at which the probability of interband transitions increases sharply. For the electric field we had the inequality (4.4.25). Proceeding to the magnetic field, we need to make the following replacement, in keeping with the expression for the Lorentz force:

$$|e|F \to |e|vH/c \approx \frac{\hbar|e|H}{mcd} = \frac{\hbar\omega_H}{d},$$

where the velocity of an electron at the boundary of the band has been estimated to be $\hbar/md$. Using this replacement, we obtain the result that the probability of a magnetic breakdown is small if

$$\hbar\omega_H \ll \frac{\Delta^2 md^2}{\hbar^2}, \tag{4.7.34}$$

with Δ being the width of the forbidden energy band. For some metals subject to fields $H \lesssim 10^5$ Oe $\approx 8 \cdot 10^6$ A/m, this inequality may have an opposite sense. For a more detailed treatment of the magnetic breakdown theory see [Ref. 4.8, Sect. 10].

4.7.2 Classical Paths

Our objective now is to solve (4.7.31, 33). Scalar multiplication of (4.7.33) by v yields

$$v \cdot \dot{k} = 0 . \tag{4.7.35}$$

Substituting (4.2.38) into (4.7.35), we find

$$\dot{E}_\zeta(\boldsymbol{k}) = 0 \ . \tag{4.7.36}$$

Consequently, when an electron moves in a magnetic field, its energy does not vary (as already stated); i.e., it is an integral of the motion. Scalar multiplication of (4.7.33) by $\boldsymbol{H}$ gives

$$\boldsymbol{H} \cdot \dot{\boldsymbol{k}} = 0 \ . \tag{4.7.37}$$

That is, the projection of $\boldsymbol{k}$ onto the field direction (z axis) is also an integral of the motion. Then the equation of the path of an electron in reciprocal space will be

$$E_\zeta(k_x, k_y, k_z^{(0)}) = E_0 \ , \tag{4.7.38}$$

where $k_z^{(0)}$ and E_0 are fixed (for paths that actually exist, E_0 is the Fermi energy), and k_x and k_y run over all the possible values that are compatible with (4.7.38).

Thus, the path of an electron in reciprocal space is the contour where the isoenergetic surface crosses the plane perpendicular to the field. Hence it is clear that a study of the various properties of metals in a magnetic field provides valuable information about the shape of the Fermi surface. The various types of isoenergetic surfaces are shown in Fig. 4.37. In analyzing the quasi-classical dynamics of an electron, it is convenient to exploit the extended-zone method. The geometric properties of the surfaces portrayed in Fig. 4.37 (single or multiple connectivity, the presence or absence of self-crossing curves, etc.) have a very substantial effect on the character of electron motion in a magnetic field. These problems are discussed in more detail by Lifshitz et al. [4.8]. Here we restrict our attention to the fundamental difference between open and closed paths. The motion in open paths is infinite, and the motion in closed paths is

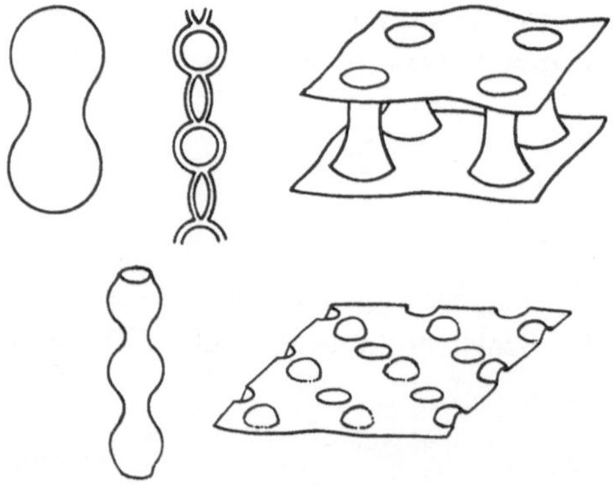

Fig. 4.37. Different types of isoenergetic band electron surface

finite. The possibility of infinite motion in a plane perpendicular to the field distinguishes an electron is a crystal from a free electron. Thus, the motion in open paths is not quantized, and, as a result, galvanomagnetic effects, particularly the behavior of the magnetoresistance, can vary appreciably [Ref. 4.8, Sect. 28].

Observe that the trajectory of an electron in reciprocal space and the trajectory of an electron in direct space are similar. This follows immediately from (4.7.33): the electron velocity vector in $\boldsymbol{k}$ space in the plane perpendicular to the field is equal to the velocity vector in $\boldsymbol{r}$ space, when the latter is multiplied by $|e|H/\hbar c$ and rotated by 90°. In consequence, the counterpart of infinite motion in $\boldsymbol{k}$ space is infinite motion in $\boldsymbol{r}$ space, and vice versa. We express (4.7.33) in terms of the components

$$\dot{k}_x = \frac{eH}{\hbar c} v_y \ , \qquad \dot{k}_y = -\frac{eH}{\hbar c} v_x \ . \tag{4.7.39}$$

We square these equations, combine them, and extract the square root

$$\frac{dk_\perp}{v_\perp} = \frac{|e|H}{\hbar c} dt \ , \tag{4.7.40}$$

where $dk_\perp = (dk_x^2 + dk_y^2)^{1/2}$, $v_\perp = (v_x^2 + v_y^2)^{1/2}$. Consequently, the period T_H of motion of an electron in a magnetic field in a closed path is

$$T_H = \frac{\hbar c}{|e|H} \oint \frac{dk_\perp}{v_\perp} \equiv \frac{2\pi}{\omega_H} \ , \tag{4.7.41}$$

with the integral being taken over the path (4.7.38). The quantity ω_H in this equation is the cyclotron frequency of a band electron; it may be represented in the ordinary form of (3.5.90):

$$\omega_H = \frac{|e|H}{m_H(E_0, k_z^{(0)})c} \ , \tag{4.7.42}$$

where, however, the cyclotron mass is a function of E_0 and $k_z^{(0)}$:

$$m_H(E_0, k_z^{(0)}) = \frac{\hbar}{2\pi} \oint \frac{dk_\perp}{v_\perp} \ . \tag{4.7.43}$$

The integral involved in (4.7.43) can be put into a more convenient form. To achieve this, we calculate the area spanned, on the one hand, by trajectory A (4.7.38) with the parameters E and $k_z^{(0)}$ and, on the other, by trajectory B with the

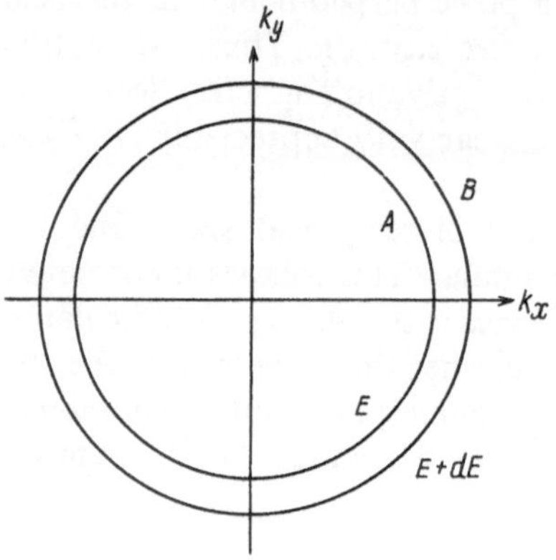

Fig. 4.38. Derivation of conditions for quantization in a magnetic field

parameters $E+dE$ and $k_z^{(0)}$ (Fig. 4.38):

$$\begin{aligned} dS &= \int\limits_{E<E_\zeta(k_x,k_y,k_z^{(0)})<E+dE} dk_x\,dk_y = \int\limits_{E<E_\zeta(k_x,k_y,k_z^{(0)})<E+dE} dk_\perp\,dk_E \\ &= \oint dk_\perp \frac{dE}{|\nabla_k E|} = \frac{1}{\hbar}\oint \frac{dk_\perp}{v_\perp}\,dE\ , \end{aligned} \tag{4.7.44}$$

where $dk_\perp$ is an element of length of contour (path) A, and dk_E is a component of the two-dimensional wave vector $\boldsymbol{k}_\perp$ and is normal to contour A. Since the gradient of E is normal to the isoenergetic surface, we have

$$dE = dk_E|\nabla_k E| = \hbar dk_E v_\perp\ , \tag{4.7.45}$$

which proves (4.7.44). Comparing (4.7.44) with (4.7.43), we find

$$m_H(E, k_z^{(0)}) = \frac{\hbar^2}{2\pi}\frac{\partial S(E, k_z^{(0)})}{\partial E}\ , \tag{4.7.46}$$

where $S(E, k_z^{(0)})$ is the area bounded by the curve (4.7.38). For a free electron, (4.7.38) circumscribes a circle

$$k_x^2 + k_y^2 = \frac{2mE}{\hbar^2} - (k_z^{(0)})^2$$

and

$$S(E, k_z^{(0)}) = \pi\left[\frac{2mE}{\hbar^2} - (k_z^{(0)})^2\right]\ . \tag{4.7.47}$$

Substituting (4.7.47) into (4.7.46), we arrive at the result that $m_H(E, k_z^{(0)}) = m$, which should be the case. In the general case, different electrons move with different periods (for open paths the period is, broadly speaking, equal to infinity).

In semiconductors with nondegenerate bands the Fermi surface has the shape of an ellipsoid

$$E = \frac{\hbar^2(k_x - k_x^{(0)})^2}{2m_1} + \frac{\hbar^2(k_y - k_y^{(0)})^2}{2m_2} + \frac{\hbar^2(k_z - k_z^{(0)})^2}{2m_3} , \tag{4.7.48}$$

where in the case of electrons, i.e., near the bottom of the conduction band, all $m_i > 0$. Furthermore,

$$m_H = \left(\frac{m_1 m_2 m_3}{m_1\alpha_1^2 + m_2\alpha_2^2 + m_3\alpha_3^2}\right)^{1/2} , \tag{4.7.49}$$

where α_i stands for the field direction cosines with respect to the major ellipsoid axes (i.e., the corresponding projection $H_x = H\alpha_1$, $H_y = H\alpha_2$, $H_z = H\alpha_3$). Here the cyclotron effective mass also is the same for all electrons, but it depends on the direction of H. Expression (4.7.49) may also be obtained straightforwardly by considering diamagnetic resonance is semiconductors [Ref. 4.50, Chap. 13].

A rule can be deduced from (4.7.33): an electron moves in such a way that a smaller-energy region lies on the right of the direction of motion at each point of the trajectory. For closed paths this signifies, according to (4.7.46), that if $m_H(E, k_z^{(0)}) > 0$, the electron revolves in the same direction as that in which a free electron does, and for $m_H(E, k_z^{(0)}) < 0$ it revolves in the opposite direction. Thus one talks of electron and hole trajectories. With the $k_z^{(0)}$ dependence of ω_H, it might seem unclear at what frequencies cyclotron resonance should occur in metals (insofar as the E dependence of ω_H goes, it is of course necessary to set $E = E_F$, since only thermally active electrons can receive energy from an externally applied ac field and participate in resonance). If the external-field frequency is

$$\omega = n\omega_H(E_F, k_z^{(0)}) , \quad n = 1, 2, 3 \ldots , \tag{4.7.50}$$

the electrons that contribute to resonance are those for which

$$|\omega_H(E_F, k_z) - \omega_H(E_F, k_z^{(0)})| \lesssim \bar{\tau}^{-1} , \tag{4.7.51}$$

with $\bar{\tau}$ being the time between collisions (Sect. 3.7.3). In the general case, such electrons occupy a layer of thickness

$$|\Delta k_z| \approx \left(\bar{\tau}\left|\frac{\partial\omega_H}{\partial k_z}\right|_{k_z = k_z^{(0)}}\right)^{-1} .$$

But if

$$\left.\frac{\partial\omega_H(E_F, k_z)}{\partial k_z}\right|_{k_z=k_z^{(0)}} = 0 , \qquad (4.7.52)$$

the number of such electrons is appreciably larger. For them,

$$|\Delta k_z| \approx \left(\bar{\tau}\left|\frac{\partial^2\omega_H}{\partial k_z^2}\right|_{k_z=k_z^{(0)}}\right)^{-1/2} .$$

Therefore, resonance will occur primarily when conditions (4.7.50, 52) are fulfilled. Apart from this, absorption anomalies can take place for minimum and maximum values of $k_z^{(0)}$. At these points the velocity is parallel to the magnetic field (Fig. 4.39).

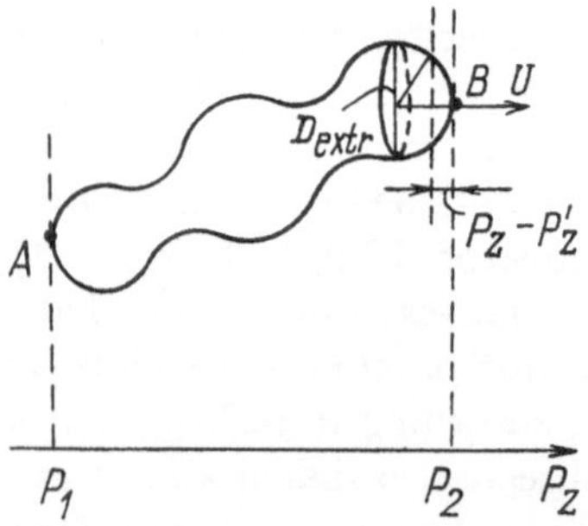

Fig. 4.39. One of the types of points in $\boldsymbol{k}$ space that contribute to cyclotron resonance

4.7.3 Quasi-Classical Energy Levels. Oscillatory Effects

The motion of an electron in a plane perpendicular to the field can also be finite. Then, according to the general principles of quantum mechanics, only certain values of the corresponding degrees of freedom will be allowed values. We quantize the effective Hamiltonian (4.7.28) according to the Bohr–Sommerfeld rule. Here it is convenient to choose a vector potential (calibration) other than that used in (4.7.3):

$$A_x = -Hy , \quad A_y = A_z = 0 . \qquad (4.7.53)$$

The effective Hamiltonian then assumes the form

$$\hat{\mathscr{H}}_{\text{ef}} = E_\zeta\left(K_x + \frac{eH}{\hbar c}y, K_y, K_z\right) . \qquad (4.7.54)$$

The variables x and z are cyclic, so we have a unique quantization condition

$$\oint K_y \, dy = 2\pi(n+\tfrac{1}{2}) \;, \qquad n = 0, 1, 2, \ldots \;, \tag{4.7.55}$$

where the integration is carried out over the closed trajectory in phase space. It follows from (4.7.53) that $K_y = k_y$, and, since x is a cyclic variable, $\dot{K}_x = 0$. Hence we obtain

$$dk_x = \frac{eH}{\hbar c} dy \;. \tag{4.7.56}$$

Substituting (4.7.56) into (4.7.55) and allowing for $\oint k_y \, dk_x = S(E, k_z)$, we find

$$S(E, k_z) = \frac{2\pi|e|H}{\hbar c}(n+\tfrac{1}{2}) \;. \tag{4.7.57}$$

Condition (4.7.57) defines the energy-band spectrum $E(n, k_z)$. Strictly speaking, this condition (like the quasi-classical approximation) is only valid for $n \gg 1$. The energy difference of two adjacent levels here is equal to

$$\Delta E = \frac{2\pi|e|H}{\hbar c}\left(\frac{\partial S(E, k_z)}{\partial E}\right)^{-1} = \hbar\omega_H(E, k_z) \;. \tag{4.7.58}$$

(see 4.7.42, 46). Thus, the correspondence principle holds: the frequency of the transition between two adjacent levels at large quantum numbers is equal to the rotation frequency.

Note that the above standard derivation of (4.7.55) applies here only to a Schrödinger equation of second order in $\partial/\partial y$. Nevertheless, it also works well in the more general case [4.51, 52].

Assuming $\hbar\omega_H$ to depend weakly on E and k_z (the dispersion relation is close to the quadratic law), the result for the number of levels under the Fermi surface will be

$$n_F \approx E_F/\hbar\omega_H \;. \tag{4.7.59}$$

The condition $n_F \gg 1$ is equivalent to (4.7.21) if $k_F \approx d^{-1}$. With nearly filled or nearly empty bands it may turn out that $n_F \approx 1$ and (4.7.57) should be inapplicable. However, the dispersion relation here is close to the quadratic law and the effective Hamiltonian is a harmonic-oscillator and free-motion Hamiltonian. For a harmonic oscillator the quasi-classical spectrum is known to coincide with the exact spectrum even when $n = 0$. Thus, quantization condition (4.7.59) may be regarded as applicable practically without restrictions (except for those required for the effective Hamiltonian itself to be applicable, Sect. 4.7.1).

The energy levels do not depend on the integral of motion K_x. We calculate the corresponding degeneracy multiplicity. The number of states with a given spin projection k_z that lie in the dk_z interval, with n being fixed, is defined as $V dk_z \Delta S/(2\pi)^3$, where V is the volume of the crystal and ΔS the difference of the areas in the k_x, k_y plane that correspond to the levels E_n and E_{n+1}. Taking into account (4.7.57), we obtain for the degeneracy multiplicity of each level

$$\frac{dq(n,k_z)}{dk_z}=\frac{V}{(2\pi)^2}\frac{|e|H}{\hbar c}=\frac{V}{(2\pi l_H)^2}\,, \tag{4.7.60}$$

as for free electrons (3.5.105). As can be seen, the degeneracy multiplicity is independent of n and k_z.

As stated in Sect. 3.5.6, quantization of the motion in the x, y plane leads to characteristic oscillations of the various physical quantities as a function of H^{-1} (for example, magnetic susceptibility, the de Haas–van Alphen effect, Sect. 3.5.6). The density of states at the Fermi level exhibits singularities when a successive Landau quantum level passes through the Fermi surface. These peculiarities, leading to oscillations, manifest themselves if the distance between adjacent Landau levels is larger than the thermal smearing of the Fermi level and the broadening of the levels themselves, which is due to collisions,

$$\hbar\omega_H \gtrsim k_B T,\ \hbar/\bar{\tau}\ . \tag{4.7.61}$$

Inequalities (4.7.61) can be fulfilled at low temperatures (on the order of several K) in pure samples and high magnetic fields ($H \lesssim 10^5$ Oe $\approx 8\cdot 10^6$ A/m). The oscillation period can be determined directly from (4.7.57)

$$\Delta(1/H)=\frac{2\pi|e|}{\hbar c}\frac{1}{S(E_F,k_z)} \tag{4.7.62}$$

since it is for this magnetic field variation that the number of the Landau level at the Fermi surface varies by unity.

Just as in the cyclotron resonance case, we may ask exactly what k_z values contribute to the oscillation period. For the same reasons as those for cyclotron resonance, the contribution is made by the intersection of the Fermi surface with the extremal area (Fig. 4.40). In any other case the oscillation periods due to the intersection of k_z and $k_z+\Delta k_z$ differ considerably, and such oscillations, on the average, suppress each other. But if $\partial S/\partial k_z = 0$, these periods differ by an

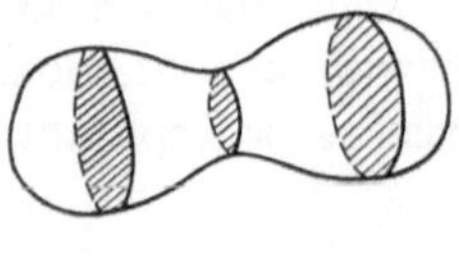

Fig. 4.40. Extreme cross sections of the Fermi surface

amount $\propto (\Delta k_z)^2$, i.e., the oscillations in the vicinity of this point occur "in phase" and enhance each other.

Thus, investigating the oscillation period of the susceptibility or magnetic moment as a function of H^{-1} enables us to determine the areas of the extreme intersections of the Fermi surface in any direction.

Important information is also contained in the magnitudes of the oscillation amplitudes. In simple cases, when the Fermi surface has a center of symmetry and any arm drawn from the center meets the surface only at one point, the form of the Fermi surface can be unambiguously reproduced from a knowledge of the oscillation periods, at arbitrary field directions. The temperature dependence of these amplitudes enables us to find the velocity distribution at the Fermi surface [4.53]. Broadly speaking—i.e., for a Fermi surface of arbitrary shape—a study of the de Haas–van Alphen effect is insufficient for an unambiguous determination of the energy spectrum of electrons near the Fermi surface. For this purpose we also make use of cyclotron resonance, the anomalous skin effect, ultrasonic measurements, so-called dimensional effects in thin metallic plates placed in a magnetic field, studies of galvanomagnetic properties, and positron annihilation. We do not linger on these very important problems; but they are elaborately treated in the literature [4.8, 44, 54]. Here we note that the initial approximation normally used in processing experimental data is a Fermi surface constructed in the nearly-free-electron approximation (Sect. 4.6.4).

We conclude this section by mentioning the very interesting phenomenon of first-order magnetic phase transitions under conditions of strong susceptibility oscillations (Shoenberg effect, see [4.8, 55]). The point is that we must take into account the magnetic field that acts on the electron not only due to the external field but also due to the other charges in the metal. Since the characteristic scale of the electron orbit r_H is large compared to the lattice period, the electron senses the medium's magnetic field, which is produced by all the external and internal charges. The averaged intensity of a field in a medium is the magnetic induction B. Thus, the electron–electron interaction in the de Haas–van Alphen effect can be adequately taken into account by the replacement $H \to B$ in the formula for the magnetization $M(H)$. At sufficiently low temperatures the $M(H)$ dependence may be very strong (this is clearly manifest in the simple two-dimensional model of the de Haas–van Alphen effect considered in Sect. 3.5.6). Consequently, upon performing the replacement $H \to B$, it is the $M(B)$ dependence that will be highly pronounced. This may result in the $H(B) = B - 4\pi M(B)$ ceasing to be a monotonic function (Fig. 4.41).

The detailed theory of the de Haas–van Alphen effect shows that at 0 K in the absence of scattering processes the amplitude of the oscillations of $M(B) \propto B^{-1/2}$ and $\partial H/\partial B = 1 - 4\pi \partial M/\partial B$ is bound to become less than zero when the B's are sufficiently small. The corresponding condition has the form

$$n_F \approx (c/v_F)^{4/3} \approx 10^3 \ , \tag{4.7.63}$$

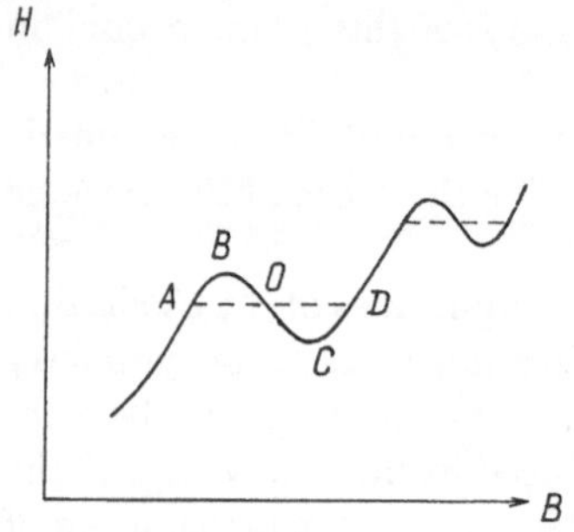

Fig. 4.41. Derivation of the Shoenberg effect

with n_F being the number of Landau levels beneath the Fermi surface. On the other hand, the thermodynamic equilibrium conditions for magnets require that the inequality

$$\left(\frac{\partial B}{\partial H}\right)_{\zeta, T} > 0 \tag{4.7.64}$$

be fulfilled [Ref. 4.50, Chap. 5]. Therefore, the BC segment in Fig. 4.41 corresponds to a thermodynamically unstable state. As is always the case in such situations, the true $H(B)$ dependence is obtained by means of the Maxwell construction [1.4]. The horizontal straight line AD should be drawn in such a fashion that the surface area of ABO is equal to that of OCD. The $H(B)$ relation thus constructed describes the first-order phase transition between states with different values of H and with the same value of B. The straight line AD is a phase coexistence line, the segments AB and CD describe a metastable state, and the segment BC describes an unstable state. This Shoenberg effect leads to alterations in the observed character of the oscillations, and also gives rise to a peculiar domain structure; i.e., it results in the metal being broken up into sections with different magnetization.

4.8 Impurity States

4.8.1 A Simple Model

Departure from ideal spatial periodicity may, just as with phonons (Sect. 2.4), lead to the occurrence of new states which are localized at a distortion-inducing defect. We restrict ourselves to point defects (impurity, vacancy, etc.). Our primary objective now is to consider a simple model problem.

Let there be a lattice of atoms, on each of which the electron is in one orbital state (tight-binding approximation, Sect. 4.6.3). The energy of the atomic level on one of the sites (we choose that site as the origin) differs from those on the other sites by U. Then the Schrödinger equation becomes

$$Ea_n = \sum_{m \neq n} \beta_{mn} a_m + U\delta_{n0} a_0 \ , \quad \beta_{mm} \equiv 0 \ , \tag{4.8.1}$$

where a_m is the amplitude of the probability that the electron will be located on the mth site, and β_{mn} is a transfer integral, that is, the matrix element of the Hamiltonian between the states of sites m, n (for simplicity, we assume that it depends only on the difference $\boldsymbol{R}_n - \boldsymbol{R}_m$; i.e., the case in which one of the sites is a zero site is not singled out in the β_{mn}). The energy E is counted from the atomic-level energy of the "host" atoms.

The function a_m is defined on a three-dimensional lattice $\boldsymbol{R}_m$ and may always be represented as a Fourier integral

$$a_m = \frac{1}{N} \sum_{\boldsymbol{k}} a(\boldsymbol{k}) \exp(\mathrm{i}\boldsymbol{k} \cdot \boldsymbol{R}_m) \ , \tag{4.8.2}$$

where the summation is carried out over an arbitrary reciprocal-lattice cell (for example, over the first Brillouin zone), and N is the number of lattice sites. Substituting (4.8.2) into (4.8.1) and using the Fourier integral expansion of β_{mn},

$$\beta_{mn} = \frac{1}{N} \sum_{\boldsymbol{q}} \beta(\boldsymbol{q}) \exp[-\mathrm{i}\boldsymbol{q} \cdot (\boldsymbol{R}_n - \boldsymbol{R}_m)] \ , \tag{4.8.3}$$

we obtain

$$E \sum_{\boldsymbol{k}} a(\boldsymbol{k}) \exp(\mathrm{i}\boldsymbol{k} \cdot \boldsymbol{R}_n) = \frac{1}{N} \sum_{\boldsymbol{kq}} \beta(\boldsymbol{q}) a(\boldsymbol{k}) \sum_m \exp[\mathrm{i}(\boldsymbol{q} - \boldsymbol{k}) \cdot \boldsymbol{R}_m] \cdot$$
$$\cdot \exp(\mathrm{i}\boldsymbol{q} \cdot \boldsymbol{R}_n) + \frac{U}{N} \sum_{\boldsymbol{q}} a(\boldsymbol{q}) \sum_{\boldsymbol{k}} \exp(\mathrm{i}\boldsymbol{k} \cdot \boldsymbol{R}_n) \ , \tag{4.8.4}$$

$$\left[\delta_{n0} = \frac{1}{N} \sum_{\boldsymbol{k}} \exp(\mathrm{i}\boldsymbol{k} \cdot \boldsymbol{R}_n) \right] .$$

Allowing for

$$\sum_m \exp[-\mathrm{i}(\boldsymbol{q} - \boldsymbol{k}) \cdot \boldsymbol{R}_m] = N\delta_{\boldsymbol{qk}} \tag{4.8.5}$$

and equating the coefficients of the orthogonal functions $\exp(\mathrm{i}\boldsymbol{k}\cdot\boldsymbol{R}_n)$, we have

$$[E-\beta(\boldsymbol{k})]\,a(\boldsymbol{k})=\frac{U}{N}\sum_{\boldsymbol{q}}a(\boldsymbol{q}) \;. \tag{4.8.6}$$

Two types of solutions exist. In the first, $\sum_{\boldsymbol{q}} a(\boldsymbol{q})=0$, i.e., $a_0=0$. The probability of finding an electron on the impurity is equal to zero, the spectrum coinciding with that of the band states, $E=\beta(\boldsymbol{k})$. In the second solution, $a_0\neq 0$. Then (4.8.6) yields

$$1=\frac{U}{N}\sum_{\boldsymbol{k}}\frac{1}{E-\beta(\boldsymbol{k})} \;. \tag{4.8.7}$$

When $\min\beta(\boldsymbol{k})<E<\max\beta(\boldsymbol{k})$ the question arises as to what the singular integral involved in (4.8.7) should imply. As with phonons, it is solved by introducing a small imaginary addition to E (Sects. 2.4, 6). Such states describe band electron scattering on an impurity.

We defer the problem of describing these states for a while and consider the case

$$E<\min\beta(\boldsymbol{k}) \tag{4.8.8}$$

or

$$E>\max\beta(\boldsymbol{k}) \;. \tag{4.8.9}$$

It follows from (4.8.7) that (4.8.8) may hold for $U<0$, and (4.8.9) for $U>0$. Further, for illustration, we consider the case of $U<0$. Introduce the notation

$$E=\min\beta(\boldsymbol{k})-\Delta \tag{4.8.10}$$

($\Delta>0$), and represent (4.8.7) as

$$1=\frac{|U|}{N}\sum_{\boldsymbol{k}}\frac{1}{\Delta+\tilde{\beta}(\boldsymbol{k})}\equiv|U|F(\Delta) \;, \tag{4.8.11}$$

with $\tilde{\beta}(\boldsymbol{k})=\beta(\boldsymbol{k})-\min\beta(\boldsymbol{k})$. Assume without loss of generality that the minimum of $\beta(\boldsymbol{k})$ is reached at the point $\boldsymbol{k}=0$ (if necessary, we displace the boundaries of the cell in reciprocal space; the case of several equal minima does not present any difficulties, either). At small $\boldsymbol{k}$

$$\tilde{\beta}(\boldsymbol{k})=\frac{\hbar^2}{2}\sum_{i=1}^{d}\frac{k_i^2}{m_i} \;, \tag{4.8.12}$$

where d is the space dimension ($d = 1, 2, 3$) and m_i is the eigenvalue of the inverse-effective-mass tensor (all $m_i > 0$). In the one- and two-dimensional cases, the value

$$F(0) = \frac{1}{N}\sum_{k}\frac{1}{\tilde{\beta}(\boldsymbol{k})} \equiv V_0 \int \frac{d\boldsymbol{k}}{(2\pi)^3}\frac{1}{\tilde{\beta}(\boldsymbol{k})} \tag{4.8.13}$$

(with V_0 being the unit-cell volume) diverges at small k (logarithmically for $d = 2$ and $\propto k^{-1}$ for $d = 1$). Since $F(\Delta)$ decreases with increasing Δ and $F(0) = \infty$, (4.8.11) here can be solved for any $|U|$. Such states are said to be localized. To understand the meaning of this term, we calculate the expression

$$a_m = N^{-1}\sum_{k} a(k)\exp(\mathrm{i}\boldsymbol{k}\cdot\boldsymbol{R}_m) = \frac{U a_0}{N}\sum_{k}\frac{\exp(\mathrm{i}\boldsymbol{k}\cdot\boldsymbol{R}_m)}{E - \beta(\boldsymbol{k})}\;, \tag{4.8.14}$$

where (4.8.2, 6) are taken into account. For $\Delta > 0$, the quantity $E - \beta(\boldsymbol{k})$ does not go to zero anywhere and, therefore, according to the familiar properties of the Fourier transformation, a_m decreases at $|\boldsymbol{R}_m| \to \infty$ more rapidly than any power of $|\boldsymbol{R}_m|^{-1}$ [for real $\beta(\boldsymbol{k})$ it decreases exponentially]. The assertion for $d = 1$ constitutes the Riemann–Lebesgue lemma [Ref. 4.17, Sect. 3.1]. The generalization for the many-dimensional case, at least at the "physical" level of rigor, is trivial enough. Consequently, the probability of detecting an electron in localized states decreases rapidly with increase in the distance from the impurity.

In the one- and two-dimensional cases localized states always arise. In the three-dimensional case they occur when the $|U|$ is sufficiently large, i.e., when

$$|U| > \left[\frac{1}{N}\sum_{k}\frac{1}{\tilde{\beta}(\boldsymbol{k})}\right]^{-1}\;, \tag{4.8.15}$$

as follows from (4.8.11) and from the decrease of $F(\Delta)$ with increasing Δ.

Localized states with $E < \min\beta(\boldsymbol{k})$ in our model correspond to donor levels in the theory of semiconductors, and with $E > \max\beta(\boldsymbol{k})$ to acceptor levels.

4.8.2 Green's Functions and the Density of States

We now wish to consider the general formulation of the problem. Let our Hamiltonian have the form

$$\hat{\mathscr{H}} = \hat{\mathscr{H}}_0 + \hat{V}\;, \tag{4.8.16}$$

with $\hat{\mathscr{H}}_0$ being the Hamiltonian of an ideal crystal, and $\hat{V}$ the perturbation. In the one-band (tight-binding) approximation, the $\hat{V}$ is given by the matrix elements V_{mn}, where m and n stand for site numbers. Thus, it was assumed in Sect. 4.8.1 that

$$V_{mn} = U\delta_{m0}\delta_{n0} \ . \tag{4.8.17}$$

We wish to determine the spectrum of the Hamiltonian $\hat{\mathscr{H}}$, or, to be more exact, the density of states

$$g(E) = \sum_{\nu} \delta(E - E_\nu) = \mathrm{Tr}\{E - \hat{\mathscr{H}}\} \ , \tag{4.8.18}$$

with E_ν being the eigenvalues of the Hamiltonian $\hat{\mathscr{H}}$. The operator $\delta(E - \hat{\mathscr{H}})$ is related to the operators

$$\hat{R}^{\pm}(E) = (E - \hat{\mathscr{H}} + \mathrm{i}\eta)^{-1}|_{\eta \to +0} \tag{4.8.19}$$

by the identity

$$\left.\frac{1}{x \pm \mathrm{i}\eta}\right|_{\eta \to +0} = \mathscr{P}\frac{1}{x} \mp \mathrm{i}\pi\delta(x) \ , \tag{4.8.20}$$

which should be understood as follows: For any "sufficiently good" function $\varphi(x)$,

$$\int dx\,\varphi(x)(x \pm \mathrm{i}\eta)^{-1} = \int dx\,\varphi(x)x(x^2 + \eta^2)^{-1} \mp \mathrm{i}\eta \int dx\,\varphi(x) \cdot (x^2 + \eta^2)^{-1} \xrightarrow[\eta \to +0]{} \mathscr{P}\int dx\,\varphi(x)x^{-1} \mp \mathrm{i}\pi\varphi(0) \ , \tag{4.8.21}$$

where $\mathscr{P}$ is the symbol of the principal value and where we have employed one of the representations of the δ function

$$\delta(x) = \lim_{\eta \to +0} \frac{\eta}{\pi(x^2 + \eta^2)} \ . \tag{4.8.22}$$

The operators $\hat{R}^{\pm}(E)$ are called resolvents, or Green's functions. The density of states can be expressed in terms of these $\hat{R}^{\pm}(E)$:

$$g(E) = \sum_{\nu} \delta(E - E_\nu) = \mp\frac{1}{\pi}\mathrm{Im}\left\{\sum_{\nu}(E - E_\nu \pm \mathrm{i}\eta)^{-1}\right\} = \mp\frac{1}{\pi}\mathrm{Im}\,\mathrm{Tr}\{\hat{R}^{\pm}(E)\} \ . \tag{4.8.23}$$

Suppose that we know the unperturbed Green's function

$$\hat{R}_0^{\pm}(E) = (E - \hat{\mathscr{H}}_0 \pm \mathrm{i}\eta)^{-1} \ . \tag{4.8.24}$$

For this function to be constructed, we need to know the eigenfunctions and the eigenvalues of the Hamiltonian $\hat{\mathscr{H}}_0$.

Equations (4.8.16, 19, 23) yield

$$[\hat{R}^{\pm}(E)]^{-1} = [\hat{R}_0^{\pm}(E)]^{-1} - \hat{V} = [\hat{R}_0^{\pm}(E)]^{-1}[1 - \hat{R}_0^{\pm}(E)\hat{V}] \ . \tag{4.8.25}$$

Transforming (4.8.25), we find

$$\hat{R}^{\pm}(E) = [1 - \hat{R}_0^{\pm}(E)\hat{V}]^{-1}\hat{R}_0^{\pm}(E) \ . \tag{4.8.26}$$

We need to prove the matrix equation

$$\det \hat{A} = \exp(\mathrm{Tr}\{\ln \hat{A}\}) \ . \tag{4.8.27}$$

Denoting $\ln \hat{A} = \hat{X}$, we obtain

$$\det \hat{A} = \det \exp \hat{X} = \lim_{n\to\infty} \det\left(1 + \frac{\hat{X}}{n}\right)^n = \lim_{n\to\infty}\left[\det\left(1 + \frac{\hat{X}}{n}\right)\right]^n$$

$$= \lim_{n\to\infty}\left[1 + \frac{\mathrm{Tr}\hat{X}}{n}\right]^n = \exp \mathrm{Tr}\hat{X} \ ,$$

which proves (4.8.27). The properties used here are

$$\det(\hat{A}\hat{B}) = \det\hat{A}\det\hat{B} \ , \quad \det(1 + \varepsilon\hat{X}) = 1 + \varepsilon\mathrm{Tr}\hat{X} + O(\varepsilon^2) \ .$$

We now transform $\mathrm{Tr}\hat{R}^{\pm}(E)$. It follows from (4.8.19) that $(\partial/\partial E)\ln\hat{R}^{\pm}(E) = -\hat{\mathrm{R}}^{\pm}(E)$ and

$$\mathrm{Tr}\{\hat{R}^{\pm}(E)\} = -\frac{\partial}{\partial E}\mathrm{Tr}\{\ln\hat{R}^{\pm}(E)\} = \frac{\partial}{\partial E}\mathrm{Tr}\ln[1 - \hat{R}_0^{\pm}(E)\hat{V}] - \frac{\partial}{\partial E}\mathrm{Tr}$$

$$\cdot \ln\hat{R}_0^{\pm}(E) = \frac{\partial}{\partial E}\mathrm{Tr}\ln[1 - \hat{R}_0^{\pm}(E)\hat{V}] + \mathrm{Tr}\hat{R}_0^{\pm}(E) \ , \tag{4.8.28}$$

where we have used (4.8.26, 27). The result for the density of states, according to (4.8.23), then is

$$g(E) = g_0(E) \mp \frac{1}{\pi}\mathrm{Im}\left\{\frac{\partial}{\partial E}\ln\det[1 - \hat{R}_0^{\pm}(E)\hat{V}]\right\} \ . \tag{4.8.29}$$

Formula (4.8.29) gives a formal solution to the problem of the variation of the density of states due to the perturbation $\hat{V}$.

We apply (4.8.28) to the problem considered in Sect. 4.7.1. The eigenfunctions and the eigenvalues of the Hamiltonian $\hat{\mathscr{H}}_0$ have the form

$$a_n(\boldsymbol{k}) = N^{-1/2}\exp(\mathrm{i}\boldsymbol{k}\cdot\boldsymbol{R}_n)\ , \qquad E(k) = \beta(\boldsymbol{k})\ , \tag{4.8.30}$$

where $\boldsymbol{k}$ runs through the first Brillouin zone. Then

$$\begin{aligned}[\hat{R}_0^-(E)]_{mn} &= \sum_{\boldsymbol{k}} a_m^*(\boldsymbol{k})a_n(\boldsymbol{k})[E-\beta(\boldsymbol{k})-\mathrm{i}\eta]^{-1}\\ &= N^{-1}\sum_{\boldsymbol{k}}\exp[\mathrm{i}\boldsymbol{k}\cdot(\boldsymbol{R}_n-\boldsymbol{R}_m)]\,[E-\beta(\boldsymbol{k})-\mathrm{i}\eta]^{-1}\ .\end{aligned} \tag{4.8.31}$$

Using (4.8.17, 31), we find the matrix $1-\hat{R}_0^-(E)\hat{V}$

$$[1-\hat{R}_0^-(E)\hat{V}]_{mn} = \delta_{mn} - U[\hat{R}_0^-(E)]_{mo}\delta_{no}\ . \tag{4.8.32}$$

This matrix differs from the unit matrix only in that it has a zero row. Therefore

$$\det[1-\hat{R}_0^-(E)\hat{V}] = 1-UF(E)\ , \tag{4.8.33}$$

where

$$F(E) = [\hat{R}_0^-(E)]_{00} = N^{-1}\sum_{\boldsymbol{k}}[E-\beta(\boldsymbol{k})-\mathrm{i}\eta]^{-1}\ . \tag{4.8.34}$$

Then

$$\begin{aligned}g(E) &= g_0(E) + \frac{1}{\pi}\frac{\partial}{\partial E}\mathrm{Im}\{\ln[1-UF(E)]\}\\ &= g_0(E) - \frac{U}{\pi}\mathrm{Im}\frac{F'(E)}{1-UF(E)}\ ,\\ F'(E) &\equiv \frac{dF(E)}{dE}\ .\end{aligned} \tag{4.8.35}$$

If $E > \max\beta(\boldsymbol{k})$ or $E < \min\beta(\boldsymbol{k})$, $F(E)$ and $F'(E)$ are real valued and the imaginary part in (4.8.35) is only nonzero provided

$$1 = UF(E)\ . \tag{4.8.36}$$

The above expression is an equation for determining the energy of localized states and coincides with (4.8.7). In this case, according to (4.8.20, 35),

$$g(E) = g_0(E) + UF'(E)\delta[1-UF(E)] = g_0(E) + \delta(E-E_l)\ , \tag{4.8.37}$$

where E_l is the localized-state energy and we have used the property of the δ function

$$\delta[\varphi(x)] = \sum_i \delta(x - x_i)/|\varphi'(x_i)| \ ,$$

with x_i being the roots of the equation $\varphi(x) = 0$. Thus a δ-function peak arises in the density of states.

Now let $\min \beta(\boldsymbol{k}) < E < \max \beta(\boldsymbol{k})$. Then

$$\operatorname{Im} F(E) = \frac{\pi}{N} \sum_{\boldsymbol{k}} \delta(E - \beta(\boldsymbol{k})) = \frac{\pi}{N} g_0(E) \ ,$$

and near E_l, which is a solution of the equation

$$1 = U \operatorname{Re} F(E_l) \ , \tag{4.8.38}$$

$$\begin{aligned} g(E) &= g_0(E) + \operatorname{Re} F'(E) \frac{\pi U}{N} g(E) \frac{1}{[1 - U \operatorname{Re} F(E)]^2 + \left[\dfrac{\pi U}{N} g_0(E)\right]^2} \\ &\approx g_0(E) + \frac{\Gamma}{\pi[(E - E_l)^2 + \Gamma^2]} \ , \end{aligned} \tag{4.8.39}$$

$$\Gamma = \frac{\pi g_0(E_l)}{N \operatorname{Re} F'(E_l)} \tag{4.8.40}$$

is what is known as the width of the quasi-local level. The density of states in this case has a maximum (Fig. 4.42). According to the uncertainty relation, the electron resides in the state with energy E_l for a time $\approx \hbar/\Gamma$. If $g_0(E_l)$ is sufficiently small, we may assume that the electron is virtually localized, thus the name "the quasilocal level".

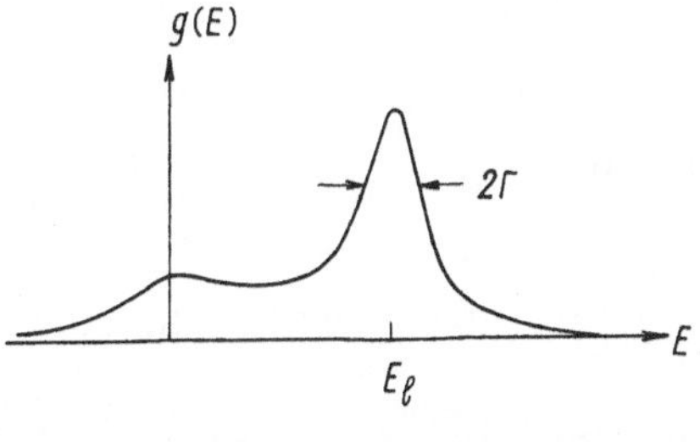

Fig. 4.42. Manifestation of the quasi-local level E_l in the density of states $g(E)$, in the electron energy band

4.8.3 Friedel Oscillations

We have omitted a number of important problems relating to the impurity states in semiconductors (see any relevant text-book), restricting ourselves to a few comments. As we have seen in Sects. 4.4, 7, the electron motion in weakly inhomogeneous and sufficiently low electric and magnetic fields may be described by some effective Hamiltonian, it being possible to neglect the mixing of states from different bands. Similar arguments are also applicable to the impurity potential $V(\boldsymbol{r})$, if, first, the characteristic scale of its variation is large compared to the lattice period and, second, the depth of the potential well is small in comparison with the width of the forbidden energy band. Then we arrive at the effective Schrödinger equation

$$[E_\zeta(-\mathrm{i}\nabla)+V(\boldsymbol{r})]\psi(\boldsymbol{r})=E\psi(\boldsymbol{r}) \ , \tag{4.8.41}$$

which was actually implied in our preliminary discussion of the impurity states of semiconductors in Sect. 4.5.1; i.e., by the replacement of the free-electron mass by the effective mass. A justification of (4.8.41) is presented by *Tsidilkovski* [4.21].

We now proceed to a description of an interesting effect that arises when an impurity is screened by a degenerate electron gas—Friedel oscillations [4.56]. Our treatment is close to that in [4.32]. We restrict our attention to the case of free electrons and a spherically symmetric potential $V(r)$.

The wave function with the orbital quantum number l can then represented as

$$\psi(\boldsymbol{r})=\frac{\chi(r)}{r}Y_{lm}(\vartheta,\varphi) \ , \tag{4.8.42}$$

where $Y_{lm}(\vartheta,\varphi)$ is a spherical harmonic and $\chi(r)$ satisfies the equation

$$\chi''(r)+\left\{\frac{2m}{\hbar^2}[E-V(r)]-\frac{l(l+1)}{r^2}\right\}\chi(r)=0 \ . \tag{4.8.43}$$

Let us write the equation for the function χ_1, which corresponds to the energy E_1:

$$\chi_1''(r)+\left\{\frac{2m}{\hbar^2}[E_1-V(r)]-\frac{l(l+1)}{r^2}\right\}\chi_1(r)=0 \ . \tag{4.8.44}$$

Multiplying (4.8.43) by $\chi_1(r)$, and (4.8.44) by $\chi(r)$, we subtract one from the other. This gives

$$\chi_1(r)\chi''(r)-\chi(r)\chi_1''(r)\equiv[\chi_1(r)\chi'(r)-\chi(r)\chi_1'(r)]'$$
$$=2m\hbar^{-2}(E_1-E)\chi_1(r)\chi(r)\ . \tag{4.8.45}$$

Integrate (4.8.45) between 0 and R, allowing for the condition $\chi(0)=0$,

$$\chi_1(R)\chi'(R)-\chi_1'(R)\chi(R)=\frac{2m}{\hbar^2}(E_1-E)\int_0^R dr\,\chi_1(r)\chi(r)\ . \tag{4.8.46}$$

The asymptote $\chi(R)$ at large distances R is known:

$$\chi(R)=A\sin[kR+\pi l/2+\delta_l(k)]\ , \tag{4.8.47}$$

where $\delta_l(k)$ is the phase shift, $k=(2mE/\hbar^2)^{1/2}$, and A is the normalization coefficient. Let k_1 tend to k: $k_1=k+\delta k$. Then we obtain

$$\chi_1(R)\approx\chi(R)+\frac{\partial\chi(R)}{\partial k}\delta k\ ,\qquad E_1-E\approx\hbar^2k\delta k/m\ . \tag{4.8.48}$$

Substituting (4.8.48) into (4.8.46) yields

$$\frac{1}{2k}\left(\frac{\partial\chi}{\partial k}\frac{\partial\chi}{\partial R}-\chi\frac{\partial^2\chi}{\partial k\partial R}\right)=\int_0^R dr\,\chi^2(r)\equiv n_l(k,R)\ . \tag{4.8.49}$$

Calculating the derivatives, we find

$$\frac{1}{2k}\left(\frac{\partial\chi}{\partial R}\frac{\partial\chi}{\partial k}-\chi\frac{\partial^2\chi}{\partial k\partial R}\right)=-\frac{1}{2k}\chi^2\frac{\partial^2}{\partial k\partial R}\ln\chi=-\frac{A^2}{2k}\sin^2\left[kR-\frac{\pi l}{2}\right.$$
$$\left.+\delta_l(k)\right]\frac{\partial}{\partial k}\left[k\operatorname{cotan}\left(kR-\frac{\pi l}{2}+\delta_l(k)\right)\right]$$
$$=\frac{A^2}{2}\left\{R+\frac{\partial\delta_l(k)}{\partial k}-\frac{1}{2k}\sin 2\left[kR-\frac{\pi l}{2}+\delta_l(k)\right]\right\}\ . \tag{4.8.50}$$

The quantity $n_l(k,R)$ represents the probability of finding a particle in the quantum state $|klm\rangle$ inside a sphere of radius R. Let it be assumed, for concreteness, that our system is placed into an impenetrable spherical box of radius L:

$$\chi(L)=0\ . \tag{4.8.51}$$

Therefore, when $R = L$, the number $n_l(k, L) = 1$, whence we obtain ($L \to \infty$)

$$A^2 = 2/L \; . \tag{4.8.52}$$

We are interested in the variation of the electron density when we turn on the impurity potential. In this case, the value for $\delta_l(k) = 0$ is subtracted from (4.8.50). Allowing for (4.8.51), we find

$$\Delta n_l(k, R) = L^{-1}\left[\frac{\partial \delta_l(k)}{\partial k} + \frac{\sin \delta_l(k)}{k} \cos(2kR - \pi l + \delta_l(k))\right] . \tag{4.8.53}$$

Using (4.8.51) and (4.8.47), the allowed values of k may vary by π/L. Thus the degeneracy multiplicity of $|kl\rangle$ is equal to $(2L/\pi)(2l+1)$, where the factor 2 is due to spin and $2l+1$ is due to the quantum number m. The result for the total electron-density variation then is

$$\Delta n(R) = \frac{2}{\pi} \sum_{l=0}^{\infty} (2l+1) \int_0^{k_F} dk \left\{ \frac{\partial \delta_l(k)}{\partial k} - \frac{\sin \delta_l(k)}{k} \cos[2kR - \pi l + \delta_l(k)] \right\} . \tag{4.8.54}$$

For $R \to \infty$ we obtain the total variation of the number of electrons, which, because of electroneutrality, should be equal to the excess effective impurity charge in comparison to the ion charge of the matrix:

$$\Delta Z = \frac{2}{\pi} \sum_{l=0}^{\infty} (2l+1) \delta_l(k_F) \tag{4.8.55}$$

(Friedel sum rule). We have omitted the quantity δ_l $(k=0)$ because if the scattering cross section is finite as $k \to 0$, then $\delta_l(0) = 0$.

The oscillatory part of the electron density is

$$\varrho(R) = \frac{1}{4\pi R^2} \frac{\partial \Delta n(R)}{\partial R} \; . \tag{4.8.56}$$

Substituting (4.8.54) into (4.8.56), performing integration by parts, and retaining the leading terms with respect to R^{-1}, we find

$$\varrho(R) = -\frac{1}{2\pi^2 R^3} \sum_{l=0}^{\infty} (2l+1)(-1)^l \sin \delta_l(k_F) \cos(2k_F R + \delta_l(k_F)) \; . \tag{4.8.57}$$

This slow and oscillatory decrease of the electron density at large distances from the charged impurity is a consequence of the Fermi–Dirac statistics. As will be seen in Chap. 5, this decrease of the electron density has important implications in the theory of the magnetic properties of transition-d metals and rare-earth metals.

4.9 The Electronic Structure of Disordered Systems

4.9.1 The Average Green's Function in the Diagonal Disorder Model

In the case of a finite impurity concentration we are dealing essentially with a disordered system (alloy). As emphasized in Sect. 1.10, a description of such a system in the language of wave functions is inadequate. This is so because each impurity configuration (or each realization of the impurity potential V) has a set of quantum states of its own. On the other hand, the density of states appears to be a self-averaging quantity [1.44]. That is, for an overwhelming majority of configurations, the density of states assumes the same value, which coincides with the mean density of states for all configurations,

$$g(E) = \frac{1}{\pi} \operatorname{Im} \{\operatorname{Tr} \langle \hat{R}(E) \rangle\} \ . \tag{4.9.1}$$

Here $\hat{R} \equiv \hat{R}^-$ is the resolvent—see (4.8.19, 23)—and the angular brackets denote averaging over impurity configurations, i.e., impurity potential realization. Therefore, when we are concerned with the determination of the electronic structure of disordered systems, we normally pose the problem of finding the average Green's function (resolvent) and the density of states that is related to it by (4.9.1). We may ask whether a more detailed description, similar to that of the dispersion relation $E(\boldsymbol{k}, \zeta)$ for crystals, can be given. It turns out that to certain approximations this is possible.

The spectrum of the eigenvalues of the operator $\hat{\mathscr{H}}$ is a set of E values for which the operator $E - \hat{\mathscr{H}}$ has no inverse; i.e., the resolvent has a pole. The spectrum of a disordered system could be defined as the poles of the mean resolvent. This definition has a correct limit if we go over to an ideal crystal. In addition, in a homogeneous alloy the mean resolvent $\langle R_{nm}(E) \rangle$ depends only on the difference $\boldsymbol{R}_n - \boldsymbol{R}_m$ (this is not true for each individual potential). Therefore, we can introduce the quasimomentum $\boldsymbol{k}$ and Brillouin zones, and carry out the Fourier transformation

$$\langle R_{nm}(E) \rangle = \frac{1}{N} \sum_{\boldsymbol{k}} \langle R(E, \boldsymbol{k}) \rangle \exp[\mathrm{i}\boldsymbol{k} \cdot (\boldsymbol{R}_n - \boldsymbol{R}_m)] \ . \tag{4.9.2}$$

If the spectrum is determined as the poles $E(\boldsymbol{k})$ of the function $\langle R(E, \boldsymbol{k}) \rangle$, it depends on quasimomentum, as in the case of an ideal crystal. But since the mean resolvent is no longer the resolvent of some Hermitian operator, the poles of the resolvent, as will be seen, are not real:

$$E = \varepsilon(\boldsymbol{k}) + \mathrm{i}\Gamma(\boldsymbol{k}) \tag{4.9.3}$$

(with ε and Γ being real). This fact, broadly speaking, signifies that the energy spectrum of a disordered system cannot be specified by the dispersion relation $E(\boldsymbol{k})$. Physically, this is due to the nonconservation of quasimomentum when the impurity potential is scattered by fluctuations. Recall that the system is assumed to be homogeneous only on the average, but not for each individual impurity. The quasimomentum thus is not a "good" quantum number and a state with a definite $\boldsymbol{k}$ cannot correspond to a definite energy, but is a wave packet and has an energy spread on the order of Γ. As is known from quantum mechanics [1.12], the occurrence of an imaginary part in the energy formally denotes that the state is nonstationary, its lifetime being

$$\tau(\boldsymbol{k}) \approx \hbar/\Gamma(\boldsymbol{k}) \ . \tag{4.9.4}$$

Now we can answer the question about the possibility of describing the electronic structure of alloys with the help of the dispersion relation $E(\boldsymbol{k})$. Evidently, such a description is reasonable if the real part of the pole of the mean resolvent $\varepsilon(\boldsymbol{k})$ is large compared to its imaginary part $\Gamma(\boldsymbol{k})$ or, put another way, the decay (damping) of states with a definite $\boldsymbol{k}$ is comparatively small. This only takes place provided that the disorder is relatively weak. It must also be emphasized that this approach obviously does not allow us to consider localized states (Sect. 4.5.4), which cannot, even approximately, be specified by quasimomenta in the one-impurity limit.

Particular methods of computing the average Green's function will be considered using the example of diagonal disorder models, which, in simplified form, describe substitutional alloys (Sect. 1.10). We assume that the energies of the atomic levels are independently distributed random quantities and the transfer integrals β_{mn} are not random but conserve the same values as those in a perfect crystal. Therefore

$$V_{mn} = U_n \delta_{mn} \ , \tag{4.9.5}$$

with all U_n being distributed with the same probability density $P(U_n)$, normalized to unity

$$\int_{-\infty}^{\infty} dU P(U) = 1 \ . \tag{4.9.6}$$

Thus, in the model of a binary alloy composed of A and B atoms with energies ε_A and ε_B and concentrations $1-c$ and c, respectively, the quantity U may assume two values ε_A and ε_B with the probabilities $1-c$ and c:

$$P(U)=(1-c)\delta(U-\varepsilon_A)+c\delta(U-\varepsilon_B) \ . \tag{4.9.7}$$

The average Green's function in such a relatively realistic model cannot be calculated exactly; approximate computational methods will be outlined later in the text. Here we consider the *Lloyd* model [4.57], in which

$$P(U)=\frac{\Gamma}{\pi}\frac{1}{\Gamma^2+(U-E_0)^2} \ . \tag{4.9.8}$$

This model is unrealistic since $P(U)$ falls off too slowly when $U\to\pm\infty$; i.e., large potential fluctuations are more than probable. On the other hand, the model enables us to calculate the function $\langle R(E,\boldsymbol{k})\rangle$ exactly and thus to demonstrate its general properties.

Before passing on to particular calculations, we wish to derive a very helpful representation for the disordered Green's function in the diagonal disorder model (4.9.5).

Let the zero-approximation Hamiltonian $\hat{\mathscr{H}}_0^{(n)}$ be chosen as the sum of the transfer Hamiltonian, i.e., the matrix β_{mn} and the perturbation matrix $\hat{V}$ in which U_n is replaced by zero (i.e., the perturbation on the nth site is eliminated). The perturbation matrix then has the form

$$V_{ml}^{(n)}=U_n\delta_{mn}\delta_{ln} \ . \tag{4.9.9}$$

We use the identity (4.8.26)

$$\hat{R}=(1-\hat{R}^{(n)}\hat{V}^{(n)})^{-1}\hat{R}^{(n)} \ , \tag{4.9.10}$$

where $\hat{R}^{(n)}$ is the resolvent corresponding to the zero approximation. Also, we use the identity

$$(1-\hat{A})^{-1}=1+\hat{A}(1-\hat{A})^{-1} \ , \tag{4.9.11}$$

which may be proved by postmultiplying by $(1-\hat{A})$. Equation (4.9.10) then becomes

$$\hat{R}=\hat{R}^{(n)}+\hat{R}^{(n)}\hat{T}^{(n)}\hat{R}^{(n)} \ , \qquad \hat{T}^{(n)}=\hat{V}^{(n)}(1-\hat{R}^{(n)}\hat{V}^{(n)})^{-1} \ . \tag{4.9.12}$$

The operators, similar to $\hat{T}^{(n)}$, play an important role in the theory of alloys. For (4.9.9), in which $\hat{V}^{(n)}$ has a unique nonzero matrix element, it is readily understood that $\hat{T}^{(n)}$ will also have a unique nonzero matrix element with the same row and column numbers. This may be shown by expanding (4.9.12) in a power series of $\hat{V}^{(n)}$. The second-order term, for example, is equal to

$$(\hat{V}^{(n)}\hat{R}^{(n)}\hat{V}^{(n)})_{ml} = \sum_{pq} V^{(n)}_{mp} R^{(n)}_{pq} V^{(n)}_{ql}$$
$$= U_n^2 \sum_{pq} \delta_{mp}\delta_{pq}\delta_{ql}\delta_{mn}\delta_{ln} R^{(n)}_{pq} = U_n^2 R^{(n)}_{nn}\delta_{mn}\delta_{ln} \; ; \quad (4.9.13)$$

i.e., the operator $\hat{V}^{(n)}$ is replaced by the number U_n, and the operator $\hat{R}^{(n)}$ by the number $R^{(n)}_{nn}$ (4.8.33).

As a result,

$$T^{(n)}_{ml} = T_n \delta_{nm}\delta_{nl} \, ,$$
$$T_n = U_n/(1 - U_n R^{(n)}_{nn}) \, . \quad (4.9.14)$$

Formulas (4.9.12, 14) permit separation of the explicit U_n dependence of the operator $\hat{R}$, since $\hat{R}^{(n)}$ is independent of U_n.

Because the quantities U_m at different m are distributed independently, expression (4.9.14) makes it possible to average over the potential of a fixed site n with the other potential being arbitrary

$$\langle T_n \rangle^{(n)} = \int_{-\infty}^{\infty} dU P(U) \frac{U}{1 - U R^{(n)}_{nn}} \, . \quad (4.9.15)$$

In the Lloyd model (4.9.8) the integral (4.9.15) is easy to calculate in the complex plane. The imaginary part of the function $R^{(n)}_{nn}$ is positive, since it is proportional to the density of states (recall that $\hat{R} \equiv \hat{R}^-$). Therefore, the function $1 - U R^{(n)}_{nn}$ does not go to zero for $\mathrm{Im}\, U > 0$, and the only singularity that remains in the upper half plane when the path of integration is closed is the pole of the function $P(U)$ at point $U = E_0 + \mathrm{i}\Gamma$ with the residue of $(2\pi \mathrm{i})$. Therefore

$$\langle T_n \rangle^{(n)} = \frac{E_0 + \mathrm{i}\Gamma}{1 - (E_0 + \mathrm{i}\Gamma) R^{(n)}_{nn}} \, , \quad (4.9.16)$$

and the averaging over U_n in the Lloyd model reduces to the replacement $U_n \to E_0 + \mathrm{i}\Gamma$. This being valid for any n, we arrive at the result that the average Green's function in the Lloyd model coincides with the Green's function of a system with a nonrandom complex perturbation, equal to $E_0 + \mathrm{i}\Gamma$ on each site, regardless of n. This permanent perturbation may be taken into account by the addition to the energy

$$\langle R(E, \boldsymbol{k}) \rangle = R_0(E - E_0 - \mathrm{i}\Gamma, \boldsymbol{k}) = \frac{1}{E - E_0 - \beta(\boldsymbol{k}) - \mathrm{i}\Gamma} \, . \quad (4.9.17)$$

Then the poles of the mean resolvent are equal to

$$E = E_0 + \beta(\boldsymbol{k}) + \mathrm{i}\Gamma \ , \tag{4.9.18}$$

which tallies exactly with (4.9.3). Only for $\Gamma \to 0$, when, from (4.9.8), $P(U) \to \delta(U - E_0)$, are the poles of the Green's function real. Allowance for potential fluctuations entails a damping of electronic states. In the Lloyd model we have come to the result that this damping is independent of E.

4.9.2 Approximate Methods of Computing the Average Green's Function in the Binary Alloy Model

In this section we consider a more realistic model with the probability distribution (4.9.7). Exact averaging is here no longer possible, and we have to employ approximate methods of calculating $\langle R(E, \boldsymbol{k})\rangle$. These methods, treated here within the framework of the simple model discussed, are used, with some alterations, also in real-band calculations of alloys [4.58, 59].

The approximations that are discussed in what follows may be viewed as resulting from the partial summation of perturbation-theory series over some small parameter. This is normally either the spread in values of the potential $\varepsilon_B - \varepsilon_A$ or the concentration of one of the constituents ($c \ll 1$ or $1 - c \ll 1$). Sometimes it is possible to use the quantity $1/z$ (with z being the number of nearest neighbors) as the small parameter. To begin with, it follows both from the general discussion of the spectrum problem (the eigenvalues are the poles of the Green's function) and from the form of the exact solution (4.9.17) in the Lloyd model that, rather than develop $\langle R(E, \boldsymbol{k})\rangle$ as a perturbation-theory series, it is more reasonable to expand $\langle R(E, \boldsymbol{k})\rangle^{-1}$, i.e., to seek the corrections to the poles straightforwardly. It is precisely this procedure that corresponds to the usual quantum-mechanical perturbation theory, in which the energy values are expressed as a power series in a small parameter.

Instead of $\langle R(E, \boldsymbol{k})\rangle$ we introduce an unknown function $\Sigma(E, \boldsymbol{k})$, called the self-energy, or the mass operator,

$$\langle R(E, \boldsymbol{k})\rangle = \frac{1}{E - \beta(\boldsymbol{k}) - \Sigma(E, \boldsymbol{k})} \ , \tag{4.9.19}$$

and seek the expansion for Σ. Equation (4.9.19) is referred to as the Dyson equation. As long as there is no algorithm for the computation of Σ, the Dyson equation is devoid of content. The computation algorithm presupposes the use of perturbation theory. Therefore, although the self-energy may always be introduced formally, this may be pointless if it is not small in comparison with the bare energy $\beta(\boldsymbol{k})$.

Thus we have to consider approximate methods of calculating $\Sigma(E, \boldsymbol{k})$. At the outset we single out from the random potential its mean value

$$\bar{U} = (1-c)\varepsilon_A + c\varepsilon_B . \tag{4.9.20}$$

Then

$$U = \bar{U} + \delta U , \tag{4.9.21}$$

where the quantity δU takes on the values

$$\begin{aligned} &\varepsilon_A - \bar{U} = c\Delta , \quad \varepsilon_B - \bar{U} = -(1-c)\Delta , \\ &\Delta \equiv \varepsilon_B - \varepsilon_A \end{aligned} \tag{4.9.22}$$

and the probability distribution is

$$P(\delta U) = (1-c)\delta(\delta U - c\Delta) + c\cdot\delta(\delta U + (1-c)\Delta) . \tag{4.9.23}$$

The simplest approximation is the virtual-crystal approximation, which consists of neglecting the fluctuating part δU. Then we have

$$\begin{aligned} &\Sigma(E, \boldsymbol{k}) = \bar{U} , \\ &\langle R(E, \boldsymbol{k})\rangle = [E - \beta(\boldsymbol{k}) - \bar{U}]^{-1} \equiv \bar{R}(E, \boldsymbol{k}) . \end{aligned} \tag{4.9.24}$$

This approximation holds true for small Δ. A point to be emphasized is that the imaginary part of Σ in the virtual-crystal approximation is equal to zero; i.e., the decay of electronic states is absent. It turns out that even in the limit $\Delta \to 0$, a certain region of $\boldsymbol{k}$ space exists in which the neglect of the decay is qualitatively incorrect. To see this, we calculate $\operatorname{Im}\Sigma(E, \boldsymbol{k})$ to the first non-vanishing order in Δ.

Let the Green's function of the unperturbed problem be $\bar{R}$, and the perturbation be δU. We expand (4.8.26) in a power series of $\hat{V}$ to the second order; this gives

$$\hat{R} = \hat{\bar{R}} + \hat{\bar{R}}\delta\hat{V}\hat{\bar{R}} + \hat{\bar{R}}\delta\hat{V}\hat{\bar{R}}\delta\hat{V}\hat{\bar{R}} + \dots , \tag{4.9.25}$$

i.e.,

$$\begin{aligned} \langle R_{nm}(E)\rangle = {} & \bar{R}_{nm}(E) + \sum_l \bar{R}_{nl}(E)\langle\delta V_l\rangle\bar{R}_{lm}(E) \\ & + \sum_{lp} \bar{R}_{nl}(E)\bar{R}_{lp}(E)\bar{R}_{pm}(E)\langle\delta V_l\delta V_p\rangle + \dots . \end{aligned} \tag{4.9.26}$$

But since the quantities U at different l are distributed independently, we have

$$\begin{aligned} &\langle \delta U_l \rangle = 0 \;, \\ &\langle \delta U_l \delta U_p \rangle = \langle (\delta U)^2 \rangle \, \delta_{lp} \;. \end{aligned} \tag{4.9.27}$$

From (4.9.23), we find

$$\langle (\delta U)^2 \rangle = c^2 \Delta^2 (1-c) + (1-c)^2 \Delta^2 c = c(1-c)\Delta^2 \;. \tag{4.9.28}$$

Substituting (4.9.27, 28) into (4.9.26) yields

$$\langle R_{nm}(E) \rangle = \bar{R}_{nm}(E) + c(1-c)\Delta^2 \bar{F}(E) \sum_l \bar{R}_{nl}(E) \bar{R}_{lm}(E) \;, \tag{4.9.29}$$

with

$$\overline{F(E)} \equiv \bar{R}_{ll}(E) = \frac{1}{N} \sum_k \frac{1}{E - \beta(\boldsymbol{k}) - \bar{U} - \mathrm{i}0} \;, \tag{4.9.30}$$

[see (4.8.34)]. Proceeding to the Fourier representation (4.9.2) and using the convolution theorem, according to which

$$\sum_{nm} \exp[\mathrm{i}\boldsymbol{k} \cdot (\boldsymbol{R}_n - \boldsymbol{R}_m)] \sum_l \bar{R}_{nl}(E) \bar{R}_{lm}(E) = \bar{R}^2(E, \boldsymbol{k}) \;,$$

we find

$$\langle R(E, \boldsymbol{k}) \rangle = \bar{R}(E, \boldsymbol{k}) + c(1-c)\Delta^2 \bar{F}(E) \bar{R}^2(E, \boldsymbol{k}) + \dots \;. \tag{4.9.31}$$

Now we recall the Dyson equation (4.9.19) and regard (4.9.31) as an expansion of the function $\bar{R}(E, \boldsymbol{k})$ as a power series in $\Sigma(E, \boldsymbol{k})$:

$$\begin{aligned} \langle R(E, \boldsymbol{k}) \rangle &= \frac{1}{E - \beta(\boldsymbol{k}) - \bar{U}} + \left(\frac{1}{E - \beta(\boldsymbol{k}) - \bar{U}} \right)^2 c(1-c)\Delta^2 \bar{F}(E) \\ &\approx \frac{1}{E - \beta(\boldsymbol{k}) - \bar{U} - c(1-c)\Delta^2 \bar{F}(E)} \;. \end{aligned} \tag{4.9.32}$$

Thus, in the second-order of perturbation in Δ, we have

$$\Sigma_2(E, \boldsymbol{k}) = c(1-c)\Delta^2 \bar{F}(E) \rightarrow \frac{c(1-c)\Delta^2}{N} \sum_q \frac{1}{\beta(\boldsymbol{k}) - \beta(\boldsymbol{q}) - \mathrm{i}0} \;, \tag{4.9.33}$$

where we have inserted the first approximation $E = \beta(\boldsymbol{k}) + \bar{U}$ into $\bar{F}(E)$ near the pole $\langle R(E, \boldsymbol{k})\rangle$. The function (4.9.33) has an imaginary part (Sect. 4.8.2):

$$\Gamma(\boldsymbol{k}) = \operatorname{Im} \textstyle\sum_2 (E, \boldsymbol{k})|_{E=\beta(\boldsymbol{k})+\bar{U}} = \pi c(1-c)\Delta^2 \frac{1}{N} g_0(\beta(\boldsymbol{k})) \; . \tag{4.9.34}$$

In the vicinity of the band edge $\beta(\boldsymbol{k}) = \min \beta(\boldsymbol{k}) + \varepsilon(\boldsymbol{k})$ (here $\varepsilon \to 0$) in the effective-mass approximation $\varepsilon(\boldsymbol{k}) = \hbar^2 k^2/2m^*$, we have (Sect. 3.5)

$$\frac{1}{N} g_0(\beta(\boldsymbol{k})) = \frac{m^*(2m^*\varepsilon)^{1/2} V_0}{2\pi^2\hbar^3} \; , \tag{4.9.35}$$

with V_0 being the unit-cell volume; recall that $g_0(\varepsilon)$ here is the density of states for one spin projection. Thus, near the band edge, the damping goes to zero as $\varepsilon^{1/2}$ and will necessarily become larger than the energy ε if ε is small. This will take place subject to the condition that

$$\frac{c(1-c)\Delta^2 m^*(2m^*\varepsilon)^{1/2}}{4\pi\hbar^3} V_0 \ll \varepsilon \; ;$$

i.e.

$$\varepsilon \ll \varepsilon_c = \frac{[c(1-c)\Delta^2]^2 m^{*3} V_0^2}{8\pi\hbar^6} \; . \tag{4.9.36}$$

Thus, even when the quantity Δ is small, near the band edges there are regions of states in which the damping is large compared with the energy. As stated in Sect. 4.9.1, the approach itself, based on the dispersion relation $E(\boldsymbol{k})$, becomes pointless for such states. The nature of these states will be considered in the next section, whereas here we limit ourselves to the statement that all of the approximations discussed are inapplicable near band edges.

If Δ is not small but the concentration is small for one of the impurities, we proceed from the exact solution of the one-impurity problem treated in Sect. 4.9.2 [see also (4.9.12)]. Notice that, in comparison with perturbation theory, allowing for electron scattering from an impurity involves the renormalization of the potential δU_n and the replacement of this potential by what is known as the one-site t matrix:

$$t_n(E) = \frac{\delta U_n}{1 - \delta U_n \bar{F}(E)} \; . \tag{4.9.37}$$

We recall that $\langle U \rangle$ is involved in the zero Hamiltonian and δU_n acts as a perturbation. In the mean-t-matrix approximation, in contrast with the virtual-crystal approximation, this replacement due to multiple electron scattering on the same site is taken into account.

We replace δU_n by an effective value $\delta\tilde{U}$, for which the t matrix is equal to the mean t matrix

$$\begin{aligned}\frac{\delta\tilde{U}}{1-\delta\tilde{U}\bar{F}(E)} = \langle t_n(E)\rangle &= \frac{(1-c)c\Delta}{1-c\Delta\bar{F}(E)} + \frac{[-c(1-c)\Delta]}{1+(1-c)\Delta\bar{F}(E)} \\ &= \frac{c(1-c)\Delta^2\bar{F}(E)}{[1-c\Delta\bar{F}(E)][1+(1-c)\Delta\bar{F}(E)]} \;.\end{aligned} \tag{4.9.38}$$

Hence we find

$$\begin{aligned}\Sigma(E,\boldsymbol{k}) = \bar{U}+\delta\tilde{U} &= \bar{U}+\frac{\langle t_n(E)\rangle}{1+\bar{F}(E)\langle t_n(E)\rangle} \\ &= \bar{U}+\frac{c(1-c)\Delta^2\bar{F}(E)}{1+(1-2c)\Delta\bar{F}(E)} \;.\end{aligned} \tag{4.9.39}$$

Formula (4.9.39) possesses a number of real advantages over the simple approximation (4.9.24). First, as will be shown, it is exact in the limit of small concentrations of one of the constituents (except for the narrow energy region near the band edges). Second, for a small Δ and an arbitrary c, it takes into account all second-order effects in Δ exactly (4.9.33). Finally, the mean-t-matrix approximation describes the important effect of the damping of electronic states in an alloy: $\operatorname{Im}\{\Sigma(E,\boldsymbol{k})\} \neq 0$.

Even better is the coherent-potential approximation (CPA). We saw in the Lloyd model that the averaged Green's function is equal to the Green's function of the "Hamiltonian" involving a homogeneous complex potential (this "Hamiltonian" is nonHermitian and fictitious). In the CPA the constant part is singled out from the fluctuating potential; in general terms, this part is complex and depends on energy

$$U_n = \Sigma(E)+\Delta U_n \;, \qquad \Delta U_n = U_n-\Sigma(E) \;. \tag{4.9.40}$$

The separation is done in such a fashion that the scattering on the "remainder" is ineffective; i.e., the mean t matrix is equal to zero:

$$\begin{aligned}\left\langle \frac{\Delta U_n}{1-\Delta U_n\mathscr{F}(E)} \right\rangle &= (1-c)\frac{\varepsilon_A-\Sigma(E)}{1-[\varepsilon_A-\Sigma(E)]\,\mathscr{F}(E)} \\ &\quad + c\frac{\varepsilon_B-\Sigma(E)}{1-[\varepsilon_A-\Sigma(E)]\,\mathscr{F}(E)} = 0 \;.\end{aligned} \tag{4.9.41}$$

Here

$$\mathscr{F}(E) = \frac{1}{N}\sum_{\boldsymbol{k}}\frac{1}{E-\beta(\boldsymbol{k})-\Sigma(E)} \;, \tag{4.9.42}$$

since the "constant" part of the potential $\Sigma(E)$ is included in the zero approximation for $R(E, \boldsymbol{k})$. Cancelling similar terms in (4.9.40), we transform this equation into

$$\Sigma(E) = \bar{U} - [\varepsilon_A - \Sigma(E)][\varepsilon_B - \Sigma(E)]\,\mathscr{F}(E)\ . \tag{4.9.43}$$

By numerically solving the system of nonlinear equations (4.9.41, 42), we can find $\Sigma(E)$ and then the average Green's function, which in the CPA is equal to

$$\langle R(E, \boldsymbol{k})\rangle = \frac{1}{E - \beta(\boldsymbol{k}) - \Sigma(E)}\ . \tag{4.9.44}$$

The CPA has been successfully employed to calculate the electronic structure of disordered systems. It should, however, be borne in mind that the states near the band edges are described incorrectly by this method (although the region of such states may be very narrow).

We conclude by making one more remark. Earlier, a rigid band model was frequently employed for alloys of metals of different valence, a procedure which neglected the variation of the dispersion relation $E(\boldsymbol{k})$ with varying concentration of the constituents, and attributed the concentration dependence of the properties of alloys to the Fermi-level motion over the energy band owing to the variation of the mean conduction-electron concentration. As is seen from the CPA equation, the effective-lattice potential $\Sigma(E)$ depends on concentration in a very complicated fashion, which may lead to serious variations of the dispersion relation $E(\boldsymbol{k})$ itself. It is only for alloys of simple metals with weak pseudopotentials—where the quasi-free-electron scheme may, in general, be used (Sect. 4.3)—that the rigid-band model is able to describe the situation qualitatively. On the other hand, adding even a small number of impurities to alloys of transition metals may alter the energy spectrum and properties of these metals appreciably [4.60].

4.9.3 The Anderson Localization

Although the problem of the average wave function is irrelevant (the wave function is not a self-averaging quantity), we can pose the problem of the most probable states. In general terms, the wave functions may be localized—i.e., they pertain to the discrete spectrum—or they may be extended and pertain to the continuous spectrum. The discrete spectrum in ordinary quantum-mechanical problems consists of a finite number of localized levels, sometimes of a countable number of levels (the harmonic oscillator problem or the hydrogen atom problem—in the latter case the discrete spectrum has a point of accumulation $E = 0$). In the periodic potential of an infinite ideal crystal, as we saw earlier, there are continuous spectrum energy bands, the discrete spectrum being absent; localized states are liable to arise in the presence of point defects. What will happen to these localized states if the concentration of such defects is increased? Will these states form a "molecular orbit," which embraces the entire crystal and is similar, in this respect, to the Bloch state, or will they remain localized in some region of the crystal? This is one of the main problems in the theory of disordered systems.

In 1958 Anderson [4.28] showed, in one of the models for disordered systems, that the probability of the electron returning to the same point during an infinite time P_∞ is nonzero when the disorder is sufficiently high. This signifies that the system has discrete spectrum states, i.e., localized states, in which, in classical terms, the electron diffuses in a finite region of space (Anderson localization). Many criteria for Anderson localization exist, and correlating them is not always possible. Thus, the equivalence of the Anderson criterion $P_\infty \neq 0$ to the existence of localized states in the usual sense is difficult to prove. Physically, however, it is rather obvious. Let it be emphasized once again that such behavior in a system with an infinite number of defects is nontrivial: after all, the density of states, calculated using the average Green's function method, here turns out to be a continuous function of energy with no split-off levels. That is, the localized states fill all the energy intervals so that their contribution to the density of states remains finite in the limit of an infinite crystal, a behavior quite unusual from the point of view of ordinary quantum-mechanical problems!

To demonstrate the Anderson localization, we will consider qualitatively the one-dimensional case [4.61]. We disorder the array of rectangular potential barriers (Sect. 4.1.2), by altering the height of the in a random fashion (Fig. 4.43). Over the interval between the nth and $n+1$th barrier the wave function has the form

$$\psi(x) = A_n e^{i\alpha x} + B_n e^{-i\alpha x} , \tag{4.9.45}$$

with $\alpha = (2mE/\hbar^2)^{1/2}$. From the requirement that the current density be constant,

$$\frac{d}{dx}\left(\psi^*(x)\frac{d\psi(x)}{dx} - \psi(x)\frac{d\psi^*(x)}{dx}\right) = 0 , \tag{4.9.46}$$

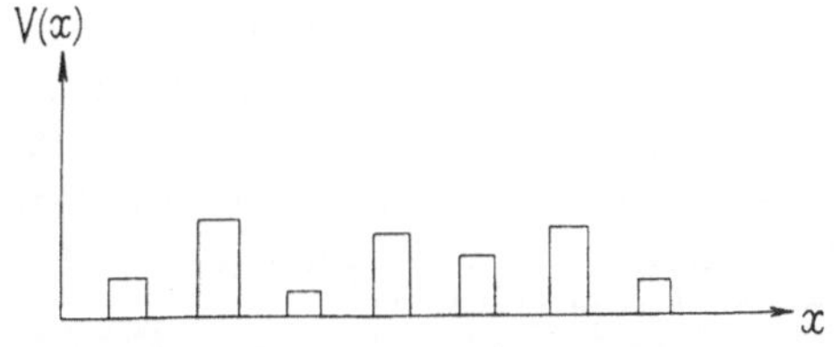

Fig. 4.43. Array of potential barriers of random height

we have

$$|A_n|^2 - |B_n|^2 = \text{const} . \tag{4.9.47}$$

Examine an array of finite length and let a wave $\exp(i\alpha x)$ fall on the array, on its left,

$$A_0 = 1, \quad B_0 = 0 .$$

Then, in virtue of (4.9.47),

$$|A_n|^2 - |B_n|^2 = 1 . \tag{4.9.48}$$

But if the system is disordered—i.e., the barrier height varies randomly—then the wave, as it undergoes random reflections, should, sooner or later, "forget" its incident direction. In other words, when the number n is sufficiently large, we have

$$|A_n| \approx |B_n| . \tag{4.9.49}$$

Conditions (4.9.47, 48) are compatible only provided that

$$|A_n|, |B_n| \gg 1 \tag{4.9.50}$$

for large n. Thus the envelope of the wave function increases with the removal from the left-hand edge of the array. Now a wave can travel from the right-hand edge of the array to meet the first wave; these two solutions can be joined somewhere (Fig. 4.44).

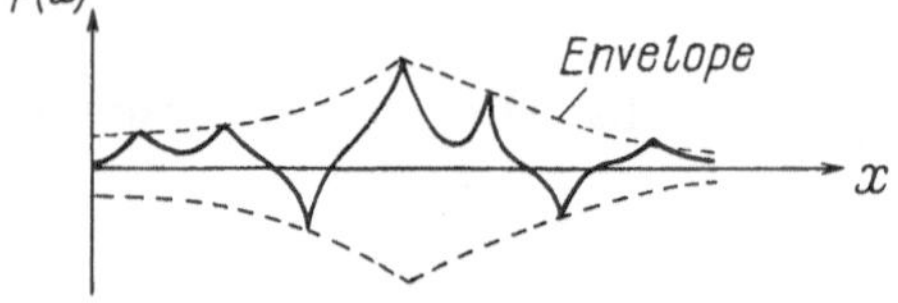

Fig. 4.44. Localization in the one-dimensional case

We will thus obtain a steady-state solution that will be localized; i.e., the electron density will have a maximum somewhere at the center of the array and will decrease rapidly toward the edges (it can be shown rigorously that this decrease is exponential). Since our reasoning involves no quantitative characteristics of disorder, the disorder may be as weak as desired. Thus we reach the unexpected conclusion that all states in a one-dimensional disordered array are always localized. Of course, we limit ourselves here to qualitative considerations; a rigorous proof is presented by *Lifshitz* et al. [1.44].

What will the situation be like in the three-dimensional case? No exact answers have been obtained thus far, but the states in the three-dimensional case are likely to be localized for some energies and delocalized for others (Sect. 4.5.4). Broadly speaking, the possibility that localized and delocalized states can coexist at the same energy may not yet be entirely dismissed. Modern treatment of the properties of disordered materials, however, is based on an alternative concept, that of the threshold mobility advanced by *Mott* (Sect. 4.5.4).

As already stated in the foregoing, the states that lie far from the energy band edge may decay weakly when the disorder is sufficiently weak. Put another way, with the quasimomentum $\boldsymbol{k}$ being fixed, the spread in E values is small. Then we should expect the true eigenstates with fixed E to be represented by a wave packet with a sufficiently weak spread in $\boldsymbol{k}$ values, i.e., to be delocalized. The states near the band edge, as emphasized above, cannot be described in terms of the dispersion relation $E(\boldsymbol{k})$ and are most likely to be localized. They are associated with relatively improbable large potential fluctuations.

We now return to the binary-alloy model (Sect. 4.9.2). Let us assume, for illustration, that $\Delta > 0$, $c = 1/2$ (the A atom concentration and the B atom concentration are equal in magnitude). The edge of the energy band $\varepsilon \equiv 0$ can then only be reached in a cluster that is composed of species A atoms (the presence of species B atoms causes the mean electron energy to increase, i.e., $\varepsilon_B > \varepsilon_A$). However, in a cluster of finite size L, the electron inevitably possesses a momentum $\approx \hbar/L$ (from the uncertainty principle) and therefore an energy

$$\varepsilon \approx \hbar^2/m^* L^2 \ . \tag{4.9.51}$$

The situation $L \to \infty$ corresponds to the band edge $\varepsilon \to 0$. But the probability, $p(L)$, that a large cluster will occur, falls off exponentially as the cluster increases in size. If the probability that a given site will be occupied by a species A atom is equal to 1/2, then the probability for a cluster containing N atoms will be

$$p(L) \approx \left(\frac{1}{2}\right)^N = \exp(-N \ln 2) \ , \tag{4.9.52}$$

but

$$N = CL^3/V_0 \ , \tag{4.9.53}$$

where C is a numerical coefficient that depends on cluster shape. Since the density of states, $g(\varepsilon)$ for $\varepsilon \to 0$, is proportional to $p(L)$, where L is related to ε by (4.9.51), we find [4.62]

$$g(\varepsilon) \propto \exp\left[-\mathrm{const}\left(\frac{\hbar^2}{m^*\varepsilon}\right)^{3/2} V_0\right] . \tag{4.9.54}$$

Thus the density of states at $\varepsilon \to 0$ is extremely small.

If the Fermi level E_F lies below the mobility threshold E_c, the conductivity σ in the region of localized states for $T=0$ should go to zero; i.e., the system should be a dielectric. It has not yet been ascertained whether σ vanishes abruptly (Mott's "minimum metallic conductivity" concept) or vanishes obeying the power law

$$\sigma \propto \left(\frac{E_F - E_c}{E_c}\right)^s . \qquad (4.9.55)$$

Recent experiments and theoretical developments provide evidence for the latter possibility [4.63].

At finite temperatures a localized electron may receive from a phonon some energy ΔE and hop over to another localized state, covering a distance R determined from the condition

$$\left(\frac{R}{a}\right)^3 \Delta E g_0 \approx 1 , \qquad (4.9.56)$$

where g_0 is the density of states at the Fermi level per atom, a is the distance between impurities, and $g_0 R^3/a^3$ is the total density of states in a cluster of size R, i.e., the inverse distance between the energy levels—this is what accounts for (4.9.56). Therefore,

$$R \approx a(\Delta E \cdot g_0)^{-1/3} . \qquad (4.9.57)$$

The probability of this hop is determined, first, by the Boltzmann factor $\exp(-\Delta E/k_B T)$, which is the number of phonons with energy $\Delta E \gg k_B T$, and, second, by the probability of quantum-mechanical tunneling between centers, $\exp(-R/\xi)$ (with ξ being the wave-function falloff radius, or the localization radius). Hopping will be accompanied preferentially by a change in energy ΔE_0 for which the hop probability, equal to

$$w \propto \exp\left(-\frac{\Delta E}{k_B T} - \frac{a}{\xi}\frac{1}{(g_0 \Delta E)^{1/3}}\right) , \qquad (4.9.58)$$

is the largest. Determining the maximum of the exponent in (4.9.58) from ΔE, we find, to within numerical factors,

$$\Delta E_0 \approx \left(\frac{a}{\xi} k_B T\right)^{3/4} \frac{1}{g_0^{1/4}} . \qquad (4.9.59)$$

(Observe that at sufficiently low temperatures the quantity ΔE_0 is really larger than $k_B T$.) Accordingly, the "optimum hop" probability is

$$w_0 \propto \exp\left[-\left(\frac{T_0}{T}\right)^{1/4}\right], \quad T_0 = \text{const}\,\frac{1}{k_B g_0}\left(\frac{a}{\xi}\right)^3 . \tag{4.9.60}$$

The static conductivity, determined at low temperatures by these optimum hops, is proportional to w_0 (the Mott $T^{1/4}$ law). For two-dimensional disordered systems, as can be readily verified, the $T^{1/4}$ law is replaced by the "$T^{1/3}$ law."

The consideration of disordered systems either here or in Sects. 1.10, 2.5, and 4.5.4 is far from complete. This area of physics is currently developing rapidly and many concepts are being revised. Specifically, electron–electron interaction effects appear to be highly important in disordered systems. However, studies of these effects have not yet yielded conclusive results. An examination of these effects calls for very complicated mathematics, and therefore we do not consider them here.

4.10 Conclusion. The Role of Many-Particle Effects

Currently, band theory forms the foundation of the entire electronic theory of metals and semiconductors, permitting a large body of experimental evidence to be systematized and accounted for. Particularly important is the fact that sufficiently reliable techniques exist not only for calculating the energy spectrum (special mention should be made of the pseudopotential method for metals) but also for determining straightforwardly the shape of the Fermi surface and the velocity distribution on it from experimental data.

The big success of band theory, however, seems surprising at first sight. It is hard to understand why allowing for the electron–electron interaction in the self-consistent field approximation—which, generally speaking, is unfounded for values of the parameters characteristic of solids—proves to be sufficient. (Incidentally, the success of band theory should not be overestimated, as often happens. In describing compounds of d and f transition metals, it is necessary to take account of the correlation explicitly—Sect. 4.5.3.) As already noted, the Pauli exclusion principle plays a decisive role here.

Consider a weakly interacting electron gas. In the absence of interactions the ground state is an electron-occupied region of k space bounded by the Fermi surface. Any interaction will smear the electronic states, giving rise to transitions between them. For pairwise collisions the energy conservation principle should hold:

$$E_1 + E_2 = E_1' + E_2' . \tag{4.10.1}$$

Here the initial states $E_{1,2}$ at the zero of temperature ($T = 0$ K) should reside beneath the Fermi surface, and the final states above it, for the Pauli exclusion principle forbids scattering to occupied states. This evidently contradicts

(4.10.1). Consequently, when $T = 0$ K, the pairwise collisions are suppressed and the Fermi surface remains well defined. Simple reasoning shows that the decay of the state with energy E is proportional to $(E - E_F)^2$; i.e., it is on the order of T^2 for thermally active electrons, and is small irrespective of whether or not the interaction is weak. However, care must be exercised here; the reasoning offered in Scct. 4.5.3 shows that a sufficiently strong interaction is nevertheless capable of destroying the Fermi surface and transforming a metal into a semiconductor. What we imply when we talk of an appreciable interaction is one that is still less than the critical value at which the Mott transition occurs.

Current carriers in metals are, of course, not quite Bloch electrons; as already stated, they are surrounded by a cloud of other electrons, with which they interact, and also by a cloud of phonons. These carriers are quasiparticles, i.e., excitations of the entire many-electron system, which, however, in many ways resemble Bloch electrons. The properties of such quasiparticles are described by Landau's Fermi-liquid theory (Sect. 5.2). One of its predictions is that the interaction affects the static properties (heat capacity, magnetic susceptibility, etc.) only through parameter renormalization (e.g., of effective mass or magnetic moment). An essential feature here is that the Fermi surface, determined, say, from the de Haas–van Alphen effect, is the Fermi surface of quasiparticles and, therefore, to some extent, its shape takes the interaction into account exactly.

Many-particle theory shows that the volume bounded by the Fermi surface does not vary when the interaction is included (Landau–Luttinger theorem). The Fermi surface in this case is defined as a surface in $\boldsymbol{k}$ space on which the electron distribution function undergoes a discontinuity at $T = 0$ K. The discontinuity persists in the many-particle system, although the magnitude of this jump is less than unity [4.64]. Figure 4.45 shows the form of the $\overline{n(\boldsymbol{k})}$ function for the isotropic case, in which $\overline{n(\boldsymbol{k})}$ depends on $|\boldsymbol{k}|$ only. As before, the quantity $\boldsymbol{k}_F$ is determined by the total density n according to (3.5.14), i.e.,

$$k_F = (3\pi^2 n)^{1/3} \ . \tag{4.10.2}$$

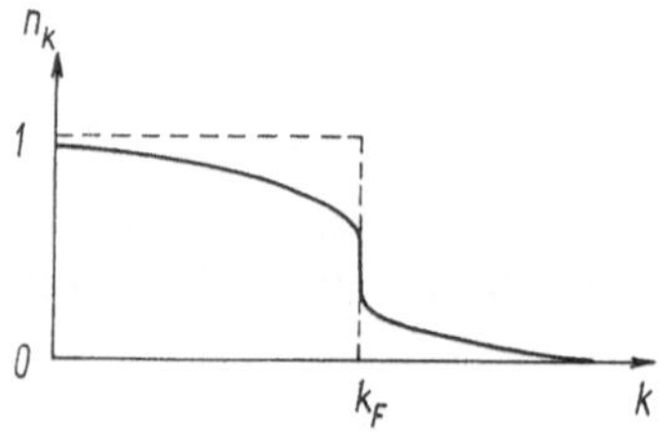

Fig. 4.45. Change of the Fermi–Dirac distribution function $\bar{n}_k$ for the mean number of occupied states with allowance made for the electron–electron correlation

We now wish to "prove" that the lifetime of the band state is long and the uncertainty of its energy therefore is small. In writing (4.10.1), we have already pointed out that each electron from the thermal smearing region $\approx k_B T$ does not collide with all n electrons of the system, but only with part of them,

$n_T \approx (k_B T/\zeta_0)n$. Apart from this, not all of these collisions are possible, even if allowed by energy and momentum conservation. Indeed, if the change in energy during collision exceeds $k_B T$, there are no vacant sites for an electron that has lost this energy. Thus the Pauli exclusion principle introduces an extra factor into the scattering probability $\approx k_B T/\zeta_0$. Therefore, the scattering probability amounts to $\sim (k_B T/\zeta_0)^2$ and the lifetime $\bar{\tau}$ of the state is $\bar{\tau} \approx (k_B T/\zeta_0)^{-2} (\hbar/\zeta_0)$, a value which is large even at room temperature. In addition, the energy uncertainty relation $\Delta E \approx \hbar/\bar{\tau}$ yields a very small quantity for ΔE; this circumstance allows us to view a Bloch electron as a stable quasiparticle.

We are now at the end of our discussion of the part of solid-state theory based on the postulate that the interaction between electrons is absent, although we have in fact taken it into account to a certain extent in the self-consistent field approximation. In the following chapter, we present a more-or-less detailed analysis of the consequences of abandoning this postulate.

5. Many-Particle Effects

This chapter deals with electron–electron and electron–phonon interaction effects. Extensive treatment is given to the plasma model of the metal, the Fermi-liquid theory of Landau, polaron theory, and other problems. Superconductivity, excitons, and magnetism are also discussed at some length. The exposition employs a minimum of mathematical machinery and will serve as an introduction to the physical ideas of contemporary quantum many-particle theory.

5.1 Plasma Phenomena. Screening

5.1.1 A Discussion of the Model

In this chapter we investigate the theory of the effects in which the electron interaction is a decisive factor. Of supreme importance are plasma phenomena, first investigated in terms of the quantum theory by *Bohm* and *Pines* in 1953 [5.1–3].

A plasma is an ionized gas. The presence of a large number of free charges bound by long-range Coulomb forces means that many properties of the plasma differ drastically from those of a normal gas (for example, the behavior in a magnetic field and screening). The gaseous plasma has been intensively investigated in connection with the problem of controlled thermonuclear fusion, as well as in astrophysical and geophysical problems (the ionosphere, the physics of the sun, etc.).

On the other hand, the definitive property of the plasma, viz., the presence of free charges, is also inherent in metals and semiconductors. Despite the quite dissimilar values of the parameters (electron density, temperature), the free electrons in a solid behave in many ways like a highly ionized gas. Thus, the electromagnetic waves in metals—plasmons, helicons, etc. (Sect. 3.7.4)—have counterparts in the gaseous plasma where they were first discovered. Sometimes we can ignore the lattice structure and the electronic arrangement of ion cores and regard the metal as an electron plasma with a uniformly smeared positive ion charge that assures electroneutrality. This is the plasma model of a metal, or the "jellium model". The model strongly emphasizes the fact that most properties of metals, even purely lattice properties (for example, the magnitudes of elastic moduli), are determined by free electrons. Frequently this approach turns out to be very profound, although, like any other model, it has only a certain scope of applicability. A presentation of the plasma model of the metal will enable us to approach some general and essential concepts and postulates, for example, the fundamentals of linear-response theory.

Different problems are associated with plasma phenomena in semiconductors; these phenomena are in many respects closer to a "classical" gaseous plasma than to the plasma in metals. First, because of the small concentration, the current carriers in semiconductors may be described by classical Maxwell–Boltzmann statistics. Second, strong electric fields are possible in semiconductors and some nonlinear plasma phenomena occur under those conditions. Although interesting, they do not have a general "solid-state" significance and are therefore not considered here.

The plasma model is exploited chiefly to describe electromagnetic properties of metals. From the standpoint of electrodynamics, a major property of plasma is its high spatial dispersion, i.e., the dependence of the electric induction at a given point in space on the field intensity in some region. This nonlocality arises from the long-range behavior of Coulomb forces and shows up dramatically in screening phenomena. To gain an insight into the properties of the plasma, we must consider the response of the electronic system to a spatially varying and time-dependent field.

5.1.2 The Equation for a Self-Consistent Plasma Potential

Let the electrons be subject to an external perturbation $U(\boldsymbol{r}, t)$, which will be regarded as weak and taken into account to the lowest order of perturbation theory. The perturbation induces in the system a charge-density redistribution which, in turn, results in an extra perturbing potential. The total potential $V(\boldsymbol{r}, t)$ can be represented as a linear functional of $U(\boldsymbol{r}, t)$ (the linearity here is due to the smallness of U):

$$V(\boldsymbol{r}, t) = \int d\boldsymbol{r}' \int\limits_{-\infty}^{t} dt' \varepsilon^{-1}(\boldsymbol{r}, \boldsymbol{r}'; t-t') U(\boldsymbol{r}', t') \ . \tag{5.1.1}$$

The operator $\hat{\varepsilon}^{-1}$ involving the kernel $\varepsilon^{-1}(\boldsymbol{r}, \boldsymbol{r}'; t-t')$ is said to be the inverse permittivity (more precisely, the longitudinal permittivity), or the inverse dielectric function. The integration over t' is carried out to t, since, by the causality principle, $U(t)$ may depend on $U(t')$ at preceding rather than subsequent instants of time. Representing U, V as Fourier integrals, for example,

$$\begin{aligned} V(\boldsymbol{r}, t) &= \int\limits_{-\infty}^{\infty} \frac{d\omega}{2\pi} \exp(-\mathrm{i}\omega t) V(\boldsymbol{r}, \omega) \ , \\ V(\boldsymbol{r}, \omega) &= \int\limits_{-\infty}^{\infty} dt \exp(\mathrm{i}\omega t) V(\boldsymbol{r}, t) \ , \end{aligned} \tag{5.1.2}$$

we obtain, subsequent to evident transformations,

$$V(\boldsymbol{r}, \omega) = \int d\boldsymbol{r}' \varepsilon^{-1}(\boldsymbol{r}, \boldsymbol{r}';\omega) U(\boldsymbol{r}', \omega) \ , \tag{5.1.3}$$

where

$$\varepsilon^{-1}(\boldsymbol{r}, \boldsymbol{r}'; \omega) = \int_0^\infty dt \exp(i\omega t) \varepsilon^{-1}(\boldsymbol{r}, \boldsymbol{r}'; t) \ . \tag{5.1.4}$$

For the spatially homogeneous case the quantity $\varepsilon^{-1}(\boldsymbol{r}, \boldsymbol{r}'; \omega)$ depends only on the difference $\boldsymbol{r}-\boldsymbol{r}'$. Then, performing the Fourier transformation with respect to the spatial coordinates, we find

$$V(\boldsymbol{q}, \omega) = U(\boldsymbol{q}, \omega)/\varepsilon(\boldsymbol{q}, \omega) \ , \tag{5.1.5}$$

where

$$V(\boldsymbol{q}, \omega) = \int d\boldsymbol{r} \exp(-i\boldsymbol{q}\cdot\boldsymbol{r}) V(\boldsymbol{r}, \omega) \ . \tag{5.1.6}$$

The same applies to U; the $1/\varepsilon(\boldsymbol{q}, \omega)$ is a Fourier transform of $\varepsilon^{-1}(\boldsymbol{r}-\boldsymbol{r}', \omega)$.

In what follows we present a general, although formal, expression for $1/\varepsilon(\boldsymbol{q}, \omega)$ and prove a number of important equations. At the outset it is, however, reasonable to calculate this quantity with comparatively simple approximations, which actually embrace nearly all of the major plasma phenomena and describe them, on the whole, correctly. To do this, we regard $V(\boldsymbol{r}, t)$ as a one-electron potential that acts on one-particle states. Using this potential, we then find the self-consistence requirements and thereby take into account its many-electron behavior. This approach for a classical plasma was proposed by *Vlasov* [5.4]. The Bohm–Pines approach, seemingly based on quite different ideas and called the random-phase approximation (RPA), is strictly equivalent to the quantum generalization of Vlasov's method.

We examine an electron system that is described in a one-particle approximation by eigenfunctions $\psi_\nu(\boldsymbol{r})$ and eigenvalues E_ν (for solids $|\nu\rangle = |\boldsymbol{k}\zeta\sigma\rangle$, with ζ being the band number, $\boldsymbol{k}$ the wave vector, and σ the spin projection). Now we introduce a one-particle density matrix

$$\hat{\varrho} = \sum_\nu w_\nu |\nu\rangle\langle\nu| \ . \tag{5.1.7}$$

This representation signifies that for any $\varphi(\boldsymbol{r})$ function,

$$\hat{\varrho}|\varphi\rangle = \sum_\nu w_\nu |\nu\rangle\langle\nu|\varphi\rangle = \sum_\nu w_\nu \psi_\nu(\boldsymbol{r}) \int d\boldsymbol{r}' \psi_\nu^*(\boldsymbol{r}')\varphi(\boldsymbol{r}') \ .$$

The quantity w_ν here is the probability of the electron being in the νth state; in equilibrium the $w_\nu = \overline{n(E_\nu)}$ coincides with the Fermi–Dirac distribution. The average for any one-particle operator $\hat{A}$ then is

$$\langle \hat{A} \rangle = \sum_\nu w_\nu \langle \nu | \hat{A} | \nu \rangle = \mathrm{Tr}(\hat{A}\hat{\varrho}) \ . \tag{5.1.8}$$

Specifically, for the electron-density operator

$$\hat{N}(\boldsymbol{r}) = \delta(\boldsymbol{r} - \boldsymbol{r}') \tag{5.1.9}$$

we obtain

$$N_{\nu\nu'} = \int d\boldsymbol{r}' \psi_\nu^*(\boldsymbol{r}') \delta(\boldsymbol{r} - \boldsymbol{r}') \psi_{\nu'}(\boldsymbol{r}') = \psi_\nu^*(\boldsymbol{r}) \psi_{\nu'}(\boldsymbol{r})$$

and

$$N(\boldsymbol{r}) = \langle \hat{N}(\boldsymbol{r}) \rangle = \sum_{\nu\nu'} N_{\nu\nu'} \varrho_{\nu'\nu} = \sum_{\nu\nu'} \varrho_{\nu'\nu} \psi_\nu^*(\boldsymbol{r}) \psi_{\nu'}(\boldsymbol{r}) \ . \tag{5.1.10}$$

The density matrix varies with the time, obeying the equation

$$\mathrm{i}\hbar \frac{\partial \hat{\varrho}}{\partial t} = [\hat{\mathscr{H}}, \hat{\varrho}]_- \tag{5.1.11}$$

obtainable from the Schrödinger equation and from the equation which is conjugate to it:

$$\mathrm{i}\hbar \frac{\partial |\nu\rangle}{\partial t} = \hat{\mathscr{H}} |\nu\rangle \ , \quad \langle \nu | \hat{\mathscr{H}} = -\mathrm{i}\hbar \langle \nu | \frac{\partial}{\partial t} \ . \tag{5.1.12}$$

The usual Dirac notation is used here. Equations (5.1.7, 12) yield

$$\begin{aligned} \mathrm{i}\hbar \frac{\partial \hat{\varrho}}{\partial t} &= \sum_\nu w_\nu \left[\left(\mathrm{i}\hbar \frac{\partial |\nu\rangle}{\partial t} \langle \nu | \right) + |\nu\rangle \left(\mathrm{i}\hbar \langle \nu \left| \frac{\partial}{\partial t} \right. \right) \right] \\ &= \sum_\nu w_\nu (\hat{\mathscr{H}} |\nu\rangle\langle\nu| - |\nu\rangle\langle\nu| \hat{\mathscr{H}}) = [\hat{\mathscr{H}}, \hat{\varrho}]_- \ . \end{aligned}$$

For the unperturbed problem, the density matrix commutes with the Hamiltonian

$$\hat{\mathscr{H}}_0 |\nu\rangle = E_\nu |\nu\rangle \ , \quad [\hat{\mathscr{H}}_0, \hat{\varrho}_0]_- = 0 \ . \tag{5.1.13}$$

We are concerned with the response of the system to the perturbation

$$V(\boldsymbol{r})\exp(-i\omega t+\eta t)|_{\eta\to+0} \;.$$

The factor $\exp(\eta t)$ is introduced to describe the slow turning-on of the potential; with transitions in the continuous spectrum, allowance for this factor is indispensable in order to avoid divergences [Ref. 1.12, Sect. 43]. Representing $\hat{\varrho}$ as $\hat{\varrho}_0 + \hat{\varrho}'$, where $\hat{\varrho}' \propto V$, and neglecting the small commutator $[\hat{\varrho}', \hat{V}] \propto V^2$, we find from (5.1.11) with allowance for (5.1.13)

$$i\hbar\frac{\partial\hat{\varrho}'}{\partial t} = [\hat{\mathscr{H}}_0, \hat{\varrho}']_- + [\hat{V}, \hat{\varrho}_0]_- \exp(-i\omega t+\eta t) \;. \tag{5.1.14}$$

It follows from (5.1.14) that $\hat{\varrho}' \propto \exp(-i\omega t+\eta t)$. Calculate the matrix elements

$$\begin{aligned}
&\langle\nu|[\hat{\mathscr{H}}_0, \hat{\varrho}']_-|\nu'\rangle = (E_\nu - E_{\nu'})\varrho'_{\nu\nu'} \;,\\
&\langle\nu|[\hat{V}, \hat{\varrho}_0]_-|\nu'\rangle = [\overline{n(E_{\nu'})} - \overline{n(E_\nu)}]\,V_{\nu\nu'} \;,\\
&\langle\nu|i\hbar\frac{\partial\hat{\varrho}'}{\partial t}|\nu'\rangle = \hbar(\omega+i\eta)\varrho'_{\nu\nu'} \;.
\end{aligned} \tag{5.1.15}$$

Insertion of (5.1.15) into (5.1.14) yields

$$\varrho'_{\nu\nu'} = \frac{\overline{n_\nu} - \overline{n_{\nu'}}}{E_\nu - E_{\nu'} - \hbar(\omega+i\eta)}V_{\nu\nu'} \quad (\overline{n_\nu} = \overline{n(E_\nu)}) \;. \tag{5.1.16}$$

According to (5.1.10), the electron-density variation induced by the potential V is

$$\begin{aligned}
\delta N(\boldsymbol{r}) &= \sum_{\nu\nu'}\psi^*_{\nu'}(\boldsymbol{r})\psi_\nu(\boldsymbol{r})\varrho'_{\nu\nu'} = \sum_{\nu\nu'}\frac{\overline{n_\nu} - \overline{n_{\nu'}}}{E_\nu - E_{\nu'} - \hbar(\omega+i\eta)}\\
&\quad\cdot\psi^*_{\nu'}(\boldsymbol{r})\psi_\nu(\boldsymbol{r})\int d\boldsymbol{r}'\psi^*_\nu(\boldsymbol{r}')V(\boldsymbol{r}')\psi_{\nu'}(\boldsymbol{r}') \;.
\end{aligned} \tag{5.1.17}$$

Finally, we write the self-consistence condition. The potential V is composed of the external potential U and the potential induced by the variation of the density $\delta N(\boldsymbol{r})$:

$$V(\boldsymbol{r}) = U(\boldsymbol{r}) + e^2\int\frac{d\boldsymbol{r}''\,\delta N(\boldsymbol{r}'')}{|\boldsymbol{r}-\boldsymbol{r}''|} \;. \tag{5.1.18}$$

Substituting (5.1.18) into (5.1.17) yields

$$U(\boldsymbol{r}) = \int d\boldsymbol{r}' \varepsilon(\boldsymbol{r}, \boldsymbol{r}'; \omega) V(\boldsymbol{r}') \ , \tag{5.1.19}$$

$$\begin{aligned} \varepsilon(\boldsymbol{r}, \boldsymbol{r}'; \omega) = \delta(\boldsymbol{r} - \boldsymbol{r}') - e^2 \sum_{\nu\nu'} \frac{\overline{n_\nu} - \overline{n_{\nu'}}}{E_\nu - E_{\nu'} - \hbar(\omega + \mathrm{i}\eta)} \\ \cdot \psi_\nu^*(\boldsymbol{r}') \psi_{\nu'}(\boldsymbol{r}') \int d\boldsymbol{r}'' \frac{\psi_{\nu'}^*(\boldsymbol{r}'') \psi_\nu(\boldsymbol{r}'')}{|\boldsymbol{r} - \boldsymbol{r}''|} \ . \end{aligned} \tag{5.1.20}$$

Solving the integral equation (5.1.19), we find the induced potential $V(\boldsymbol{r})$ from a given external perturbation $U(\boldsymbol{r})$. The equation

$$\int d\boldsymbol{r}' \varepsilon(\boldsymbol{r}, \boldsymbol{r}'; \omega) V(\boldsymbol{r}') = 0 \tag{5.1.21}$$

defines the spectrum of natural plasma oscillations in the absence of an external effect.

In the following we restrict our attention to the free-electron gas model. Therefore

$$\psi_\nu(\boldsymbol{r}) = (2\pi)^{-3/2} \mathrm{e}^{\mathrm{i}\boldsymbol{k}\cdot\boldsymbol{r}} \ ; \tag{5.1.22}$$

here a normalization to the δ function is used, $\sum_\nu \to \int d\boldsymbol{k}$. With allowance for

$$\int d\boldsymbol{r}'' \frac{\exp(\mathrm{i}\boldsymbol{f}\cdot\boldsymbol{r}'')}{|\boldsymbol{r} - \boldsymbol{r}''|} = \frac{4\pi}{\boldsymbol{f}^2} \exp(\mathrm{i}\boldsymbol{f}\cdot\boldsymbol{r}) \ , \tag{5.1.23a}$$

$$\int d\boldsymbol{r} \exp(\mathrm{i}\boldsymbol{f}\cdot\boldsymbol{r}) = (2\pi)^3 \delta(\boldsymbol{f}) \tag{5.1.23b}$$

a little manipulation yields

$$\begin{aligned} \varepsilon(\boldsymbol{r}, \boldsymbol{r}'; \omega) = \int \frac{d\boldsymbol{q}}{(2\pi)^3} \varepsilon(\boldsymbol{q}, \omega) \exp[\mathrm{i}\boldsymbol{q}\cdot(\boldsymbol{r} - \boldsymbol{r}')] \ , \\ \varepsilon(\boldsymbol{q}, \omega) = 1 - \frac{e^2}{\pi^2 \boldsymbol{q}^2} \int d\boldsymbol{k} \frac{\overline{n(\boldsymbol{k} + \boldsymbol{q})} - \overline{n(\boldsymbol{k})}}{E(\boldsymbol{k} + \boldsymbol{q}) - E(\boldsymbol{k}) + \hbar(\omega + \mathrm{i}\eta)} \ . \end{aligned} \tag{5.1.24}$$

We have allowed for the fact that

$$\langle \nu | V | \nu' \rangle = \delta_{\sigma\sigma'} \langle \boldsymbol{k} | V | \boldsymbol{k}' \rangle$$

$$\sum_{\nu\nu'} \to \int d\boldsymbol{k}\, d\boldsymbol{k}' \sum_{\sigma\sigma'} \delta_{\sigma\sigma'} = 2 \int d\boldsymbol{k}\, d\boldsymbol{k}'$$

(the factor 2 occurs because of the summation over spin). Expression (5.1.24) is sometimes referred to as the Lindhard formula. Using the value of $\overline{n(\boldsymbol{k})}$ for $T=0\,\mathrm{K}$ (3.5.20),

$$\overline{n(\boldsymbol{k})} = \begin{cases} 1, & |\boldsymbol{k}| < k_F \\ 0, & |\boldsymbol{k}| > k_F \ , \end{cases} \tag{5.1.25}$$

substituting $E(\boldsymbol{k}) = \hbar^2 \boldsymbol{k}^2/2m$, and carrying out the integration, we obtain

$$\begin{aligned} \varepsilon(q,\omega) = 1 - \frac{2me^2}{\pi\hbar^2 q^3}\Bigg\{ &-qk_F + \frac{1}{2}\left[k_F^2 - \left(\frac{m\omega}{\hbar q} - \frac{q}{2}\right)^2\right] \\ &\cdot \ln\left(\frac{\omega + i\eta - \hbar q^2/2m + \hbar q k_F/m}{\omega + i\eta - \hbar q^2/2m - \hbar q k_F/m}\right) - \frac{1}{2}\left[k_F^2 - \left(\frac{m\omega}{\hbar q} + \frac{q}{2}\right)^2\right] \\ &\cdot \ln\frac{\omega + i\eta + \hbar q^2/2m + \hbar q k_F/m}{\omega + i\eta + \hbar q^2/2m - \hbar q k_F/m}\Bigg\} \ . \end{aligned} \tag{5.1.26}$$

Note that the factor $\exp(\eta t)$ has proved to be essential—its absence would have led to the occurrence of diverging integrals. A somewhat unexpected effect arises here: an imaginary part occurs in $\varepsilon(q,\omega)$ when

$$\omega < \frac{\hbar}{m}\left(qk_F + \frac{q^2}{2}\right) \tag{5.1.27}$$

according to the formula

$$\ln(x \pm i\eta) = \ln|x| \pm i\pi\vartheta(-x) \ , \tag{5.1.28}$$

where $\vartheta(x)=1$, $x>0$, $\vartheta(x)=0$, and $x<0$. In this case, the major logarithmic branch is implied throughout. The occurrence of $\mathrm{Im}\,\varepsilon(q,\omega)$ reflects a very important physical phenomenon, the Landau damping (Sect. 5.1.4).

5.1.3 Static Screening

We start by considering the response of the electronic system to the static potential. For $\omega = 0$, (5.1.26) yields

$$\begin{aligned} &\varepsilon(q,0) = 1 + \frac{2me^2 k_F}{\pi\hbar^2 q^2}\left(1 + \frac{1-\xi^2}{2\xi}\ln\left|\frac{1+\xi}{1-\xi}\right|\right) \ , \\ &\xi = q/2k_F \ , \end{aligned} \tag{5.1.29}$$

the imaginary parts of the logarithms in (5.1.26) canceling out. For $q \ll k_F$, we find

$$\varepsilon(q, 0) \approx 1 + \frac{1}{\lambda^2 q^2} , \tag{5.1.30}$$

where

$$\lambda = \left(\frac{\pi \hbar^2}{4me^2 k_F} \right)^{1/2} . \tag{5.1.31}$$

If $U(q)$ is the potential of a point charge Ze, then

$$U(q) = \frac{4\pi Ze^2}{q^2} , \tag{5.1.32}$$

$$V(q) = \frac{4\pi Ze^2}{q^2 \varepsilon(q, 0)} \approx \frac{4\pi Ze^2}{q^2 + \lambda^{-2}} \tag{5.1.33}$$

at $q \ll k_F$. Inversion of the Fourier transform in (5.1.33) gives

$$V(r) = \frac{Ze^2}{r} \exp\left(-\frac{r}{\lambda} \right) . \tag{5.1.34}$$

Therefore, λ has the meaning of a screening radius.

If $\varepsilon(q, 0)$ were a "good" function, the behavior of $V(r)$ with $r \to \infty$ would be determined by the Fourier transform of $V(q)$ with $q \to 0$, i.e., by (5.1.34). In classical statistics, this is true of electrons (with a different expression for λ, which we will see later). For the Fermi statistics, however, as can be seen from (5.1.29), the quantity $\varepsilon(q, 0)$ with $q = 2k_F$ has a singularity of the form

$$\varepsilon(q, 0) \propto (q - 2k_F) \ln(q - 2k_F) . \tag{5.1.35}$$

This can be shown [4.32] to entail the presence of an oscillatory part in $V(r)$,

$$V(r) \approx \text{const} \frac{\cos 2k_F r}{r^3} , \qquad r \to \infty , \tag{5.1.36}$$

which is at least qualitatively consistent with the treatment of Friedel oscillations in Sect. 4.8.3. Recall that the occurrence of a slowly decreasing oscillatory "tail" in the screened potential is due to the singularities (5.1.35) in $\varepsilon(q, 0)$ and these singularities are, in turn, due to the existence of the Fermi surface.

Collisions or finite temperature, since they smear this surface, lead to a decay of the "tail," but over a length that is much larger than λ.

To elucidate the physical sense of (5.1.34), which describes the smooth part of $V(r)$, we derive this equation in another way. If we assume that $V(r)$ varies in space slowly, the electron density at point r can be found by the formula

$$N(r) = \int d\varepsilon g(\varepsilon) \Bigg/ \left(\exp\left(\frac{\varepsilon - \zeta + V(r)}{k_B T}\right) + 1\right) = N(\zeta - V(r)) \ , \tag{5.1.37}$$

where the notation adopted is the same as that used in Chap. 3. If $V(r)$ is small (which is undoubtedly true for large distances), the result for the density variation at point r in comparison with the unperturbed case is

$$\delta N(r) \approx -\frac{\partial N}{\partial \zeta} V(r) \ . \tag{5.1.38}$$

Then the self-consistent potential $V(r)$ can be found from the Poisson equation

$$\Delta V = -4\pi e^2 \delta N \ . \tag{5.1.39}$$

Taking account of the fact that for spherically symmetric functions

$$\Delta V = \frac{1}{r}\frac{\partial^2}{\partial r^2}(rV) \tag{5.1.40}$$

and substituting (5.1.38, 40) into (5.1.39), we obtain

$$V(r) = \frac{C}{r}\exp\left(-\frac{r}{\lambda}\right) , \tag{5.1.41}$$

where C is a constant that has yet to be defined (it may be found from the condition $V(r) \approx Ze^2/r$ with $r \to 0$, when the screening is unimportant; hence $C = Ze^2$), and

$$\lambda = \left(4\pi e^2 \frac{\partial N}{\partial \zeta}\right)^{-1/2} . \tag{5.1.42}$$

For Fermi statistics $\partial N/\partial \zeta = g(\zeta_0)$; in the free-electron model this gives (5.1.31). In classical statistics

$$N \propto \exp(\zeta/k_B T) \ , \quad \frac{\partial N}{\partial \zeta} = \frac{N}{k_B T} \ ,$$
$$\lambda = (k_B T/4\pi N e^2)^{1/2} \ . \tag{5.1.43}$$

Formulas (5.1.34, 43) were derived by *Debye* and *Hückel* [5.5]; for this reason, the quantity λ in (5.1.43) is frequently referred to as the Debye screening radius. The λ involved in (5.1.31) is called the Thomas–Fermi radius. The latter term is due to (5.1.37, 39) underlying the Thomas–Fermi statistical theory of the atom.

Static screening plays an important, decisive role in the physics of metals. This screening results in the Coulomb ionic binding energy weakening and the metallic structures being stabilized preferentially by band electron energy (Sect. 4.5.2); i.e., it is reasonable in general to talk of a metallic bond. *Mott* was the first to note the decisive importance of screening in the metal–insulator criterion [4.24]. The point is that there always exist an infinite number of bound states in the Coulomb potential. Therefore, an electron that has left an atom would immediately be bound again to the hole that has been created, and thus that electron could not become a conduction electron. With an increase in electron concentration λ decreases, and at some critical value of λ/a_B, where $a_B = \hbar^2/me^2$ is the Bohr radius, there are no longer any bound states in the potential (5.1.34) (allowance for Friedel oscillations does not alter this conclusion). The corresponding concentration is

$$N^{1/3} a_B \approx 0.4 \ . \tag{5.1.44}$$

Thus, after Mott, if we bring atoms with one valence electron closer together, then, as long as the electron density is less than the critical value (5.1.44), the system will be an insulator. When the concentration reaches the value (5.1.44) the electrons become free and a first-order metal–insulator transition takes place. Qualitatively, the criterion (5.1.44) is similar to that considered in Sect. 4.5.3, where we talked of inner-incomplete-shell electrons which interact with short-range forces.

An important characteristic is the number of electrons in a sphere of radius λ:

$$Q = N\lambda^3 \propto (k_F a_B)^{3/2} \ . \tag{5.1.45}$$

If $Q \gg 1$, the fluctuations of the potential they produce are small, the self-consistent-potential (random-phase) approximation is justified. An inspection of (5.1.45) shows that this corresponds to high densities. When $Q \ll 1$, the correlations of individual particles play a decisive role and the plasma description is ineffective.

A comparison of (5.1.44, 45) shows that the criterion for the applicability of the plasma description will approximately coincide with the Mott metal–insulator criterion.

5.1.4 Plasmon

Our objective now is to consider the high-frequency properties of the plasma, to be more exact, the case

$$\omega \gg q v_{\mathrm{F}} = \frac{\hbar q k_{\mathrm{F}}}{m} . \tag{5.1.46}$$

We will be concerned with the natural oscillation spectrum, defined by the condition (5.1.21) (see above), which in the one-dimensional case assumes the form

$$\varepsilon(q,\omega) = 0 . \tag{5.1.47}$$

Expanding (5.1.26) in a power series of ω^{-1} with allowance for (5.1.46) and substituting the result of the expansion into (5.1.47) yield

$$\omega^2 = \omega_{\mathrm{P}}^2(q) = \omega_{\mathrm{P}}^2 + \tfrac{3}{5} q^2 v_{\mathrm{F}}^2 + \frac{\hbar^2 q^4}{4m^2} + \ldots , \tag{5.1.48}$$

where

$$\omega_{\mathrm{P}}^2 = \frac{4\pi N e^2}{m} . \tag{5.1.49}$$

For the threshold frequency of natural electron-density oscillations (plasmons) we obtain the same expression as in Sect. 3.7.4. Allowance for the spatial dispersion, however, leads to the plasmon frequency depending on the wave vector (Fig. 5.1). At some value of $k_{\mathrm{c}} \approx \omega_{\mathrm{P}}/v_{\mathrm{F}} \approx \lambda^{-1}$, (5.1.27) ceases to hold, an imaginary part occurs in $\varepsilon(q, \omega)$, resulting in an imaginary part for the solution of (5.1.47) (Landau damping). For small $\mathrm{Im}\{\varepsilon(q, \omega)\}$ the imaginary frequency part may be found from the formula

$$\gamma = \omega \left. \frac{\mathrm{Im}\{\varepsilon(q, \omega)\}}{\dfrac{\partial}{\partial \omega} \mathrm{Re}\{\varepsilon(q, \omega)\}} \right|_{\omega = \omega_{\mathrm{P}}(q)} \tag{5.1.50}$$

which results from the replacement $\omega \to \omega + \mathrm{i}\gamma$ in (5.1.47) and from first-order expansion in γ.

Landau damping arises from the energy being transferred from a collective excitation (plasmon) to one-particle excitations. This follows from the expression for $\mathrm{Im}\,\varepsilon(\boldsymbol{q}, \omega)$, which can be derived from (5.1.24) through the identity

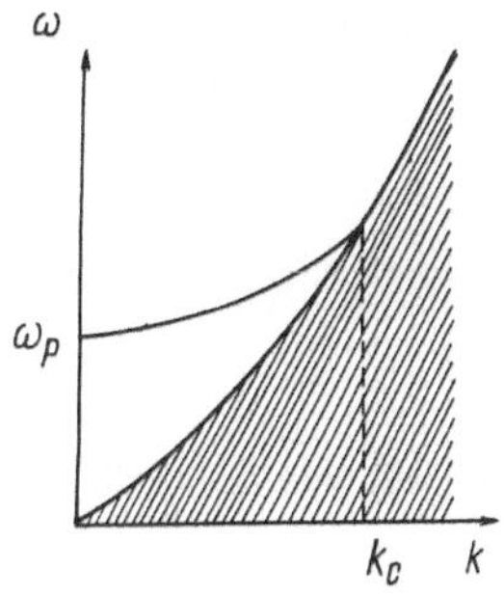

Fig. 5.1. Dispersion relation of a plasmon: the shadowed region refers to Landau damping

$$\mathrm{Im}\{(x+\mathrm{i}\eta)^{-1}\} = -\pi\delta(x):$$

$$\mathrm{Im}\{\varepsilon(\boldsymbol{q},\omega)\} = \frac{e^2}{\pi \boldsymbol{q}^2}\int d\boldsymbol{k}[\overline{n(\boldsymbol{k}+\boldsymbol{q})} - \overline{n(\boldsymbol{k})}]\delta[E(\boldsymbol{k}+\boldsymbol{q}) - E(\boldsymbol{k}) + \hbar\omega] \ . \tag{5.1.51}$$

It then follows that $\mathrm{Im}\{\varepsilon(\boldsymbol{q},\omega_{\mathrm{P}}(\boldsymbol{q}))\}$ is different from zero, provided the energy and momentum conservation principles hold:

$$E(\boldsymbol{k}+\boldsymbol{q}) = E(\boldsymbol{k}) + \hbar\omega_{\mathrm{P}}(\boldsymbol{q}) \ . \tag{5.1.52}$$

The term $\overline{n(\boldsymbol{k}+\boldsymbol{q})} - \overline{n(\boldsymbol{k})}$ in (5.1.51) is nonzero at $T = 0$ K if the transitions are made from below the Fermi surface to free states, as required by the Pauli exclusion principle. The postulate that Landau damping is the transfer of energy to one-particle degrees of freedom can be corroborated by straightforwardly calculating the effect of the potential $V(\boldsymbol{r},t)$ on the one-electron energy [5.6].

Assume that the system is subject to a plasma potential $V(\boldsymbol{r})\exp(-\mathrm{i}\omega t) + V^*(\boldsymbol{r})\exp(-\mathrm{i}\omega t)$. The latter initiates transitions between states ψ_ν whose probabilities are determined from nonstationary perturbation theory. The transition probability per unit time to the lowest order in V is

$$w_{\nu\rightleftarrows\nu'} = \frac{2\pi}{\hbar}|V_{\nu\nu'}|^2[\delta(E_\nu - E_{\nu'} - \hbar\omega) + \delta(E_\nu - E_{\nu'} + \hbar\omega)] \ . \tag{5.1.53}$$

Then the total rate of change of the one-electron energy is

$$\begin{aligned} dE_{\mathrm{e}}/dt &= \sum_{\nu\nu'} \omega_{\nu\rightleftarrows\nu'}\bar{n}_\nu(1-\overline{n_{\nu'}})(E_{\nu'} - E_\nu) = \tfrac{1}{2}\sum_{\nu\nu'} w_{\nu\rightleftarrows\nu'}(E_{\nu'} - E_\nu)(\overline{n_\nu} - \overline{n_{\nu'}}) \\ &= \pi\omega\sum_{\nu\nu'}|V_{\nu\nu'}|^2(\overline{n_\nu} - \overline{n_{\nu'}})[\delta(E_\nu - E_{\nu'} + \hbar\omega) \\ &\quad - \delta(E_\nu - E_{\nu'} - \hbar\omega)] \ . \end{aligned} \tag{5.1.54}$$

The term $n_\nu(1-\overline{n_{\nu'}})$ denotes that only transitions from occupied to free states are possible. We have also taken into account that $x\delta(x\pm\omega)=\mp\,\omega\delta(x\pm\omega)$. For free electrons $V(\boldsymbol{r})=V_q\exp(\mathrm{i}\boldsymbol{q}\cdot\boldsymbol{r})$, and we obtain

$$\frac{dE_e}{dt}=2\pi\omega|V_q|^2\int\frac{d\boldsymbol{k}}{(2\pi)^3}[\overline{n(\boldsymbol{k}+\boldsymbol{q})}-\overline{n(\boldsymbol{k})}][\delta(E(\boldsymbol{k}+\boldsymbol{q})-E(\boldsymbol{k})+\hbar\omega)$$
$$-\delta(E(\boldsymbol{k}+\boldsymbol{q})-E(\boldsymbol{k})-\hbar\omega)]=\frac{\omega q^2|V_q|^2}{2\pi e^2}\,\mathrm{Im}\{\varepsilon(\boldsymbol{q},\omega)\}\ , \tag{5.1.55}$$

where we have allowed for (5.1.51) as well as the oddness of the function $\mathrm{Im}\{\varepsilon(\boldsymbol{q},\omega)\}=-\mathrm{Im}\{\varepsilon(\boldsymbol{q},-\omega)\}$.

On the other hand, according to the familiar formula from the electrodynamics of continuous media [2.17], the energy stored in a weakly decaying plasma wave is equal to

$$E_\mathrm{P}=\frac{1}{4\pi e^2}\frac{d}{d\omega}[\omega\,\mathrm{Re}\{\varepsilon(\boldsymbol{q},\omega)\}]\bigg|_{\omega=\omega_\mathrm{P}(\boldsymbol{q})}\int d\boldsymbol{r}|\nabla V(\boldsymbol{r})|^2$$
$$=\frac{\omega q^2|V_q|^2}{4\pi e^2}\frac{\partial}{\partial\omega}\mathrm{Re}\{\varepsilon(\boldsymbol{q},\omega)\}\bigg|_{\omega=\omega_\mathrm{P}(\boldsymbol{q})}\ , \tag{5.1.56}$$

where $-e^{-1}\nabla V$ is the electric wave field intensity and allowance is made for the fact that $\mathrm{Re}\{\varepsilon(\boldsymbol{q},\omega)\}=0$ when $\omega=\omega_\mathrm{P}(\boldsymbol{q})$. If the wave field amplitude decays as $\exp(-\gamma t)$, the wave energy loss per unit time is equal to $2\gamma E_\mathrm{P}$. Using (5.1.50, 55, 56) yields

$$2\gamma E_\mathrm{P}=\frac{dE_e}{dt}\ , \tag{5.1.57}$$

which proves the statement about the nature of the Landau damping

Plasmons manifest themselves most dramatically in experiments on the characteristic energy losses of high-speed electrons (Fig. 5.2, [5.7]). The scattering of these high-speed electrons is determined by the dynamics of the conduction-electron density, i.e., the plasmons or quanta of electron-density fluctuation. By analogy with the scattering of phonons (Sect. 2.7), the result for

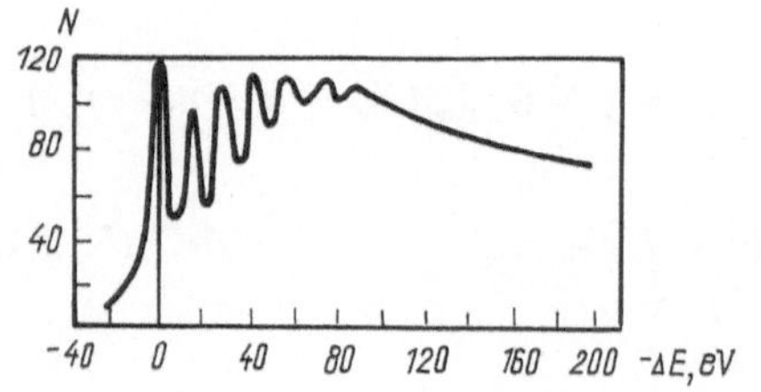

Fig. 5.2. Experimental curves of characteristic losses for Al (N here stands for the number of scattered electrons)

one-plasmon scattering as obtained from the laws of conservation of energy and momentum is

$$E(\boldsymbol{k}+\boldsymbol{q})-E(\boldsymbol{k})=-\hbar\omega_{\mathrm{P}}(\boldsymbol{q})\ ,\qquad E(\boldsymbol{k})=\frac{\hbar^2 \boldsymbol{k}^2}{2m}\ , \tag{5.1.58}$$

with $\boldsymbol{q}$ being a scattering vector. The latter is related to the scattering angle ϑ (Fig. 5.3) by the equations that follow from (5.1.58):

$$\boldsymbol{q}^2=\frac{2m}{\hbar^2}\left[E(\boldsymbol{k})+E(\boldsymbol{k}+\boldsymbol{q})-2\sqrt{E(\boldsymbol{k})E(\boldsymbol{k}+\boldsymbol{q})}\cos\vartheta\right]\ . \tag{5.1.59}$$

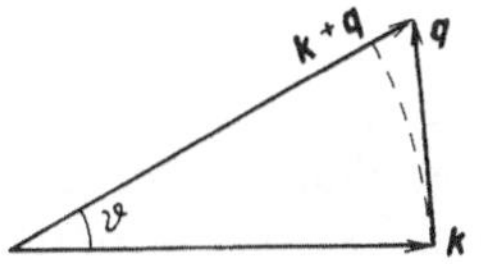

Fig. 5.3. Determination of the scattering angle

For small ϑ—i.e., when $q \ll k$ and $\hbar\omega_{\mathrm{P}} \ll E(\boldsymbol{k})$—we obtain

$$q^2 \approx k^2\left[\vartheta^2+\left(\frac{\hbar\omega_{\mathrm{P}}}{2E(\boldsymbol{k})}\right)^2\right]\ . \tag{5.1.60}$$

Thus, by studying the dependence of the energy loss on the scattering angle, we can determine the plasmon spectrum $\omega_{\mathrm{P}}(\boldsymbol{q})$. Experimental data give a fairly good fit to both (5.1.48) and the existence of a limiting scattering angle (related to $\boldsymbol{k}_{\mathrm{c}}$, as shown in Fig. 5.1) which, when reached, is responsible for the smearing of the spectrum of characteristic losses.

We will return to the theory of characteristic losses in Sect. 5.1.6. We note here that these losses occur not only in metals but also in semiconductors, and N implies the total number of valence electrons, in spite of the fact that all of them are in a wholly occupied band. As is seen from Table 5.1, the values of ω_{P} for metals and semiconductors are comparatively close in magnitude. This is so because $\hbar\omega_{\mathrm{P}}$ in these materials is much larger than the forbidden bandwidth G, and the electrons respond to a high-frequency interaction in the same way as free electrons do. The m involved in (5.1.49) is therefore the mass of a free electron. But if $\hbar\omega_{\mathrm{P}} \ll G$, we may use the effective-mass approximation and

Table 5.1. Experimental plasmon energy values for different metals [5.8]

Element	Al	Be	Mg	Si	Ge	Sb	Na
ΔE [eV]	15.0	18.9	10.5	16.9	16.0	15.3	5.7

neglect the interband transitions. The N here should be understood as the number of free carriers, N_c, and m as the effective mass m^*; e^2 should be replaced by e^2/ε_0, with ε_0 being the static permittivity (Sect. 4.5.1).

Thus, along with a high-frequency plasmon, a low-frequency plasmon exists in semiconductors. The frequency of this low-frequency plasmon, which manifests itself in, for example, optical properties, is

$$\omega_P^* = \left(\frac{4\pi N_c e^2}{m^* \varepsilon_0}\right)^{1/2} . \tag{5.1.61}$$

5.1.5 Phonons in the Plasma Model

Now we wish to make a general survey of the lattice properties of the plasma model. Ions were replaced by a compensating positive-charge phonon of density N/Z (with $Z|e|$ being the ion charge). Apparently they should oscillate with the frequency

$$\omega_{Pi} = \left(\frac{4\pi N(Ze)^2}{MZ}\right)^{1/2} = \left(\frac{4\pi N Z e^2}{M}\right)^{1/2} \tag{5.1.62}$$

(M stands for the ion mass), and therefore this model would not permit description of the acoustic character of the spectrum. However, a decisive factor in this case is screening. The spatial dispersion of ions may be neglected for

$$\omega \gg q v_T , \tag{5.1.63}$$

where $v_T = (3k_B T/M)^{1/2}$ is the thermal velocity of ions (for which we adopt the classical statistics). Then the ionic contribution to permittivity is equal to $-\omega_{Pi}^2/\omega^2$. But if, at the same time,

$$\omega \ll q v_F , \tag{5.1.64}$$

a static limit should be used for the electronic contribution. Inequalities (5.1.63, 64) are simultaneous because $v_F \gg v_T$. Thus the ion motion is screened by electrons as a static motion (this is what actually accounts for the adiabatic approximation which was often mentioned above). The total permittivity takes the form

$$\varepsilon_{tot}(q,\omega) = \varepsilon(q,0) - \frac{\omega_{Pi}^2}{\omega^2} , \tag{5.1.65}$$

with (5.1.47) yielding

$$\omega^2 = \frac{\omega_{Pi}^2}{\varepsilon(q,0)} . \tag{5.1.66}$$

Employing (5.1.30) for $\varepsilon(q, 0)$ with $q \ll k_F$ gives

$$\omega^2 = s^2 q^2 \ , \quad s = \lambda\omega_{Pi} = \left(\frac{Zm}{3M}\right)^{1/2} v_F \ , \tag{5.1.67}$$

with (5.1.31, 62) and the equality $N = k_F^3/3\pi^2$ substituted into (5.1.30). Expression (5.1.67) [5.9] gives an estimate of the sound velocity in metals, which is in reasonable agreement with experimental data. The plasma model is insensitive to the difference between solid and liquid metals and is therefore incapable of describing shear waves characteristic of solids. Nevertheless, it provides an excellent illustration of the physics of the phonon spectrum formation in metals and clears up the meaning of the adiabatic approximation. According to (5.1.35, 66), the phonon spectrum should exhibit a weak singularity at $q = 2k_F$ (Migdal–Kohn anomaly [5.10, 11]). We assume that by measuring the dispersion of short-wave phonons in the different crystallographic directions we can reproduce the form of the Fermi surface. Unfortunately, these anomalies are rather slight and the method itself cannot compete with other, primarily magnetic, techniques.

5.1.6 Fluctuation-Dissipation Theorem

We conclude this section by exploring some general properties of the function $\varepsilon(\boldsymbol{q}, \omega)$. To this end, we consider rigorously the problem of the response of a many-electron system to a weak external perturbation. We introduce into the system a trial charge described by the density

$$eN_{ext}(\boldsymbol{r}, t) = ev_{\boldsymbol{q}} \exp[\mathrm{i}(\boldsymbol{q}\cdot\boldsymbol{r} - \omega t) + \eta t] \ . \tag{5.1.68}$$

The Hamiltonian describing the interaction of the electronic system with the trial charge will then be

$$\hat{V} = e^2 \int \frac{d\boldsymbol{r}\, d\boldsymbol{r}'\, \hat{N}(\boldsymbol{r}) N_{ext}(\boldsymbol{r}')}{|\boldsymbol{r} - \boldsymbol{r}'|} = \frac{4\pi e^2}{q^2} \hat{N}_{-\boldsymbol{q}} v_{\boldsymbol{q}} \exp[\mathrm{i}(\boldsymbol{q}\cdot\boldsymbol{r} - \omega t) + \eta t] \ , \tag{5.1.69}$$

with $\hat{N}_{\boldsymbol{q}}$ being the Fourier transform of the electron-density operator. The Fourier transformation is carried out allowing for (5.1.6, 22, 68). We wish to find the variation of the density matrix of the entire electronic system $\hat{\varrho}$ by the action of perturbation $\hat{V}$ to the lowest order. This variation can be obtained from (5.1.16) by making the replacement $|v\rangle \rightarrow |n\rangle$, where $|n\rangle$ stands for the eigenstates of the many-electron system. Here

$$w_n = Z^{-1}\exp\left(-\frac{E_n}{k_B T}\right) \tag{5.1.70}$$

is the Gibbs distribution function, or the eigenstate of the total unperturbed density matrix, and Z is the partition function. Thus,

$$\varrho'_{nm} = \frac{w_n - w_m}{E_n - E_m - \hbar(\omega + i\eta)} V_{nm} = \frac{4\pi e^2}{q^2} v_q \frac{(w_n - w_m)(N_{-q})_{nm}}{E_n - E_m - \hbar(\omega + i\eta)} . \tag{5.1.71}$$

The result for the variation of the mean value of $\langle \hat{N}_q \rangle$ due to the perturbation is

$$\langle \hat{N}_q \rangle = \sum_{mn} (N_q)_{mn} \varrho'_{nm} = \frac{4\pi e^2}{q^2} v_q \sum_{mn} \frac{(w_n - w_m)|(N_q)_{mn}|^2}{E_n - E_m - \hbar(\omega + i\eta)} , \tag{5.1.72}$$

where we have taken account of the fact that $(N_{-q})_{nm} = (N_q)^*_{mn}$. The dielectric function (permittivity) is the ratio of the external charge density v_q to the total charge density $v_q + \langle \hat{N}_q \rangle$. Since $\operatorname{div} \boldsymbol{D} = 4\pi v$, $\operatorname{div} \boldsymbol{E} = 4\pi(v + \langle \hat{N} \rangle)$, this definition coincides with the usual expression $D_q = \varepsilon(\boldsymbol{q}, \omega) E_q (D_q = -4\pi i \boldsymbol{q} e v_q / q^2$, $E_q = -4\pi i \boldsymbol{q} e(v_q + \langle \hat{N}_q \rangle)/q^2)$.

Equation (5.1.72) yields

$$\frac{1}{\varepsilon(\boldsymbol{q}, \omega)} = 1 + \frac{\langle \hat{N}_q \rangle}{v_q} = 1 + \frac{4\pi e^2}{q^2} \times \sum_{mn} \frac{w_n - w_m}{E_n - E_m - \hbar(\omega + i\eta)} |(N_q)_{mn}|^2 . \tag{5.1.73}$$

Expression (5.1.73) is formally exact, but has little use in particular computations, inasmuch as it requires a knowledge of the exact eigenfunctions and the energies of the many-electron system. On the other hand, the formula allows us to derive some important general relations. We calculate $\operatorname{Im}\{1/\varepsilon(\boldsymbol{q}, \omega)\}$

$$\begin{aligned} \operatorname{Im}\{1/\varepsilon(\boldsymbol{q}, \omega)\} &= \frac{4\pi^2 e^2}{q^2} \sum_{mn} (w_n - w_m)|(N_q)_{mn}|^2 \\ &\times \delta(E_n - E_m - \hbar\omega) = -\frac{4\pi^2 e^2}{Z q^2} \sum_{mn} \exp\left(-\frac{E_m}{k_B T}\right) \\ &\times \left[1 - \exp\left(-\frac{E_n - E_m}{k_B T}\right)\right] |(N_q)_{mn}|^2 \delta(E_n - E_m - \hbar\omega) \\ &= -\frac{4\pi^2 e^2}{Z q^2} \left[1 - \exp\left(-\frac{\hbar\omega}{k_B T}\right)\right] \sum_{mn} \exp\left(-\frac{E_m}{k_B T}\right) \\ &\times |(N_q)_{mn}|^2 \delta(E_n - E_m - \hbar\omega) . \end{aligned} \tag{5.1.74}$$

The expression on the right-hand side of (5.1.74) can be related to the density correlator that was introduced in Sect. 2.7.1:

$$S(\boldsymbol{q}, \omega) = \int_{-\infty}^{\infty} dt e^{i\omega t} \langle \hat{N}_{\boldsymbol{q}}(t) \hat{N}_{-\boldsymbol{q}}(0) \rangle$$

$$= \int_{-\infty}^{\infty} dt e^{i\omega t} \sum_{mn} w_m (N_{\boldsymbol{q}})_{mn} \exp\left[\frac{i}{\hbar}(E_m - E_n)t\right] \quad (5.1.75)$$

$$\times (N_{-\boldsymbol{q}})_{nm} = 2\pi\hbar \sum_{mn} w_m |(N_{\boldsymbol{q}})_{mn}|^2 \delta(E_n - E_m - \hbar\omega) ,$$

$$S(\boldsymbol{q}, \omega) = -\frac{\hbar q^2}{2\pi e^2} \frac{1}{1 - \exp(-\hbar\omega/k_B T)} \operatorname{Im}\{1/\varepsilon(\boldsymbol{q}, \omega)\} . \quad (5.1.76)$$

Equation (5.1.76) expresses the content of what is known as the fluctuation-dissipation theorem [5.12].

Now we can express the scattering probability of fast electrons in terms of permittivity. Substituting $v_{\boldsymbol{q}} = 4\pi e^2/\boldsymbol{q}^2$ and (5.1.76) into (2.7.15) yields

$$R = \left(\frac{4\pi e^2}{\hbar \boldsymbol{q}^2}\right)^2 S(\boldsymbol{q}, \omega)$$

$$= -\frac{8\pi e^2}{\hbar \boldsymbol{q}^2} \frac{1}{1 - \exp(-\hbar\omega/k_B T)} \operatorname{Im}\{1/\varepsilon(\boldsymbol{q}, \omega)\} , \quad (5.1.77)$$

$$\omega = [E(\boldsymbol{k}) - E(\boldsymbol{k} + \boldsymbol{q})]/\hbar .$$

In the absence of damping, we write $\varepsilon(\boldsymbol{q}, \omega) \to \varepsilon(\boldsymbol{q}, \omega) + i\delta$, $\delta \to +0$,

$$\operatorname{Im}\{1/\varepsilon(\boldsymbol{q}, \omega)\} = -\pi\delta(\varepsilon(\boldsymbol{q}, \omega)) . \quad (5.1.78)$$

Substituting (5.1.78) into (5.1.77), we find

$$R = \frac{8\pi^2 e^2}{\hbar q^2} (\overline{n_\omega} + 1)\delta(\varepsilon(q, \omega)) , \quad (5.1.79)$$

where

$$\overline{n_\omega} = \frac{1}{\exp(\hbar\omega/k_B T) - 1} \quad (5.1.80)$$

is the Planck distribution function. The occurrence of the δ function (5.1.79) indicates that the scattering is accompanied by creation of a plasmon and conforms to (5.1.58); the occurrence of the representative term $\overline{n_\omega} + 1$ (the same as that for phonons) signifies that plasmons obey Bose–Einstein statistics. Incidentally, since the quantity $\hbar\omega_{\mathrm{P}}$ is normally approximately equal to 10 eV, we may always assume that $\overline{n_\omega} \approx 0$.

A very important relationship exists between the permittivity and the free-energy of a many-electron system. We represent the electron–electron interaction Hamiltonian as

$$\begin{aligned}\hat{\mathscr{H}}_{\mathrm{int}} &= \tfrac{1}{2}\sum_{ij}{}' \frac{e^2}{|\boldsymbol{r}_i - \boldsymbol{r}_j|} = \frac{2\pi e^2}{q^2} \sum_{q \neq 0} \hat{N}_{\boldsymbol{q}} \hat{N}_{-\boldsymbol{q}} \\ &= \frac{2\pi e^2}{q^2} \sum_{\boldsymbol{q}} (\hat{N}_{\boldsymbol{q}} \hat{N}_{-\boldsymbol{q}} - N) ,\end{aligned} \tag{5.1.81}$$

where we use the expression $\hat{N}_{\boldsymbol{q}} \hat{N}_{-\boldsymbol{q}} = N^{-1} \sum_{ij} \exp[\mathrm{i}\boldsymbol{q}\cdot(\boldsymbol{r}_i - \boldsymbol{r}_j)] \xrightarrow[q\to 0]{} N$. The free energy is equal to

$$F = -k_{\mathrm{B}} T \ln Z = -k_{\mathrm{B}} T \ln \mathrm{Tr} \exp\left(-\frac{\hat{\mathscr{H}}_0 + \hat{\mathscr{H}}_{\mathrm{int}}}{k_{\mathrm{B}} T}\right), \tag{5.1.82}$$

with $\hat{\mathscr{H}}_0$ being the Hamiltonian of noninteracting electrons. Making the replacement $e^2 \to \lambda e^2$ in (5.1.81), we calculate $\partial F(\lambda)/\partial\lambda$:

$$\begin{aligned}\frac{\partial F}{\partial \lambda} &= -\frac{k_{\mathrm{B}} T}{Z(\lambda)} \frac{\partial Z(\lambda)}{\partial \lambda} = \frac{1}{Z(\lambda)} \mathrm{Tr}\left\{ \hat{\mathscr{H}}_{\mathrm{int}} \exp\left(-\frac{\hat{\mathscr{H}}_0 + \lambda \hat{\mathscr{H}}_{\mathrm{int}}}{k_{\mathrm{B}} T}\right)\right\} \\ &= \langle \hat{\mathscr{H}}_{\mathrm{int}} \rangle_\lambda ,\end{aligned} \tag{5.1.83}$$

where $\langle \dots \rangle_\lambda$ is the averaging over the Gibbs ensemble with the replacement $e^2 \to \lambda e^2$. We then substitute (5.1.81) into (5.1.83), allowing for the fact that

$$\langle N_{\boldsymbol{q}} N_{-\boldsymbol{q}} \rangle = \int_{-\infty}^{\infty} \frac{d\omega}{2\pi} S(\boldsymbol{q}, \omega) \tag{5.1.84}$$

Here we have to invert the Fourier transform (5.1.75) and set $t = 0$. In turn, $S(\boldsymbol{q}, \omega)$ is expressed in terms of $\mathrm{Im}\{1/\varepsilon(\boldsymbol{q}, \omega)\}$ according to (5.1.76). As a result, we find

$$\begin{aligned}\frac{\partial F}{\partial \lambda} = -\hbar \sum_{\boldsymbol{q}} \Bigg(&\int_{-\infty}^{\infty} \frac{d\omega}{2\pi} [1 - \exp(-\hbar\omega/k_{\mathrm{B}} T)]^{-1} \\ &\times \mathrm{Im}\{1/\varepsilon_\lambda(\boldsymbol{q}, \omega)\} + \frac{2\pi N e^2}{\hbar q^2} \Bigg),\end{aligned} \tag{5.1.85}$$

where the replacement $e^2 \to \lambda e^2$ should be made in the calculation of $\varepsilon_\lambda(\boldsymbol{q}, \omega)$. For $\lambda = 0$ the quantity F is known. Calculating the permittivity within some approximation (for example, the random-phase approximation) and integrating (5.1.85) over λ between 0 and 1, we find the free energy of a many-electron system and thus all its thermodynamic properties.

5.2 The Fermi-Liquid Theory

5.2.1 Major Postulates of the Landau Theory

As is clear from Sect. 5.1, the random-phase approximation provides a sufficiently complete description of the electronic system of a metal in the high-density limit

$$k_F a_B \gtrsim 1 \ , \tag{5.2.1}$$

with a_B being the Bohr radius. It is possible to calculate the permittivity, to find the free energy of the system according to (5.1.85), and thereby to determine the heat capacity, compressibility, and other thermodynamic properties. Effects of spin and of an externally applied field can be taken into account. Also, high-frequency and "short-wave" properties can be described. However, the quantity k_F in metals is on the order of a_B^{-1} and so, even qualitatively, the random-phase approximation is dubious (although qualitatively embracing nearly all of the major features of many-particle coupling).

Landau proposed a phenomenological approach to the description of a system of interacting electrons—the Fermi-liquid theory [5.13]—applicable also when the binding energy is on the order of the kinetic energy of the particles. When applied to electrons in metals, this means that $k_F a_B \approx 1$. On the other hand, Fermi-liquid theory deals only with sufficiently low-frequency and long-wave excitations

$$\hbar\omega \ll \zeta_0 \ , \qquad q \ll k_F \tag{5.2.2}$$

and static properties at

$$k_B T \ll \zeta_0 \tag{5.2.3}$$

so that, for example, the influence of Fermi-liquid effects on plasmons ($\hbar\omega_P \gtrsim \zeta_0$) cannot be treated correctly. However, high-frequency properties are described fairly well in the random-phase approximation. Landau's theory enjoys wide use not only in solid-state physics but also in nuclear theory, the theory of neutron stars, liquid He^3, and other Fermi systems. Originally, Landau proposed his theory for neutral Fermi liquids (He^3) [5.13]. A generalization of the Fermi-liquid theory for charged Fermi liquids (electrons in a metal) has been given by Silin [5.14]—see also [5.15].

In the foregoing we repeatedly emphasized the importance of the concept of quasiparticles. Now we choose electrons as quasiparticles. The electrons interact with other quasiparticles and, consequently, are "clad" with a cloud of electron-hole pairs, plasmons, etc. (a mobile electron perturbs the state of the system, admixing different excitations to the ground state of the Fermi gas; this, in turn, affects the proper motion of the electron). With the interaction "coming into play," the ground state of the system, generally speaking, readjusts. This may entail, for example, a metal–insulator transition, structural changes in the lattice (Sect. 4.5) or a transition to the superconducting state (see below), i.e., radical changes in the ground state and in the character of the current-carrying states.

If, however, the interaction is not very strong (and the system is non-superconductive), the character of the spectrum will virtually not change. For instance, no gap will arise between the ground-state level and the first excited level, in contrast to superconductors. The energy levels will certainly shift, but they will be specified by the same quantum numbers (band number, quasi-momentum, spin projection, etc.) as those by which the states are characterized in the absence of an interaction. In other words, an unambiguous correspondence persists between the states of band electrons and the states of the quasiparticles introduced, which are also described by Fermi–Dirac statistics, obeying the Pauli exclusion principle. This is the first and fundamental postulate of Landau's theory. If this postulate holds, one talks of a normal Fermi liquid. Experiment shows that a system of conduction electrons may usually be regarded as a normal Fermi liquid.

Quasiparticles occupy reciprocal-space states bounded by the Fermi surface. According to the Landau–Luttinger theorem (Sect. 4.10), the volume bounded by the Fermi surface does not vary when the interaction is included, but the shape of the Fermi surface may, broadly speaking, be distorted. We will disregard the latter circumstance, limiting ourselves to the case of an isotropic dispersion relation and a spherical Fermi surface. Since quasiparticles obey the Pauli exclusion principle, the arguments given in Sect. 4.10 are applicable to them, and the damping of these quasiparticles on the Fermi surface goes to zero. If conditions (5.2.2, 3) are satisfied, only excitations with an energy close to ζ_0 are essential, and a quasiparticle description is applicable. Far from the Fermi surface the quasiparticles are poorly defined and the magnitude of their damping is on the order of the excitation energy (provided that the interaction itself is not small).

Another postulate of Landau's theory is that the energy of the system, $\mathscr{E}$, is a functional of the single-particle distribution function of quasiparticles $\overline{n_\nu}$. The energy of a quasiparticle, ε_ν, is defined as the variation of the total energy of the system when a particle is added to the state $|\nu\rangle$ (or withdrawn from this state), i.e., as the first functional derivative of $\mathscr{E}$,

$$\varepsilon_\nu = \frac{\delta\mathscr{E}[\overline{n_\nu}]}{\delta\overline{n_\nu}} . \tag{5.2.4}$$

In a system of interacting quasiparticles the ε_ν depends on the state of all the other quasiparticles. Of supreme importance is Landau's Fermi-liquid interaction function, which is equal to the second functional energy derivative

$$f_{\nu\nu'} = \frac{\delta\varepsilon_\nu}{\delta\bar{n}_{\nu'}} = \frac{\delta^2\mathscr{E}}{\delta\bar{n}_\nu\delta\bar{n}_{\nu'}} . \tag{5.2.5}$$

It follows from (5.2.5) that $f_{\nu\nu'} = f_{\nu'\nu}$. In the simple case of a free-electron gas the ν is the momentum and the projection of the spin $\sigma = \uparrow, \downarrow$. The functions $\overline{n_{p\sigma}}$ may be viewed as eigenvalues of the density matrix in spin space $\bar{n}_{\sigma'\sigma}(\boldsymbol{p})$. Then we obtain

$$\begin{aligned}\delta\mathscr{E} &= \sum_{p} \varepsilon_{\sigma\sigma'}(\boldsymbol{p})\delta\bar{n}_{\sigma'\sigma}(\boldsymbol{p}) \\ &\quad + \tfrac{1}{2}\sum_{pp'} f_{\sigma_1\sigma_2;\sigma'_1\sigma'_2}(\boldsymbol{p},\boldsymbol{p}')\delta\bar{n}_{\sigma'_2\sigma'_1}(\boldsymbol{p}')\delta\bar{n}_{\sigma_2\sigma_1}(\boldsymbol{p}) + \dots ,\end{aligned} \tag{5.2.6}$$

where a summation over the double spin subscripts is implied. In the isotropic case the most general form of the matrix $\hat{f}$ is

$$f_{\sigma_1\sigma_2;\sigma'_1\sigma'_2}(\boldsymbol{p},\boldsymbol{p}') = \varphi(\boldsymbol{p},\boldsymbol{p}')\delta_{\sigma_1\sigma_2}\delta_{\sigma'_1\sigma'_2} + \psi(\boldsymbol{p},\boldsymbol{p}')\boldsymbol{\sigma}_{\sigma_1\sigma_2}\boldsymbol{\sigma}_{\sigma'_1\sigma'_2} ,$$

or, in a contracted representation,

$$f(\boldsymbol{p},\boldsymbol{p}') = \varphi(\boldsymbol{p},\boldsymbol{p}') + \psi(\boldsymbol{p},\boldsymbol{p}')\hat{\boldsymbol{\sigma}}\hat{\boldsymbol{\sigma}}' , \tag{5.2.7}$$

where $\hat{\boldsymbol{\sigma}}$ stands for the Pauli matrices, φ is called a direct-interaction function, and ψ is an exchange-interaction function. The latter is a consequence of the Pauli exclusion principle; its genesis (the spin dependence of energy) was elucidated in Sect. 4.6.1 in connection with the Hartree–Fock–Slater method. As follows from the arguments presented in that section, an energetically favorable situation is a parallel orientation of electronic spins (this weakens the Coulomb repulsion); i.e., normally

$$\psi(\boldsymbol{p},\boldsymbol{p}') < 0 . \tag{5.2.8}$$

The spin–orbit interaction leads to the occurrence of terms such as $\boldsymbol{p}\hat{\boldsymbol{\sigma}}$ in (5.2.7).

This interaction, however, is much [a factor of $(v_F/c)^2$] weaker than the exchange interaction and may therefore be neglected.

An equilibrium distribution function can only be derived from combinatorial considerations, associated with the Pauli exclusion principle. We write the expression for the entropy $S[\bar{n}_\nu]$ and employ the principle of minimum thermodynamic potential.

$$\frac{\delta}{\delta \bar{n}_\nu}(\mathscr{E} - TS - \zeta N) = 0 \ . \tag{5.2.9}$$

The form of $S[\bar{n}_\nu]$ here does not depend on the interaction at all (we simply have to calculate the statistical weight of a particular state), and it is precisely $\delta\mathscr{E}/\delta\bar{n}_\nu$, i.e., (5.2.4), that plays the role of the energy of a quasiparticle. Therefore we can write for the equilibrium distribution function

$$\bar{n}_\nu^{(0)} = \frac{1}{\exp(\varepsilon_\nu - \zeta/k_B T) + 1} , \tag{5.2.10}$$

where, however, ε_ν is a functional of $\bar{n}_\nu^{(0)}$.

Further, we consider the isotropic case. From the Landau–Luttinger theorem (or simply because of the unambiguous correspondence of particles and quasiparticles), the quantity p_F is defined by the ordinary formula (3.5.16). In the vicinity of $p = p_F$ the energy, counted from ζ_0, is represented as

$$\varepsilon(\boldsymbol{p}) - \zeta_0 = v(|\boldsymbol{p}| - p_F) \equiv \frac{p_F}{m^*}(|\boldsymbol{p}| - p_F) , \tag{5.2.11}$$

where v has the meaning of velocity of quasiparticles on the Fermi surface and m^*, by definition, is the effective mass

$$m^* = p_F \left| \frac{\partial \varepsilon}{\partial \boldsymbol{p}} \right|^{-1}_{|\boldsymbol{p}| = p_F} . \tag{5.2.12}$$

This definition of m^* differs somewhat from those given earlier. It is this effective mass, though, that enters into the density of states at the Fermi level

$$g(\zeta_0) = 2V_0 \int \frac{dS}{(2\pi\hbar)^3} v^{-1} = \frac{m^* p_F V_0}{\pi^2 \hbar^3} , \tag{5.2.13}$$

with dS being an element of area of the Fermi surface, and V_0 the unit-cell volume. Only the values $|\boldsymbol{p}| = p_F$ are important to us. In the isotropic case the quantity $f(\boldsymbol{p}, \boldsymbol{p}')_{|\boldsymbol{p}| = |\boldsymbol{p}'| = p_F}$ depends only on the angle ϑ between the vectors $\boldsymbol{p}$

and $\boldsymbol{p}'$. This function can be expanded in Legendre polynomials $p_l(\cos\vartheta)$, which form a complete system of functions in ϑ, according to the formulas

$$g(\zeta_0)\varphi(\boldsymbol{p},\boldsymbol{p}') = \sum_{l=0}^{\infty}(2l+1)A_l P_l(\cos\vartheta)\,, \tag{5.2.14}$$

$$g(\zeta_0)\psi(\boldsymbol{p},\boldsymbol{p}') = \sum_{l=0}^{\infty}(2l+1)B_l P_l(\cos\vartheta)\,.$$

5.2.2 Thermodynamic Properties

Since the distribution function is defined by (5.2.10) the thermodynamic properties of a Fermi liquid are derived nearly in the same way as for a Fermi gas. The only difference is that, in itself, the spectrum ε_ν is a functional of $\bar{n}_\nu$ and therefore depends on T, ζ, and the external magnetic field. This should be taken into account in calculating the derivatives of the thermodynamic potential.

The heat capacity is equal to

$$C_v = T\left(\frac{\partial S}{\partial T}\right)_{V,N}\,. \tag{5.2.15}$$

But the entropy S is a functional of $\bar{n}_\nu$ [Ref. 1.4, Sect. 55]. Then, allowing for (5.2.4, 9), we find

$$\begin{aligned} C_v &= T\sum_\nu \frac{\delta S}{\delta \bar{n}_\nu}\frac{\partial \bar{n}_\nu}{\partial T} = \sum_\nu \frac{\delta(\mathscr{E}-\zeta N)}{\delta \bar{n}_\nu}\frac{\partial \bar{n}_\nu}{\partial T} \\ &= \sum_\nu(\varepsilon_\nu-\zeta)\frac{\partial \bar{n}_\nu}{\partial T} = \int d\varepsilon\, g(\varepsilon)(\varepsilon-\zeta)\frac{\partial \bar{n}}{\partial T}\,. \end{aligned} \tag{5.2.16}$$

Next, we calculate $\partial\bar{n}/\partial T$:

$$\frac{\partial \bar{n}}{\partial T} = \frac{\partial \bar{n}}{\partial \varepsilon}\frac{1}{k_B}\frac{\partial}{\partial T}\left(\frac{\varepsilon-\zeta}{T}\right) = -\frac{1}{k_B T}\frac{\partial \bar{n}}{\partial \varepsilon}\left[\frac{\varepsilon-\zeta}{T} - \frac{\partial}{\partial T}(\varepsilon-\zeta)\right]. \tag{5.2.17}$$

Using the calculations of the Sommerfeld integrals (Sect. 3.5.2), the result for the heat capacity is

$$\begin{aligned} C_v &= \int d\varepsilon\, g(\varepsilon)\left(-\frac{\partial \bar{n}(\varepsilon)}{\partial \varepsilon}\right)\left[\frac{(\varepsilon-\zeta)^2}{T} - \frac{1}{2}\frac{\partial(\varepsilon-\zeta)^2}{\partial T}\right] \\ &\approx \frac{\pi^2}{3}(k_B T)^2\left[\frac{g(\zeta_0)}{T} - \frac{1}{2}\frac{\partial g(\zeta_0)}{\partial T}\right]. \end{aligned} \tag{5.2.18}$$

If $\partial g(\zeta_0)/\partial T$ is finite for $T \to 0$, the second term can be neglected. Taking account of (5.2.13), the result for the renormalization of the heat capacity in comparison with the Fermi gas is

$$\frac{C_v}{C_v^{(0)}} = \frac{g(\zeta_0)}{g^{(0)}(\zeta_0)} = \frac{m^*}{m} . \tag{5.2.19}$$

The effective mass is determined by means of the kinetic equation in (5.2.3). Thus, to the lowest order in $k_B T/\zeta_0$, the effect of the Fermi-liquid interaction on the heat capacity reduces to the renormalization of the effective mass. To higher orders in $k_B T/\zeta_0$, these effects are more substantial and lead to a nonanalytical dependence [5.16]:

$$C_v = \gamma T - BT^3 \ln(T/T_0) . \tag{5.2.20}$$

Further, we find the renormalization of the compressibility κ. Since ζ is the Gibbs potential per particle, we have for $T = \text{const}$

$$d\zeta = \frac{V}{N}\, dp ; \tag{5.2.21}$$

here p is the pressure, and N the total number of particles in the system. Then, since ζ depends on N and V only through the N/V ratio, we obtain

$$\frac{1}{\kappa} = -\frac{\partial p}{\partial V} = \frac{N}{V}\left(-\frac{\partial \zeta}{\partial V}\right) = \frac{N^2}{V^2}\frac{\partial \zeta}{\partial N} . \tag{5.2.22}$$

If $T = 0$ K, the chemical potential coincides with the energy of the last filled level. With the removal of one particle, first, the p_F varies and, second, the correlational energy correction due to the variation of $\delta \bar{n}$ arises:

$$\delta\zeta_0 = \frac{\partial \zeta_0}{\partial p_F}\,\delta p_F - \tfrac{1}{2}\,\mathrm{Tr}_\sigma \mathrm{Tr}_{\sigma'} \int \frac{dp'\, V_0}{(2\pi\hbar)^3} f(p\sigma, p'\sigma')\,\delta\bar{n}_{p'\sigma'} . \tag{5.2.23}$$

According to the equation $N = Vp_F^3/3\pi^2\hbar^3$,

$$\delta N = \frac{V p_F^2 \delta p_F}{\pi^2 \hbar^3} , \tag{5.2.24}$$

it follows from (5.2.12) that

$$\frac{\partial \zeta_0}{\partial p_F}\,\delta p_F = \frac{\pi^2\hbar^3}{V p_F^2}\frac{\partial \zeta_0}{\partial p_F}\,\delta N = \frac{\pi^2\hbar^3}{m^* V p_F}\,\delta N . \tag{5.2.25}$$

In the second term we set

$$\delta \bar{n}_{p'\sigma'} = \frac{1}{g(\zeta_0)} \delta(\varepsilon_{p'} - \zeta_0)\delta N \ , \tag{5.2.26}$$

where the factor $1/g(\zeta_0)$ is determinable from the condition

$$\mathrm{Tr}_{\sigma'} \int \frac{d\boldsymbol{p}' V_0}{(2\pi\hbar)^3} \delta n_{p'\sigma'} = \delta N \ , \tag{5.2.27}$$

$f(\boldsymbol{p}, \boldsymbol{p}')$ being replaced by its own value, averaged over the angles of the vectors $\boldsymbol{p}, \boldsymbol{p}'$ and over the spins σ, σ'; i.e., $[g(\zeta_0)]^{-1} A_0$ (5.2.14). Substitution of (5.2.13, 25, 26) into (5.2.23) yields

$$\frac{\partial \zeta}{\partial N} = \frac{\pi^2 \hbar^3 (1 + A_0)}{m^* V p_\mathrm{F}} \tag{5.2.28}$$

or, for the compressibility renormalization,

$$\frac{\kappa}{\kappa_0} = \frac{m^*}{m} \frac{1}{1 + A_0} \ . \tag{5.2.29}$$

We carry out a similar calculation for the paramagnetic susceptibility. When an external magnetic field $\boldsymbol{H}$ is applied, the energy variation of a particle is equal to

$$\begin{aligned} \delta \varepsilon_{p\sigma} &= -\mu \boldsymbol{H}\boldsymbol{\sigma} + \mathrm{Tr}_{\sigma'} \int \frac{d\boldsymbol{p}' V_0}{(2\pi\hbar)^3} \delta n_{p'\sigma'} f(\boldsymbol{p}\sigma, \boldsymbol{p}'\sigma') \\ &= -\mu \boldsymbol{H}\boldsymbol{\sigma} + \mathrm{Tr}_{\sigma'} \int \frac{d\boldsymbol{p}' V_0}{(2\pi\hbar)^3} \frac{\partial n_{p'\sigma'}}{\partial \varepsilon_{p'\sigma'}} \delta \varepsilon_{p'\sigma'} f(\boldsymbol{p}\sigma, \boldsymbol{p}'\sigma') \\ &= -\mu \boldsymbol{H}\boldsymbol{\sigma} - \mathrm{Tr}_{\sigma'} \int \frac{d\boldsymbol{p}' V_0}{(2\pi\hbar)^3} \delta(\varepsilon_{p'\sigma'} - \zeta_0) \delta \varepsilon_{p'\sigma'} f(\boldsymbol{p}\sigma, \boldsymbol{p}'\sigma') \ , \end{aligned} \tag{5.2.30}$$

with μ being the magnetic moment of a free electron. We try solution of (5.2.30) in the form

$$\delta \varepsilon_{p'\sigma'} = -\gamma \boldsymbol{H}\boldsymbol{\sigma}' \ . \tag{5.2.31}$$

Substituting $\hat{f}$ from (5.2.7) and allowing for the fact that

$$\mathrm{Tr}_{\sigma'} \boldsymbol{\sigma}' = 0 \ , \quad \mathrm{Tr}_{\sigma'} \{(\boldsymbol{\sigma}\boldsymbol{\sigma}')\boldsymbol{\sigma}'\} = 2\boldsymbol{\sigma} \ , \tag{5.2.32}$$

we find $\gamma = \mu - \gamma B_0$, or

$$\gamma = \frac{\mu}{1 + B_0} . \tag{5.2.33}$$

The susceptibility χ is determined from the expression for the magnetic moment

$$\begin{aligned} \chi \boldsymbol{H} &= \mu \mathrm{Tr}_\sigma \int \frac{d\boldsymbol{p}\, V_0}{(2\pi\hbar)^3} \boldsymbol{\sigma} \delta n_{\boldsymbol{p}\sigma} = \mu \mathrm{Tr}_\sigma \left\{ \int \frac{d\boldsymbol{p}\, V_0}{(2\pi\hbar)^3} \boldsymbol{\sigma} \frac{\partial \bar{n}_{\boldsymbol{p}\sigma}}{\partial \varepsilon_{\boldsymbol{p}\sigma}} \delta\varepsilon_{\boldsymbol{p}\sigma} \right\} \\ &= \frac{\mu^2 g(\zeta_0) \boldsymbol{H}}{1 + B_0} . \end{aligned} \tag{5.2.34}$$

Thus the renormalization here is

$$\frac{\chi}{\chi_0} = \frac{m^*}{m} \frac{1}{1 + B_0} . \tag{5.2.35}$$

By virtue of (5.2.8), the exchange interaction enhances the spin paramagnetism by $(1 + B_0) < 1$. This conclusion has important implications for the theory of the magnetism of transition metals (Sect. 5.6), in which case the exchange interaction may become so strong that $B_0 = -1$.

5.2.3 Kinetic Equation for Quasiparticles

Nonstationary processes in a Fermi liquid, such as different types of waves, electrical conduction, etc., are described on the basis of a kinetic equation [5.15, 17]. The energy $\varepsilon(\boldsymbol{p}, \boldsymbol{r}, \sigma)$ acts as a one-particle Hamiltonian (in the inhomogeneous case the $\boldsymbol{r}$ dependence should be taken into account). For simplicity, the spin effects will be disregarded. The quasi-classical kinetic equation will then assume the form

$$\frac{\partial \bar{n}(\boldsymbol{p}, \boldsymbol{r}, t)}{\partial t} + \dot{\boldsymbol{r}} \cdot \frac{\partial \bar{n}(\boldsymbol{p}, \boldsymbol{r}, t)}{\partial \boldsymbol{r}} + \dot{\boldsymbol{p}} \cdot \frac{\partial \bar{n}(\boldsymbol{p}, \boldsymbol{r}, t)}{\partial \boldsymbol{p}} = I[\bar{n}] , \tag{5.2.36}$$

with $I[\bar{n}]$ being the collision integral. Allowance for the Hamiltonian equations

$$\dot{\boldsymbol{r}} = \frac{\partial \varepsilon}{\partial \boldsymbol{p}} \qquad \dot{\boldsymbol{p}} = -\frac{\partial \varepsilon}{\partial \boldsymbol{r}} \tag{5.2.37}$$

yields

$$\frac{\partial \bar{n}}{\partial t} + \frac{\partial \varepsilon}{\partial \boldsymbol{p}} \cdot \frac{\partial \bar{n}}{\partial \boldsymbol{r}} - \frac{\partial \varepsilon}{\partial \boldsymbol{r}} \cdot \frac{\partial \bar{n}}{\partial \boldsymbol{p}} = I[\bar{n}] \ . \tag{5.2.38}$$

In the absence of Fermi-liquid correlations, (5.2.38) would take the form

$$\frac{\partial \bar{n}}{\partial t} + \frac{\partial \varepsilon}{\partial \boldsymbol{p}} \cdot \frac{\partial \bar{n}}{\partial \boldsymbol{r}} + e\left(\boldsymbol{E} + \frac{1}{c}\left(\frac{\partial \varepsilon}{\partial \boldsymbol{p}} \times \boldsymbol{H}\right)\right) \cdot \frac{\partial \bar{n}}{\partial \boldsymbol{p}} = I[\bar{n}] \ , \tag{5.2.39}$$

with $\boldsymbol{E}$ and $\boldsymbol{H}$ standing for the intensities of the electric and magnetic fields acting on a particle. If the long-range character of the forces between electrons in metals is taken into account explicitly, $\boldsymbol{E}$ and $\boldsymbol{H}$, (or, to be more precise, $\boldsymbol{B}$) should involve not only the fields of the external sources but also the self-consistent field, associated with the motion of the electrons themselves (Sect. 5.1). The fields $\boldsymbol{E}$, $\boldsymbol{B}$ are determinable from Maxwell equations, where the current density and charge density comprise both external and induced sources. The latter are determined by the distribution function $\bar{n}$, as will be shown later.

The Fermi-liquid interaction gives correlational corrections to the self-consistent field approximation. This, according to (5.2.5), results in the occurrence in $\partial\varepsilon/\partial\boldsymbol{r}$ of an extra term

$$F_c(\boldsymbol{p}, \boldsymbol{r}, \sigma) = \left(\frac{\partial \varepsilon}{\partial \boldsymbol{r}}\right)_{\text{corr}} = \mathrm{Tr}_{\sigma'}\left\{\int \frac{d\boldsymbol{p}' V_0}{(2\pi\hbar)^3} f(\boldsymbol{p}\sigma, \boldsymbol{p}'\sigma') \frac{\partial \bar{n}(\boldsymbol{p}', \boldsymbol{r}, \sigma')}{\partial \boldsymbol{r}}\right\} . \tag{5.2.40}$$

Since the correlation is short-range and we deal with a weakly inhomogeneous state—this is, from (5.2.2), the only thing which we may claim—the possible nonlocality of the coupling (5.2.40), i.e., the dependence of $\delta\varepsilon(\boldsymbol{r})$ on $\delta\bar{n}(\boldsymbol{r}')$ at $\boldsymbol{r} \neq \boldsymbol{r}'$, is neglected.

We restrict our attention to the simple case $\boldsymbol{H} = 0$ and omit the spin variables. To first order in $\boldsymbol{E}$ and $\delta\bar{n} = \bar{n} - \bar{n}_0$, the quantity $\bar{n}$ in the term $\partial\varepsilon/\partial\boldsymbol{r}$ $\partial\bar{n}/\partial\boldsymbol{p}$ may be replaced by $\bar{n}_0$. Then adding (5.2.40) to (5.2.39) yields

$$\begin{aligned} &\frac{\partial}{\partial t}\delta\bar{n}(\boldsymbol{p}, \boldsymbol{r}, t) + \frac{\partial \varepsilon_p}{\partial \boldsymbol{p}} \cdot \frac{\partial}{\partial \boldsymbol{r}}\delta\bar{n}(\boldsymbol{p}, \boldsymbol{r}, t) - \frac{\partial \bar{n}_0(\boldsymbol{p})}{\partial \boldsymbol{p}} \\ &\quad \cdot \sum_{\boldsymbol{p}'} f(\boldsymbol{p}, \boldsymbol{p}') \frac{\partial}{\partial \boldsymbol{r}} \delta\bar{n}(\boldsymbol{p}', \boldsymbol{r}, t) + e\boldsymbol{E} \cdot \frac{\partial \bar{n}_0(\boldsymbol{p})}{\partial \boldsymbol{p}} = I[\bar{n}] \ . \end{aligned} \tag{5.2.41}$$

Let us analyze the structure of the collision integral. Similar to (3.6.15), we obtain for elastic isotropic impurity scattering

$$I[\bar{n}] = \sum_{p'} W_{pp'}\delta(\varepsilon_p - \varepsilon_{p'})[\bar{n}_p(1-\bar{n}_{p'}) - \bar{n}_{p'}(1-\bar{n}_p)]$$
$$= \sum_{p'} W_{pp'}\delta(\varepsilon_p - \varepsilon_{p'})(\bar{n}_p - \bar{n}_{p'}) \; . \tag{5.2.42}$$

Here ε_p is the total energy in the pth state; it is equal to the sum of the equilibrium value ε_p^0 and its change due to the variation of $\delta\bar{n}_p$:

$$\delta\varepsilon_p = \sum_{p'} f(\boldsymbol{p}, \boldsymbol{p}')\delta\bar{n}_{p'} \; . \tag{5.2.43}$$

For states that differ insignificantly from the equilibrium state,

$$I[\bar{n}] = \sum_{p'} W_{pp'}\delta(\varepsilon_p^0 + \delta\varepsilon_p - \varepsilon_{p'}^0 - \delta\varepsilon_{p'})(\bar{n}_{p'}^0 + \delta\bar{n}_{p'} - \bar{n}_p^0 - \delta\bar{n}_p) \; . \tag{5.2.44}$$

We represent the Fermi-equilibrium function as

$$\bar{n}_p^0 \approx \bar{n}_0(\varepsilon_p^0 + \delta\varepsilon_p) - \frac{\partial\bar{n}_p^0}{\partial\varepsilon_p^0}\delta\varepsilon_p \tag{5.2.45}$$

and take account of the fact that, because of the presence in (5.2.44) of the δ function, the terms $\bar{n}_0(\varepsilon_p^0 + \delta\varepsilon_p) - \bar{n}_0(\varepsilon_{p'}^0 + \delta\varepsilon_{p'})$ do not contribute to $I[\bar{n}]$. The collision integral to the lowest order in $\delta\bar{n}_p$ then has the form

$$I[\bar{n}] = \sum_{p'} W_{pp'}\delta(\varepsilon_p^0 - \varepsilon_{p'}^0)(\delta\tilde{n}_{p'} - \delta\tilde{n}_p) \; , \tag{5.2.46}$$

where

$$\delta\tilde{n}_p = \delta\bar{n}_p - \frac{\partial\bar{n}_0(\boldsymbol{p})}{\partial\varepsilon_p^0}\delta\varepsilon_p$$

or

$$\delta\tilde{n}_p = \delta\bar{n}_p - \frac{\partial\bar{n}_0(\boldsymbol{p})}{\partial\varepsilon_p^0}\sum_{p'} f(\boldsymbol{p}, \boldsymbol{p}')\delta\bar{n}_{p'} \; . \tag{5.2.47}$$

The kinetic equation (5.2.41) becomes

$$\frac{\partial}{\partial t}\delta\bar{n}_p + \frac{\partial\varepsilon_p}{\partial\boldsymbol{p}}\cdot\frac{\partial}{\partial\boldsymbol{r}}\delta\tilde{n}_p + e\boldsymbol{E}\cdot\frac{\partial\bar{n}_p^0}{\partial\boldsymbol{p}} = I[\delta\tilde{n}_p] \; . \tag{5.2.48}$$

It follows from (5.2.48) that, in those cases where the time derivative may be neglected (e.g., for static electrical, thermo-magnetic, and galvano-magnetic phenomena, for the anomalous skin effect), the kinetic equation has the usual form with the replacement $\delta\bar{n}_p \to \delta\tilde{n}_p$. The latter denotes a renormalization of the constants (effective mass, magnetic moment, etc.).

This is what underlies band theory.

For high-frequency properties the Fermi-liquid correlation manifests itself in the occurrence of new wave types and in a variation of the dispersion relation for original waves. For lack of space, we do not consider these problems here; see [5.2, 14].

Now we calculate the renormalization of the effective quasiparticle mass. In the absence of collisions and external fields, the kinetic equation (5.2.48) together with (5.2.47) yields the continuity equation

$$\frac{\partial \delta\varrho}{\partial t} + \operatorname{div}\boldsymbol{j} = 0 \ , \tag{5.2.49}$$

where $\delta\varrho = \sum_p \delta\bar{n}_p$ has the significance of quasiparticle density variation and, consequently, the current density has the form

$$\boldsymbol{j} = \sum_p \boldsymbol{v}_p \delta\tilde{n}_p = \sum_p \boldsymbol{v}_p \left(\delta\bar{n}_p - \frac{\partial \bar{n}_p^0}{\partial \varepsilon_p} \sum_{p'} f(\boldsymbol{p},\boldsymbol{p}')\,\delta\bar{n}_{p'} \right) , \tag{5.2.50}$$

with $\boldsymbol{v}_p = \partial\varepsilon_p/\partial\boldsymbol{p}$. The current transferred by one quasiparticle is equal to

$$\boldsymbol{j}_p = \frac{\delta \boldsymbol{j}}{\delta\bar{n}_p} = \boldsymbol{v}_p - \sum_{p'} \frac{\partial \bar{n}_{p'}^0}{\partial \varepsilon_{p'}} f(\boldsymbol{p},\boldsymbol{p}')\,\boldsymbol{v}_{p'} \ . \tag{5.2.51}$$

On the other hand, because of the unambiguous correspondence between the states of particles and quasiparticles, the flux due to one quasiparticle being added to the state $\boldsymbol{p}$ is equal to $\boldsymbol{p}/m$. Substitution of $\boldsymbol{v}_p = \boldsymbol{p}/m^*$ yields

$$\frac{\boldsymbol{p}}{m} = \frac{\boldsymbol{p}}{m^*} - \sum_{p'} \frac{\partial \bar{n}_{p'}^0}{\partial \varepsilon_{p'}} f(\boldsymbol{p},\boldsymbol{p}')\,\frac{\boldsymbol{p}'}{m^*} \ , \tag{5.2.52}$$

with $\boldsymbol{p}, \boldsymbol{p}'$ lying on the Fermi surface. Multiplying the left-hand and right-hand sides by $\boldsymbol{p}/p_F^2$, introducing the angle ϑ between $\boldsymbol{p}$ and $\boldsymbol{p}'$, and substituting expansion (5.2.14), we obtain

$$\frac{1}{m} = \frac{1}{m^*}\left[1 + \tfrac{1}{2}\int_0^\pi d\vartheta \sin\vartheta \cos\vartheta \sum_{l=0}^{\infty} (2l+1)\,A_l P_l(\cos\vartheta) \right] . \tag{5.2.53}$$

Allowing for

$$\int_0^\pi d\vartheta \sin\vartheta \cos\vartheta \, P_l(\cos\vartheta) = 0 \ , \quad l \neq 1$$

$$\int_0^\pi d\vartheta \sin\vartheta \cos\vartheta \, P_1(\cos\vartheta) = \int_0^\pi d\vartheta \sin\vartheta \cos^2\vartheta = 2/3 \ ,$$

we find

$$m^* = m(1 + A_1) \ . \tag{5.2.54}$$

In conclusion we remind the reader of the way the quasiparticle current density is determined in terms of $\delta\tilde{n}_p$,

$$j = \sum_p v_p \, \delta\tilde{n}_p \ , \tag{5.2.55}$$

which agrees with the usual definition (v_p is the velocity of a quasiparticle, and $\delta\tilde{n}_p$ satisfies the standard kinetic equation at $\partial\tilde{n}_p/\partial t = 0$).

Thus, Fermi-liquid theory substantiates band theory, leading in many cases to results that are analogous to those obtained in the noninteracting electron model with an arbitrary dispersion relation. New effects should be expected, first, in the high-frequency properties and, second, in cases where, per se, the theory of a normal Fermi liquid needs modification (superconductivity, possibly some disordered systems, and compounds of transition-*d* metals and rare-earth metals).

Bear in mind, however, that even in the static properties of a normal Fermi liquid, where the correlation effects reduce to quasiparticle parameter renormalization, these effects could lead to qualitatively new results. The renormalization factors are nonanalytical functions of the external parameters if narrow density of electronic states peaks are present in the vicinity of ζ_0 [5.18], a situation which is rather typical of *d* and *f* metals, their alloys, and their compounds (Sect. 5.6). This may lead to various anomalies in the properties of metals and alloys and even entail electronic phase transitions [5.18].

The theory of a normal Fermi liquid may be shown [5.19] to hold in the isotropic case if

$$1 + A_l > 0 \ , \quad 1 + B_l > 0 \ , \quad l = 0, 1, 2 \ldots . \tag{5.2.56}$$

If one of the conditions (5.2.56) is violated, a phase transition occurs in the electronic subsystem. For instance, when $1 + B_0 \to 0$, the spin susceptibility suffers a divergence, i.e., a ferromagnetic transition takes place, and when $1 + A_0 \to 0$, the compressibility diverges leading to an abrupt increase in volume, etc. It is not quite clear where the metal–insulator transition, due to correlation, should be placed in this scheme and whether it can be described in this fashion. In general, the problem as to the scope of applicability of Landau's theory is one of the most important and complicated issues in many-body theory. Sufficiently general and reliable results have not as yet been obtained.

5.3 Electron–Phonon Interaction

5.3.1 Formulation of the Problem

Up to this point we have neglected the interaction of two major types of elementary excitation in solids—electrons and phonons. In reality, this interaction is essential and is responsible for many properties of solids. The major effects are as follows:

1. *Phonon spectrum renormalization* and even a rearrangement of lattice structure because of the interaction of phonons with electrons (Sect. 4.5.2). Even the "bare" acoustic-phonon spectrum in metals is due to electrons (Sect. 5.1.5), and we should be very cautious here not to allow for the electron–phonon interaction twice.

2. *Electron spectrum renormalization.* Whereas Landau quasiparticles are electrons "clad" with a cloud of electron-hole excitations, a real current carrier is also "clad" with phonons. An example of such a carrier is the polaron (Sect. 5.3.4).

3. *The electron, as it moves, perturbs the phonon subsystem,* the perturbation, in turn, affecting other electrons. This gives rise to an electron–electron interaction which, owing to its special properties, results in the important phenomenon of superconductivity (Sect. 5.3.5).

4. *The electron, as it is scattered on phonons, imparts to the latter its momentum and energy.* These processes determine the temperature trend of resistivity and other kinetic coefficients in metals (Sect. 5.3.3).

To calculate these effects, we must write the electron–phonon interaction Hamiltonian. In the adiabatic approximation (Sects. 1.9, 5.3.2) the electronic subsystem is described by the Hamiltonian $\mathscr{H}(\hat{\boldsymbol{p}}_i, \hat{\boldsymbol{r}}_i, \boldsymbol{R}_l)$, with $\hat{\boldsymbol{p}}_i$ and $\hat{\boldsymbol{r}}_i$ being the electron momenta and coordinates, and $\boldsymbol{R}_l$ the ion coordinates on which $\hat{\mathscr{H}}$ depends parametrically. Regarding the deviations of these $\boldsymbol{R}_l$ from the equilibrium position $\boldsymbol{R}_l^0$ as small, we leave in $\hat{\mathscr{H}}$ only linear terms in displacements $\boldsymbol{u}_l = \boldsymbol{R}_l - \boldsymbol{R}_l^0$.

For simplicity, each lattice cell is assumed to contain one atom. Since we consider only acoustic branches of the vibrational spectrum, l is the cell subscript. Further, if $\boldsymbol{u}_l$ is independent of $\boldsymbol{R}_l$, the interaction Hamiltonian should go to zero, for the energy does not change when the lattice is displaced as a whole. Consequently, the Hamiltonian involves only the derivatives $\partial u_\alpha/\partial R_\beta$ ($\alpha, \beta = x, y, z$). The Hamiltonian may not contain the antisymmetric tensor $\partial u_\alpha/\partial R_\beta - \partial u_\beta/\partial R_\alpha$, which is nonzero when the lattice as a whole is rotated through an angle $\delta\varphi$ about the axis prescribed by the unit vector $\boldsymbol{n}$:

$$\delta \boldsymbol{u}_l = \delta\varphi(\boldsymbol{n} \times \boldsymbol{R}_l) \tag{5.3.1}$$

(we leave the proof as an exercise). Therefore, the electron–phonon interaction Hamiltonian should be linear in the deformation tensor

$$u_{\alpha\beta}(\boldsymbol{R}) = \frac{1}{2}\left(\frac{\partial u_\alpha}{\partial R_\beta} + \frac{\partial u_\beta}{\partial R_\alpha}\right) , \tag{5.3.2}$$

$$\hat{\mathscr{H}}_{\text{int}} = \sum_l \Lambda_{\alpha\beta}(\hat{\boldsymbol{p}}_i, \hat{\boldsymbol{r}}_i, \boldsymbol{R}_l^0) u_{\alpha\beta}(\boldsymbol{R}_l^0) \tag{5.3.3}$$

(the recurrent tensor subscripts here imply summation). The operator $\hat{\Lambda}$ is called the deformation potential; the methods of computing this potential are not discussed here (a somewhat obsolete but very lucid physical treatment is given by *Ziman* [5.20]).

We now wish to take account of the expression for $\boldsymbol{u}(\boldsymbol{R}_l^0)$, which may be obtained from (2.7.22):

$$\boldsymbol{u}(\boldsymbol{R}_l^0) = \mathrm{i}\sum_{\boldsymbol{q}j}\left(\frac{\hbar}{2MN\omega_{\boldsymbol{q}j}}\right)^{1/2} \exp(\mathrm{i}\boldsymbol{q}\cdot\boldsymbol{R}_l^0)\boldsymbol{e}_{\boldsymbol{q}j}(b_{\boldsymbol{q}j}^+ - b_{-\boldsymbol{q}j}) , \tag{5.3.4}$$

$$\omega_{\boldsymbol{q}j} = s_j|\boldsymbol{q}| , \tag{5.35}$$

with s_j being the sound velocity. The quantity $u_{\alpha\beta}$ is easy to calculate from (5.3.2). We write the matrix element of the Hamiltonian (5.3.3); this matrix element corresponds to the scattering of an electron from state $|\boldsymbol{k}'\rangle$ to state $|\boldsymbol{k}\rangle$ with simultaneous absorption or emission of a phonon $|\boldsymbol{q}j\rangle$ (for simplicity, the band subscript is omitted, and the interband scattering is neglected). The operator $\hat{\Lambda}$ acts only on the electronic variables, and b^+, b act only on the phonon variables. From the definition of b^+, b in Sect. 2.1.4 we see that the matrix elements of the latter variables are

$$\begin{aligned} \langle N_{\boldsymbol{q}j}|b_{\boldsymbol{q}j}^+|N_{\boldsymbol{q}j} - 1\rangle &= \sqrt{N_{\boldsymbol{q}j}} , \\ \langle N_{\boldsymbol{q}j}|b_{\boldsymbol{q}j}|N_{\boldsymbol{q}j} + 1\rangle &= \sqrt{N_{\boldsymbol{q}j} + 1} , \end{aligned} \tag{5.3.6}$$

where $N_{\boldsymbol{q}j}$ is the number of phonons in the corresponding state. As a result, we obtain

$$\langle \boldsymbol{k}', N_{\boldsymbol{q}j} | \hat{\mathscr{H}}_{\text{int}} | \boldsymbol{k}, N_{\boldsymbol{q}j}+1 \rangle = \left(\frac{\hbar (N_{\boldsymbol{q}j}+1)}{2MN\omega_{\boldsymbol{q}j}} \right)^{1/2} \tfrac{1}{2} (q_\alpha e_{\boldsymbol{q}j;\beta} + q_\beta e_{\boldsymbol{q}j;\alpha}) \\ \times \sum_l \exp(\mathrm{i}\boldsymbol{q}\cdot\boldsymbol{R}_l^0) \int d\boldsymbol{r} \psi_{\boldsymbol{k}'}^*(\boldsymbol{r}) \Lambda_{\alpha\beta}(\hat{\boldsymbol{p}}, \boldsymbol{r}; \boldsymbol{R}_l^0) \psi_{\boldsymbol{k}}(\boldsymbol{r}) \; . \tag{5.3.7}$$

Now we exploit Bloch's theorem (4.2.1). Under the integration sign in (5.3.7) we make the replacement $\boldsymbol{r} \to \boldsymbol{r} - \boldsymbol{R}_l^0$ and allow for the fact that, due to spatial homogeneity, $\Lambda(\hat{\boldsymbol{p}}, \boldsymbol{r}; \boldsymbol{R}_l^0)$ depends only on $\boldsymbol{r} - \boldsymbol{R}_l^0$. Then

$$\int d\boldsymbol{r} \psi_{\boldsymbol{k}'}^*(\boldsymbol{r}) \Lambda_{\alpha\beta}(\hat{\boldsymbol{p}}, \boldsymbol{r}; \boldsymbol{R}_l^0) \psi_{\boldsymbol{k}}(\boldsymbol{r}) = \exp[\mathrm{i}(\boldsymbol{k}'-\boldsymbol{k})\cdot\boldsymbol{R}_l^0] \Lambda_{\boldsymbol{k}'\boldsymbol{k}}^{\alpha\beta} \; . \tag{5.3.8}$$

Further

$$\sum_l \exp[\mathrm{i}(\boldsymbol{k}'-\boldsymbol{k}+\boldsymbol{q})\cdot\boldsymbol{R}_l^0] = \sum_g \delta_{\boldsymbol{k}'-\boldsymbol{k}+\boldsymbol{q},\boldsymbol{b}_g^*} \; , \tag{5.3.9}$$

with $\boldsymbol{b}_g^*$ being reciprocal lattice vectors (1.3.3, 1.5.12). Substitution of (5.3.8, 9) into (5.3.7) yields

$$\langle \boldsymbol{k}', N_{\boldsymbol{q}j} | \hat{\mathscr{H}}_{\text{int}} | \boldsymbol{k}, N_{\boldsymbol{q}j}+1 \rangle = \frac{1}{2} \left[\frac{\hbar (N_{\boldsymbol{q}j}+1)}{2MN\omega_{\boldsymbol{q}j}} \right]^{1/2} \\ \times (q_\alpha e_{\boldsymbol{q}j;\beta} + q_\beta e_{\boldsymbol{q}j;\alpha}) \Lambda_{\boldsymbol{k}'\boldsymbol{k}}^{\alpha\beta} \sum_g \delta_{\boldsymbol{k}'-\boldsymbol{k}+\boldsymbol{q},\boldsymbol{b}_g^*} \; . \tag{5.3.10}$$

For the process accompanied by emission of a phonon we need to make the replacement $N_{\boldsymbol{q}j}+1 \to N_{\boldsymbol{q}j}$, according to (5.3.6). It follows from (5.3.5) that, as $\boldsymbol{q} \to 0$, the probability of this process, proportional to the square of the modulus of the matrix element, goes to zero $\propto q$.

Now we are able to write the electron–phonon collision integral (Sect. 3.6.1). The number of electrons scattered from state $\boldsymbol{k}$ and into this state per unit time can be represented as

$$I[\bar{n}_{\boldsymbol{k}}] = b_+ - b_- = \sum_{\boldsymbol{k}\boldsymbol{q}\boldsymbol{k}'} \{ w(\boldsymbol{k}', \boldsymbol{q}; \boldsymbol{k}) [\bar{n}_{\boldsymbol{k}'}(1-\bar{n}_{\boldsymbol{k}}) N_{\boldsymbol{q}} \\ - \bar{n}_{\boldsymbol{k}}(1-\bar{n}_{\boldsymbol{k}'})(N_{\boldsymbol{q}}+1)] \delta(\varepsilon_{\boldsymbol{k}} - \varepsilon_{\boldsymbol{k}'} - \hbar\omega_{\boldsymbol{q}}) \sum_g \delta_{\boldsymbol{k},\boldsymbol{k}'+\boldsymbol{q}+\boldsymbol{b}_g^*} \\ - w(\boldsymbol{k}'; \boldsymbol{k}, \boldsymbol{q}) [\bar{n}_{\boldsymbol{k}'}(1-\bar{n}_{\boldsymbol{k}})(N_{\boldsymbol{q}}+1) - \bar{n}_{\boldsymbol{k}}(1-\bar{n}_{\boldsymbol{k}'}) N_{\boldsymbol{q}}] \\ \times \delta(\varepsilon_{\boldsymbol{k}} - \varepsilon_{\boldsymbol{k}'} + \hbar\omega_{\boldsymbol{q}}) \sum_g \delta_{\boldsymbol{k}',\boldsymbol{k}+\boldsymbol{q}+\boldsymbol{b}_g^*} \} \tag{5.3.11}$$

(for simplicity, the phonon number subscript j is omitted). Here $\omega(\boldsymbol{k}', \boldsymbol{q}; \boldsymbol{k})$ and $\omega(\boldsymbol{k}'; \boldsymbol{k}, \boldsymbol{q})$ stand for the quantum-mechanical probabilities of the corresponding transitions with emission and absorption of phonons, calculated in the first order of perturbation theory (the final states are indicated first).

5.3.2 Conditions for the Applicability of the Adiabatic Approximation

Before applying the results obtained in Sect. 5.3.1 to an examination of particular effects of the electron–phonon interaction in crystals, we discuss the scope of applicability of the underlying adiabatic approximation and consider possible variations of this approximation in solids.

The total Hamiltonian of the crystal can be represented as

$$\hat{\mathscr{H}}_{\text{tot}} = \hat{\mathscr{H}}(\hat{\boldsymbol{r}}_i, \hat{\boldsymbol{p}}_i) + \hat{V}(\hat{\boldsymbol{r}}_i, \hat{\boldsymbol{R}}_l) + \hat{T}(\hat{\boldsymbol{\mathscr{P}}}_l) \;, \tag{5.3.12}$$

where $(\hat{\boldsymbol{r}}_i, \hat{\boldsymbol{p}}_i)$ are the coordinates and momenta of the electrons; $(\hat{\boldsymbol{R}}_l, \hat{\boldsymbol{\mathscr{P}}}_l)$ are the coordinates and momenta of the nuclei; $\hat{\mathscr{H}}(\hat{\boldsymbol{r}}_i, \hat{\boldsymbol{p}}_i)$ describes the kinetic energy of the electrons and the electron–electron interaction; $\hat{V}(\boldsymbol{r}_i, \boldsymbol{R}_l)$ involves the Coulomb electron–nucleus interaction energy and also the Coulomb nuclear repulsive interaction energy; and $\hat{T}(\hat{\boldsymbol{\mathscr{P}}}_l)$ is the kinetic nuclear energy operator

$$\hat{T}(\hat{\boldsymbol{\mathscr{P}}}_l) = \sum_l \frac{\hat{\boldsymbol{\mathscr{P}}}_l^2}{2M_l} \tag{5.3.13}$$

with M_l being the mass of the lth nucleus.

To start with, we neglect the operator $\hat{T}(\hat{\boldsymbol{\mathscr{P}}}_l)$, which is small with respect to what is called the adiabatic parameter $\kappa \equiv (m/M)^{1/4}$ where M is the characteristic nuclear mass and $\hat{T}(\hat{\boldsymbol{\mathscr{P}}}_l) \propto \kappa^4$. The quantity $\hat{V}(\hat{\boldsymbol{r}}_i, \hat{\boldsymbol{R}}_l)$ depends only on nuclear coordinates but not on nuclear momenta. Therefore, neglecting the operator (5.3.13), the eigenfunctions of the operator (5.3.12), which are the solutions of the Schrödinger equation

$$[\mathscr{H}(\boldsymbol{r}_i, \hat{\boldsymbol{p}}_i) + V(\boldsymbol{r}_i, \boldsymbol{R}_l)]\psi_\alpha(\boldsymbol{r}_i, \boldsymbol{R}_l) = E_\alpha(\boldsymbol{R}_l)\psi_\alpha(\boldsymbol{r}_i, \boldsymbol{R}_l) \;, \tag{5.3.14}$$

depend on $\boldsymbol{R}_l$ parametrically. Using the functions $\psi_\alpha(\boldsymbol{r}_i, \boldsymbol{R}_l)$ as the basis for expanding the total wave function of the crystal, we have

$$\psi(\boldsymbol{r}_i, \boldsymbol{R}_l) = \sum_\alpha \chi_\alpha(\boldsymbol{R}_l)\psi_\alpha(\boldsymbol{r}_i, \boldsymbol{R}_l) \;. \tag{5.3.15}$$

Operation of the Hamiltonian (5.3.12) on the function (5.3.15) and allowance for (5.3.13 and 14) yield

$$\hat{\mathscr{H}}_{\text{tot}}\psi(\boldsymbol{r}_i,\boldsymbol{R}_l)=\sum_\alpha\{[\mathscr{H}(\boldsymbol{r}_i,\hat{\boldsymbol{p}}_i)+V(\boldsymbol{r}_i,\boldsymbol{R}_l)]\psi_\alpha(\boldsymbol{r}_i,\boldsymbol{R}_l)\}$$
$$\cdot\chi_\alpha(\boldsymbol{R}_l)+\sum_{\alpha l}\frac{\hat{\mathscr{P}}_l^2}{2M_l}\psi_\alpha(\boldsymbol{r}_i,\boldsymbol{R}_l)\chi_\alpha(\boldsymbol{R}_l)\tag{5.3.16}$$
$$=\sum_\alpha\{E_\alpha(\boldsymbol{R}_l)\psi_\alpha(\boldsymbol{r}_i,\boldsymbol{R}_l)\chi_\alpha(\boldsymbol{R}_l)+\sum_l[\psi_\alpha(\boldsymbol{r}_i,\boldsymbol{R}_l)$$
$$\cdot\frac{\hat{\mathscr{P}}_l^2}{2M_l}\chi_\alpha(\boldsymbol{R}_l)+\frac{1}{2M_l}(\hat{\mathscr{P}}_l^2\psi_\alpha(\boldsymbol{r}_i,\boldsymbol{R}_l)$$
$$+2\hat{\mathscr{P}}_l\psi_\alpha(\boldsymbol{r}_i,\boldsymbol{R}_l)\hat{\mathscr{P}}_l)\chi_\alpha(\boldsymbol{R}_l)]\}\ .$$

Substituting (5.3.16) into the Schrödinger equation

$$\hat{\mathscr{H}}_{\text{tot}}\psi(\boldsymbol{r}_i,\boldsymbol{R}_l)=E\psi(\boldsymbol{r}_i,\boldsymbol{R}_l)\ ,\tag{5.3.17}$$

premultiplying it by $\psi_\beta^*(\boldsymbol{r}_i,\boldsymbol{R}_l)$, and integrating over all $\boldsymbol{r}_i$ with allowance for the orthonormalization of the functions $\psi_\alpha(\boldsymbol{r}_i,\boldsymbol{R}_l)$, we have

$$[T(\hat{\mathscr{P}}_l)+E_\beta(\boldsymbol{R}_l)]\chi_\beta(\boldsymbol{R}_l)+\sum_\alpha C_{\beta\alpha}(\boldsymbol{R}_l,\hat{p}_l)\chi_\alpha(\boldsymbol{R}_l)=E\chi_\beta(\boldsymbol{R}_l)\ .\tag{5.3.18}$$

The notation introduced here is

$$C_{\beta\alpha}(\boldsymbol{R}_l,\hat{p}_l)=\sum_l[A_{\beta\alpha}(\boldsymbol{R}_l)\hat{\mathscr{P}}_l+B_{\beta\alpha}(\boldsymbol{R}_l)]\ ,$$
$$A_{\beta\alpha}(\boldsymbol{R}_l)=\frac{1}{M_l}\int\prod_i d\boldsymbol{r}_i\psi_\beta^*(\boldsymbol{r}_i,\boldsymbol{R}_l)\hat{\mathscr{P}}_l\psi_\alpha(\boldsymbol{r}_i,\boldsymbol{R}_l)\ ,\tag{5.3.19}$$
$$B_{\beta\alpha}(\boldsymbol{R}_l)=\frac{1}{2M_l}\int\prod_i d\boldsymbol{r}_i\psi_\beta^*(\boldsymbol{r}_i,\boldsymbol{R}_l)\hat{\mathscr{P}}_l^2\psi_\alpha(\boldsymbol{r}_i,\boldsymbol{R}_l)\ .$$

If we neglect the terms $C_{\beta\alpha}$ in the exact equation (5.3.18), one term may be retained in the sum (5.3.15), and we come to the adiabatic approximation

$$\psi(\boldsymbol{r}_i,\boldsymbol{R}_l)=\psi_\beta(\boldsymbol{r}_i,\boldsymbol{R}_l)\chi_\beta^{(\mu)}(\boldsymbol{R}_l)\ ,\tag{5.3.20}$$

where $\psi_\beta(\boldsymbol{r}_i,\boldsymbol{R}_l)$ is determined by the solution of the Schrödinger equation for fixed nuclei (5.3.14), and $\chi_\beta^{(\mu)}(\boldsymbol{R}_l)$ by the equation

$$[T(\hat{\mathscr{P}}_l)+E_\beta(\boldsymbol{R}_l)]\chi_\beta^{(\mu)}(\boldsymbol{R}_l)=E_\mu\chi_\beta^{(\mu)}(\boldsymbol{R}_l)\ ,\tag{5.3.21}$$

in which $E_\beta(\boldsymbol{R}_l)$ plays the part of the potential energy of the nuclear subsystem, and μ numbers the eigenstates of this subsystem with a fixed β. Expand the energy $E_\beta(\boldsymbol{R}_l)$ as a function of $\boldsymbol{R}_l$ near the equilibrium position $\boldsymbol{R}_l^0$ to within quadratic terms

$$E_\beta(\boldsymbol{R}_l) \cong E_\beta(\boldsymbol{R}_l^0) + \tfrac{1}{2}\sum_{l;ij}\left(\frac{\partial^2 E_\beta}{\partial R_l^i \partial R_l^j}\right)_{\boldsymbol{R}_l=\boldsymbol{R}_l^0} u_l^i u_l^j \tag{5.3.22}$$

$(i, j = x, y, z)$. Then we see that (5.3.21) is the Schrödinger equation for a crystal to a harmonic approximation (Chap. 2).

In this case

$$E_\mu = E_\beta(\boldsymbol{R}_l^0) + \sum_i (n_i + 1/2)\hbar\omega_i\ , \qquad n_i = 0, 1, 2, \dots\ , \tag{5.3.23}$$

where ω_i stands for phonon frequencies. Note that $\omega_i \propto M^{-1/2} \propto \kappa^2$. The representative values of the displacements u_l^i here will be on the order of $(\hbar/M\omega_i)^{1/2} \propto \kappa$, which is what justifies the use of (5.3.22) for $\kappa \ll 1$. Note also that the representative values of the nuclear momenta $\mathcal{P}_l \approx (M\hbar\omega_i)^{1/2} \propto \kappa^{-1}$. Then, since $A_{\beta\alpha}$ and $B_{\beta\alpha}$ are, according to (5.3.19), on the order of M^{-1}—i.e., on the order of κ^4—the perturbation operator $C_{\alpha\beta}$ is estimated as $\kappa^4 \cdot \kappa^{-1} = \kappa^3$; that is, it is less than the bare spectrum energy difference, which is on the order of $[E_\alpha(\boldsymbol{R}_l^0) - E_\beta(\boldsymbol{R}_l^0)] \propto \kappa^0$ for $\alpha \neq \beta$ and constitutes approximately $\hbar\omega_i \propto \kappa^2$ when $\alpha = \beta$, $\mu \neq \mu'$. This justifies the neglect of the terms with $C_{\beta\alpha}$ in (5.3.18) and thus the adiabatic approximation (5.3.20). Moreover, since

$$\psi_\beta(\boldsymbol{r}_i, \boldsymbol{R}_l) - \psi_\beta(\mathrm{r}_i, \boldsymbol{R}_l^0) \approx \sum_{jl}\frac{\partial\psi_\beta}{\partial R_l^j} u_l^j \propto \kappa\ ,$$

(5.3.20) may be replaced by a simpler expression

$$\psi(\boldsymbol{r}_i, \boldsymbol{R}_l) = \chi_\beta^{(\mu)}(\boldsymbol{R}_l)\psi_\beta(\boldsymbol{r}_i, \boldsymbol{R}_l^0)\ , \tag{5.3.24}$$

the Born–Oppenheimer approximation. Thus we have rigorously validated the underlying assumption of band theory that electronic states can be considered not only for fixed but also for equilibrium values of nuclear coordinates.

From the foregoing treatment it follows that the adiabatic approximation may be violated only when the electronic states are degenerate:

$$E_\alpha(\boldsymbol{R}_l^0) = E_\beta(\boldsymbol{R}_l^0) \tag{5.3.25}$$

occurs for some $\boldsymbol{R}_l^0$, since otherwise the smallness of the adiabatic parameter κ allows a rigorous validation of the approximation. Initially, metals would seem to be unlikely candidates for condition (5.3.25), since it is always possible there to construct a state of the entire system that is as close in energy to the ground state as desired. To do this, it suffices to promote the electron from the Bloch state immediately below the Fermi surface into the state immediately above the Fermi surface. However, the phase volume occupied by such "dangerous" electrons is very small (being a layer of thickness $\omega_D \propto \kappa^2$ near E_F). It can be shown [5.21], that this suffices to enable calculation of the phonon spectra and elastic properties of the metal in the adiabatic approximation. On the other hand, the nonadiabatic effects in the spectrum of electronic states in the immediate vicinity of the Fermi energy E_F may be considerable. Specifically, the electron mass renormalization on the Fermi surface due to the electron–phonon interaction is not small with respect to the adiabatic parameter and may be very appreciable [5.10].

Condition (5.3.25) may hold also for some ion or atom in the crystal if that ion or atom contains a partially filled shell. For instance, a fivefold degenerate d level in a cubic symmetry field, as may be shown using group-theory methods, splits into threefold and twofold degenerate states t_{2g} and e_g, i.e., the degeneracy is not completely removed. In an octahedral environment the t_{2g} level lies lower than e_g, and in a tetrahedral environment the situation is reversed (Fig. 5.4). Therefore, ions of configuration d^1 (Ti^{3+}, V^{4+}) in a tetrahedral environment or ions $d^9(Cu^{2+})$ in an octahedral environment, for example, have a doubly degenerate configuration. In the former case, the single d electron, and in the latter case the d hole, may be in either of the two e_g states.

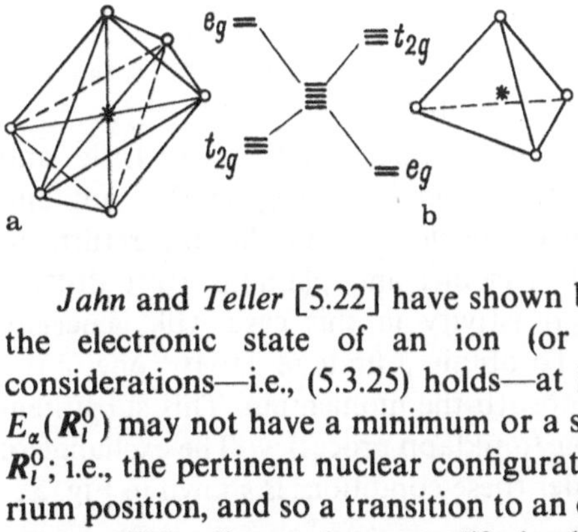

Fig. 5.4. Relative position of t_{2g} and e_g levels in the crystalline field of octahedral (**a**) and tetrahedral (**b**) environment

Jahn and *Teller* [5.22] have shown by use of group-theory methods that, if the electronic state of an ion (or atom) is degenerate by symmetry considerations—i.e., (5.3.25) holds—at least one of the "adiabatic potentials" $E_\alpha(\boldsymbol{R}_l^0)$ may not have a minimum or a stationary point at all at these values of $\boldsymbol{R}_l^0$; i.e., the pertinent nuclear configuration may not correspond to the equilibrium position, and so a transition to an equilibrium position may occur, i.e., the lattice will be distorted so as to lift the degeneracy of the electronic states (static Jahn–Teller effect). A detailed study of the problem [5.23] shows that several symmetry-equivalent equilibrium positions exist separated by potential barriers; i.e., there is a many-minimum potential such as that considered in Sect. 2.5. If the barriers are sufficiently high, the probability of quantum-mechanical tunneling of an ion from one minimum to another during the time of observation is negligibly small; the ion will remain in its shifted equilibrium position. If the tunneling probability is not negligibly small, the ion will sample all the new equilibrium positions (dynamic Jahn–Teller effect—the symmetry of the system in this case, on average, is not lowered). Under the conditions of the Jahn–Teller effect the adiabatic approximation is not applicable, and the system is described by what are known as vibron states [5.23], in which the electron motion cannot be separated from the motion of the atomic nuclei. The study of crystals that contain Jahn–Teller ions is a rapidly developing area of solid-state physics and chemistry [5.23, 24].

5.3.3 Temperature Dependence of the Electrical Conductivity in Metals

Now we employ the collision integral obtained and evaluate the contribution of the electron–phonon scattering to the electrical resistivity of metals. Rather than solve the kinetic equation, we wish to exploit simple semiquantitative considerations. Here we follow, in the main, [5.25]; see also [3.19, 5.20, 26].

We start noting that, in contrast to neutron scattering from the lattice, scattering from zero-point vibrations does not contribute to the resistivity. It can be readily verified that for $N_{\boldsymbol{q}} = 0$ the collision integral (5.3.11) is equal to zero. For this purpose we make the replacement $\boldsymbol{k} \rightleftarrows \boldsymbol{k}'$ in b_- and use the equality

$$w(\boldsymbol{k};\boldsymbol{k}',\boldsymbol{q}) = w(\boldsymbol{k}',\boldsymbol{q};\boldsymbol{k}) \ , \tag{5.3.26}$$

which is valid in the lowest order of perturbation theory [Ref. 1.12, Sect. 126].

Bloch's pioneering calculation of the electrical conductivity [5.27] was based on the assumption that the phonons are in equilibrium, i.e., that (2.1.48) could be used in (5.3.11) for $N_{\boldsymbol{q}}$. This certainly calls for justification. If no impurities and no other defects are present in the system and if we neglect the umklapp processes (below), the momentum imparted by electrons to phonons returns in part to the electronic subsystem, and electrons and phonons start drifting together at a constant velocity. The resistivity in this case will, generally speaking, be equal to zero (Sect. 2.4.3). To obtain a finite resistivity, one of the subsystems, being rigidly fixed, has to absorb the momentum. This subsystem can be impurities or the lattice as a whole (umklapp processes). The exchange of momentum between the subsystems under these conditions is shown in Fig. 5.5.

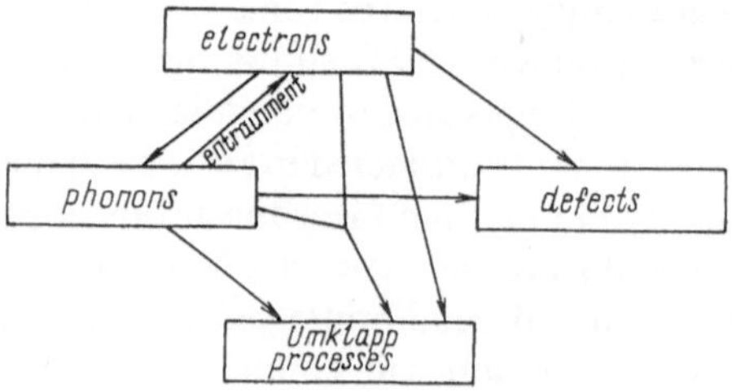

Fig. 5.5. Diagram of momentum transfer during scattering

The effect of the phonon subsystem not being in equilibrium is called entrainment. Entrainment is substantial if the impurity scattering is sufficiently weak and umklapp processes are unfavorable. Thus, if the Fermi surface does not touch the Brillouin-zone boundaries (Fig. 5.6a, b) (or the boundaries of the arbitrarily chosen reciprocal-space unit cell), the umklapp processes are allowed only for sufficiently large phonon wave vectors $\boldsymbol{q}$. Also, at low temperatures the number of phonons is exponentially small:

$$N_{\boldsymbol{q}} \approx \exp(-\hbar\omega_{\boldsymbol{q}}/k_{\mathrm{B}}T) \ . \tag{5.3.27}$$

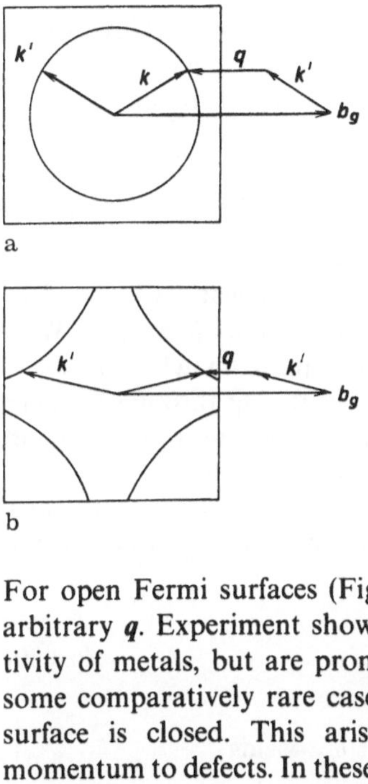

Fig. 5.6. Umklapp processes for closed (**a**) and open (**b**) Fermi surfaces

For open Fermi surfaces (Fig. 5.6b) umklapp processes are allowed with any arbitrary $\boldsymbol{q}$. Experiment shows that entrainment effects are slight in the resistivity of metals, but are pronounced in the thermal emf [5.20, 26]. There are some comparatively rare cases where entrainment is observed and the Fermi surface is closed. This arises from the fact that phonons transfer their momentum to defects. In these cases, the distance between the Fermi surface and the Brillouin-zone boundaries appears to be small and, therefore, the effect of the exponentially small $N_{\boldsymbol{q}}$ in (5.3.27) may only show up at very low temperatures. In the following, we assume the phonons to be in equilibrium.

Now we evaluate typical values of w at high and low temperatures ($T \gg \vartheta_{\mathrm{D}}$) and ($T \ll \vartheta_{\mathrm{D}}$). With $T \gg \vartheta_{\mathrm{D}}$, all phonons are excited and the mean transferred momentum $\boldsymbol{q}$ is on the order of the reciprocal-lattice constant. All representative electron energies in metals, including deformation potentials, are on the order of ζ_0. The quantity $\hbar/M\omega_{\boldsymbol{q}}$, involved in (5.3.10), is on the order of $\hbar^2/Mk_{\mathrm{B}}\vartheta_{\mathrm{D}}$. According to (5.1.67),

$$\frac{m}{M} \approx \left(\frac{S}{v_{\mathrm{F}}}\right)^2 \approx \left(\frac{k_{\mathrm{B}}\vartheta_{\mathrm{D}}}{\zeta_0}\right)^2 ,$$

whence

$$\frac{\hbar^2}{Mk_B\vartheta_D} \approx \frac{k_B\vartheta_D}{\zeta_0}\frac{1}{k_F^2} .$$

Since w has the dimensions of inverse time, we write

$$w \approx \frac{k_B\vartheta_D}{\hbar} . \tag{5.3.28}$$

When $T \gg \vartheta_D$, the energy of a phonon is small compared to the thermal smearing of the "Fermi step," so we can neglect this energy in the δ functions involved in (5.3.11) and regard the scattering as elastic. The collision frequency is proportional to w and the number of phonons $N_q \approx T/\vartheta_D$. For the specific electrical resistivity we write

$$\varrho = \sigma^{-1} = \frac{m}{Ne^2}\frac{1}{\bar{\tau}} , \quad \frac{1}{\bar{\tau}} \approx \frac{k_B T}{\hbar} , \quad T \gg \vartheta_D . \tag{5.3.29}$$

Thus, at high temperatures the resistivity increases linearly with T. This result was already obtained in the itinerant electrons model (Sect. 3.4). With $T \gg \vartheta_D$, the scattering may be regarded as elastic and approximately isotropic—phonons with all values of $\boldsymbol{q}$ are excited and the scattering through different angles is approximately equiprobable. Thus all the conclusions drawn in Sect. 3.6, in particular the Wiedemann–Franz relation, are valid.

Of greater interest is the low-temperature case. The characteristic quantities $\omega_{\boldsymbol{q}}$ are on the order of $k_B T/\hbar$, and the characteristic quantities $\boldsymbol{q}$ are on the order of

$$q \approx \frac{k_B T}{\hbar s} \approx \frac{k_B T}{\hbar v_F}\left(\frac{M}{m}\right)^{1/2} . \tag{5.3.30}$$

An estimate for w, similar to that made above for $\omega_{\boldsymbol{q}}$, yields

$$w \approx \frac{k_B T}{\hbar} , \quad T \ll \vartheta_D , \tag{5.3.31}$$

(we leave the calculations as an exercise: calculate carefully the powers of the temperature and dimensionless parameter m/M and exploit dimension considerations). The phonon energy is on the order of the magnitude of the thermal Fermi-surface smearing so that, in terms of energy relaxation, each collision is effective. Momentum relaxation is quite different. For $T \ll \vartheta_D$, (5.3.30) yields

$$\frac{q}{k_F} \approx \frac{k_B T}{\zeta_0}\left(\frac{M}{m}\right)^{1/2} \approx \frac{T}{\vartheta_D} \ll 1 . \tag{5.3.32}$$

Consequently, the momentum transferred is small and only small-angle scattering is involved. The scattering process can be represented as a slow diffusion of an electronic angular momentum, imparted by an electric field, on the Fermi surface. The scattering probability W is proportional to the number of collisions with phonons and proportional to w, which specifies the efficiency of collisions. The phonons fill a sphere in $\boldsymbol{q}$ space of radius (5.3.30). The law of conservation of momentum determines $\boldsymbol{k}'$ for each particular $\boldsymbol{k}$; the law of conservation of energy

$$\varepsilon_{\boldsymbol{k}} = \varepsilon_{\boldsymbol{k}+\boldsymbol{q}} \pm \hbar\omega_{\boldsymbol{q}} , \tag{5.3.33}$$

fixes the direction of the vector $\boldsymbol{q}$ at its prescribed value. For free electrons

$$\varepsilon_{\boldsymbol{k}+\boldsymbol{q}} - \varepsilon_{\boldsymbol{k}} \approx \frac{\hbar^2 \boldsymbol{k} \cdot \boldsymbol{q}}{m} .$$

By virtue of the condition $s \ll v_{\mathrm{F}}$, (5.3.33) yields $\boldsymbol{q} \cdot \boldsymbol{k} \approx 0$; i.e., the vector $\boldsymbol{q}$ should be perpendicular to the fixed vector $\boldsymbol{k}$. This condition singles out the cross section of the phonon-filled sphere in $\boldsymbol{q}$ space. The collision number is proportional to the cross-sectional area, i.e., $(T/\vartheta_{\mathrm{D}})^2$. Since W and w should have the same dimensions, we obtain

$$W \approx w \left(\frac{T}{\vartheta_{\mathrm{D}}} \right)^2 \approx \frac{k_{\mathrm{B}} T^3}{\hbar \vartheta_{\mathrm{D}}^2} . \tag{5.3.34}$$

The total momentum that the electronic subsystem receives from the field within a given time span is exactly compensated by its scattering of electrons in $\boldsymbol{k}$ space during diffusion. The particle flux per unit time is proportional to the product of the diffusion coefficient D by the gradient of the distribution function in momentum space; the mean value of the latter is proportional to the total current $\boldsymbol{j}$. On the other hand, the change in momentum due to the electric field per unit time is proportional to electric field intensity $\boldsymbol{E}$. Thus

$$\boldsymbol{E} \propto D\boldsymbol{j} \quad \text{or} \quad \varrho \propto D . \tag{5.3.35}$$

The proportionality factor characterizes not the scattering process but the electronic subsystem. Specifically, for a degenerate case, the proportionality factor cannot depend on temperature since the characteristic electron velocity is independent of temperature.

We now determine the T dependence of the diffusion coefficient. By analogy with the normal diffusion in gases, we write for the mean variation of the wave vector during the time Δt

$$\overline{(\Delta \boldsymbol{q})^2} \approx D \Delta t . \tag{5.3.36}$$

During the time W^{-1} the quantity q^2 varies approximately as the square of the quantity (5.3.30), whence

$$D \approx W\left(\frac{k_B T}{\hbar s}\right)^2 \propto \frac{T^5}{\vartheta_D^4} . \tag{5.3.37}$$

In view of (5.3.36, 37) and by dimension considerations, we have for ϱ

$$\varrho \approx \frac{m}{Ne^2} \frac{k_B T}{\hbar}\left(\frac{T}{\vartheta_D}\right)^4 . \tag{5.3.38}$$

Thus, the resistivity of metals at low temperatures, which is due to electron–phonon scattering, decreases with T as T^5. For energy relaxation processes (thermal conductivity) the role of the relaxation time is played simply by W^{-1}, but not $W^{-1}(\vartheta_D/T)^2$, as with momentum relaxation. The Wiedemann–Franz law should therefore be violated at low temperatures, certainly provided that the number of defects is small and electron–phonon scattering still plays a dominant role.

The arguments presented in Sect. 4.10 show that the electron–electron scattering makes a contribution to the relaxation frequency. This contribution is on the order of

$$\bar{\tau}_{ee}^{-1} \approx (k_B T)^2/\hbar\zeta_0 . \tag{5.3.39}$$

(but only provided that the Fermi surface is open and umklapp processes are allowed). Comparison of (5.3.38) with (5.3.39) shows that the contribution (5.3.39) becomes equal to the momentum relaxation frequency only with phonon scattering at low temperatures:

$$T \lesssim \vartheta_D\left(\frac{k_B\vartheta_D}{\zeta_0}\right)^{1/3} .$$

In fact, however, the contribution of the electron–electron scattering to $\bar{\tau}^{-1}$ is much less than that predicted by (5.3.39), and apparently contains small numerical factors. At any rate, the contribution $\propto T^2$ to the resistivity of metals is difficult to single out experimentally because the interpretation of results is often ambiguous.

5.3.4 Polarons

Now we consider another effect of electron–phonon coupling: the current carrier spectrum renormalization. This effect is particularly significant in ionic crystals, oxides, and chalcogenides of d and f elements, and in some other semiconductors. The electron interacts preferentially with longitudinal optical lattice vibration modes, because they are accompanied by a dipole moment and considerable Coulomb forces. (In Sect. 5.3.3 we could disregard optical phonons since the number of these phonons at low temperatures is exponentially small; here, as will be seen, virtual phonons are essential and the interaction with them is very important even at $T=0$ K). Our treatment here is close to that in [5.28–30].

We neglect, for simplicity, the dispersion of the frequency of optical phonons and write their Hamiltonian in the form

$$\hat{\mathscr{H}}_{\rm ph}=\hbar\omega_0\sum_{\boldsymbol{q}} b_{\boldsymbol{q}}^+ b_{\boldsymbol{q}} \ . \tag{5.3.40}$$

The polarization vector $\boldsymbol{P}$ is proportional to the displacements of the atomic sublattices and may be represented as

$$\boldsymbol{P}=eF\sum_{\boldsymbol{q}} \boldsymbol{e}_{\boldsymbol{q}}[b_{\boldsymbol{q}}\exp(\mathrm{i}\boldsymbol{q}\cdot\boldsymbol{r})+b_{\boldsymbol{q}}^+\exp(-\mathrm{i}\boldsymbol{q}\cdot\boldsymbol{r})] \ , \tag{5.3.41}$$

where $\boldsymbol{e}_{\boldsymbol{q}}$ is the polarization vector of a phonon, F is a real-valued constant (due to the hermiticity of the operator $\boldsymbol{P}$) constant, which will be defined in what follows. As is known from electrodynamics, the polarization charge density is equal to

$$\varrho=-\operatorname{div}\boldsymbol{P}=\mathrm{i}Fe\sum_{\boldsymbol{q}} \boldsymbol{e}_{\boldsymbol{q}}\boldsymbol{q}[b_{\boldsymbol{q}}\exp(\mathrm{i}\boldsymbol{q}\cdot\boldsymbol{r})-b_{\boldsymbol{q}}^+\exp(-\mathrm{i}\boldsymbol{q}\cdot\boldsymbol{r})] \ . \tag{5.3.42}$$

Equation (5.3.42) shows that the contribution to the charge-density perturbation is made only by the longitudinal mode, for which $\boldsymbol{e}_{\boldsymbol{q}}\parallel\boldsymbol{q}$ (the other two optical modes in a lattice with two atoms per unit cell are transverse modes, $\boldsymbol{e}_{\boldsymbol{q}}\perp\boldsymbol{q}$).

The "lattice" charge-density operator in this case is

$$\varrho = e\sum_{\boldsymbol{q}}[v_{\boldsymbol{q}}\exp(\mathrm{i}\boldsymbol{q}\cdot\boldsymbol{r})+v_{-\boldsymbol{q}}^+\exp(-\mathrm{i}\boldsymbol{q}\cdot\boldsymbol{r})], \qquad v_{\boldsymbol{q}}=\mathrm{i}F|\boldsymbol{q}|b_{\boldsymbol{q}} \ . \tag{5.3.43}$$

Then, similar to (5.1.69), the potential energy of an electron at point $\boldsymbol{r}$ is the electrostatic potential multiplied by e, and is equal to

$$\begin{aligned} V(\boldsymbol{r})&=4\pi e^2\sum_{\boldsymbol{q}}\frac{1}{q^2}[v_{\boldsymbol{q}}\exp(\mathrm{i}\boldsymbol{q}\cdot\boldsymbol{r})+v_{-\boldsymbol{q}}^+\exp(-\mathrm{i}\boldsymbol{q}\cdot\boldsymbol{r})] \\ &=4\pi\mathrm{i}Fe^2\sum_{\boldsymbol{q}}\frac{1}{|\boldsymbol{q}|}[b_{\boldsymbol{q}}\exp(\mathrm{i}\boldsymbol{q}\cdot\boldsymbol{r})-b_{\boldsymbol{q}}^+\exp(-\mathrm{i}\boldsymbol{q}\cdot\boldsymbol{r})] \ . \end{aligned} \tag{5.3.44}$$

The interaction of electrons with optical phonons, in particular, gives rise to an effective attraction between electrons. We should consider the case of electrons spaced a large distance apart, when the effect of interest can be treated by perturbation theory. The first electron carries (virtually) the system over into the excited state $|\boldsymbol{q}\rangle$, which corresponds to the creation of an optical phonon. The optical phonon then is absorbed (virtually) by the second electron. Mathematically, this is expressed by the familiar formula for the ground-state energy in second-order perturbation theory:

$$\begin{aligned}\Delta E(\boldsymbol{r}_1,\boldsymbol{r}_2)=&-\frac{1}{\hbar\omega_0}\sum_{\boldsymbol{q}}[\langle 0|V(\boldsymbol{r}_1)|\boldsymbol{q}\rangle\langle\boldsymbol{q}|V(\boldsymbol{r}_2)|0\rangle\\&+\langle 0|V(\boldsymbol{r}_2)|\boldsymbol{q}\rangle\langle\boldsymbol{q}|V(\boldsymbol{r}_1)|0\rangle]\\=&-\frac{2}{\hbar\omega_0}\langle 0|V(\boldsymbol{r}_1)V(\boldsymbol{r}_2)|0\rangle\ ,\end{aligned}\tag{5.3.45}$$

where $\hbar\omega_0$ is the excitation energy and we have made use of the completeness condition

$$\sum_{\boldsymbol{q}}\langle\varphi|\boldsymbol{q}\rangle\langle\boldsymbol{q}|\psi\rangle=\langle\varphi|\psi\rangle\ .\tag{5.3.46}$$

We now recall that, when averaged over the ground state,

$$\langle 0|b_{\boldsymbol{q}}b^{+}_{\boldsymbol{q}'}|0\rangle=\delta_{\boldsymbol{q}\boldsymbol{q}'}\ ,\qquad\langle 0|b^{+}_{\boldsymbol{q}}b_{\boldsymbol{q}'}|0\rangle=0\ .\tag{5.3.47}$$

Substituting (5.3.44) into (5.3.45), a little manipulation yields

$$\Delta E(\boldsymbol{r}_1,\boldsymbol{r}_2)=-\frac{8\pi e^2F^2}{\hbar\omega_0}\,\frac{e^2}{|\boldsymbol{r}_1-\boldsymbol{r}_2|}\ .\tag{5.3.48}$$

We wish to ascertain to which case this expression corresponds. In the absence of lattice polarization, the binding energy of two electrons is equal to $e^2/\varepsilon_\infty|\boldsymbol{r}_1-\boldsymbol{r}_2|$, with ε_∞ being the permittivity at high frequencies, when the lattice does not have enough time to respond to a perturbation. But in reality the binding energy has the form $e^2/\varepsilon_0|\boldsymbol{r}_1-\boldsymbol{r}_2|$, where ε_0 is the static permittivity. This difference represents an attraction between electrons (5.3.48):

$$\frac{1}{\varepsilon_0}=\frac{1}{\varepsilon_\infty}-\frac{8\pi e^2F^2}{\hbar\omega_0}\ .\tag{5.3.49}$$

The formula allows the coupling constant to be expressed in terms of observables.

We now calculate the electron-spectrum renormalization for a weak interaction in second-order perturbation theory:

$$\Delta E_{\boldsymbol{k}} = -\sum_{\boldsymbol{q}} \frac{|\langle \boldsymbol{k}, 0|V|\boldsymbol{k}-\boldsymbol{q}, \boldsymbol{q}\rangle|^2}{E_{\boldsymbol{k}-\boldsymbol{q}} - E_{\boldsymbol{k}} + \hbar\omega_0} \ , \tag{5.3.50}$$

with $E_{\boldsymbol{k}}$ being the unperturbed energy of an electron, $|\boldsymbol{k}, 0\rangle$ the state with an electron with wave vector $\boldsymbol{k}$ without phonons, and $|\boldsymbol{k}-\boldsymbol{q}, \boldsymbol{q}\rangle$ the state with a phonon $|\boldsymbol{q}\rangle$ and, according to the momentum conservation principle, with an electron in the state $|\boldsymbol{k}-\boldsymbol{q}\rangle$. The matrix element in (5.3.50) is, according to (5.3.44), equal to $4\pi \mathrm{i} F e^2/|\boldsymbol{q}|$. Therefore

$$\begin{aligned} \Delta E_{\boldsymbol{k}} &= -(4\pi F e^2)^2 \int \frac{d\boldsymbol{q} V_0}{(2\pi)^3} \frac{1}{q^2} \frac{1}{\dfrac{\hbar^2}{2m}[(\boldsymbol{k}-\boldsymbol{q})^2 - \boldsymbol{k}^2] + \hbar\omega_0} \\ &= -\frac{8mF^2e^4V_0}{\pi\hbar^2} \int_0^\infty dq \int_{-1}^{1} dx \frac{1}{2kqx + q^2 + 2m\omega_0/\hbar} \ . \end{aligned} \tag{5.3.51}$$

Calculating the integral, the result obtained for slow electrons with $k \ll (m\omega_0/\hbar)^{1/2}$, allowing for (5.3.51) is

$$\begin{aligned} \tilde{E}_{\boldsymbol{k}} &= E_{\boldsymbol{k}} + \Delta E_{\boldsymbol{k}} = \frac{\hbar^2 \boldsymbol{k}^2}{2m^*} - \alpha\hbar\omega_0 \ , \\ m^* &= m(1-\alpha/6)^{-1} \approx m(1+\alpha/6) \ , \end{aligned} \tag{5.3.52}$$

where

$$\alpha = \frac{e^2}{2\hbar\omega_0} \left(\frac{2m\omega_0}{\hbar}\right)^{1/2} \left(\frac{1}{\varepsilon_\infty} - \frac{1}{\varepsilon_0}\right) . \tag{5.3.53}$$

is a dimensionless constant used in the theory of polarons, and we have set $\alpha \lesssim 1$ (the easiest way of calculating this integral is by a second-order expansion in $\boldsymbol{k}$). Thus, the electron–phonon interaction leads to an increase in current carrier effective mass [similar to the electron–electron coupling (5.2.54)].

We have considered what is known as the weak-coupling polaron [5.31]. When $1 \lesssim \alpha \lesssim 10$, we talk of intermediate coupling, and when $\alpha \gtrsim 10$, of a strong coupling. The strong coupling polaron is the self-trapped state of the lattice electron. The existence of such states was propounded by *Landau* [5.32] and *Frenkel* [5.33], and the strong and intermediate polaron coupling theory was developed by *Pekar*, *Feynman*, *Bogolyubov*, *Tyablikov*, and many others. For a detailed treatment of the theory, we refer the reader to [5.29, 30, 34].

We consider briefly *Pekar*'s approach [5.34], applicable in the strong coupling case. The electron brings about a lattice displacement, thereby causing the total energy of the system to decrease. The total energy of the system consists of the following components: the kinetic energy of the electron

$$T[\psi]=\frac{\hbar^2}{2m}\int d\boldsymbol{r}|\nabla\psi(\boldsymbol{r})|^2 \ , \tag{5.3.54}$$

with $\psi(\boldsymbol{r})$ being the wave function; the electrostatic energy of the distorted lattice [2.17]

$$U[\boldsymbol{P}]=2\pi\varepsilon^*\int d\boldsymbol{r}\boldsymbol{P}^2(\boldsymbol{r}) \ , \tag{5.3.55}$$

with ε^* being the effective dielectric permittivity

$$\frac{1}{\varepsilon^*}=\frac{1}{\varepsilon_\infty}-\frac{1}{\varepsilon_0} \ , \tag{5.3.56}$$

which is involved in the expression for F (5.3.49) and, consequently, for $\boldsymbol{P}$ (5.3.41); and the energy of the interaction of the lattice polarization with the electron-initiated electric induction $\boldsymbol{D}[\psi,\boldsymbol{r}]$:

$$V=-\int d\boldsymbol{r}\boldsymbol{P}(\boldsymbol{r})\boldsymbol{D}[\psi,\boldsymbol{r}] \ , \tag{5.3.57}$$

$$\boldsymbol{D}[\psi,\boldsymbol{r}]=e\int\frac{d\boldsymbol{r}'(\boldsymbol{r}-\boldsymbol{r}')}{|\boldsymbol{r}-\boldsymbol{r}'|^3}|\psi(\boldsymbol{r}')|^2 \ . \tag{5.3.58}$$

The Hamiltonian of the problem has the form

$$\mathscr{H}=T+U+V \ . \tag{5.3.59}$$

and is valid in the adiabatic approximation (the displacements are assumed to be static). This is true if $\hbar\omega_0$ is much less than the characteristic electron energies.

Minimization of (5.3.59) with respect to $\boldsymbol{P}$ for $\psi=$const yields

$$\boldsymbol{P}(\boldsymbol{r})=\frac{1}{4\pi\varepsilon^*}\boldsymbol{D}[\psi,\boldsymbol{r}] \ . \tag{5.3.60}$$

Substituting (5.3.60) into (5.3.59), we obtain

$$\mathscr{H}[\psi]=\frac{\hbar^2}{2m}\int d\boldsymbol{r}|\nabla\psi|^2-\frac{1}{8\pi\varepsilon^*}\int d\boldsymbol{r}\boldsymbol{D}^2[\psi,\boldsymbol{r}] \ . \tag{5.3.61}$$

If we denote by l the characteristic size of the region which "traps" the electron (in reality, this region may move over the crystal as a whole), the first term is on the order of $\hbar^2/ml^2$, and the second term on the order of $e^2/\varepsilon_\infty l$. The first term may be evaluated from the uncertainty relation: if the uncertainty of the coordinate of the electron in the well is $\approx l$, then the momentum is $\approx \hbar/l$ and the kinetic energy is $\approx \hbar^2/ml^2$. The second term is the electrostatic energy of the electron in the region of the excessive positive charge of size l. Minimization with respect to l gives

$$\frac{\partial}{\partial l}\left(\frac{\hbar^2}{ml^2}-\frac{e^2}{\varepsilon^* l}\right)\Bigg|_{l=l_0}=0 \ , \tag{5.3.62}$$

$$l_0 \approx \frac{\hbar^2\varepsilon^*}{me^2} \ .$$

Then,

$$E(l_0) \approx -\text{const}\frac{me^4}{\hbar^2\varepsilon^{*2}} \approx -\text{const}\,\hbar\omega_0\alpha^2 \ . \tag{5.3.63}$$

Pekar has found the ground-state energy of the Hamiltonian (5.3.61) by the direct variational method (Sect. 4.6.1), using a trial function of the form

$$\psi(\boldsymbol{r})=(\beta_1+\beta_2 r+\beta_3 r^2)\mathrm{e}^{-\beta_4 r} \ . \tag{5.3.64}$$

The result is as follows:

$$E_0=-0.1088\,\alpha^2\hbar\omega_0 \ . \tag{5.3.65}$$

Considering a moving polaron—i.e., assuming ψ, $\boldsymbol{D}$, and $\boldsymbol{P}$ to be functions of $\boldsymbol{r}-\boldsymbol{v}t$—and adding the kinetic energy of ions, we can determine its effective mass. The minimum energy at a fixed small velocity $\boldsymbol{v}$ has the form

$$E=E_0+\frac{m^*v^2}{2} \ , \tag{5.3.66}$$

where it is natural to regard m^* as the polaron effective mass.

The effective mass obtained by Landau and Pekar [5.35] at $\alpha \approx 10$ (the case of NaCl) is hundreds of times larger than the free-electron mass. Thus, the strong coupling polaron is a very slow, heavy quasiparticle. On account of this, the transfer processes in which it takes part may change in character. The point is that the usual kinetic equation is applicable when the mean free time $\bar{\tau}$ is much larger than the impact time. For very heavy quasiparticles and strong coupling, the mean free time is on the order of the impact time, and the approach based on the kinetic equation ceases to be applicable, or at least needs modification. Polaron transfer is a complicated problem, far from being understood perfectly. It is solved in different ways, depending on whether the polaron radius l_0 is large or small compared with the lattice period d. All of our calculations so far have related to the case of large-radius polarons, $l_0 \gg d$. The kinetic properties of large-radius intermediate and strong coupling polarons seem to be the most complicated.

We now consider briefly the theory of small-radius polarons where $l_0 \leqslant d$ [5.30]. The Hamiltonian of the system may be represented as

$$\hat{\mathscr{H}} = \hat{\mathscr{H}}_e + \hat{\mathscr{H}}_{ph} + \hat{V} , \tag{5.3.67}$$

where $\hat{\mathscr{H}}_{ph}$ is the Hamiltonian of a phonon (5.3.40); $\hat{\mathscr{H}}_e$ is the Hamiltonian of a band electron, for which we adopt the tight-binding approximation

$$(\hat{\mathscr{H}}_e)_{mn} = \begin{cases} \beta & \text{with } m, n \text{ being nearest neighbors} \\ 0 & \text{otherwise, } m \text{ and } n \text{ are site numbers} \end{cases} \tag{5.3.68}$$

and $\hat{V}$ is the Hamiltonian of the electron–phonon interaction (5.3.44), written in the site representation with allowance for diagonal matrix elements only (this is justified for small-radius polarons):

$$V_{nn} = \sum_q \frac{4\pi i F e^2}{|q|} \langle n|e^{iq\cdot r}|n\rangle (b_q^+ - b_{-q}) . \tag{5.3.69}$$

But in the tight-binding approximation (Sect. 4.6.3) we have

$$\langle n|e^{iq\cdot r}|n\rangle = \int_\Omega dr |\chi(r-R_n)|^2 e^{iq\cdot r} = e^{iq\cdot R_n} \int_\Omega dr |\chi(r)|^2 e^{iq\cdot r} . \tag{5.3.70}$$

Substituting (5.3.70) into (5.3.69), we write V_{nn} in the form

$$V_{nn} = i \sum_q M_q e^{iq\cdot R_n} (b_q^+ - b_{-q}) ,$$

$$M_q = \frac{4\pi F e^2}{|q|} \int_\Omega dr |\chi(r)|^2 e^{iq\cdot r} . \tag{5.3.71}$$

We solve the problem by the canonical transformation method; i.e., we find a unitary operator $\hat{U}$ such that the transformed Hamiltonian

$$\hat{\mathscr{H}}' = \hat{U} \hat{\mathscr{H}} \hat{U}^+ , \tag{5.3.72}$$

becomes as simple in form as possible. In particular, by use of canonical transformation we try to eliminate the electron–phonon interaction term $\hat{V}$. The $\hat{U}$ will be sought in the form

$$\hat{U}=\exp\left[\mathrm{i}\sum_{q}\hat{A}(q)(b_q^+ + b_{-q})\right], \tag{5.3.73}$$

where $\hat{A}(q)$ is a matrix that depends on electronic variables, with $A_{nm}(q)=0 (n\neq m)$; $A_{nn}(q)$ has to be found from the maximum simplicity condition for $\mathscr{H}'$. We proceed to the calculation

$$\hat{V}'=\hat{U}\hat{V}\hat{U}^+ . \tag{5.3.74}$$

Obviously, the matrix $\hat{U}$ will be unitary if $A_{nn}(q)$ is real valued. Then, since the matrix $\hat{A}$ is diagonal, we have

$$U_{nm}=0 , \quad n\neq m$$
$$U_{nn}=\exp\left[\mathrm{i}\sum_{q}A_{nn}(q)(b_q^+ + b_{-q})\right] . \tag{5.3.75}$$

Therefore $V'_{nm}=0$ at $n\neq m$, and

$$V'_{nn}=\mathrm{i}\sum_{q}M_q\exp(\mathrm{i}q\cdot R_n)\exp\left[\mathrm{i}\sum_{p}A_{nn}(p)(b_p^+ + b_{-p})\right]$$
$$\cdot(b_q^+ - b_{-q})\exp\left[-\mathrm{i}\sum_{p}A_{nn}(p)(b_p^+ + b_{-p})\right] . \tag{5.3.76}$$

Vary the operator

$$B_q^+=\exp\left[\mathrm{i}\sum_{p}A_{nn}(p)(b_p^+ + b_{-p})\right]b_q^+\exp\left[-\mathrm{i}\sum_{p}A_{nn}(p)(b_p^+ + b_{-p})\right], \tag{5.3.77}$$

involved in (5.3.76), with respect to $A_{nn}(p)$ (analogous computations were performed in Sect. 2.7):

$$\frac{\delta B_q^+}{\delta A_{nn}(p)}=\mathrm{i}\exp\left[\mathrm{i}\sum_{p}A_{nn}(p)(b_p^+ + b_{-p})\right](b_p^+ + b_{-p})b_q^+$$
$$\cdot\exp\left[-\mathrm{i}\sum_{p}A_{nn}(p)(b_p^+ + b_{-p})\right]-\mathrm{i}\exp\left[\mathrm{i}\sum_{p}A_{nn}(p)\right.$$
$$\left.\cdot(b_p^+ + b_{-p})\right]b_q^+(b_p^+ + b_{-p})\exp\left[-\mathrm{i}\sum_{p}A_{nn}(p)(b_p^+ + b_{-p})\right] \tag{5.3.78}$$
$$=\mathrm{i}U_{nn}[b_p^+ + b_{-p}, b_q^+]_- U_{nn}^+=\mathrm{i}\delta_{-q,p} ,$$

using the commutation relations for the operators b_q^+ and b_q (Sect. 2.8). Since $B_q^+ = b_q^+$ for $A_{nn}(\boldsymbol{p}) = 0$, (5.3.78) yields

$$B_q^+ = b_q^+ + \mathrm{i}A_{nn}(-\boldsymbol{q}) \ . \tag{5.3.79}$$

Similarly,

$$\exp\left[\mathrm{i}\sum_p A_{nn}(\boldsymbol{p})(b_p^+ + b_{-p})\right] b_{-q} \exp\left[-\mathrm{i}\sum_p A_{nn}(\boldsymbol{p})(b_p^+ + b_{-p})\right] = b_{-q} - \mathrm{i}A_{nn}(\boldsymbol{q}) \ . \tag{5.3.80}$$

Substituting (5.3.79, 80) into (5.3.76) yields

$$V'_{nn} = V_{nn} - 2\sum_q M_q A_{nn}(-\boldsymbol{q}) \exp(\mathrm{i}\boldsymbol{q} \cdot \boldsymbol{R}_n) \ . \tag{5.3.81}$$

The calculations of the quantity

$$(\mathscr{H}'_{\mathrm{ph}})_{nn} = U_{nn} \mathscr{H}_{\mathrm{ph}} U_{nn}^+ \ . \tag{5.3.82}$$

are quite analogous, and are left as an exercise for the reader. As a result, we find

$$(\hat{\mathscr{H}}'_{\mathrm{ph}})_{nn} = \hat{\mathscr{H}}_{\mathrm{ph}} + \mathrm{i}\hbar\omega_0 \sum_q A_{nn}(\boldsymbol{q})(b_{-q} - b_q^+) + \sum_q A_{nn}(-\boldsymbol{q}) M_q \exp(\mathrm{i}\boldsymbol{q} \cdot \boldsymbol{R}_n) \ . \tag{5.3.83}$$

Then, if we choose

$$A_{nn}(-\boldsymbol{q}) = \frac{M_q}{\hbar\omega_0} \exp(-\mathrm{i}\boldsymbol{q} \cdot \boldsymbol{R}_n) \ , \tag{5.3.84}$$

the term in the Hamiltonian which is linear with respect to the phonon operators cancels out, since, from (5.3.71, 81, 83), we have

$$(\hat{\mathscr{H}}'_{\mathrm{ph}})_{nn} + \hat{V}_{nn} = \hat{\mathscr{H}}_{\mathrm{ph}} - \frac{1}{\hbar\omega_0} \sum_q M_q^2 \ . \tag{5.3.85}$$

The result is

$$
(\hat{\mathscr{H}}')_{nm} = \begin{cases} \beta \exp\left[\mathrm{i} \sum_{q} \frac{M_q}{\hbar\omega_0} \mathrm{e}^{\mathrm{i} q \cdot R_n} (b_q^+ + b_{-q}) \right] \cdot \\ \quad \cdot \exp\left[-\mathrm{i} \sum_{q} \frac{M_q}{\hbar\omega_0} \mathrm{e}^{\mathrm{i} q \cdot R_n} (b_q^+ + b_{-q}) \right] , & n, m \text{ are nearest neighbours,} \\ \hat{\mathscr{H}}_{\mathrm{ph}} - \Delta , & n = m, \\ 0 & \text{otherwise} \end{cases}
$$

5.3.86)

where

$$
\Delta = \frac{1}{\hbar\omega_0} \sum_q M_q^2 . \tag{5.3.87}
$$

Thus, the canonical transformation leads us to the result that the effect of the electron–phonon interaction boils down, first, to a decrease in electron energy at each site by an amount Δ (known as the polaron shift) and, second, to the occurrence of a complicated dependence of the transfer integral β on phonon operators. To understand the consequences of this, we average $\hat{\mathscr{H}}'_{nm}$ over the phonon subsystem for $\boldsymbol{R}_n - \boldsymbol{R}_m = \boldsymbol{\delta}$ (the nearest-neighbor vector). This quantity is quite similar to the one used in Sect. 2.7.3, where we calculated the Debye–Waller factor (2.7.53, 55). The two quantities differ from each other only in the prefactors of the phonon operators. Without repeating all those calculations, we immediately put down the result (2.7.55, 56):

$$
\tilde{\beta} = \langle \hat{\mathscr{H}}'_{nm} \rangle = \beta \exp(-S_T) , \tag{5.3.88}
$$

$$
\begin{aligned} S_T &= 2 \sum_q \frac{M_q^2}{(\hbar\omega_0)^2} [1 - \cos \boldsymbol{q} \cdot \boldsymbol{\delta}] [N(\omega_0) + \tfrac{1}{2}] \\ &= \frac{1}{(\hbar\omega_0)^2} \operatorname{cotanh}\left(\frac{\hbar\omega_0}{2k_{\mathrm{B}}T} \right) \sum_q M_q^2 (1 - \cos \boldsymbol{q} \cdot \boldsymbol{\delta}) . \end{aligned} \tag{5.3.89}
$$

Taking account of (5.3.89), we find the order of magnitude of S_T to be

$$
S_T \approx \frac{\Delta}{\hbar\omega_0} \operatorname{cotanh}\left(\frac{\hbar\omega_0}{2k_{\mathrm{B}}T} \right) \approx \begin{cases} \dfrac{\Delta}{\hbar\omega_0} , & k_{\mathrm{B}}T \ll \hbar\omega_0 \\ 2k_{\mathrm{B}}T \left(\dfrac{\Delta}{\hbar\omega_0} \right)^2 , & k_{\mathrm{B}}T \gg \hbar\omega_0 . \end{cases} \tag{5.3.90}
$$

Thus, with $S_T \gg 1$, the transfer integral decreases exponentially and, consequently, the conduction band narrows (5.3.88). In addition, the band decreases in width as the temperature is increased; i.e., the effective mass increases. In replacing $\hat{\mathscr{H}}'_{nm}$ by its mean value we have disregarded the fluctuation scattering processes in the phonon subsystem. As may be shown [5.30], this scattering is rather slight at very low temperatures. However, with increase in T, on the one hand, these scattering processes themselves build up and, on the other, the spin-polaron effective mass $m^* \propto |\beta|^{-1}$. As a result, the narrow-band motion at some temperature gives place to thermally activated jumps having the character of a random diffusion process.

5.3.5 The Cooper Phenomenon

We saw in Sect. 5.3.4 that the electron–phonon interaction leads to an effective attraction between electrons. This attraction brings about a drastic readjustment of the ground state of the electronic subsystem: a transition to a superconducting state. Our objective here is to consider a problem whose solution is indicative of this instability [5.36].

We explore the interaction of two electrons. The electron–electron interaction potential has the form $V(\boldsymbol{r}_1 - \boldsymbol{r}_2)$; both electrons are in Fermi-surface states. The Schrödinger equation for an electron pair has the form

$$-\frac{\hbar^2}{2m}(\Delta_1 + \Delta_2)\psi(\boldsymbol{r}_1, \boldsymbol{r}_2) + V(\boldsymbol{r}_1 - \boldsymbol{r}_2)\psi(\boldsymbol{r}_1, \boldsymbol{r}_2) = \left(E + \frac{\hbar^2 k_F^2}{m}\right)\psi(\boldsymbol{r}_1, \boldsymbol{r}_2) , \quad (5.3.91)$$

where the energy E is counted from the doubled Fermi energy. Expand $\psi(\boldsymbol{r}_1, \boldsymbol{r}_2)$, $V(\boldsymbol{r}_1 - \boldsymbol{r}_2)$ in a double Fourier series

$$\psi(\boldsymbol{r}_1, \boldsymbol{r}_2) = \sum_{\boldsymbol{k}_1 \boldsymbol{k}_2} \exp[\mathrm{i}(\boldsymbol{k}_1 \cdot \boldsymbol{r}_1 + \boldsymbol{k}_2 \cdot \boldsymbol{r}_2)]\psi_{\boldsymbol{k}_1 \boldsymbol{k}_2} . \quad (5.3.92)$$

Transform the second sum in (5.3.91):

$$\sum_{\boldsymbol{q}\boldsymbol{k}_1\boldsymbol{k}_2} V_{\boldsymbol{q}}\psi_{\boldsymbol{k}_1\boldsymbol{k}_2}\exp[\mathrm{i}(\boldsymbol{k}_1 + \boldsymbol{q})\cdot\boldsymbol{r}_1 + \mathrm{i}(\boldsymbol{k}_2 - \boldsymbol{q})\cdot\boldsymbol{r}_2] = \sum_{\boldsymbol{q}\boldsymbol{k}'_1\boldsymbol{k}'_2} V_{\boldsymbol{q}}\psi_{\boldsymbol{k}'_1 - \boldsymbol{q}, \boldsymbol{k}'_2 + \boldsymbol{q}}\exp[\mathrm{i}(\boldsymbol{k}'_1 \cdot \boldsymbol{r}_1 + \boldsymbol{k}'_2 \cdot \boldsymbol{r}_2)] . \quad (5.3.93)$$

Equating the coefficients of equal exponents (possible because they are orthonormal), we obtain

$$\frac{\hbar^2}{2m}[(k_1^2 - k_F^2) + (k_2^2 - k_F^2)]\psi_{k_1 k_2} + \frac{1}{N}\sum_q V_q \psi_{k_1 - q, k_2 + q} = E\psi_{k_1 k_2} . \qquad (5.3.94)$$

Introduce new variables—the center-of-mass quasimomentum $\boldsymbol{\kappa}$ and the quasimomentum with respect to the motion $\boldsymbol{K}$—using the formulas

$$\begin{aligned} \boldsymbol{\kappa} &= \boldsymbol{k}_1 + \boldsymbol{k}_2 , \quad \boldsymbol{K} = \tfrac{1}{2}(\boldsymbol{k}_1 - \boldsymbol{k}_2) , \\ \boldsymbol{k}_1 &= \frac{\boldsymbol{\kappa}}{2} + \boldsymbol{K} , \quad \boldsymbol{k}_2 = \frac{\boldsymbol{\kappa}}{2} - \boldsymbol{K} . \end{aligned} \qquad (5.3.95)$$

Equation (5.3.94) involves only ψ with one value of $\boldsymbol{\kappa}$. We choose the value $\boldsymbol{\kappa} = 0$. This corresponds to a pair at rest; a moving pair can be considered simply by passing to a moving frame of reference. Denoting

$$\varepsilon_K = \frac{\hbar^2}{2m}(K^2 - k_F^2) , \quad \varphi_K = \psi_{K,-K} , \qquad (5.3.96)$$

We obtain from (5.3.94)

$$2\varepsilon_K \varphi_K + \frac{1}{N}\sum_q V_q \varphi_{K-q} = E\varphi_K , \qquad (5.3.97)$$

analogous to the equation considered in Sect. 4.9.1.

To solve this equation, we need to analyze the form of V_q. Before we do so, one more circumstance must be noted. Up to this point nothing has been said about the orientation of the spins of the electron pair selected. The pair should have minimum energy, and for the two-electron system this is achieved with the spins being antiparallel, due to the antisymmetry of the electron wave function and due to some general theorems of quantum mechanics. It follows from the fact that antiparallel spins correspond to an antisymmetric spin-wave function and a symmetric coordinate wave function. The latter does not vanish anywhere and can therefore describe the ground state [Ref. 1.12, Sects. 62, 63]. Thus, we consider a pair of electrons with antiparallel spins and with total momentum equal to zero.

Now we consider the potential. The electron–electron attraction is effective only in a layer of thickness $\hbar\omega_D$ near the Fermi surface, ω_D being the Debye frequency.

Using perturbation theory, we see that the magnitudes of the corresponding energy contributions of virtual excited states fall off rapidly with increasing excitation energy. The energy $\hbar\omega_D$ is the minimum characteristic binding energy and, therefore, it determines the variation scale of V_q.

Further, we assume that scattering occurs only in states that lie above the Fermi surface, since scattering below the Fermi surface is prohibited by the Pauli exclusion principle. For this to be taken into account, we must introduce the dependence of V_q in (5.3.97) to include not only $\boldsymbol{q}$ but also $\boldsymbol{K}$. By doing this it certainly ceases to be a Fourier transform of the pairwise interaction potential, but its physical significance is retained completely. A more rigorous treatment confirms this simple, although not altogether correct, mathematical approach. Thus, we choose the phonon contribution to V_q as

$$V_q^{ph} = \begin{cases} -V & \text{for } 0 \leqslant \varepsilon_K, \varepsilon_{K-q} \leqslant \hbar\omega_D \\ 0 & \text{otherwise}, \end{cases} \tag{5.3.98}$$

where $V > 0$. Furthermore, we have to take into account, in one way or another, the screened Coulomb repulsion between electrons, which acts more or less uniformly over the entire energy band. We model this repulsion by the following contribution to the potential:

$$V_q^e = \begin{cases} U & \text{for } 0 \leqslant \varepsilon_K, \varepsilon_{K-q} \leqslant \hbar\omega_c \\ 0 & \text{otherwise}, \end{cases} \tag{5.3.99}$$

where $\hbar\omega_c$ is an energy on the order of the bandwidth.

We now set $V_q = V_q^{ph} + V_q^e$. From (5.3.97) it is easy to obtain

$$\begin{aligned} \varphi_K &= \frac{V\Delta_1 - U\Delta_2}{2\varepsilon_K - E}, \quad 0 < \varepsilon_K < \hbar\omega_D, \\ \varphi_K &= -\frac{U\Delta_2}{2\varepsilon_K - E}, \quad \hbar\omega_D < \varepsilon_K < \hbar\omega_c, \end{aligned} \tag{5.3.100}$$

where

$$\Delta_1 = \frac{1}{N}\sum_{k;0<\varepsilon_k<\hbar\omega_D} \varphi_k; \quad \Delta_2 = \frac{1}{N}\sum_{k;0<\varepsilon_k<\hbar\omega_c} \varphi_k. \tag{5.3.101}$$

Substitution of (5.3.100) into (5.3.101) yields

$$\begin{aligned} \Delta_1 &= (V\Delta_1 - U\Delta_2)\Lambda_1, \\ \Delta_2 &= \Delta_1 - U\Delta_2(\Lambda_2 - \Lambda_1), \end{aligned} \tag{5.3.102}$$

where the notation introduced is

$$\Lambda_1 = \frac{1}{N} \sum_{k;0<\varepsilon_k<\hbar\omega_D} (2\varepsilon_k - E)^{-1} , \tag{5.3.103}$$

$$\Lambda_2 = \frac{1}{N} \sum_{k;0<\varepsilon_k<\hbar\omega_c} (2\varepsilon_k - E)^{-1} .$$

The condition for the system of equations (5.3.102) to have solutions is

$$\begin{vmatrix} 1 - V\Lambda_1 & U\Lambda_1 \\ -1 & 1 + U(\Lambda_2 - \Lambda_1) \end{vmatrix} = 0 , \tag{5.3.104}$$

or, if we calculate the determinant,

$$1 - V\Lambda_1 + U\Lambda_2 + UV(\Lambda_2 - \Lambda_1)\Lambda_1 = 0 , \tag{5.3.105}$$

We transcribe (5.3.105) in the form

$$1 - [V - U + VU(\Lambda_2 - \Lambda_1)]\Lambda_1 + U(\Lambda_2 - \Lambda_1) = 0 ,$$

or

$$1 = V_{\mathrm{ef}}\Lambda_1 , \tag{5.3.106}$$

where

$$V_{\mathrm{ef}} = V - U \frac{1}{1 + U(\Lambda_2 - \Lambda_1)} . \tag{5.3.107}$$

The density of states $g(\varepsilon)$ in a layer $0 < \varepsilon < \hbar\omega_D$ and, to a lesser accuracy, in a layer $0 < \varepsilon < \hbar\omega_c$ may be regarded as constant. Then

$$\Lambda_1 = g \int_0^{\hbar\omega_D} \frac{d\varepsilon}{2\varepsilon - E} = \frac{g}{2} \ln \left| \frac{2\hbar\omega_D - E}{E} \right| , \tag{5.3.108}$$

$$\Lambda_2 = \frac{g}{2} \ln \left| \frac{2\hbar\omega_c - E}{E} \right| \quad (E < 0) .$$

We set $|E| \ll \hbar\omega_D, \hbar\omega_c$. Therefore

$$\Lambda_2 \approx \Lambda_1 + (g/2)\ln(\omega_c/\omega_D) , \tag{5.3.109}$$

and

$$V_{ef} = V - \frac{U}{1 + (gU/2)\ln \omega_c/\omega_D} \, . \tag{5.3.110}$$

Since $\ln(\omega_c/\omega_D) \gg 1$, the contribution of the Coulomb interaction decreases abruptly [5.37]. Therefore, the quantity V_{ef} is normally greater than zero. Equation (5.3.106) here has the solution

$$E = -\hbar\omega_D \exp\left(-\frac{2}{gV_{ef}}\right) , \tag{5.3.111}$$

which, as is easy to understand from a comparison with Sect. 4.8, describes the bound state of two electrons (a Cooper pair). The size ξ of the pair may be evaluated from both (5.3.100) and simple dimensional considerations.

The formation of a bound state near the Fermi surface results in the uncertainty of the electron momenta δp being such that

$$\frac{p_F \delta p}{m} \approx |E| \, . \tag{5.3.112}$$

On the other hand, it follows from the uncertainty relation that $\delta p \simeq \hbar/\xi$, whence we obtain

$$\xi \approx \frac{\hbar p_F}{m|E|} \gg d \, . \tag{5.3.113}$$

Thus, a Cooper pair cannot be conceived of as some point entity, since it is large in size compared to the lattice period. Nevertheless, a Cooper pair is a stable entity and moves as a single unit.

The condition $|E| \ll \hbar\omega_D$ is fulfilled for $gV_{ef} \lesssim 1$. Due to the existence of the Fermi surface, the quantity g is finite (for the one-electron problem the g in the three-dimensional case goes to zero at the band edges). For this reason a bound state here arises with an arbitrarily small $V_{ef} > 0$. Such an instability, representing the energetic favorableness of the binding of electrons into Cooper pairs, leads to a radical change in the properties of the system. The Cooper phenomenon underlies the microscopic Bardeen–Cooper–Schrieffer (BCS) theory of superconductivity [5.38, 39]. We do not elaborate on the BCS theory here, but concentrate on some simple results of the phenomenological theory, based on the concept of superconductivity as a macroscopic quantum phenomenon. The foundation of this treatment was laid by F. *London* and G. *London* [5.40], and the theory itself created by *Landau* and *Ginzburg* [5.41].

5.4 Superconductivity

To start with, we wish to enumerate some of the major experimental facts. The phenomenon of superconductivity was first discovered by *Kamerlingh-Onnes* in 1911 in mercury. Later, it was found to be inherent in a large number of metals and compounds. The crux of the phenomenon is that when a metal is cooled to some temperature T_c, its electrical resistivity goes abruptly to zero. Later on it appeared that at the transition point the heat capacity changed discontinuously (Fig. 5.7) and, therefore, the transition to the superconducting state is a second-order phase transition. It was also found that a superconductor can be brought into the normal state not only by heating but also by placing it in a sufficiently strong magnetic field H_c. The magnitude of this critical field decreases with increasing temperature and vanishes smoothly at $T = T_c$ (Fig. 5.8).

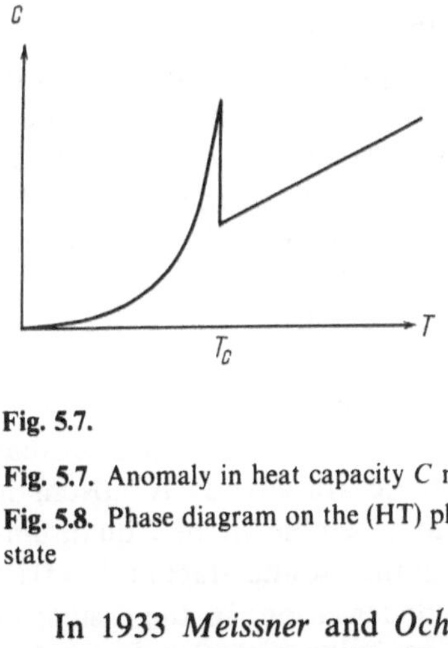

Fig. 5.7.

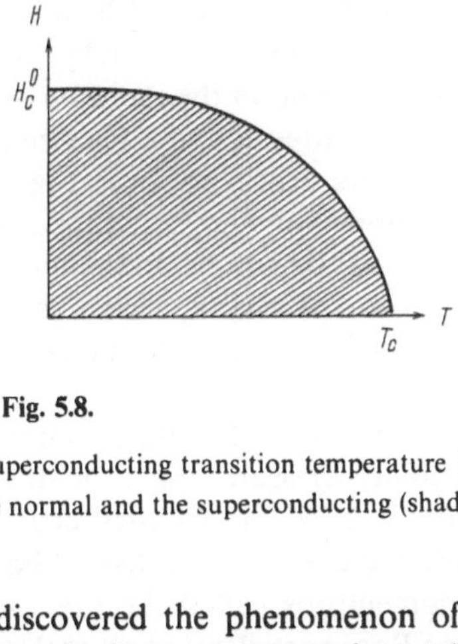

Fig. 5.8.

Fig. 5.7. Anomaly in heat capacity C near the superconducting transition temperature T_c

Fig. 5.8. Phase diagram on the (HT) plane of the normal and the superconducting (shaded region) state

In 1933 *Meissner* and *Ochsenfeld* discovered the phenomenon of the expulsion of a magnetic field of intensity $H < H_c$ from a superconductor (Fig. 5.9). In the bulk of a superconductor the magnetic induction B is equal to zero. The induction curve of an infinitely long superconducting cylinder in a field parallel to its axis is portrayed in Fig. 5.10a.

Subsequently, it was found that along with this behavior (superconductors of the first kind), a more complicated type of behavior occurs (Fig. 5.10b), in which the field over some interval $H_{c1} < H < H_{c2}$ penetrates partially into the superconductor (a superconductor of the second kind). In sufficiently weak fields any superconductor behaves like an ideal diamagnetic material. We should point out that if the superconductor were a "normal" metal but exhibited zero conductivity, Ohm's law would give $\boldsymbol{E} = 0$ in the bulk of the superconductor and it thus would follow from Maxwell's equation

$$\operatorname{curl} \boldsymbol{E} = -\frac{1}{c}\frac{\partial \boldsymbol{B}}{\partial t}$$

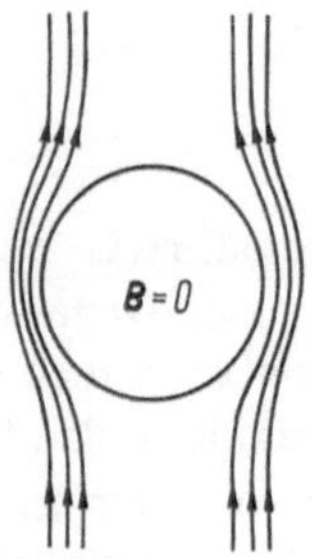

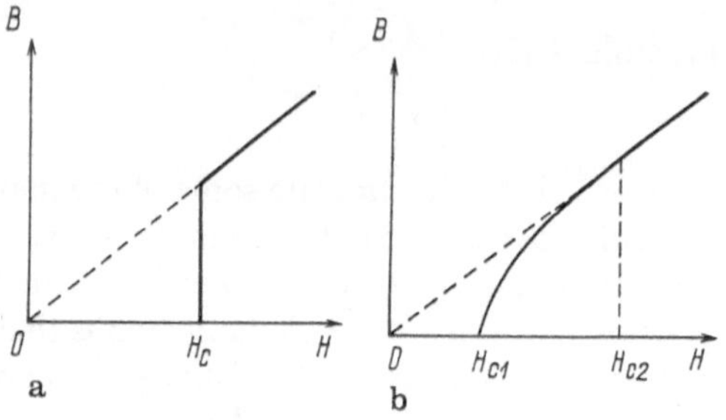

Fig. 5.9.

Fig. 5.10.

Fig. 5.9. Meissner effect
Fig. 5.10. Magnetization curves for first-order (**a**) and second-order (**b**) superconductors

that $\partial \boldsymbol{B}/\partial t = 0$; i.e., during the superconducting transition in a sufficiently weak magnetic field, the field would be frozen rather than expelled from the sample.

We have outlined, very briefly, the major properties of superconductors. Without dwelling upon the thermodynamics of the superconducting transition, we center our attention on the unusual properties of the ground state in superconductors. Note, however, that if superconductivity does arise from the Cooper phenomenon and, consequently, vanishes because of the annihilation of Cooper pairs, T_c should be on the order of $|E|/k_B$, $|E|$ being the binding energy (5.3.111). Normally, real T_c values range approximately from 1 to 10 K.

Thus, what is the cause of superconductivity? The Cooper pairs that form are no longer fermions, but are bosons, i.e., the wave function of two Cooper pairs does not reverse sign when their coordinates are exchanged. We saw earlier that the probability of scattering to a state where there are already N bosons is proportional to $N+1$, so that bosons tend to assemble in one quantum-mechanical state. In an ideal gas all bosons are in the ground state at $T = 0$ K; this phenomenon is called the Bose–Einstein condensation. In the system of interacting Bose particles, not all of the particles are in the ground state. Cooper pairs cannot supply their momentum to scattering centers singly, because the motion of the pairs is highly correlated and there are no unbound electrons at $T = 0$ K (to get unbound electrons, an energy on the order of $|E|$ is required to destroy a pair).

Since the macroscopic number of Cooper pairs moves in a correlated fashion, a wave function of the entire system is formed which has a macroscopic meaning. This is impossible for unpaired electrons because, according to the Pauli exclusion principle, they cannot be in one and the same state. What we imply here is not the wave function $\psi(\boldsymbol{r}_1, \ldots, \boldsymbol{r}_N)$ in configuration space, which may be introduced in all the cases, but the function $\psi(\boldsymbol{r})$, which depends on three spatial variables. The $\psi(\boldsymbol{r})$ may be visualized as the wave function of any Cooper pair from among those performing the correlated motion. More precisely, it is necessary to introduce the creation and annihilation operators of a Cooper pair at the point $\boldsymbol{r}$ (these are to some extent analogous to the phonon operators that are familiar to us). Then $\psi(\boldsymbol{r})$ is the mean value of the annihilation operator. The function $\psi(\boldsymbol{r})$ may be represented as

$$\psi(\boldsymbol{r}) = [\tfrac{1}{2}N_s(\boldsymbol{r})]^{1/2}\exp[\mathrm{i}\vartheta(\boldsymbol{r})] \,, \tag{5.4.1}$$

with $N_s(\boldsymbol{r})$ and $\vartheta(\boldsymbol{r})$ being real-valued functions ($N_s > 0$).

We assume that the superconducting current $\boldsymbol{j}$ is related to ψ by the usual quantum-mechanical formula

$$\boldsymbol{j} = \frac{e\hbar}{2m\mathrm{i}}(\psi^* \nabla\psi - \psi\nabla\psi^*) \,, \tag{5.4.2}$$

where $2e$ is the charge of a Cooper pair, $2m$ the mass of Cooper pair, $N_s/2$ the density of pairs, and N_s the density of paired electrons. Substituting (5.4.1) into (5.4.2) yields

$$\boldsymbol{j} = \frac{e\hbar}{2m}N_s\nabla\vartheta \,. \tag{5.4.3}$$

The magnetic field enters into both the Hamiltonian and the expression for the current (5.4.2) through the replacement

$$-\mathrm{i}\hbar\nabla \rightarrow -\mathrm{i}\hbar\nabla - \frac{2e}{c}\boldsymbol{A} \quad \text{or} \quad \nabla \rightarrow \nabla - \frac{2\mathrm{i}e}{\hbar c}\boldsymbol{A} \,,$$

by acting on ψ, and through $\nabla \rightarrow \nabla + (2\mathrm{i}e/\hbar c)\boldsymbol{A}$ by acting on ψ^*, where $\boldsymbol{A}$ is the vector potential (pair charge $2e$). Then we have, instead of (5.4.3),

$$\boldsymbol{j} = \frac{e\hbar}{2m}N_s\left(\nabla\vartheta - \frac{2e\boldsymbol{A}}{\hbar c}\right) . \tag{5.4.4}$$

In a homogeneous superconductor, $N_s = \text{const}$. Applying the curl operation to (5.4.4), we find, with allowance for the fact that $\operatorname{curl}\boldsymbol{A} = \boldsymbol{B}$ [5.40],

$$\operatorname{curl}\boldsymbol{j} = -\frac{N_s e^2}{mc}\boldsymbol{B} \,. \tag{5.4.5}$$

Equation (5.4.5) stands for Ohm's law for superconductors. In turn, the induction $\boldsymbol{B}$ is related to the current by the Maxwell equation

$$\operatorname{curl}\boldsymbol{B} = \frac{4\pi}{c}\boldsymbol{j} \,. \tag{5.4.6}$$

Allowing for (5.4.5) and the identity $\operatorname{curl}\operatorname{curl}\boldsymbol{B} = \operatorname{grad}\operatorname{div}\boldsymbol{B} - \Delta\boldsymbol{B} = -\Delta\boldsymbol{B}$ ($\operatorname{div}\boldsymbol{B} = 0$), application of the curl operation in (5.4.6) yields

$$\Delta\boldsymbol{B} - \frac{4\pi N_s e^2}{mc^2}\boldsymbol{B} = 0 \ . \tag{5.4.7}$$

In the one-dimensional case where $\boldsymbol{B}$ depends only on x, the solution to (5.4.7) for a sample that occupies the half-space ($x > 0$) has the form

$$\boldsymbol{B}(x) = \boldsymbol{B}(0)\exp(-x/\delta_\mathrm{L}) \ , \tag{5.4.8}$$

where

$$\delta_\mathrm{L} = \left(\frac{mc^2}{4\pi N_s e^2}\right)^{1/2} \ , \tag{5.4.9}$$

is said to be the London penetration depth. Thus, the field and current in a superconductor are concentrated in a skin of thickness δ_L. If N_s is on the order of the valence electron density, then $\delta_\mathrm{L} \approx 3\cdot 10^{-5}$ cm. As distinct from a diamagnetic material, where the magnetic induction decreases because of molecular currents in the bulk of the material, the field "expulsion" in a superconductor is assured by the surface current that flows in a skin about δ_L thick.

The Londons equation (5.4.5), which disregards the spatial dispersion (i.e., the nonlocal relationship between $\boldsymbol{j}$ and $\boldsymbol{B}$) can be compared with the skin effect in metals in the case of a smoothly varying field. Thus the free path length in metals compares with the spatial extent of the Cooper pair ξ (5.3.113), which assigns the characteristic scale of delocalization.

An opposite case is possible,

$$\delta_\mathrm{L} \ll \xi \ , \tag{5.4.10}$$

which is said to be Pippardian, similar to the hyperanomalous skin effect in metals (3.7.2). Repeating *Pippard*'s qualitative arguments [5.42] based on the concept of effective pairs with the density

$$N_s^{\mathrm{ef}} = N_s\frac{\delta}{\xi} \ , \tag{5.4.11}$$

where δ is the penetration depth (only this part of the effective pairs interacts with the field), we obtain, similar to (3.2.67),

$$\delta \approx (\delta_\mathrm{L}^2\xi)^{1/3} \ . \tag{5.4.12}$$

We now return to the usual London case. A striking phenomenon, associated with the expulsion of the magnetic field from the superconductor, is the quantization of the magnetic flux [5.43]. We place a ring, whose thickness considerably exceeds δ_L, in a magnetic field and then bring it into the superconducting state. The field lines will be pushed out of the sample and, after the external field has been turned off, they will close (Fig. 5.11). The field lines are incapable of crossing the sample, so a certain magnetic flux ϕ will be captured by the ring.

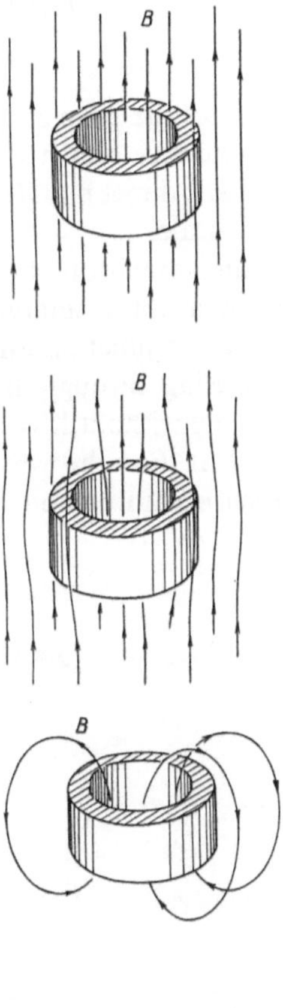

Fig. 5.11. Capture of the magnetic flux by a superconducting loop

In the bulk of the ring $\boldsymbol{j}=0$, and integration of (5.4.4) over any closed contour inside the ring gives

$$\Delta\vartheta = \frac{2e}{\hbar c}\oint \boldsymbol{A}\cdot d\boldsymbol{l} = \frac{2e\phi}{\hbar c}\,, \tag{5.4.13}$$

where $\Delta\vartheta$ is the phase change due to tracing around a path, and we have exploited Stokes theorem relating the circulation of the vector $\boldsymbol{A}$ to the flux of its curl $\boldsymbol{B}$. But since in tracing around a closed path we have returned to the same point and $\psi(\boldsymbol{r})$ is a single-valued function, we should have

$$\Delta\vartheta = 2\pi n \;, \qquad n = 0, \pm 1, \pm 2 \ldots . \tag{5.4.14}$$

Substitution of (5.4.14) into (5.4.13) yields

$$\phi = \phi_0 n \;, \qquad \phi_0 = \frac{\pi\hbar c}{|e|} \approx 2\cdot 10^{-7}\,\mathrm{Gs}\cdot\mathrm{sm}^2 \;. \tag{5.4.15}$$

The quantity ϕ_0 is called the flux quantum. Thus, we arrive at the conclusion that the ring can capture only a specific amount of magnetic flux, which is a multiple of ϕ_0. This effect was observed experimentally and is a direct manifestation of the laws of quantum mechanics on a macroscopic scale.

Another example of a macroscopic quantum effect is the Josephson effect [5.44]. We consider two superconductors that are joined by a thin insulating layer (Fig. 5.12) [5.45]. We apply a potential difference V to the junction and denote the probability amplitude of Cooper pairs penetrating through the insulator because of the tunneling effect by K (we may assume without loss of generality that the penetration probability amplitude is real). For the wave functions on either side of the barrier, ψ_1 and ψ_2, we then obtain combined equations

$$\begin{aligned} i\hbar\frac{\partial\psi_1}{\partial t} &= U_1\psi_1 + K\psi_2 \;, \\ i\hbar\frac{\partial\psi_2}{\partial t} &= K\psi_1 + U_2\psi_2 \;, \end{aligned} \tag{5.4.16}$$

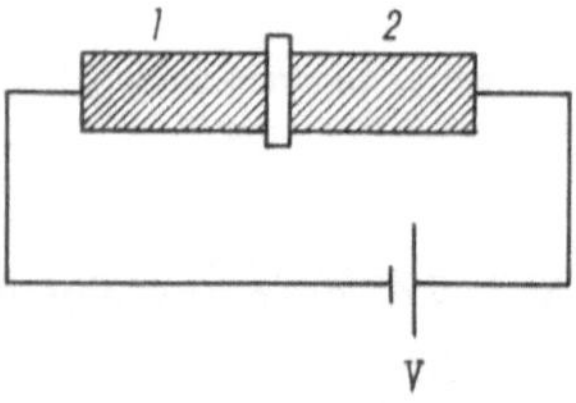

Fig. 5.12. Josephson effect

$U_1 - U_2 = 2eV$. We choose the quantity $1/2(U_1 + U_2)$ as the zero of energy. Then $U_1 = eV$, $U_2 = -eV$. Using the substitutions

$$\psi_{1,2} = (\tfrac{1}{2}N_{s(1,2)})^{1/2}\exp(i\vartheta_{1,2}) \;,$$

and equating the real and imaginary parts, a little manipulation yields

$$\begin{aligned}
\frac{\partial N_{s1}}{\partial t} &= \frac{2K}{\hbar}(N_{s1}N_{s2})^{1/2}\sin(\vartheta_2-\vartheta_1)\ ,\\
\frac{\partial N_{s2}}{\partial t} &= -\frac{2K}{\hbar}(N_{s1}N_{s2})^{1/2}\sin(\vartheta_2-\vartheta_1)\ ,\\
\frac{\partial \vartheta_1}{\partial t} &= \frac{K}{\hbar}(N_{s1}N_{s2})^{1/2}\cos(\vartheta_2-\vartheta_1)-\frac{eV}{\hbar}\ ,\\
\frac{\partial \vartheta_2}{\partial t} &= \frac{K}{\hbar}(N_{s1}N_{s2})^{1/2}\cos(\vartheta_2-\vartheta_1)+\frac{eV}{\hbar}\ ,
\end{aligned} \tag{5.4.17}$$

Hence it follows that $\partial N_{s1}/\partial t = -\partial N_{s2}/\partial t$. This quantity is related to the current flowing through the junction

$$J = 2e\frac{\partial N_{s1}}{\partial t} = \frac{2eK}{\hbar}(N_{s1}N_{s2})^{1/2}\sin(\vartheta_2-\vartheta_1) \equiv J_0\sin(\vartheta_2-\vartheta_1)\ . \tag{5.4.18}$$

From the second pair of equations (5.4.17) it follows that

$$\frac{\partial}{\partial t}(\vartheta_2-\vartheta_1) = \frac{2eV}{\hbar} \equiv \omega_V\ , \quad \vartheta_2-\vartheta_1 = \delta_0+\omega_V t\ . \tag{5.4.19}$$

Equations (5.4.18, 19) signify that when a constant potential difference is applied, an alternating current of frequency ω_V flows through the junction. The effect may be observed also if an alternating voltage $v\cos\omega t$ is applied in addition to the direct voltage V, with $v \ll V$. Then

$$\vartheta_2-\vartheta_1 = \delta_0+\frac{2eVt}{\hbar}+\frac{2ev}{\hbar\omega}\sin\omega t\ ,$$

and we have, up to linear terms in v,

$$\begin{aligned}
J &\approx J_0\left[\sin(\delta_0+\omega_V t)+\frac{2ev}{\hbar\omega}\sin\omega t\cos(\delta_0+\omega_V t)\right]\\
&= J_0\left\{\sin(\delta_0+\omega_V t)+\frac{ev}{\hbar\omega}[\sin(\delta_0+(\omega+\omega_V)t)\right.\\
&\quad \left.+\sin(\delta_0+(\omega-\omega_V)t)]\right\}\ .
\end{aligned} \tag{5.4.20}$$

When $\omega=\omega_V$, the mean value of the last term is other than zero and a resonance increase in direct current through the junction occurs. The Josephson effect, in particular, allows very accurate voltage and magnetic field measurements [5.45].

We conclude this section by returning to the problem of the difference in the properties of first-order and second-order superconductors (type I and type II). For superconductors of the second kind the surface tension at the interface between the normal and the superconducting phase turns out to be negative [Ref. 5.46, Chap. 5]. The increase in free energy when the field penetrates into the superconductor and thus gives rise to a normal-phase region may thus be compensated, at $H > H_{c1}$, by a surface energy gain due to the formation of an interface. Similarly, starting with the normal phase at high fields, the formation of a superconducting phase is favorable at $H < H_{c2}$. Thus, in the range $H_{c1} < H < H_{c2}$, the sample is in a mixed state. In this state, it is advantageous to have a phase boundary surface that is as large as possible. To achieve this the regions of normal phase have the form of filaments, such that there is a maximum surface for a given volume. Each normal-phase filament should contain a flux of ϕ_0 and not more, since it is energetically favorable to decrease the energy of the magnetic field of each filament. The interaction between filaments results in their being ordered into a two-dimensional lattice [5.47].

5.5 Excitons

Here we examine very briefly one more type of elementary excitation that plays an important role in the energy spectrum of semiconductors and insulators. Here we are concerned with electrically neutral (currentless) electron–hole excitations, called excitons. First we illustrate this (undoubtedly somewhat primitive) definition with particular examples.

Consider a molecular crystal in which, by definition, the binding energy of the electrons with each other and with the nuclei in a molecule is large compared with the intermolecular binding energy. Such a crystal may consist only of molecules with saturated chemical bonds, or of atoms of rare gases. Suppose that one of the molecular electrons on the mth site has been transferred from the ground state $|0\rangle$ into the excited state $|1\rangle$ and, accordingly, a hole has remained in the state $|0\rangle$ (Fig. 5.13a). The excited electron and hole can propagate independently to produce current-carrying states (Fig. 5.13b) or conjointly as

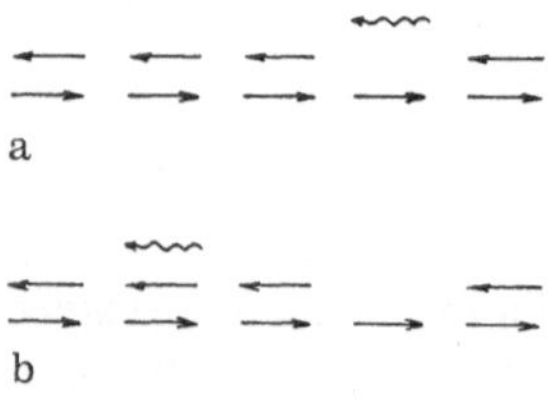

Fig. 5.13. Exciton state types: Frenkel electron–hole currentless exciton (**a**), electron–hole current-carrying excitation (**b**) (straight arrow refers to the excited state $|0\rangle$; undulating arrow refers to the state $|1\rangle$)

the excited state of the site to produce currentless states (Fig. 5.13a). In the latter case we get an energy gain equal to the magnitude of the Coulomb repulsion of electrons on the site in the state $|1\rangle$ and in the state $|0\rangle$ (to a zero approximation, the intermolecular contribution to the energy may be neglected).

This bound state of the electron and hole on one site propagating over the crystal as a whole is said to be a Frenkel exciton [5.48]. To estimate its energy we use the Hamiltonian of a molecular crystal has the form

$$\hat{\mathscr{H}} = \sum_m \hat{\mathscr{H}}_m + \tfrac{1}{2}\sum_{mn}{}' \hat{V}_{mn} \, . \tag{5.5.1}$$

In this expression $\hat{\mathscr{H}}_m$ is the Hamiltonian of the electrons in the mth molecule; $\hat{V}_{mn}$ is the Hamiltonian of the intermolecular, normally dipole–dipole, interaction

$$\hat{V}_{mn} = \frac{\hat{\boldsymbol{d}}_n \cdot \hat{\boldsymbol{d}}_m}{|\boldsymbol{R}_m - \boldsymbol{R}_n|^3} - \frac{3(\hat{\boldsymbol{d}}_{\dot{n}}(\boldsymbol{R}_m - \boldsymbol{R}_n))(\hat{\boldsymbol{d}}_{\dot{m}}(\boldsymbol{R}_m - \boldsymbol{R}_n))}{|\boldsymbol{R}_m - \boldsymbol{R}_n|^5} \, , \tag{5.5.2}$$

where $\hat{\boldsymbol{d}}_m$ is the dipole moment operator of the mth molecule, and $\boldsymbol{R}_m$ is the vector of the same site (for simplicity, we consider the case of one molecule per unit cell).

To a zero approximation with respect to $\hat{V}$, the wave function of a crystal with an excited mth molecule may be chosen as the product of wave functions of individual molecules

$$\psi_m = \varphi_m^{(1)} \prod_{n \neq m} \varphi_n^{(0)} \, , \tag{5.5.3}$$

with $\varphi_m^{(0,1)}$ being the wave function of the mth molecule in the state (0, 1). The antisymmetry effects (Pauli exclusion principle) are insignificant in the problem of interest and will therefore be neglected. This is certainly justified for excitations without spin flip (singlet excitons). By contrast, exchange effects are important for excitations with spin flip known as triplet excitons. The matrix elements of the Hamiltonian (5.5.1) for the functions (5.5.3) are as follows (see similar computations in Sect. 4.6.1):

$$\langle \psi_m | \hat{\mathscr{H}} | \psi_l \rangle = \Delta + \langle \varphi_m^{(1)} \varphi_l^{(0)} | \hat{V}_{ml} | \varphi_m^{(0)} \varphi_l^{(1)} \rangle \equiv \Delta + v_{ml} \, , \tag{5.5.4}$$

where

$$\Delta = \langle \varphi_m^{(1)} | \hat{\mathscr{H}}_m | \varphi_m^{(1)} \rangle - \langle \varphi_m^{(0)} | \hat{\mathscr{H}}_m | \varphi_m^{(0)} \rangle \, ,$$

is the excitation energy in the molecule. The zero of energy is the energy of the state $|0\rangle$, i.e., the ground state. The dipole–dipole interaction matrix element v_{ml}

depends only on the difference $\boldsymbol{R}_m - \boldsymbol{R}_l$. Then, just as in the electron-spectrum problem in the LCAO method (Sect. 4.5.3), we can immediately write the spectrum and eigenfunctions

$$E(\boldsymbol{k}) = \Delta + v(\boldsymbol{k}) \; ,$$

$$\psi_{\boldsymbol{k}} = \sum_m \psi_m \exp(\mathrm{i}\boldsymbol{k}\cdot\boldsymbol{R}_m) \; , \tag{5.5.5}$$

where $\boldsymbol{k}$ is the quasimomentum running over the Brillouin zone, and $v(\boldsymbol{k})$ the Fourier transform of v_{ml}. Thus, the interaction between molecules expands the excited level into an exciton band. The exciton velocity is, as usual, equal to $(1/\hbar)\partial E(\boldsymbol{k})/\partial \boldsymbol{k}$. Excitons do not transport current but transfer energy and may show up in optical spectra [5.49–51].

Another limiting case is the bound state of an electron and hole with a radius that is much larger than the lattice constant. This case occurs frequently in semiconductors (Wannier–Mott excitons). If the electron and hole are described in the effective-mass approximation and their interaction is described with the help of the static dielectric permittivity, the effective Hamiltonian will be

$$\hat{\mathscr{H}} = -\frac{\hbar^2}{2m_\mathrm{e}}\Delta_\mathrm{e} - \frac{\hbar^2}{2m_\mathrm{h}}\Delta_\mathrm{h} - \frac{e^2}{\varepsilon_0 |\boldsymbol{r}_\mathrm{e} - \boldsymbol{r}_\mathrm{h}|} \; , \tag{5.5.6}$$

where the subscripts e and h refer to electrons and holes respectively. This is analogous to the problem of the hydrogen atom with the substitutions $m \to m_\mathrm{e}$, $M \to m_\mathrm{h}$ (M being the nuclear mass), $e^2 \to e^2/\varepsilon_0$. We can thus immediately write the expression for the spectrum

$$E_n(\boldsymbol{K}) = -\frac{\mu e^4}{2\hbar^2\varepsilon_0^2}\frac{1}{n^2} + \frac{\hbar^2 \boldsymbol{K}^2}{2(m_\mathrm{e} + m_\mathrm{h})} \; , \qquad n = 1, 2, 3, \dots \; , \tag{5.5.7}$$

where $\mu = m_\mathrm{e} m_\mathrm{h}/(m_\mathrm{e} + m_\mathrm{h})$ is the reduced mass, and $\boldsymbol{K}$ is the center-of-mass quasimomentum. The radius of the bound state at $n = 1$,

$$a \approx \frac{\hbar^2 \varepsilon_0}{\mu e^2} \; , \tag{5.5.8}$$

is large for the same reasons as in the treatment of impurity states in semiconductors (Sect. 4.5.1), i.e., the smallness of m_e and, consequently, μ (normally $m_\mathrm{h} \gg m_\mathrm{e}$ and $\mu \approx m_\mathrm{e}$) and the large values of ε_0. The fulfillment of the inequality $a \gg d$ justifies allowance for the crystalline environment by introducing static dielectric permittivity (Sect. 4.5.1).

A more complicated type of elementary excitation is the biexciton, the bound state of two Wannier–Mott excitons in analogy to the H_2 molecule. Biexcitons are quite stable [5.52]. However, since holes are much lighter than protons, the energy of the zero-point oscillations of such a "molecule" is large, and therefore it is only loosely bound. In fact, the binding energy of a biexciton, expressed in units of $\mu e^4/2\hbar^2\varepsilon_0^2$, is appreciably less than that of H_2, expressed in units of $me^4/2\hbar^2$. The use of lasers permits creation of large exciton concentrations $n \approx 10^{16}\,\mathrm{cm}^{-3}$ in semiconductors; this is feasible when

$$na^3 \approx 1 \ . \tag{5.5.9}$$

In this case the excitons form a liquid rather than a gas. The liquid could be similar to liquid hydrogen (i.e., it could consist of biexcitons), or to liquid alkali metals (i.e., it could form a two-constituent electron–hole plasma). Because of the small binding energy of the biexciton, the latter alternative is the one actually found [5.53]. When the condition (5.5.9) is satisfied, the bound state of the electron and hole vanishes because of Coulomb potential screening, and an electron–hole plasma is formed (cf. the discussion of the Mott transition in Sect. 5.1.3 and formula (5.1.44)). This formation of an electron–hole plasma results in a metal–insulator transition in a system of light-excited electric current carriers. The metallic electron–hole liquid can also form drops that can be observed experimentally. The study of this type of liquid is a rapidly developing area of the physics of semiconductors.

5.6 Transition Metals and Their Compounds

5.6.1 Properties of *d* and *f* States

An interesting group of solids is the family of materials containing transition-group elements whose atoms have incomplete *d* or *f* subshells (Chap. 1). These materials can be metals, semiconductors, or insulators, and most of the solid-state concepts introduced earlier are applicable to them subject to the important restriction that many-electron effects play a much larger role in these compounds. Many-electron effects are manifest most dramatically in the phenomenon of magnetic ordering, which actually has not ever been detected in substances not containing transition elements. Even when such materials possess no magnetic order, they still exhibit unusual thermal, magnetic, optical, electrical, and even mechanical properties. The nature of these anomalies is due to the peculiar behavior of *d* and *f* states (to be discussed in Sect. 5.6.3).

We start by considering the atom of a transition element. There are several groups of elements that have unfilled *d* and *f* subshells. The electronic configurations of these atoms are specified in Table 1.10 and in Sect. 3.1. We have to pay attention to the small radius of *d*-electron and particularly *f*-electron subshells in comparison with characteristic solid-state distances. By characteristic spacing we mean the distance between nearest ions in the metallic state of a relevant element. Another interesting feature is that the filling of d and f subshells proceeds in jumps at the middle and end of each series. With the atomic number changing by unity, the occupation number changes by two electrons. The situation at the end of a series is clear—this is the usual effect of the elevated stability of a filled shell; the effect of the stability of a half-filled shell occurs, for example, in chromium, where the configuration of the ground state is $3d^5 4s^1$. Here we are confronted with a fact that is impossible from the one-electron standpoint since two states ($3d$ and $4s$) are partially filled. It would seem that either the energy of $4s$ states should be lower than that of $3d$ states—then the configuration of the ground state should be $3d^4 4s^2$—or vice versa—then we should have $3d^6 4s^0$. This indicates that the one-electron approach is inadequate to describe the atoms of transition metals. In particular, we have to take account of the exchange correlation interaction (Sect. 4.6.1), which leads to the formation of an atomic magnetic moment. As has been stated (Sect. 4.6.1), the effect comes about because electrons with parallel spins are at larger mean distances than are electrons with antiparallel spins, because of the Pauli exclusion principle, and, therefore, repel more weakly. If this interaction is sufficiently large compared with the characteristic one-electron excitation energies (for example, the transition from the $3d^5$ to the $4s^1$ state), which are relatively small for transition elements, then it is more favorable energetically to have two partially filled states rather than have electrons with antiparallel spins.

Thus, isolated atoms can possess magnetic moments. This applies not only to transition-element atoms but also to atoms with partially filled s and p states. The magnitude of the magnetic moments may be determined from Hund's rules, according to which the total spin moment S is maximum in the ground state with a given configuration and the orbital moment L is maximum when S is fixed [1.12]. The total atomic moment J arises from the spin–orbit coupling and is equal to $L+S$ for a more than half-filled subshell and is equal to $|L-S|$ in the opposite case. For a subshell filled exactly by half, $L=0$ and $J=S$.

What happens to electronic states as atoms are united into a crystal? We are concerned primarily with the states of unfilled subshells, for these states are immediately responsible for all the properties of the crystal, apart from the ultrahigh frequency properties. As stated in Sect. 4.5.3, when atomic states form a band we have a gain in kinetic energy—the stronger the overlap of wave functions the larger the gain—and a loss in Coulomb repulsive energy. If the radius of a relevant electron subshell exceeds the nearest atom or ion distance, the gain in kinetic energy turns out to be decisive and Bloch states arise. The interaction here should be approached in terms of the Fermi-liquid theory or similar approaches. This situation occurs for outer s- and p-shell electrons. These electrons are either shared between all the atoms, as in metals, or between pairs of adjacent sites to form covalent bonds as in silicon-type crystals, or they fill the shells of only some of the atoms as in ionic crystals (Sect. 1.8).

A different situation occurs for the states of f electrons. These electrons maintain their atomlike character and do not form a band (except for the so-called intermediate valence cases [5.54]). In fact they only participate in chemical bonding indirectly, through their influence on valence electrons. The magnetic moments of f elements in compounds or in the metallic state are normally close to those of the corresponding atoms.

Both cases may be realized for d states. The atom-like behavior of d states persists in many semiconducting compounds, for example NiO. In other cases metal–insulator transitions occur due to changes in the behavior of d states and due to band formation. In pure d metals, as will be seen, we deal with a "band" limit, although the latter shows some features of the atom-like localized behavior of d states.

Attempts at providing a unified description of the "localized itinerant" behavior of d electrons date back to *Shubin* and *Vonsovsky* [5.55], who developed what is known as the polar model of the crystal. Despite the subsequent effort by many physicists, the problem of constructing such a description is still far from being resolved.

We start by considering localized d (or f) states. Each transition atom then possesses a magnetic moment. Electron motion, associated with a change in the number of electrons on given lattice sites, is energetically unfavorable and the only type of low-lying energy excitations is due to the rotation of moments relative to each other (known as spin waves). This situation is described in a model by Heisenberg [1.25], which we consider briefly in what follows.

5.6.2 The Heisenberg Model

To start with, we briefly recall some of the results discussed in Sect. 1.7. We examine a system of two electrons that interact via the potential $V(\boldsymbol{r}-\boldsymbol{r}')$, which will be viewed as a perturbation. In the absence of an interaction, the electrons were in the states ϕ_1 and ϕ_2. The total wave function of two noninteracting electrons should be antisymmetric with respect to the permutation of spin and spatial coordinates; on the other hand, this function is the product of the coordinate function and spin-wave function

$$\psi(\boldsymbol{r}_1, \sigma_1; \boldsymbol{r}_2, \sigma_2) = \phi(\boldsymbol{r}_1, \boldsymbol{r}_2)\chi(\sigma_1, \sigma_2) \ . \tag{5.6.1}$$

Permuting σ_1 and σ_2 for parallel spins, with total spin $S=1$, alters nothing, so

$$\chi(\sigma_1, \sigma_2) = \chi(\sigma_2, \sigma_1) \ ,$$
$$\phi(\boldsymbol{r}_1, \boldsymbol{r}_2) = -\phi(\boldsymbol{r}_2, \boldsymbol{r}_1) \ .$$

Then

$$\phi_{S=1}(\boldsymbol{r}_1,\boldsymbol{r}_2)=\frac{1}{\sqrt{2}}[\varphi_1(\boldsymbol{r}_1)\varphi_2(\boldsymbol{r}_2)-\varphi_1(\boldsymbol{r}_2)\varphi_2(\boldsymbol{r}_1)]\ . \tag{5.6.2}$$

For antiparallel spins, $S=0$, $\chi(\sigma_1,\sigma_2)=-\chi(\sigma_2,\sigma_1)$, as is proved in any textbook on quantum mechanics. Consequently, $\phi(\boldsymbol{r}_1,\boldsymbol{r}_2)=\phi(\boldsymbol{r}_2,\boldsymbol{r}_1)$ and

$$\phi_{S=0}(\boldsymbol{r}_1,\boldsymbol{r}_2)=\frac{1}{\sqrt{2}}[\varphi_1(\boldsymbol{r}_1)\varphi_2(\boldsymbol{r}_2)+\varphi_1(\boldsymbol{r}_2)\varphi_2(\boldsymbol{r}_1)]\ . \tag{5.6.3}$$

When there is no interaction, the energy does not depend on the total spin. In first-order perturbation theory the energy correction due to an interaction is

$$\Delta E_S=\int d\boldsymbol{r}_1 d\boldsymbol{r}_2|\phi_S(\boldsymbol{r}_1,\boldsymbol{r}_2)|^2\,V(\boldsymbol{r}_1-\boldsymbol{r}_2)=A\pm J\ , \tag{5.6.4}$$

where

$$\begin{aligned}A&=\int d\boldsymbol{r}_1 d\boldsymbol{r}_2|\varphi_1(\boldsymbol{r}_1)|^2|\varphi_2(\boldsymbol{r}_2)|^2\,V(\boldsymbol{r}_1-\boldsymbol{r}_2)\ ,\\ J&=\int d\boldsymbol{r}_1 d\boldsymbol{r}_2\varphi_1^*(\boldsymbol{r}_1)\varphi_2(\boldsymbol{r}_1)\varphi_1(\boldsymbol{r}_2)\varphi_2^*(\boldsymbol{r}_2)V(\boldsymbol{r}_1-\boldsymbol{r}_2)\ .\end{aligned} \tag{5.6.5}$$

The minus sign in (5.6.4) corresponds to $S=1$ and the plus sign to $S=0$. We introduce the total-spin operator

$$\hat{\boldsymbol{S}}=\hat{\boldsymbol{s}}_1+\hat{\boldsymbol{s}}_2\ ,$$

the square of which is equal to $S(S+1)$:

$$\begin{aligned}S(S+1)&=\hat{\boldsymbol{s}}_1^2+\hat{\boldsymbol{s}}_2^2+\ +2\hat{\boldsymbol{s}}_1\cdot\hat{\boldsymbol{s}}_2=2\cdot\tfrac{1}{2}(\tfrac{1}{2}+1)+2\hat{\boldsymbol{s}}_1\cdot\hat{\boldsymbol{s}}_2\\ &=\tfrac{3}{2}+2\hat{\boldsymbol{s}}_1\cdot\hat{\boldsymbol{s}}_2\ .\end{aligned} \tag{5.6.6}$$

Hence the operator with the eigenvalues -1 for $S=1$ and $+1$ for $S=0$ may be represented as

$$\hat{P}=1-S(S+1)=-\tfrac{1}{2}(1+4\hat{\boldsymbol{s}}_1\cdot\hat{\boldsymbol{s}}_2)\ . \tag{5.6.7}$$

Expression (5.6.4) may then be modified to

$$\Delta E=A-J/2-2J\hat{\boldsymbol{s}}_1\cdot\hat{\boldsymbol{s}}_2\ . \tag{5.6.8}$$

Thus, the antisymmetry of the total wave function for the system of two interacting electrons leads to the energy depending on the mutual orientation of spins of the form (5.6.8) [1.26].

This coupling, called the exchange interaction, considerably exceeds the magnetic dipole–dipole interaction of spin moments, which is proportional to $(v/c)^2$ (with v being the mean velocity of electrons, and c the velocity of light in vacuum). The hypothesis as to the exchange nature of the forces responsible for magnetic ordering was propounded independently by Frenkel [5.56] and Heisenberg [1.25]. For the many-electron system the spin-dependent part of the interaction Hamiltonian is represented in the form

$$\hat{\mathscr{H}}_{\mathrm{ex}} = -\sum_{ik}{}' J_{ik}\hat{s}_i \cdot \hat{s}_k \ , \qquad (5.6.9)$$

with the sum being carried out over all electrons. Expression (5.6.9) is the most general representation that can be constructed out of spin operators by taking into account only pairwise spin interactions for spin 1/2, and $\hat{s}_i^2 = 3/4$ and no other functions, except for a linear function, can be constructed out of $\hat{s}_i \cdot \hat{s}_k$.

As for expressions like (5.6.5), we should not take them too seriously. First, it is not always possible to allow for the interaction by using perturbation theory and, second, contributions to J can exist which are of a quite different nature. We briefly consider, for example, what is known as kinetic exchange.

Let two electrons be on two neighboring sites in the nondegenerate orbital state (Fig. 5.14). The repulsive interaction energy of two electrons on the same site is larger than that of two electrons on different sites (the former exceeds the latter by an amount U), and the matrix element of the Hamiltonian, β, which corresponds to transitions of an electron from site to site, in the excited state, is much less than U. For the antiparallel spin orientation (Fig. 5.14a) this virtual process, taken into account in the second-order perturbation theory, then causes the energy of the system to decrease by $2\beta^2/U$. The factor 2 here has appeared because of the contribution of two electronic transitions. For the parallel orientation (Fig. 5.14b) this process is prohibited. The corresponding contribution to the exchange integral is equal to the energy difference for differing spin orientations

$$J = E_{\uparrow\uparrow} - E_{\uparrow\downarrow} = -\frac{2\beta^2}{U} \ . \qquad (5.6.10)$$

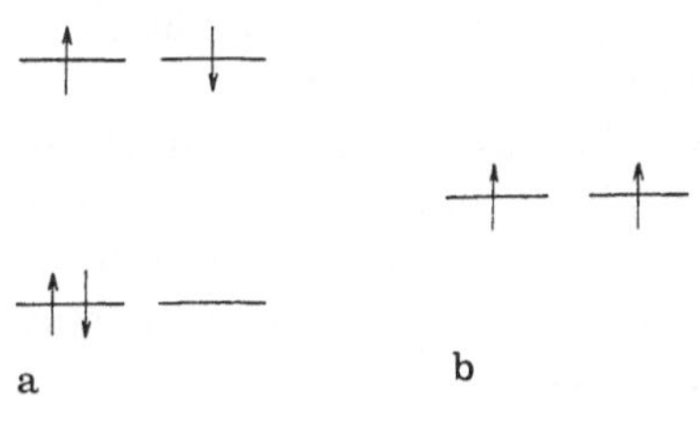

Fig. 5.14. Indirect exchange: lowering of energy as a result of virtual transitions for antiparallel spin orientation (**a**), in comparison with the parallel orientation (**b**)

Indeed, the exchange integral is half the energy difference of the singlet and the triplet state

$$J = \tfrac{1}{2}(E_{S=1} - E_{S=0}) \ .$$

$E_{S=1}$ is the energy of the state with parallel spins $E_{\uparrow\uparrow}$. The state with antiparallel spins, may with equal probability be either a singlet or a triplet state with $S_z = 0$. So

$$E_{\uparrow\downarrow} = \tfrac{1}{2}(E_{S=1} + E_{S=0}) \ ,$$

and therefore

$$J = E_{S=1} - \tfrac{1}{2}(E_{S=1} + E_{S=0}) = E_{\uparrow\uparrow} - E_{\uparrow\downarrow} \ .$$

Actually, an approximately similar exchange interaction, due to the excited states of anions that separate transition-element ions, the indirect Kramers–Anderson exchange also plays a leading role in transition-metal compounds. The resultant exchange integral may be both positive (5.6.5) and negative (5.6.10).

It is normally believed that the exchange integral between $d(f)$ electrons on different sites is independent of the orbital states in which these electrons reside. Thus, in (5.6.9) we may then carry out the sum over all electrons on each site and omit the intrasite exchange energy to obtain

$$\hat{\mathscr{H}} = -\sum_{ij}{}' J_{ij}\hat{S}_i \cdot \hat{S}_j \ , \tag{5.6.11}$$

where the sum now is taken over the sites and $\hat{S}_i$ is the total spin of the $d(f)$ shell. The model described by the Hamiltonian (5.6.11) is referred to as the Heisenberg model.

Another simplification that is used frequently and which we will employ, is the nearest-neighbors approximation:

$$J_{ij} = \begin{cases} J, & \text{if } i \text{ and } j \text{ are nearest neighbors} \\ 0, & \text{in all other cases} \ . \end{cases} \tag{5.6.12}$$

If $J > 0$ in the model (5.6.11), then parallel ordering of all spins is the most favorable energetically. With $J < 0$, if the lattice can be broken up into two sublattices, A and B, so that all nearest neighbors for any sublattice A atom are sublattice B atoms and vice versa, an energetically favorable situation arises when the spins in A are parallel to each other and antiparallel to the spins of sublattice B atoms (Fig. 5.15). Such an ordering is said to be antiferromagnetic, or more precisely, a collinear antiferromagnetic structure.

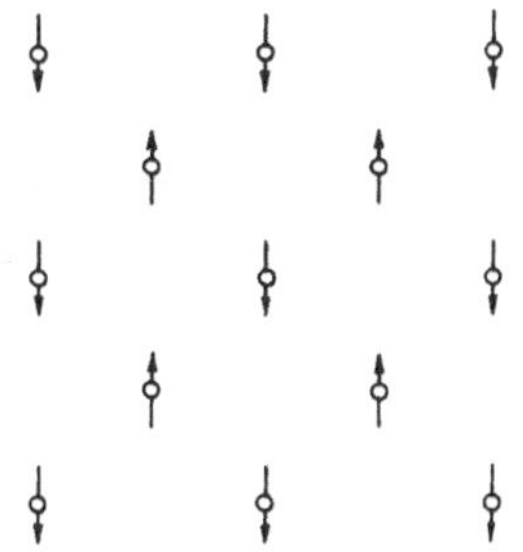

Fig. 5.15. Collinear antiferromagnetic structure in a plane analog of the bcc lattice

If the atoms in A and those in B are different, it can happen that the total magnetic moment in such an ordering is nonzero. Then one talks of ferrimagnetism. Lattices that can be separated into two sublattices are said to be alternating. Thus, the simple and the bcc lattices are alternating, whereas the fcc lattice is not. What is essential is the type of sublattice formed by the transition-element atoms.

Bear in mind, however, that the sublattice spin $\sum_{i \in A} \hat{S}_i$, as can be readily verified, does not commute with the Hamiltonian (5.6.11), so the state pictured in Fig. 5.15 cannot be a ground state in the quantum case and is not even an eigenstate of the Hamiltonian. *Anderson* [5.57] has proved, however, that the mean value of the Hamiltonian in this state differs from the ground-state energy by an amount that is small with respect to the parameter $(zS)^{-1}$, where z is the number of nearest neighbors and S the magnitude of spin in a site [5.58]. In the one-dimensional case, however, the quantum effects are so substantial that the ground state is not antiferromagnetic at all [5.59]. Note that long-range antiferromagnetic order in the three-dimensional case can be described without introducing the concept of magnetic sublattices [5.60].

Now we restrict our attention to the ferromagnetic case ($J > 0$). We start our analysis of the Hamiltonian (5.6.11) with the molecular-field approximation which was introduced by *Weiss* long before the creation of quantum mechanics. In doing so, we replace the interaction energy of two spins by the energy of their interaction with the self-consistent field, which is determined by the mean values of these spins:

$$2\hat{S}_i \cdot \hat{S}_j \rightarrow \hat{S}_i \cdot \langle \hat{S} \rangle + \hat{S}_j \cdot \langle \hat{S} \rangle \ . \tag{5.6.13}$$

Then

$$\hat{\mathscr{H}} = -2zJ \langle \hat{S} \rangle \cdot \sum_i \hat{S}_i \ . \tag{5.6.14}$$

The problem of calculating the partition function Z of the Heisenberg Hamiltonian has been reduced to a calculation of the partition function of a spin in an external field

$$Z = \mathrm{Tr}\{\exp(-\hat{\mathscr{H}}/k_B T)\} = Z_1^N \ , \tag{5.6.15}$$

where

$$Z_1 = \mathrm{Tr}\{\exp(-\boldsymbol{x}\cdot\hat{\boldsymbol{S}}/S)\} = \sum_{M=-S}^{S} \exp(-xM/S)$$
$$= \frac{\exp(-x) - \exp(x(S+1)/S)}{1-\exp(x/S)} = \frac{\sinh[x(2S+1)/2S]}{\sinh(x/2S)} . \tag{5.6.16}$$

Here we have denoted

$$\boldsymbol{x} = \frac{2JzS}{k_B T}\langle\hat{\boldsymbol{S}}\rangle \equiv \boldsymbol{n}x , \tag{5.6.17}$$

where $\boldsymbol{n}$ is a unit vector in the direction of the vector $\langle\hat{\boldsymbol{S}}\rangle$, and have employed the formula for the sum of terms in a geometric progression. The mean value of $\hat{\boldsymbol{S}}$ is found from (5.6.16):

$$\langle\hat{\boldsymbol{S}}\rangle = Z^{-1}\mathrm{Tr}\{\hat{\boldsymbol{S}}\exp(-\boldsymbol{x}\cdot\hat{\boldsymbol{S}}/S)\} = -\boldsymbol{n}S\frac{1}{Z}\frac{\partial Z}{\partial x}$$
$$= -\boldsymbol{n}S\frac{\partial \ln Z}{\partial x} = \boldsymbol{n}S\,B_S(x) , \tag{5.6.18}$$

where

$$B_S(x) = \frac{2S+1}{2S}\,\mathrm{cotanh}\left(\frac{2S+1}{2S}x\right) - \frac{1}{2S}\,\mathrm{cotanh}\left(\frac{x}{2S}\right) , \tag{5.6.19}$$

is called the Brillouin function. Plots of this function for different S are presented in Fig. 5.16. Allowing for (5.6.17, 19), we can obtain an equation for

$$\frac{k_B T}{2JzS^2}x = B_S(x) , \tag{5.6.20}$$

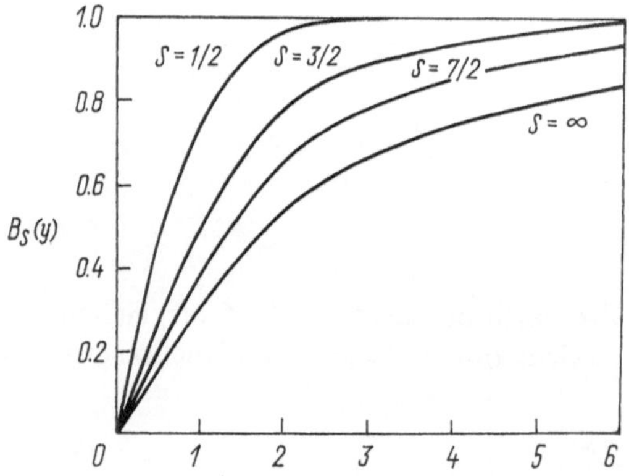

Fig. 5.16. Plots of the Brillouin function $B_s(x)$ for different values of spin S

that has a nontrivial solution $x \neq 0$, provided that

$$1 < \frac{2JzS^2}{k_B T} B'_S(0) = \frac{2JzS(S+1)}{3k_B T} \tag{5.6.21}$$

(Fig. 5.17) or

$$T < T_C \equiv \frac{2JzS(S+1)}{3k_B} . \tag{5.6.22}$$

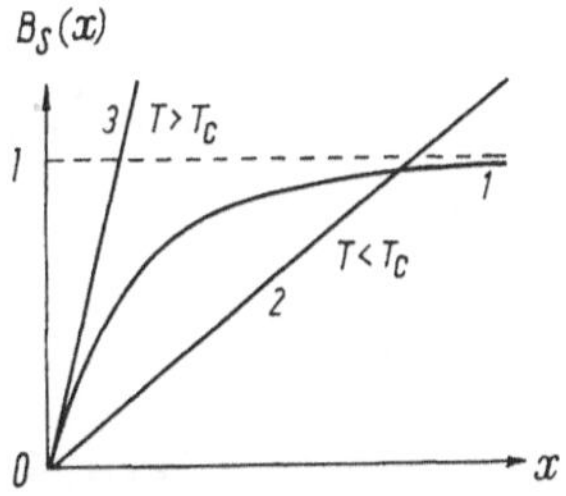

Fig. 5.17. Graphic solution of (5.6.20): (*1*) right-hand side of (5.6.20), (*2*) left-hand side of (5.6.20) for $T < T_c$, (*3*) left-hand side of (5.6.20) for $T > T_c$

Apart from this, there exists the solution $x = 0$, but it does not yield the minimum energy, which, according to (5.6.14), is equal to

$$\langle \hat{\mathscr{H}} \rangle = -2zJN\langle \hat{S} \rangle^2 . \tag{5.6.23}$$

Thus, at temperatures that are lower than some temperature, called the Curie point T_C, there exists in the system a spontaneous magnetization

$$\boldsymbol{M} = g\mu_B \langle \hat{\boldsymbol{S}} \rangle , \tag{5.6.24}$$

(with g being the g factor), which disappears at high temperatures. This is a second-order ferromagnet–paramagnet phase transition.

We now calculate the magnetic susceptibility χ at $T > T_C$. In an external magnetic field $\boldsymbol{h}$ the Hamiltonian will involve an extra term

$$\hat{\mathscr{H}}_h = -g\mu_B \boldsymbol{h} \cdot \sum_i \hat{\boldsymbol{S}}_i , \tag{5.6.25}$$

it being necessary in (5.6.18) to make the replacement $\boldsymbol{x} \to \boldsymbol{x} + \boldsymbol{y}$, where

$$\boldsymbol{y} = g\mu_B \frac{S\boldsymbol{h}}{k_B T} \equiv y\boldsymbol{n} . \tag{5.6.26}$$

Here we have taken account of the fact that $\langle \hat{S} \rangle \parallel \boldsymbol{h} = h\boldsymbol{n}$. Then

$$x = \frac{2JzS^2}{k_B T} B_S(x+y) \approx \frac{2zJS^2}{k_B T} B'_S(0)(x+y) = \frac{T_C}{T}(x+y) ,$$
$$x = \frac{y}{T/T_C - 1} , \tag{5.6.27}$$

since in sufficiently weak fields $x, y \ll 1$ and we may set $B_S(x+y) \approx B'_S(0)(x+y)$.

Allowing for (5.6.17, 22, 26), the result for the magnetization will be

$$\boldsymbol{M} = g\mu_B \langle \hat{S} \rangle = g\mu_B \frac{k_B T \boldsymbol{x}}{JzS} = \frac{g\mu_B T}{JzS} \frac{\boldsymbol{y}}{T/T_C - 1}$$
$$= \frac{g^2 \mu_B^2 S(S+1)\boldsymbol{h}}{3k_B(T - T_C)} , \tag{5.6.28}$$

or

$$\chi = \frac{g^2 \mu_B^2 S(S+1)}{3k_B(T - T_C)} , \tag{5.6.29}$$

i.e., the Curie–Weiss law. When $T \to T_C$, the magnetic susceptibility diverges.

On the whole, the self-consistent field approximation describes the properties of the Heisenberg model fairly well. However, by giving up a detailed consideration of the spin-exchange interaction, this approximation disregards the correlation in spin direction and spin motion, which is particularly significant, first, near the Curie point and, second, at low temperatures. The problem of allowing for spin correlations near T_C relates to the theory of phase transitions. Neither features peculiar to solids nor the quantum character of spins manifest themselves here (for very subtle reasons), so we do not elaborate on this problem, referring the reader to the literature on statistical physics and phase transition theory. Insofar as the correlation of spin motion at $T \ll T_C$ goes, an exhaustive description is possible in terms of elementary excitations (quasiparticles), that is, spin waves. We now focus our attention on this problem.

We introduce, as is normally done in quantum mechanics, the operators $\hat{S}_i^z$, $\hat{S}_i^{\pm} = S_i^x \pm \mathrm{i} S_i^y$ satisfying the permutation relations

$$[\hat{S}_i^+, \hat{S}_j^-]_- = 2\hat{S}_i^z \delta_{ij}, \quad [\hat{S}_i^z, \hat{S}_j^{\pm}]_- = \pm \hat{S}_i^{\pm} \delta_{ij} , \tag{5.6.30}$$

which can be readily derived from the input equations for the spin components

$$[\hat{S}_i^x, \hat{S}_j^y]_- = \mathrm{i}\hat{S}_i^z \delta_{ij} , \quad [\hat{S}_i^y, \hat{S}_j^z]_- = \mathrm{i}\hat{S}_i^x \delta_{ij} ,$$
$$[\hat{S}_i^z, \hat{S}_j^x]_- = \mathrm{i}\hat{S}_i^y \delta_{ij} . \tag{5.6.31}$$

The operator $\hat{S}_i^+$ increases the spin projection on site i by unity, and the operator $\hat{S}_i^-$ decreases it by unity.

Consider the state with one spin changed:

$$\psi = \sum_i c(i)\hat{S}_i^- |0\rangle \ , \tag{5.6.32}$$

where $|0\rangle$ is the ground state of the ferromagnet, and the spin projection on each site is equal to S. The equation of motion for the operator $\hat{S}_i^-$ has the form

$$i\hbar \frac{d\hat{S}_i^-}{dt} = [\hat{S}_i^-, \mathscr{H}]_- = -\sum_{jk}{}' J_{jk}[\hat{S}_i^-, \hat{\boldsymbol{S}}_j \cdot \hat{\boldsymbol{S}}_k] \ . \tag{5.6.33}$$

Calculate the commutator in (5.6.33), allowing for the equations

$$\hat{\boldsymbol{S}}_j\hat{\boldsymbol{S}}_k = \tfrac{1}{2}(\hat{S}_j^+\hat{S}_k^- + \hat{S}_j^-\hat{S}_k^+) + \hat{S}_j^z\hat{S}_k^z \ , \tag{5.6.34}$$

and

$$[\hat{A}, \hat{B}\hat{C}]_- = [\hat{A}, \hat{B}]_-\hat{C} + \hat{B}[\hat{A}, \hat{C}]_- \ . \tag{5.6.35}$$

As a result, we obtain from (5.6.30, 34, 35)

$$\begin{aligned}
[\hat{S}_i^-, \hat{\boldsymbol{S}}_j \cdot \hat{\boldsymbol{S}}_k]_- &= [\hat{S}_i^-, \hat{\boldsymbol{S}}_j]_-\hat{\boldsymbol{S}}_k + \hat{\boldsymbol{S}}_j[\hat{S}_i^-, \hat{\boldsymbol{S}}_k]_- \\
&= \tfrac{1}{2}[\hat{S}_i^-, \hat{S}_j^+]_-\hat{S}_k^- + [\hat{S}_i^-, \hat{S}_j^z]_-\hat{S}_k^z + \tfrac{1}{2}\hat{S}_j^-[\hat{S}_i^-, \hat{S}_k^+]_- \\
&\quad + \hat{S}_j^z[\hat{S}_i^-, \hat{S}_k^z]_- = -\delta_{ij}\hat{S}_i^z\hat{S}_k^- + \delta_{ij}\hat{S}_i^-\hat{S}_k^z \\
&\quad - \delta_{ik}\hat{S}_j^-\hat{S}_i^z + \delta_{ik}\hat{S}_j^z\hat{S}_i^- = -\delta_{ij}\hat{S}_k^-(\hat{S}_i^z - \delta_{ik}) \\
&\quad + \delta_{ij}\hat{S}_i^-\hat{S}_k^z - \delta_{ik}\hat{S}_j^-\hat{S}_i^z + \delta_{ik}\hat{S}_i^-(\hat{S}_j^z - \delta_{ij}) \ .
\end{aligned} \tag{5.6.36}$$

Acting on the state $|0\rangle$, the operator $\hat{S}_i^z$ gives the number S at any value of i. Here we are only interested in the case of $j \neq k$. Then

$$\begin{aligned}
i\hbar \frac{d\hat{S}_i^-}{dt}|0\rangle &= S\sum_{jk}{}' J_{jk}[\delta_{ij}(\hat{S}_i^- - \hat{S}_k^-)|0\rangle + \delta_{ik}(\hat{S}_i^- - \hat{S}_j^-)|0\rangle] \\
&= 2S\sum_j J_{ij}(\hat{S}_i^- - \hat{S}_j^-)|0\rangle \ .
\end{aligned} \tag{5.6.37}$$

$(J_{ij} = J_{ji})$. The Schrödinger equation for the function ψ then becomes

$$\begin{aligned}
i\hbar \frac{\partial\psi}{\partial t} &= E\psi = 2S\sum_{ij} c(i)J_{ij}(\hat{S}_i^- - \hat{S}_j^-)|0\rangle \\
&= 2S\sum_{ij} J_{ij}[c(i) - c(j)]\hat{S}_j^-|0\rangle \ .
\end{aligned} \tag{5.6.38}$$

Expression (5.6.38) is satisfied identically if

$$Ec(i) = 2S\sum_j J_{ij}[c(i) - c(j)] \ . \qquad (5.6.39)$$

Since J_{ij} depends only on $\boldsymbol{R}_i - \boldsymbol{R}_j$, we, as usual, try solutions of the equation in the form of travelling waves

$$c(i) = \exp(\mathrm{i}\boldsymbol{k}\cdot\boldsymbol{R}_i) \ . \qquad (5.6.40)$$

Here

$$E(\boldsymbol{k}) = 2S\sum_j J_{ij}[1 - \exp(\mathrm{i}\boldsymbol{k}\cdot(\boldsymbol{R}_i - \boldsymbol{R}_j))] \ . \qquad (5.6.41)$$

In the nearest-neighbors approximation

$$E(\boldsymbol{k}) = 2JS\sum_{\boldsymbol{\delta}}[1 - \exp(\mathrm{i}\boldsymbol{k}\cdot\boldsymbol{\delta})] \ , \qquad (5.6.42)$$

with $\boldsymbol{\delta}$ being nearest-neighbor vectors. In the isotropic case for small $\boldsymbol{k}$

$$2\sum_{\boldsymbol{\delta}}[1 - \exp(\mathrm{i}\boldsymbol{k}\cdot\boldsymbol{\delta})] \approx \sum_{\boldsymbol{\delta}}(\boldsymbol{k}\cdot\boldsymbol{\delta})^2 \ ,$$

$$\sum_{\boldsymbol{\delta}}(\boldsymbol{k}\cdot\boldsymbol{\delta})^2 = \frac{z}{3}k^2d^2$$

and

$$E(\boldsymbol{k}) = Dk^2d^2 \ ,$$

$$D = \tfrac{1}{3}zJS \approx \frac{k_\mathrm{B}T_\mathrm{C}}{2(S+1)} \ . \qquad (5.6.43)$$

The D is called the spin-wave stiffness constant. The vanishing of $E(\boldsymbol{k})$ with $\boldsymbol{k}\to 0$ results from the symmetry of the model (see acoustic branches of vibrations in Sect. 2.1.3). Indeed, for all spins to be rotated through the same angle, no energy needs to be expended; therefore, to rotate site spins through angles that differ as little as desired [$\boldsymbol{k}\to 0$ when $c(i)\approx$ const], it is necessary to spend an arbitrarily small amount of energy.

In an external magnetic field or in taking account of a weak spin–orbit interaction (i.e., when the energetically most favorable spin directions with respect to crystallographic axes arise—magnetic anisotropy), the symmetry is perturbed and a gap emerges in the spin-wave spectrum. Normally the gap is small compared to $k_B T$ and will therefore be disregarded.

At low temperatures ($T \ll T_C$) the number of spin waves is small and on the right-hand side of (5.6.30) we may set $\hat{S}_i^z \approx S$. Then $(2S)^{-1/2}\hat{S}_i^+$, $(2S)^{-1/2}\hat{S}_i^-$ satisfy Bose permutation relations, just as for phonons and, according to (5.6.34), spin waves are created by quasi-Bose operators. At low temperatures the spin-wave distribution function is therefore defined by the Planck formula (2.1.48, 2.7.36):

$$\bar{N}_k = \frac{1}{\exp(E(\boldsymbol{k})/k_B T) - 1} . \tag{5.6.44}$$

The total number of spin waves defines the deviation of the magnitude of magnetization from its saturation value at $T = 0$ K:

$$\frac{\langle \hat{S}^z \rangle}{S} - 1 = -\frac{1}{NS}\sum_k \bar{N}_k = -\frac{d^3}{(2\pi)^3 S}\int d\boldsymbol{k}\, \bar{N}_k . \tag{5.6.45}$$

At low temperatures ($D \gg k_B T$) small k are essential, so we may exploit (5.6.43) and extend the integration over all k. Then

$$\begin{aligned} d^3 \int d\boldsymbol{k}\, \bar{N}_k &= 4\pi \int_0^\infty (kd)^2 d(kd)[\exp(D(kd)^2/k_B T) - 1]^{-1} \\ &= 4\pi \left(\frac{k_B T}{D}\right)^{3/2} \int_0^\infty \frac{dx\, x^2}{\exp x^2 - 1} . \end{aligned} \tag{5.6.46}$$

The integral involved in (5.6.46) is equal to

$$\begin{aligned} \int_0^\infty \frac{dx\, x^2}{\exp x^2 - 1} &= \frac{1}{2}\int_0^\infty \frac{dy\, y^{1/2}}{e^y - 1} = \frac{1}{2}\sum_{n=1}^\infty \int_0^\infty dy\, y^{1/2} e^{-ny} \\ &= \frac{1}{2}\int_0^\infty dy\, y^{1/2} e^{-y} \sum_{n=1}^\infty n^{-3/2} = \frac{\sqrt{\pi}}{4}\zeta(3/2) , \end{aligned}$$

with $\zeta(x)$ being the Riemann ζ function. Allowance for (5.6.45, 46) yields

$$\frac{\langle \hat{S}^z \rangle}{S} - 1 = -C\left(\frac{T}{T_C}\right)^{3/2} , \tag{5.6.47}$$

where C is a numerical factor that depends on S. This formula was first derived by *Bloch* [5.27], and gives a fairly good fit to experimental data for many ferromagnets. The self-consistent field approximation, as is seen from the solution to (5.6.2), would have given an exponentially small value for the quantity (5.6.47) at $T \to 0$ K, contradicting experiment.

5.6.3 *d* Metals

We have stated previously that electrons in *d* metals occupy, in a way, an intermediate position between localized *d* electrons in magnetic insulators and normal itinerant electrons. On the whole, the problem as to the nature of *d* states is solved rather in favor of the band model [5.58, 62]. First, direct X-ray, electron, and optical spectroscopy data that provide straightforward information on energy-spectrum structure indicate the existence of a d band in *d* metals. Second, *d* metals normally exhibit large values for the heat-capacity, a fact which may be explained only by taking into account the large density of *d* electron states. Therefore, we have to assume that *d* electrons form a Fermi gas or, equally, a Fermi liquid, since the heat capacity of a localized subsystem is not proportional to the first power of temperature, but is proportional to the number of spin waves, i.e., $T^{3/2}$. Third, direct studies of the Fermi surface in *d* metals with the help of the de Haas–van Alphen effect and other methods indicate that *d* metals possess a highly complicated structure, which is very difficult to understand unless allowance is made for the delocalization of *d* states.

The arguments given above apply to all *d* metals, both to those which possess a magnetic order and to those which do not. Of the 24 pure *d* elements, two elements (Cr and Mn) are antiferromagnets and three elements (Fe, Co, Ni) are ferromagnets. These prominent properties will be discussed later, whereas at this point we note that the paramagnetism of the other *d* metals is definitely inconsistent with the localized model. At the same time, this paramagnetism is often dissimilar to Pauli paramagnetism, indicating that a purely band description of *d* bands is, nevertheless, not exhaustive.

The *d* band and other corresponding energy bands exhibit a number of specific features. First and foremost, because of the comparatively weak overlap of *d*-type wave functions, the band is rather narrow. Accordingly, the density of *d* states is very large. A typical feature is that the energy dependence of the density of states is a two-peaked curve, portrayed in Fig. 5.18 together with the corresponding curve for *s* states. This two-hump dependence is explained with the help of the following simple considerations [5.63] (see also Sect. 1.7).

We consider a diatomic molecule with a potential such as that pictured in Fig. 5.19a. If the potential barrier between two wells is sufficiently large, the

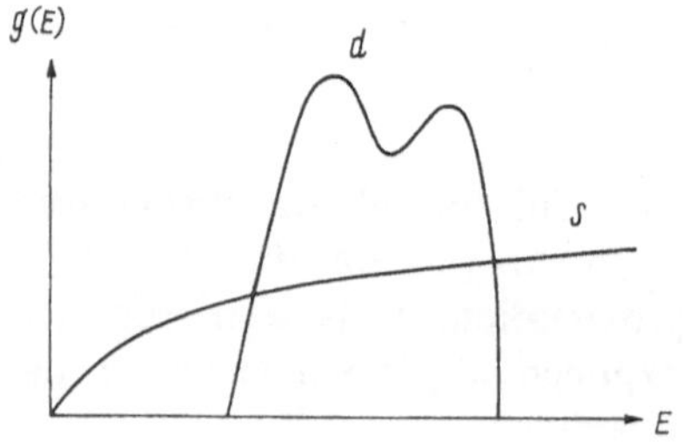

Fig. 5.18. Density-of-electron-states function $g(\varepsilon)$ in transition-*d* metals, for *d* and *s* bands

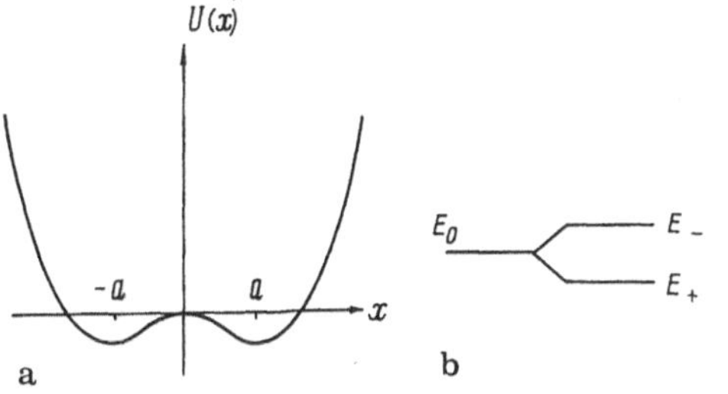

Fig. 5.19. Definition of bonding and antibonding states

zero-approximation function may be chosen in the form $\varphi_{1,2}(x) = \varphi(x \pm a)$, with $\varphi(x)$ being the wave function of an electron with energy E_0 in a well. Allowance for the overlap leads to the level E_0 being split up into two levels (Fig. 5.19b). The lower level has a symmetric wave function with respect to spatial coordinates and a maximum between the atomic centers (the bonding orbital). To the upper level corresponds an antisymmetric function (the antibonding orbital).

The separation of states into bonding and antibonding ones is rather typical of narrow bands, which are described by the LCAO approximation (Sect. 4.6.3). In the crystal the levels $E_\pm$ (Fig. 5.19b) smear into two density-of-states peaks. As regards the splitting of *d* states in the crystalline field, which is sometimes propounded as an explanation of the two-peaks, band calculations show that this splitting is small and is much less important than the separation of states into bonding and antibonding ones. The filling of the first peak (bonding states) is accompanied by an energy gain compared to isolated atoms, and the filling of the second peak involves a loss in energy. This accounts for the well-known fact that the bonding energy (and, accordingly, refractive index) is increased in the middle of each *d* transition series. Incidentally, such a correlation between *d*-band filling and bonding energy also indicates that the contribution of *d* electrons to chemical bonding is large and *d* electrons are itinerant.

The complicated structure of the density of *d* states is influenced also by the energy overlap of these states with the *s* states of conduction electrons. As we saw in connection with the weak coupling approximation (Sect. 4.3), an abrupt spectrum readjustment, or hybridization of states, occurs near the intercept of the energy curves (Fig. 5.20). This readjustment has a large effect on the density of states. As an example, Fig. 5.21 presents the calculated density of states for molybdenum [5.64].

We now discuss the problem of the temperature dependence of the magnetic susceptibility χ in *d* transition metals. This dependence is very unusual. For a number of *d* metals the magnetic susceptibility decreases rather appreciably with increasing temperature, for other *d* metals it increases, and for Pd it even exhibits a maximum at some temperature.

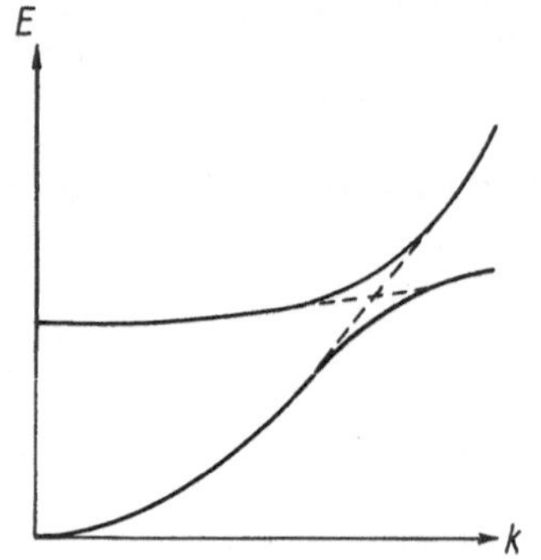

Fig. 5.20. Hybridization of s and d states

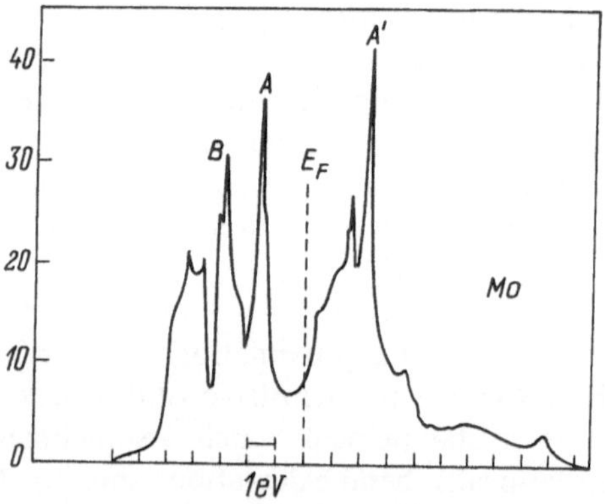

Fig. 5.21. Density of electron states as a function of energy $g(\varepsilon)$ in Mo according to band calculation

Some of the features peculiar to the χ versus T curve may be understood on the basis of the formula (3.5.87) for the Pauli susceptibility with an arbitrary dispersion relation. Here we rewrite the Pauli susceptibility in a somewhat different form:

$$\chi_P(T) = \chi_P(0)\left[1 - \frac{\pi^2}{6}(k_B T)^2 \left.\frac{\partial^2 \ln g(\varepsilon)}{\partial \varepsilon^2}\right|_{\varepsilon = \zeta_0}\right]. \tag{5.6.48}$$

Since the function $g(\varepsilon)$ is large and varies abruptly with varying ε, the second term in (5.6.48) may be very large indeed and may be both positive and negative. Yet this formula cannot account for, say, the non-monotonic $\chi(T)$ relation in Pd because the correlation of d electrons plays a leading part here. In the simplest version it is taken into account in terms of Fermi-liquid enhancement (5.2.35), by introducing the temperature dependence in the numerator and denominator of this formula:

$$\chi(T) = \chi_P(T)\frac{1}{1 + B_0\,\chi_P(T)/\chi_P(0)}. \tag{5.6.49}$$

When $1 + B_0 \ll 1$ (as with Pd), the temperature dependence may become very strong and non-monotonic. Nevertheless, (5.6.49) does not embrace all cases, either. Some alloys and compounds of transition metals, as well as pure Fe, Co, and Ni, which are ferromagnets at sufficiently low temperatures, exhibit a Curie–Weiss $\chi(T)$ dependence in the paramagnetic region rather than a Pauli or Fermi-liquid dependence, yet the d electrons are undoubtedly itinerant. This is explained by the following circumstance. The quasi–single-particle description is believed to be fairly good for the ground state of the d subsystem. As far as thermodynamics is concerned, however, the free energy is determined not only by the ground-state energy but also by entropy. The entropy of single-particle degrees of freedom is small, because of the Pauli exclusion principle, since the entropy is proportional to the number of thermally active electrons, i.e., $k_B T/\zeta_0$. On the other hand, in a system of strongly interacting particles there exist collective excitations whose energy is comparatively low—spin density fluctuations. In ferromagnets these are spin waves, which are due to the rotation of moments with a "dephasing" of the form $\exp(i\boldsymbol{q}\cdot\boldsymbol{R})$ and these waves exist whether or not the moments are described by the Heisenberg model. For any model invariant under rotation of all spins through the same angle, the spin frequency will tend to zero if $\boldsymbol{q} \to 0$. In a paramagnet the spin waves are known as paramagnon. They attenuate strongly and are therefore not quasiparticles in the usual sense, but they do contribute to the thermodynamics. According to (5.1.85), the contribution to the free energy is made by the singularities $1/\varepsilon(\boldsymbol{q}, \omega)$. In spin systems it is the $\boldsymbol{q}$- and ω-dependent susceptibility that plays a similar role. If the singularity is not a pole on the real axis of ω and no quasiparticle corresponds to it, then it will contribute to the thermodynamics.

It is these paramagnons that are associated with the peculiarities of the susceptibility χ, which has a large imaginary part of ω. Since the contribution of the one-electron degrees of freedom to the entropy is suppressed, the entropy is determined chiefly by the spin fluctuations and these are responsible, in particular, for the Curie–Weiss behavior. For a more detailed treatment of the various theoretical approaches to the temperature dependence of the susceptibility in d metals, see [5.65, 66].

What has been said above applies equally to ferromagnetic d metals. The condition of instability with respect to ferromagnetism may be obtained by use of the Fermi-liquid theory, from the divergence of the susceptibility

$$1 + B_0 < 0 \quad \text{or} \quad 1 < g(\zeta_0)I \ , \tag{5.6.50}$$

where I is the modulus of the exchange interaction ψ averaged over angles ($\psi < 0$). Formula (5.6.50) is called the Stoner criterion. The highest values of the density of states $g(\zeta_0)$ near the Fermi surface actually occur in Fe, Co, and Ni, of all the pure d transition elements. Stoner constructed a theory of ferromagnetism in d transition metals [5.67], based on the thermodynamics of a Fermi gas with an energy that depends on spin direction

$$\varepsilon_{\uparrow,\downarrow}(\boldsymbol{k}) = \varepsilon(\boldsymbol{k}) \mp \Delta \;, \tag{5.6.51}$$

where Δ was determined in a self-consistent fashion as the product of I by the total spin $(2N)^{-1}(N_\uparrow - N_\downarrow)$ [by analogy with (5.6.14)—the mean field is proportional to the magnetization]. The total spin is found from the ordinary one-electron formulas

$$N_{\uparrow,\downarrow} = \int d\varepsilon\, g(\varepsilon) \frac{1}{\exp((\varepsilon \mp \Delta)/k_B T) + 1} \;. \tag{5.6.52}$$

The criterion that holds for the ground state is (5.6.50). At finite temperatures Stoner's theory is inadequate for reasons stated in the foregoing. The theory has currently been superseded by various versions of the so-called spin-fluctuation theory of the magnetism of itinerant electrons [5.65].

We conclude by noting that the strongly correlated character of d electrons may, in our view (which, incidentally, is not a generally accepted point of view), manifest itself in the Fermi-liquid approach being inadequate even for the ground state. Elsewhere we have set forth relevant arguments, based on a treatment of a simple model problem [5.68].

5.6.4 Magnetism in the 4*f* Metals

Insofar as 4*f* (RE) metals go, the 4*f* state there, as already stated, is usually well localized and maintains the atomlike character. The magnetic moment for the atom of the 4*f* element in the metal is nearly the same as that for isolated atoms. The magnetic moment is determined by the total angular momentum of the atom rather than the spin angular momentum. In d metals the orbital moments collapse in going from atomic states to Bloch states, and the magnetism of Fe, Co, and Ni is almost a pure spin magnetism (with a small admixture of the polarization of the orbital moment due to the spin–orbit interaction). In f metals nearly all intrashell interactions, including the spin–orbit interaction, are stronger than the energy corrections of f states in the crystal relative to their atomic energies, so one has to take as a zero approximation the classification of the states of the f shell as in an isolated atom (according to LSJ).

Thus it appears that the f subsystem can be described by the Heisenberg model (with the replacement $\hat{\boldsymbol{S}}_i \to \hat{\boldsymbol{J}}_i$). What then are the magnetic structures that occur in rare-earth metals (REM)? Experimental facts can be summarized as follows:

1. *All "heavy" REMs from* Gd *to* Tm *(except for* Yb*) are ferromagnetic at low temperatures.* As the temperature is raised, ferromagnetism gives way to antiferromagnetism (except for Gd) rather than paramagnetism, and only as a result of one more phase transition as the temperature is increased further is it replaced

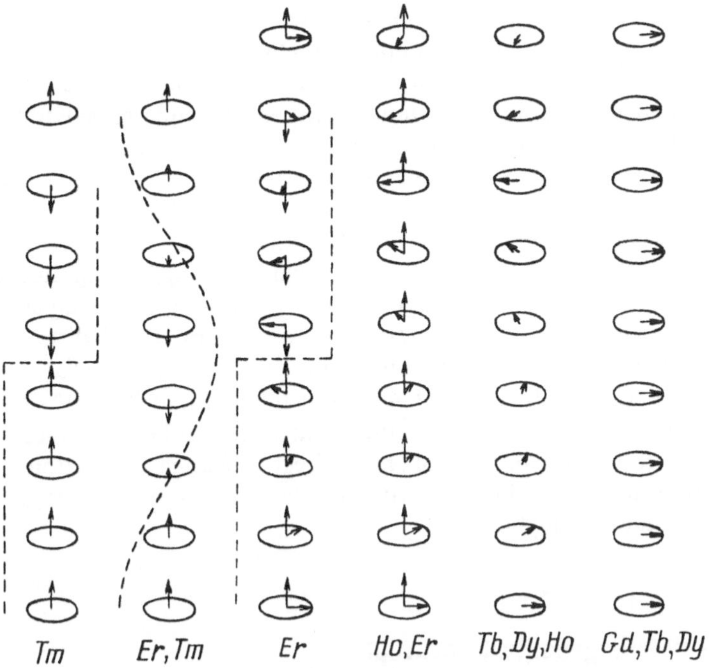

Fig. 5.22. Noncollinear atomic magnetic structures in heavy REMs

by the Langevin paramagnetism [i.e., a paramagnetism with the Curie–Weiss dependence of the susceptibility, $\chi(T)$].

2. *All "light"* REMs *from* Ce to Eu *are apparently antiferromagnetic*. The element Yb is a Pauli paramagnet. The various types of antiferromagnetic ordering in REMs are not similar to the familiar collinear structure (Fig. 5.15), being some complicated system of standing spin-density waves. In some ferromagnets (for example, Ho, Er) a helical ordering of spins in the perpendicular plane is superposed on the ferromagnetism [5.58].

This complicated magnetic behavior comes from the specific nature of the exchange interaction between the magnetic moments of the f subsystem.

The overlap of the wave functions of f states on different sites is negligible, even for nearest neighbors. Much more important here is the exchange interaction of f electrons with conduction electrons (the s–d (f) exchange model [5.69, 70]). The f moment causes the conduction-electron spin density to polarize. This polarization $\sigma(r)$ falls off with distance, obeying the same law as that for the total density around a charged impurity (Sect. 4.8.3):

$$\sigma(r) \propto \frac{\cos(2k_F r)}{r^3} . \quad (5.6.53)$$

In turn, the polarized spin density, which is separated from the first f moment by a distance r, tends, because of the s–f coupling, to align the second moment in a specific direction. Therefore, the effective f–f exchange integral decreases with distance obeying (5.6.53). This interaction is known as the Ruderman–Kittel–Kasuya–Yosida (RKKY) indirect exchange interaction. Since the aforementioned process corresponds to the energy contribution in second-order perturbation theory with respect to the s–f exchange integral I and occurs via the excited states of conduction electrons, for which the characteristic energy is the Fermi energy, the effective exchange interaction between neighboring sites is on the order of I^2/ζ_0 ($|I| \ll \zeta_0$). A large variety of complicated magnetic structures results because the RKKY interaction decreases slowly with distance in an oscillatory manner and, what is most important, the period of oscillations is incommensurate with the lattice period. Also bear in mind that (5.6.53) holds for the isotropic dispersion relation of s electrons. Since the real Fermi surface is anisotropic, the period of oscillations of the indirect exchange interaction is different in different directions, thus rendering the character of magnetic structures in REMs still more complicated.

5.7 Anderson's "Orthogonality Catastrophe"

A very interesting class of many-particle effects is associated with the response of the electron gas to external dynamic space-localized effects. Examples of these may be the Coulomb field of an X-radiation-induced hole in the ion core shell, the field of a charged particle diffusing in the metal, and other effects. At first sight, with neglect of the electron–electron interaction, the effect of such a field reduces to a redistribution of conduction electrons in single-particle states. Anderson [5.71] has demonstrated that the actual situation is much more complicated and even a weak perturbation may cause a radical reconstruction in the ground state of the system of a macroscopically large number of conduction electrons. This "orthogonality catastrophe" is an essentially many-particle effect, associated with the Pauli exclusion principle.

Consider the following problem [5.71]. Assume that at the time zero the system of noninteracting electrons is in the ground state $|\phi_0\rangle$, which is known to be described by the determinant (4.6.3) of single-particle states $\psi_\nu^{(0)}(\boldsymbol{r}, s)$, with s being the spin projection. As the set of quantum numbers ν, we choose the orbital quantum numbers l and m, the spin projection σ, and the principal quantum number n, specifying the energy:

$$\psi_\nu^{(0)}(\boldsymbol{r}, s) = \varphi_{nl}^{(0)}(r)\, Y_{lm}(\vartheta. \varphi)\chi_\sigma(s) \ , \qquad (5.7.1)$$

[see (4.8.42)]. Just as in Sect. 4.8.3, we assume for illustration that the system is in a large sphere of radius L with impermeable walls. Then, similar to (4.8.42, 47, 52), we have at large r

$$\varphi_{nl}^{(0)}(r) \approx \left(\frac{2}{L}\right)^{1/2} \frac{1}{r} \sin\left[\frac{\pi n}{L} r - \frac{\pi l}{2}\left(1 - \frac{r}{L}\right)\right], \tag{5.7.2}$$

$\varphi_{nl}^{(0)}(L) \equiv 0$, the energy levels are $E_{nl}^{(0)} = \pi^2\hbar^2/2mL^2(n + l/2)^2$. Assume that then a spherically symmetric perturbation V is applied, giving rise to scattering phases δ_{nl}. The new eigenfunctions satisfying the condition $\varphi_{nl}(L) = 0$, are of the form

$$\varphi_{nl}(r) \approx \left(\frac{2}{L}\right)^{1/2} \frac{1}{r} \sin\left[\frac{\pi n}{L} r - \left(\frac{\pi l}{2} - \delta_{nl}\right)\left(1 - \frac{r}{L}\right)\right] \tag{5.7.3}$$

with r being much larger than the range of the potential V. The corresponding energy levels are

$$E_{nl} = \frac{\pi^2\hbar^2}{2mL^2}\left(n + \frac{l}{2} - \frac{\delta_{nl}}{\pi}\right)^2 .$$

The reconstruction of single-particle states under the influence of the perturbation V may be characterized by the overlap integral of the "old" and "new" functions:

$$\begin{aligned} \Sigma_{\nu\nu'} &= \sum_s \int d\boldsymbol{r}\, \phi_{nlm\sigma}^{*}(\boldsymbol{r}, s)\, \psi_{n'l'm'\sigma'}(\boldsymbol{r}, s) \\ &= \delta_{ll'}\, \delta_{mm'}\, \delta_{\sigma\sigma'}\, S_{nn'}^{l} , \end{aligned} \tag{5.7.4}$$

where

$$S_{nn'}^{l} = \int_0^L dr\, r^2\, \varphi_{n'l}^{(0)}(r)\, \varphi_{nl}(r) \approx \frac{\sin\delta_{nl}}{\pi} \frac{1}{n' - n + \delta_{nl}/\pi} . \tag{5.7.5}$$

In calculating (5.7.5), we have allowed for Eqs. (5.7.2) and (5.7.3) and assumed n, n' to be much larger than unity (but $n - n'$ may be arbitrary). For weak V we have $\delta_{nl}/\pi \ll 1$, and

$$S_{nn}^{l} = \frac{\sin\delta_{nl}}{\delta_{nl}} \approx 1 .$$

Thus a small perturbation induces an equally small change in a single-particle state. The "orthogonality catastrophe" arises because this is no longer true for a many-particle state $|\phi\rangle$. We calculate

$$S = \langle \phi_0 | \phi \rangle . \tag{5.7.6}$$

Proceeding in a way similar to (4.6.4), we find

$$S = \det \Sigma_{\nu\nu'} = \prod_l [\det S^l_{nn'}]^{2(2l+1)} \equiv \prod_l S_l^{2(2l+1)} , \tag{5.7.7}$$

where S_l is the determinant of the $N_F \times N_F$ matrix ($N_F = k_F L/\pi$):

$$\begin{aligned} S_l &= \det \left\| \frac{\sin \pi \tilde{\delta}_n}{\pi} \frac{1}{n' - n + \tilde{\delta}_n} \right\| = \prod_{n=1}^{N_F} \frac{\sin \pi \tilde{\delta}_n}{\pi} \Delta_{N_F} , \\ \Delta_N &= \det \left\| \frac{1}{n' - n + \tilde{\delta}_n} \right\|_{n', n = 1, \ldots, N} . \end{aligned} \tag{5.7.8}$$

Here we have allowed for (5.7.5), introduced the notation $\tilde{\delta}_n = \delta_n/\pi$, and omitted for brevity the subscript l. The determinant involved in (5.7.8) has the following structure:

$$D_N = \det \left\| \frac{1}{a_{n'} + b_n} \right\| \equiv \begin{vmatrix} \frac{1}{a_1 + b_1} & \frac{1}{a_1 + b_2} & \cdots & \frac{1}{a_1 + b_N} \\ & \cdots\cdots & & \\ \frac{1}{a_N + b_1} & \frac{1}{a_N + b_2} & \cdots & \frac{1}{a_N + b_N} \end{vmatrix} \tag{5.7.9}$$

($a_{n'} = n'$, $b_n = -n + \tilde{\delta}_n$). On subtracting the last row from the preceding ones and placing the common multipliers outside, we have

$$D_N = \frac{\prod_{n=1}^{N-1} (a_N - a_n)}{\prod_{m=1}^{N} (a_N + b_m)} \begin{vmatrix} \frac{1}{a_1 + b_1} & \frac{1}{a_1 + b_2} & \cdots & \frac{1}{a_1 + b_N} \\ & \cdots\cdots & & \\ \frac{1}{a_{N-1} + b_1} & \frac{1}{a_{N-1} + b_2} & \cdots & \frac{1}{a_{N-1} + b_N} \\ 1 & 1 & \cdots & 1 \end{vmatrix} \tag{5.7.10}$$

Subtracting now the last column from the preceding ones gives

$$D_N = \frac{\prod_{n=1}^{N-1} (a_N - a_n)(b_N - b_n)}{\prod_{m=1}^{N} (a_N + b_m)(b_N + a_m)} D_{N-1} , \tag{5.7.11}$$

whence it is easy to prove by induction that

$$D_N = \frac{\prod\limits_{m>n} (a_m - a_n)(b_m - b_n)}{\prod\limits_{m',n'} (a_{m'} + b_{n'})} . \tag{5.7.12}$$

Substitution of (5.7.12) into (5.7.8) yields

$$\begin{aligned} S_l &= \prod_{n=1}^{N_F} \frac{\sin \pi \tilde{\delta}_n}{\pi} \frac{\prod\limits_{m>m'}^{N_F} \prod\limits_{n>n'}^{N_F} (m-m')(n'-n+\tilde{\delta}_n - \tilde{\delta}_{n'})}{\prod\limits_{m,n}^{N_F} (m-n+\tilde{\delta}_n)} \\ &= \prod_{n=1}^{N_F} \frac{\sin \pi \tilde{\delta}_n}{\pi \tilde{\delta}_n} \prod_{m>m'} \prod_{k>k'} \frac{(m-m')(k'-k+\tilde{\delta}_k - \tilde{\delta}_{k'})}{(m-m'+\tilde{\delta}_{m'})(k-k'+\tilde{\delta}_{k'})} . \end{aligned} \tag{5.7.13}$$

Calculating $\ln |S_l|$ with allowance for the identity

$$\frac{\sin \pi x}{\pi x} = \prod_{m=1}^{\infty} \left(1 - \frac{x^2}{m^2}\right) , \tag{5.7.14}$$

we find

$$\begin{aligned} \ln |S_l| = \sum_{n=1}^{N_F} \Bigg\{ \sum_{m=1}^{\infty} \ln\left(1 - \frac{\tilde{\delta}_n^2}{m^2}\right) + \sum_{m=1}^{n-1} \Bigg[\ln\left(1 + \frac{\tilde{\delta}_m - \tilde{\delta}_n}{n-m}\right) \\ - \ln\left(1 + \frac{\tilde{\delta}_m}{n-m}\right) - \ln\left(1 - \frac{\tilde{\delta}_n}{n-m}\right) \Bigg] \Bigg\} . \end{aligned} \tag{5.7.15}$$

We expand ln in a power series of $\tilde{\delta}_n$ up to terms $\sim \tilde{\delta}_n^2$:

$$\begin{aligned} \ln |S_l| &\approx - \sum_{n=1}^{N_F} \tilde{\delta}_n^2 \sum_{m=1}^{\infty} \frac{1}{m^2} - \frac{1}{2} \sum_{n=1}^{N_F} \sum_{m=1}^{n-1} \left(\frac{\tilde{\delta}_m - \tilde{\delta}_n}{n-m} \right)^2 \\ &\quad + \frac{1}{2} \sum_{n=1}^{N_F} \sum_{m=1}^{n-1} \frac{\tilde{\delta}_n^2 + \tilde{\delta}_m^2}{(n-m)^2} = - \sum_{n=1}^{N_F} \tilde{\delta}_n^2 \sum_{m=1}^{\infty} \frac{1}{m^2} - \frac{1}{2} \sum_{n=1}^{N_F} \sum_{p=1}^{N_F - n} \\ &\quad \times \left(\frac{\tilde{\delta}_{n+p} - \tilde{\delta}_n}{P} \right)^2 + \frac{1}{2} \sum_{m=1}^{N_F} \tilde{\delta}_m^2 \sum_{p=1}^{N_F - m} \frac{1}{p^2} + \frac{1}{2} \sum_{n=1}^{N_F} \tilde{\delta}_n^2 \sum_{k=1}^{n} \frac{1}{k^2} \\ &= - \frac{1}{2} \sum_{n=1}^{N_F} \tilde{\delta}_n^2 \left(\sum_{m=N_F-n+1}^{\infty} \frac{1}{m^2} + \sum_{m=n+1}^{\infty} \frac{1}{m^2} \right) \\ &\quad - \frac{1}{2} \sum_{n=1}^{N_F} \sum_{p=1}^{N_F-n} \left(\frac{\tilde{\delta}_{n+p} - \tilde{\delta}_n}{p} \right)^2 . \end{aligned} \tag{5.7.16}$$

At $n \gg 1$

$$\sum_{m=n}^{\infty} \frac{1}{m^2} \approx \int_{n}^{\infty} \frac{dm}{m^2} = \frac{1}{n} ; \quad \sum_{m=1}^{n} \frac{1}{m} \approx \ln n . \tag{5.7.17}$$

Allowing for the fact that $\tilde{\delta}_{N_F} \equiv \delta_l(k_F)/\pi$ is finite and $\delta_0 = 0$ (Sect. 4.8.3), we find that the first sum in (5.7.16) diverges at $N_F \to \infty$ as $-1/2[\delta_l(k_F)/\pi]^2 \ln N_F$. The second sum remains finite. It may be shown by a straightforward calculation that all the contributions to (5.7.15) by the higher-order terms of the series expansions of $\tilde{\delta}_n$ starting from $\tilde{\delta}_n^3$) are also finite. As a result, allowing for (5.7.7), we find

$$|S| \sim N_F^{-\alpha} , \quad \alpha = \sum_{l=0}^{\infty} (2l+1)\left[\frac{\delta_l(k_F)}{\pi}\right]^2 . \tag{5.7.18}$$

Thus when $N_F \to \infty$ and the perturbation is as small as desired but finite $S = 0$, that is, the perturbation brings about a radical reconstruction of the ground state of the many-electron system. For the generalization of the formula (5.7.18) for the case of a nonspherically symmetric perturbation, we refer the reader to [5.72]. The divergence that has led to Eq. (5.7.18) arises from the presence of a sharp boundary between occupied and empty states and from the finite density of states at the Fermi level (Sect. 5.3.5, where the instability of the ground state of the Fermi gas in the case of an arbitrary small attraction between particles arose essentially for the same reasons).

This is an example of the infrared catastrophe. The name derives from quantum field theory and implies peculiarities in the quantities calculated which arise due to the contribution of low-energy excitations. Such excitations in metals are electron-hole pairs, i.e., the passage of an electron from beneath the Fermi surface to an empty state with an energy close to the initial energy. As has already been stated in the foregoing, the Anderson orthogonality catastrophe manifests itself in the structure of X-ray absorption spectra [5.30, 5.73], and also in the temperature dependence of the quantum diffusion coefficient (Sect. 1.8.7) of a heavy particle in metals (for example, a muon) [5.74, 5.75].

A key feature in the occurrence of infrared catastrophes in the physics of metals, in addition to the Pauli exclusion principle, is the dynamic character of the perturbation. Thus in our case the perturbation was switched on gradually. Infrared catastrophes occur also when electrons are scattered in quantum systems with internal degrees of freedom. The most important and most studied example is the Kondo effect [5.76] which arises in the case of scattering from magnetic impurities, where the internal degrees of freedom are spin coordinates. Unfortunately, the theory of this effect calls for sophisticated mathematical methods which lie beyond the scope of this book; we refer the reader to reviews [5.77, 5.78]. Infrared catastrophes arise also in problems of the interaction of electrons with two-level centers (Sect. 2.5) in metallic glasses [5.79] and with local phonons [5.80]. These same catastrophes are responsible for anomalies in the properties of metals and alloys exhibiting narrow peaks in the density of electron states near the Fermi level [5.18, 5.81].

5.8 Conclusion

The treatment of many-electron effects concludes our introduction to the quantum theory of solids and, in some cases, brings the reader close to topical problems of this theory in its modern form. The greatest achievement of the many-electron theory is the concept of quasiparticles, which develops and generalizes the usual one-electron band theory. At the same time, the introduction of correlation effects leads to the formation of collective elementary excitations (plasmons, excitons, spin waves, etc.) and, in some cases, results in a radical readjustment of the ground state (metal–insulator transition, superconductivity, magnetic ordering, etc.) or the properties of current carriers (for example, in the case of polarons). The interaction of the various elementary excitations is responsible for scattering processes and, consequently, the kinetic properties of solids.

We hope that the material presented here will adequately prepare the reader for a further, more profound, and mathematically more rigorous study of the problems enumerated. Methods of quantum field theory can be employed for this purpose. These methods simplify calculations to a considerable extent and, when skillfully used, introduce clarity into the presentation. In the final analysis, however, they do not bring us outside the scope of the general concepts about the role of many-particle effects with which we have familiarized ourselves in this book.

Addenda: Recent Developments

We wish to make some comments and minor additions to the book in the light of recent publications that appeared after the main text was written.

1. Section 4.5.6 treats a two-dimensional model of the electron gas in a quantizing magnetic field. In this context it is worth noting that such a system may in fact be realized at the surface of semiconductors, to which the electrons are "pushed" by an externally applied electric field (for example in metal-oxide-semiconductor structures). These entities have recently been the focus of attention in connection with the discovery by K. von Klitzing of a so-called quantized Hall effect (1985 Nobel Prize for Physics). On this topic we refer the reader to K. von Klitzing's Nobel lecture [A.1] and the literature cited therein.

2. In Sect. 4.2.5 we considered topological electronic transitions due to the emergence of van Hove singularities of the density of electron states $g(E)$ on the Fermi level E_F. If there is a sufficiently narrow and sharp $g(E)$ peak (a situation which is typical of transition metals, their alloys and compounds), then, as shown in [5.18], it may affect the properties of metals even when separated from E_F by a distance that is much larger than the thermal smearing $k_B T$. In this case electronic phase transitions of non-topological type are possible. The authors of [A.2] report an experimental discovery of an electronic transition of this type in titanium-based alloys.

3. In Sect. 4.6 we treated very briefly methods of band calculations of the energy spectrum. The contemporary development of computational solid state physics is characterized primarily by successful attempts to calculate sufficiently complicated physicochemical properties of solids on the basis of the density functional method without invoking any fitting parameters. Specifically, a study has been made of the causes for the relative stability of the various crystalline structures [A.3–5]. Calculations have been carried out to determine the lattice spacings of not only pure metals but also compounds [A.6], the formation energy of chemical compounds [A.7], the elastic moduli of transition metals [A.8], and the various magnetic characteristics of ferromagnets [A.9, 10].

4. In Sect. 4.9 it is state that the theory of disordered systems deals chiefly with self-averaging quantities. Recently there has arisen and has been developing rapidly a theory of so-called mesoscopic fluctuations of the conductivity, the density of electron states, and other quantities in disordered systems with small, but not microscopic, linear dimensions in some direction. It is the non-self-

averaging effects that are in the foreground here. For these problems see, for example, [A.11–13].

5. A remark to Sect. 5.4. Doubtless, one of the most interesting events in solid state physics has recently been the discovery of high-temperature superconductivity in the systems $La_{2-x}(Sr, Ba)_xCuO_4$ and in other similar compounds [A.14]. The nature of this phenomenon remains to be ascertained.

6. A remark to Sect. 5.6. One of the most rapidly developing areas of solid state physics has now become the investigation of the so-called heavy-fermion or heavy-electron systems. These are, for example, cerium or uranium compounds ($CeAl_3$, $CeCu_6$, UBe_{13}, UPt_3), characterized by giant values of the density of states at the Fermi level with effective masses of the order of 10^3 electron masses, and by many other unusual properties such as exotic superconductivity [A.15, 16, 17]. In [A.17] one can find a lot of interesting material on the magnetism of compounds of uranium and transuranium elements and on the problem of intermediate valence.

7. A very interesting area of condensed matter physics, not touched upon in the book, is the physics of systems with charge and spin density waves [A.18, A.19].

8. In recent years, applications of methods and phenomena of the physics of elementary particles such as positron annihilation, muon propagation, properties of positrons and mesoatoms in solids have begun to develop rapidly in solid state physics [A.20, A.21].

References

Chapter 1

1.1 J. Frenkel: *Kinetic Theory of Liquids* (Clarendon Press, Oxford 1946)
1.2 G. Tamman: Ann. d. Phys. **82**, 240 (1927)
1.3 M. Hansen, K. Anderko: *Constitution of Binary Alloys* (McGraw Hill, New York 1958)
1.4 L.D. Landau, E.M. Lifshitz: *Statistical Physics*, Part 1 (Pergamon Press, Oxford 1980)
1.5 A.V. Bushman, V.E. Fartov: Usp. Fiz. Nauk **140**, 174 (1983)
1.6 G. Friedel: Ann. Phys. **18**, 273 (1922)
1.7 P.G. de Gennes: *The Physics of Liquid Crystals* (Clarendon Press, Oxford 1974)
1.8 E.S. Fedorov: *Symmetry and Crystal Structure* (Izd. AN SSSR, Moskva 1949 (in Russian)
1.9 A. Schoenflies: *Theorie der Kristalstruktur* (Berlin 1923)
1.10 M.J. Buerger: *Elementary Crystallography* (Wiley, New York 1963)
1.11 R.W. Wickoff: *Crystal Structures* (Interscience, New York 1963)
1.12 L.D. Landau, E.M. Lifshitz: *Quantum Mechanics* (Pergamon Press, Oxford 1977)
1.13 a) H. Streitwolf: *Gruppentheorie in der Festkörperphysik* (Akademie-Verlag, Leipzig 1967)
b) V. Heine: *Group Theory in Quantum Mechanics* (Pergamon Press, Oxford 1960)
c) J.P. Elliott, P.G. Dawber: *Symmetry in Physics* (Macmillan Press, London 1979)
d) M. Hammermesh: *Group Theory and its Application to Physical Problems* (Pergamon Press, Oxford 1962)
1.14 A.P. Cracknell: *Applied Group Theory* (Pergamon Press, Oxford 1968)
1.15 V.M. Goldschmidt: Skrifter der Norsre Videnskaps **11**, 11 (1926)
1.16 L. Pauling: Proc. Roy. Soc. **114**, 181 (1927)
1.17 *International Tables of* X-*Ray Spectroscopy* (Birmingham 1965)
1.18 P.W. James: *The Optical Principles of the Diffraction of* X-*Rays* (London 1950)
1.19 P. Ewald, ed.: *Fifty Years of* X-*Ray Diffraction* (Utrecht 1962)
1.20 R. Nathans, S. Pickart: "Magnetism", Vol. 3, ed. by G.T. Rado, H. Suhl (Acad. Press, New York 1963)
1.21 A.J. Freeman, R.E. Watson: Phys. Rev. **127**, 2058 (1962)
1.22 M. Born, J.R. Oppenheimer: Ann. d. Phys. **84**, 457 (1927)
1.23 J.C. Slater: *Quantum Theory of Molecules and Solids*, Vol. 1, *Electronic Structure of Molecules* (McGraw Hill, New York 1963)
1.24 J.J. Rehr, E. Zaremba, W. Kohn: Phys. Rev. **B12**, 2062 (1975)
1.25 W. Heisenberg: Z. Phys. **49**, 619 (1928)
1.26 P.A.M. Dirac: Proc. Roy. Soc. **A123**, 714 (1929)
1.27 L. Pauling: *The Nature of the Chemical Bond* (Cornell Univ. Press, New York 1960)
1.28 H. Eyring, J. Walter, G.E. Kimball: *Quantum Chemistry* (Cornell Univ. Press, New York 1946)
1.29 P. Ewald: Ann. Phys. **64**, 251 (1921)
1.30 J.M. Ziman: *Principles of the Theory of Solids* (Cambridge Univ. Press, London 1972)
1.31 V. Heine, D. Weaire: "Solid State Physics", Vol. 24, ed. H. Ehrenreich, F. Seitz, D. Turnbull (Acad. Press, New York 1970)

1.32 P.S. Rudman, J. Stringer, R.I. Jaffee (eds.): *Phase Stability in Metals and Alloys* (McGraw Hill, New York 1967)
1.33 A.I. Kitaigorodskii: *Molecular Crystals* (Nauka, Moskva 1971 (in Russian)
1.34 J.D. Watson: *Molecular Biology of the Gene* (Benjamin, Massachusetts 1965), Chap. 4
1.35 L. Pauling: J. Am. Chem. Soc. **57**, 2680 (1935)
1.36 V.G. Vaks, V.I. Zinenko, B.E. Shneider: Usp. Fiz. Nauk **141**, 629 (1983)
1.37 J. de Boer: Physica **14**, 139 (1948)
1.38 I.M. Khalatnikov: *Theory of Superfluidity* (Nauka, Moskva 1971) (in Russian)
1.39 G.A. Vardanyan: Usp. Fiz. Nauk **144**, 113 (1984)
1.40 J.C. Slater: *Quantum Theory of Molecules and Solids*, Vol. 3, *Insulators, Semiconductors and Metals* (McGraw Hill, New York 1967)
1.41 S.P. Shubin: *Phys. Z. Sowjetunion*, **5**, 81 (1934)
1.42 L.T. Chadderton: *Radiation Damage in Crystals* (Cambridge Univ. Press, London 1964)
1.43 M.W. Thompson: *Defects and Radiation Damage in Metals* (Cambridge Univ. Press, London 1969)
1.44 I.M. Lifshitz, S.A. Gredeskul, L.A. Pastur: *Introduction to the Theory of Disordered Systems* (Nauka, Moskva 1982) (in Russian)
1.45 J.D. Bernal: Proc. Roy. Soc. **A280**, 299 (1964)
1.46 J.C. Wright: IEEE Trans. Magnetics, Mag. -12, No 2, 95 (1976)
1.47 C.H. Bennett: J. Appl. Phys. **43**, 2727 (1972)
1.48 H.-J. Günterodt, H. Beck (des.): *Glassy Metals 1. Ionic Structure, Electronic Transport and Crystallization* (Springer, Berlin 1981)
1.49 J.M. Ziman: *Models of Disorder* (Cambridge Univ. Press, London 1979)
1.50 S.V. Vonsovsky, E.A. Turov: Izvestiya AN SSSR (ser. fiz.) **42**, 1570 (1978)

Chapter 2

2.1 A.A. Maradudin, E.W. Montroll, G.H. Weiss: *Theory of Lattice Dynamics in the Harmonic Approximation* (Acad. Press, New York 1971)
2.2 M. Born, Th. v. Karman: Z. Phys. **13**, 297 (1912); **14**, 15, 65 (1913)
2.3 P.P. Debye: Ann. d. Phys. **39**, 789 (1912)
2.4 J. Goldstone: Nuovo Cimento **19**, 154 (1961)
2.5 W.C. Owerton: 7th Intern. Conf. Low Temp. Phys. (National Research Council of Canada, Toronto 1960), p. 677
2.6 L. van Hove: Phys. Rev. **89**, 1189 (1953)
2.7 R.E. Peierls: *Quantum Theory of Solids* (Clarendon Press, Oxford 1955)
2.8 W.H. Keesom, D.W. Dobrzynski: Physica **1**, 1085 (1934)
2.9 R. Peierls: Ann. d. Phys. **93**, 1055 (1929)
2.10 I.M. Lifshitz: Zh. Eksp. Teor. Fiz. **38**, 1569 (1960)
2.11 P.A.M. Dirac: *The Principles of Quantum Mechanics* (Clarendon Press, Oxford 1958)
2.12 L.D. Kudryavtsev: *Mathematical Analysis* (Vysshaya Shkola, Moskva 1970), Chap. 7 (in Russian)
2.13 L.D. Landau, E.M. Lifshitz: *Theory of Elasticity* (Pergamon Press, Oxford 1970)
2.14 P.W Anderson, B.I. Halperin, C.M. Varma: Phil. Mag. **25**, 1 (1972)
2.15 W.A. Phillips: J. Low Temp. Phys. **7**, 351 (1972)
2.16 V.G. Karpov, M.I. Klinger, F.N. Ignatjev: Zh. Eksp. Teor. Fiz. **84**, 760 (1983)
2.17 L.D. Landau, E.M. Lifshitz: *Electrodynamics of Continuous Media* (Pergamon Press, Oxford 1984)
2.18 L. van Hove: Phys. Rev. **95**, 249, 1374 (1954)
2.19 S.V. Tyablikov: *Methods of the Quantum Theory of Magnetism* (Nauka, Moskva 1965) (in Russian)

2.20 G.C. Wick: Phys. Rev. **80**, 268 (1950)
2.21 C. Bloch, C. de Dominicis: Nucl. Phys. **7**, 459 (1958)
2.22 L.I. Mandelstamm, G.S. Landsberg: Compt. Rend. **189**, 109 (1928)
2.23 Ch. Raman, K.S. Krishnan: Nature **121**, 3048 (1928)
2.24 R. Mössbauer: Z. Phys. **151**, 124 (1958)
2.25 A. Abragam: "Low Temperature Physics", ed. by C. DeWitt (Plenum Press, New York 1962)
2.26 H. Frauenfelder: *The Mössbauer Effect* (Benjamin, Massachusetts 1962)
2.27 H.J. Lipkin: *Quantum Mechanics* (North-Holland, Amsterdam 1973)
2.28 L.D. Landau: Zh. Eksp. Teor. Fiz. **11**, 592 (1941); J. Phys. USSR **5**, 91 (1941)
2.29 M. Born, K. Huang: *Dynamical Theory of Crystal Lattices* (Clarendon Press, Oxford 1954)
2.30 F. Jona, G. Shirane: *Ferroelectric Crystals* (Pergamon Press, Oxford 1962)
2.31 A.H. Cottrell: *Dislocations and Plastic Flow in Crystals* (Clarendon Press, Oxford 1953)
2.32 H.G. van Bueren: *Imperfections in Crystals* (North-Holland, Amsterdam 1960)
2.33 A.M. Kosevich: *Physical Mechanics of Real Crystals* (Naukova Dumka, Kiev 1981) (in Russian)
2.34 N.W. Ashcroft, N.D. Mermin: *Solid State Physics* (Holt, Rinehart and Winston, New York 1976)

Chapter 3

3.1 R. Tolman, T. Stewart: Phys. Rev. **8**, 97 (1916); **9**, 164 (1917)
3.2 I.M. Tsidilkovski: Usp. Fiz. Nauk **115**, 321 (1975)
3.3 J. Frenkel: Z. Phys. **26**, 117 (1924)
3.4 L.D. Landau, E.M. Lifshitz: *Mechanics* (Pergamon Press, Oxford 1976)
3.5 H. Fröhlich: Physica **6**, 406 (1937)
3.6 J. Dorfmann: Z. Phys. **23**, 286 (1924)
3.7 W. Pauli: Z. Phys. **41**, 81 (1927)
3.8 J. Frenkel: Z. Phys. **49**, 31 (1928)
3.9 L.D. Landau: Z. Phys. **64**, 629 (1930)
3.10 J.A. van Leeven: J. Phys. et Radium (6) **2**, 361 (1921)
3.11 Ya. P. Terletsky: Zh. Eksp. Teor. Fiz. **9**, 796 (1939)
3.12 N. Bohr: Dissertation, Copenhagen (unpublished)
3.13 J.E. Mayer, M. Goeppert Mayer: *Statistical Mechanics* (Wiley, New York 1977)
3.14 E. Stoner: Proc. Roy. Soc. **A152**, 672 (1935)
3.15 R.E. Peierls: Z. Phys. **81**, 186 (1933)
3.16 F. Seitz: *Modern Theory of Solids* (McGraw Hill, New York 1940)
3.17 J. Frenkel: Phys. Z. Sowjetunion **7**, 452 (1935)
3.18 L.E. Gurevitch: J. Phys. USSR **9**, 4 (1945); **10**, 67 (1946)
3.19 F.J. Blatt: *Physics of Electronic Conduction in Solids* (McGraw Hill, New York 1968)
3.20 S.R. de Groot, P. Mazur: *Non-Equilibrium Thermodynamics* (North-Holland, Amsterdam 1962)
3.21 R. Gans: Ann. d. Phys. **20**, 293 (1906)
3.22 I.K. Kikoin (ed.): Tables of Physical Quantities (Atomizdat, Moskva 1976), p. 470 (in Russian)
3.23 P.L. Kapitsa: Proc. Roy. Soc. **A123**, 292 (1929)
3.24 A.B. Pippard: Rep. Prog. Phys. **23**, 176 (1960)
3.25 G.E.H. Reuter, E.H. Sondheimer: Proc. Roy. Soc. **A195**, 336 (1948)
3.26 M. Born, E. Wolf: *Principles of Optics* (Pergamon Press, Oxford 1968)
3.27 R.W. Wood: Phys. Rev. **44**, 353 (1933)
3.28 R.L. Kronig: Nature **133**, 211 (1934)

3.29 M. Ya. Azbel, E.A. Kaner: Zh. Eksp. Teor. Fiz. **32**, 896 (1957)
3.30 J. Dorfmann: Doklady AN SSSR **81**, 765 (1951)
3.31 R.B. Dingle: Proc. Roy. Soc. **A212**, 38 (1952)
3.32 O.V. Konstantinov, V.I. Perel: Zh. Eksp. Teor. Fiz. **38**, 161 (1960)
3.33 P. Aigrain: Proc. Intern. Conf. Semiconductors Phys. (1960), p. 224.

Chapter 4

4.1 J. Frenkel: *Principles of Wave Mechanics* (Moskva, Leningrad 1933), Sect. 34 (in Russian)
4.2 F. Bloch: Z. Phys. **52**, 555 (1928)
4.3 R.L. Kronig, W.G. Penney: Proc. Roy. Soc. **180**, 199 (1931)
4.4 V. Rojansky: *Introductory Quantum Mechanics* (Acad. Press, New York 1938)
4.5 H. Jones: *The Theory of Brillouin Zones and Electron States in Crystals* (North-Holland, Amsterdam 1962)
4.6 H.A. Kramers: Physica **2**, 483 (1935)
4.7 L. Brillouin: J. Phys. Rad. **1**, 377 (1930)
4.8 I.M. Lifshitz, M. Ya. Azbel, M.I. Kaganov: *Electron Theory of Metals* (Nauka, Moskva 1971) (in Russian)
4.9 I.M. Lifshitz: Zh. Eksp. Teor. Fiz. **38**, 1569 (1960)
4.10 V.G. Vaks, A.V. Trefilov, S.V. Fomichev: Zh. Eksp. Teor. Fiz. **80**, 1613 (1981)
4.11 D.R. Overcash, T. Davis, J.W. Cook, M.J. Skove: Phys. Rev. Lett. **46**, 287 (1981)
4.12 R.E. Peierls: Ann. d. Phys. **4**, 121 (1930)
4.13 S.P. Shubin: *Lectures in the Quantum Theory of Solids* (Sverdlovsk, 1933–1935, unpublished)
4.14 W. Shockley: Phys. Rev. **52**, 866 (1937)
4.15 W.V. Houston: Phys. Rev. **57**, 184 (1940)
4.16 S.V. Vonsovsky: Zh. Eksp. Teor. Fiz. **9**, 154 (1939)
4.17 M.V. Fedorjuk: *The Saddle Point Method* (Nauka, Moskva 1977) (in Russian)
4.18 L.D. Landau: Phys. Z. Sowjetunion **1**, 88 (1932)
4.19 A.H. Wilson: Proc. Roy. Soc. **A133**, 458 (1931)
4.20 R.A. Smith: *Semiconductors* (Cambridge Univ. Press, London 1964)
4.21 I.M. Tsidilkovski: *Electrons and Holes in Semiconductors* (Nauka, Moskva 1972); *Band Structure of Semiconductors* (Nauka, Moskva 1978) (in Russian)
4.22 M. Ya. Azbel: Phys. Rev. Lett. **43**, 1954 (1979)
4.23 E.H. Lieb, F.Y. Wu: Phys. Rev. Lett. **20**, 1445 (1968)
4.24 N.F. Mott: Proc. Phys. Soc. (London) **A62**, 416 (1949)
4.25 M.S. Svirsky, S.V. Vonsovsky: Fiz. Met. Metalloved. **4**, 392 (1957)
4.26 M.I. Katsnelson, V. Yu. Irkhin: J. Phys. **C17**, 4291 (1984)
4.27 N.F. Mott: *Metal–Insulator Transitions* (Taylor and Francis, London 1974)
4.28 P.W. Anderson: Phys. Rev. **109**, 1492 (1958)
4.29 H.A. Bethe: *Intermediate Quantum Mechanics* (Benjamin, Massachusetts 1964)
4.30 D.R. Hartree: Proc. Camb. Phil. Soc. **24**, 89 (1928)
4.31 V.A. Fock: Z. Phys. **61**, 126 (1930)
4.32 W.A. Harrison: *Solid State Physics* (McGraw Hill, New York 1970)
4.33 J.C. Slater: *Quantum Theory of Molecules and Solids*, Vol. 4, *The Self-Consistent Field in Molecules and Solids* (McGraw Hill, New York 1974)
4.34 V.L. Moruzzi, J.F. Janak, A.R. Williams: *Calculated Electronic Properties of Metals* (Pergamon Press, Oxford 1978)
4.35 O. Gunnarsson, B.I. Lundqvist: Phys. Rev. **B13**, 4274 (1976)
4.36 J.P. Perdew, A. Zunger: Phys. Rev. **B23**, 5048 (1981)
4.37 A.K. Rajagopal: "Advances in Chem. Phys.", Vol. 41, ed. by I. Prigogine, S.A. Rice (Wiley, New York 1980), p. 59

4.38 C. Herring: Phys. Rev. **57**, 1169 (1940)
4.39 V. Heine, M.H. Cohen: "Solid State Physics", Vol. 24, ed. by H. Ehrenreich, F. Seitz, D. Turnbull (Acad. Press, New York, 1970)
4.40 J.M. Ziman: "Solid State Physics", Vol. 26, ed. by H. Ehrenreich, F. Seitz, D. Turnbull (Acad. Press, New York 1971)
4.41 J.C. Slater: Phys. Rev. **51**, 846 (1937)
4.42 D.D. Koelling: Rep. Prog. Phys. **44**, 139 (1981)
4.43 H.L. Skriver: *The* LMTO *Method* (Springer, Berlin 1984)
4.44 A.P. Cracknell, K.C. Wong: *The Fermi Surface* (Clarendon Press, Oxford 1973)
4.45 D. Shoenberg, P.J. Stiles: Proc. Roy. Soc. **A281**, 62 (1964)
4.46 G.A. Burdick: Phys. Rev. **129**, 138 (1963)
4.47 W. Kohn: Phys. Rev. **115**, 1460 (1959)
4.48 E.I. Blount: Phys. Rev. **126**, 1636 (1962)
4.49 R.E. Peierls: Z. Phys. **80**, 763 (1933)
4.50 S.V. Vonsovsky: *Magnetism*, Vol. 1 (Wiley, New York 1974)
4.51 A.I. Baz', Ya. B. Zeldovich, A.M. Perelomov: *Scattering, Reactions and Decays in the Nonrelativistic Quantum Mechanics* (Nauka, Moskva 1971), Chap. 5 (in Russian)
4.52 V.P. Maslov, M.V. Fedorjuk: *Quasiclassical Approximation for the Quantum Mechanics Equations* (Nauka, Moskva 1976) (in Russian)
4.53 I.M. Lifshitz, A.V. Pogorelov: Dokl. AN SSSR **96**, 1143 (1954)
4.54 D. Shoenberg: "The Physics of Metals", Vol. 1, ed. by J.M. Ziman (Cambridge University Press, London 1961); Proc. 9th Intern. Conf. Low Temp. Phys. (Plenum Press, New York 1965)
4.55 A. Pippard: "The Physics of Metals", Vol. 1, ed. by J.M. Ziman (Cambridge University Press, London 1961)
4.56 J. Friedel: Phil. Mag. **43**, 153 (1952)
4.57 P. Lloyd: J. Phys. **C2**, 1717 (1969)
4.58 H. Ehrenreich, L. Schwartz: "Solid State Physics", Vol. 31, ed. by H. Ehrenreich, F. Seitz, D. Turnbull (Acad. Press, New York 1976)
4.59 J.S. Faulkner: Prog. Mater. Sci. **27**, 1 (1982)
4.60 M.I. Katsnelson, A.S. Shcherbakov: Phil. Mag. **B46**, 357 (1982)
4.61 N.F. Mott, W.D. Twose: Adv. Phys. **10**, 107 (1961)
4.62 I.M. Lifshitz: Usp. Fiz. Nauk **83**, 617 (1964)
4.63 N.F. Mott: Phil. Mag. **B49**, L75 (1984)
4.64 A.B. Migdal: Zh. Eksp. Teor. Fiz. **32**, 399 (1957)

Chapter 5

5.1 D. Bohm, D. Pines: Phys. Rev. **92**, 609 (1953)
5.2 P.M. Platzmann, P.A. Wolf: *Waves and Interactions in Solid State Plasmas* (Acad. Press, New York 1973)
5.3 V.P. Silin, A.A. Rukhadze: *Electromagnetic Properties of Plasmas and Plasma-like Media* (Atomizdat, Moskva 1961), Chap. 4 (in Russian)
5.4 A.A. Vlasov: Zh. Eksp. Teor. Fiz. **8**, 291 (1938)
5.5 P.P. Debye, E. Hückel: Z. Phys. **24**, 185 (1923)
5.6 B. Kh. Ishmukhametov, M.I. Katsnelson: Fiz. Met. Metalloved. **45**, 484 (1978)
5.7 G. Ruthemann: Ann. d. Phys. **2**, 113 (1948)
5.8 H. Raether: "Springer Tracts in Modern Physics", Vol. 38 (Springer, Berlin 1965), p. 84
5.9 D. Bohm, T. Staver: Phys. Rev. **84**, 836 (1951)
5.10 A.B. Migdal: Zh. Eksp. Teor. Fiz. **34**, 1438 (1958)
5.11 W. Kohn: Phys. Rev. Lett. **2**, 393 (1959)

5.12 H.B. Callen, T.R. Welton: Phys. Rev. **83**, 34 (1951)
5.13 L.D. Landau: Zh. Eksp. Teor. Fiz. **30**, 1058 (1956)
5.14 V.P. Silin: Zh. Eksp. Teor. Fiz. **33**, 405, 1282 (1957); **34**, 707 (1958)
5.15 D. Pines, P. Noziéres: *The Theory of Quantum Liquids*, Vol. 1 (Benjamin, Massachusetts 1966)
5.16 C.J. Pethick, G.M. Carneiro: Phys. Rev. **A7**, 304 (1973)
5.17 V.P. Silin: Fiz. Met. Metalloved. **29**, 681 (1970)
5.18 M.I. Katsnelson, A.V. Trefilov: Pisma Zh. Eksp. Teor. Fiz. **40**, 303 (1984); Phys. Lett. **109A**, 109 (1985)
5.19 I. Ya. Pomeranchuk: Zh. Eksp. Teor. Fiz. **35**, 524 (1958)
5.20 J.M. Ziman: *Electrons and Phonons* (Oxford University Press, Oxford 1960)
5.21 E.G. Brovman, Yu. M. Kagan: Usp. Fiz. Nauk **112**, 369 (1974)
5.22 H.E. Jahn, E. Teller: Proc. Roy. Soc. **A161**, 200 (1937)
5.23 I.B. Bersuker, V.Z. Polinger: *Vibron Interactions in Molecules and Crystals* (Nauka, Moskva 1983) (in Russian)
5.24 K.I. Kugel, D.I. Khomsky: Usp. Fiz. Nauk **136**, 621 (1982)
5.25 L.D. Landau, A.S. Kompaneetz: *Conductivity of Metals* (ONTI, Kharkov 1935) (in Russian)
5.26 E.M. Lifshitz, L.P. Pitaevsky: *Physical Kinetics* (Pergamon Press, Oxford 1981)
5.27 F. Bloch: Z. Phys. **59**, 208 (1930)
5.28 C. Kittel: *Quantum Theory of Solids* (Wiley, New York 1963)
5.29 J. Appel: "Solid State Physics", Vol. 21, ed. by H. Ehrenreich, F. Seitz, D. Turnbull (Acad. Press, New York 1967), p. 193
5.30 G. Mahan: *Many-Particle Physics* (Plenum Press, New York 1981)
5.31 H. Fröhlich, H. Pelzer, S. Zienau: Phil. Mag. **41**, 221 (1950)
5.32 L.D. Landau: Sow. Phys. **3**, 664 (1933)
5.33 J. Frenkel: Sow. Phys. **9**, 158 (1936)
5.34 S.I. Pekar: *Investigations in the Electron Theory of Crystals* (GITTL, Moskva-Leningrad 1951) (in Russian)
5.35 L.D. Landau, S.I. Pekar: Zh. Eksp. Teor. Fiz. **18**, 419 (1948)
5.36 L.N. Cooper: Phys. Rev. **104**, 1189 (1956)
5.37 N.N. Bogoliubov, V.V. Tolmachev, D.V. Shirkov: *A New Method in the Theory of Superconductivity* (Izd. AN SSSR, Moskva 1958) (in Russian)
5.38 J. Bardeen, L. Cooper, J. Schrieffer: Phys. Rev. **108**, 1175 (1957)
5.39 N.N. Bogoliubov: Zh. Eksp. Teor. Fiz. **34**, 58 (1958); Nuovo Cimento **7**, 794 (1958)
5.40 F. London, H. London: Proc. Roy. Soc. **A155**, 71 (1935)
5.41 V.L. Ginzburg, L.D. Landau: Zh. Eksp. Teor. Fiz. **20**, 1064 (1950)
5.42 A.B. Pippard: Proc. Roy. Soc. **A216**, 547 (1953)
5.43 F. London: *Superfluids*, Vol. 1 (Wiley, New York 1950)
5.44 B.D. Josephson: Phys. Lett. **1**, 251 (1962)
5.45 R.P. Feynman, R.B. Leighton, M. Sands: *The Feynman Lectures on Physics*, Vol. 3 (Addison-Wesley, Reading 1963)
5.46 E.M. Lifshitz, L.P. Pitaevsky: *Statistical Physics*, Part 2 (Pergamon Press, Oxford 1980)
5.47 A.A. Abrikosov: Zh. Eksp. Teor. Fiz. **32**, 1442 (1957)
5.48 J. Frenkel: Phys. Rev. **37**, 17, 1276 (1931)
5.49 R.S. Knox: *Theory of Excitons* (Acad. Press, New York 1963)
5.50 V.M. Agranovich: *Theory of Excitons* (Nauka, Moskva 1968) (in Russian)
5.51 F. Bassani, G. Pastori Parravicini: *Electron States and Optical Transitions in Solids* (Pergamon Press, Oxford 1975)
5.52 T.M. Rice, J.C. Hensel, T.G. Phillips, G.A. Thomas: "Solid State Physics", Vol. 32, ed. by H. Ehrenreich, F. Seitz, D. Turnbull (Acad. Press, New York 1977)
5.53 L.V. Keldysh: Proc. 9th Intern. Conf. Semicond. (Moskva 1968); "Electrons in Semiconductors" (Nauka, Moskva 1971), p. 5 (in Russian)

5.54 J.M. Lawrence, P.S. Riseborough, R.D. Parks: Rep. Prog. Phys. **44**, 1 (1981)
5.55 S.P. Shubin, S.V. Vonsovsky: Proc. Roy. Soc. **A145**, 149 (1934)
5.56 J. Frenkel: Z. Phys. **49**, 31 (1928)
5.57 P.W. Anderson: Phys. Rev. **83**, 1260 (1951)
5.58 S.V. Vonsovsky: *Magnetism*, Vol. 2 (Wiley, New York 1974)
5.59 L. Hülten: Archiv. Mat. Ast. Phys. **26A**, 11 (1938)
5.60 V. Yu. Irkhin, M.I. Katsnelson: Z. Phys. **B62**, 201 (1986)
5.61 F. Bloch: Z. Phys. **61**, 206 (1931)
5.62 C. Herring: "Magnetism", Vol. 4, ed. by G.T. Rado, H. Suhl (Acad. Press, New York 1966)
5.63 N.F. Mott: Proc. Phys. Soc. (London) **47**, 571 (1935)
5.64 D.D. Koelling, F.M. Mueller, A.J. Arko, J.B. Ketterson: Phys. Rev. **B10**, 4889 (1974)
5.65 T. Moriya: J. Magn. Magn. Mater. **14**, 1 (1979)
5.66 K.L. Liu, A.H. Macdonald, J.M. Daams, S.H. Vosko, D.D. Koelling: J. Magn. Magn. Mater. **12**, 43 (1979)
5.67 E. Stoner: Proc. Roy. Soc. **A154**, 656 (1936); Rep. Prog. Phys. **11**, 43 (1948)
5.68 S.V. Vonsovsky, M.I. Katsnelson: J. Phys. **C12**, 2055 (1979); "Problems of Modern Physics", ed. by A.P. Aleksandrov (Nauka, Leningrad 1980), p. 233 (in Russian)
5.69 S.P. Shubin, S.V. Vonsovsky: Phys. Z. Sowjetunion, **7**, 292 (1935)
5.70 S.V. Vonsovsky: J. Phys. USSR, **10**, 468 (1946)
5.71 P.W. Anderson: Phys. Rev. Lett. **18**, 1049 (1967); Phys. Rev. **164**, 352 (1967)
5.72 K. Yamada, K. Yosida: Prog. Theor. Phys. **68**, 1504 (1982)
5.73 P. Noziéres, C.T. de Dominicis: Phys. Rev. **178**, 1097 (1969)
5.74 K. Yamada: Prog. Theor. Phys. **72**, 195 (1984)
5.75 Yu. M. Kagan, N.V. Prokofiev: Zh. Eksp. Teor. Fiz. **90**, 2176 (1986)
5.76 J. Kondo: Prog. Theor. Phys. **32**, 37 (1964)
5.77 J. Kondo: "Solid State Physics", Vol. 23, ed. by F. Seitz, D. Turnbull, H. Ehrenreich (Acad. Press, New York 1969), p. 184
5.78 A.M. Tsvelick, P.B. Wiegmann: Adv. Phys. **32**, 453 (1983)
5.79 A. Zawadovsky: Phys. Rev. Lett. **45**, 211 (1980)
5.80 V. Yu. Irkhin, M.I. Katsnelson: Fiz. Tverd. Tela **28**, 3648 (1986); Z. Phys. **B70**, 371 (1988)
5.81 M.I. Katsnelson, A.V. Trefilov: Pisma Zh. Eksp. Teor. Fiz. **42**, 393 (1985)

Addenda

A.1 K. von Klitzing: Rev. Mod. Phys. **58**, 519 (1986)
A.2 A.S. Shcherbakov, M.I. Katsnelson, A.V. Trefilov, E.N. Bulatov, N.V. Volkenshtein, E.G. Valiulin: Pisma Zh. Eksp. Teor. Fiz. **44**, 393 (1986)
A.3 H.L. Skriver: Phys. Rev. **B31**, 1909 (1985)
A.4 V.L. Moruzzi, P. Oelhafen, A.R. Williams: Phys. Rev. **B27**, 7194 (1983)
A.5 N.E. Christensen: Phys. Rev. **B29**, 5547 (1984)
A.6 A.P. Malozemoff, A.R. Williams, V.L. Moruzzi: Phys. Rev. **B29**, 1620 (1984)
A.7 M. Methfessel, J. Kübler: J. Phys. **F12**, 41 (1982)
A.8 N.E. Christensen: Solid State Commun. **49**, 701 (1984)
A.9 B.L. Gyorffy, A.J. Pindor, J. Staunton, G.M. Stocks, H. Winter: J. Phys. **F15**, 993 (1985)
A.10 A.I. Liechtenstein, M.I. Katsnelson, V.P. Antropov, V.A. Gubanov: J. Magnetism and Magn. Mater. **67**, 65 (1987)
A.11 B.L. Altshuler: Pisma Zh. Eksp. Teor. Fiz. **41**, 530 (1985)
A.12 P.A. Lee, A.D. Stone: Phys. Rev. Lett. **55**, 1622 (1985)
A.13 A.I. Larkin, D.E. Khmelnitskii: Zh. Eksp. Teor. Fiz. **91**, 1815 (1986)
A.14 J.C. Bednorz, K.A. Müller: Z. Phys. **B64**, 189 (1986)
A.15 G.R. Stewart: Rev. Mod. Phys. **56**, 755 (1984)

A.16 P.A. Lee, T.M. Rice, J. W. Serene, L.J. Sham, J.W. Wilkins: Comments Condens. Matter Phys. **12**, 99 (1986)
A.17 Proceedings of the International Conference on Anomalous Rare Earths and Actinides (Grenoble, July 1986): J. Magnetism and Magn. Mater **63–64** (1986)
A.18 Charge Density Waves in Solids: Lecture Notes in Physics, **217** (Springer-Verlag, Heidelberg 1985), ed. by Gy. Hutiray, J. Solyom
A.19 R.L. Withers, J.A. Wilson: J. Phys. **C19**, 4809 (1986)
A.20 Positrons in Solids (Springer-Verlag, Heidelberg 1979), ed. by P. Hantojärvi
A.21 Muons and Pions in Materials Research (North-Holland, Amsterdam 1984), ed. by J. Chappert, R.I. Grynszpan

Subject Index